Advances in Intelligent Systems and Computing

Volume 1163

The series “Advances in Intelligent Systems and Computing” contains publications on theory, applications, and design methods of Intelligent Systems and Intelligent Computing. Virtually all disciplines such as engineering, natural sciences, computer and information science, ICT, economics, business, e-commerce, environment, healthcare, life science are covered. The list of topics spans all the areas of modern intelligent systems and computing such as: computational intelligence, soft computing including neural networks, fuzzy systems, evolutionary computing and the fusion of these paradigms, social intelligence, ambient intelligence, computational neuroscience, artificial life, virtual worlds and society, cognitive science and systems, Perception and Vision, DNA and immune based systems, self-organizing and adaptive systems, e-Learning and teaching, human-centered and human-centric computing, recommender systems, intelligent control, robotics and mechatronics including human-machine teaming, knowledge-based paradigms, learning paradigms, machine ethics, intelligent data analysis, knowledge management, intelligent agents, intelligent decision making and support, intelligent network security, trust management, interactive entertainment, Web intelligence and multimedia.

The publications within “Advances in Intelligent Systems and Computing” are primarily proceedings of important conferences, symposia and congresses. They cover significant recent developments in the field, both of a foundational and applicable character. An important characteristic feature of the series is the short publication time and world-wide distribution. This permits a rapid and broad dissemination of research results.

**** Indexing: The books of this series are submitted to ISI Proceedings, EI-Compendex, DBLP, SCOPUS, Google Scholar and Springerlink ****

More information about this series at http://www.springer.com/series/11156

P. Suresh · U. Saravanakumar ·
Mohammed Saleh Hussein Al Salameh
Editors

Advances in Smart System Technologies

Select Proceedings of ICFSST 2019

Springer

Editors
P. Suresh
Vel Tech Rangarajan Dr. Sagunthala
R&D Institute of Science and Technology
Chennai, Tamil Nadu, India

U. Saravanakumar
Vel Tech Rangarajan Dr. Sagunthala
R&D Institute of Science and Technology
Chennai, Tamil Nadu, India

Mohammed Saleh Hussein Al Salameh
Jordan University of Science
and Technology
Irbid, Jordan

ISSN 2194-5357 ISSN 2194-5365 (electronic)
Advances in Intelligent Systems and Computing
ISBN 978-981-15-5028-7 ISBN 978-981-15-5029-4 (eBook)
https://doi.org/10.1007/978-981-15-5029-4

This Springer imprint is published by the registered company Springer Nature Singapore Pte Ltd.
The registered company address is: 152 Beach Road, #21-01/04 Gateway East, Singapore 189721, Singapore

Preface

FSST 2019, the first edition of International Conference on Frontiers in Smart System Technologies, was held during April 3–5, 2019, in Vel Tech Rangarajan Dr. Sagunthala R&D Institute of Science and Technology, Chennai, India. This conference provides a lively platform for delivering the innovative ideas in the research and application of smart systems. We are pleased to present the proceedings of the conference as its published record.

The conference was opened by the President Dr. R. Rangarajan and Foundress President Dr. Sagunthala Rangarajan of Vel Tech who cordially welcomed the chief guest Dr. Mylswamy Annadurai (Former Director—ISRO Satellite Centre and Vice President, Tamil Nadu State Council for Science and Technology), guest of honor Professor Dr. Virendrakumar C. Bhavsar (Honorary Research Professor and Professor Emeritus, University of New Brunswick, Canada) and the participants from all over the world.

FSST 2019 represents the efforts of many people. We thank Dr. Mylswamy Annadurai for inaugurating the conference and sharing his experience in the area of smart structures. FSST 2019 organizers extend sincere thanks to the eminent speakers: Dr. Virendrakumar C. Bhavsar shared his experience in high-performance computing; Dr. Chua-Chin Wang of National Sun Yat-sen University, Taiwan, delivered his research contributions in VLSI circuit fabrications for smart systems; Dr. Po-Ming Lee, Professor of Southern Taiwan University of Science and Technology, Taiwan, shared his experience in EDA computing environment; Dr. Alvaro Manuel Reis Da Rocha, Professor of University of Coimbra, Portugal, for sharing his insights into the effective business management using ICT; and Mr. Sundar Dandapani, Director of Capgemini, delivered on Security Considerations for Connected Devices. It is our immense pleasure to thank all the International and National Advisory Board members for great support and valuable suggestions to make this event success.

FSST 2019 received 376 submissions in the areas of smart systems, connected devices, connecting mechanisms and cloud architectures, and the conference accepted 77 papers (20.47%) from various countries. The conference would not be possible without the excellent papers contributed by the authors. We thank all the authors for their contributions and their participation in FSST 2019! We sincerely appreciate and thank the reviewers for their hard work in reviewing submissions.

We gratefully acknowledge all support from Vel Tech to make this event success. We would also like to thank all the academic leaderships of the Vel Tech, faculty and students for their continuous support and encouragements which made this event a grand success.

We hope that this program will further stimulate research in smart systems and provide better techniques, algorithms and hardware for deployment. We feel honored and privileged to provide the best recent developments in this field to you through this great event.

Chennai, India — Dr. G. Vairavel
Chennai, India — Dr. P. Suresh
Chennai, India — Dr. U. Saravanakumar
Irbid, Jordan — Dr. Mohammed Saleh Hussein Al Salameh

Acknowledgements

It has been our pleasure to host all the participants of the FSST 2019 at Vel Tech. We are thankful to all the participants for coming to Vel Tech to attend the FSST 2019. We have been fortunate to have some of the eminent persons from academia, industry and utility working in the area of smart systems. We are assured that the participants must have benefitted by attending this conference.

We are very much thankful to the management for giving us the opportunity to hold FSST 2019 at Vel Tech Rangarajan Dr. Sagunthala R&D Institute of Science and Technology, Avadi, Chennai. We thank our academic leaders, Dr. Beela Sathyanarayana, Chancellor; Dr. V. S. S. Kumar, Vice-Chancellor; Dr. E. Kannan, Registrar; Dr. Koteeswara Rao Anne, Director Academics; Dr. P. Sarasu, Director, Industry Relations and R&D; Dr. A. Abudhahir, Director, Quality Assurance; and Dr. S. Sivaperumal, Director, International Relations, for their suggestions and supports.

We sincerely thank the members of the advisory committee for their valuable suggestions.

We sincerely thank our chief guest Dr. Mylswamy Annadurai (Former Director—ISRO Satellite Centre and Vice President, Tamil Nadu State Council for Science and Technology) and the guest of honor Dr. Virendrakumar C. Bhavsar (Honorary Research Professor and Professor Emeritus, University of New Brunswick, Canada) for kindly accepting our invitation and gracing the inaugural function.

We also thank the Vel Tech administration, in general, for providing us with all possible logistic support toward organizing this event.

We thank all the plenary speakers, the tutorial speakers and the delegates for their enthusiastic participation in this conference. We take this opportunity to thank all the reviewers and technical committee members for providing their valuable comments in time and help toward the improvement of quality of papers presented in the conference. We once again thank everyone whose contribution has made this conference a successful one.

Contents

About the Editors

Dr. P. Suresh (SM'10–M'13–SM'19) research interests are in the field of Optical Engineering, Nano Photonics, Nano Optics, System on Chip, Reconfigurable computing and Embedded Systems. He is associated with many professional societies like IEEE from 2010 and obtained Senior Member recognition from IEEE in 2019. He obtained Chartered Engineer a Member of Institution of Engineers, India from 2019. He is a Fellow of Optical Society of India from 2018. He is associated with OSA Community from 2013 and designated in different membership grade. He delivered his research contribution in various technical meetings and conferences. He has published more than 50 journal articles in reputed journals. He has edited technical books for Springer, IGI Global, CRC Press (Taylor and Francis Group), etc. He has edited several technical issues, research books, special thematic issues, etc.

Dr. U. Saravanakumar research interests span both VLSI Design and Silicon Photonics. Much of his work has been on improving the understanding, design, and performance of on chip communication, mainly through the application of network concept in to System on Chip. He has given numerous invited talks, tutorials and demos on VLSI / FPGA. Dr. Saravanakumar is an editor of Advances in Intelligent Systems (Springer) and is the author of over thirty papers on VLSI and Silicon Photonics. He is an editorial board member of International Journal of Electronics, Communications, and Measurement Engineering (IJECME) (IGI Global Publishers) and American Journal of Electrical and Computer Engineering (Science Publishing Group, USA). He is a reviewer of various publishers such as IEEE Access, Springer, Inderscience and various conferences. He is a Senior Member of IEEE, Member of IE and Member of SIGDA.

Dr. Mohammed Saleh Hussein Al Salameh research interests include electromagnetic compatibility EMI/EMC, bioelectromagnetics, minimization of human exposure to fields and radiations, EMP interaction, coupling and shielding, crosstalk, satellite communications, neural networks, optical fibers and integrated optical waveguides, unconventional microstrip circuits, dielectric resonator antennas

(DRAs), radar sensing, printed circuit boards, VLSI interconnects, and computer modeling for real-world problems. He is also developing numerical methods, such as the finite element method, finite difference method, and method of moments, for practical EMI/EMC applications. Dr. Al Salameh is an expert in the field of the health effects of electromagnetic energy.

AMI and Its Wireless Communication Security Aspects with QOS: A Review

Priyanka D. Halle and S. Shiyamala

Abstract Advanced metering infrastructure (AMI) is a vital part of smart grid (SG).The communication infrastructure is a key element of AMI. This paper focuses on wireless communication security for AMI. Also, this paper reviews research on various different schemes for security of wireless communication infrastructure with the aim—to examine the newly emerged wireless communication technology; to identify the different kinds of attack on wireless communication infrastructure; to investigate research direction for security in AMI; to identify the different algorithms to overcome different kinds of attack; and to identify the different simulation platform for analyzing the different parameters. The survey of this paper tries to provide information the possibility of different kinds of attacks on communication infrastructure with their solution. This paper focuses on the security of wireless communication infrastructure of the AMI by providing routing algorithm, by avoiding the previous algorithm's problem. The proposed algorithm is dynamic source destination sequenced distance vector routing (DS2DVR) which improves the security of wireless communication infrastructure. Different kinds of attack are considered for security. For simulation, network simulator 2(NS2) is used.

Keywords Wireless communication security · DS2DVR algorithm · Attacks · AMI

1 Introduction

Smart meters are nothing but advanced metering infrastructure which is a part of smart grid which provides bidirectional communication between suppliers and consumers. As a result of bidirectional communication, it manages smartly by controlling the system. Communication infrastructure is very important part of the

P. D. Halle · S. Shiyamala (✉)
Vel Tech Rangarajan Dr. Sagunthala R &D Institute of Science and Technology, Avadi, Chennai, Tamilnadu, India
e-mail: drshiyamala@veltech.edu.in

P. Suresh et al. (eds.), *Advances in Smart System Technologies*,
Advances in Intelligent Systems and Computing 1163,
https://doi.org/10.1007/978-981-15-5029-4_1

AMI. By providing wireless communication security for communication infrastructure, system becomes more reliable [1].This paper focuses for security aspects for home area network (HAN) of AMI infrastructure. In the communication of smart meter, Quality of service (QOS) will help to improve the performance of AMI [2]. AMI is the next generation of power system which provides strong wireless communication infrastructure [3]. AMI is significant part of smart SG infrastructure and involves smart meters, communication networks, data management systems abusing bidirectional communication from the utilities through to customers for information interchange as well as mixing of collected data into software application stages for suitable control response, less pricing and rules [4]. AMI provides different services to consumers [5] —(1) AMI provides security and privacy. (2) Consumers can check prizing of their energy usage from anywhere. (3) Consumers will alert for their energy usage. (4) Consumers will plan to usage of electricity.

AMI provides cyber security. From the beginning of electricity market, there is a problem of electricity theft; for this, AMI is the best option [6]. For social development, AMI performs vital role. The demand of electricity rapidly increases day by day for automation of industrial as well as social development. For this, the use of energy will be smartly. AMI tries to provide smart electricity market [7]. This paper focuses by providing smart meters for HAN; the utilization of smart meter allows gaining daily basis information like their daily energy consumption and billing. Without any wasting time, AMI provides information according to their request. And, this will help to improve electricity market [8]. AMI has a number of challenging aspects like communication infrastructure, cost, wireless communication technology, routing and networking and security. The number of researchers had worked on these challenges. Still there is a lot of problems related to performance of the AMI. This big problem is security. The number of researchers tries to provide security by considering different attacks. Still there are different kinds of problems occurring like speed, range, cost, flexibility, QOS parameters and communication infrastructure [9]. Smart meter is a key element of SG. The number of researchers is going to reduce the cost of AMI hardware by improving the security of meter data management system (MDMS) means information security [10]. MDMS is the heart of the AMI. By considering security parameters like integrity, confidentiality, availability and accountability, we can make more secure AMI. There are different protocols and algorithms for security parameters. Many researchers are going to work on security parameters, but still AMI degrades its performance due to security. AMI has three important components, namely smart home, communication device and utility pole or utility office [11]. This paper tries to provide different constraints regarding secure wireless communication infrastructure. AMI needs to maintain the data because unauthorized person can access the data, and because of this, the AMI will degrade the performance. Security is a big challenge of AMI because different kinds of attacker attack on the AMI [12]. AMI has established a network of smart meters. It consists different networks like home area network (HAN), neighborhood area network (NAN), wireless sensor network (WSN), wide area network (WAN). These networks communicated with each other by using wireless technology. For security purpose of AMI, there are different key management schemes. Researchers have

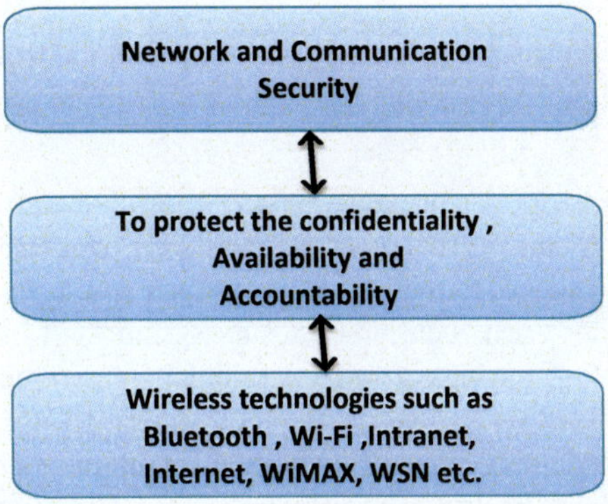

Fig. 1 Security aspects for AMI

done work on different key management schemes for security purpose of different kinds of networks. AMI can work with different types of communication like unicast, multicast, broadcast [13]. AMI provides smart utilities [14].

Usually, a utility works with vendors and integrators to design AMI infrastructure. The stability, range and difficulty of the infrastructure are combined with its endless growth in terms of quality, use of wireless technology, different kinds of network, topology, hardware/firmware/software, functionality and security controls for different kinds of attack [15]. Figure 1 gives the idea about the different security aspects for wireless communication infrastructure of the AMI.

AMI supports number of features so stakeholders, utility companies, energy markets, security researchers and regulators attracts towards to the AMI [16]. For AMI infrastructure, the smart communication infrastructure and intelligent subsystem are required. AMI measures the consumption of energy and other related billing parameters in the defined time interval. The measured data are modulated according to the different algorithms and protocols. And then showed to the management system [17]. AMI needs to do energy management accordingly—(1) utility side management, (2) demand side management and (3) generation planning and management. AMI transfers real-time data but the number of problems arises in bidirectional communication like fast speed to transfer the data. The number of researchers is going to work on wireless technology by considering parameters of QOS [18]. The security requirement for AMI is based on two stages—primary security services (functional requirement) and secondary security services (supportive for functional). This paper focuses on primary security services by providing QOS [19]. Primary security services try to provide security in terms of confidentiality, integrity, availability (CIA) CIA model and privacy. The performance of AMI is totally based on communication infrastructure. Attacker has lot of options to do attack on AMI. That is why security is a big challenge on each step of the AMI [20].

2 Role of Quality of Service (QOS) in the Performance of AMI

QOS tries to increase the performance of any smart system. The parameters of QOS are classified basically as two types—quantitative requirements and qualitative requirements.

By considering these two requirements, any system will increase their performance [21]. Table 1 gives idea about quantitative requirements and qualitative requirements.

For quantitative requirements and qualitative requirements have different solutions. Many researchers are trying to provide different solutions for the same. In these, security is a big issue for the performance of the AMI. Table 2 gives brief idea for the solution of quantitative requirements and qualitative requirements.

Interoperable, secure and flexible bidirectional networking technologies that encounter each SG component QoS necessities are vital for the fruitful awareness

Table 1 Classification of QOS [36]

Quantitative requirements	Qualitative requirements
Latency	Data accuracy
Bandwidth	Data validity
Data rates	Accessibility
Throughput	Interoperability
Reliability	Security

Table 2 Solution for quantitative requirements and qualitative requirements for the QOS

QOS parameters	Different algorithms/methodology/protocols for the solution
Latency	Rule based auto-focus algorithm, Conjugate Gradient algorithm, algorithm combining interior penalty function with D.C,OLSR protocol
Bandwidth	A novel polling algorithm, multi-objective genetic algorithm is proposed based on the elitist archive and the K-means approaches
Data rates	PPG signal motion artifacts correction algorithm based on feature estimation
Throughput	secure multiparty computation (MPC), homomorphic encryption (FHE) [22]
Reliability	Hop count routing: a routing algorithm
Data accuracy	Quantum evolutionary algorithm
Data validity	Cellular networks with opportunistic scheduling using queuing theory and stochastic geometry
accessibility	Iterative resource efficient power allocation (IREPA) algorithm
interoperability	Emotional-based strategic communications (EBSC), null encryption algorithm, time-based onetime password algorithm
Security	Greedy algorithm, topology reconfiguration algorithm, real-time distributed state estimation algorithm, automatic network reconfiguration algorithm, etc.

of the AMI [21]. By integrating Wireless Fidelity (Wi-Fi) and the long-term evolution (LTE), we can improve the QOS. Class-based weighted fair queuing (CBWFQ) and round robin (RR), deficit weighted round robin queuing (DWRRQ) techniques are used for calculation of jitter, end-to-end delay and traffic received. The Optimized Network Engineering Tools (OPNET) Modeler 17.2 are used for simulation to examine the QoS structure in LTE and Wi-Fi networks [23]. A lot of research has been conducted on the topic of QoS counting several surveys that have concentrated on the QoS problems for explicit wireless networks such as WSN, wireless mesh networks (WMNs) and IEEE802.11-based WLANs. QOS metrics are divided into three parts—timeliness, bandwidth and reliability, stability and availability [24]. The significant functionality of the routing algorithm is to choose a route with enough capitals that satisfies the QoS necessity of the particular application [25]. End-to-end delay is one of the important QOS metrics for the performance calculation of the network. In WSNs, enhancing end-to-end delay and energy consumption must be accomplished in demand to confirm QoS wants, since routing choices can impact the whole network [26]. For performance analysis of WSN, QOS metrics are considered like network lifetime, total energy consumption, average residual energy, throughput, packet delivery ratio (PDR), normalized overhead, end-to-end delay [27]. Many researchers are trying to search routing protocol to fulfill these QOS metrics. QOS metrics for video communication are different than routing algorithm [28]. Applications can choose among different soft QoS data delivery services by using multimodal reinforcement learning-based routing [29]. QOS tries to reduce overload of the system by increasing or decreasing QOS metrics [30]. Novel Quality of Service tries to solve complex control problems [31]. QOS provides the information according to service availability, service responsiveness and budgets. Researchers will get choice to develop particular application [32]. SDOWN controller helps in improving the QoS and QoE of the end users' in terms of Q-factor, SNR, OSNR, latency, jitter and BER [33].

3 Security Requirement Parameters

Security is based on many parameters. Communication protocols try to solve security issue. Basically, security is calculated by considering integrity, confidentiality, availability, accountability, authorization and authentication. Confidentiality is extra significant than the integrity of data and availability of capitals in the scenario of home applications [34]. Many researchers are trying to provide different algorithms for security in terms of integrity, confidentiality, availability, accountability [35].

Table 3 provides information about security of wireless communication infrastructure. Security basically based on CIA model.

Table 3 Security services or security levels for communication in AMI

Ref. No.	Confidentiality	Integrity	Availability	Accountability or non-repudiation	Authorization	Authentication
[34]	✓	✓	NC	NC	NC	✓
[35]	✓	✓	✓	✓	NC	NC
[36]	✓	✓	✓	NC	NC	✓
[37]	✓	✓	✓	✓	NC	NC
[38]	✓	✓	✓	✓	NC	NC
[39]	✓	✓	✓	NC	NC	NC
[40]	NC	✓	✓	NC	NC	NC
[41]	NC	✓	✓	NC	NC	✓
[42]	✓	✓	NC	NC	NC	✓
[43]	✓	✓	✓	✓	✓	✓
[44]	✓	✓	✓	NC	NC	NC
[45]	✓	✓	✓	NC	NC	NC
[46]	NC	✓	✓	NC	NC	NC
[47]	NC	NC	✓	NC	NC	NC
[48]	NC	✓	NC	NC	NC	NC
[49]	✓	✓	✓	✓	✓	✓

✓ Considered for security providing solution
NC not considered

4 Wireless Sensor Network (WSN)

In this paper, authors are using WSN for data gathering from the AMI. This section will discuss different types of attacks occurring on wireless communication infrastructure of the AMI and their solution. Table 4 provides this information.

The WSN is nothing but the combination of different sensors located on different places for collecting the physical data. In this paper, researcher is going to use WSN for collecting the meter reading from different smart meters [50]. These collected data will be send to the MDMS or base station or sink node for gathering the data. Wireless communication will support for this process [59]. The number of attackers will attack on wireless communication infrastructure. This paper will provide the solution for this. The number of researchers has done work on this, but still there is a big problem. WSN has a big issue of network performance because of different kinds of attacks occurred on this. For this, different researchers are going to do work [51]. In WSN, overcoming of clustering is a big challenge. As a result of clustering, the network of WSN reduces performance [52]. By considering security parameters like confidentiality and integrity, we can boost the security of the WSN [53]. WSN provides smart infrastructure for automation. WSN uses broadcast medium for

Table 4 Different parameters considered for security of communication infrastructure of AMI

Ref. No.	Considered attack/considered methodology for security for WSN	Algorithm	Simulation software results
50.	Detection rate of hybrid DoS attack, node energy consumption analysis technique, intrusion detection method, flood attack	Energy consumption prediction algorithm, energy trust algorithm, IDSET algorithm	NS2 simulation tool
51.	Energy optimized secure routing (EOSR) routing protocol, Sybil attack, black hole attack, malicious attack, tampering attack	Routing algorithm	NS2 simulation tool
52.	Cluster-based protocol named sec-LEACH Sybil, HELLO flood, selective forwarding, spoofing or replaying the route information, sinkhole and wormhole attack	Routing algorithm	
53.	Strong encryption unpredictable cryptographic primitives key recovery attack, chop-chop attack, FMS attack, stream cipher algorithm	Robust stream cipher algorithm, AES encryption algorithm, block cipher algorithm	
[54].	PSO-RBF, malicious attacks	K means algorithm	
55.	Context attacks, DOS attack	Online algorithm	TOSSIM (version 2.1.2
56.		Distance-vector hop (DV-Hop) algorithm and weighted centroid DV-Hop-based algorithms, a weighted centroid DV-Hop algorithm, EWCL algorithm	MATLAB 2007
50.	Detection rate of hybrid DoS attack, node energy consumption analysis technique, intrusion detection method, flood attack	Energy consumption prediction algorithm, energy trust algorithm, IDSET algorithm	NS2 simulation tool

(continued)

Table 4 (continued)

Ref. No.	Considered attack/considered methodology for security for WSN	Algorithm	Simulation software results
51.	Energy optimized secure routing(EOSR) routing protocol, Sybil attack, black hole attack, malicious attack, tampering attack	Routing algorithm	NS2 simulation tool
52.	Cluster-based protocol named sec-LEACH Sybil, HELLO flood, selective forwarding, spoofing or replaying the route information, sinkhole and wormhole attack	Routing algorithm	
53.	Strong encryption unpredictable cryptographic primitives key recovery attack, chop-chop attack, FMS attack, stream cipher algorithm	Robust stream cipher algorithm, AES encryption algorithm, block cipher algorithm	

communication. The different kinds of attacker try to attack on wireless communication medium. This is a big problem which occurs in WSN communication media [55]. National security application is provided by WSN. For the development for WSN, different localization schemes are needed [56]. In WSN, different sensors will leakage the important information. The researchers are trying to solve the problem by using different techniques [57]. WSNs are frequently subject to numerous security threats and attacks like impersonation, intentionally providing false information, eavesdropping, data modification and sensor node detention attacks [58]. WSN can increase efficiency and stability of network. In the SG, WSN gathers and processes the exact and valuable data in board area and monitors control devices permitting bidirectional information exchange, monitoring, control and maintenance in real time [60].

5 Secure Routing Algorithms/Protocols for AMI

Routing protocols are required for best performance of AMI. It delivers the data efficiently. The number of researchers tries to deliver different varieties of routing protocols for the growth of AMI communication infrastructure. This research tries to give information that WSN is the best solution for communication purpose in AMI. For this, different categories of routing protocols are required for effective delivery of data. The performance of routing protocols is estimated through throughput, end-to-end delay, and hop count and packet delivery ratio. The collection of routing protocol is based on the type of a message in the AMI. The performance of the routing protocol is based on the type of a message. These performance characteristics are calculated with help of simulation platform like NS2.

From Table 5, DSR is the best solution for security purpose. For throughput, there is a big problem. Combining the DSDV and DSR, we can improve the performance and security. The proposed algorithm is DS2DVR. Cognitive radio (CR) is the one of the best choices for AMI communication. CR improves the energy efficiency. From the literature survey, it is clear that latency and energy efficiency have more problems in AMI. Routing protocol for low power and Lossy networks (RPL) is a primary solution to improve the performance of communication infrastructure of AMI. CRB-RBL (cognitive receiver based RPL) protocol tries to improve the performance in terms of latency and energy efficiency. It is receiver-based routing protocol. It is not suitable for multiple networks. The researchers have scope to do work on this [61]. OFQS overcomes the problems of RPL. OFQS tries to provide the QOS in terms of end-to-end delay, network lifetime, load balancing and packet delivery ratio [62].

6 Future Scope

From Table 5, get the information, the researcher has many areas to do work on routing protocol for the development of different parameters like delay, network lifetime, load balancing and packet delivery ratio, throughput. From Table 3, it is clear that many researchers are going to do work on security parameters in terms of CIA model and accountability, authorization and QOS for routing protocols. Still, the issue of security is not solved yet. The researchers have to do work on wireless communication security for AMI. Electricity sector faces lot of problems of different kinds of attacks. By providing different secure routing algorithms and different simulation platforms, we can develop electricity sector. By making smart electricity sector, we can make social development.

7 Conclusion

Ultimately, by providing the security for wireless communication of AMI, we are going to develop electricity sector. We are proposed new routing protocol for communication infrastructure of AMI by considering QoS metrices. NS2 is one of the best platforms for simulation of networking. It provides efficient results for researchers

Table 5 Security routing protocol with simulation result [29]

Sr. No.	Category of a message	Throughput (KbPS)	End-to-end delay (ms)	Packet transfer ratio	Routing protocol	Comment
1.	Control message [29]	10.94	518.71	99.78	AODV	All four routing protocols have some drawbacks with respect to different metrics. A researcher has to do work on this. It is a future scope
		10.31	617.046	99.71	AOMDV	
		9.81	135.58	99.59	DSR	
		7.53	519.527	99.31	DSDV	
2.	Billing message [29]	179.66	553.59	99.92	AODV	DSR is best solution for delivery of billing message, but still there is a problem of throughput. It is a future scope.
		181.65	672.665	99.75	AOMDV	
		146.14	140.597	99.56	DSR	
		117.2	432.54	99.29	DSDV	
3.	Notification message [29]	52.87	445.11	99.71	AODV	DSR is best solution for delivery of notification message, but still there is a problem of throughput. It is a future scope
		51.34	456.09	99.68	AOMDV	
		45.81	190.19	99.56	DSR	
		39.26	464.61	99.49	DSDV	
4.	Request message [29]	53.18	424.79	99.85	AODV	DSR is best solution for delivery of notification message, but still there is a problem of throughput. It is a future scope
		51.29	433.58	99.82	AOMDV	
		45.26	181.36	99.65	DSR	
		42.62	437.69	99.56	DSDV	

to develop any system. Eventually, the secure wireless communication makes AMI more remarkable. Ultimately, we support for social enlargement.

References

1. Ahmad, M.W., Mourshed, M., Mundow, D., Sisinni, M., Rezguia, Y.: Building Energy Metering and Environmental Monitoring—A State-of-the-Art Review and Directions for Future Research, vol. 120, pp. 85–102, 15 May 2016
2. Pau, M., Patti, E., Barbierato, L., Estebsari, A., Pons, E., Ponci, F., Monti, A.: A Cloud-Based Smart Metering Infrastructure for Distribution Grid Services and Automation, vol. 15, pp. 14–25, September 2018
3. Hasan, M.R., Zhao, Y., Luo, Y., Wang, G., Winter, R.M.: An Effective AODV-based Flooding Detection and Prevention for Smart Meter Network, vol. 129, pp. 454–460 (2018)
4. Otuoze, A.O., Mustafa, M.W., Larik, R.M.: Smart Grids Security Challenges: Classification by Sources of Threats, February 2018
5. Meister, J., Ihle, N., Lehnhoff, S., Uslar, M.: Smart Grid Digitalization in Germany by Standardized Advanced Metering Infrastructure and Green Button, pp. 347–371 (2018)
6. Yip, S.-C., Tan, W.-N., Tan, C.K., Gan, M.-T., Wong, K.S.: An Anomaly Detection Framework for Identifying Energy Theft and Defective Meters in Smart Grids, vol. 101, pp. 189–203, October 2018
7. Wen, L., Zhou, K., Yang, S., Li, L.: Compression of Smart Meter Big Data: A Survey, vol. 91, pp. 59–69, August 2018
8. Sofana Reka, S., Dragicevic, T.: Future Effectual Role of Energy Delivery: A Comprehensive Review of Internet of Things and Smart Grid, vol. 91, pp. 90–108, August 2018
9. Abbasinezhad-Mood, D., Nikooghadam, M.: Design of an Enhanced Message Authentication Scheme for Smart Grid and its Performance Analysis on an ARM Cortex-M3 Microcontroller, vol. 40, pp. 9–19, June 2018
10. Baloglu, U.B., Demir, Y.: Lightweight Privacy-Preserving Data Aggregation Scheme for Smart Grid Metering Infrastructure Protection, vol. 22, pp. 16–24, September 2018
11. Ferrag, A.M., Maglaras, L.A., Janicke, H., Jiang, J., Shu, L.: A Systematic Review of Data Protection and Privacy Preservation Schemes for Smart Grid Communications, vol. 38, pp. 806–835, April 2018
12. Ford, V., Siraj, A., Rahman, M.A.: Secure and Efficient Protection of Consumer Privacy in Advanced Metering Infrastructure Supporting Fine-grained Data Analysis, vol. 83, issue 1, pp. 84–100, February 2017
13. George, N., Nithin, S., Kottayil., S.K.: Hybrid Key Management Scheme for Secure AMI Communications, pp. 862–869 (2016)
14. Habibzadeh, H., Soyata, T., Kantarci, B., Boukerche, A., Kaptan, C.: A Survey of the Sensing, Communication, and Security Planes in Smart City System Design, vol. 144, pp. 163–200, 24 October 2018
15. Hansen, A., Staggs, J., Shenoi, S.: Security Analysis of an Advanced Metering Infrastructure, vol. 18, pp. 3–19, September 2017
16. Iang, R., LuKim-Kwang, R., Choo, R.: Security Analysis of an Advanced Metering Infrastructure, vol. 18, pp. 3–19, September 2017. Achieving High Performance and Privacy-Preserving Query Over Encrypted Multidimensional Big Metering Data, vol. 78, Part 1, pp. 392–401, January 2018
17. Kabalci, Y.: A Survey on Smart Metering and Smart Grid Communication, vol. 57, pp. 302–318, May 2016
18. Kakran, S., Chanana, S.: Smart Operations of Smart Grids Integrated with Distributed Generation: A Review, vol. 81, Part 1, pp. 524–535, January 2018

19. Leszczyna, R.: A Review of Standards with Cyber security Requirements for Smart Grid, vol. 77, pp. 262–276, August 2018
20. Leszczyna, R.L.: Standards on Cyber Security Assessment of Smart Grid, vol. 22, pp. 70–89, September 2018
21. Faheem, M., Shah, S.B.H., Butt, R.A., Waqar Ashraf, M.: Smart Grid Communication and Information Technologies in the Perspective of Industry 4.0: Opportunities and Challenges, vol. 30, pp. 1–30, November 2018
22. Samet Tonyali, Kemal Akkayaa, Nico Saputro, A. Selcuk Uluagac, Mehrdad Nojoumian. Privacy-preserving protocols for secure and reliable data aggregation In IoT-enabled Smart Metering systems; Volume 78, Part 2, January 2018, pp 547–557
23. Saeed, A.T., Esmailpour, A., Nasser, N.: Performance analysis for the QoS support in LTE and WiFi. In: 2016 IEEE Wireless Communications and Networking Conference Workshops (WCNCW)
24. Malik, A., Qadi, J., Ahmad, B., Alvin Yau, K.-L., Ullah, U.: QoS in IEEE 802.11-Based Wireless Networks: A Contemporary Review
25. Zaheeruddin, Lobiyal, D.K., Prasad, S.: Antbased Pareto Optimal Solution for QoS aware Energy Efficient Multicast in Wireless Networks, vol. 55, pp. 72–81, June 2017
26. Yahiaoui, S., Omar, M., Bouabdallah, A., Natalizio, E., Challal, Y.: An Energy Efficient and QoS Aware Routing Protocol for Wireless Sensor and Actuator Networks, vol. 83, pp. 193–203, January 2018
27. Deepa, O., Suguna, J.: An Optimized QoS-Based Clustering with Multipath Routing Protocol for Wireless Sensor Networks, 5 December 2017
28. Chen, X., Zhao, Y., Li, Y., Chen, X., Zhao, Y., Li, Y.: QoE-Aware Wireless Video Communications for Emotion-Aware Intelligent Systems: A Multi-layered Collaboration Approach, vol. 47, pp. 1–9, May 2019
29. Basagni, S., Di Valerio, V., Gjanci, P., Petrioli, C.: MARLIN-Q: Multi-Modal Communications for Reliable and Low-latency Underwater Data Delivery, vol. 82, pp. 134–145, January 2019
30. Armentia, V., Gangoiti, U., Orive, D., Marcos, M. Dynamic QoS Management for Flexible Multimedia Applications, pp. 5920–5925, July 2017
31. Qiu, C., Cui, S., Yao, H., Xu, F., Richard Yu, F., Zhao, C.: A Novel QoS-Enabled Load Scheduling Algorithm Based on Reinforcement Learning in Software-Defined Energy Internet, September 2018
32. Wu, Z., Lu, Z., Hung, P.C.-K., Huang, S.-C., Tong, Y., Wang, Z.: QaMeC: A QoS-driven IoVs Application Optimizing Deployment Scheme in Multimedia Edge Clouds, vol. 92, pp. 17–28, March 2019
33. Singh, S., Jha, R.K.: SDOWN: A Novel Algorithm for Better Quality of Service and Experience in Software Defined Optical Wireless Network, vol. 176, pp. 662–684, January 2019
34. Rubio, J.E., Alcaraz, C., Lopez, J.: Recommender System for Privacy-Preserving Solutions in Smart Metering, vol. 41, pp. 205–218, October 2017
35. Rohokale, V., Prasad, R.: Cyber Security for Smart Grid—The Backbone of Social Economy, January 2016
36. Otuozea, A.O., Mustafaa, M.W., Larik, R.M: Review Smart Grids Security Challenges: Classification by Sources of Threats, 7 February 2018
37. El Mrabet, Z., Kaabouch, N., El Ghazi, H., El Ghazi, H.: Cyber-Security in Smart Grid: Survey and Challenges, vol. 67, pp. 469–482, April 2018
38. Benmalek, M., Challal, Y., Derhab, A., Bouabdallah, A.: VerSAMI: Versatile and Scalable key management for Smart Grid AMI Systems, vol. 132, pp. 161–179, 26 February 2018
39. Guan, X., Yang, B., Chen, C., Dai, W., Wang, Y.: A Comprehensive Overview of Cyber-Physical Systems: From Perspective of Feedback System, vol. 3, issue 1, 10 January 2016
40. Rinaldi, S., Giustina, D.D., Ferrari, P., Flammini, A., Sisinni, E.: Time Synchronization over Heterogeneous Network for Smart Grid Application: Design and Characterization of a Real Case, vol. 50, pp. 41–57, 1 November 2016
41. Tonyali, S., Akkayaa, K., Saputro, N., Selcuk Uluagac, A., Nojoumian, M.: Privacy-preserving protocols for secure and reliable data aggregation. In: IoT-enabled Smart Metering systems, vol. 78, Part 2, pp. 547–557, January 2018

42. Mahmood, K., Chaudhry, S.A., Naqvi, H., Kumari, S., Li, X., Sangaiah, A.K.: An Elliptic Curve Cryptography Based Lightweight Authentication Scheme for Smart Grid Communication, vol. 81, pp. 557–565, April 2018
43. Le, T.N., Chin, W.-L., Chen, H.-H.: Standardization and Security for Smart Grid Communications Based on Cognitive Radio Technologies—A Comprehensive Survey, vol. 19, issue 1, First quarter 2017
44. Wang, K., Du, M., Maharjan, S., Sun, Y.: Strategic Honeypot Game Model for Distributed Denial of Service Attacks in the Smart Grid, vol. 8, issue 5, Sept. 2017
45. He, H., Yan, J.: Cyber-Physical Attacks and Defences in the Smart Grid: A Survey, vol. 1, issue 1, p. 12 (2016)
46. Emmanuel, M., Rayudu, R.: Communication Technologies for Smart Grid Applications: A Survey, vol. 74, pp. 133–148, October 2016
47. Renewable and Sustainable Energy Reviews. Journal Homepage. www.elsevier.com/locate/rser
48. Asplund, M., Nadjm-Tehrani, S.: Attitudes and Perceptions of IoT Security in Critical Societal Services, 29 April 2016
49. Mustafa, M.A., Zhang, N., Kalogridis, G., Fan, Z.: DEP2SA: A Decentralized Efficient Privacy-Preserving and Selective Aggregation Scheme in Advanced Metering Infrastructure, 07 December 2015
50. Jinhui, X., Yang, T., Feiyue, Y., Leina, P., Juan, X., Yao, H.: Intrusion Detection System for Hybrid DoS Attacks using Energy Trust in Wireless Sensor Networks, vol. 131, pp. 1188–1195 (2018)
51. Yang, T., Xiangyang, X., Peng, L., Tonghui, L., Leina, P.: A Secure Routing of Wireless Sensor Networks Based on Trust Evaluation Model, vol. 131, pp. 1156–1163 (2018)
52. Tanwar, S., Thakkar, K., Thakor, R., Singh, P.K.: M-Tesla-Based Security Assessment in Wireless Sensor Network, vol. 132, pp. 1154–1162 (2018)
53. Asimi, Y., Asimi, A., Guezzaz, A., Tbatou, Z., Sadqi, Y.: Unpredictable Cryptographic Primitives for the Robust Wireless Network Security, vol. 134, pp. 316–321 (2018)
54. Wu, H., Xian, J., Wang, J., Khandge, S., Mohapatra, P.: Missing Data Recovery Using Reconstruction in Ocean Wireless Sensor Networks, September 2018
55. Bradbury, M., Jhumka, A., Leeke, M.: Hybrid Online Protocols for Source Location Privacy in Wireless Sensor Networks, vol. 115, pp. 67–81, May 2018
56. Kaur, A., Kumar, P., Gupta, G.P.: A Weighted Centroid Localization Algorithm for Randomly Deployed Wireless Sensor Networks, 9 February 2017
57. Li, Y., Wang, X., Kim, D.W., Zhang, J., Dai, R.: Designing Self-destructing Wireless Sensors with Security and Performance Assurance, vol. 141, pp. 44–56, 4 August 2018
58. Zhang, J., Li, H., Li, J.: Key Establishment Scheme for Wireless Sensor Networks Based on Polynomial and Random Key Predistribution Scheme, vol. 71, pp. 68–77, 15 March 2018
59. Yoldaş, Y., Önen, A., Muyeen, S.M., Vasilakos, A.V., Alan, İ.: Enhancing Smart Grid with Micro Grids: Challenges and Opportunities, vol. 72, pp. 205–214, May 2017
60. Yang, Z., Ping, S., Aghvam, H.S.A.H.: CRB-RPL: A Receiver-based Routing Protocol for Communications in Cognitive Radio Enabled Smart Grid, vol. 66, issue 7, July 2017
61. Nassar, J., Berthomé, M., Dubrulle, J., Gouvy, N., Mitton, N., Quoitin, B.: Multiple Instances QoS Routing. RPL: Application to Smart Grids 18(8), 30 Jul 2018. pii: E2472. https://doi.org/10.3390/s18082472

Improving Energy Efficiency in Cooperative MIMO with HARQ-CC Mechanism

P. Bhuvaneswari and L. Nithyanandan

Abstract This paper proposes energy efficiency for the physical layer and MAC layer in cooperative MIMO with hybrid automatic repeat request (HARQ) scheme. By this scheme, the retransmission count changes based on the significance of scheduling, and it reduces the packet collision. The transmit power at the physical layer determines the contending nodes and outage probability per unit area. Regarding delay and energy, accessing the channel in the MAC layer is very expensive because it requires extra time and energy to obtain the channel, and also, the antenna selection scheme in slow fading scenario is addressed. Finally, the proposed system lessens the contending nodes and delay and improves the network performance.

Keywords Cooperative · MIMO · PHY and MAC layer · Energy efficiency

1 Introduction

Most of the wireless networks and devices are battery powered and have a minimum lifetime. The most obvious technique to reduce energy consumption at wireless nodes is diversity. The diversity technique can be achieved by cooperative communication and MIMO. This technique has been used for energy efficiency (EE) in clustering and routing algorithms. One of the critical performance metrics in a wireless network is energy consumption, which reduces the lifetime of the system which can be solved by cooperative MIMO. In cooperative MIMO, the transmitter and receiver work as a virtual antenna where the transmitter and receiver are formed by cooperative nodes, which improve the lifetime of the network, reduce delay and improve the coverage. The scenarios of various types of MIMO are shown in Fig. 1 where the users in cooperative MIMO (CMIMO) have more benefit than SU-MIMO and MU-MIMO.

P. Bhuvaneswari (✉) · L. Nithyanandan
Department of Electronics and Communication Engineering, Pondicherry Engineering College, Puducherry 605014, India
e-mail: bhuvana.ece06@pec.edu

L. Nithyanandan
e-mail: nithi@pec.edu

P. Suresh et al. (eds.), *Advances in Smart System Technologies*,
Advances in Intelligent Systems and Computing 1163,
https://doi.org/10.1007/978-981-15-5029-4_2

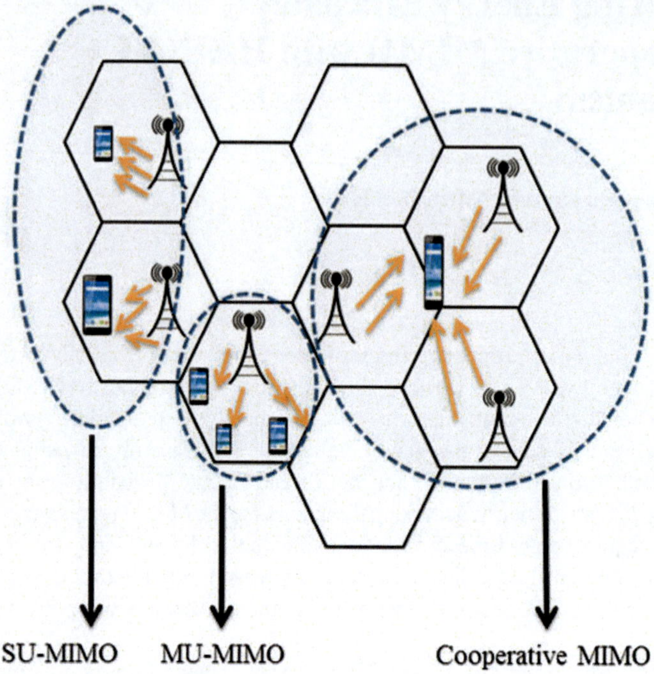

Fig. 1 Scenario of different types of MIMO

The author in [1] proposed cooperative MIMO in the wireless sensor network (WSN) with Alamouti code and showed that it reduces the energy consumption when the transmission distance is long. The power allocation between inter- and intracluster in energy-efficient CMIMO has been addressed in [2]. To improve the network performance and energy efficiency in CMIMO, singular value-based adaptive modulation is proposed [3], and randomly distributed nodes are considered. MIMO interference channel is considered [4] for energy efficiency from every individual node. But this technique is mainly used for limited energy networks because it cannot maintain the energy efficiency in every specific node. Max-min precoding is investigated [5] to improve energy efficiency in all base stations with various transmit powers, and the sum rate is achieved at low SNR. For the given SNR, MIMO-OFDM decode and forward relay scenario is considered which is responsible for higher information rate and energy efficiency [6] than MIMO decode and forward scenario. The performance of the relay scenario is improved by enabling multiple antennas to every node. The author in [7] proposed the relay selection methods which maximize the SNR at poorest UE node and reduce the events of outage, and also, it shows that the multiplexing gain is increased when the diversity order is kept constant. CMIMO-spatial modulation is proposed in anad-hoc network for the multihop scenario where BER and the number of hops were investigated [8], and for different BER, the energy consumption is reduced by selecting the number of hops.

FEC calculates the sum of required retransmission, and it adjusts the total energy desired to deliver an individual message to various nodes from a single node. To reduce the retransmission and to counter fading channel, hybrid retransmission techniques are used. HARQ schemes increase the network operation. For a long-range retransmission, HARQ-CC and low code rate are used [9] which ensure low error rates with low power, and for a short-range retransmission, simple HARQ technique and high code rate are used to minimize the computational load which reduces the time consumed in data bit. CMISO and CMIMO in WSN are addressed [10] over a quasi-static fading channel, where the CMIMO has better performance in long distance than CMISO, but it produces more transmission delay and consumes extra energy on the receiver side. The author in [11] investigated data gathering for each hop in CMIMO mechanism, minimizing energy consumption by keeping the neighbor sensors in a small distance. HARQ-IR is addressed [12] to maximize energy efficiency by considering the constant consumption of power by the least amount of HARQ rounds in massive MIMO. In [13] distributed MIMO with ZF receiver in large-scale fading channel is proposed to perform a detailed approximation of instantaneous SNR with PDF for the system and the number of antennas and user location parameter is newly derived for the spectral efficiency. The transmission of information with decode and forward protocol [14] is proposed to reduce the error rate which improves the accuracy in the relay scenario of the received signal. However, it consumes more time and power for determining the nodes. Similarly, author in [15] addressed incremental decode and forward protocol with different MIMO transmission schemes in cross-layer approaches of two different layers known as PHY and MAC layer. Here, the EE in all MIMO schemes is gradually lowered when the amount of nodes increases. Therefore, the energy consumption problem can be solved by HARQ-CC in cooperative MIMO with multiple relays in the slow fading scenario. HARQ-CC improves energy efficiency, increases throughput and reduces the average delay when the number of contending nodes decreases.

2 System Model

A cooperative MIMO network for a communication system consists of source S, relay R and the destination D with the number of antennas having the same power, HARQ-CC scheme and MAC protocol as shown in Fig. 2, and the system parameters are listed in Table 1. The multiple antennas for S, R and D are denoted as M_S, M_R and M_D, and the active antennas are denoted as L_S, L_R and L_D. The distance between S and D is denoted by X_{SD}, the distance between S and R is denoted by X_{SR} and the distance between R and D is denoted by X_{RD}. Initially, the S broadcasts information to R and D and the rebroadcasted information is forwarded to the destination by relay from the source.

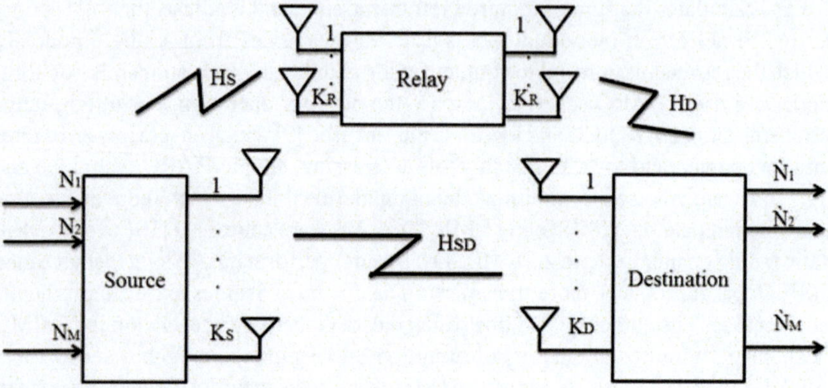

Fig. 2 System model

Table 1 System parameters

Parameter	Value
Bandwidth	20 MHz
Frequency	2.4 GHz
Header/payload	36/2000 Bytes
ACK	15 Bytes
RTS/CTS	20/16 Bytes
Path loss exponent	4

2.1 *Physical Layer*

HARQ-CC retransmission mechanism is considered in the physical layer. In this mechanism, the past erroneous error is not canceled, but they are combined at the receiver. The transmission mechanism increases the possibility of correct decoding by authorizing *N* transmission of the same packet in the event of the following outage. Then, at the receiver side, the multiple packets are combined as in maximum ratio combining method. The outage probability is given as [8]

$$O_P(N) = 1 - e^{\frac{-\Upsilon_o}{\ddot{\Upsilon}}} \sum_{x=1}^{N} \frac{(\frac{-\Upsilon_o}{\ddot{\Upsilon}})^{x-1}}{(x-1)!} \tag{1}$$

where $\ddot{\Upsilon}$ is the SNR at destination and $_o = 2^{Rb/B} - 1$. The number of appropriate transmission *L* per data packet is given by

$$L(N) = (x + a)^n = \sum_{x=0}^{N-1} (O_P(x)) \tag{2}$$

The O_P is nonzero for a limited amount of allowed transmissions due to slow fading; the essential data rate (R_D) is written as,

$$R_D(N) = R_b \sum_{x=1}^{N} \left(O_P \frac{(x-1) - O_P(x)}{x} \right) \quad (3)$$

MAC Layer

A handshake mechanism in MAC protocol is established when the source transmits a packet (RTS), and the destination acknowledges with clear to send (CTS). Then, the source sends the null data packet (NDP), authorizes the terminal to evaluate the CSI and responds for CFR to support the channel estimation by MIMO schemes. To allow cooperation, MAC mechanism is introduced to simultaneously accomplish the channel valuation and the choice of relay in the case of erroneous decoding by the destination. Subsequently, a request message from the destination for retransmission (RFT) after packet transmission from the source and the CSI is valuated with "helper." Then, the cooperation is established with "helper" and earlier to acknowledge by an authorized packet called helper RTS. In a MAC layer, there is an effective connection among contending nodes and the transmission power. By network allocation vector (NAV), the nodes in the transmission ranges of source and destination are closed, and from the transmission side, when the medium is chosen for rebroadcasting, the cooperating nodes in the range of transmission of a relay are closed. The blockage takes place when the contending nodes listen to the packets of RTS/CTS/hRTS. The overhearing node for the minimum received power is denoted as P_{RX} in a circular radius. For the channel access, other nodes have contended with the transmitter which is denoted as C_{CT} with the density ρ, and the contending nodes C_N with the path loss exponent (α) are given by [15],

$$C_N = \rho\pi\, C^2_{CT} \quad (4)$$

$$= [\rho\pi\, (\frac{P_S}{P_{RX}})^{\frac{2}{\alpha}}] \quad (5)$$

where P_S is the transmit power. At a random time, the probability (Q_r) of transmission at one node and the probability (Q_c) of transmission which occupied the successful channel are given by

$$Q_r = -(1-\Phi)^a + 1 \quad (6)$$

$$Q_c = \frac{a\Phi(1-\Phi)^{a-1}}{-(1-\Phi)^a + 1} \quad (7)$$

The probability of transmission of packets (Φ) which are initially established by a node is calculated as,

$$\Phi = \frac{2(1-2Q)}{(V+1)(1-2Q) + QV\left(1-(2Q)^z\right)} \tag{8}$$

where Q is the collide transmitted packet, z and v are contention window for maximum and minimum size. The messages in MAC layer are delivered to the rate control which maintains transmission messages at error-free control, whereas in PHY layer, higher data transmission is employed.

2.2 Power Consumption

During transmission and reception, a considerable amount of energy consumption in the RF circuit and the consumption of power when broadcasting [15] is expressed as

$$W_{Brd} = (1+\acute{\eta})\, W_S + L_S\, W_{TXN} + (R+D)\, W_{RXN} \tag{9}$$

where $\acute{\eta}$ is efficiency loss, W_{TXN} and W_{RXN} are active antenna at the transmitter and receiver. The power consumption for MIMO cooperation is given by

$$W_{Coop} = W_{Brd} + W_{Rtxn} \tag{10}$$

$$W_{Rtxn} = (1+\acute{\eta})\, W_i + L_i\, W_{TXN} + (D)\, W_{RXN} \tag{11}$$

where W_{Rtxn} is the total power consumption of the retransmission phase.

3 Transmission Schemes

The energy consumption and instantaneous SNR for cooperative SISO and antenna selection are considered as a transmission scheme for the proposed work.

3.1 SISO

In SISO, let us assume that the antennas for M_S, M_R and M_D are 1. For the retransmission and broadcast phase in a network, the consumption of power is given by,

$$W_{Brd,\,SISO} = (1+\acute{\eta})\, W_i + W_{TXN} + 2\, W_{RXN} \tag{12}$$

$$W_{Rtxn,\ SISO} = (1+\acute{\eta})\ W_x + W_{TXN} + W_{RXN} \tag{13}$$

At the link a-b, the instantaneous SNR is

$$\psi_{ab,SISO} = |H_{ab}|^2 \psi_{ab}^{*} \tag{14}$$

3.2 Antenna Selection

In this transmission scheme, one antenna is chosen which is triggered at every cooperating node; hence, L_S, L_R and $L_D = 1$. The antenna selection is based on SNR at the receiver, and in the transmitter side, it makes a feedback coming out of the receiver side which notifies the significance of the preferred antenna; for instance, it is considered to form by this route a failure-free feedback medium which given as,

$$W_{Brd,\ AS} = (1+\acute{\eta})\ W_i + W_{TXN} + 2\ W_{RXN} \tag{15}$$

$$W_{Rtxn,\ AS} = (1+\acute{\eta})\ W_x + W_{TXN} + W_{RXN} \tag{16}$$

The instantaneous SNR is

$$\psi_{ab,AS} = \max_{a,b} |Hab|^2 \psi\prime_{a,b} \tag{17}$$

The selected antennas are best for source and destination link because the link between source and destination has full diversity order and the mutual information in the antenna selection scheme is similar to SISO, whereas, in relay case, the antenna selection scheme is entirely random at the source. Thus, the source to relay link is SISO, because the source has only one antenna.

3.3 System Throughput

The throughput of the system depends on the transmission delay in both PHY and MAC layer. In PHY layer, the delay occurs due to packet transmission, and in the MAC layer, the lag is due to control packet transmissions and channel access. The transmission delay in PHY layer is given as [15],

$$D_{PHY} = \frac{A_{dr}}{B_N} \tag{18}$$

where A_{dr} is the average data rate and B_N is the number of bits per packets. The delay in MAC layer is the amount of time consumption by collision $[T_C]$, protocol overhead and backoff count $[C_b]$ which is expressed as

$$D_{MAC} = V_{MAC} + [T_C][C_b] + \frac{t}{1-K} \tag{19}$$

where V_{MAC} is the overhead and t is the collision node time period. The MAC overhead is defined by the sum of time consumption by RTS and CTS messages with respect to propagation delay.

3.4 Energy Consumption

In PHY layer, the energy consumed on channel access is not considered, but it has measured from the target outage probability. Due to the data transmission delay, the network consumes high energy which is subject to the amount of bits successfully transmitted. Hence,

$$E_C = (P_S + P_{RX})D_{PHY} \tag{20}$$

The energy consumption for MAC layer is written as,

$$E_C = E_{\mathrm{Ctime}} + E_{\mathrm{Caccess}} \tag{21}$$

where E_{Ctime} is the waiting time interval consumed in the direction of backoff counter and E_{Caccess} is the time consumption of obtaining the channel. If the time period of backoff counter is expired, then successful transmission, unsuccessful transmission and no transmission possibilities occur.

4 Simulation Results

The proposed HARQ-CC mechanism at PHY and MAC layer in cooperative MIMO is simulated, and the transmission schemes are compared with the existing and proposed method in the tool Network Simulator version 2 (NS2). The average delay is reduced due to efficient retransmission, and the system performance is improved when the channel attempts increase. The contending nodes with different transmission trail are taken on an account, and the energy consumption of both the layers is shown with respect to the power.

In Fig. 3, the contending nodes are reduced when the retransmissions enabled the appropriate relay power to encounter the outage probability. In MAC layer, the delay is entirely dependent on the contending nodes, and delay is high compared

Fig. 3 Contending nodes

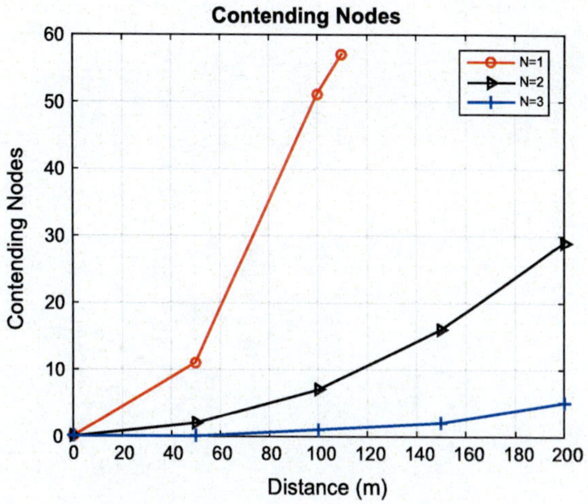

to PHY layer. The overall system throughput in MAC layer is also affected when the system performance increases. The retransmission for short distances provides high power in PHY layer to control the higher number of transmissions; however, it provides achievable throughput in large distances.

The retransmission gain for throughput and EE for different transmission trails are shown in Figs. 4 and 5. From the figure, for the long distances, the throughput is increased, and the gain above 0 dB shows a network improvement compared without retransmission $N = 1$. When $N = 2$, 0 dB exceeds the margin, and the gain at the initial stage decreases because of the increasing number of N. Hence, the decreasing contending node gives higher gain.

The average delay in PHY layer is reduced as shown in Fig. 6, while it is increased in MAC layer. This is due to the channel access, and the transmit power determines the

Fig. 4 Throughput

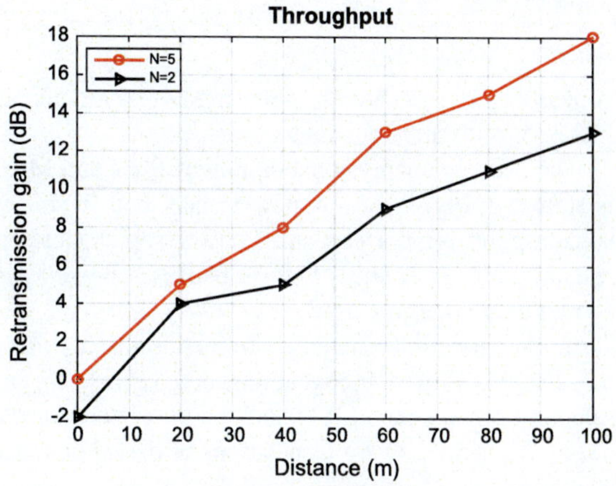

Fig. 5 Energy efficiency gain

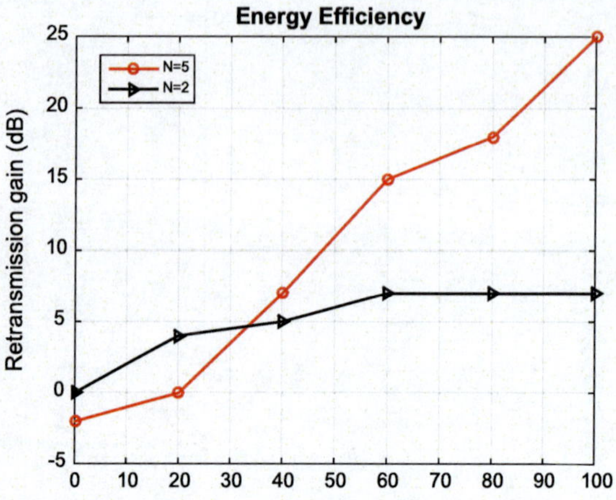

Fig. 6 Average delay

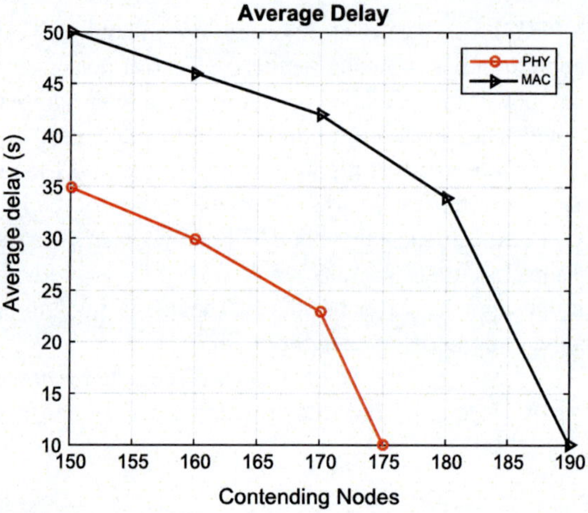

contending nodes. Thus, the delay in MAC layer increases when collision probability increases with more contending nodes.

The average energy consumption of PHY and MAC layer with respect to power is shown in Fig. 7. In low power state, both layers consume nearly equal energy consumption after a particular power region, and the MAC layer increases the consumption of energy. This is because the relay power increases when more cooperating nodes are used.

In the cooperative scheme, the medium between S and D is commonly close to the channel position. In Fig. 8, dSR at 0.3 gives best energy efficiency in the antenna selection scheme than SISO, and the outperformance in AS is better in all cases of dSR. This is due to various positions of the relay and also the number of successful

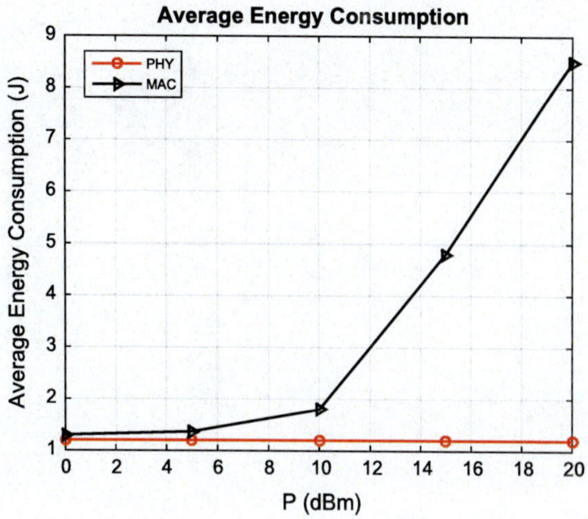

Fig. 7 Average energy consumption

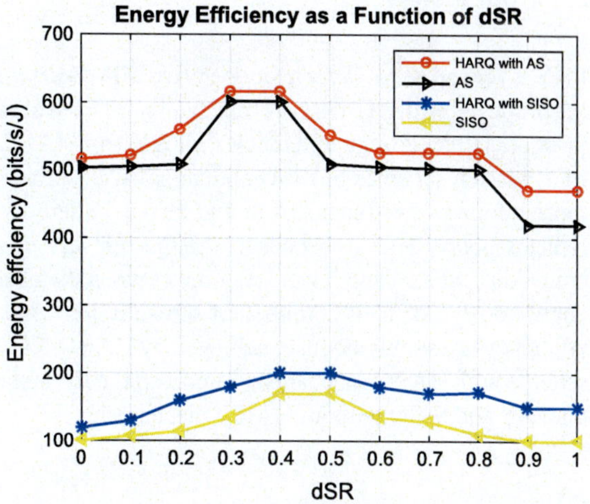

Fig. 8 Energy efficiency

retransmission in HARQ-CC with a number of channel attempts than the existing method.

The performance of outage probability for AS and SISO is shown in Fig. 9. AS with HARQ outperforms SISO and AS schemes. It can be noted from the figure, for AS and SISO with and without HARQ mechanism, the minimum required power varies around −3 to 4 dBm when the outage probability is 0.3. The spatial correlation in MIMO schemes increases the outage probability which leads to increase power for AS and SISO schemes.

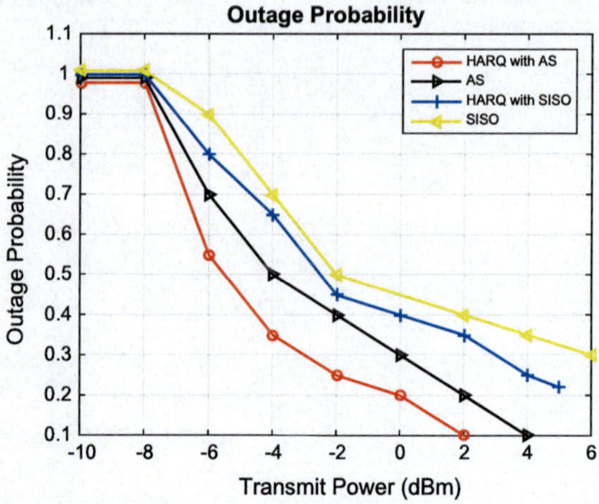

Fig. 9 Outage probability

5 Conclusion

PHY layer and MAC layer in cooperative MIMO network in the slow fading scenario are proposed with a HARQ-CC mechanism. The average delay, energy consumption, throughput and energy efficiency for PHY and MAC layer are investigated with the transmission schemes of SISO and antenna selection. With HARQ-CC mechanism, more number of retransmission and channel attempts gives higher throughput and reduces delay. This is because HARQ with less transmitting power reduced the contending nodes, and it has less impact when the channel access attempts increase continuously. With the number of transmission trials, the throughput and energy efficiency are simultaneously achieved by HARQ-CC. Finally, the antenna selection scheme with HARQ mechanism improved the EE in the cooperative MIMO network than AS and SISO without HARQ mechanism.

References

1. Cui, S., Goldsmith, A.J., Bahai, A.: Energy-efficiency of MIMO and cooperative MIMO techniques in sensor networks. IEEE J. Selected Areas Commun. **22**(6), 1089–1098 (2004)
2. Zhou, Z., Zhou, S., Cui, S., Cui, J.: Energy-efficient cooperative communication in a clustered wireless sensor networks. IEEE Trans. Wireless Commun. **57**(6), 3618–3627 (2008)
3. Jiang, J., Thopson, J.S., Sun, H.: A singular-value-based adaptive modulation and cooperation scheme for virtual-MIMO systems. IEEE Trans. Vehicular Technol. **60**(6), 2495–2504 (2011)
4. He, S., Huang, Y., Yang, L., Ottersten, B.: Coordinated multicell multiuser precoding for maximizing weighted sum energy efficiency. IEEE Trans. Signal Process. **62**(3), 741–751 (2014)
5. Li, Y., Fan, P., Beaulieu, N.C.: Cooperative downlink max-min energy efficient precoding for multicell MIMO network. IEEE Trans. Vehicular Technol. **60**(11), 9425–9430 (2016)

6. Anunchit: Realizing cooperative beam-forming scenario with MIMO-OFDM DF relay channels and performance comparison based on modulation scheme and energy efficiency. In: Proceedings of International Conference on Advances in Computing, Communications and Informatics, pp. 1762–1768, IEEE (2015)
7. Silva, S., Amarasuriya, G.: Relay selection strategies for MIMO two-way relay networks with spatial multiplexing. IEEE Trans. Commun. **63**(12), 4694–4710 (2015)
8. Peng, Y., Al-Hazemi, F.: Design and optimization for energy-efficient cooperative MIMO transmission in Ad Hoc networks. IEEE Trans. Vehicular Technol. **66**(1), 710–719 (2017)
9. Rosas, F., Souza, R.D.: Optimizing the code rate of energy-constrained wireless communications with HARQ. IEEE Trans. Wireless Commun. **15**(1), 191–206 (2016)
10. Grira, L., Bouallegu, R.: Energy efficiency of Cooperative MIMO in wireless sensor networks over Rayleigh fading Channel. In: 31st International Conference on Advanced Information Networking and Applications, pp. 107–112, IEEE (2017)
11. Li, X., Tao, X.: Energy-efficient cooperative MIMO-based random walk routing for wireless sensor networks. IEEE Commun. Lett. **20**(11), 2280–2283 (2016)
12. Kim, S., Yu, H.: Energy-efficient HARQ-IR for massive MIMO systems. IEEE Trans. Commun. **66**(9), 3892–3902 (2018)
13. Almelah, B., Hamdi, K.A.: Spectral efficiency of distributed large-scale MIMO systems with ZF receivers. IEEE Trans. Vehicular Technol. **66**(6), 4834–4844 (2017)
14. Sri Harsha Varma, U., Sai Nikhil, M.V.: Cooperative MIMO with relay selection for LTE-advanced system. In: Proceedings in International Conference on circuits Power and Computing Technologies, pp. 1–6 (2017)
15. Peron, G., Brante, G.: Physical and MAC cross-layer analysis of energy-efficient cooperative MIMO networks. IEEE Trans. Commun. **66**(5), 1940–1954 (2018)

Implementation of Adaptive Huffman Codes for PAPR Reduction in OFDM Systems

S. P. Vimal, N. Sathish Kumar, and M. Kasiselvanathan

Abstract Orthogonal Frequency Division Multiplexing (OFDM) is one of the upcoming techniques for achieving large data rates in 4G/5G wireless communication. In OFDM system, the data are transmitted over sizable subcarriers and modulated with low symbol rate. High-peak-to-average-power ratio (PAPR) is the major disadvantage in OFDM systems that results in out of band energy distortion, intermodulation among the subcarriers, and saturation in power amplifiers. This paper deals with the analyzing the PAPR reduction in OFDM systems using digital communication codes. Adaptive Huffman codes (AHC) is selected and applied in OFDM transmitter section for reducing the high peak signals. The encoding part selected in AHC is Vitter's algorithm. QAM16 and PSK mapping schemes are used for modulating the data. Simulation is carried out for the proposed work using MATLAB software, and PAPR results are showed in Complementary Cumulative Distribution Function (CCDF) plot. Further, the PAPR results are compared with Huffman coding and using no codes. Simulation results show that for all subcarriers, the PAPR reduction of adaptive Huffman codes dominates the Huffman codes and using no codes.

Keywords PAPR · Adaptive huffman codes (AHC) · QAM · PSK · CCDF

S. P. Vimal (✉) · N. Sathish Kumar · M. Kasiselvanathan
Department of Electronics and Communication Engineering, Sri Ramakrishna Engineering College, Coimbatore, Tamil Nadu 641022, India
e-mail: vimal.sp@srec.ac.in

N. Sathish Kumar
e-mail: sathishkumar.n@srec.ac.in

M. Kasiselvanathan
e-mail: kasiselvanathan.m@srec.ac.in

P. Suresh et al. (eds.), *Advances in Smart System Technologies*,
Advances in Intelligent Systems and Computing 1163,
https://doi.org/10.1007/978-981-15-5029-4_3

1 Introduction

The emerging domain in communication research area is wireless communication. The necessary goal of the wireless communication is to enhance intersymbol interference, bandwidth reduction, intercarrier interference, and complexity of the system. However, OFDM is one of the latest techniques for achieving high data rates in 4G/5G wireless communication systems [1]. In multipath fading environment, OFDM systems can be operated in high speed. Spectral overlapping is possible in OFDM systems. Due to spectral overlapping, more than fifty percentage of bandwidth have been saved in OFDM systems when compared to ordinary FDM systems. The application in OFDM systems leads to less energy loss in frequency domain. The main disadvantage in OFDM systems is high PAPR. The utilization of large number of subcarriers in OFDM systems will result in high PAPR. High PAPR makes the amplifier side to work in nonlinear region which causes bit error rate degradation and adjacent channel interference [2]. In this paper, the adaptive Huffman encoding algorithm is applied in the OFDM transmitter section, so that the high peak powers are eliminated before the transmission starts [3, 4]. There are various PAPR reduction techniques such as tone reservation, tone injection, selective scrambling, coding schemes, envelope scaling, and random phase updating [5]. The coding scheme of encoding is selected in this work. The idea behind the coding scheme is to reduce the PAPR of OFDM signal with constant modulus constellation points combined with modulation scheme. The coded OFDM signal not only reduces the peak power but also it reduces the data rate of the system when compared to other schemes. The coding schemes have a advantage of error correcting capability which results in BER and PAPR reduction. The utilization of Huffman and adaptive Huffman codes (AHC) plays a very important role in this paper [6]. The encoding procedure is carried out for adaptive Huffman codes. The encoding procedure selected in this work is Vitter's algorithm [7]. PAPR results are calculated for AHC and compared with Huffman codes [8, 9]. Simulated PAPR results are displayed in CCDF plot. Simulation results show that the contribution of AHC performs well in PAPR reduction when compared to Huffman codes.

2 Proposed System

Figure 1 shows the block diagram of the proposed OFDM transmitter system using adaptive Huffman encoder. In an OFDM system, data are carried over in many subcarriers. OFDM is a combination of multiplexing and modulation. The transmission behind OFDM systems is said to be multicarrier transmission. The subcarriers use the frequency spectrum, and it is necessary to convert a serial data into several parallel data that are to be divided among the individual carriers. The generated inputs are in random from S/P converter. They may include characters (both upper case A-Z and lower case a-z), special characters, and numbers. The output of S/P converter

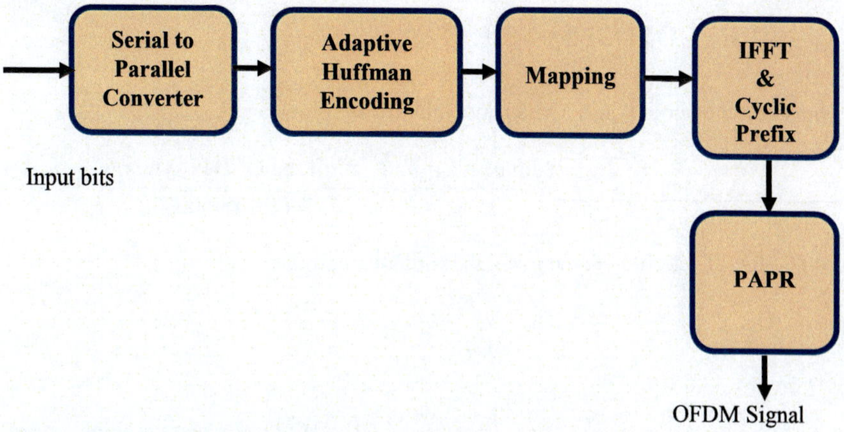

Fig. 1 Transmitter block diagram of the proposed system

shows the serial data streams placed in individual parallel subcarrier. The serial-to-parallel converter output is encoded using adaptive Huffman codes [10]. AHC is a type of adaptive coding techniques where the encoding procedures are similar to Huffman coding. Initial knowledge of source distribution is not needed in AHC where building the code as a symbol is possible. Huffman encoding is based on two-pass procedure. To construct the Huffman tree, first pass defines the identification of each character followed by file compression specified by second pass [11]. AHC requires one-pass procedure that is initialized by Vitter's algorithm. The input bits are encoded based on Vitter's algorithm [7]. Further, the encoded bits are mapped by conventional modulation schemes such as phase shift keying (PSK) and quadrature amplitude modulation (QAM) [12]. The IFFT operation specifies the conversion of signals from frequency domain to time domain. After IFFT cyclic prefix is carried out, the cyclic prefix ensures that the subcarriers are orthogonal to each other. Adding cyclic prefix will reduce the intercarrier interference (ICI) in the OFDM transmitter section. After taking IFFT, the symbols in subcarriers are transmitted. In OFDM, the information symbols are not directly transmitted. IFFT operation is performed over information symbols, and only the IFFT samples are transmitted. The PAPR is significantly high in OFDM systems. The PAPR rises as the number of subcarriers gets increased. The reason for increase in PAPR is because of IFFT preprocessing. In single-carrier system, there is no IFFT preprocessing, where in OFDM, the application of IFFT preprocessing enhances the instantaneous swing of the transmitted symbols over the average level [13]. Data symbols across subcarriers are added to produce a high peak value signal. The different symbols that are loaded in the subcarriers are random, and depending upon the nature, they are occasionally all add up across the subcarriers to produce a high peak value when compared to the normal value. After reducing the high peak signal, the OFDM signal is transmitted to the receiver side.

3 PAPR of OFDM System

The signal power average is determined by the relation

$$\text{Average Power} = \frac{\text{Sum of magnitude of all the OFDM symbols}}{\text{Total Number of OFDM symbols}}$$

The PAPR relation can be given in mathematically as

$$\text{PAPR} = \frac{\max|x(t)|^2}{E\left[|x(t)|^2\right]} \tag{1}$$

where $\max|x(t)|^2$ is the peak signal power and $E\left[|x(t)|^2\right]$ is the average signal power.

4 Algorithm for Adaptive Huffman Coding

The algorithm for AHC and Vitter's algorithm is given below:

Step1: Apply the input characters.
Step2: Parallel data are generated using S/P converter.
Step3: Encode the bits using Vitter's algorithm.
Step4: Zero padding is done before mapping.
Step5: Mapping is done using QAM16 and PSK16 modulation techniques.
Step6: Before adding cyclic prefix, compute IFFT for the mapped sequence.
Step7: Calculate PAPR.
Step8: Graph is plotted between threshold and CCDF values.

4.1 Vitter's Algorithm

Step1: A tree structure is created to represent a code. In a code, every node was initialized by unique number and corresponding weight.
Step2: Number follows downward subjected from right to left.
Step3: Distribution of weights is followed by sibling property where the nodes must be displayed in the preference of decreasing weight with each node adjacent to each other.
Step4: The weight denotes the count of symbols transmitted.
Step5: Here a block specifies set of nodes with same weight.
Step6: Every node in a code is assigned in a binary tree.
Step7: Binary number transmission for every symbol in alphabet is done by "Not Yet Transmitted" (NYT) node.

Step8: Transmit the code for NYT node, for that we have to transmit the NYT symbol.
Step9: Transmit the code for every leaf node, where every symbol is within a tree.
Step11: Execution of update procedure is followed for symbol in both transmitter and receiver sides.
Step11.1: Add two child nodes to NYT if current symbol is NYT. The one will be the leaf node of the symbol, and other is our new NYT node. Increase weight for old NYT and new leaf node, then go to Step 11.4. If not, go to leaf node symbol.
Step11.2: Except if that node is its parent, swap it with the node having the highest number if this node does not have the highest number in a block,
Step11.3: Weight is increased for current node.
Step11.4: Go to parent node if this is not the root node, then go to Step 11.2.

5 Simulation and Results

In this work, MATLAB software is selected for simulation. The input bits chosen for encoding are randomly selected. QAM 16 and PSK mapping scheme are selected for modulation purpose. The coding is constructed for any number of subcarriers (N). Here, $N = 32$, 64, 128, and 256 subcarriers are selected for simulation. Simulation results are viewed in CCDF plot. CCDF plot shows relative power level of a signal versus probability of occurrence. CCDF plot shows how often the PAPR of OFDM symbol exceeds the threshold values. The ratio between power average and the power level of a signal is expressed in decibels (dB). The relation for CCDF is given below

$$\text{CCDF} = \text{Probability}(\text{PAPR} > \text{PAPR}_0)$$

Graph is plotted for CCDF and threshold values, and the CCDF plot specifies the power signals. Threshold value fixation ranges from zero to maximum value. Calculation of threshold value formula is given as

$$\text{Threshold} = 0 : \frac{\text{Max PAPR} - \text{Min PAPR}}{\text{Max PAPR} : \text{Min PAPR}}$$

Figures 2, 3, 4, 5, 6, 7, 8 and 9 show the simulation results of CCDF plot for PAPR reduction using adaptive Huffman code, Huffman code, and PAPR with no coding.

Figure 2 represents the CCDF plot for QAM 16 modulation with $N = 32$. By applying adaptive Huffman codes, there is a 5.7 dB PAPR reduction. There is a difference of 0.6 dB PAPR reduction between AHC and Huffman codes. The CCDF plot for QAM 16, $N = 64$ is shown in Fig. 3. The plot shows a reduction in PAPR of 10.5 dB for Huffman codes, and for AHC, there is an 8.2 dB. There is a 2.3 db difference between Huffman coding and AHC. The CCDF plot for QAM 16, $N = 128$ is shown in Fig. 4. The utilization of AHC shows a nice improvement when compared to Huffman and no coding. The application of AHC in PAPR reduction

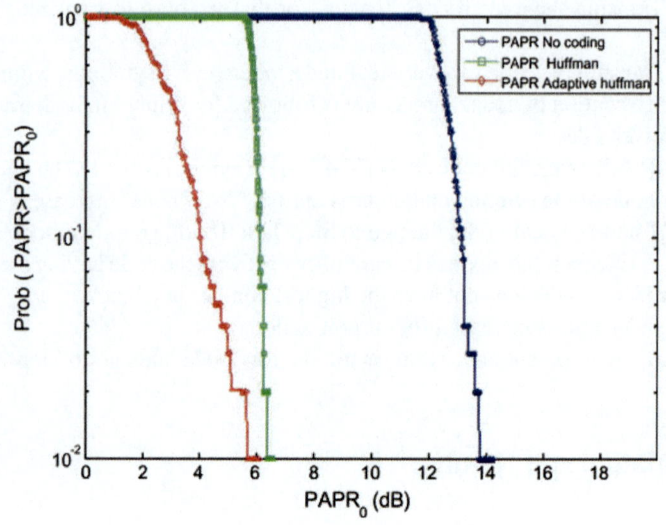

Fig. 2 CCDF plot, PAPR of QAM 16 modulation for $N = 32$

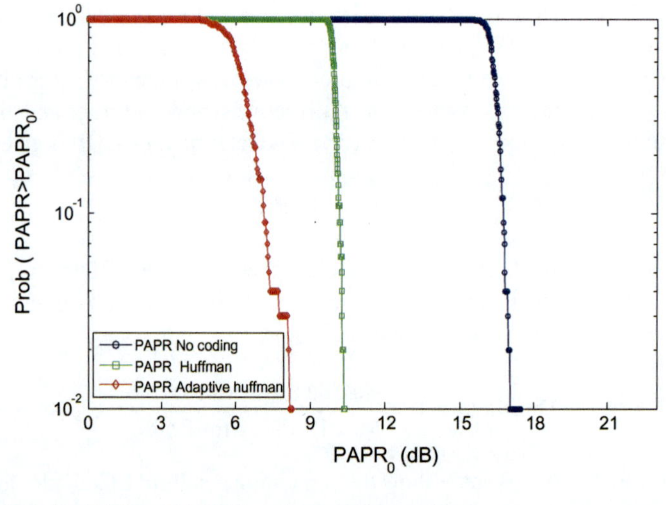

Fig. 3 CCDF plot, PAPR of QAM 16 modulation for $N = 64$

shows a good result for higher subcarriers also. For $N = 256$, QAM 16 in Fig. 5 there is a 4 db reduction between AHC and Huffman codes. For all the subcarriers used in this study, it is shown that if N increases, PAPR also increases. The CCDF plot for PSK, $N = 32$ in Fig. 6 shows a 0.5 db PAPR difference between AHC and Huffman codes. For all subcarriers, $N = 64$, 128, and 256 are shown in Figs. 7, 8 and 9; the usage of AHC shows a good improvement. The PAPR comparison with different subcarriers for QAM 16 and PSK mapping is shown in Table 1.

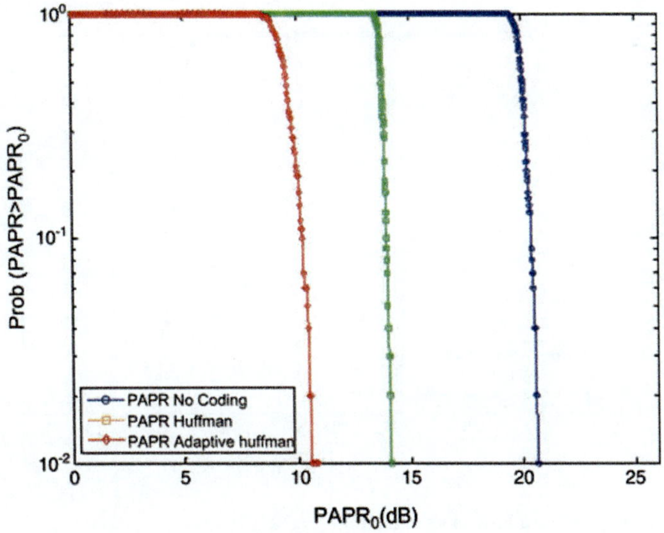

Fig. 4 CCDF plot, PAPR of QAM 16 modulation for $N = 128$

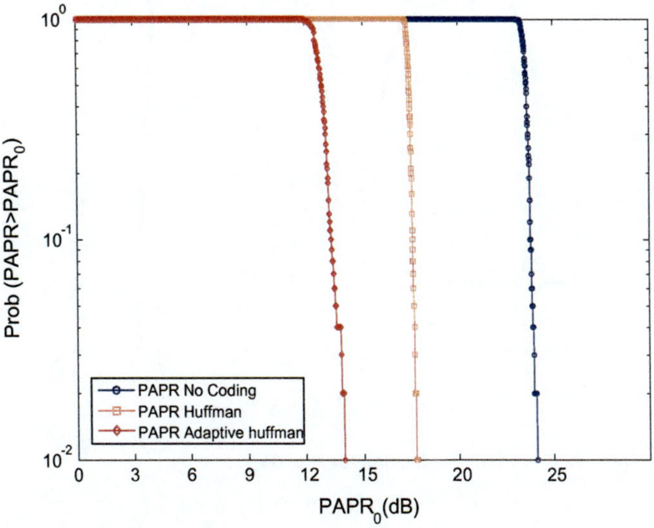

Fig. 5 CCDF plot, PAPR of QAM 16 modulation for $N = 256$

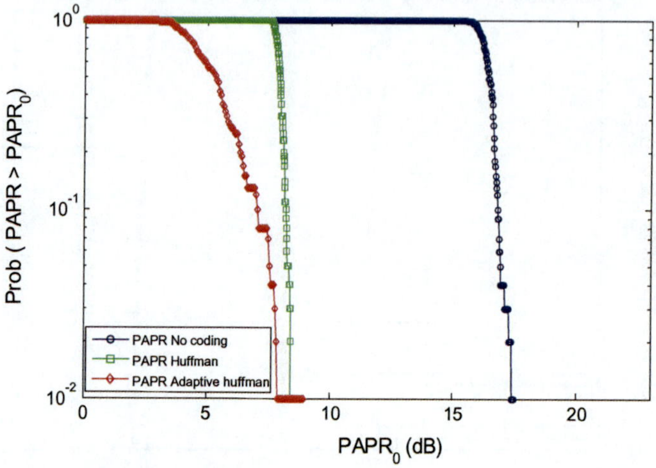

Fig. 6 CCDF plot, PAPR of PSK modulation for $N = 32$

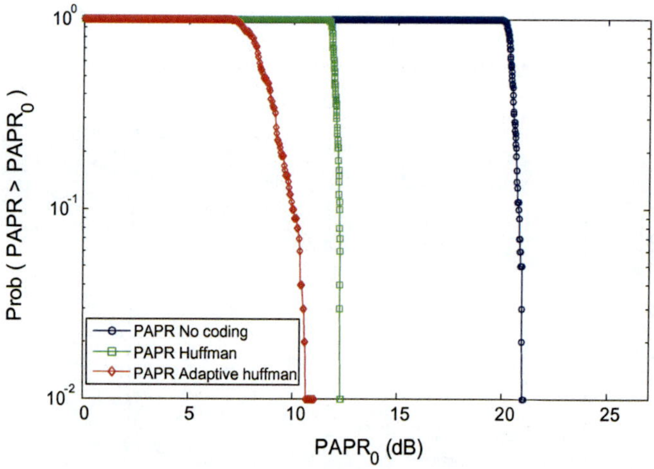

Fig. 7 CCDF plot, PAPR of PSK modulation for $N = 64$

6 Conclusion

The PAPR reduction comparison table shows that Huffman code shows a good improvement when compared to no coding. But the utilization of adaptive Huffman codes dominates the Huffman and no coding for all the subcarriers used in this study. Further, the usage of QAM 16 shows a good PAPR reduction results for all coding schemes and "*N*" values used in this paper when compared to PSK modulation schemes.

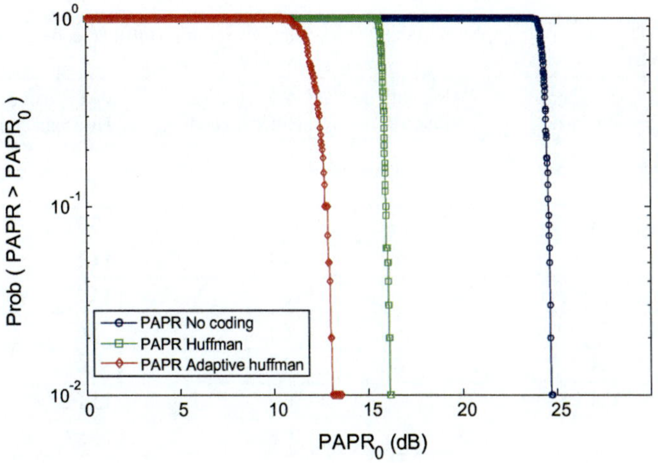

Fig. 8 CCDF plot, PAPR of PSK modulation for $N = 128$

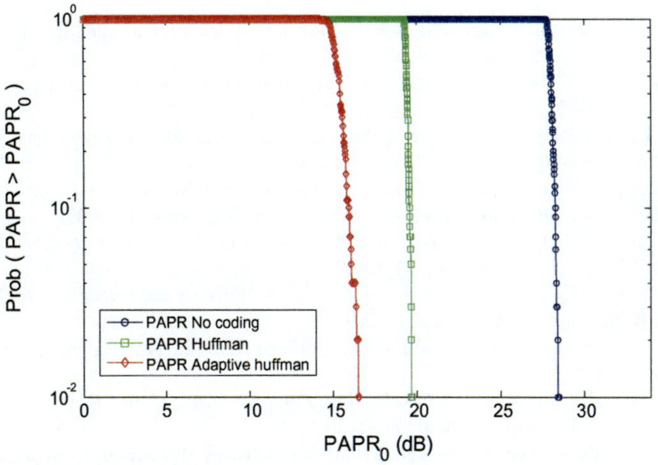

Fig. 9 CCDF plot, PAPR of PSK modulation for $N = 256$

Table 1 PAPR reduction comparison table values using no coding, Huffman codes, andladaptive Huffman codes

Mapping	Number of subcarriers	PAPR with no coding (dB)	PAPR using Huffman coding (dB)	PAPR using adaptive Huffman coding (dB)
16 QAM	32	13.8	6.3	5.7
	64	17.7	10.5	8.2
	128	21.1	14.1	11.1
	256	24	18.3	14.3
PSK	32	17.22	8.1	7.6
	64	21	12.1	10.91
	128	24.1	16.2	13.12
	256	28.3	19.8	15.87

References

1. Tellado, J.: Peak to Average Power Reduction for Multicarrier Modulation, Ph.D. thesis, Stanford university, September (1999)
2. Foomooljareon, P., Fernando, W.A.C.: PAPR Reduction in OFDM Systems. Thammasaltn. J. Sc. Tech. **7**, 70–79 (2002)
3. Nandi, U., Mandal, J.K.: Comparative study and analysis of adaptive region based Huffman compression techniques. Int. J. Inf. Technol. Convergence Serv. (IJITCS) **2**(4) (2012)
4. Tharini, C., Vanaja Ranjan, P.: Design of modified adaptive Huffman data compression algorithm for wireless sensor network. J. Comput. Sci. **5**(6), 466–470 (2009)
5. Jiang, T., Wu, Y.: An Overview: peak-to-average power ratio reduction techniques for OFDM signals. IEEE Trans. Broadcasting **54**(2) (2008)
6. Lin, J.-Y., Liu, Y., Yi, K.-C.: Balance of 0, 1 bits for Huffman and reversible variable-length coding. IEEE Trans. Commun. **52**(3) (2004)
7. Vitter, J.S.: Design and analysis of dynamic Huffman codes. J. Assoc. Comput. Mach. **34**(4), 825–845 (1987)
8. Kavousianos, X., Kalligeros, E., Nikolos, D.: Optimal selective Huffman coding for test-data compression. IEEE Trans. Comput. **56**(8) (2007)
9. Ashraf Eltholth, A., Adel Mikhail, R., Elshirbini, A., Moawad Moawad, I., Abdelfa tah, A.I.: Peak-to-average power ratio reduction in OFDM systems using Huffman coding. World Acad. Sci. Eng. Technol. **43** (2008)
10. Gadbail, R.N., Gulhane, V.S.: Effective compression technique by using adaptive Huffman coding algorithm for XML database. Int. J. Latest Trends Comput. **1**(2) (2010)
11. Suresh. P.: Creation of optical chain in the focal region of high NA lens of tightly focused higher order Gaussian beam. J. of Opt. **46**, 225–230 (2017)
12. Latif, A., Gohar, N.D.: On the PAPR reduction properties of Hybrid QAM-FSK (HQFM) OFDM transiever. J. Appl. Sci., 1061–1066 (2008)
13. Tarokh, V., Jafarkhani, H.: On the computation and reduction of the peak-to-average power ratio in multicarrier communications. IEEE Trans. Commun. **48**(1), 37–44 (2000)

An Effective Network Connectivity Model for Dynamic Channel Allocation in Wireless Communication

S. P. Vimal, A. Sivasakthivel, and L. Gururajan

Abstract The advent of smart devices requires effective communication between distinct devices which gains desire attention of communication. Those devices require effective communication performance in order to meet the challenges associated with wireless network devices. This paper examined the problem related to cognitive wireless network devices for channel allocation since effective allocation of channel is complex task. To construct the collaborative channel distribution method which is introduced, the utmost network connectivity will act as an optimal aim through this paper. This developed methodology considers the channel inert status as well as the transmission sweep of psychological clients. This designates right off the bat the inert channel to bottleneck hub in addition to dole out the suited channels to entire other psychological clients as could be allowed. In such a way, the above-mentioned current three channel designation methodologies along with the greatest system availability through the omnidirectional as well as bearing receiving wires. It additionally proposes the comparing calculation to play out the channel distribution intended for every single intellectual client in subjective remote systems. Reenactment results demonstrate that the proposed methodology is promising.

Keywords Network connectivity · Channel allocation · Collaborative model

---

S. P. Vimal (✉)
Department of Electronics and Communication Engineering, Sri Ramakrishna Engineering College, Coimbatore, Tamil Nadu 641022, India
e-mail: vimal.sp@srec.ac.in

A. Sivasakthivel · L. Gururajan
Department of Electronics and Communication Engineering, Arunai Engineering College, Tiruvannamalai, Tamil Nadu 606603, India
e-mail: shivasakthi17@gmail.com

L. Gururajan
e-mail: gururagu@gmail.com

P. Suresh et al. (eds.), *Advances in Smart System Technologies*,
Advances in Intelligent Systems and Computing 1163,
https://doi.org/10.1007/978-981-15-5029-4_4

1 Introduction

The quantity of D2D enhancement is the most important dispute in D2D underlay networks as well as combining the links also CSI feedback necessities. Here, we introduced a new resolution for the purpose of D2D channel distribution in which it needs the adjacent data of D2D communicating hubs only. Also, to exploit the depended D2D is the major objective of this paper as well as the channel distribution issue will be fixed as a mixed-integer programming (MIP) [1].

Furthermore, there are two techniques which have been introduced for the intention of joint resource optimization athwart wireless links as well as core systems such as non-orthogonal multiple access (NOMA) method is used on behalf of the concert increase in 5 g networks along with for resource distribution in core networks, software-defined network (SDN) will be introduced. The second method is joint system utility optimization algorithm along with multipath has been proposed. To calculate the performance of the above-mentioned two algorithms, extensive simulation has been used [2].

An algorithm is used for the purpose of enhancing the efficiency as well as diminishes the computation difficulty, so it is called as distributed iteration resource distribution algorithm which is going on the basis of alternation direction method of multipliers (ADMM). The local as well as double variables are updated through each base station (BS) in corresponding as well as separately. The global variables updating may have the chance to failure due to the noise leads to improper data exchange between BSs. Here, a robust distributed algorithm has been introduced to overcome the above-mentioned problem for the updating of data exchange that will be more successful in global variables [3].

The following factors are included in joint resource allocation such as development of subcarrier distribution, subcarrier pairing (SCP), as well as power distribution in a cooperative secure orthogonal frequency division multiple access (OFDMA) communication network along in the midst of untrusted users. The joint power distribution as a common curved dilemma will be proved next to the detection of optimal subcarrier allocation for the purpose of amplify and forward (AF) relayed networks. Hence, the optimal SCP pair subcarriers will be proved in which the effective channel gains will be diminished [4].

The one among the methodology for 5G mobile cellular systems is massive MIMO (multiple-input-multiple-output). It is used for the purpose of enhancing the channel capability as well as usage of spectrum along with; it will turn out to be a hotspot in the region of wireless communications. To overcome the low difficulty of channel calculation algorithm intended for massive MIMO system, we introduced an algorithm which is on the basis of intrinsic sparsity of wireless communication channel [5] (Figs. 1 and 2).

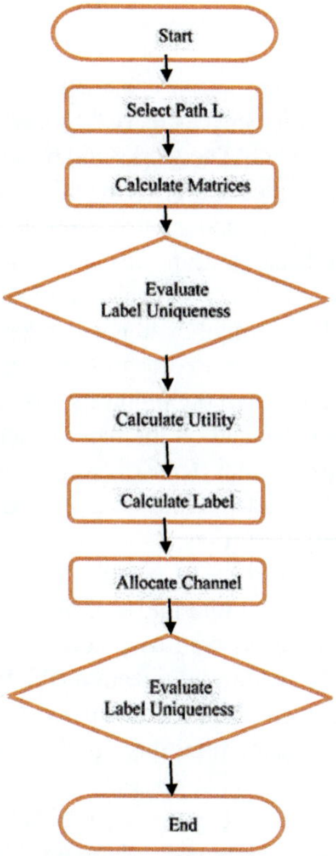

Fig. 1 Flow chart of the proposed approach

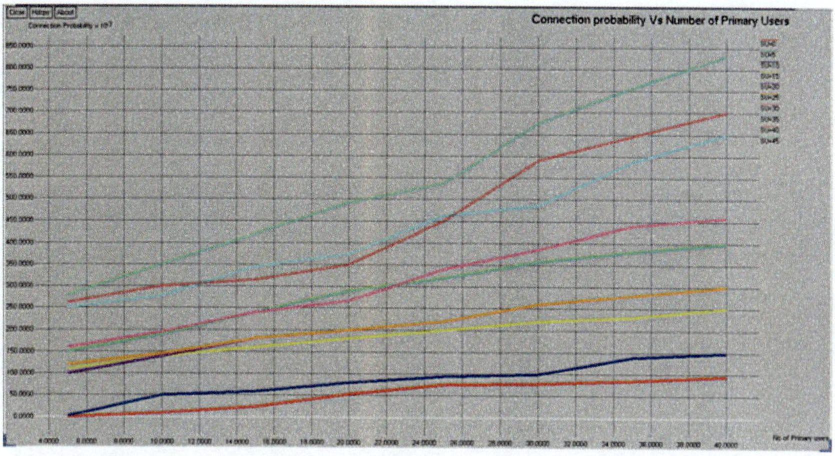

Fig. 2 Number of users versus connection probability

2 Literature Survey

Here, joint optimization of component carrier (CC) chosen as well as resource distribution in 5G carrier aggregation (CA) networks will be determined. The first thing is the quantity of CC will be computed as well as the throughput optimization dilemma is NP-hard. The second is, to overcome the first problem, a greedy-based algorithm will be introduced. Through this algorithm, we can able to perform ½ of the optimal performance. Finally, throughput as well as calculation difficulties will be done [6].

On behalf of increasing the potential of ad hoc networks, we introduced an algorithm named as Hyper-Erlang Channel Allocation Factor-based QoS Enhancement Mechanism (HECAFQEM). The two important factors are used for the purpose of the amount of refusal of user demand on the basis of queuing delay, broadcasting delay such as channel allocation factor (CAF) as well as expected channel allocation factor (ECAF). Finally, the QoS adaptations are executed by HECAFQEM through the estimation of queuing delay [7] (Figs. 3 and 4).

Here, the study has been taken based on two joint channel allocation as well as power control problems along in the midst of dissimilar limitation, in addition with a joint channel distribution as well as power control method will be introduced on the basis of particle swarm optimization (PSO) which is used to competently manage interference as well as enhance the network throughput. The algorithm evades trapping into infeasible resolutions through the scheming of fitness ranges of the 2 dilemmas [8].

Here, the new methodology has been introduced which is named as deadline-driven link quality aware channel assignment (LACA). Along with vital deadlines as well as heavy collisions, the link will be prioritized by LACA. Except, the utilization of the spare slots meant for rebroadcasting on lossy links will be permitted by LACA.

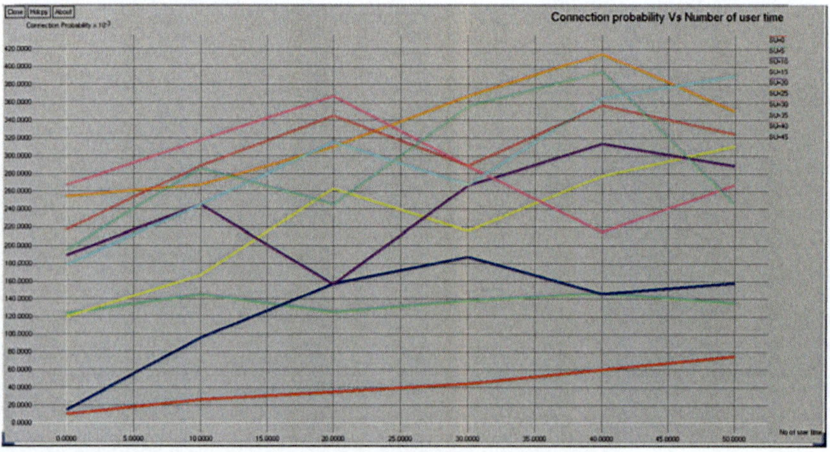

Fig. 3 Number of users versus connection probability

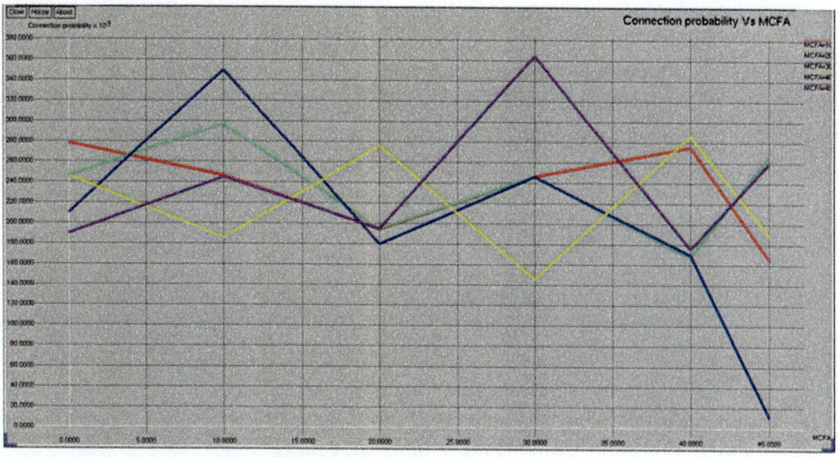

Fig. 4 Number of users versus MCFA

Furthermore, the comparison has been taken to existing systems and it is proved that LACA can achieve more packet delivery ratio [9].

Here, the survey has been taken based on the trendy progression in multichannel wireless mesh networks, focal point on wireless intrusion methods as well as channel allocation algorithms along in the midst of the aim of enhancing the system concert. Also, in together omnidirectional as well as directional antenna systems, the channel distribution algorithms having dissimilar approaches have been reviewed. Finally, there are both merits and demerits in static as well as dynamic channel distribution, and moreover, the proposed method of channel allocation algorithms will be more reliable on the intrusion technology [10].

3 Proposed Methods

Now, this paper discusses the proposed channel distribution scheme in the below division in detail.

3.1 Maximum Network Connectivity

The linking of the cognitive user system will be termed as utmost system connectivity. Consider that the obtainable spectrum is separated into a series of bands, also, these bands are said to be as channels. Every channel has dissimilar bandwidth as well as coverage; moreover, every channel is entirely orthogonal. The divergence as well as intrusion may occur at the time of the similar channel that will be utilized by the numerous users subsequently. The system topology will not vary in excess of time as well as the position of the entire users is stable. A graph coloring problem will be illustrated through the channel assignments. Several dissimilar colors will be utilized to indicate dissimilar channels, for the cognitive wireless systems along with n cognitive users as well as m channels. The channel obtainable matrix, channel intrusion matrix, channel capability matrix as well as channel assignment matrix will be included in the graph coloring model.

$$V = \left\{ v_{n,m} \middle| v_{n,m} \in (0, 1) \right\}_{k*h} \tag{1}$$

where
V—Channels cognitive users can exploit,
n—Cognitive users,
k and h—Number of cognitive and primary users,
m—Licensed channels.

If channel m will be used by cognitive user n, then $v_{n\ m},= 1$. The channel intrusion matrix equation will be given below as

$$I = \left\{ i_{n,z,m} \middle| i_{n,z,m} \in (0, 1) \right\}_{k*k*h} \tag{2}$$

where
I—The interference between the primary as well as cognitive users that utilize the same channel,
n and z—Cognitive users,
m—Licensed channel.

If the interference exists as soon as cognitive users n as well as z utilize the channel m at the same time, then $i_{n^*k^*m} = 1$.

The channel utility matrix can be represented as:

$$U = \left\{ u_{n,m} \right\}_{k*h} \tag{3}$$

where
U—Illustrates about the utility, while channel m will be utilized by cognitive user n effectively.

The channel allocation matrix can be given below in equation as:

$$A = \left\{ a_{n,m} \middle| a_{n,m} \in (0, 1) \right\}_{k*h} \tag{4}$$

where
A—Channels can be assigned to which cognitive users,
k and h—the number of cognitive and primary users,
n—Cognitive users,
m—Licensed channels.

If channel m is allocated to cognitive user n, then $a_{n,m} = 1$. The components of A meet the below limitations:

$$a_{n,m} * a_{n,m} = 1 if i_{n,m} = 0 \tag{5}$$

$N, z < k, m < h$

The most extreme system network is to give increasingly psychological clients a chance to embrace the accessible channel as well as make them ready to participate in sending information bundles. To accomplish the most extreme system availability, this paper expects that the channel utility grid VU is bigger, more components of channel portion framework. A one framework, as well as less components of Channel obstruction network I is one. In the accompanying, these papers are to talk about the issue.

3.2 Channel Allocation Scheme

To evade the intrusion, dissimilar coverage regions will be managed by the cognitive users at the time of utilizing dissimilar channels in the multihop cognitive wireless network. At the time of computing channel obtainable matrix V, there is no determination of distances among primary users as well as cognitive users as compared to prior methodologies, but also acquire into account individuals among cognitive users. The distance among cognitive users will be determined for the channel intrusion matrix I, but also it will be determined the time at the channel is utilized. The distance among primary as well as cognitive users will be regarded by the channel utility matrix U. In addition, the case that essential clients utilize the channel ought to likewise be considered. In the event that there is a bigger separation among essential as well as subjective clients, the comparing component in U holds bigger esteem. On the off chance that essential client does not utilize the channel, this paper let the comparing component in U be greatest esteem. Since the channel accessible grid V of each psychological client in the multibounce subjective remote system is extraordinary, this paper ought to specially allot the channel to intellectual client along with less accessible diverts so as to ensure arrange network.

On the basis of maximal connectivity as well as fairness, a channel distribution algorithm will be introduced to cognitive wireless network users. Here, cooperative multihop path also broken to aid the channel distribution next to considering maximal connectivity in multihop circumstances.

3.3 Channel Allocation for Cognitive Users

Cognitive users will have dissimilar utmost broadcasting distances for the intention of dissimilar channels at dissimilar moments for the purpose of transmitting the information packets. Every cognitive user manages an utmost broadcasting distance matrix in the above case. Based on the above-mentioned matrix, this paper could able to achieve an accessible channel matrix V1. The channel accessible matrix V will be given below in equation as:

$$V = V_1 * V_2 \tag{6}$$

where V_2—another available channel matrix ascertained by the distance among primary as well as cognitive users. Moreover, while figuring the impedance network, this paper consolidates the time, channel, and subjective client. Provided that this is true, this paper can construct the suited obstruction grid that depicts which psychological client has the impedance when utilizing which channel at which minute. At the point when the essential client does not utilize its authorized channel, the psychological client can use the channel by the most extreme transmission separate. Furthermore, in such a case, it additionally holds the bigger utility esteem. Typically, a few channels are not utilized by essential clients in a more extended time. Assuming this is the case let psychological client uses it right off the bat to guarantee organize topology is generally steady. Moreover, while dispensing the channel to psychological clients, it offers need to the intellectual client more remote from the essential client. For this situation, the impedance to the essential client can be additionally diminished. It fabricates the channel utility grid as:

$$U = T_m * n * D \tag{7}$$

where
T_m—The maximum transmission distance of cognitive users,
N—Total number of the time slots that primary users do not develop the channel,
D—The distance among primary as well as cognitive user.

The channels can be successfully distributed in the omnidirectional antenna pattern with respect to the above-mentioned 3 matrixes as well as assignment approaches. There is a collision on primary as well as some other cognitive users in a particular direction at the time of cognitive users developed the directional antenna to broadcast the signals.

3.4 Algorithm

1. Maximum connectivity fairness channel allocation algorithm.
2. Input: N cognitive users as well as M primary users information to get
3. Source hub s along with destination node d
4. Output: channel available matrix V as well as path L
5. % building path L from s and hub s along with destination hub d and
6. % distributing the channel for every hub in path L

7. Procedure:
8. Construct path L from source hub s as well as destination hub d
9. Select channel m
10. Compute utility matrix U
11. Compute obtainable matrix V
12. Analyze interfering matrix I
13. Estimate the least transmission power
14. Distribute channel m to cognitive user c
15. Else
16. Distribute channel m to cognitive user c_j
17. Else
18. Determine utility matrix U
19. Assign channel m to cognitive user c_u
20. End
21. Modernize accessible matrix V
22. If neighbor matrix of channel m is not empty
23. Again back to Line 16 as well as replicate the above procedure
24. End
25. If there survive a few channel m that is not allocated
26. End
27. Save channel allocation result as well as exit.

4 Results and Discussion

In this part, the future system conducts a series of tests as well as simulations to authenticate the proposed algorithm MCFA. Reenactment situations are a square region by 5–50, incorporating 5 essential clients in the midst of various channels, subjective clients' most extreme transmission sweep are 10, 15, as well as 25, individually, and along with the quantities of intellectual clients are 10, 15, and 25. This paper let subjective clients use omnidirectional receiving wires as well as directional radio wires to do reproduction investigation individually. This proposed effort is connected by way of synergistic max-whole transmission capacity rule.

Here, the primary users are represented as PUs as well as cognitive users are as SUs, and furthermore, this is the relation between primary users, cognitive users as well as the possibility of system linking, respectively. Also, if the quantity of primary users turns out to be more, there is an enhancement in the possibility of system connections. The above-mentioned process will be equivalent to cognitive users also. On behalf of MCFA, the more the quantity of psychological clients as well as (or) essential clients is, the more the chance of constructing the association between hubs. This demonstrates the quantity of intellectual clients along with essential clients that have an imperative effect on system availability.

The dynamics of network connectivity changes the amount of primary users in a network. This paper can clearly see that system connectivity turns out to be terrible in excess of time whereas adding the more primary users as well as it can achieve the better network connectivity. This implies that network connectivity embraces the dynamic nature.

Furthermore, evaluate the achievement of MCFA as well as a comparison is also taken along with MCAA. Here, prior to 60 time slot, there is no usage of channels, as well as the channels will able to utilize after time slot 60. The above-mentioned figure shows that there is no usage of channels at the primary users, to develop the utmost broadcasting radius R; it will be found out by the cognitive users who are used to construct the linking between hubs. MCFA as well as MCCA manages the high possibility of linking, and at the time of broadcasting, radius R is smaller which will be sought out by the above-mentioned paper. As time passes, their likelihood of associations step by step diminishes, whereas they acquire the expanding likelihood of associations as soon as utilizing the greater transmission sweep. After availability 70, their likelihood of associations stays stable. Rather than MCAA, MCFA holds the bigger likelihood of system associations. This shows that MCFA holds the better execution.

5 Conclusion

To enhance the cognitive wireless network achievement intended for medical applications, an utmost connectivity channel distribution algorithm has been proposed. By utilizing omnidirectional as well as directional antennas, the introduced algorithm will be discussed as well as determined in the above-mentioned paper. The need portion of channels to bottleneck hub is performed here. By distributing the suitable channel to each intellectual client, the proposed calculation can adequately keep the most extreme availability for psychological remote systems. Reproduction results demonstrate that this proposed methodology is compelling as well as promising. In this proposed future work, it will additionally investigate the constant property of this proposed methodology, along with this paper applies the above-mentioned proposed channel distribution plan to the vast scale subjective remote systems to approve the execution of this proposed methodology.

References

1. Wang, F., et al.: Joint scheduling and resource allocation for device-to-device underlay communication. In: Wireless Communications and Networking Conference (WCNC), 2013 IEEE (2013)
2. Huang, X., Yuan, T., Zhang, Y.: Utility-optimized bandwidth and power allocation for non-orthogonal multiple access in software defined 5G networks. J. Netw. Comput. Appl. **113**, 75–86 (2018)
3. Li, W., et al.: Energy efficiency maximization oriented resource allocation in 5G ultra-dense network: Centralized and distributed algorithms. Comput. Commun. **130**, 10–19 (2018)

4. Saini, R., Mishra, D., De, S.: Subcarrier pairing as channel gain tailoring: Joint resource allocation for relay-assisted secure OFDMA with untrusted users. Phys. Commun. **32**, 217–230 (2019)
5. Khan, I., Singh, D.: Efficient compressive sensing based sparse channel estimation for 5G massive MIMO systems. AEU-Int. J. Electron. Commun. **89**, 181–190 (2018)
6. Gao, W., Ma, L., Chuai, G.: Joint optimization of component carrier selection and resource allocation in 5G carrier aggregation system. Phys. Commun. **25**, 293–297 (2017)
7. Amuthan, A., et al.: Hyper-Erlang channel allocation factor-based QoS enhancement mechanism for mobile ad hoc networks. Alexandria Eng. J. **57**(2), 799–811 (2018)
8. Xu, J., Guo, C., Zhang, H.: Joint channel allocation and power control based on PSO for cellular networks with D2D communications. Comput. Netw. **133**, 104–119 (2018)
9. Gao, W., et al.: Link quality aware channel allocation for multichannel body sensor networks. Pervasive Mobile Comput. **41**, 354–364 (2017)
10. Ding, Y., Xiao, L.: Channel allocation in multi-channel wireless mesh networks. Comput. Commun. **34**(7), 803–815 (2011)

Analysis of Brushless DC Motor Using Deep Neural Network and BAT Algorithm

K. Balamurugan and R. Mahalakshmi

Abstract Brushless DC (BLDC) motors are specially used in actuation process control, automation industry, spacecraft industry, and military appliances. The model specifications and efficiency parameter depending upon its analysis and design of the BLDC motor are necessary to collect consistent data and significant information. The parameter identification is derived by mathematical analysis via optimization techniques using the Deep Neural Network (DNN) and BAT algorithm. The torque, voltage, current, speed, and temperature of Brushless DC motor are analyzed using sensors with PID controller. The analyzed data is to monitor and control through the Internet of things (IOT) using GPRS.

Keywords Brushless DC motor · Internet of things (IoT) · BAT algorithm · Deep neural network · PID controller

1 Introduction

The monitoring and control of Brushless DC motor drive can be measured in sensorless method, but to scale down the total cost of energizing devices, sensorless methods are used. The merit of sensorless mode of Brushless DC motor is that sensing components and the device can be eliminated, and thus the cost of overall machine drive can be minimized. The demerits of this particular type of sensorless method are most requirements for optimization control algorithms. [1–3]. The Brushless DC motor

K. Balamurugan (✉)
Department of Electrical and Electronics Engineering, Sri Ramakrishna Engineering College, Coimbatore, India
e-mail: kb.sreceee@gmail.com

R. Mahalakshmi
Department of Electrical and Electronics Engineering, Sri Ramakrishna Institute of Technology, Coimbatore, India

P. Suresh et al. (eds.), *Advances in Smart System Technologies*,
Advances in Intelligent Systems and Computing 1163,
https://doi.org/10.1007/978-981-15-5029-4_5

drive has back EMF trapezoidal voltage and it is commonly agitated by 3Φ sinusoidal current, and consistently mechanized by currents having a square waveform [4–7].

In this paper, speed control and Simulink modeling of Brushless DC motor are presented and the parameter evaluation is done. The primitive strategy of parameter model estimation might be to measure optimal data for a Simulink model which expects reliant variable results or occurrence determined by independent variable parameter inputs. For a certain variable observation, independent variable parameter inputs are classified into input vector and dependent variable parameter outputs are classified into output vector [8]. The optimization techniques like BAT optimization algorithm and Deep Neural Network (DNN) methods are designed for the estimation and control parameters of the Brushless DC motor using Internet of Things.

As a new emerging technology said about recent advances in cloud computing, big data analysis, Internet of Things (IoT), modern wireless or sensorless communication has created more attention to a wide number of industrial applications. Internet of Things (IoT) is giving aid to attain the automation process through remote operation. It will help to increase productivity by communicating with remotely operating devices [9].

The theoretical analysis of the Brushless DC motor consisting of the mechanical and electrical system is discussed in the second chapter, to be followed by the estimation of the parameter in the third chapter. The parameter estimation, as quoted earlier comprises the recent methodologies viz Deep Neural Network based PID controllers and BAT algorithm analysis. The fourth and fifth chapters comprise of Simulink model, Experimental results, and conclusion, respectively.

2 Mathematical Analysis of Brushless DC Motor

Figure 1 shows that the mathematical model of BLDC motor drive. Some of the

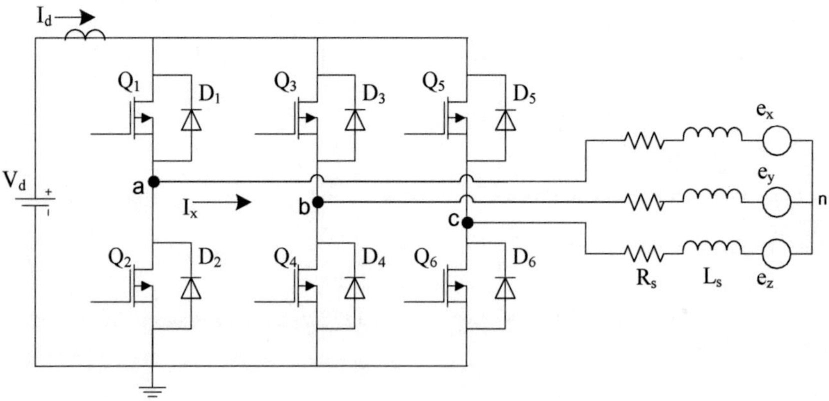

Fig. 1 Analysis circuit of Brushless DC motor

assumptions are discussed to reduce the analytical issues in DC motor are uniform air-gap, armature reaction, no magnetic saturation, eddy current losses, no hysteresis, symmetrical three-phase winding is ignored and mutual inductance is also eliminated. The mathematical analysis of armature winding is

$$V_x = i_x R + L\frac{\mathrm{d}i_x}{\mathrm{d}t} + e_x \tag{1}$$

$$V_y = i_y R + L\frac{\mathrm{d}i_y}{\mathrm{d}t} + e_y \tag{2}$$

$$V_z = i_z R + L\frac{\mathrm{d}i_z}{\mathrm{d}t} + e_z \tag{3}$$

where
V_x, V_y, V_z—phase voltage in volts
i_x, i_y, i_z—Stator phase current in ampere
e_x, e_y, e_z—back EMF in volts
L—Self-inductance per phase of armature in Henry
R—Resistance per phase of armature in Ohms.

The back emf equation are written by 120 electrical degrees from one phase to another, and it can be written as

$$e_x = K_{\mathrm{emf}} f(\theta_{ea})\omega_m \tag{4}$$

$$e_y = K_{\mathrm{emf}} f\left(\theta_{ea} - \frac{2\pi}{3}\right)\omega_m \tag{5}$$

$$e_z = K_{\mathrm{emf}} f\left(\theta_{ea} + \frac{2\pi}{3}\right)\omega_m \tag{6}$$

where
ω_m—rotor speed in rad/s
K_{emf}—constant back emf in V/rad/s
f—frequency in Hz
θ_{ea}— rotor electrical angle in rad.

Subtract the Eq. (2) from Eqs. (1) and (3) from Eq. (2)

$$V_{xy} = R(i_x - i_y) + L(\frac{\mathrm{d}i_x}{\mathrm{d}t} - \frac{\mathrm{d}i_y}{\mathrm{d}t}) + (e_x - e_y) \tag{7}$$

$$V_{yz} = (i_y - i_z) + L(\frac{\mathrm{d}i_y}{\mathrm{d}t} - \frac{\mathrm{d}i_z}{\mathrm{d}t}) + (e_y - e_z) \tag{8}$$

For a three-phase winding, by using Kirchhoff's (KCL) Current Law, the total phase current is zero, and the equation can be expressed as

$$i_x + i_y + i_z = 0 \tag{9}$$

Substituting i_c from (9) to (8) gives

$$V_{yz} = R(i_x + 2i_y) + L(\frac{di_x}{dt} + 2\frac{di_y}{dt}) + (e_y - e_z) \tag{10}$$

The torque equation of Brushless DC motor drive can be given as

$$T_{em} = \frac{e_x i_x + e_y i_y + e_z i_z}{\omega_m} \tag{11}$$

or

$$T_{em} = K_t[f(\theta_e)i_a + f(\theta_e - \frac{2\pi}{3})i_b + f(\theta_e + \frac{2\pi}{3})i_c] \tag{12}$$

where
T_{em}—Torque in Nm
K_t—torque constant in Nm/A.

Under steady-state condition, electromagnetic torque (T_{em}) equation can be expressed in terms of friction torque, load torque and inertia torque.

$$T_{em} = T_L + J\frac{d\omega_m}{dt} + \beta\omega_m \tag{13}$$

where
T_L—load torque in Nm
J—moment of inertia in kgm^2
β—rotor friction in Nm s/rad.

From the above Eqs. 7, 10, and 13 can be converted into Laplace transformation, which can be written in terms of state-space expression provides (MIMO) a multiple-input–multiple-output. State-space expression as

$$\begin{bmatrix} i_x \\ i_y \\ i_z \end{bmatrix} = \begin{bmatrix} -\frac{R}{L} & 0 & 0 \\ 0 & -\frac{R}{L} & 0 \\ 0 & 0 & -\frac{\beta}{J} \end{bmatrix} \begin{bmatrix} i_x \\ i_y \\ \omega_m \end{bmatrix} + \begin{bmatrix} \frac{2}{3L} & \frac{1}{3L} & 0 \\ -\frac{1}{3L} & \frac{1}{3L} & 0 \\ 0 & 0 & \frac{1}{J} \end{bmatrix} \begin{bmatrix} V_{xy} - e_{xy} \\ V_{yz} - e_{yz} \\ T_{em} - T_L \end{bmatrix} \tag{14}$$

$$\begin{bmatrix} i_x \\ i_y \\ i_z \end{bmatrix} = \begin{bmatrix} 1\,0\,0 \\ 0\,1\,0 \\ 0\,0\,1 \end{bmatrix} \begin{bmatrix} i_x \\ i_y \\ \omega_m \end{bmatrix} \tag{15}$$

3 Implementation of BAT and DNN Technique Using IoT

Figure 2 shows the parameter estimation assign to the parameters of a system to an input. Thus, in this process, the analysis of the machine parameter and the mathematical model is of very importance. The issue of torque ripple estimation applies to the type of "reverse problems" in which the familiarity of the dynamic system is calculated from the input and output system.

Mathematical model Simulink of parameter analysis is to understand the system's characteristics equation. These estimation parameters describe the speed control and stability behavior of the system. The BAT algorithm technique is used for the elimination of torque, speed parameters as shown in (Fig. 3).

The interfacing architecture is shown in Fig. 4 classified into two parts: one on the processor running Linux—streaming data onto the IoT platform and another running on equipment processor—collecting various data from the sensors. The interfacing architecture of BLDC motor supports multi-axis application running multi-drive on the constant device. On the electrical device, it can be used to interface to control the

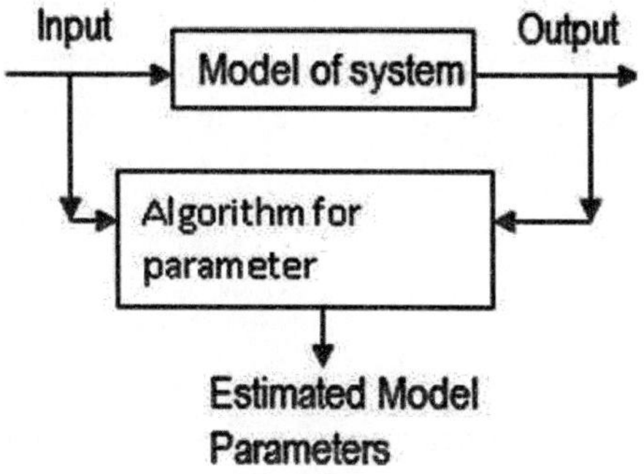

Fig. 2 Block diagram of BLDC estimation parameters

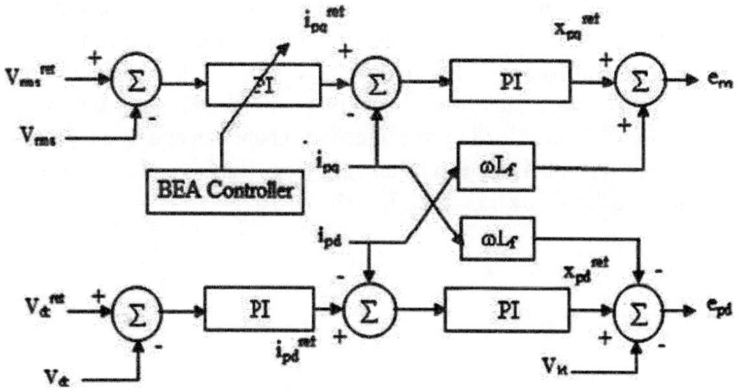

Fig. 3 Parameter estimation of BAT algorithm

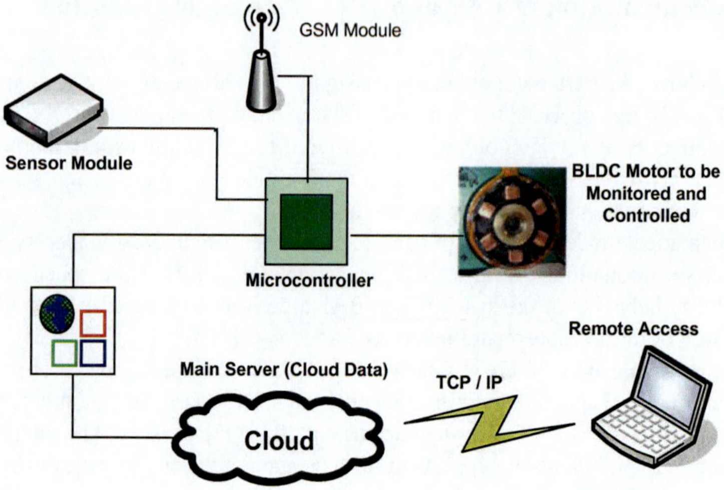

Fig. 4 Interfacing architecture of Brushless DC motor using Internet of Things(IoT)

software. For the prototype module, Internet of Things (IoT) interface with control application to Initial State platform with the help of Raspberry PI interface with UART port module of the controller.

4 Simulation Result

The Simulink model for deep neural network analysis optimization is shown in (Fig. 5).

The Simulink model for BAT algorithm optimization is shown in (Fig. 6).

The BLDC motor is simulated as shown in the figure and different values speed and torque are simulated with two different optimization techniques with MATLAB/SIMULINK using trial and error method. Then, the optimization technique of Deep Neural Network and BAT Echo optimization techniques are simulated and executed to estimate the parameters of the BLDC motor with the inputs as electric torque, voltages, load torque, currents, change in speed and current with respect to time in second. The Deep Neural Network (DNN) and BAT optimization of torque and speed are shown in Figs. 7, 8, 9, and 10.

5 Conclusion

In this proposed article, the parameter identification methods are implemented to the BLDC motor model. With these estimated parameters, torque and speed of the

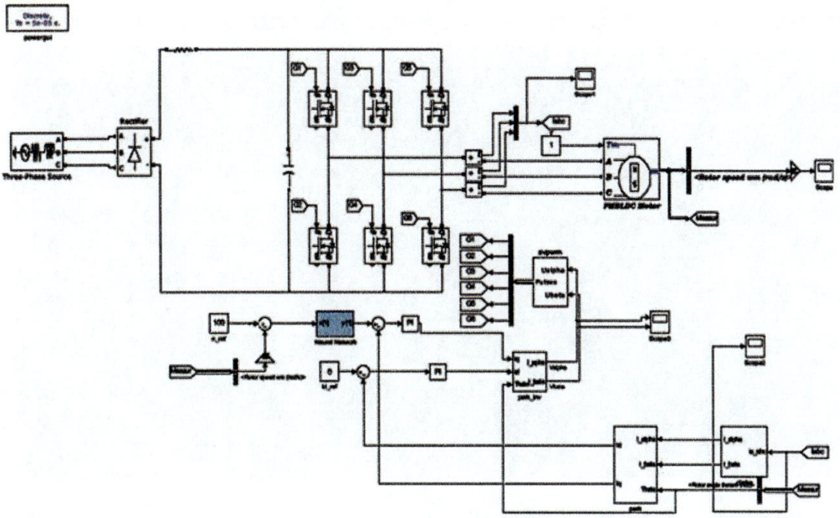

Fig. 5 Simulink model for deep neural network analysis

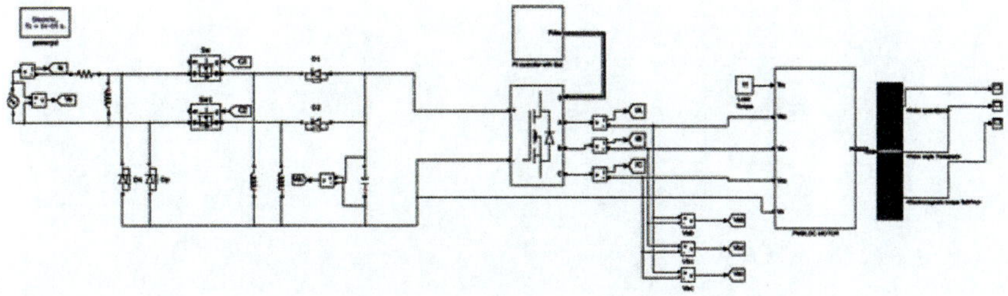

Fig. 6 Simulink model for BAT algorithm optimization

Brushless DC motor can be varied with less settling time as shown in the results. Therefore, control of the DC motor drive with the measured parameters is more efficient than with the parameters determined with the trial and error method. The algorithms also give the parameters in less time with more accuracy as compared to the conventional methods. The analysis, design, and implementation of Internet of Things (IoT) for controlling and monitoring of BLDC motor in automation industries with help of optimization technique. The main idea of this proposed model is to give controllable and remote area connectivity between the user and industrial environment.

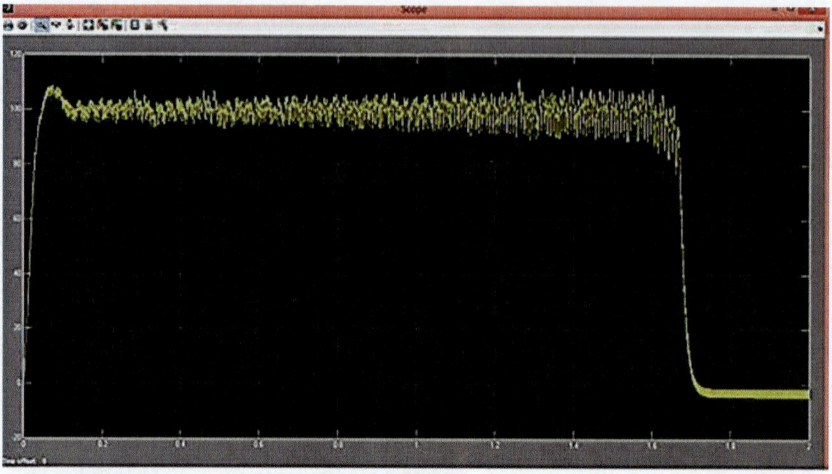

Fig. 7 Motor speed for deep neural network analysis

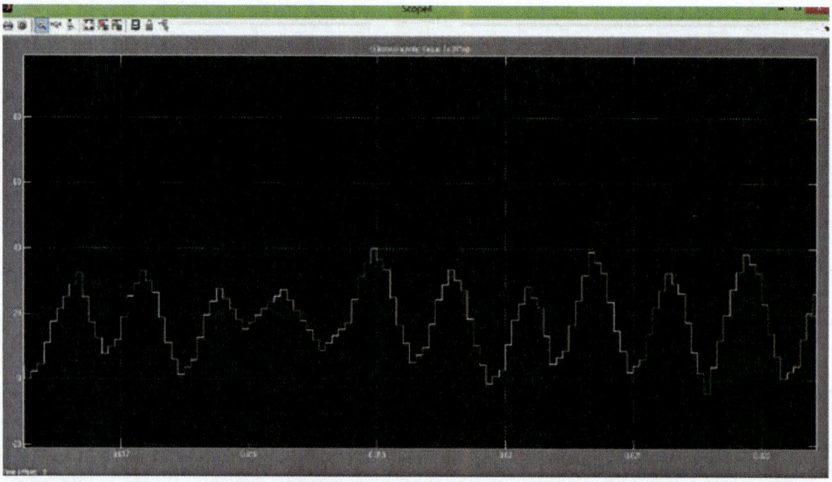

Fig. 8 Motor torque for deep neural network analysis

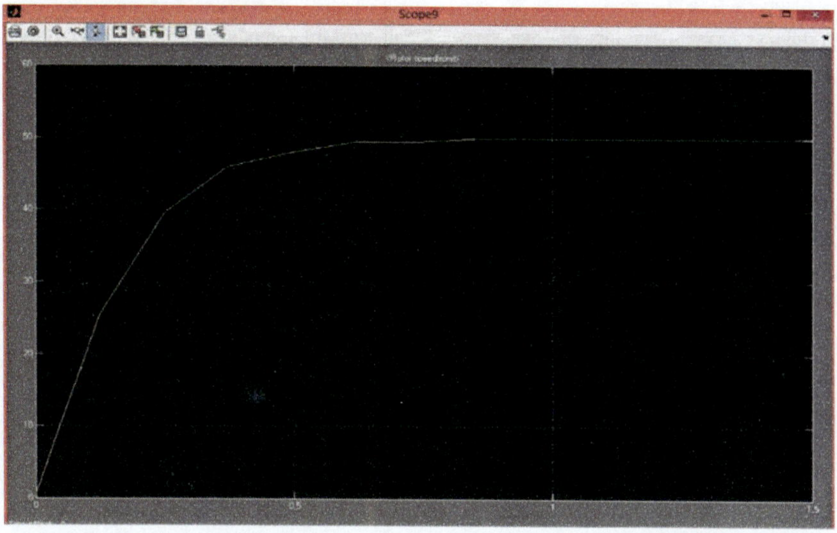

Fig. 9 Motor speed for BAT algorithm

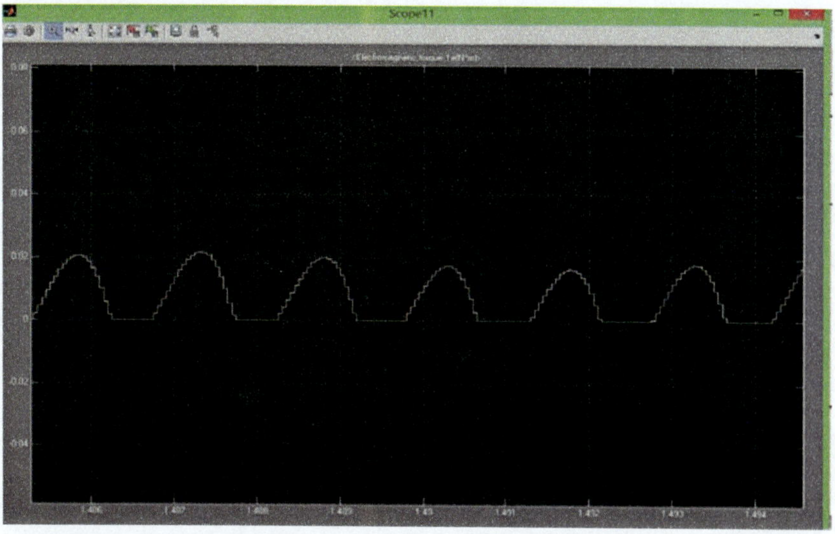

Fig. 10 Motor torque for BAT algorithm

References

1. Iizuka, K., Uzuhashi, H., Kano, M., Endo, T., Mohri, K.: Microcomputer control for sensorless brushless motor. IEEE Trans. Ind. Appl. **IA**(21), 595–601 (1985)
2. Becerra, R.C., Ehsani, M.: High-speed torque control of brushless permanent magnet motors. IEEE Trans. Ind. Electron. **1**(35), 402–406 (1988)

3. Hubik, V., Sveda, M., Singule, V.: On the development of BLDC motor control run-up algorithms for aerospace application. In: Proceedings of the 13th Power Electronics and Motion Control Conference (EPE-PEMC 2008), Poznan, Poland, pp. 1620–1624 (2008)
4. Bonfe, M., Bergo, M.A.: Brushless motor drive with sensorless control for commercial vehicle hydraulic pumps. In: Proceedings of the IEEE International Symposium on Industrial Electronics (ISIE 2008), Cambridge, England, pp. 612–617 (2008)
5. Bianchi, N., Bolognani, S., Jang, J.H., Sul, S.K.: Comparison of PM motor structures and sensorless control techniques for zero-speed rotor position detection. IEEE Trans. Power Electron., 2466–2475 (2007)
6. Su, G.J., McKeever, J.W.: Low-cost sensorless control of brushless DC motors with improved speed range. IEEE Trans. Power Electron. **19**, 296–302 (2004)
7. Damodharan, P., Vasudevan, K.: Indirect back-EMF zero crossing detection for sensorless BLDC motor operation. In: Proceedings of the International Conference on Power Electronics and Drives Systems (PEDS 2005), Kuala Lumpur, Malaysia, pp. 1107–1111 (2008)
8. Premkumar, K., Manikandan, B.V.: Bat algorithm optimized fuzzy PD based speed controller for brushless direct current motor. Eng. Sci. Technol. Int. J. **19**, 818–840 (2016)
9. Thompson, H.A.: Wireless and Internet communications technologies for monitoring and control. Control Eng. Pract. **12**, 781 (2004)

Performance Comparison of Various Machine Learning Algorithms for Ultrasonic Fetal Image Classification Problem

N. Sathish Kumar, M. Kasiselvanathan, and S. P. Vimal

Abstract The machine learning uses statistical algorithms to provide the ability to learn with and without data programming. Image classifier is used to categorize a subject or an object present in an image into predefined classes. A statistical algorithm used in numerous fields like pattern classification, regression, control systems, identification, and prediction. This research paper presents a novel approach performance comparison of machine learning classifiers used for the classification of ultrasonic fetal images. It presents Gabor feature extraction and various classification techniques to classify three different classes of ultrasound images. The proposed method is presented as follows. Initially, the Gabor features are obtained from raw images. To remove redundancy in features and dimensionality reduction principal component analysis (PCA) is applied. Finally, the features obtained from PCA are fed to the various machine learning classifiers and its performances are evaluated. The simulations results are carried out using MATLAB image processing toolbox. From the results, it is observed that decision tree (DT) algorithm and multi-layer perceptron (MLP) perform to be closer and this classifier outperforms all other classifiers.

Keywords Machine learning · Principal component analysis (PCA) · Decision tree classifier (DT) · Classifier · Multi-layer perceptron (MLP)

N. Sathish Kumar (✉) · M. Kasiselvanathan · S. P. Vimal
Department of Electronics and Communication Engineering, Sri Ramakrishna Engineering College, Coimbatore, Tamilnadu 641022, India
e-mail: nsk20022002@gmail.com

M. Kasiselvanathan
e-mail: kasiselvanathan@gmail.com

S. P. Vimal
e-mail: srecdoc@gmail.com

P. Suresh et al. (eds.), *Advances in Smart System Technologies*, Advances in Intelligent Systems and Computing 1163,
https://doi.org/10.1007/978-981-15-5029-4_6

1 Introduction

Machine learning algorithm is for identifying patterns that can be applied to real-world applications. It is very useful tool that can help in interpretation of medical images [1]. Machine learning uses machine learning algorithm which used to compute the image features that are supposed to be of importance in making the prediction or final conclusion regarding the diagnosis of disease. Multi-modal biometric recognition systems have greater advantages that provide good anti-spoofing abilities. Gaikwad proposed a system that uses contourlet transform for analyzing the features present in palm print and palm vein images [2]. Feature detection may be used to find the image information and provide a local decision at every image point to check whether there is an image feature of the given type presented in that point. This feature detection technique provides robust to image transformations [3, 4]. A near-infrared (NIR) imaging of Palm Dorsa Subcutaneous Vein Pattern(PDSVP) system has been introduced for data acquisition, PDSVP extraction [5].

The machine learning algorithm identifies the most excellent grouping of these image features for classifying the image or computing some metric for the given images. It is measured a branch of artificial intelligence (AI) since it provides the extraction of meaningful signal components from the image. The process of assigning objects to one of several predefined types is known as classification. The classification model has set of input attributes '*x*' and output labels '*y*'. The classification of a given task of mapping is as shown in Fig. 1.

2 Image Classifier Algorithms

Image classification is defined as the process of extracting information classes from a multi-band image [6]. Figure 2 shows the flowchart of image classifier. It is broadly classified in to two types, namely structural and statistical. Specific classifiers are considered in this paper, namely DT classifier, k-nearest neighbor (kNN) classifier, logical regression (LR), MLP, linear discriminant analysis (LDA) and Naive Bayes (NB) classifier [7].

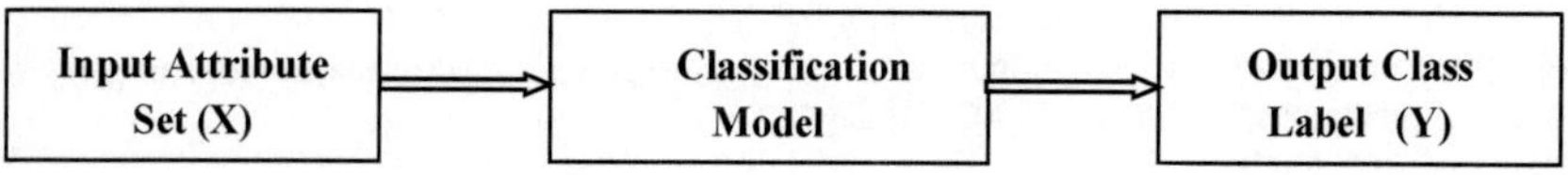

Fig. 1 Classification model

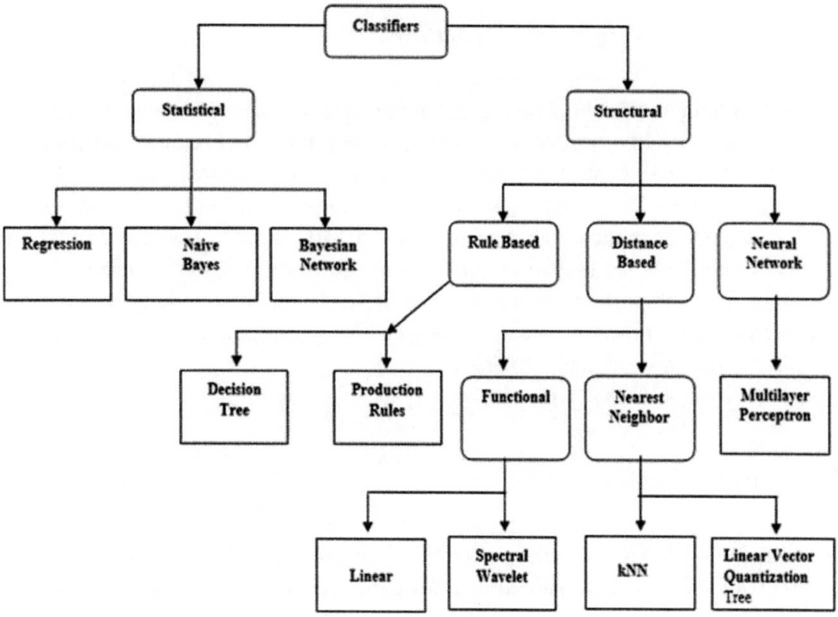

Fig. 2 Classification of classifiers

2.1 Decision Trees in Machine Learning

DT classifier can be used to observe and to clearly represent decisions and decision making. DT uses a tree-like model of decisions and supervised learning algorithm for classification problems. DT is related equally to continuous, categorical output and input variables [8]. A tree has three nodes named as root node, internal node, and leaf node. A root has more outgoing edges and has no incoming edges. An internal node has more than two outgoing edges and has one incoming edge. Leaf node has no outgoing edges and has one incoming edge.

A decision tree is made with its root at the top. Actual dataset will have an enormous features and this has a branch in a much bigger tree. DT algorithm is also called as learning decision tree from data. Regression trees (RT) are also predicted continuous values. Generally, DT algorithms are also called as 'CART' and RT. This algorithm can handle both numerical and unconditional data and also multi-output cases. It requires little effort from users for data preparation. They also have disadvantages such as instability when small variation in data, it can also create complex tress resulting in overfitting.

2.2 k-Nearest Neighbors Algorithm

kNN algorithm is a classification algorithm in machine learning and is used in many other applications like intrusion detection, pattern recognition, and data mining. It is a simplest classification algorithm and is also known as non-parametric, lazy learning algorithm. The data points are spitted into several classes to endow with a new sample points in the database. It can be used for classification and regression predictive problems for industrial applications. To assess the performance of this algorithm, three important aspects are considered, namely simplicity to interpret output, calculation time, and predictive power. The given training data classifies the coordinates which is divided into groups.

2.3 Discriminant Analysis

The two classic classifiers namely quadratic discriminant analysis (QDA) and linear discriminant analysis (LDA) are discussed. These classic classifiers provide closed-form solutions which can be computed easily. QDA is more flexible since it can learn quadratic boundaries. Both these classifiers can be derived from simple probabilistic models. In LDA, it is assumed that all classes have the equivalent estimated covariance. Rescaling of the data is given in Eq. (1).

$$X^{*} = D^{-1/2}U^{t}X \text{ with } \Sigma = UDU^{t} \tag{1}$$

It can be noted that a data point after scaling is equivalent to finding the estimated class mean $\mu k*$ which is nearer to the data point in the Euclidean distance. It is done after projecting on the $K-1$ subspace generated for all classes. From this, it is confirmed that by linear projection, LDA can provide a dimensionality reduction onto a $K-1$ dimensional space. The variance of the $\mu k*$ after projection is maximized because dimensions are still further reduced by projecting onto the linear subspace.

2.4 Logistic Regression (LR)

LR classification algorithm is used to predict the probability of a categorical dependent variable. In LR, the dependent variable contains data coded as 1 (True) or 0 (False) and LR model predicts $P(Y{=}1)$ as a function of X. LR is a predictive algorithm that is used when the 'Y' variable is binary categorical. It takes only two values namely '1' or '0'. The main goal is to determine a mathematical equation that can be used to predict the probability of event 1. The following assumptions are made in LR model. Binary LR requires the dependent variable to be binary in which the factor level 1 of the dependent variable must describe the desired outcome. In LR model,

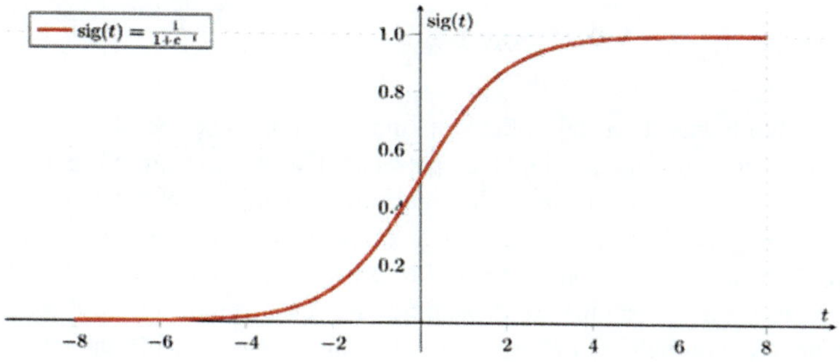

Fig. 3 Sigmoid function

the predicted output is the probability that the input sample belongs to a targeted class which is digit '1.'The LR hypothesis model is described in Eqs. (2) and (3).

$$P(y = 1|x) = H_w(x) = \left[1/1 + \exp\left(-W^T x\right)\right] = \text{Sigmoid}\left(W^T x\right) \quad (2)$$

$$P(y = 0|x) = 1 - p(y - 1|x) = 1 - H_w(x) \quad (3)$$

The sigmoid function transforms the predicted output into probability space in which the values are in range [0, 1]. Figure 3 shows the sigmoid function.

2.5 *Multi-layer Perceptron*

MLP algorithm is a feed-forward network called as artificial neural network (ANN) model that is used to map the sets of input data onto a set of appropriate outputs [9]. A MLP has three layers named as a hidden layer, an input layer, and an output layer. All the nodes use a nonlinear activation function except input nodes.

The structure of MLPis shown in Fig. 4. The parameters such as the number of input layer features, number of training iterations, number of hidden layer array, learning rate are to be focused in the design of MLP. MLP is widely used in nonlinear models.

Fig. 4 MLP structure

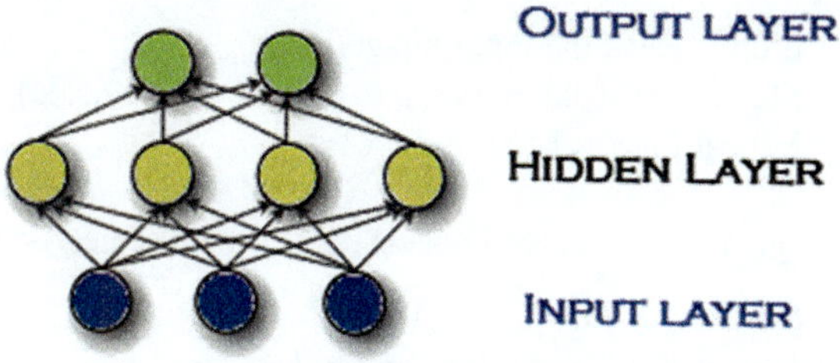

2.6 Naive Bayes (NB) Algorithm

NB algorithm is based on Bayes theorem and has an assumption of independence among predictors. It is assumed that the presence of a special feature is unrelated to other features presented in a class. NB algorithm provides better performance with respect to categorical input variables when compared to numerical variables. Let us consider an example, an apple if it is red or green, round, and about 3.5 inches in diameter. Even these features are depending on each other, all of these properties contribute to the probability independently that this fruit is an apple and that is why it is known as 'Naïve.' Bayes theorem is used to calculate the posterior probability $P(a|b)$ from $P(a)$, $P(b)$ and $P(b|a)$. Mathematically, it can be modeled as in Eq. (4).

Posterior Probability = (likely hood probability). (Class Prior Probability)/Predictor Prior probability.

$$P(a|b) = P(a|b)\,P(a)\,/\,P(b) \tag{4}$$

where,
$P(a|b) = P(b_1|a).\ P(b_2|a).\ P(b_n|a).P(a)$
$P(a|b)$ represents the posterior probability of class
$P(a)$ represents the prior probability of class
$P(b|a)$) represents the likelihood which is the probability of predictor given class
$P(b)$) represents the prior probability of predictor.

2.7 Proposed Method

The proposed method contains Gabor features extraction and the outputs are given to the PCA block. The machine learning classifier classifies and labels the output in different classes. PCA is used to find a low dimension set of axes that summarize dataset composed by a set of properties. PCA will select features from the dataset and constructs new set of properties based on combination of the previous set. This classifier performs a linear transformation and moves the original features to a new features collected by principal component. These new features are algebraic. The flowchart of the proposed method is as shown in Fig. 5.

GABOR–PCA Algorithm

Step 1: Read the Input imageImage.
Step 2: Set scale = [0.3, 0.6, 0.9]; m = size(scale)
Repeat for $j = 1$ to m.

Image resize = image Imageis resized with scale(j).

Set Ø= [0, 45, 90, 135, 180]
k = size(Ø) Repeat for $i = 1$ to k.

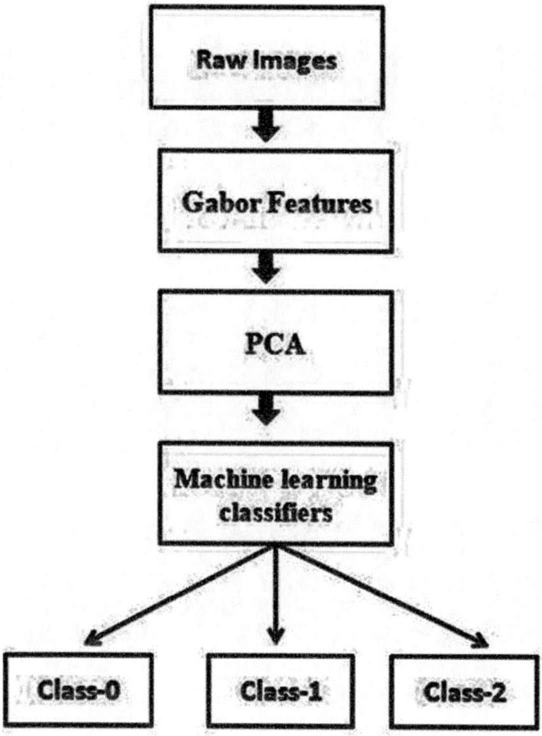

Fig. 5 Flowchart of the proposed method

Apply Gabor transform with resized image Imageresize and angle thet $k(i)$.end loop

First Level Optimization: Positional features of Gabor matrices are taken to calculate the arithmetic average. It is a reduced matrix by factor of 5.

Second Level Optimization: Then PCA is applied for on each average Gabor feature matrix and eigenvectors are extracted for the corresponding significant eigenvalues. End loop.

3 Results and Discussion

The machine learning is used statistical algorithms to provide the ability to learn with and without data programming. Image classifier is used to categorize a subject or an object present in an image into predefined classes. The proposed method and existing methods are evaluated using machine learning and its performance is analyzed. 540 ultrasound images are considered as the dataset with features. The algorithms use fetal images. Simulations results are carried out using MATLAB image processing

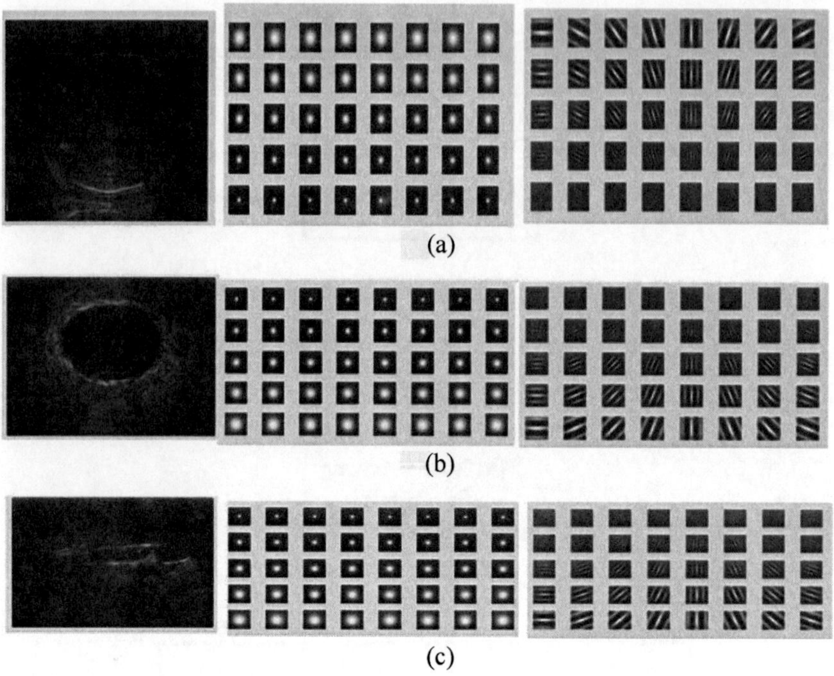

Fig. 6 a–c Feature extraction processes

toolbox. The feature extraction process and related outputs are shown in Fig. 6.From the results, it is observed that DT algorithm and MLP algorithm perform to be closer and this classifier outperforms all other classifiers. Table 1 shows the performance comparison of metric of various Classifiers and it is noted that the accuracy, precision, recall, and F1 score of DT algorithm are superior compared to other algorithms that are considered in this paper.

Table 1 Comparison of performance metric of various classifiers

Classifier	Accuracy	Precision	Recall	F1 Score
Proposed method (Gabor–PCA)	70.55	0.75	0.75	0.75
DT	69.55	0.695	0.695	0.69
MLP	68.5	0.68	0.68	0.68
kNN	45.33	0.45	0.43	0.45
LDA	37.4	0.37	0.37	0.37
LR	35.0	0.35	0.34	0.35
NB	35.4	0.35	0.35	0.35

4 Conclusion

There has been tremendous progress in machine learning technology since this algorithm was first invented and tested. Machine learning algorithms are used in various applications and providing solutions to real-world problems and are benefited through the learning process. The proposed method provides significantly better performance than the existing algorithms. The performance metric has been significantly improved by increasing the number of iterations, particulars of more features, and other related parameters. From the simulation results, DT algorithm and MLP algorithm perform to be closer and this classifier outperforms all other classifiers. Convolution neural networks can also provide better performance. The proposed method accuracy is higher with 70.55% compared to the classical methods. Hence, more hybrid algorithms in combination with Gabor–PCA will yield high accuracy. Hybrid algorithms like neuro-fuzzy can also yield superior results compared to the existing classical algorithms.

References

1. Petrov, D., Gutman, B.A., (Julie) Yu, S.-H., Alpert, K., et al.: Machine learning for large scale quantity of 3D shape models in neuro imaging. In: 8th International Workshop Machine Learning in Medical Imaging, MLMI 2017, vol. 10541, pp. 371–378. LNCS Springer publications (2017)
2. Gaikwad, D.P., Narote, S.P.: Multi-modal biometric system using palm print and palm vein features. Annual IEEE India Conference (INDICON) (2014). https://doi.org/10.1109/INDCON.2013.6726010
3. Moghaddam, B., Nastar, C., Pentland, A.: A Bayesian similarity measure for deformable image matching. Image Vision Comput. **19**(5), 235–244 (2001)
4. Shan, B.: A novel image correlation matching approach. J. Multimedia **5**(3), 268–275 (2010)
5. Joardar, S., Chatterjee, A., Rakshit, A.: Real-time NIR imaging of palm dorsa subcutaneous vein pattern based biometrics: an SRC based approach. IEEE Instrumentation Meas. Mag., 13–19 (2016)
6. Laimeche, L., Farida Merouani, H., Mazouzi, S.: A New Feature Extraction Scheme in Wavelet Transform for Stego Image Classification. Springer Evolving Systems, pp. 1–14 (2017)
7. Chapelle, O., Sindhwani, V., Keerthi, S.S.: Optimization techniques for semi- supervised support vector machines. J. Mach. Learn. Res. **9**, 203–233 (2013)
8. Moradi, M., Syeda-Mahmood, T., Hor, S.: Tree-based transforms for privileged learning. In: Proceeding of International Workshop on Machine Learning in Medical Imaging MLMI-2016, pp. 188–195 (2016)
9. Bianchini, M., Scarselli, F.: On the complexity of neural network classifiers: a comparison between shallow and deep architectures. IEEE Trans. Neural Netw. Learn. Syst. **25**(8) (2014)

A Three-Pronged Approach to Mitigate Web Attacks

Ganga Rama Koteswara Rao and R. Satya Prasad

Abstract In recent days, one of the popular Web servers used by the majority of Web sites in the globe is Apache. Furthermore, lack of proper tuning of these Web servers causes delay in the network by 40% approximately. The present digital world depends upon Web applications to deliver its functionality. Economies, governments, organizations, and ultimately people rely heavily on the performance of the Web applications. Hence, in real time, the practical problem arises in tuning of the Web servers. The Web servers and poor coding of Web applications are the source for the vulnerabilities in the Web. In this, paper focuses on defending the vulnerabilities in Web applications deployed on Apache Web server through protecting the HTTP header and implementing the secured configuration of Apache Web server and ModSecurity Web application firewall.

Keywords Vulnerabilities · ModSecurity · Apache · Web attacks · Web browser · Misconfiguration

1 Introduction

A huge number of Web sites in the world use Apache Web server. Configuring the Apache Web server is a critical task, and it is a major challenging task in the field of cyber security. The protected communication is provided on World Wide Web using HTTPS protocol [1]. The main objective of HTTPS is to provide validation of the Web site visited by the user, shielding the privacy and providing there liability of information exchange between the client and the server. The information exchange

G. R. K. Rao (✉)
Department of Computer Science & Engineering, Koneru Lakshmaiah Education Foundation, Vaddeswaram, Vijayawada, India
e-mail: drgrrao@kluniversity.in

R. Satya Prasad
Department of Computer Science, Acharya Nagarjuna University, Nagarjuna Nagar, India
e-mail: profrsp@gmail.com

P. Suresh et al. (eds.), *Advances in Smart System Technologies*,
Advances in Intelligent Systems and Computing 1163,
https://doi.org/10.1007/978-981-15-5029-4_7

is encrypted both at from-end and to-end to prevent from man-in-middle attacks, tampering, and eavesdropping.

Web sites typically use at least a single database for their functionality. The databases of the Web site can be easily accessed by the unauthorized user due to vulnerabilities in Web applications. SSL and its successor TLS provide encrypted association between the browser and the Web server. SSL guarantees the information exchange among Web server, and browser remains encrypted [2]. The most dangerous Web security attacks are XSS, SQLI, CSRF, command inoculation, password cracking, security misconfiguration, and exposure of sensitive data. The following are the risk areas for Web server safety.

- Vulnerabilities in different versions of the Web servers
- Configurations of Web servers
- Unencrypted message transmission between Web server and browser
- Vulnerable Web sites hosted on the Web server
- Hackers may change or damage the information hosted on the Web server
- Unauthorized accessing of resources of the Web server.

2 Literature Survey

Heshman Abusaimeh and Mohammad Shkoukani focused on discussing the safety of networking practices. They studied on development of the safety of the high speed Internet.

Merve Bas Seyyar et al. proposed a technique to identify the assault-oriented inspection and to differentiate them from additional kinds of visits. The assaults are detected depending on the information presented at the log data of the Web servers such as Apache or IIS. The rules are defined to acknowledge the Web attacks such as XSS and SQL injection.

Fernando Roman Munoz et al. focused on sensing the Web vulnerabilities on the Web sites and fixed them on the Web sites related to banking, media, etc. The latest Web applications are examined to identify the novel additional susceptibilities and observed that the penetration testing tools cannot examine all types of Web application susceptibilities.

Yao-Wen Huang et al., examined the weaknesses in the Web application that helps the hacker to perform Web attacks due to poor coding and suggested the testing techniques to identify the vulnerabilities in the Web applications.

3 Proposed System

Knowledge of the default configuration of Web browsers, Apache Web server and ModSecurity Web application firewall (WAF) helps the hackers to hack the Web sites. This paper focuses on securely configuring the Apache Web server and ModSecurity WAF to defend themselves from various types of Web attacks [3]. The assaults connected with the Web applications are code injection, losing of sensitive information, damage of information, malware, worms, phishing, link spams, information warfare, account invasion, click scam, etc. The majority of the assaults occur due to misconfiguration in the security [4]. The safety of the Web applications depends on the right configuration of the system. The system is vulnerable when there is a weakness in the configuration settings. The proposed system focuses on protecting the vulnerabilities in the Web applications by performing the following steps

- Defending HTTP headers
- Secure configuration of the Apache Web server
- Secure configuration of the ModSecurity.

The architecture diagram of the proposed system is as follows

In the proposed system, Apache Web server is securely configured, and ModSecurity is configured with blocking rules to obstruct both the incoming and outgoing malicious Web traffic and to alert the administrator through messages [5].This in effect leads to securing the client browser as well as the Web server.

4 Defending HTTP Header

Because of the incessant utilization of online Web applications for different everyday exercises, Web applications are ending up most appropriate focus for assailants. Cross-site scripting otherwise called XSS assault is a standout among the most conspicuous electronic assault which can prompt trade-off of entire program as opposed to simply the real Web application, from which assault has begun. Verifying Web applications utilizing server-side arrangements is not productive as designers are not really security aware [6]. In this manner, program merchants have endeavored to advance customer-side channels to guard against these assaults. This paper demonstrates that even the premier winning XSS channels conveyed by most recent renditions of most generally utilized Internet browsers do not give suitable safeguard. We assess three programs—Internet Explorer, Chrome, Safari and Firefox—for reflected XSS assault against various kind of vulnerabilities [7]. We locate that none of the above is totally ready to safeguard against all conceivable kind of reflected XSS vulnerabilities. Further, we assess Firefox in the wake of introducing an extra named XSS-Me, which is broadly utilized for testing the reflected XSS vulnerabilities. Exploratory outcomes demonstrate that this customer-side arrangement can shield against more noteworthy level of vulnerabilities than different programs. It is seen to be progressively favorable if this extra is coordinated inside the program rather being authorized as an expansion.

The various Web sites are compromised by the attacker due to failure in the configuration of the Web server. The Web site can be protected by configuring the behavior of the browsers while handling the malicious Web sites to mitigate the assaults and safety weaknesses [8]. Below are some of the most prominent headers that can be set to bolster the security of their Web applications. The following HTTP protection headers can be used to improve the security features of the Web applications, since defending the HTTP headers are the foundation for securing a Web site.

4.1 X-XSS-Protection

The reflected cross-site scripting attack can be moderated in the modern Chrome and Safari browsers by using the XSS auditor or XSS filter. The XSS auditor blocks the request containing the malicious Javascript injected by the hacker into the server response [9]. The X-XSS-protection header is used to enable the XSS auditor in these modern browsers to block the Web page containing the malicious code. Usage of this header is also recommended by Web security experts. The header can be configured in different ways such as disabling the cross-site scripting filter, enabling the cross-site scripting filter, and sterilizing the page, if the assault identified; enabling the cross-site scripting filter and blocking the page, if assault is identified; and enabling the cross-site scripting filter and blocking the account destruction, if assault is identified.

4.2 HTTP Strict Transport Security

The HSTS policy enables the Web browsers to access response headers through secure connections. The threats such as passive and active network assaults, SSL protocol assaults, and cookie hijacking can be prevented using HSTS. The HSTS helps the Web browser to switch from insecure connections to secured connections and recommend the users to use TLS/SSL while connecting to the Web sites [10]. It forces the Web server to respond to browsers through HTTPS connection. The user agent will contact the server for a period of time in a safe mode. If the hacker tries to connect to Web site through HTTP version, the HSTS policy will routinely convert the HTTP version to HTTPS version to block the stealing of sensitive information by the hacker such as credit card information and session cookies and also prevents the man-in-middle attack executed by the hacker.

4.3 X-Frame-Options

All the exiting browsers are the victims of the click jacking. The click jacking depends on the fake buttons with the attractive titles that appears to be authenticated. These fake buttons will be attractive to the user to click on it which contains hidden malicious links. The hacker employs iframes in the site, and when the user selects the malicious buttons, the hidden code in the button will be executed which drives the data to the iframe pointing to malicious sites [11], and allows the users to visit unexpected Web sites or making the sensitive information of the user visible to all or using Twitter/Facebook accounts without user consent. The click jacking assault can be defended with the help of x-frame-option. The x-frame-option can deny the loading of Web page in a frame or Web page can be loaded in a frame from specific origin.

4.4 X-Content-Type-Options

MIME sniffing allows the browsers to deny the Web sites testing the content exhibited on the Web sites, which causes severe safety vulnerabilities. The hacker can perform operations that cannot be detected by the Web administrator or clients. MIME sniffer helps the hacker to execute XSS assaults easily on the Web sites [12]. The safety of the Web site can be improved using HTTP safety headers. These headers halt the functionality of the MIME sniffing on the Web browsers. The assets cannot be accessed by the user unless the type of the file is declared by the server. The browser denies the MIME types not defined by the source server and ignores exhibiting the assets to the users.

4.5 Access-Control-Allow-Origin

The Web page permits the user to load the images from different location from World Wide Web. The resources are pooled from diverse sources by the Web pages. The risks are involved with the appealing resources hosted on diverse Web server. These risks can be minimized on the browsers using safety policies. The Access-Control-Allow-Origin allows the resources hosted on the server will interact with other resources hosted on the same server, and these resources cannot interact with resources hosted on different servers. This policy uses protocol, host, and port number.

This policy provides safety to Web browsers to block the Javascript code building the requests against a diverse origin. It also allows genuine communications between a server and clients of an acknowledged and dependence source.

4.6 *HTTP Public Key Pinning*

It allows Web sites to avoid mimic by hackers using fake certificates. It supplies public keys to the browsers. The man-in-middle assaults can be identified and prevented using HTTP public key pinning. The Web site sends the public keys pins to the client's browser via HTTP header when the client selects the Web page of the Web site. The browser checks the validity of the public pin with the certificate chain. The certificate chain must contest with the pinned public key.

When someone visits the Web site, it returns the public key pins to the visitor's browser via HTTP headers. The browser checks at least one of the pins which is valid for the certificate chain and caches the pins for the next visit. In order to validate the chain, at least one of the certificate chain's public keys needs to match a pinned public key.

The whole mechanism tells an application that the certificate chain from a server must contain at least one of the pinned certificates and to sound the alarm if it finds none. In doing so, users can be more certain that there is no-one interfering with communication over a network, which is especially important for Web sites that handle sensitive information like healthcare or financial records.

5 Securing the Apache HTTP Server

The most popularly used Web server is the Apache, and it is an open-source cross-platform server for Web. Being most extensively used, Web server becomes most susceptible service to assault and having more than 200 vulnerabilities. If is very difficult to configure Apache Web server, the following steps are followed to minimize the risk on Apache Web server.

- Block all recourses that are accessed by default.
- Malicious activities are identified by logging all Web requests.
- Authenticate the uniqueness of the server by the client.
- Providing secure communications between the client and the server.

The Web site will be the victim for attacking when the Web server is misconfigured or setup by default. The penetrating testing needs to be performed on the Web application frequently to identify online cyber threats. The scans need to be done automatically using penetration testing tools like Nessus to prevent the Web applications and Web servers from hackers.

The default configured Web server can be easily hacked by the hacker easily. Almost all Web application is having one or more vulnerabilities. The attackers try to perform following operations such as examining the Apache version and operating system, stealing of cookie information, hijacking the session to steal the sensitive information of the user, making use of cross-site scripting assaults to steal cookie information, and examining the document root directory in the lack of index file.

The algorithm is implemented as shown below for securing the Apache Web server to block the above-specified operations executed by the hacker.

The port scanning tools help the hacker to discover the version of the Apache server and its operating system. The default configuration of the Apache server causes the hacker to exploit the Web vulnerabilities on the Web site hosted on the server.

6 ModSecurity

ModSecurity is the open-source Web application firewall used to mitigate the assaults and susceptibility on Apache, IIS, and Ngnix Web servers, and it is placed between users and the Web server. All the Web applications and Web servers are the basis of Web security weaknesses and are open for attacker to exploit Web vulnerabities on the Web sites. The vulnerabilities occur in Web applications and Web servers due to poor technical skills, outsourced code, lack of resources, and third-party code. The Web attacks on Web applications and Web servers can be minimized by using ModSecurity WAF. This paper focuses on ModSecurity firewall rules to examine both inbound and outbound data streams.

The ModSecurity core rules are used to perform following operations

- Identifying the destructions in the HTTP protocol
- Identifying the malwares based on the Web.
- Mitigating HTTP overflowing and HTTP denial of service assaults.
- Identifying bots and crawlers.
- Identifying the uploading of malevolent files on the Web sites.
- Identifying the trojan horses
- Identifying misconfiguration on the Web servers.
- Appends security layer to the Web application.
- Examines the HTTP traffic efficiently compared with the traditional firewalls.

The top security risks such as XSS, SQLI, CSRF, password cracking, security misconfiguration, and exposure of sensitive data in Web applications are blocked using ModSecurity.

Challenging tasks involved in Web application firewall is configuring and maintaining it. The ModSecurity firewall cannot mitigate Web vulnerabilities by default, and it can mitigate based on the firewall rules defined by the administrator [13]. The research paper focuses on the configuring the Web application firewall rules to mitigate the Web security vulnerabilities and unknown vulnerabilities by understanding the behavior of the Web applications to improve the safety features of the applications of Web. The Web attacks are detected and logged in both Web application firewall log file and in Apache Web server (Table 1).

Table 1 Algorithm to secure configuring the Web server

Set the server token value to "prod".
Set ServerSignature option to "Off".
Disable TraceEnable option.
Set all pages provided by the web server with X-Frame Options header .
Locate all cookies with HttpOnly and Secure flag.
Set the X-XSS-Protection
header. Set options to None.
Set FileETag option to None.
Create user and group for authorization to run the web server from
non- privileged account.
Defend the directory permissions of binary and configuration file.
Set AllowOverride option to None.
Add **deny from all** statement in LimitExcept directory. Set TraceEnable option to off.
Append the following directive Header edit Set-Cookie ^(.*)$ $1;HttpOnly;Secure.
Alter SSLCipherSuite directive to recognize barely advanced encryption algorithms.

7 Results

The frequent weaknesses such reflected cross-site scripting and command injection assaults are tested on DVWA Web site. First, the above-said assaults are executed on the Web site by disabling the ModSecurity defending rules, and the results are examined by comparing wireshark traffic, review log files of Apache2 and ModSecurity [14]. Finally, the assaults are blocked by defining the appropriate rules in the ModSecurity firewall, and the audit log files are examined to find whether the assault is stopped (Fig. 1).

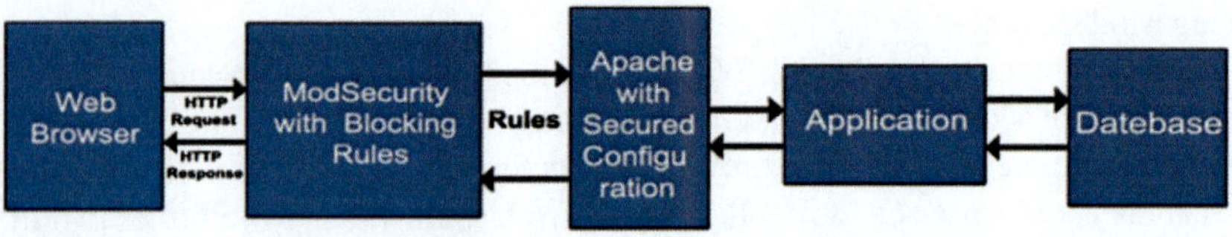

Fig. 1 Proposed system diagram

7.1 *Examining the Command Injection Attack*

In this paper, the command "cat/etc./passwd" is injected into the name "Enter an IP address" as exposed in Fig. 2. The browser displays the each user account information used in the Linux device as depicted in Fig. 2.

Figure 3 demonstrates the traffic captured by the wireshark tool during the command injection assault. It illustrates the POST request with command injection in the packet #4. The packet #13 demonstrates that it accepted the request.

Figure 4 is access log file of the Apache2 access log, and it demonstrates that it acknowledged the appeal to the Web server.

Fig. 2 Command injection attack by disabling the ModSecurity blocking rules

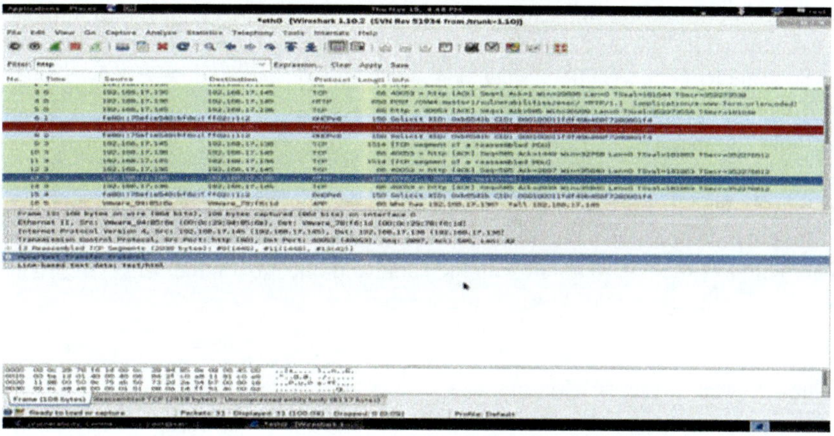

Fig. 3 Traffic captured by wireshark tool during the command injection assault

Fig. 4 Access log file of Apache2 before enabling ModSecurity WAF

7.2 *Obstructing Command Injection Assault*

The command injection uses the keywords of the operating system on which the Web server is running. The attack can be blocked by scrutinizing the text at URI. For instance, the following line can added in the/etc./modsecurity.conf file to mitigate the command injection assault.

SecRule REQUEST_URI "cat|etc.|passwd|ls|grep|tr"

To stop, more advanced command injection assaults by adding strings are given in Table 2 to the existing rules.

When the similar appeal was sent after inserting the defending rule, the browser responded with a "Forbidden" error code, as exposed in Fig. 5.

Figure 6 demonstrates the traffic captured by the wireshark tool. It demonstrates that the POST request was sent with the command injection in the packet #25. The packet #27 illustrates that it forbids the appeal.

Table 2 Common command injection strings

tail	sort	uniq	dpkg	less	ping	nslookup	ping
kill	Netcat	cron	chown	nohup	traceroute	wget	df
dd	find	mount	pwd	useradd	userdel	rm	md5sum

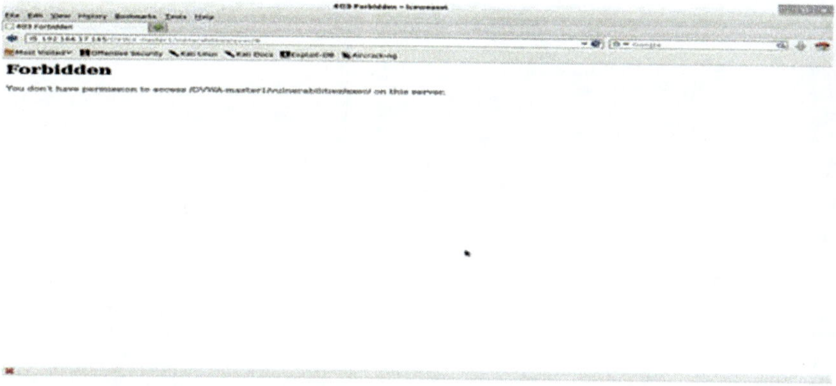

Fig. 5 Browser reply with the command injection rule

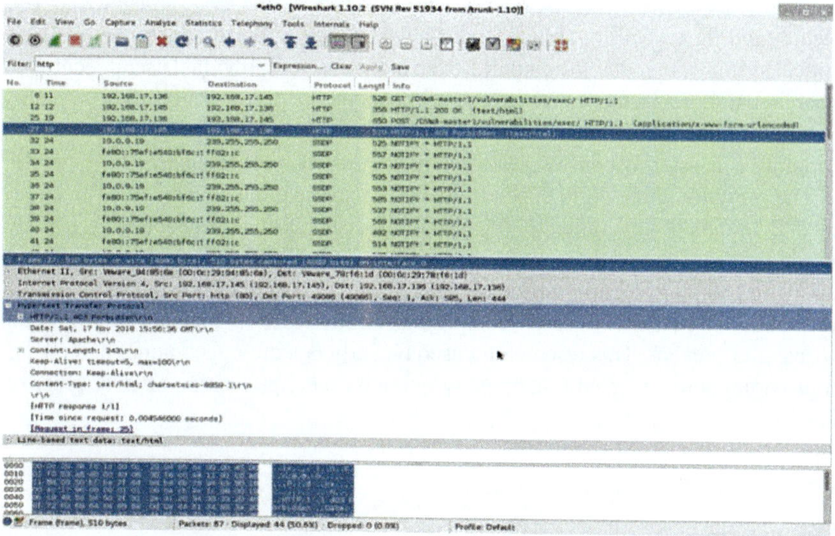

Fig. 6 Blocking of command injection attack with ModSecurity rules

Figure 7 illustrates that audit log file of Apache2 denies the requests of the Web server after implementing the ModSecurity defending rules.

8 Conclusions

Most of the administrators and users use the default security policies created at the time of initial deployment of the systems. These systems have to be implicitly highly

Fig. 7 Apache2 error log with the command injection defending rule

secure, and thus, security assumes the highest concern for administrators of these critical systems. In today's digital life, even in the case of individual users, security is an area that, if neglected, causes a severe impact on the performance of the systems, leading to huge losses, financial or critical, for instance, a denial of service attack crippling a disaster response service having a disastrous consequence.

The present digital world depends upon Web applications to deliver its functionality. Economies, governments, organizations, and ultimately people rely heavily on the performance of these Web uses. XSS, SQLI, CSRF, command inoculation, password cracking, security misconfiguration, and exposure of sensitive data are the most common methods to easily compromise Web security which are very difficult to detect or prevent. This research focused on early detection techniques and means of providing uninterrupted functionality through timely prevention of these types of attacks.

References

1. Jiang, S., Smith, S., Minami, K.: Securing Web servers against insider attack. In: Seventeenth Annual Computer Security Applications Conference, New Orleans, LA, USA, ISBN 0-7695-1405-7
2. Fong, E., Gaucher, R., Okun, V., Black, P.E.: Building a test suite for web application scanners. In: Proceedings of the 41st Hawaii International Conference on System Sciences, 07–10 January 2008, IEEE Computer Society Washington, DC, USA
3. Abusaimeh, H., Shkoukani, M.: Survey of web application and internet security threats. Int. J. Comput. Sci. Netw. Security **12**(12), December 2012
4. Joshi, C., Singh, U.K.: Performance evaluation of web application security scanners for more effective defense. Int. J. Sci. Res. Publications **6**(6), June 2016, ISSN 2250-3153
5. Dalai, A., Jena, S.K.: Evaluation of web application security risks and secure design patterns. In: Proceedings of the 2011 International Conference on Communication, Computing & Security, 12–14 February 2011, ACM. 978-1-4503-0464-1/11/02

6. Munoz, F.R., Cortes, I.I.S., Villalba, L.J.G.: Enlargement of vulnerable web applications for testing. J. Supercomputing **74**(12), 6598–6617 (2018) (Springer)
7. Adhyaru, R.P.K.: Techniques a For attacking web application security. Int. J. Inf. Sci. Techniques (IJIST) **6**(1/2), March 2016
8. Rao, G.R.K., Satya Prasad, V: Combating cross-site scripting assaults without proprietary software. Int. J. Appl. Eng. Res. **12**(17), 6788–6796 (2017). ISSN 0973-4562
9. Foster, E.: Testing Web Application Security Scanners against a Web 2.0 Vulnerable WebApplication, 11 October 2018, Copy right SAN Institute
10. Eshete, B., Villafiorita, A., Weldemariam, K.: Early detection of security misconfiguration vulnerabilities in web applications. In: 2011 Sixth International Conference on Availability, Reliability and Security, Vienna, Austria. https://doi.org/10.1109/ares.2011.31
11. Medeiros, I., Neves, N.F., Correia, M.: Automatic detection and correction of web application vulnerabilities using data mining to predict false positives. In: International World Wide Web Conference Committee (IW3C2), 7–11 April 2014, Seoul, Korea, ACM. 978-1-4503-2744-2/14/04, http://dx.doi.org/10.1145/2566486.2568024
12. Bacudio, A.G., Yuan, X., Bill Chu, B.-T., Jones, M.: An overview of penetration testing. Int. J. Netw. Security Appl. (IJNSA) **3**(6) November 2011
13. Erturk, E., Rajan, A.: Web Vulnerability Scanners: A Case Study, Submitted on 25 Jun 2017, Cornel University Library
14. Black, P.E., Fong, E., Okun, V., Gaucher, R.: Software assurance tools: web application security scanner functional specification version 1.0. In: Software Diagnostics and Conformance Testing Division Information Technology Laboratory National Institute of Standards and Technology Gaithersburg, MD 20899

Mitigation of ARP Cache Poisoning in Software-Defined Networks

A. Vishnu Priya and Hiran Kumar Singh

Abstract Software-defined networking (SDN) is an emerging centralized network paradigm that solves the major network management issues in the modern computer networks. The security of the central controller network still is an open challenge. In that, address resolution protocol (ARP) cache poisoning is one of the major threats to controller. Since SDN is a programmable network, we could resolve this issue by an application programs. This paper suggests the mitigation of ARP cache poisoning attack using sFlow monitor. We implemented and tested the procedure in floodlight controller. This method can reduce the controller overhead and improve the network performance.

Keywords RP cache poisoning · sFlow · Software-defined networks · Floodlight controller

1 Introduction

ARP is the stateless protocol which used to receive the MAC address of any devices in local area network (LAN). Each Layer 2 switch has a cache known as content addressable memory (CAM) table which stores information of node's IP and MAC address. A node wants the MAC address of any node that sends the ARP request to switch then the switch will forward that to all nodes via broadcasting. The intended machine that matches with IP sends the response with MAC address.

ARP is the type of active sniffing attack and is the process that the attacker can monitor and capture all data packets passing through the network. Many enterprise switches are open, so anyone can easily plug into the network using ethernet cable. Generally, this attack operates at the second layer of OSI model. The major threats of

A. Vishnu Priya (✉) · H. K. Singh
Department of Electronics & Communication Engineering, Vel Tech Rangarajan Dr. Sagunthala R&D Institute of Science and Technology, Chennai, India
e-mail: vishnu.priya@veltech.edu.in

H. K. Singh
e-mail: hksingh@veltech.edu.in

P. Suresh et al. (eds.), *Advances in Smart System Technologies*,
Advances in Intelligent Systems and Computing 1163,
https://doi.org/10.1007/978-981-15-5029-4_8

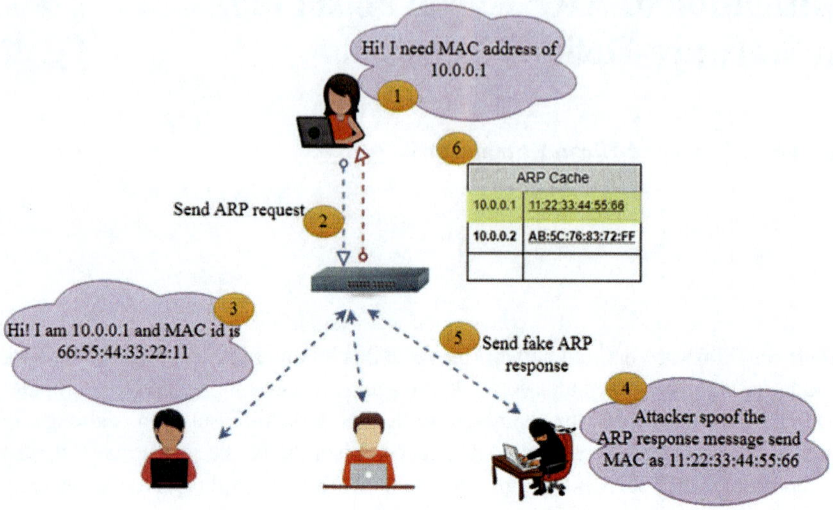

Fig. 1 ARP cache poisoning attack

ARP poisoning is packet sniffing, session hijacking, denial of service (DoS), man-in-the-middle attack, and stealing of passwords. Cain and Abel are the very popular common attacking tools toward ARP poisoning.

ARP poisoning attack is depicted in Fig. 1, and in that, the attacker spoofs the ARP packets and alters the MAC address with its own MAC value. Then, he floods the ARP packets with fake entries to all nodes. It alters the MAC address of attacker who has replaced with spoofed IP address. This will lead the attacker who can divert all data packets to his machine.

In traditional networks, many approaches had proposed like XARP that allows administrator to monitor whole ARP packets. Dynamic host configuration protocol (DHCP) is another way to resolve the IP/MAC translation issue without broadcasting, but it will not support in case of static IP [1]. In this paper, we proposed the anti-ARP cache poisoning in SDN networks using sFlow monitors, since our approach is lightweight to mitigate the attack without controller overhead.

2 Related Works

Mohammad Z. Masoud et al., have discussed ARP poisoning attack prevention in software-defined networks [2]. They proposed the algorithm for both static (SDN_STA) and dynamic IP address (SDN-DYN) scenario.

Ahmed M. AbdelSalam et al., have discussed ARP mitigation in spoofing attacks in SDN [3]. The algorithm detects the ARP request, ARP reply, and DHCP packets and starts to count that, and if large number of packets arrived, then it stops them by changing of OpenFlow rules.

Ajay Nehra et al., have proposed traffic pattern-based solution to ARP-related threats (FICUR) that finds the ARP poison attack by traffic pattern analysis [4]. They created IP-MAC binding logs and found the violation of received traffic. They segregated the doubtful packets and detected the flooding attack based on threshold. They have created and tested the man-in-the-middle (MitM) module in POX controller.

In ONOS security report discussed the different defense measures for ARP poisoning attack includes ARP alert, packet filtering, and encryption. They also suggested SSL implementation on OpenFlow protocol [5]. Ghazi Al Sukkar has described attack implementation and defense tool using "Auto detect IP" method. It verifies the IP and MAC addresses in the status, and if any violation, it will reset the ARP table. They also suggested prevention and detection systems [6]. In all the approach, controller detects the attack and modifies the flow rules. Here, sFlow monitor will collect the metrics of all traffic and detects the attack.

3 Background Study

3.1 Software-Defined Networks

SDN controller is the foremost component in the global view network. It composed of three planes such as data plane, control plane, and application plane. Infrastructure layer holds SDN-enabled switches. Since the user applications are providing the new management tasks and policies to the data plane through the controller [7], the controller is responsible to convert these tasks into flow rules. The switches are just like a forwarding element that act according the flow rules derived by SDN controller.

The switches maintain the flow table that determines the path of the packet. Communication between the controller and the switch will reinforce by southbound protocol as in Fig. 2. OpenFlow is a southbound protocol, standardized by open networking foundation (ONF) [8]. If any new packet arrived in the network, the data plane elements will check the corresponding flow rules in the flow table. If it is not mentioned in the flow table, then the switches forward that packet into the controller. Controller will modify the flow rules based on application commands.

3.2 Floodlight Controller

Floodlight controller is a SDN controller is an open-source package works with OpenFlow protocol. The controller maintains all network rules and necessary instructions to the infrastructure on how the network traffic must be handled. Floodlight controller

Fig. 2 SDN architecture

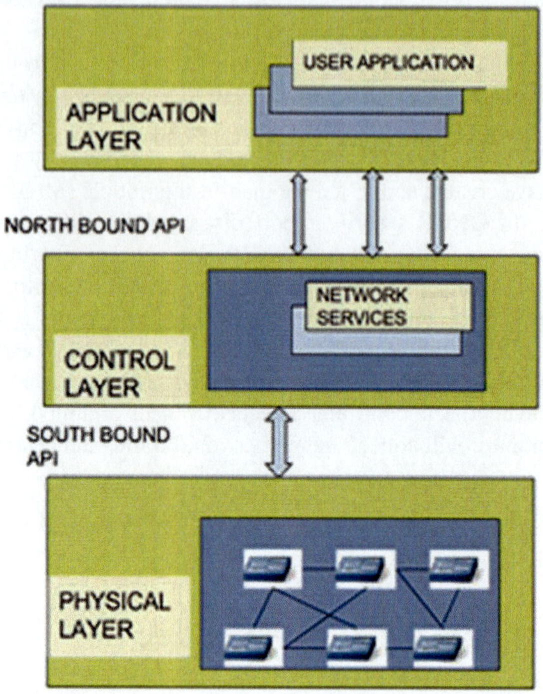

can be advantageous because it offers the ability to easily adapt to the software and develop applications. The modules are written in core Java.

Floodlight has its own floodlight GUI implementation using REST APIs that makes it comfortable for the user to understand the program interface [9]. The GUI provides read and write access to the controller, meaning it allows one to modify the network state in addition to query it. It is GUI supported that allows user to access state information and devices on the network and install the flow rules in flow table.

3.3 sFlow-RT

It is open source network monitoring tool that operates in layer 2 of OSI model. It has the global view of network that helps to defense against network attacks [10]. The sFlow-RT services are real-time, scalable, and programmable. It receives continuous telemetry updates on network devices, hosts, and applications. This information is then used to configure the network flow by changing the controller flow. It is highly adoptable with application that allows them to communicate with network devices for retrieving metrics. JavaScript API allows sFlow-RT to be customized to support a variety of cases, in our case to modify the flow of the controller.

3.4 Mininet

Mininet is a network emulator that could operate in virtual platform. User can create any network topology and easily communicate with it using Mininet CLI. It helps to customize and share the environment to others or real hardware. Topology can create using Mininet GUI interface and can convert it as Python script. In this work, we created network topology and initiated attacks on that same using C codes. Mininet is compatible with different controllers that can connect it to some default port [11].

4 Proposed Algorithm

In the proposed algorithm, initially create the IP-MAC bindings all hosts present in the network. Attacker replaces the table entries by ARP-reply messages. So, we inspect all ARP-messages using sFlow-RT agent which was installed in the switches of the network. Every agent collect the IP address (ip_rtv), MAC address (mac_rtv), switch DPID (OpenFlow data path identifier), port, and attachment point values. Each IP-MAC values of ARP replies is compared with log values. If any violation, sFlow agent will send the block message to static flow pusher module of floodlight controller. Floodlight controller will modify the flow rule. The detailed algorithm and flowchart are shown in Fig. 3.

ARP Cache Poisoning mitigation procedure

Network consists of Si switches, Controller C and Ni nodes where i=1,2,3…..n
Start Controller C
From each switches Si, where i=1, 2, 3….n
Retrieve IP and MAC and create IP & MAC binding ip_log, mac_log
Install Agent Aj in switches where j=1, 2, 3…..m
From every Agent Aj
Inspect IP & MAC address of network devices ip_rtv, mac_rtv, switch DPID, port, and attachment point
For each ip_rtv
Compare mac_rtv=mac_log
If else
Send block event message to staticflowpusher module of Floodlight controller

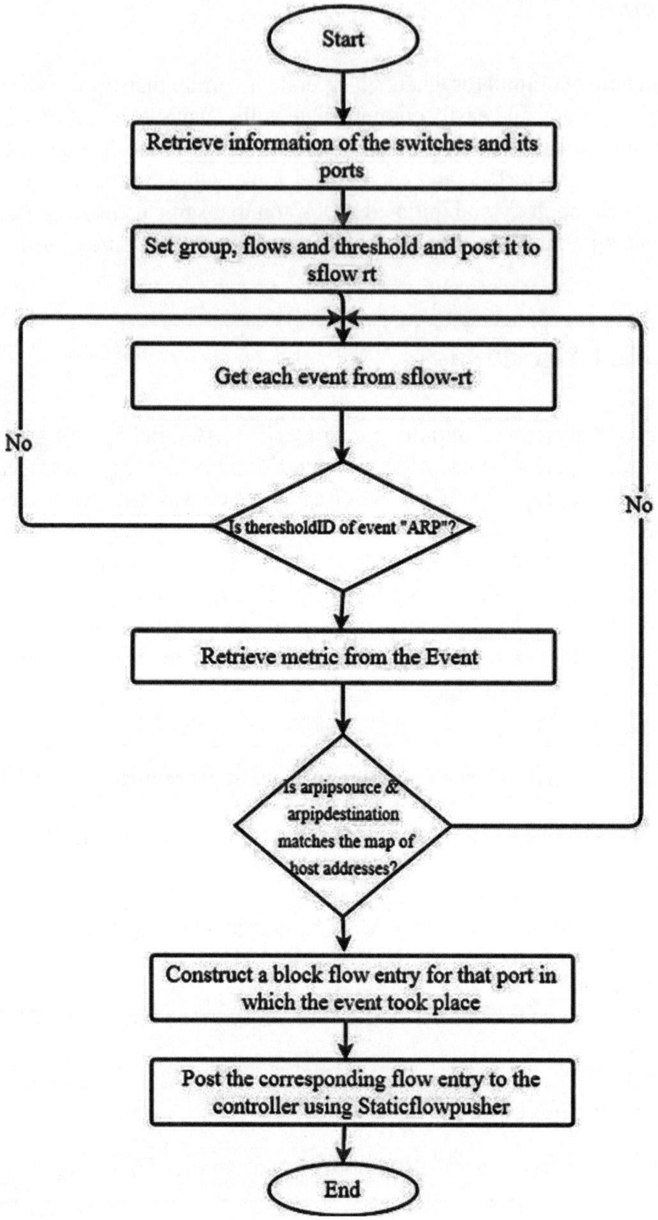

Fig. 3 ARP spoofing mitigation flowchart

5 Implementation and Results

The experiment conducted on windows 8.1 PC and has i5-4210U CPU with installed RAM of 6.00 GB. A network topology with one switch and six hosts is created in Mininet installed in Ubuntu 14.10 Linux virtual machine as shown in Fig. 4. This network is connected with floodlight controller which is running in 10.10.10.2:6633 IP address and port number. ARP attack has been induced through Ettercap tool and collect the information of the packets through Wireshark and pass them to the sFlow-rt and the controller [12].

The network metrics such as host IP and host MAC has retrieved from floodlight controller GUI through Python script. This data is passed as arguments to the script to verify the log details with IP-MAC bindings. If any modification found in the retrieved metrics, the updated flow rule will send to static flow pusher module in floodlight. The final result after mitigation is shown in Fig. 5. The attack code is run on host 1(10.0.0.1) and tries to poison the host 2 (10.0.0.2) cache. After attack, target MAC address changes by Host 1 MAC address. When host 3 (10.0.0.3) sends a message to Host 2, the message could be hijacked by Host 1. But the received sFlow metrics could easily identify the ARP cache poison. Figure 6 shows the sFlow ARP mitigation output once it identifies the attack and it simply blocks the particular vulnerable port and sends the warning message to network administrator.

This method is a simple and effective solution for SDN network without direct controller monitoring. The controller's throughput has evaluated using iperf tool in three different scenarios like normal TCP flow, TCP flow during attack, and TCP flows after mitigation. The readings are collected over 10 ms, and the graph is plotted

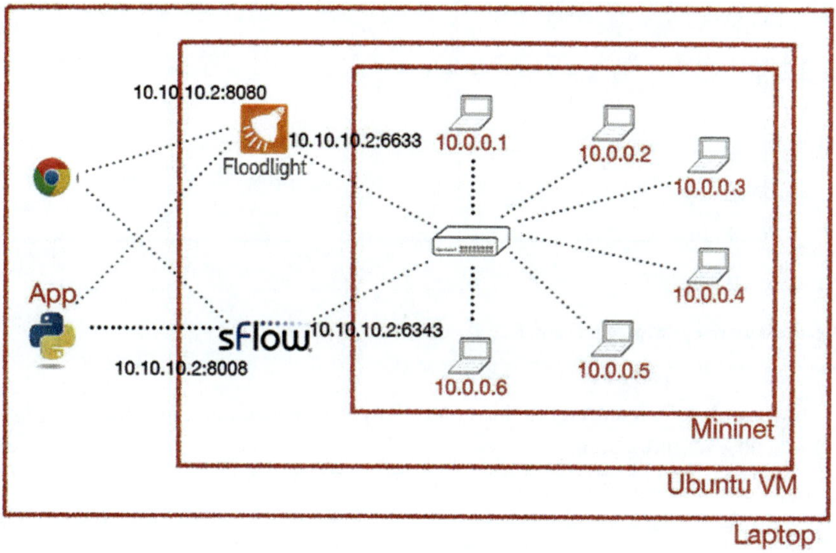

Fig. 4 Network topology

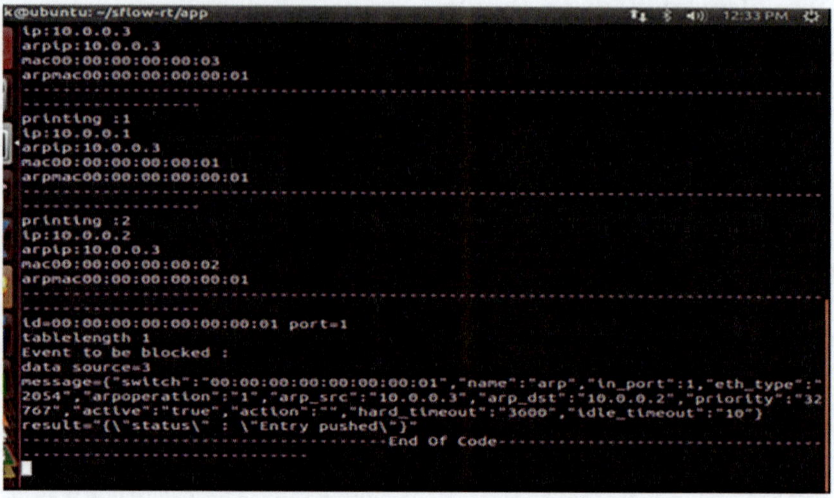

Fig. 5 Flow modification message after mitigation from sFlow agent

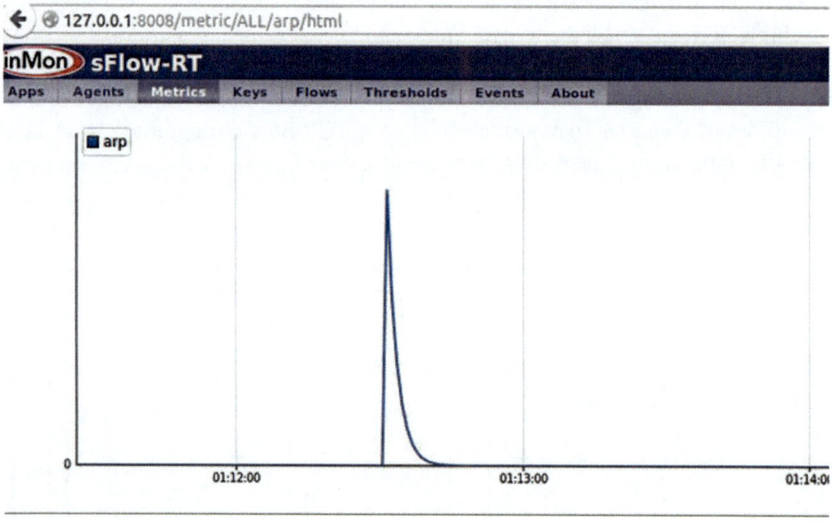

Fig. 6 Retrieved metric from sFlow-RT

as shown in Fig. 7. It shows that the controller performance remains almost similar before and after mitigation.

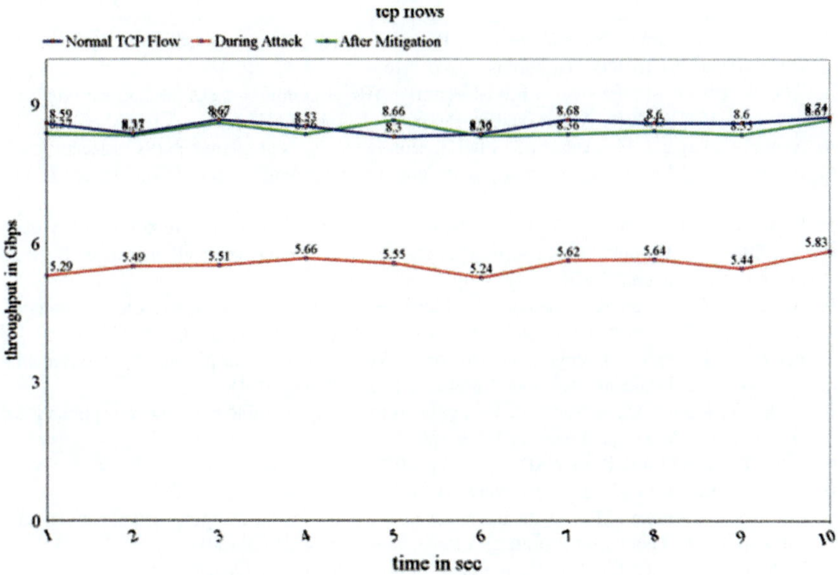

Fig. 7 Controller performance

6 Conclusions

ARP flooding attack which is exploited by sending a storm of spoofed ARP requests to poison the target's ARP cache. In this, we proposed a lightweight mitigation procedure for software-defined networks. This scans all ARP request and reply messages and detects the attack. This will be the effective and a recommended method when the network is scalable up to 100 K flows. As SFlow detects the spoofed traffic autonomously, it makes the mitigation procedure simpler. The mitigation scripts of ARP poisoning further helps the network from undergoing a few more LAN-based attacks like man-in-the-middle attack, IP spoofing. The ARP cache poisoning mitigation ensures the strong security in layer 2 is promises for preventing from the consequence higher layer attacks.

The proposed mitigation algorithm improves the controller performance. It identifies the spoofed flow and blocks the corresponding port. The algorithm considered two parameters to mitigate the attack, whereas in our future work, we will inspect the entire flow details and other entities.

References

1. Cox, J.H., Clark, R.J., Owen, H.L.: Leveraging SDN for ARP security. In: Conference Proceedings—IEEE SOUTHEASTCON, vol. 2016–July, 2016

2. Masoud, M.Z., Jaradat, Y., Jannoud, I.: On Preventing ARP Poisoning Attack Utilizing Software Defined Network (SDN) Paradigm pp. 0–4 (2015)
3. Abdelsalam, A.M., El-sisi, A.B.: Mitigating ARP spoofing attacks in software-defined networks, ICCTA 2015, At Alexandria, Egypt, no. October, 2015
4. Nehra, A., Tripathi, M., Gaur, M.S.: FICUR: Employing SDN programmability to secure ARP. In: 2017 IEEE 7th Annual Computing and Communication Workshop and Conference CCWC 2017
5. Hong, S., Xu, L., Wang, H., Gu, G.: Poisoning network visibility in software-defined networks: new attacks and countermeasures. In: Proceedings of 2015 Network and Distributed System Security Symposium, Feb 2015, pp. 8–11
6. Al Sukkar, G., Saifan, R., Khwaldeh, S., Maqableh, M., Jafar, I.: Address resolution protocol (ARP): spoofing attack and proposed defense. Commun. Netw. (August), 118–130 (2016)
7. Feamster, N., Rexford, J., Zegura, E.: The road to SDN: an intellectual history of programmable networks. ACM Sigcomm Comput. Commun. **44**(2), 87–98 (2014)
8. Braun W., Menth, M.: Software-defined networking using OpenFlow: protocols, applications and architectural design choices 302–336 (2014)
9. Izard, R.: Static Entry Pusher API. pp. 1–11 (2017)
10. Systems, B.C.: Brocade sFlow for Network Traffic Monitoring, p. 12 (2009)
11. Demirci, M., Ammar, M.: Design and analysis of techniques for mapping virtual networks to software-defined network substrates. Comput. Commun. **45**, 1–10 (2014)
12. Vishnupriya, A.: ICCCE 2018 Committees. Springer, Singapore (2018)

Performance Investigation of Fiber to the Home (FTTH) Ingress Network Based on GPON with Optisystem

T. Kavitha, K. Prabhakar Reddy, and J. Sravani

Abstract Nowadays in our day to day life very high-speed data rate is very exigent. It can be increased according to the users of Internet with the applications like smart phone, video surveillance, live broadcasting of news, video conference, etc. Fiber to the home (FTTH), also called "fiber to the premises" (FTTP), is the installation of optical fiber from a central point directly to individual buildings such as residences, apartment buildings and businesses to provide unprecedented high-speed Internet access. FTTH dramatically increases the connection speeds available to computer users compared with other technologies. This paper investigates the performance of the FTTH, and the performance is analyzed by using the optisystem software. The proposed architecture minimizes the Bit error rate (BER) and increases the Q-factor. The minimum BER attained for upstream configuration with a distance of 15 km is $1.01489e^{-073}$ and for downstream configuration is $1.24672e^{-026}$. Maximum Q-factor for upstream configuration at a length 15 km is 18.1225 and for downstream configuration is 10.6166. The Q-factor for both the cases meets the standards because both the configuration values are above 6. The standard sensitivity of this configuration is −28 dBm, and the calculation values using optisystem is −26.205 dBm. So, the proposed system is feasible.

Keywords Bit error rate · Fiber to the home · Q-factor · Optisystem

1 Introduction

Fiber optical communication plays a key role in transmission of signals in the form of light. Optical fiber is an advanced transmission media capable of supporting the next-generation services [1]. The need for high transmission rate, larger bandwidth, and

T. Kavitha · K. Prabhakar Reddy (✉) · J. Sravani
Veltech Rangarajan Dr. Sagunthala R&D Institute of Science and Technology, Chennai, India
e-mail: Prabhakarpavan1234@gmail.com

T. Kavitha
e-mail: drkavitha@veltech.edu.in

P. Suresh et al. (eds.), *Advances in Smart System Technologies*,
Advances in Intelligent Systems and Computing 1163,
https://doi.org/10.1007/978-981-15-5029-4_9

long distance is achieved by using the optical fiber communication technology. Data, voice, and video signals are transmitted with large transmission speed to the office or home via optical fiber, called the technology as FTTH, and it is a broadband network technology [2]. Various communication and multimedia services are provided by the FTTH technology such as video conferencing, digital cable television, carrier class telephony, and high-speed Internet access.

Compared with copper-based technologies, the proposed technology will deliver a higher capacity signals [3, 4]. Broad band passive optical network and gigabit capable passive optical networks are used to eliminate the bandwidth bottleneck problem as well as to satisfy the customer bandwidth requirements and increase bandwidth requirements in metro networks [5]. To increase the performance of the system, the network provider adopts the FTTH with GPON system. It is a cost-effective and reasonable solution to give high capacity optical access [6].

To satisfy the requirements of customer, the proposed GPON architecture is simulated in optisystem software and measures the parameters BER, *Q*-factor, and power budget. This paper is organized as follows: Simulation configuration is explained in Sect. 2. Simulation results are discussed in Sect. 3 with conclusion in Sect. 4.

2 Simulation Configuration

2.1 Network Analyzing Standards

The performance of FTTH network can be analyzed by using optisystem software, in which the configuration is given as per standards. Both the calculation for the PLB and simulation parameters standards are collected from PT.Telkom. The device enumeration in FTTH implementation is given in Table 1.

Table 1 Device characteristics

S. No.	Component	Enumeration/specification
1	FOC	Maximum 0.35 dB/km
2	Rx sensitivity	−28 dB
3	Splitter 1:8	10.38 dB
4	Connector loss	0.25 dB (approx)
5	Distance	Maximum 15 km
6	Slitter 1:4	Maximum 7.25 dB

2.2 Design Investigation

The below given table shows the requirement of list of materials that can be used specifically in downstream configuration table (Table 2).

The length of the cable is 15 km. SC-UPC connector is used per element. In aerial systems, concrete or steel poles are used with accessories. FTTH configuration consists of optical line termination, optical distribution cabinet, optical distribution point, and optical network unit. OLT provides interface between PON systems and service providers. ODC serves as a splitter or distribution point of the feeder cable into a distribution cable. Using 1:8 splitters, the ODP unit distributes the cable into a drop cable. ONU is installed at the customer side. The upstream and downstream configurations are shown in Figs. 1 and 2.

Table 2 List of materials (LoM)

Segment	Type of material	Component
Feeder segment	Termination material	Optical distribution frame
		Optical distribution cabinet
	Cable	Optical fiber cables
	Other material	Slitter 1:4
		Optical power meter
		Connector
Distribution segment	Termination material	Optical distribution point
	Cable	Optical fiber cables
	Other	Slitter 1:N (depends on the users)
		Connector
Drop segment	Termination material	OTP
	Cable	Optical fiber cables
	Other	Connectors
Indoor segment	Termination material	ONT/ONU
	Cable	Indoor cable
	Other	Connector BER analyzer

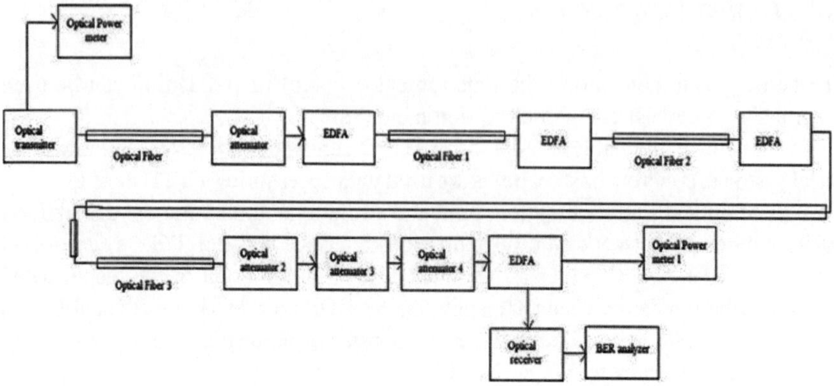

Fig. 1 Upstream configuration

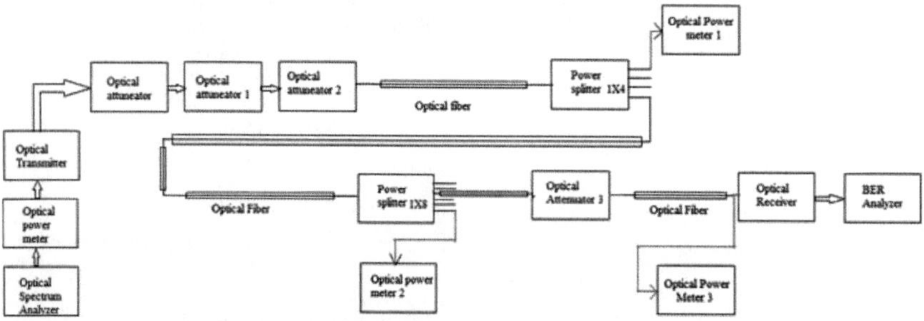

Fig. 2 Downstream configuration

3 Results and Discussions

3.1 BER for Upstream Configuration

BER analyzer in the optisystem simulator is used to display the eye diagram. Eye diagram analysis method is used to analyze the digital transmission system accurately. Eye diagram measurement is done in time domain, and distortion effects are visualized by the oscilloscope. Uplink configuration eye diagram is shown in Fig. 3. The analysis is performed for a length of 15 km. We got the BER as $1.01489e^{-073}$ and Q-factor as 18.1255 for the length 13 km (Table 3).

3.2 BER for Downstream Configuration

See Fig. 4 and Table 4.

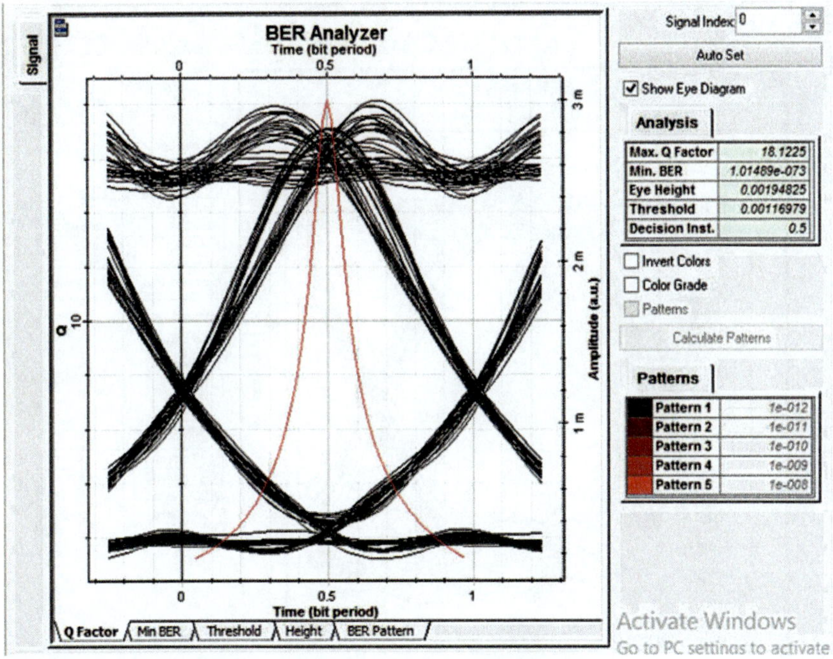

Fig. 3 Solution for uplink configuration from BER analyzer

Table 3 Comparison results for upstream configuration

S. No.	Distance (km)	BER	Q-factor
1	7	$1.07121e^{-068}$	17.4609
2	8	$3.01251e^{-025}$	17.2653
3	9	$9.12411e^{-036}$	17.34
4	10	$1.12512e^{-032}$	17.2993
5	11	$1.80104e^{-089}$	17.1701
6	12	$4.14722e^{-099}$	17.7836
7	13	$1.01489e^{-073}$	18.1255
8	14	$1.24121e^{-070}$	17.726
9	15	$1.60426e^{-070}$	17.71

The downstream configuration of the proposed system with same distance as upstream is simulated. Eye pattern opening is clear. This shows that there is no error in transmission. The simulation results prove that the network is worked in good condition up to 15 km.

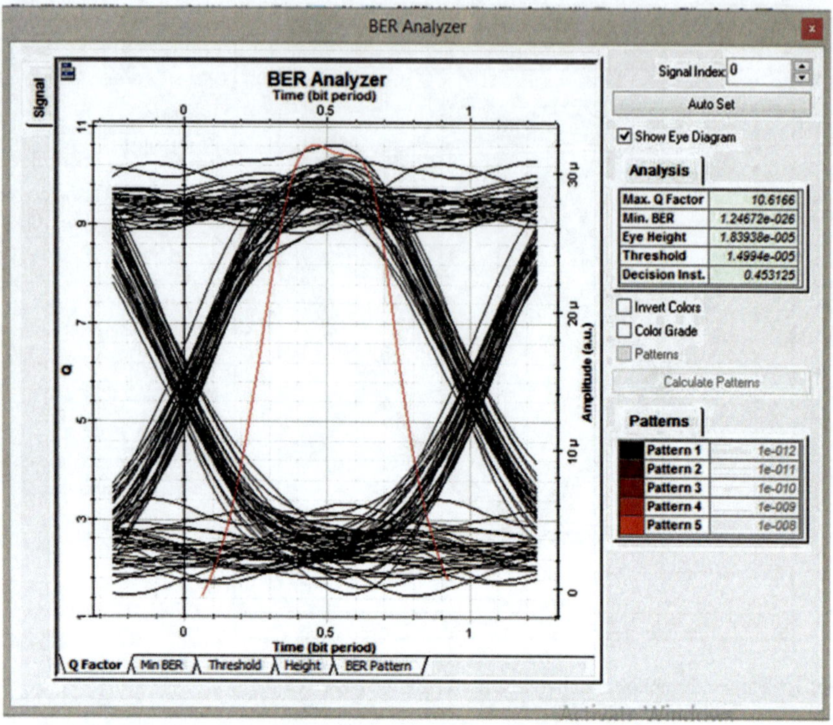

Fig. 4 Results of BER analyzer for downstream configuration

Table 4 Comparison results for downstream configuration

S. No.	Distance (km)	BER	*Q*-factor
1	7	$2.11155e^{-036}$	10.5451
2	8	$3.56761e^{-044}$	9.8914
3	9	$2.75999e^{-033}$	11.0624
4	10	$3.1161e^{-028}$	9.05024
5	11	$8.51758e^{-023}$	9.7578
6	12	$5.21093e^{-021}$	9.33147
7	13	$1.20064e^{-026}$	10.6202
8	14	$1.73008e^{-028}$	11.0088
9	15	$1.24672e^{-026}$	10.6166

4 Conclusion

The performance of the system was evaluated using various methods like eye diagram, bit error rate, and quality factor. We have found all the parameter performances at a distance of 15 km, and a wavelength for downstream configuration is

1490 nm and for upstream configuration is 1310 nm. Based on the simulation in optisystem, the maximum *Q*-factor for downstream and upstream configurations are 11.0624 and 18.1255, respectively. Both the values are above the threshold value, i.e., minimum *Q*-factor is 6. At a distance of 13 km we got the BER as 1.01489e−073 for upstream configuration and for downstream configuration is 1.20064e−026, So, it can be concluded that both values can meet the minimum value of acceptance that has been set by PT.Telkom.

References

1. Palais, J.C.: Fiber Optic Communications. Prentice Hall, Englewood Cliffs (1988)
2. Wang, Y., Guan Y.: Performance simulations for a high-speed optical transmission system based on OptiSystem. In: IEEE 7th International Congress on Image and Signal Processing, pp. 907–911 (2014)
3. Verma, S., Kakati, A., Bhulania, P., Performance analysis of *Q*-factor and polarization for GPON network using optisystem. In: IEEE International Conference on Information Technology, pp, 138–141 (2016)
4. Olasoji, Y.O.: Design of a wavelength division multiplexing fiber to the home network: university campus as a case study. Int. J. Innov. Eng. Technol. 41–54 (2016)
5. Al-Quzwini, M.M.: Design and implementation of a fiber to the home FTTH access network based on GPON. Int. J. Comput. Appl. **92**(6) (2014)
6. Suresh, P.: Creation of optical chain in the focal region of high NA lens of tightly focused higher order Gaussian beam. J. of Opt. **46**(3), 225–230 (2017)

Secure Ciphertext Policy Using Blowfish Hybridized Weighted Attribute-Based Encryption

S. Porkodi and D. Kesavaraja

Abstract There are lots of application which are been used by the people where a user's data has to be protected from the attackers who steal the data by using fake identities which makes difficult to find the real culprit. The main issue is to give strong security to the data and to minimize computational cost with an efficient key management system. We propose a new encryption method in this paper, based on user attributes and access control policy using blowfish which is hybridized and weighted encryption which is a combination of features from two different algorithms to provide a high security to user data and to generate key efficiently. The two algorithms ciphertext policy attribute encryption with access control policy and the hierarchical identity-based encryption with weighted attributes are included. Here, based on the priority, the weights are assigned to the attributes and then the message is combined with access control policy given by the user and encrypted. This algorithm is used to find the culprit who is misusing the application. Also we can add that person to the blacklist, and all the accounts associated with that identity can also be blacklisted. The proposed system provides multi-authority security, efficiency, collusion resistant, secure against chosen plain text attack and reliable.

Keywords Blowfish · Ciphertext policy attribute-based encryption (CP-ABE) · Data transfer · Hierarchical identity-based encryption (HIBE) · Multi-authority security · Secure hash algorithm-512

S. Porkodi (✉) · D. Kesavaraja
Department of Computer Science and Engineering, Dr. Sivanthi Aditanar College of Engineering, Tiruchendur, Tamil Nadu, India
e-mail: Ishwaryaporkodi6296@gmail.com

D. Kesavaraja
e-mail: dkesavraj@gmail.com

P. Suresh et al. (eds.), *Advances in Smart System Technologies*,
Advances in Intelligent Systems and Computing 1163,
https://doi.org/10.1007/978-981-15-5029-4_10

1 Introduction

To secure the user data when transferred over the Internet, cryptography algorithms such as encryption based on the user attributes are used. Many application uses encryption for the secure transfer of data from one place to another. In today's world, there is lot of hackers and attackers using some software as medium to spread malware to be installed on to the mobile device which in turn sends data from our mobile to the hacker. Even some malware is used to hack the mobile's front camera to get detailed information, photos and videos of the users. These hackers are hard to find since they use several different numbers and they do not have a unique user ID to be tracked. To avoid these types of hacking, the usage of CP-ABE encryption with user attributes is recommended. So that when data is been encrypted and transferred from one user to another, only they both can see the actual text and others cannot hack the data from the applications. The same technique can be used in mobile too, which ensures that the data from the mobile can be only seen to the person to whom the data is transferred. As the CP-ABE encryption uses the attributes of the user to encrypt data, only that particular user will have the permission to access the data by matching their attribute. Whereas others who try to hack the data will not have the attribute of that particular user, they cannot get the actual data and it is hard to hack. This algorithm can also be used to deduct the person who is illegally using the applications in a wrong way, and we can also blacklist them. As unique ID is an attribute, the different accounts using the same identity can also be blacklisted.

In this proposed work, to protect the data of the user, the ciphertext policy assigns weights on user attributes which is hybridized with the blowfish and the message is encrypted which is referred as (CP-BHWABE); it is constructed to protect user's message file by encrypting with the user's real identity and to protect user privacy with which fine-grained access control can be achieved. In the encryption based on attributes, increase in the authority is directly proportional to the increase of the suitability of the data access control system policy. Encryption based on user attributes and weight of the attributes is hybridized using blowfish to build a hierarchical tree-like structure in which the attributes of the user are arranged in terms of their priority and used for encryption. An admin may have multiple authorities, or many admin may belong to the same department to handle the same files; all these functions are handled with hierarchical identity-based encryption. The blowfish is used for generating the keys, encrypting the message and decrypting the message.

Encryption of the user identity, security of the data transacted by using encryption, for managing the keys efficiently to reduce computational costs is the important purpose of the paper. Then, to blacklist the user when the user is found to be in link with the harmful activities, admin can reveal the identity of the user and blacklist them. Security and performance analyses are done so that the system can be safe with more efficiency.

2 Related Works

Some existing systems of ciphertext policy using hybridized weighted attribute-based encryption with access control policy are listed below:

ABE is an encryption where key values and ciphertext depend on the attributes of the user which was proposed first by Goyal et al. [1]. Lots of applications use encryption to protect data where the data is protected but the identity of the illegal or problematic user cannot be deducted since the accounts in the application or registered mostly with mobile numbers or email ids. These details are not unique identities of the users and thus the culprits cannot be found. The challenges faced in attribute-based encryption are key coordination, key escrow and key revocation and thus different types of attribute-based encryptions were proposed. The key policy attribute-based encryption and ciphertext policy attribute-based encryption are the two types of attribute-based encryption.

In 2006, Sahai and Waters [2] described attribute-based encryption. Then, it was classified into two where in KP-ABE, the encrypted data depends on the set of user attributes but access rules which supports mono access tree where in CP-ABE encryption [3], the access policy written by the encryptors are used to encrypt the plaintext and the private key of the user is based on the attribute set of the user. The decryption is only done when the user attributes satisfy the access control policy of the ciphertext. The CP-ABE encryption is adopted in this project since the data owner can create their access policy in their own way.

In 2011, for the flexible and scalable access, a system was proposed by Wan et al. [4] with access control scheme which supports DNF which is disjunction normal form and access control policy by CNF which is conjunctive normal form. But this system cannot provide flexible access. In the same year, Brent Waters proposed an efficient, expressive and secure realization of CP-ABE encryption system [5]. In 2014, for more flexible access, Hua Deng and team [6] proposed a system in which LSSS is enriched with the expression of the access control policy but this system is with less efficiency in the receiver's decryption side. In 2014, for multiple authorities, Kai He and team [7] proposed a system which is based on hierarchical identity-based encryption where priority is added to each attribute and hierarchical structure is created.

Some relevant works which are implemented during recent times are shown below:

In phones, the files are uploaded and downloaded where the files are collected in the cloud but the server has no idea about the cloud security within the cloud and the user. The water-marking algorithm consists of the user file which uses classical access security and lightweight security which was developed by Wang et al. [8] to give security for the data within user and cloud by system authentication method. Data transaction error can be decreased by combining water-marking method with the Solomon and Reed code.

Li et al. proposed that the cryptography has an ability to check the versatility of access control policy was important which was then expanded by attribute-based encryption [9]. In ABE encryption, the cost of computation, management of the key and decryption are highly expensive and produce them with high quality. So to get a clear solution, the third part is used to do computational tasks and verify the results from the third party.

Li et al. proposed the personal health record framework [10] in which files of all the patients are encrypted and the data can be accessed. By this, clear and stable data can be obtained, but it is different from securing the data by ABE encryption. The complexity of the management of the key in the multiple security domains is degraded due to the multiple divisions of data. The efficiency, scalability and security are attained by using access control policy.

In this proposed work, ciphertext policy using blowfish hybridized weighted attribute-based encryption (CP-BHWABE) to encrypt with the user's real identities with which fine-grained access control can be achieved. Based on the user attributes, the priority/weight for the user attributes is generated by the system. For example, if user1 is from the block list department and user2 is from message regulation department, the system generates priority/weight values based on user1 attributes for user1 and user2 attributes for user2. According to their priority/weight, they decrypt the files they have access to, whereas user2 cannot decrypt the documents of user1 even though both the users are admin.

If a user is reported to involve in cyberbullying or cybercrime, then the account of that user and the accounts with the same identity can also be blacklisted. Since the same user can have different accounts but their identity be the same, it is easy to blacklist them by avoiding threats to the common people. As well as if any action is to be taken, then that can also be done as now the identity of the user can be found if needed.

3 Hierarchical Attribute-Based Encryption

The features of ciphertext policy attribute-based encryption [12] (CP-ABE) and hierarchical identity-based encryption (HIBE) are merged to get attribute-based encryption with hierarchical structure (HABE). This system is all about the fine access control policy, and it generates key delegation with authority attributes to achieve full delegation. When comparing others to other systems, this system has hierarchical structure to represent the importance of the attributes.

CP-ABE: The inverse system of key policy attribute-based encryption (KP-ABE) is CP-ABE. Here, based on the user's attribute set, the secret key of the user is obtained, and by using the access tree policy, the data is encrypted. The access control policy is given by the data producer in which user has the data control power and the data is confidential. The steps involved in the CP-ABE algorithm are,

1. Setup (): This algorithm is to randomize and the input is unstated security parameters; after the setup, the public key P_{k} and master secret key M_{k} are obtained as the output.

2. Keygen (M_k, AS): The inputs of this algorithm are master secret key M_k and non-monotonic access policy structure AS; then the keygen function is executed to give the output as secret key S_k based on attributes to the user.
3. Encrypt (P_k, S_a, m): The inputs of the encrypt function are public key P_k, the descriptive attribute S_a and message m. The encryption process is carried out to obtain ciphertext C_t as the output.
4. Decrypt (C_t, S_k): The input of the decrypt function includes ciphertext C_t where t is the access control policy tree and S_k is the secret key of the user which depend on the attribute set of the user S_a, and the output generated is the original message m which is sent by the data producer. This can only be done if t is satisfied by S_a.

In the CP-ABE encryption, the access control policy structure is attached with ciphertext till the attribute set is interpreted by the key for decryption. The monotonic access is given with the approximate threshold value to the encryption system. When the access control policy is fulfilled by the decryption key, then ciphertext can be decrypted. The CP-ABE approach is better by the fine encryption access control of the data than KP-ABE. The only drawback in CP-ABE encryption is the necessities of the access control policy which cannot be fulfilled as it requires flexible and efficient performance.

HIBE: The identity-based encryption (IBE) is later extended into identity-based encryption with hierarchical structure (HIBE). The private key along with public key is generated by solo private key generator (PKG). The PID "primitive ID," it is also known as 1-HIBE in ID-based encryption system. The main drawback is heavy key management. So to overcome this, 2-HIBE was found, in which the security is added. It consists of root and domain of PKG, with the help of random PID which is string and is connected with the users, where private key is produced by domain PKG to provide domain secret key as per the request, which is then acquired by root PKG. The root certificate authority is also involved in the cryptosystem, where certificates hierarchy is permitted. So with the implementation of different HIBE levels, allotment of key escrow and workload for the server can be reduced.

HABE: The combination of CP-ABE encryption along with HIBE encryption yields HABE; that is, ciphertext access policy is hybridized with ABE encryption. The steps involved in HABE algorithm are

1. Setup (k) $\geq P_k$, M_{k0}: The k is the security parameter which is taken as input to give master secret key M_{k0} and public key P_k as outputs.
2. Keygen (P_k, M_{k1}, S_a) $\geq S_k$: This function takes public key P_k, domain authority's master secret key M_{k1} and the attribute set S_a as the input, which provides private key S_k as output.
3. Delegate (P_k, M_{k1}, S_a) $\geq M_{k1+1}$: The inputs are public key P_k, secret key of domain authority for attribute set S'_a and the attribute set S_a in which $S_a \varepsilon S'_a$. The output is master secret key M_{k1+1} belongs to domain authority.
4. Encrypt (P_k, m, t) $\geq C_t$; The encrypt function's input is public key P_k, access control policy tree t, a message m and the ciphertext C_t is obtained as output.

5. Decrypt (t, C_t, S_k) $\geq m$: The secret key of the attribute S_k, access control tree policy t and ciphertext C_t are given as input to the decrypt function which generated the output as original message m.

In HABE, the similar user attribute can be managed by multiple domain owners, but this method cannot be implemented in the practical situation. Also in the area of the compound attributes, HABE cannot work and also it degrades the performance of multi-value tasks. So to make these limitations possible in real life, a new approach is proposed, ciphertext policy using blowfish hybridized weighted attribute-based encryption (CP-BHWABE).

4 Ciphertext Policy Using Blowfish Hybridized Weighted Attribute-Based Encryption

Data security and user privacy are the major aspects while transferring data through the Internet. The main objectives can be done with CP-BHWABE. In many ABE algorithms, single authority is used to handle both public key and secret key. But most of the time, user holds attribute set in multiple authorities while the holder of the data shares the content of message file with associated user, managed by the relevant authority. Lots of multiple authority-based attribute encryption is developed. In the access policy tree, to update the ciphertext, the user should be present in online all times, and the attributes are also given similar characters. In this system, blowfish method is used to give the weights for the attributes to provide data security and privacy of the user. The system involves five steps,

1. Data sender/holder: A person, who creates the data, encrypts it and sends it via the Internet.
2. Server: It stores data.
3. Weight Attributes Authority (WAA): It is mainly for authentication, to update the values of the users and to validate the values of the user attributes, assigned with respective weights for the user priority.
4. Central Authority (CA): For every user, a unique identity is allocated to user and according to weight attribute, public key is generated.
5. Data receiver: A person, who decrypts the ciphertext to obtain the original data.

The proposed system, ciphertext policy using blowfish hybridized weighted ABE encryption (see Table 1).

In CP-BHWABE system, for encryption and decryption of data, blowfish is used and also randomly key is generated. Eventually, the WAA generates priority based on attributes of user. Secure hash algorithm-512 is used to digest the larger data to the smaller data. Then, with access key policy and key values, the messages are encrypted and sent to the receiver. CP-BHWABE is reliable and has high security. In real time, the system is applicable. CP-BHWABE encryption ensures fine-grained access control, collusion resistant and supports multiple authorities. The proposed

Table 1 CP-BWABE system

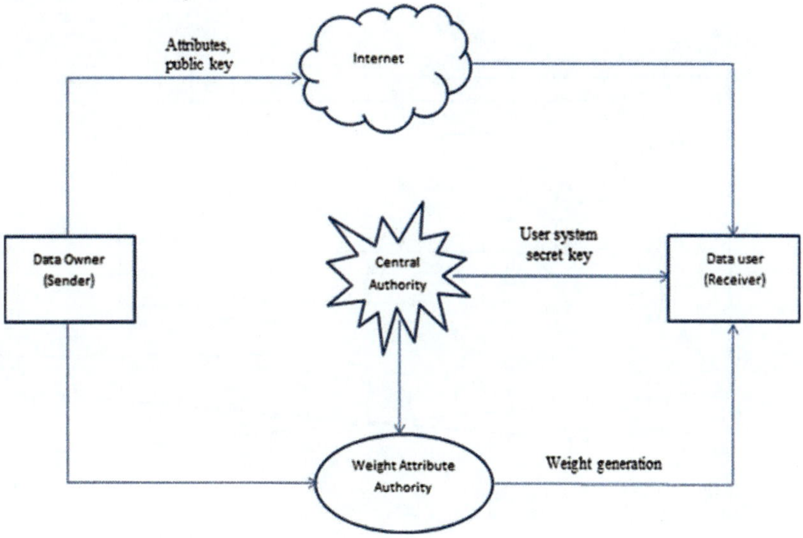

system can be described in two levels; they are algorithmic level and systemic level. In algorithmic level, blowfish defines all operations in the system. Whereas at the systemic level, only higher level operations are explained, like setup the system, registration of user, collection of user details, file creation, accessing file and file transfer.

4.1 Algorithmic Level

Blowfish

One of the encryptions with symmetric key is blowfish; it uses only one key for both encrypt and decrypt algorithms [11]. In blowfish, fixed block size is 64 bits, variable length of key is 32–448 bit size, maximum key range can be 448 bits, and for which at least 2448 groups are needed to define every key. The ciphertext is 16 round Feistel cipher that uses S-boxes dependent on large keys. The message is separated as 64 bits of block and encrypted individually.

Algorithmic level consists of five sub-keys, four 256-entry S-boxes and one 18-entry P-array. Each round is termed as r, which consists of four actions to perform nonlinear functions. The total rounds in blowfish are shown in Fig. 1. f is a function that splits 32 bits input into four 8 bits quarters (chunks) and gives it as the S-box's input. The output in S-box is given as the carry for MOD 232 addition, after that it performs XOR operation. Decryption is the reverse technique and is done by

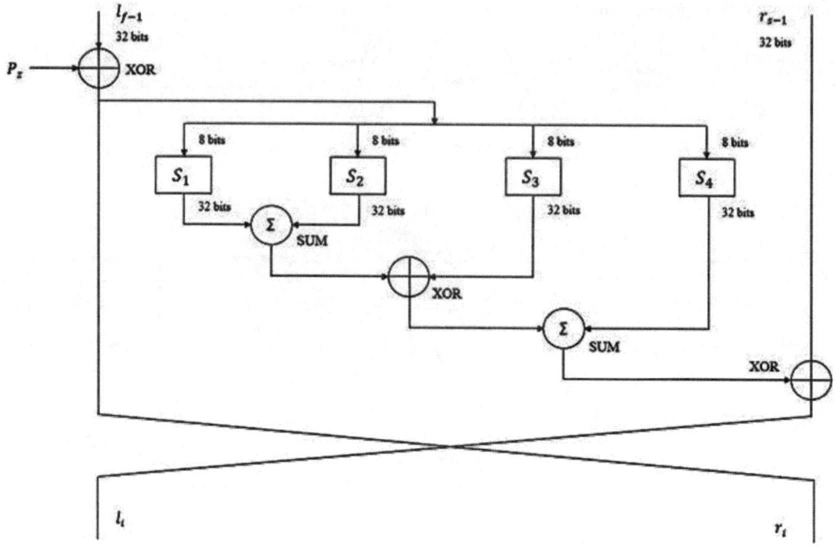

Fig. 1 One round of blowfish

inverting P_{17} and P_{18} blocks and giving P-box entries reversely. Key expansion and data encryption are the phases of blowfish.

Key expansion: Key of 448 bits is transformed into 4168 bytes of sub-key groups, with 18 P-array of 32 bits sub-key from P_1 to P_{18} and four 32 bits 256 entries of S-box.

Steps of key expansion,

1. Setup and values of P-box and S-box are initialized.
2. The user input key's variable length is XOR^{ed} with entries of P till all arrays of P are XOR^{ed} to bits of input key.
3. Encryption is done to a block consisting only of zeros; in P_1 and $P_{2,}$ the results are applied.
4. Then, again encrypt the ciphertext which is got from the encrypted zero blocks. Use it to P_3 and P_4.
5. Repeat same until the exchange of every S-box and P-box. Altogether there is a need of 521 iterations to get necessary sub-keys.

Encryption of data is completed with 16v rounds of process. Two operations are performed for each round of encryption, they are key permutation and XOR operator is used for key and data replacement and then addition is done for 32-bit blocks. Blowfish adds everything into dominant, exhalent process as given below.

Blow - fish Algorithm for Encryption
Input: Z
Output: Recombination of Z
Splits Z as, two 32 – bits: Z_l, Z_r
For i = 1 to 16
  $Z_l = z_l$ XOR P_i
  $Z_r = F_{(yl)}$ XOR Z_r
  Swap z_l and Z_r
Swap z_l and Z_r
  $Z_r = Z_r$ XOR P_{17}
  $Z_l = Z_l$ XOR P_{18}
Recombine Z_l and Z_r
End

Higher optimism and larger assurance are offered by the blowfish algorithm.

4.2 Systemic Level

The systemic level process of the CP-BHWABE system includes the following,

Setup Admin calls setup function to get public parameters. Data sender chooses the security parameters and then requests admin to yield the private key S_k components and sends the encrypted data to CA, whereas the CA verifies user attributes, and if the attributes are valid then CA utilizes P_k and M_k of the system to create P_k and S_k to a newly registered user. The weight attribute authority allocates weight priority to the attributes according to their importance.

Keygen While a user logs in for the first time into server, a unique user ID is been generated for user by CA. The attribute set of users is sent to WAA; the attributes are authenticated by attribute authority. If the attributes are valid, then secret key is generated for user who logs in for the first time, based on their importance of the attributes by WAA. Then, the receiver's secret key of system and attribute secret key of the user is transferred separately by WAA and CA. Both authority and central authority setup are initiated to get the keys.

Encrypt Before the plain text message file is uploaded to the Internet, data sender first has to log in through uniquely generated ID for them. Then, encryption key is randomly generated to encrypt the file to be sent. Data sender also defines a threshold weighted access structure (W) for receivers of the data, and then, the message data file is encrypted with W and sent over the Internet as illustrated in Fig. 2.

Decrypt The received files are been downloaded by the data receiver, and the message file is decrypted. Only if the attribute secret key of the receiver is authenticated, system generates priority values to its attributes based on the level of attributes.

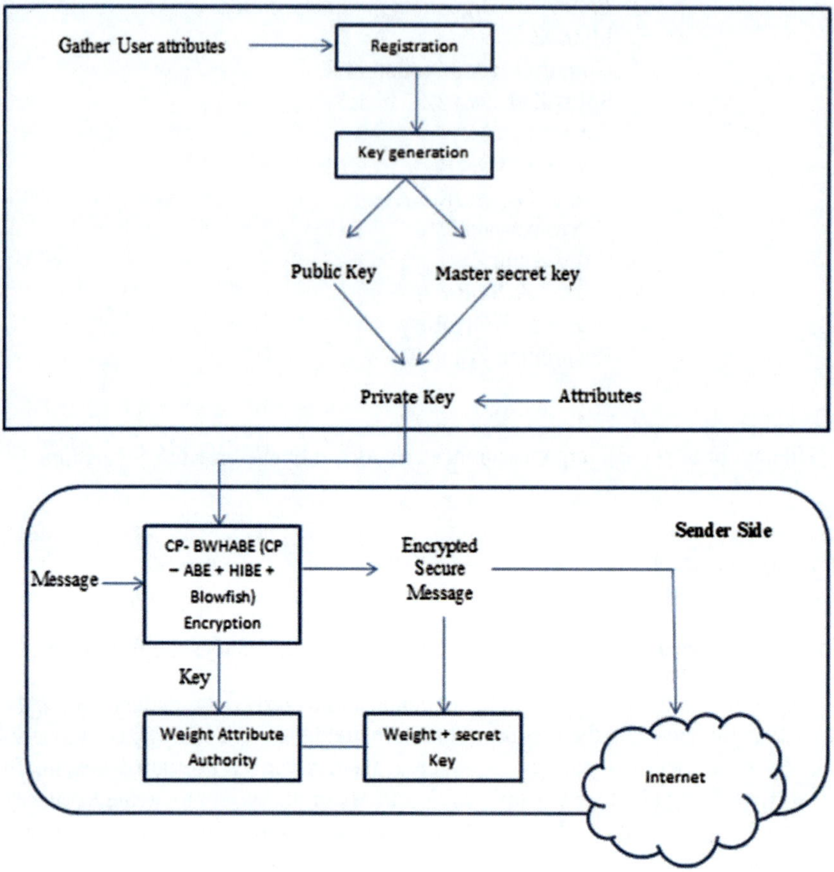

Fig. 2 Sender side (data encryption)

Then with respect to W, the receiver decrypts the file. If a user is found to be invalid, user is not allowed to decrypt as illustrated in Fig. 3. The invalid user is reported. If they are found to be the attacker, then they are added to the blacklist.

5 Results and Comparison

Experimental results are analyzed to get the evaluation of performance. The CP-BHWABE system is implemented in Java platform (J Creator), with an Intel i3 core with 4 GB RAM. The performance analysis of computational cost in encryption and decryption is computed. This system not only provides security for data, also ensures the control of accessing the ciphertext. While comparing CP-ABE and HIBE, the

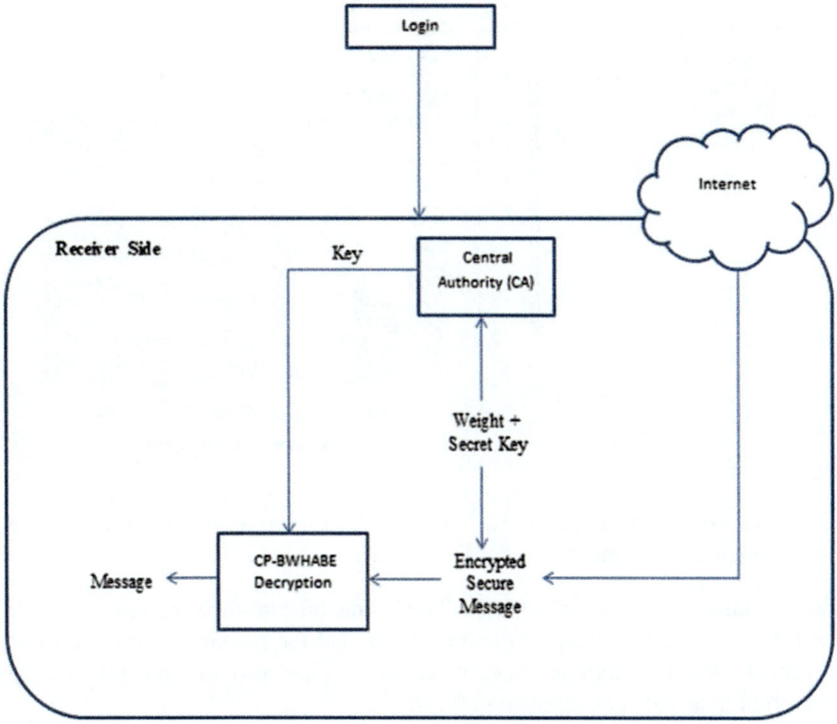

Fig. 3 Receiver side (data decryption)

proposed CP-BHWABE performs with full delegation, less workload and lightweight management of key for large number of users.

In CP-BHWABE system, many files are taken in various different sizes (in kb and mb) and several different file formats (txt, pdf, jpg, png, mp3, mp4, mov) as input, which are encrypted, then decrypted by CP-BHWAB. Generating necessary keys and weights is done with CP-BHWABE. This system is mainly designed for encrypting user identity, data security and privacy and also to do these in the lesser execution time. The final result of this system is discussed in Table 2. The time cost of the

Table 2 Performance analysis of computational cost of encryption, decryption and throughput for CP-BHWABE

Input (MB)	Size of file	Time taken to encrypt data (s)	Time taken to decrypt data (s)	Calculated throughput (MB)
100	21	7	0.00476	100
200	41	18	0.00487	200
300	62	26	0.00483	300

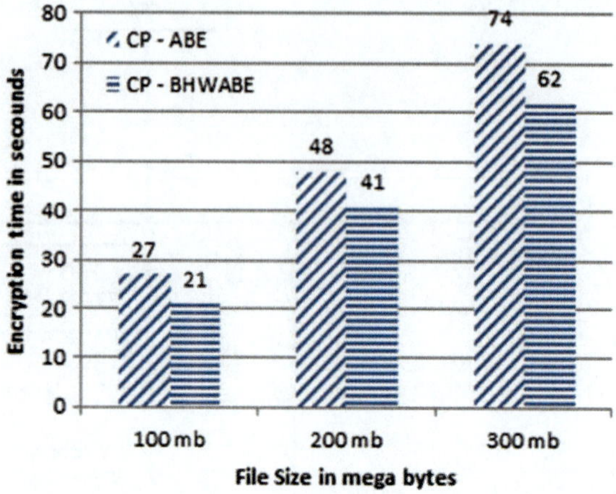

Fig. 4 Comparison of computational cost of encryption between CP-ABE and CP-BHWABE

CP-BHWABE is compared with CP-ABE system by considering the performance metrics of encryption and decryption.

Encryption time Encryption time is defined as the time taken to encrypt the data. It can further be used to estimate the throughput and the system speed to generate encrypted data. The duration taken to convert a plain text to ciphertext defines encryption time and is represented in Fig. 4.

Decryption time Decryption is the reverse of encryption. The time taken to decrypt is estimated to be the time taken to convert ciphertext into plain text. The time to decrypt data for CP-ABE and CP-BHWABE is calculated and analyzed which is represented in Fig. 5.

Calculated throughput Ratio between file size and data encrypted time is defined as throughput. Throughput is shown in Fig. 6 can be calculated as

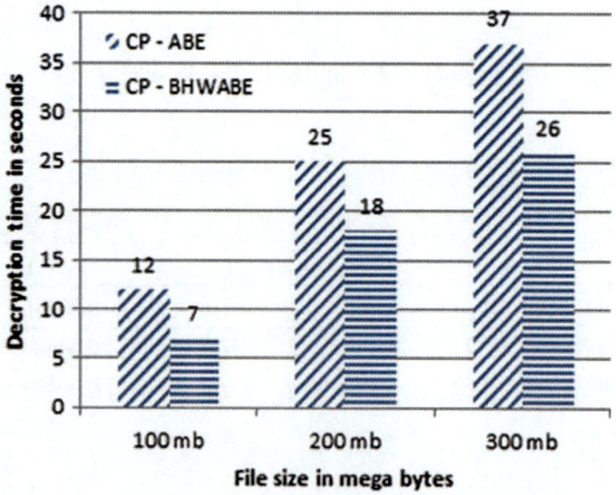

Fig. 5 Comparison of computation of decryption time between CP-ABE and CP-BHWABE

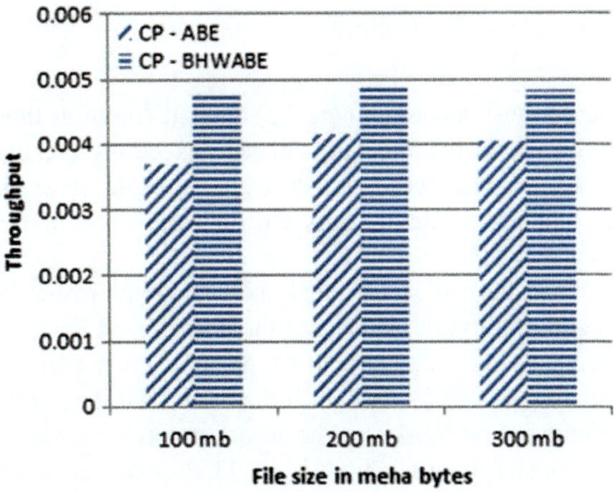

Fig. 6 Comparison of throughput between CP-ABE and CP-BHWABE

$$\text{Throughput} = \frac{\text{File size(mb)}}{\text{Encryption time}} \tag{1}$$

5.1 Analysis of Security

Data shared over the Internet is encrypted with CP-BHWABE that provided high security from the attack of "chosen plain text" and Diffie–Hellman attack. The analyses of security property of the system are

Confidentiality of data The confidentiality of the data is obtained against a user who does not have the valid attribute set that satisfies the access policy; for this data is encrypted with access policy. Even though server handles the process of data encryption, still it cannot decrypt the data since they do not have the attribute for decryption of the ciphertext generated in the encryption process. In decryption process, if the attributes set is not able to satisfy the access control policy in the ciphertext, the server cannot get the sender user weights. So that only users who have attributes that are valid can satisfy the access control policy and can do decryption on received data file.

Access Control The access control is mainly for the flexibility in specifying many different access control policies of each and every individuals. So to bring this type of access control policy, CP-BHWABE is utilized. In encrypting data, sender's access control policy is flexible and generates key for encrypting data. Importantly, access tree consists of access control policy for the data to be encrypted. It performs complex operations which include OR and AND gates; these are used to represent data in any needed access condition.

5.2 Data Collusion

Data collusion is combining the same files multiple times or different files altogether to form a new file; this can be done by executing a set of operations. In the Internet, an attacker can use such methods to guess private key and hack inside system, accessing the other user's files illegally. CP-BHWABE system solves this problem by increased security.

Consider two users, User1 and User2 with respective identities user1 and user2. In addition to this, User1 has the attribute set Set1; Authority Auth1 monitors the elements. The access structure f (N1, User1) can only be satisfied by Set1 but it cannot satisfy (N2, user2). Also, the other User2 has the attribute set Set2; Authority Auth2 monitors their elements. The access structure f (N2, User2) can only be satisfied by Set2 but it cannot satisfy (N1, user1). Both obtain keys from Auth1 and Auth2. But if user is not found in encryption list, (N2, user2) cannot be obtained in the ciphertext. So, the decryption algorithm will not work and it fails. So from this, understand that the user who is allowed to access the data can only access it.

6 Conclusion

In the data transfer between two applications from sender to receiver via the Internet, the user authentication, data security, user privacy and key management are main issues. So, an algorithm with efficiency and scalable access control policy is proposed. The system comprises the combination of ciphertext policy attribute-based encryption and identity-based encryption which is hybridized and weighted with blowfish algorithm. It provides data security even in the multiple authority as hierarchical structure is maintained with the importance of each attributes; also it provides lightweight key management for large number of users. This system also reduces the computational cost. The CP-ABE and blowfish are used for encryption and decryption of the data and are transferred securely. When the authenticated user wants to send a data file, it is encrypted with the values of the user attributes and sent over the Internet. With valid key generated by blowfish, the receiver has to decrypt the ciphertext. From the result, we can see that the CP-BHWABE system is reliable, secure, efficient, data confident and flexible access control policy.

The extension of this work in future can probably be protection—saving property basis information sharing with re-encryption and quality-based encryption.

References

1. Goyal, V., Pandey, O., Sahai, A., Waters, B.: Fuzzy identity-based encryption. In: EUROCRYPT'05, pp. 457–473. Springer (2005)
2. Goyal, V., Pandey, O., Sahai, A., Waters, B.: Attribute based encryption for fine-grained access control of encrypted data. In: ACM Conference on Computer and Communication Safety, pp. 89–98 (2006)
3. Bethencourt, J., Sahai, A., Waters, B.: Ciphertext-policy attribute-based encryption. In: Rado, G.T., Suhl, H. (eds.) Magnetism, vol. III, pp. 271–350. Academic, New York
4. Wan, Z., Liu, J., Deng, R.H.: HASBE: a hierarchical attribute-based solution for flexible and scalable access control in cloud computing. In: IEEE Trans. Inf. Forensics Secur. **7**(2) (2012)
5. Waters, B.: Ciphertext-policy attribute-based encryption: an expressive, efficient, and provably secure realization. In: International Association for Crytologic Research, PKC 2011, LNCS 6571, pp. 53–70, (2011)
6. Deng, H., Wu, Q., Qin, B., Domingo-Ferrer, J., Zhang, L., Liu, J., Shi, W.: Ciphertext-policy hierarchical attribute-based encryption with short ciphertexts. Int. Sci. **275**, 370–384 (2014)
7. He, K., Chen, M.-R., Mao, Y., Zhang, X., Zhan, Y.: Efficient hierarchical identity–based encryption for mobile ad hoc network. In: Mobile Information System, pp. 407–425
8. Wang, H., Wu, S., Chen, M., Wang, W.: Security protection between users and mobile media cloud. IEEE Commun. Mag. **52**, 73–79 (2014)
9. Li, J., Huang, X., Li, J., Chen, X., Xiang, Y.: Securely outsourcing attribute-based encryption with checkability. IEEE Trans. Parallel Distrib. Syst. **25**, 2201–2210 (2014)
10. Li, M., Yu, S., Zheng, Y., Ren, K., Lou, W.: Scalable and secure sharing of personal health records in cloud computing using attribute-based encryption. IEEE Trans. Parallel Distrib. Syst. **24**, 131–143 (2013)
11. Mousa, A.: Data encryption performance based on Blowfish. In: Proceedings of 74th International Symposium ELMAR, pp 131–134 (2015)
12. Wang, Y., Gao, J.: A regulation scheme based on the ciphertext policy hierarchical attribute—based encryption in bitcoin system. In: IEEE Access vol. 6, pp 16267–16278 (2018)

Reconfigurable Antenna Using RF MEMS Switches Issues and Challenges: A Survey

Mallikharjuna Rao Sathuluri and G. Sasikala

Abstract The reconfigurable antenna provides on-demand services over the present and future communication applications with the help of a RF switches. The features of reconfigurable antenna are offering multiple resonant frequencies depending on the mode of the antenna. The switching is not only possible from one resonant frequency to another; it is also possible to switch from one polarization to other polarization, i.e., circular polarization to linear polarization. This paper discusses about the basic features of microstrip antenna re-configurability, power efficiency, polarization, multipath, inference, fading, and RF distortion. Additionally, the paper describes several key topics related to RF MEMS switch-based reconfigurable patch antenna material analysis. This paper surveys the role and challenges of RF MEMS switches based reconfigurable patch antennas. A little discuss is extended on the PIN diode-based reconfigurable antennas.

Keywords Patch antenna · Re-configurability · RF MEMS switch · PIN diode · FET · Polarization · Bandwidth · Gain · Directivity

1 Introduction

In the history, antennas would take enough space in the communication applications. But still, the role of antenna is significant and expecting the antenna with smart features like re-configurability in terms of frequency and polarization. The switch-based reconfigurable antennas can be designed using different switches like PIN

M. R. Sathuluri (✉) · G. Sasikala
Department of ECE, Vel Tech Rangarajan Dr. Sagunthala R&D Institute of Science and Technology (Deemed to be University), Chennai, Tamil Nadu 600062, India
e-mail: smr.aliet@gmail.com

G. Sasikala
e-mail: gsasikala@veltech.edu.in

M. R. Sathuluri
Department of ECE, Andhra Loyola Institute of Engineering and Technology, Vijayawada, Andhra Pradesh, India

P. Suresh et al. (eds.), *Advances in Smart System Technologies*, Advances in Intelligent Systems and Computing 1163,
https://doi.org/10.1007/978-981-15-5029-4_11

diodes, varactor diodes, MEMS switches, Field effect transistors, optical controlled switches. The simple reconfigurable antenna can be design by interfacing the two antennas with RF switches. From the literature, it is clear that RF switches based reconfigurable antennas are offering best performance compared with other switches. In this reconfigurable antenna using RF switches, paradigm is facing some real-time challenges like multipath, inference, fading and RF distortion is the major issues will rise, when we try to interface two antennas. PDMS-based high-Q microstrip patch antennas may not offer high radiation efficiency. The antenna radiation efficiency is also an important factor in antenna optimization. Generic algorithms will help to optimize the reconfigurable pixel antennas [1]. Without using the switches also we can design reconfigurable antennas, but the tuning frequency range is very low; the bandwidth offered by this type of antenna is also very low [2]. The general antenna performance characterization parameters are gain, directivity, efficiency, directivity gain, bandwidth, VSWR, spillover loss, taper loss, cross polarization loss, maximum element reflective loss, insertion and isolation losses of RF MEMS switch [3]. In the design of hybrid antenna designs, use of unwanted couplings majorly impacts the radio frequency performance [4]. Different types of switches are used to design electrically and optically reconfigured antennas, i.e.,

- RF MEMS: The switching in the transmission lines is created by placing micro/nano-mechanical structures with the incorporation of electrostatic (or) magneto static (or) thermal (or) piezoelectric actuation methods.
- PIN Diodes: These switches are fabricated sing GaAs technology. No need of mechanical actuation for switching. Offers high switching speed and low isolation compared to MEMS switches.
- Varactors: These are designed using p–n junction diode switches. The varactor capacitance can be change by varying the switching of p–n junction diodes.
- Photoconductive Elements: These switches are optical based. The switch ON and OFF activity is achieved by light of appropriate wavelength. For light, laser diodes are useful.

At low operating frequency PIN diode, varactor diode, photoconductive elements are offering the good performance. But, the operating frequency is above 100 GHz; these switches are not able to reach the expectation. MEMS switches are giving solution for high-frequency operating reconfigurable antennas.

2 Related Work

The re-configurability paradigm is most widely discussed area in both industry person and researcher. The re-configurability in terms of frequency and polarization is the smart sign to fulfill the requirement in present-day communication applications. In this paper, we discuss few challenges and future direction of RF MEMS switches based reconfigurable antennas. In this section, several survey papers are included which is the best of our knowledge. By redistributing antenna, currents with respect to electromagnetic fields can help to reconfigurable antenna in terms of frequency, polarization or radiation characteristics. Modifications in geometry, electrical behavior are required to design reconfigurable antennas in terms of bandwidth enhancement, change in operating frequency, polarization and radiation pattern.

2.1 RF MEMS Switches

RF MEMS technology is a multidomain concept; the design requires the knowledge on radio frequency, mechanical and electrical domains [2, 3, 5–7]. In 1991, Larson proposed the concept of RF MEMS switches, and later, so many researchers proposed different RF MEMS switches for wide range of communication applications.

The paper [8] proposes a new structure to achieve best insertion losses, multiband RF MEMS switches with minimum actuation voltage. When the switch is in ON state, it is offering an insertion loss of −0.1 dB and return losses of −36.8 dB up to 25 GHz. When the switch is in OFF state, it offers two isolation peaks, i.e., −48. dB at 4.5 GHz and −54.56 dB at 9.7 GHz. A novel structure is proposed with two cantilever structures.

The paper [9] elaborated clamped–clamed structure RF MEMS switch. GaAs is used as a substrate. The air gap is 2.5 μm which is requiring 30 V actuation voltages.

2.2 Reconfigurable Antennas

The paper [2] presents a detailed study about the E-shaped reconfigurable patch antenna using two RF MEMS switches. The paper is concluded that the E-shape antennas will offer more bandwidth compared with other shapes. The proposed switch is for cognitive radio application. The switch off state frequency is in the range 2–2.6 GHz, and the switches on state frequency are in the range 2.6–3.2 GHz range.

The studies of the paper [5] over viewed circular/linear polarization reconfigurable antenna. Circular shape is chosen for the patch antenna. Overall, one RF MEMS switch is used for re-configurability in terms of polarization. The antenna is designed for satellite applications. The antenna frequency is in the K-band region.

The authors [6] presented pixel slot antenna using RF MEMS switch Radant SPST-RMSW101. Overall, the design is incorporated with four switches. Total number of modes is five. Tuning frequency range is 1.56–3 GHz.

The paper [7] proposes dual band circular polarized antenna. The antenna is reflect array (RA) type. The antenna can switch from K-band to Ka-band. The reflection phase of circularly polarized waves is able to control by the switches at 24.4 and 35.5 GHz independently.

The paper [3] overviews reflect array antenna monolithically oriented with ninety RF MEMS switches. Aperture coupled microstrip patch antenna elements are used to pattern a 10 × 10 element reconfigurable reflect array antenna radiates at 26.5 GHz.

The authors [10] presented slot array antenna with RF MEMS. Antenna is designed for X-Band applications. Overall, four switches are used in the design.

The paper [11] proposes three states of polarization with pneumatically controlled reconfigurable antenna. Four PIN diodes are used for design of antenna.

3 Reconfigurable Antenna Using RF MEMS Switches Framework

An antenna with re-configurability feature would enable change of radiation pattern, impedance bandwidth, polarization, operating frequency. The simple method to make single antenna as reconfigurable is by extending its length using matching switches. The switch mismatching may effect impedance bandwidth and operating frequency.

3.1 Polarization

Antenna is an bidirectional energy converter. It converts electrical to electromagnetic and electromagnetic to electrical. The electromagnetic wave consist E-plane and H-plane. E-plane helps to find the type of polarization. There exist different polarizations like circular, elliptical and linear polarization. In RFID application, transmitting antenna needs circular polarization and receiving antenna needs linear polarization.

Linear polarization

If the electromagnetic wave is propagating either horizontal plane or vertical plane, then it comes in the category of linear polarization. If the E-field in the electromagnetic wave is parallel with the earth surface, then we can call it as horizontal linear polarization, and if the E-field is maintaining 90° with earth surface, then it is vertical linear polarization. Linear polarization has demand in C-band and Ku-band applications (Fig. 1).

Circular polarization

Satellite communication highly demands this type of polarization. In this type of polarization, the wave propagates with same frequency and magnitude but the electric field is in the circular shape. The limitation of this polarization is limited range of the RF lose because they splits power across two different planes (Fig. 2).

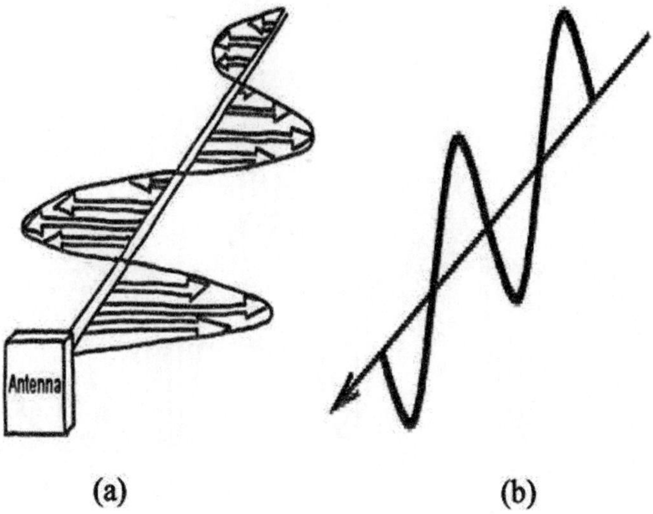

Fig. 1 Linear polarization **a** horizontal and **b** vertical

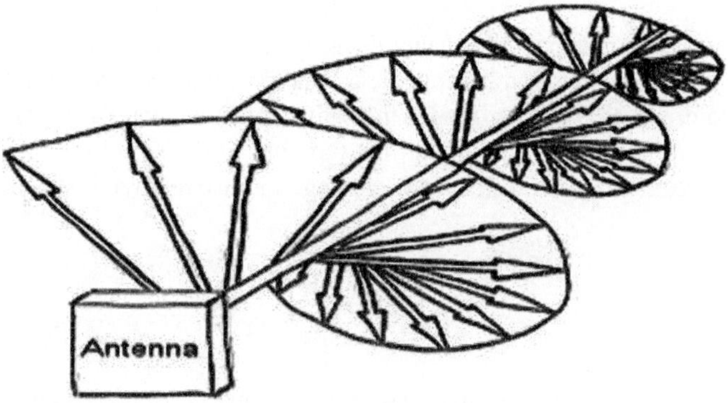

Fig. 2 Circular polarization

Elliptical polarization

It is one mixed verity of circular and linear polarization. In this polarization, the EM radiation is in elliptical shape if we relate with propagation direction. This polarization precisely requires in satellite communication as similar to circular polarization, but the linear polarization is used in regular simple antenna applications (Fig. 3).

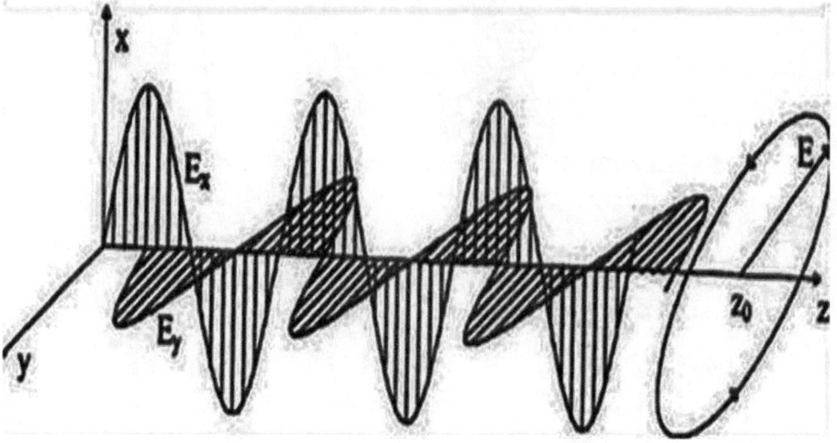

Fig. 3 Elliptical polarization

3.2 *Fading*

Fading is one of the antenna performance deciding factors. It indicates how the signal is attenuated by depending on time, geographical position and radio frequency. By using random process, we can model the fading phenomenon of the antenna. In practical fading because of multipath, this type of fading is referred as multipath fading. Bad weather is also one reason for fading; it is also referred is shadow fading.

3.3 *Interference*

Transmitters and electrical systems used in the communication cause interference. It may affect the simultaneous multiple signal reception; sometime, it leads to signal loss. The quality of the signal also depends on the interference property of the antenna.

3.4 *Multipath*

Because of reflections, the signal receives at antenna in multiple paths. These signal reflections are because of obstacles present in the path of propagation. A communication with high line of sight is preferable; here line of sight refers the single direct path between transmitter and receiver without displacement error as shown in Fig. 4.

If the signal receives with some reflections, it is known as multipath. Achieving the high line of sight, especially in metropolitan cities, is really a challenging task.

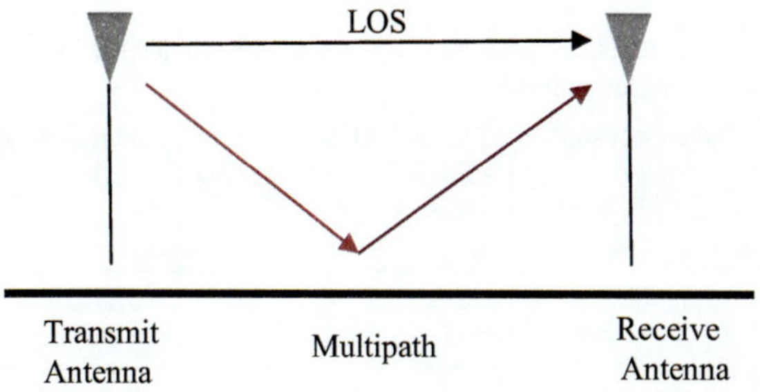

Fig. 4 Multipath in antenna

Signal loss and some phase errors will occur because of multipath phenomenon (Tables 1, 2, 3 and 4).

4 Antenna Basic Measurements

An antenna is basically a transducer. It converts radio frequency (RF) electrical current into an electromagnetic (EM) wave of the same frequency. Any antenna can be characterized with antenna parameters.

4.1 Antenna Impedance

It is defined as the ratio of input voltage to input current,

$$Z_a = \frac{V_i}{I_i}\Omega \tag{1}$$

Here, 'Z_a' is a complex quantity, and it is written as

$$Z_a = R_a + jX_a \tag{2}$$

where the reactive part 'X_a' results from fields surrounding the antenna. The resistive part R_a is given as

$$R_a = R_l + R_r. \tag{3}$$

Here, 'R_l' represents losses in the antenna, and 'R_r' represents radiation resistance.

Table 1 Comparison of the related work with the survey based on RF MEMS switches, actuation voltage, insertion losses, isolation losses

Survey paper	Year	Topic discussed	RF MEMS switch type	Gap (μm)	Pull in voltage (V)	Isolation losses	Insertion losses
Ramana et al. [12]	2018	Serpentine structure, pull in voltage	Series metal contact	3	4	−67.9714 dB at 12 GHz	−0.094 dB at 12 GHz
Pertin et al. [13]	2018	Bridge structure, pull in voltage	Shunt capacitive	0.6	19	−12 dB	−0.05 dB
Angira et al. [14]	2016	Dual cantilever, insertion losses, pull in voltage	Shunt capacitive	2	12.75	5–17 dB (1–25 GHz)	0.02–0.1 dB (1–25 GHz)
Zhu et al. [8]	2014	High isolation, high power applications	Series metal contact and shunt capacitive	10	62	−50 dB at 5 GHz, −45 dB at 20 GHz and −60 dB at 35 GHz	0.2–0.6 dB (1–35 GHz)
Persano et al. [15]	2015	Bridge structure, GaAs substrate	Shunt capacitive	2.5	30	–	–
Mulloni et al. [9]	2013	Clamped-clamped structure, Influence of temperature on the actuation voltage	Shunt capacitive	2.9	65	–	–
Ghodsian et al. [16]	2008	Cantilever beam, contact force, contact resistance, CPW transmission line	DC contact series	3	48	−39 dB at 5 GHz,−32 dB at 10 GHz	0.1–0.5 dB (1–40 GHz)

Table 2 Extensive survey on effect of antenna shape on bandwidth

Survey paper	Antenna shape	Frequencies	Impedance band width of antenna
Kona et al. [17]	Stacked	Radar application-1.26 GHz Radiometer application-1.4 GHz	±10 MHz ±10 MHz
Jung et al. [5]	Circular with stub	16.9–22.5 GHz	22%
Luk et al. [18]	L-shaped	3.76–5.44 GHz	35%
Huynh et al. [19]	U-slotted	812 MHz to 1.2 GHz	45%
Yang et al. [20]	E-shaped	1.9 and 2.4 GHz	30.3%

Table 3 Comparison of the related work with the survey based on the reconfigurable antenna overview, antenna materials, parametric analysis, antenna features, RF switches

Survey paper	Year	Topic discussed	Antenna shape	Number of switches	Switches	Switches state and resonant frequency	Antenna application
Rajagopalan et al. [2]	2014	RF MEMS switches, Particle swarm optimization, impedance bandwidth, antenna shape	E-shape	2	RMSW 100HP—SPST, high power (10 W), DC-12 GHz	S1 and S2: ON-2–2.6 GHz S1 and S2: OFF-2.6–3.2 GHz	Cognitive radio
Jung et al. [5]	2012	linear and circular polarization, switch packaging, impedance bandwidth	Circle	1	DC contact RF MEMS	S1: OFF-circular polarization (17.4–21.9 GHz) S1: ON-linear polarization(16.9–22.5 GHz)	Satellite
Chiu et al. [6]	2012	Re-configurability with RF MEMS switches	Pixel slot and loop	4	RMSW101—SPST, high power (10 W), DC-12 GHz	All switches: ON: 3.3 GHz S1: ON: 2.5 GHz S1 + S2: ON:2 GHz S1 + S2 + S3: ON: 1.56 GHz S1 + S2 + S3 + S4: ON: 2.42 GHz	Wi-Fi
Guclu et al. [7]	2012	Circularly polarization, reflect array (RA), sequential rotational principal	Split rings	6	Series DC contact RF MEMS switch	24.4 GHz and 35.5 GHz independently	K-band and Ka-band
Bayraktar et al. [3]	2012	Reflect array (RA), phase shift, ACMPA	Aperture coupled microstrip patch antenna	90	Series RF MEMS switch	26.5 GHz	Ka-band
Sánchez-Escuderos et al. [10]	2011	RF MEMS switches, antenna gain	Slot array design	4	RMSW101—SPST, High Power (10 W), DC-12 GHz	10.25 GHz	X-band

(continued)

Table 3 (continued)

Survey paper	Year	Topic discussed	Antenna shape	Number of switches	Switches	Switches state and resonant frequency	Antenna application
Cheng et al. [4]	2009	Antenna-filter-antenna (AFA), Electronically scanned arrays(ESAs), RF MEMS switches	Programmable lens-array antenna	5 switches, 4 modes	Capacitive RF MEMS switches	35 GHz	Ka-band
Jung et al. [21]	2008	RHCP, dual frequency, high-frequency ratio, VSWR perturbation stub	Square shape	1	Series RF MEMS switch	S1 ON—4.8 GHz S1 OFF—7.6 GHz	C-band
This survey		Overview of reconfigurable antenna, design issues, parametric issues, future challenges, material analysis	–	–	–	–	–

Table 4 Extensive survey on reconfigurable antenna with PIN diode

Survey paper	Topic discussed	Antenna shape	Number of PIN diodes	Number of operating modes	Mode and operating frequency	Application
Wu et al. [11]	LP/LHCP/RHCP and LP/orthogonal-LP/LHCP/RHCP	Circular	4	2	2.42 GHz	Cognitive radio
Wang et al. [22]	Magnetic current	Series of rectangular antennas	2	3	M1:D1-ON,D2-OFF:Directinal status 1, M2:D1-OFF,D2-ON-:Directional status 1, M3:D1-OFF, D2-OFF-Bidirectional	WiMax
Alam et al. [23]	Reconfigurable band rejection antenna	Stub and slot/diodes	3	4	*M1:D1-ON,D2-OFF,D3-OFF* = *3.5 GHz* *M2:D1-ON,D2-ON,D3-ON* = *5.5 GHz* *M3:D1-OFF,D2-OFF,D3-OFF* = *7.5 GHz* *M4:D1-OFF,D2-ON,D3-ON* = *9.5 GHz*	WiMax wide local area network (WLAN) and ETSI HiperLAN/2

4.2 Directivity

The directivity (D) of the antenna is the ratio of the amount of energy transmitted in the direction with the strongest power to total amount energy transmitted in all directions.

$$D = 4\Pi \frac{U_{\max}}{P_{\text{rad}}} \tag{4}$$

4.3 Efficiency

For antenna, the complex impedance is $Z_{\text{ant}} = R_{\text{ant}} + JX_{\text{ant},}$ where R_{ant} is the antenna resistance; it is a combination of radiation resistance (R_{rad}) and dissipation resistance (R_{diss}). The antenna phasor current $I_{\text{o}} = I_{\text{s}}e^{\text{j}}$, and the radiation power is $P_{\text{rad}} = (1/2)I_{\text{o}}^{2}R_{\text{rad},}$ the dissipation power is $P_{\text{diss}} = (1/2)I_{\text{o}}^{2}R_{\text{diss}}$. From the radiated power and the dissipated power, we can write the expression for antenna efficiency i.e.,

$$E = \frac{P_{\text{rad}}}{P_{\text{rad}} + P_{\text{diss}}} \tag{5}$$

4.4 Gain

The ability of direct the input power into radiation in a particular direction is measured with the antenna gain. It can be expressed in terms of antenna efficiency and directivity as follows:

$$G = \text{Efficiency} * \text{Directivity} \tag{6}$$

4.5 Bandwidth

The antenna bandwidth can be expressed as the range of frequencies over which an antenna meets a published performance requirement.

$$\text{Bandwidth} = f_2 - f_1 \tag{7}$$

where 'f_1' is the lower cutoff frequency and 'f_2' is the uppercut of frequency. The impedance bandwidth is nothing but the fractional bandwidth; the expression for impedance bandwidth is

$$\text{Impedance Bandwidth} = \frac{2(f_{\max} - f_{\min})}{(f_{\max} + f_{\min})} \times 100 \tag{8}$$

4.6 VSWR

Voltage standing wave ratio (VSWR) is one of the antenna performance indices which will help to measure the antenna efficiency in the transmission of radio frequency power, by using transmission line, into a load. If the VSMR value is one, it indicates that there are no signal reflections. If the VSWR value is above one, it indicates that signal reflections are significantly increasing. The VSWR can be expresses as

$$\text{VSWR} = \frac{1 + \Gamma}{1 - \Gamma} \tag{9}$$

where 'Γ' is the reflection coefficient.

4.7 Return Losses

The return loss is the ratio of the input signal power to reflected signal power. It is usually expressed as a ratio in decibels (dB). Antenna with return loss below −10 dB is acceptable. The antenna return loss can be expressed as (Table 5)

$$\text{ReturnLosses} = 10 \log_{10} \frac{P_{\text{input}}}{P_{\text{reflected}}} \tag{10}$$

5 Material Selection

Materials or medium that are homogenous, linear and isotropic are called simple materials or media. There are three different types of materials which are of much use in EM theory which are conductor, dielectric and magnetic. A good conductor is defined as a material having electrical parameters $\varepsilon_{0,}\ \mu_{0,}\ \alpha \gg \omega\varepsilon_0$ at operating frequency $f = \omega/2\pi$ below optical region. In general, we can use Au or Cu as conductors. In antenna, substrate acts as dielectric material. The substrate material properties

Table 5 An extensive survey on frequency bands and respective applications

Frequency band	Frequency range	Application
L-band (1–2 GHz)	1–2 GHz	Long-range air traffic control and surveillance (radar), GPS
S-band (2–4 GHz) and C-band (4–8 GHz)	3.3–3.6 GHz; 5.25–5.825 GHz	IEEE 802.16-WiMax Cognitive radio
C-band (4–8 GHz)	5.15–5.35 GHz; 5.725–5.825 GHz	IEEE 802.11a—wide local area network (WLAN)
C-band (4–8 GHz)	5.15–5.35 GHz, 5.47–5.725 GHz	IEEE 802.11a—ETSI Hiper LAN/2
X-band (8–12 GHz)	7.25–7.75 GHz	Fixed and mobile satellites
	7.75–7.9 GHz	Fixed satellite
	7.9–8.4 GHZ	Fixed earth exploration satellites
	8.5–10.5 GHz	Radio location
	8–12 GHz	Synthetic-aperture radar (SAR)
Ku-band (12–18 GHz)	12–18 GHz	Direct broadcast satellite, DHT television
K-band (18–27 GHz)	18–27 GHz	Surface movement radars
Ka-band	27–40 GHz	Surface movement radars
V-band	40–75 GHz	High attenuation radar applications
W-band	75–110 GHz	High-frequency radar

will show major impact on the performance indices of the antenna in terms of operating frequency, efficiency and bandwidth. So selection of the antenna materials is also very important. List of different substrate materials is listed in Table 6. The feed line width is proportional to substrate thickness and inversely proportional to dielectric constant. The antenna impedance locus is more inductive when the thickness of the substrate material and inductance of the feed line is increased.

6 Conclusion

The survey on the reconfigurable antenna using the RF MEMS switches is presented in this paper. The challenges in the design of reconfigurable antenna are fading, multipath, inference, switch mismatching. By using RF MEMS switches, we can design antennas with reconfigurable polarization or reconfigurable operating frequency. The reconfigurable antennas can be designed without using the switches, but the frequency is in the low range. The RF MEMS switches based antennas will offer high tuning frequency. RF MEMS switches can operate at high power, and the power consumption is very low; this feature attracts the future research on reconfigurable antennas using RF MEMS switches.

Table 6 Substrate materials for patch antenna

Substrate	Dielectric constant	Size radiation	Bandwidth
Benzocyclobutane2	2.6	Medium	Medium
Duroid 6010	10.7	Lowest	Minimum
Nylon fabric	3.6	Medium	Medium
Roger 4350	3.48	Medium	Medium
RT-Duroid	2.2	Medium	Medium
Foam	1.05	Highest	Maximum
FR-4	4.4	Medium	Medium
Nylon 610	2.84	Medium	Medium
Fused Quartz	3.78	Medium	Medium
Beryllia	6.4	Lowest	Minimum
TMM 10	9.2	Lowest	Minimum
Silicon	11.9	Lowest	Minimum

References

1. Song, S., Murch, R.D.: An efficient approach for optimizing frequency reconfigurable pixel antennas using genetic algorithms. IEEE Trans. Antennas Propag. **62**(2), 609–620 (2014)
2. Rajagopalan, H., Kovitz, J.M., Rahmat-Samii, Y.: MEMS reconfigurable optimized E-shaped patch antenna design for cognitive radio. IEEE Trans. Antennas Propag. **62**(3), 1056–1064 (2014)
3. Bayraktar, O., Civi, O.A., Akin, T.: Beam switching reflectarray monolithically integrated with RF MEMS switches. IEEE Trans. Antennas Propag. **60**(2), 854–861 (2012)
4. Cheng, C.-C., Lakshminarayanan, B., Abbaspour-Tamijani, A.: A programmable lens-array antenna with monolithically integrated MEMS switches. IEEE Trans. Microw. Theory Tech. **57**(8), 1874–1884 (2009)
5. Jung, T.J., Hyeon, I.-J., Baek, C.-W., Lim, S.: Circular/linear polarization reconfigurable antenna on simplified RF-MEMS packaging platform in K-band. IEEE Trans. Antennas Propag. **60**(11), 5039–5045 (2012)
6. Chiu, C.-Y., Li, J., Song, S., Murch, R.D.: Frequency-reconfigurable pixel slot antenna. IEEE Trans. Antennas Ropaga. **60**(10), 4921–4924 (2012)
7. Guclu, C., Perruisseau-Carrier, J., Civi, O.A.: Proof of concept of a dual-band circularly-polarized RF MEMS beam-switching reflectarray. IEEE Trans. Antennas Propag. **60**(11), 5451–5455 (2012)
8. Zhu, Y., Han, L., Qin, M., HuangKey, Q.: Novel DC-40 GHz MEMS series-shunt switch for high isolation and high power applications. Sens. Actuators A: Phys. **214**, 101–110 (2014)
9. Mulloni, V., Solazzi, F., Ficorella, F., Collini, A., Margesin, B.: Influence of temperature on the actuation voltage of RF-MEMS switches. Microelectron. Reliab. (2013)
10. Sánchez-Escuderos, D., Ferrando-Bataller, M., Baquero-Escudero, M., Herranz, J.I.: Reconfigurable slot-array antenna with RF-MEMS. IEEE Antennas Wirel. Propag. Lett. **10**, 721–725 (2011)
11. Wu, B., Okoniewski, M., Hayden C.: A pneumatically controlled reconfigurable antenna with three states of polarization. IEEE Trans. Antennas Propag. **62**(11), 5474–5484 (2014)
12. Ramana, R., Shanmugananthamb, T., Sindhanaiselvic, D.: Analysis of RF MEMS series switch with serpentine spring shaped cantilever beam for wireless applications. Mater. Today: Proc. **5**, 1890–1896 (2018)

13. Pertin, O., Kurmendra, Pull-in-voltage and RF analysis of MEMS based high performance capacitive shunt switch. Microelectron. J. **77**, 5–15 (2018)
14. Angira, M., Rangra, K.: A novel design for low insertion loss, multi-band RF-MEMS switch with low pull-in voltage. Eng. Sci. Technol. Int. J. (2015)
15. Persano, A., Quaranta, F., Martucci, M.C., Siciliano, P., Cola, A.: On the electrostatic actuation of capacitive RF MEMS switches onGaAs substrateAnna. Sens. Actuators A: Phys. A, 232, 202–207 (2015)
16. Koszewskia, A., Souchona, F., Levya, D.: Mechanical properties of MEMS switches: modeling and characterization. Procedia Chem. 626–629 (2009)
17. Kona, K.S., Bahadori, K., Rahmat-Samii, Y.: Stacked microstrip-patch arrays as alternative feeds for spaceborne reflector antennas. IEEE Antennas Propag. Mag. **49**(6), (2007)
18. Luk, K.M., Mak, C.L., Chow, Y.L., Lee, K.F.: Broadbaand microstrip patch antenna. Electron. Lett. **34**(15) (1998)
19. Huynh T., Lee, K.-F.: Single-layer single-patch wideband microstrip antenna. Electron. Lett. **31**(16) (1995)
20. Yang, F., Zhang, X.-X., Ye, X., Rahmat-Samii, Y.: Wide-band E-shaped patch antennas for wireless communications. IEEE Trans. Antennas Propag. **49**(7) (2001)
21. Jung, C.W., Lee, M.-J., De Flaviis, F.: Reconfigurable dual-band antenna with high frequency ratio (1.6:1) using MEMS switches. Electron. Lett. **44**(2) (2008)
22. Wang, R., Wang, B.-Z., Gao, G.-F., Ding, X., Wang, Z.-P.: Low-profile pattern-reconfigurable vertically polarized end-fire antenna with magnetic-current radiators. IEEE Antennas Wirel. Propag. Lett. 1–4 (2018)
23. Alam, M.S., Abbosh, A.: Reconfigurable band-rejection antenna form ultra-wideband applications. IET Microw., Antennas Propag., **12**(2), 195–202 (2018)

Transformer Oil Health Monitoring Techniques—An Overview

R. Rengaraj, G. R. Venkatakrishnan, Pranav Moorthy, Ravi Pratyusha, Ritika, and K. Veena

Abstract A transformer is an electrical static machine used for changing the level of voltage as a ratio of the primary to secondary turns. At the distributor side of the power system network, transformers are required to step down the voltage from distributor voltage (high voltage) to the level required by industries and consumers (230 V). Oil is used in oil-core transformers for the purpose of cooling as well as insulation. The maintenance of transformer oil is essential as chemical reactions and dielectric breakdown occur due to the presence of anomalies and even at nominal operating conditions with ageing. This paper deals with various offline and online transformer oil monitoring methods which have been used until recent times. The pros and cons of the various methods and interpretation techniques for fault detection and fault-type prediction have been presented.

Keywords Power system network · Distribution transformer · Oil-core · Transformer oil monitoring methods

R. Rengaraj · G. R. Venkatakrishnan · P. Moorthy · R. Pratyusha · Ritika · K. Veena (✉)
Department of Electrical and Electronics Engineering, SSN College of Engineering, Chennai, Tamil Nadu, India
e-mail: veena16118@eee.ssn.edu.in

R. Rengaraj
e-mail: rengarajr@ssn.edu.in

G. R. Venkatakrishnan
e-mail: venkatakrishnangr@ssn.edu.in

P. Moorthy
e-mail: pranavmoorthy16068@eee.ssn.edu.in

R. Pratyusha
e-mail: ravipratyusha16078@eee.ssn.edu.in

Ritika
e-mail: ritika16079@eee.ssn.edu.in

P. Suresh et al. (eds.), *Advances in Smart System Technologies*,
Advances in Intelligent Systems and Computing 1163,
https://doi.org/10.1007/978-981-15-5029-4_12

1 Introduction

In recent times, electricity has become a part of our lives and is not a luxury but rather a necessity. Electricity has amassed so much importance that the development of nations is influenced by the amount of electric power produced by the said nation. This is because various industries are directly dependent on the production of electricity, and every other industry is dependent on electricity in one form or the other. The responsibility of producing power in India falls upon the Ministry of Power.

Providing electric power to various consumers is composed of three stages—generation, transmission and distribution. Hydro, coal, nuclear fission are some of the sources of energy which are used to drive a turbine, thereby converting mechanical energy to electrical energy which constitutes the generating system. The voltage of the generated energy is stepped up with the help of the power transformer as transmission of electrical power at higher voltage levels leads to the reduction of losses of power in the form of heat and also reduces the cost of copper required for the purpose of the transmission line. According to paper [1], the electric power after being transmitted over large distances in high voltage levels such as 11 kV is stepped down with the help of a step down transformer and then distributed through distribution transformers to the various consumers who may be industrial consumers or domestic consumers.

The transformer is an integral part of the power system as it is used to step up the voltage level of the electric power at the generation stage and is used to step down the voltage level after transmission of electric power over large distances. Without the transformer, the electric power would have to be transmitted at the voltage level at which it was generated. This would lead to the loss of huge amounts of power (I2R loss) because of which extra power would need to be generated. With an increase in demand, transformers are being loaded beyond their rated capacity. Generator transformers are heavily loaded to maximize profit and meet the demand. As stated by [2, 3] loss of a transformer in a utility can cost many lakhs of rupees to the economy.

All faults that occur in a transformer, viz. short-circuit fault, partial discharge (PD), degradation due to gases dissolved in transformer oil, are to be recorded for every transformer to prevent breakdown. Replacing a transformer is an expensive and tedious process. Hence, monitoring of the health of the transformer is a very important part of power system maintenance. Online methods using spectrometer chips are currently in testing.

Although there are numerous kinds of faults possible in a transformer, the root cause of maximum faults (over voltage, temperature rise, discharge, etc.) is the gases evolved that get dissolved in insulation oil under working conditions, especially at peak loads. Since the 1900s, these faults due to oil degradation have been watched and dissolved gas analysis (DGA) has been developing since the 1950s. Many techniques, offline and online, are being used to monitor transformers in order to prevent catastrophic failure and minimize outages.

2 Functions of Transformer Oil

The four functions of insulating oil in a transformer according to findings of papers [4] are:

Cooling In a transformer, due to the exchange of power through induction and the losses incorporated by the core, heat is generated as a result of eddy current loss based on the type of material used. Thus, to carry the heat away from the core to avoid failure of equipment, insulating oil is required.

Insulation The oil acts as an insulation between the core windings.

Protecting against chemical attacks In an air core transformer, due to heat production, there are chances for chemical reaction and formation of compounds. Few compounds can affect the paper insulation, causing ageing of the transformer. Thus, in oil core, the heat gets carried away and gases get dissolved in the oil and are safe for a certain rated limit after which the oil can be replaced for better maintenance and reduce ageing.

Prevention of sludge build-up Reduction in compound formation and a dense liquid medium prevents the formation of sludge (waste) at the bottom of the transformer.

Other uses of oil in the core of transformer which is given by authors of [5] are

- They provide an arc quenching medium.
- Reduces humming noise in the transformer, hence reduced vibrations.
- The paper insulation failure forms sludge due to the presence of oil, instead of settling down to the tank. Removal of sludge is an easy process.
- It acts as an indicator for Buckholtz relay.

Most of the faults that occur in the transformer will be a consequence of heating of insulation failure, a sudden burst which is linked to the oil characteristics of the transformer. (viz. The insulation failure is caused due to the heating of oil beyond the rated temperature as a result of gases dissolved. Hence, the need for transformer oil testing is indispensable and has to be revolutionized for optimizing and obtaining better efficiency from the constructed power system.

3 Properties of Transformer Oil

Properties of transformer oil based on IEC protocols have been used to differentiate between healthy and faulty condition. There are various standards set for transformer healthy conditions of which IEC standards are more commonly used.

By the data presented by authors of [6] and views of writers of paper [4, 5, 7], Table 1 lists the characteristics and standard testing protocols of transformer oil. The cause of fault and its effect has also been depicted.

Table 1 Characteristics of transform oil and standard test protocols

Characteristics of transformer	Standard protocol for testing	Description	Cause and effects
Electrical properties			
Breakdown voltage or dielectric strength	IEC 60156	The ability to withstand electrical stress	Lowering of breakdown voltage is due to particles in oil particularly large water content
Dissipation factor of dielectric	IEC 60247	The measure of dielectric loss of oil	Due to conductivity of Oil Indicates the presence of dissolved metal ions and acids
Chemical Properties			
Water content	IEC 60814	Buildup of water due to ageing and leakage Water redistributes between insulation and oil at different temperatures	Due to ageing and leakage Excess water increases the rate of ageing of paper (solid insulation)
Acidity	IEC 62021	Amount of potassium hydroxide (KOH) needed to neutralize the acid in oil is called acidity	Due to oxidation of oil acids are produced Indicator of degree of ageing
Inhibitor content	IEC 60666	Inhibitors are compounds added to oil in order to arrest or slow down oxidation process leading to ageing	Prevents chain reaction of oxidation products (thus ageing) by decomposing in intermediate sinhibitor content to prevent rapid ageing
Physical properties			
Colour	ASTM D1500	Colour of insulating oil	Indicator of ageing Along with chemical analysis, it has some diagnostic values
Interfacial tension (IFT)	ASTM D971-99	Interfacial tension between oil and water is a measure of amount of polar contaminants	Influenced by even non-acidic oxidation product Indicator of ageing

(continued)

Table 1 (continued)

Characteristics of transformer	Standard protocol for testing	Description	Cause and effects
Corrosive sulphur	IEC 62535	Formation of copper sulphide in cellulosic insulation detection	Realized that copper strip (ASTM D1275) and silver strip (DIN 51353) tests were not adequate Causes failure in extreme conditions

4 Causes of Transformer Failure

By the authors of [8, 9], the various reasons for failure and their contribution as a percentage of total fault occurred in the year 1998 accumulated by HSB survey are listed below:

Lightning Surge: Lighting contains high voltage charges, which cause high voltage generation to electric power cables or communication cables at that instance. It accounts for 12.4% as a cause for failure.

External and internal short circuits Low resistance path across the tabs of transformer results in external short circuit. Failure of insulation layer between electrodes results in internal short circuits. The connections at the outside of the transformer core get into contact, resulting in fault which is 21.5% (maximum) of the total faults recorded.

Power workmanship The errors in the efforts of workers put into designing and building of power transformer is a cause of 2.9% of faults, which has been decreasing since the year 1975.

Deterioration of insulation When the insulation of power transformer is not in the same condition as when installed, the insulation is said to be deteriorated by a certain percentage. The cause of deterioration is primarily heat produced due to dielectric loss and conduction currents in electrically stressed insulation. This deterioration accounted for 14% for faults in 1998.

Overloading Transformers are overloaded when economic demand increases. Beyond a certain rated limit, overloading can cause failure of the transformer. 2.4% of transformers failed due to overloading in records.

By reports collected by authors of [10], the other types of loads are as follows:

Inadequate maintenance Various physical, visual, testing and oil analysis methods are available for detection of the transformer for maintenance. The period of maintenance varies from place to place. In case the testing is not carried out at proper intervals or is inadequate, the transformer may fail accounting for 11.3% (the second largest reason for failure). Thus, oil testing must be carried out at smaller intervals to detect fault scenario.

Loose connection 6.0% of failures occur due to a loose connection in transformers and connection with the power system.

Other causes All other causes like atmospheric interaction, human or animal intervention account to 24.4% of faults. They are not traceable, hence let us focus on the rest.

Apart from visual faults, the external short circuit, maintenance of transformer at regular and adequate intervals is very important.

Maintenance requires for the transformer to be shut down for a few days to analyse the insulation and oil characteristics which are sent as samples for testing in laboratories. The time-consuming process is eliminated by formulating a method of online monitoring on oil and its characteristics and only shutting down when a fault is predicted. To monitor the operation of power transformers, many devices have evolved, such as Buchholz relays or differential relays. But the main shortcoming of these devices is that they only respond to the severe power failures which require removal of the equipment from the service. Thus, techniques for early detection of the malfunction would be very valuable to avoid outages. Among the existing methods for identifying the incipient faults, dissolved gas analysis (DGA) is commonly used.

5 Dissolved Gas Analysis

The detection of abnormalities in power transformers is made unfussy by dissolved gas analysis [11] (DGA). Suggestions on identifying defective transformers and the types of faults are given in well-informed guides for interpretation of gases. Based on [12], about 70% of the most common failures in power transformers can be detected using DGA.

The operating conditions of the power transformer are monitored using DGA. It is used to identify the incipient faults in the power transformers originated by the electrical and thermal stresses prevalent. The oil and paper insulation degrading due to faults such as overheating, partial discharge and constant arcing generates gases which include hydrogen (H_2), methane (CH_4), ethylene (C_2H_4), acetylene (C_2H_4) and ethane (C_2H_6), carbon monoxide (CO) and carbon dioxide (CO_2) whose concentrations helps in identification of faults and estimation of their severity. The method gas chromatogram is the most primitive method of dissolved gas analysis issued by ASTM as ASTM standard D3612-02. The complexity of equipment requirement of gas chromatogram makes it be conducted only a laboratory environment [10]. Extracted gases from insulating oil by vacuum extraction, stripper extraction and headspace sampling are analysed using gas chromatogram. The author James Dukram has suggested that with the knowledge of technique like key gas, Dornenberg ratio, Roger's ratio, IEC ratio and Duval triangle interpretation DGA data are made possible.

6 Faults Detectable by DGA

The author in [13] has discussed in detail, the various incipient faults detectable by using DGA. Out of one hundred and seventy cases of faults in the transformer in service by visually inspecting the power transformers and 19 of them are simulated in laboratories. There are about six categories of faults accounted for by DGA.

Thermal faults In [14], the faults are classified based on the development of hotspots either in oil or paper insulation. At temperature ($T < 300$ °C), carbonized paper was found in seven cases upon inspection, and at the temperature ($T > 300$ °C), five cases have been recorded of identification of brownish paper. In paper [15], concentration of CO_2 and CO has increased with the prevalence of such faults. The other case of development of hotspots in oil was a result of circulating currents in tanks or clamps or bad contacts in weld and winding.

Low energy discharges According to paper [14], low energy discharges having 25 cases have been recorded due to small arcing, sparking, carbonized pinhole in paper or particles of carbon in oil. Relatively more amount of C_2H_2 is produced during a small breakdown in oil. These low energy discharged are easily notable in transformers in service due to the formation of a huge concentration of dissolved gases. But there are possibilities that these faults may not be detectable unless the partial discharge activity is intense or at have a longer period. The low to medium partial discharge activity during high-voltage factory tests on transformers are approached using acoustic or other electrical technique. About 1–10 mJ is the minimum electrical energy discharge required to cause carbonized paper insulation. Visible damage to paper insulation requires about 100,000–1,000,000 mJ of electric discharge. Monitoring of damage to paper insulation due to partial discharges by acoustic and electrical methods proves to be more economical. Production of free bubbles in oil due to partial discharge is harmful.

Partial discharges of corona type In [15], partial discharges of corona type on inspection are identified by the formation of x-wax deposits which is a saturated hydrocarbon. These on further degradation produce increase dissolved hydrogen. So far, there has been a detection of over nine such faults in the transformer in service. In spite of their low amplitude from 10 to 30 pC, they are produced over a longer period making them easily observable using DGA.

High energy discharges From paper [15], these faults are identified by the presence of corrosion in the windings, laminations, ducts, etc., and clogging of molten copper due to high-density current and power follow.

7 Methods of Dissolved Gas Analysis

As suggested by the authors in the paper [14], there are a few methods currently available for determining the concentration of dissolved gases present on faulty transformers. These are as follows.

7.1 Gas Chromatography

By the writings of paper [16], for nearly sixty years, the method of gas chromatogram introduced by James and Martin has been into existence for identifying the dissolved gases in transformer oil. As this technique quantifies all the dissolved gases such as total dissolved gases (TDG), individual gases in gas blanket (IGp) and the individual dissolved gases (IDG), it is acknowledged as one of the best methods of DGA [9]. Due to the complexity in conduction of this test, the procedures involved in extraction of oil samples from the transformer in service, its transportation to the laboratories and identification of dissolved gases following certain grade, the ASTM D923 suggests various methods of extraction of gas such as headspace sampling, vacuum extraction, shake test and stripper column extraction.

Basic gas chromatography has the following components, a carrier gas supply which is argon, helium, nitrogen or hydrogen as they do not interfere or react with the extracted gases. The components in the gaseous state are separated in a column from volatilized oil sample in the injection port and travel with the carrier gases or condense in a stationary state which is a thin film of polymer that can withstand high temperatures at the walls of the column. Sometimes, the condensed gases in the stationary state may get mobilized and travel along with the carrier gas or vice versa. Out of packed and capillary columns, the latter is preferred as the stationary phase is coated on the column walls and has higher separation capabilities due to its small diameter but longer column length. The temperature is essentially kept constant to have better separation capacity. By papers [14, 17] the time of diffusion, the gases are removed on after the other. The detectors such as flame ionization, flame photometry, nitrogen-phosphorus, electron capture, atomic emission, thermal conductivity and electrolytic conductivity generate electrical signals when heated up by the contact of the gaseous components. Measurement of complex gases is made possible with mass spectrometer detection with the help of electron ionization detectors [16]. High sensitivity is the major factor which makes flame ionization detectors suitable for hydrocarbon and carbon dioxide gases. Hydrogen, oxygen, nitrogen could be detected using thermal conductivity detectors. Gas chromatography can also be used to quantify other components such as methyl acetate, furan, methanol, acetone, isopropyl alcohol, 2-methylfuran, phenol, methyl ethyl ketone and methyl formate.

7.2 Hydrogen Oil Monitoring

The disadvantages of gas chromatogram such as complexity of extraction and laboratory environment are overcome by hydrogen online monitoring which has proven to be one of the most recent, low cost and robust techniques. Introduced by Syprotec and functionality further enhanced by the Institute de Recherched' Hydro Québec, this technique is widely accepted as hydrogen was the by-product of the majority of the faults associated with transformer insulating oil. Based on [18], in addition to hydrogen and carbon monoxide, other forms of hydrocarbons such as ethylene and acetylene are also detected but with less accuracy. The hydrogen concentration accuracy was measured to be $+10\%$ to -10% for the temperature ranging from 20 to 40 °C.

The basic functional unit of the hydrogen online monitoring is a sensor in contact with the insulating oil and an electronic unit. The sensor is housed in a brass case which consists of a temperature sensor, fuel cell and Teflon membrane. It is corrosion-resistant. The sensor is mainly installed in the transformer in between the cooling unit and the transformer bank. Atmospheric oxygen present in between the brass case facilitates the permeation of hydrogen gas through the Teflon membrane to the fuel cell where it is electrolytically burnt to produce current proportional to the hydrogen gas concentration. The amplified current, when passed through a load resistor between the electrodes of the fuel cell, yields a potential voltage drop whose value when exceeded triggers an alarm. It is unable to identify the type of fault, but the detection percentage of hydrogen is 100%, followed by carbon monoxide at 15%, acetylene at 8% and ethylene at 1%.

7.3 Photo-Acoustic Spectroscopy

The phenomenon of thermal expansion of object caused by the absorption of light was first identified by Alexander Graham Bell in the late eighteenth century [19, 20]. Spectroscopy is a technique utilizing electromagnetic effect for the estimation of the energy level and the atomic structure of the compounds. Electrical, rotational and vibrational transition of molecules upon absorption of electromagnetic radiation causes deformation in the covalent bond and changes in the nuclear magnetic moment. The different absorption characteristics within the electromagnetic spectrum were due to the transition caused by different absorption at different energy levels. Absorption of near-infrared to infrared radiation causes vibrational and rotational transition, especially for colourless gas as ultra-violet radiations, penetrates the molecules. From the atomic spectra database of high-resolution transmission molecular absorption and the National Institute Of Standards and Technology, USA, the NIR spectrum can be employed to detect the concentration of CO, CO_2, CH_4, C_2H_2 and IR spectrum traces the concentration of C_2H_4,C_2H_6.

In [17], the principle of operation of the spectrometer is infrared radiation absorbed by the dissolved gas molecules which causes the transition of states, and this kinetic energy is in the form of sound waves. A microphone detects the waves and transforms them into an electrical signal. The intensity of pressure wave produces an absorption spectrum of fault gases which are selected using optical fibres to get the individual concentration of dissolved gases [19]. Even though this method is capable of determining the incipient faults, it is unable to predict the concentration of H_2 in the transformer oil [15]. Selection of centre wavelength is the most crucial task as the fault gases have different specific wavelength and the detection accuracy varies with respect to environmental conditions.

The authors of [21] have presented an offline and an online method for monitoring of power transformers. Sweep frequency response analysis (SFRA) is a progressive and well-known offline method for short circuits detection in transformers. In this method, a low and constant potential difference is applied to the terminals of the transformer, with frequency varying from 0.1 Hz to 5 MHz. The response of the transformer to a broad range of frequencies is studied, and the resonant frequencies of the winding are found and are compared with resonant frequencies of a healthy transformer. The occurrence of a fault results in the change of the number of poles in the response. This method cannot identify the section that has been short-circuited.

Maintenance requires for the transformer to be shut down for a few days to analyse the insulation and oil characteristics which are sent as samples for testing in laboratories [22]. This time-consuming process is eliminated by formulating a method of online monitoring on oil and its characteristics and only shutting down when a fault is predicted. These factors have led to a shift from offline methods of monitoring to online methods of monitoring power transformers in service.

The authors of [23] express that the life cycle of power transformers can be enriched by supervising the voltages, currents and the temperature of the transformer oil. The health of transformers diminishes when the transformer insulation deteriorates due to various factors. Transformer insulation system consists of paper and oil, which are prone to diverse types of ageing namely electrical ageing, thermal ageing, mechanical ageing and, environmental ageing. It is found that the rate of ageing is proportional to temperature, moisture and presence of air. Also, good quality oil should be preferred for prolonging the life of the transformer. Numerous methods such as dissolved gas analysis, degree of polymerization, Furan analysis, recovery voltage measurement, tangent delta, insulation resistance test, partial discharge, etc., are used for diagnosis of faults. Among these methods, the dissolved gas analysis is used most prominently worldwide. Gases like hydrogen, carbon dioxide, carbon monoxide, nitrogen, oxygen and hydrocarbons are detectable using this method. The quality of paper insulation is deduced by measuring the degree of polymerization and Furan analysis of the insulation. Recovery voltage measurement checks the content of moisture and ageing of the insulation. Tangent delta is an estimate of dielectric losses in the insulation system. A high value of tangent delta denotes the deterioration of insulation in the transformer. Insulation resistance test examines the accumulation of polarizable material. And, partial discharge examines the decay of insulation due to ionization.

Condition-based monitoring [24, 25] comprises of continuous surveillance of the transformer and monitoring of changes in the operational characteristics. Predictive maintenance of transformer is done through a real-time data acquisition setup that is composed of a signal conditioning board, an AD converter and, a multiple sensor package. Initially, a healthy transformer is taken as reference, and its frequency and harmonic components are noted. Thereupon, diverse faults are introduced into the transformer and its frequency and harmonic components are recorded again, hence acquiring the distinctive signature of each fault. This data are stored in the embedded system [26]. Live data are acquired from the transformer when it is under operation, and faults are analyzed through wavelet decomposition and realized with FFT, using the reference data obtained initially.

The authors of [27] have presented the implementation of online health monitoring of transformers using GSM module. The main feature of this system is automatic decision making. The system includes the SIM 900GSM module, microcontroller, ACS712 current sensor and a mobile phone. The various units of the system are the data collector unit, data processor unit and a communicating unit. The values of voltage, current, oil level and oil temperature are collected. Data are secured continuously from the transformer, and when there is any irregularity, the microcontroller sends a message to the controller cell by establishing serial communication with the modem. Once communication is established successfully between the microcontroller and GSM modem, data are sent to the mobile phone from the GSM module. The system is profitable, easy to install and can be notified remotely. However, it requires a lot of connections and apparatus. Hence, skilled personnel is mandatory for installation.

In [28], the authors have developed a reliable mobile embedded system, to monitor the parameters of a distribution transformer. The monitoring system consists of remote terminal unit (RTU), the transmission network and the monitoring unit stations. The RTU is composed of an MCU (microchip unit-PIC18F4550), a peripheral circuit (liquid crystal display, RS232 communications, analog to digital converter and GSM-GPRS wireless communication module (SIM808). The RTU is designed to collect data from sensors connected to the MCU and to send data to the monitoring node via GSM-GPRS module. The monitoring unit receives data from the RTU and displays it. Current sensor, temperature sensor, oil level sensor and vibration sensor are used. They are connected to the transformer, and the digital output is given to the microcontroller. When a fault occurs, alert messages are sent to mobile phones as well as monitoring units using the GSM/GPRS module. Data transmission using GPRS is convenient, highly efficient and has a low cost. Hence, it provides a feasible solution for the monitoring of distribution transformers.

In [29], the authors have explained the design of a solid oxide fuel cell (SOFC) detector. The SOFC detector is fabricated in a mould. For calibration purpose, double-logarithmic model (DLM) and linearity model (LM), are two mathematical models which are compared. The SOFC detector exhibits good reliability and linearity with DLM. Experimental results suggest that DLM is suitable for H_2 and other hydrocarbons, whereas LM is suitable for CO. Measurement of H_2, CH_4, CO, C_2H_4, C_2H_6 and C_2H_6 can be determined using the SOFC detector with high precision.

The authors of [8] have presented a monitoring system for fault diagnosis of power transformers, which is designed using P89V51RD2 microcontroller. Critical parameters such as current, voltage, frequency, oil level and temperature are monitored using sensing units and the values are updated from time to time. A liquid crystal display connected to the system shows the values of these parameters. The system consists of various units such as the voltage and current monitoring unit, frequency counter, voice player and the auto-dialing unit. The voltage and current monitoring and frequency counter units monitor the voltages, currents and frequency, respectively. The voice player stores up to 100 min of recorded voice. And, the auto-dialing system dials a pre-stored number and plays a pre-recorded voice, in case of an abnormality. Access to the parameter data is available 24/7. The system requires less maintenance and has a long lifetime. The system is compact and productive in the monitoring of power transformers.

The monitoring system MS2000 in [30] uses field bus technology with intelligent bus terminals. The system monitors oil characteristics like gas-in-oil content, moisture in oil, oil temperature and oil level in conservator, bushings, cooling unit and tap changer of the power transformer. Using short cables, the sensors are connected to bus terminal boxes, which consist of a bus coupler as a header station and electronic bus terminal strips. The signals from the sensors are transmitted to an industrial PC for processing via a measuring transducer. The control PC can be installed anywhere on the site. It is advantageous that using a single control PC, it is possible to monitor several transformers in close vicinity. The design is compact and simple.

The online method of monitoring of transformers according to the authors of [21] involves the determination of voltages and currents in primary and secondary windings at a frequency of 50 Hz. A healthy transformer is simulated with an input of 230 V, 50 Hz, and a constant load, as a reference. When a fault occurs, the voltages/currents drift from the operating values. Simulations of faulty transformer models indicate that primary current can be considered as the prime indication for primary winding short circuit, whereas secondary faults are indicated by both input current and output voltage. And a short circuit between the windings can be registered with the aid of output voltage and the input and output currents. Also, it is problematic to figure the fault locations except if a short circuit occurs between windings.

8 Interpretation of Faults

The results obtained from various offline and online methods are interpreted using different algorithms in various papers [31]. Fuzzy logic and regressions are pioneer among them. The authors of [32] have stated the following methods for interpretation of faults.

8.1 Key Gas Method

The key gas method is one of the methods used to analyse the results obtained after performing the dissolved gas analysis on the oils used in the transformers. It detects when faults or anomalies occur in the oil using the fact that the gases that are present during the fault require a very high amount of heat energy in order for the chemical bonds, which are present in the transformer oil to break. The concentration of six gases is studied, and the obtained data are interpreted on whether or not faults are present. The six gases being carbon monoxide, hydrogen, methane, ethane, ethene and ethyne. The percentage of various gases for fault concentration limitations is specified by the experience of various experts. This results in this method of interpretation not being reliable and only has an accuracy of 42% as found out by studies based on an IEC data bank.

8.2 Doernenburg Ratio Method (DRM)

The ratio of the concentration of various gases is used in this method to infer the kind of fault that has occurred in the transformer. Values of the gas concentration ratios of CH_4/H_2, C_2H_2/CH_4 and C_2H_6/C_2H_2 ratios are predefined, and these ratios are used in order to interpret the result obtained from the dissolved gas analysis of the transformer oil. The requirement for this method to be accurate is that the strength or concentration of at least one of the key gases (H_2, C_2H_4, CH_4 and C_4H_4) exceed twice the value of their L1 concentration which is given in the table below and the at least one gas exceeds its L1 concentration. Table 2 shows the concentration of L1 for major gases in oil. In Table 3, the limits for ratios of gases in oil and gas space for analysis by Doernenburg ratio method are listed.

Table 2 Concentrations of L1 for Doernenburg ratio method

Key gases in oil	Concentration of L1 (parts per million)
Carbon monoxide (CO)	350
Methane (CH_4)	120
Hydrogen (H_2)	100
Ethane (C_2H_6)	65
Ethylene (C_2H_4)	50
Acetylene (C_2H_2)	35

Table 3 Ratio of key gases—Doernenburg ratio method

Fault diagnosis	Ratio 1 CH_4/H_2		Ratio 2 C_2H_2/C_2H_4		Ratio 3 C_2H_2/CH_4		Ratio 4 C_2H_6/C_2H_2	
	Oil	Gas space	Oil	Gas space	Oil	Gas space	Oil	Gas space
Thermal decomposition	>1.0	>1.0	<0.75	<1.0	<0.3	<0.1	>0.4	>0.2
Corona (low intensity PD)	<0.1	<0.01	Not significant		<0.3	<0.1	>0.4	>0.2
Arching (high intensity PD)	>0.01 <0.1	>0.001 <0.1	>0.75	>1.0	>0.3	>0.1	<0.4	<0.2

Table 4 Rogers ratio method diagnoses

Suggested fault diagnosis	CH_4/H_2	C_2H_2/C_2H_4	C_2H_6/C_2H_2
Normal	>0.1 <1.0	<0.1	<1.0
Low energy density PD	<0.1	<0.1	<1.0
Arching (high intensity PD)	0.1–1	0.1–3	>3.0
Low temperature thermal fault	>0.1 <1.0	<0.1	1–3
Thermal fault <700 °C	>1.0	<0.1	1–3
Thermal fault >700 °C	>1.0	<0.1	>3.0

8.3 Rogers Ratio Method

The Rogers ratio method (RRM) bears similarity to that of the Doernenburg ratio method in many ways. However, one aspect of the difference between the two methods is that the Doernenburg ratio method (DRM) requires substantial concentration of fault gases, whereas the only requirement for RRM method is that the concentration of the fault gases exceeds the L1 value of the particular gas. Four concentrations ratios were initially used by the RRM for detection of the fault. The concentration ratios being C_2H_6/CH_4, C_2H_2/C_2H_6, CH_4/H_2 and C_2H_4/C_2H_6. These concentration ratios lead to a total of 12 possible diagnoses. However, this method is highly inconsistent with an accuracy rate of a mere 58.9%. Fault interpretations based on gas ratio values by RRM are listed in Table 4.

8.4 IEC Ratio Method

The same three ratios of gas concentration as used in Doernenburg ratio method and Rogers method are used in the [33] IEC ratio method (IRM). However, their

Table 5 Suggested IEC ratio method diagnoses

Case	Characteristic fault	CH_4/H_2	C_2H_2/C_2H_4	C_2H_4/C_2H_6
PD	Partial discharges	<1.0	Not significant	<0.2
D1	Discharges of low energy	0.1–0.5	>1.0	>1.0
D2	High energy discharges	0.1–1	0.6–2.5	>2.0
T1	Thermal fault <300 °C	>1.0 but *t* Not significant	Not significant	<1.0
T2	Thermal fault >300 °C and <700 °C	>1.0	<0.1	1–4
T3	Thermal fault >700 °C	>1.0	<0.2	>4.0

difference lies in the ratio ranges and interpretation of the dissolved gas analysis. The ethyne/hydrogen ratio has been introduced to detect contamination of on-load tap-changer compartments. This method provides the result in form of a 3D graphical representation which improves the reliability of the diagnoses improving the accuracy. Prediction of various faults by IEC ratios is listed in Table 5.

8.5 Duval Triangle Method

In the Duval triangle method, the data obtained from the results of the dissolved gas analysis are interpreted in the form of a triangle whose three sides are given by the concentration of methane, ethene and ethyne. The obtained triangle is divided into seven zones based on the type of fault that is occurring with the few of the various types being partial discharge, thermal faults and electric arcing. Despite providing a more accurate diagnosis of the occurring faults as compared to the above-discussed methods, interpretation is to be done cautiously as an incorrect interpretation of the resultant triangle would lead to incorrect diagnosis. Tables 6, 7 and 8 list the faults interpreted using the Duval triangles DRM1, DRM2, DRM3, respectively.

The authors in the paper [34] have compared the fault diagnoses of transformer oil using artificial neural network and wavelet neural network. In this proposed method, a hybrid genetic algorithm is used in order to optimize decision variables and in order to detect faults in the transformer. The WNN consists of an input layer, a hidden layer

Table 6 Faults in Duval triangle 1

PD	Partial discharge
D1	Discharges of low energy
D2	Discharges of high energy
DT	Combination of thermal fault and discharges
T1	Thermal faults <300 °C
T2	Thermal faults >300 °C and <700 °C
T3	Thermal faults >700 °C

Table 7 Faults in Duval triangle 2

PD	Partial discharge
S	Stray gassing of mineral oil
C	Hotspots with carbonization of paper ($T > 300$ °C)
O	Overheating ($T < 250$ °C)

Table 8 Faults in Duval triangle 3

PD	Partial discharge
S	Stray gassing of mineral oil
C	Hotspots with carbonization of paper ($T > 300$ °C)
O	Overheating ($T < 250$ °C)
T3	Thermal faults of very high temperature (T > 700 °C)

and an output layer. The purpose of the mother function has been fulfilled by the Gauss function, which has been defined as $(x) = 2a/2\ (2a\ x - b)2\ e{-}0.5(2^{a-b})^2$ where a and b are the dilation coefficient and translation coefficient, respectively. The input data have been real encoded in order to avoid losses due to discretization. This algorithm follows a kind of elitist strategy by ensuring the survival of strings whose fitness value is high in their respective generation. The concentration of the six gases, namely hydrogen, methane, ethane, ethene, ethyne and total hydrocarbon are assigned as the six input vectors after being fuzzified in order to distill the data information effectively. The membership values of the output vectors are calculated using the sigmoid function. The output vector yields six single fault modes. The paper [35], WNN_GA, WNN is trained with the help of 300 input sample data, and the mean error curve obtained is compared with that of an ANN system and is observed to have a faster rate of convergence and better stability. The ANN system reached a mean error of 107.8264 after a total of 1000 iterations which is unacceptable for fault detection purposes as compared to the 0.3132 mean error of WNN_GA-based system after 420 iterations. The WNN_GA system is found to have an accuracy of close to 92% which is greater than both WNN-based and ANN-based systems. Table 9 compares ANN and WNN methods of integrating algorithm.

The authors in [36] have proposed a model using genetic programming which generates a random search space in which every search unit is assigned a particular fitness value, thereby producing the next search space unit using which at the end we obtain an optimal solution to the given problem. The concentration of the gases hydrogen, methane, ethane, ethene, ethyne, carbon monoxide and carbon dioxide is used to primarily diagnose the fault occurring in the transformer oil. The root relative squared error is used as a fitness function. It has been experimentally found out that the error percentage is as low as 1.53% and the posterior average error is 1.13%.

The author in [17] has proposed an algorithm which has taken inspiration from the human immune system. The fault samples are mapped into shape space as antigens, and random antibodies are generated. The affinity between the antigen and antibody

Table 9 Comparison of ANN and WNN

	Parameter	Settings
WNN	The mother wavelet	$g(t) = t\mathrm{e}^{-t^2/2}$
	Number of input layer nodes	6
	Number of output layer nodes	6
	Largest number of hidden layer nodes	15
	Ranges of dilation	[−5, 5]
	Ranges of translation	[−5, 5]
	Ranges of weighting values	[−1, 1]
GA	Population size	50
	Maximal generation	100
	Adjustment factor	9.9

is calculated, and the memory antigen and antibodies are purified. The fault samples consist of normalized concentrations of hydrogen, ethene, ethane and ethyne. The diagnoses result can further be improved by adding the fault type F to the input sample data. In the learning algorithm, a memory set is built which recognizes the input data. The antibodies with high affinity to the antigens are cloned, and those clones whose affinity value is greater than that of a fixed value are eliminated. After training, an improved KNN classifier is implemented in which the Euclidean distance between the memory antibody and the test antigen the fault type of the power transformer is diagnosed. This method is not effective in case of more than one transformer fault because In that case, the distance of the antibody between any of the fault antigens almost become equal. This method is found to have diagnoses accuracy greater than that of IEC ratio method and backpropagation neural network.

The author of [37] has compared the accuracy of ANFIS network with that of an artificial neural network. The input vector consisted of the concentration of the gases hydrogen, methane, ethyne, ethane carbon dioxide and ethane. The data are processed with the help of the sigmoid function and are input to both the ANN and ANFIS network. The results of ANN with back propagation algorithm and the ANFIS network are compared and training data are loaded in both the networks. In paper [38] ANFIS, network is found to have greater accuracy than that of the ANN network with backpropagation. The ANFIS network also has a faster rate of convergence than that of the ANN network

The authors of [39] have proposed an algorithm combining ANN using backpropagation algorithm along with various particle swarm algorithms in order to detect the incipient faults in the power transformer. The accuracies of all the particle swarm algorithms have been experimentally determined and compared with one another. In ANN, the data are trained using the compositions of the gas in the transformer oil. Hundred input data which consisted of a concentration of the six types of gases were input to the ANN of which 70% was the training set and 15% was the test set to confirm that the simulated output agrees with the target output. Then a particle swarm algorithm is used in which a swarm is initiated along with the PSO parameters, and the personal best and global best values are recorded updating its position and velocity until it meets the convergence. The evolutionary PSO employs the concepts of reproduction, duplication and mutilation. The Iteration particle swarm algorithm modifies the velocity function, in order to improve the accuracy of the PSO. The accuracy of ANN, ANN-PSO, ANN-IPSO and ANN-EPSO is compared and is found that ANN-EPSO has the highest accuracy of 98% followed by ANN-IPSO with an accuracy of 97%. The ANN-PSO has an accuracy of 96%, and ANN has the lowest accuracy of 95%. The authors of [40] have proposed the use of a support vector machine in which the parameters were optimized with the help of several particle swarm algorithms to determine the transformer fault. This method uses stepwise regression which allows the user to determine concentrations of which gas are important in order to detect the fault in the transformer. The particle swarm algorithms which are used are conventional PSO, iteration PSO, evolutionary PSO, modified PSO-time varying acceleration coefficient (TVAC) and modified EPSO-TVAC. Out of all these algorithms, SVM-MEPSO-TVAC has the highest accuracy of 99.50% using stepwise regression.

9 Conclusion

The paper presents the need for monitoring transformers used in power systems. It deals with the numerous methods of offline and online health monitoring of transformer oil. The disadvantage of using the offline method of fault detection and analysis is that the transformer has to be shut down and the sample collected has to be transported which may lead to error in analysis owing to further degradation of oil with time. Hence, it is inferred that the shift from offline to online monitoring provides more accurate results and is practicable. Various algorithms for analysing fault types have been introduced briefly.

References

1. Copiun, M.K.: Coieitioy Monitoring of HV Transformers Using Oil Analysis Techniques, pp. 1–2
2. Xianyong, L., Fangjie, Z., Fenglei, H.: Research on on-line DGA using FTIR. In: Proceedings International Conference on Power System Technology 2002, pp. 1875–1880 (2002)
3. Kita, D.M., Lin, H., Agarwal, A., Richardson, K., Luzinov, I., Gu, T., Hu, J.: On-chip infrared spectroscopic sensing: redefining the benefits of scaling. IEEE J. Sel. Top. Quantum Electroni. **23**(2), 1–10 (2017)
4. Gray, I.A.R.: Transformer Chemistry Services A Guide to Transformer Oil Analysis, pp. 1–13
5. Guide for the Interpretation of Gases Generated in Oil-Immersed Transformers IEEE Standard, C57.104 (2008)
6. Product Information Dissolved gas analysis and supervision of oil condition in transformers and reactors, ABB company. http://www.abb.com

7. Sun, C., Ohodnicki, P.R., Stewart, E.M.: Chemical sensing strategies for real-time monitoring of transformer oil: a review. IEEE Sens. J. **17**(18), 5786–5806 (2017)
8. Thangam, G.K., Rajasaranya, T.: TRANS-INFORMER—an integrated system for health monitoring of power transformers. Int. J. Comput. Sci. Mobile Comput., IJCSMC **2**(4), 105–110 (2013)
9. Schwarz, R., Muhr, M.: Diagnostic methods for transformers. In: International Conference on Condition Monitoring and Diagnosis, Beijing, China (2008)
10. Kim, Y.M., Lee, S.J., Seo, H.D., Jung, J.R., Yang, H.J.: Development of dissolved gas analysis (DGA) expert system using new diagnostic algorithm for oil—immersed transformers. In: IEEE International Conference on Condition Monitoring and Diagnosis, pp. 365–369, Bali, Indonesia (2012)
11. Hamrick, L.: Dissolved Gas Analysis for Transformers Neta World (2010). http://www.netaworld.org
12. Life Management Techniques for Power Transformers. Technical Brochure 227—CIGRE (2013)
13. Duval, M., Dukarm, J.: Improving the Reliability of transformer gas-in-oil diagnosis. IEEE Electr. Insul. Mag. **21**(4), 21–27 (2005)
14. Abu-Elanien, A.E.B., Salama, M.M.A.: Survey on the transformer condition monitoring. In: Large Engineering Systems Conference on Power Engineering, pp. 187–191 (2007)
15. Chang, W., Hao N., (2011) Prediction of dissolved gas content in transformer oil based on genetic programming and DGA. In: International Conference on Transportation, Mechanical, and Electrical Engineering (TMEE), pp. 1133–1136, Changchun, China (2011)
16. Abu Bakar, N., Abu-Siada, A.: High voltage power transformer dissolved gas analysis. In: Measurement and Interpretation Techniques IDC Technologies, pp. 1–17 (2017)
17. Abu Bakar, N., Abu-Siada A., Islam, S.: A review of dissolved gas analysis measurement and interpretation techniques. IEEE Electr. Insul. Mag. **30**(3), 39–49 (2014)
18. Makky, M.: Review of Transformer On-Line Condition Monitoring (2014)
19. Abu Bakar, N., Abu-Saida, A.: A new method to detect dissolved gases in transformers oil using NIR-IR spectroscopy. IEEE Trans. Dielectr. Electr. Insul. **24**(1), 409–419 (2017)
20. Geibler D., Jaya, M., Leibfried, T.: Analysis of frequency domain spectroscopy measurements on power transformers by the use of a finite element based model. In: IEEE International Conference on Condition Monitoring and Diagnosis, pp. 206–210, Bali, Indonesia (2012)
21. Aburaghiega, E., Farrag, M.E., Hepburn Glasgow, D.M., Garcia, B.: Power transformer health monitoring: a shift from off-line to on-line detection. In: 50th International Universities Power Engineering Conference (UPEC) (2015)
22. Redding, B., Liew, S.F., Sarma, R., Cao, H.: On-chip random spectrometer. In: Conference on Lasers and Electro-Optics Europe and International Quantum Electronics Conference CLEO EUROPE/IQEC (2013)
23. Arvind, D., Khushdeep, S., Deepak, K.: Condition monitoring of power transformer: a review. In IEEE/PES Transmission and Distribution Conference and Exposition (2008)
24. Hussain, M., Salman, M., Rohit, Subhan, A., Khalid, H., Zaidi, S.S.H.: Condition based health monitoring of transformers. In: International Conference on Computing, Mathematics and Engineering Technologies—iCoMET (2018)
25. Chen, G., Xu, M., Liu, T., Ni, J., Xie, D., Zhang, Y.: Research on condition monitoring and evaluation method of power transformer. In: Third International Conference on Intelligent System Design and Engineering Applications, pp. 1155–1158 (2013)
26. Avinash Nelson, A., Jaiswal, G.C., Ballal, M.S., Tutakne, D.R.: Remote condition monitoring system for distribution transformer. In: IEEE Transaction on Instrumentation and Measurement (2014)
27. Rahman, S., Dey, S.K., Bhawmick, B.K., Das, N.K.: Design and implementation of real time transformer health monitoring system using GSM technology. In: International Conference on Electrical, Computer and Communication Engineering (ECCE), pp. 258–261, Bangladesh (2017)

28. Pawar, R.R., Deosarkar, S.B.: Health condition monitoring system for distribution transformer using internet of things (IoT). In: IEEE International Conference on Computing Methodologies and Communication (ICCMC), pp. 117–122 (2017)
29. Fan, J., Wang, F., Sun, Q., Bin, F., Ding, J., Ye, H.: SOFC detector for portable gas chromatography: high-sensitivity detection of dissolved gases in transformer oil. IEEE Trans. Dielectr. Electr. Insul. **24**(5), 2854–2863 (2017)
30. Tenbohlen, S., Figel, F.: On-line condition monitoring of power transformers. In: IEEE Winter Meeting, pp. 1–5., Singapore (2002)
31. Malik, H., Singh, S., Mantosh, K.R., Jarial, R.K.: UV/VIS response based fuzzy logic for health assessment of transformer oil. In: International Conference on Communication Technology and System Design, vol. 30, 905–912 (2011)
32. de Pablo, A., Fergusan, W., Mudryk, A., Golovan, D.: On-line condition monitoring of power transformers—a case history. In: Electrical Insulation Conference, pp 285–288. Annapolis, Maryland (2011)
33. Mineral oil-impregnated equipment in service- interpretation of dissolved and free gases analysis IEC 60599 (2007)
34. Pan, C., Chen, W., Yun, Y.: Fault diagnostic method of power transformers based on hybrid genetic algorithm evolving wavelet neural network .IET Electr. Power Appl. **2**(1), 71–76 (2008)
35. Hao, X., Sun, C.: Artificial immune network classification algorithm for fault diagnosis of power transformer, IEEE Trans. Power Deliv. **22**(2), 930–935 (2007)
36. Duval, M.: A review of faults detectable by gas-in-oil analysis in transformers. IEEE Insul. Mag., 2002 **18**(3), 8–17 (2002)
37. Geetha, R.: Fault diagnosis of power transformer using duval triangle based artificial intelligence techniques. Int. J. Recent Trends Eng. Res. (IJRTER) **2**(10), 78–88 (2016)
38. Muhamad, N.A., Phung, B.T., Blackburn, T.R.: Comparative study and analysis of DGA methods for mineral oil using fuzzy logic. In: 8th International Power Engineering Conference, pp. 1301–1306 (2007)
39. Illias, H.A., Chai, X.R. Abu Bakar, H., Mokhlis, H.: Transformer Incipient Fault Prediction Using Combined Artificial Neural Network and Various Particle Swarm Optimization Techniques, pp. 1–16 (2015)
40. AzilIllia H., Zhao Liang, W.: Identification of transformer fault based on dissolved gas analysis using hybrid support vector machine-modified evolutionary particle swarm optimization, pp. 1–15 (2018)

Investigations on Pipeline Optimized Adaptive Fir Filter Architecture for Audio De-noising

R. Sangeeta and S. Padmapriya

Abstract This project proposes a high-speed retimed de-noising adaptive filter on different FPGA platforms using FIR filter. Retiming technique is the minimization of the clock period in a circuit. Using retiming, an adaptive filter can have a low critical path, low power consumption, and high throughput. In an adaptive filter, the least mean square (LMS) is the utmost familiar adaptive algorithm by virtue of its simple structure. LMS algorithm is easy to execute in the real-time systems; to improve its critical path, it is crucial to correct the LMS adaptive algorithm. The retimed VLSI architecture lifts up the operational speed of an adaptive filter through limiting the critical path using delay components and also has a faster convergence. A fine-grain pipelined LMS method is used to further increase the maximum operating speed which provides pipelining at a computational component level. The critical path delay in the Virtex-5 series FPGA platform for direct form implementation of retimed LMS is found to be 9.104 ns, whereas that of traditional un-retimed LMS structure is 24.283 ns, thereby minimization of 37% critical path delay is achieved. A fine-grain retimed adaptive filter on Virtex-5 series FPGA platform achieves 8% improvement in clock frequency compared to a retimed structure.

Keywords Adaptive filter · Least mean square (LMS) · Retiming · Fine-grain pipelining

R. Sangeeta (✉)
VLSI Design, Sri Ramakrishna Engineering College, Coimbatore, India
e-mail: sangeeta17696@gmail.com

S. Padmapriya
Department of Electronics and Communication Engineering, Sri Ramakrishna Engineering College, Coimbatore, India
e-mail: padmapriya.s@srec.ac.in

P. Suresh et al. (eds.), *Advances in Smart System Technologies*, Advances in Intelligent Systems and Computing 1163,
https://doi.org/10.1007/978-981-15-5029-4_13

1 Introduction

In modern communication systems, the level of noise has increased because of the noisy environment of urban life due to exposure of electrical and electronic products and with the rapid industrialization. When the signal transmitted from transmitter to receiver, the desired speech signal is distorted by some noises which are due to time-varying processes that are unknown and unpredictable. This variation can only be tracked by an adaptive filter, and it increases the sensitivity of receiver end of the communication system. Some of the applications of adaptive filtering in digital signal processing are de-noising, echo cancellation, channel estimation, system identification, etc [1]. That may include noise reduction from a speech signal, hearing aid, and noise reduction in electrocardiography (ECG) signals [2].

Digital filters are the fundamental element in the domain of signal processing, biomedical engineering, communication engineering, embedded system, and various industrial electronics. Digital filters are to be categories in two different classes, finite impulse response (FIR) and infinite impulse response (IIR). In digital signal processing (DSP), FIR filter is one with an output response or behavior of the system which is finite. An adaptive filter can be designed by using both FIR and IIR filter. Because of the recursive nature of IIR filter, the pole can deviate from the unit circle causing the instability in the filter, whereas the FIR filters have exact linear phase property and provide the guaranteed stability to the digital filter [3]. In the adaptive filter, the least mean square (LMS) algorithm is utilized for suppressing the noise by continuously adapting the weight factor. An LMS adaptive filter structure can be designed using the retimed technique to enhance the speed of a filter. The LMS algorithm is a typically used adaptive algorithm due to its simple structure but it suffers from the large critical path delay for combinational circuit and also is affected from slower convergence rate. The retiming technique is the most useful technique to reduce the clock period or to increase the operation clock frequency by relocating the delay in an optimized way and also improve the rate of convergence. The retimed structure employed with fine-grain pipelining which results in speed enhancement for the critical path in the filter at each element and operator level.

2 Related Works

An adaptive filter is one of the classes of the digital filter, which modifies its weight coefficient or adapts its weight according to changes in the surrounding. However, some work focus on optimization of the adaptive algorithm. Dhal et al. [4] proposed **a** variable-step block LMS (BLMS) algorithm which dynamically changes the block length. A variable step-size feedback filtered-x LMS (FxLMS) algorithm is used for removal of acoustic noise that provides the better convergence than FXLMS [5]. Modified sign-error LMS (MSLMS) adaptive algorithm employs a three-level quantization strategy which provides less area delay product (ADP) and high convergence rate as presented in [6]. A high-performance feedback filtered-x LMS (FxLMS) for active noise cancellation which attenuates the noises in narrow bandwidth is presented in [7]. Basant et al. [8] proposed the delay optimized implementation for 2-parallel delayed LMS adaptive FIR filter which consists of fine-grained dot-product unit, fine-grained fused multiply-add unit and a multiple-input-addition unit which is used to reduce the maximum critical path delay in the circuit. A logarithmic number system is used to realize an adaptive filter with error nonlinearities in hardware presented in [9]. Mula et al. [10] propose a delayed FxLMS (DFxLMS) and delayed FsLMS (DFsLMS) algorithm to solve the problem of air–electrical interface of active noise cancellation (ANC) and has low-complexity realization by using the technique called resource sharing [3]. Proposed IIR LMS adaptive filter with accuracy compared to FIR but suffer from stability problem.

3 Adaptive Filter

An adaptive filter is based on the FIR digital filter; in this gradient descent algorithm is used to adjust filter coefficients automatically [11]. An adaptive filter has two inputs, i.e., primary and reference inputs, which are shown in Fig. 1. The original signal s_k corrupts with noise signal n_{0k}. Noise n_{1k} is the reference input. The noise n_{1k} is correlated with n_{0k} with some unknown way, but it is uncorrelated with s_k which signifies the following cases:

$$d_k = s_k + n_{0k} \tag{1}$$

$$E[s_k * n_{0k}] = 0 \text{ for all } k \tag{2}$$

$$E[s_k * n_{1k}] = 0 \text{ for all } k \tag{3}$$

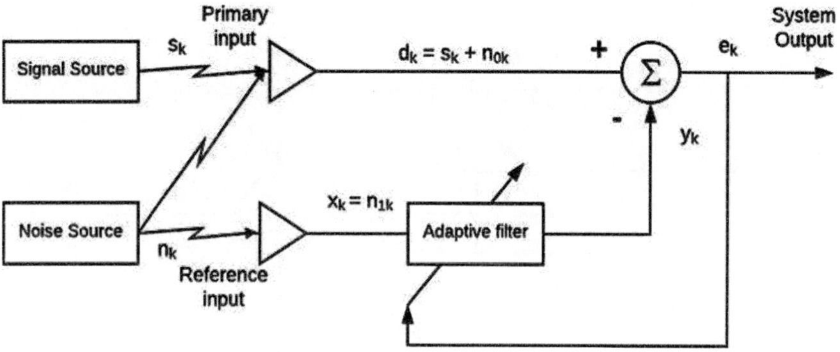

Fig. 1 Adaptive filter block diagram

$$E[n_{0k} * n_{1k}] = p_k \text{ for all } k \tag{4}$$

where d_k is the desired signal, p_k is probability for the correlation between n_{1k} and n_{0k}, and $E[*]$ is the mean [15, 20]. The reference input n_{1k} is processed in the adaptive filter in which it tries to minimize the error between the output y_k and the primary input d_k by adapting its impulse response iteratively. The primary inputs signals are subtracted with the output of filter y_k to produce system output.

$$e_k = s_k + n_{0k} - y_k \tag{5}$$

3.1 LMS Algorithm

The most widely used adaptive control algorithm is LMS algorithm due to its simplicity feature and also robustness to the signal statistic [12]. The error signal is fed to LMS algorithm which updates the individual tap weight of the filter for several iterations until the signal e_k is sufficiently small which is shown in Fig. 2. This algorithm can be described in three parts:

The adaptive filter output

$$y_k = \sum_{i=0}^{N-1} w_{i,k}\, x_{k-i} \tag{6}$$

The feedback error signal

$$e_k = d_k - y_k \tag{7}$$

The filter taps coefficient expression

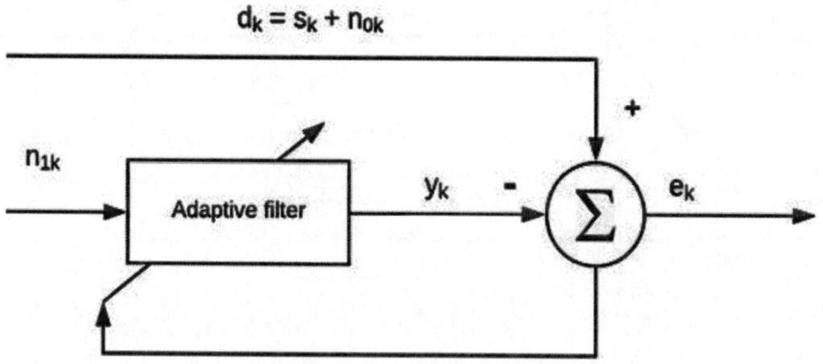

Fig. 2 LMS adaptive filter algorithm

$$W_{k+1} = W_k + 2\mu e_k X_k \tag{8}$$

where $W_{k+1} = \left[w_{0,k}, w_{1,k}, \ldots, w_{N-1,k}\right]^T$ is the weight, i.e. filter coefficients at time n, $X_k = \left[x_k, x_{k-1}, \ldots, x_{k-N+1}\right]^T$ is the reference input of N samples at time n and the means of controls for convergence rate and stability of filter is μ, i.e., step size of filter. A smaller value of μ ensures better stability, and larger value of μ improves the convergence speed [4, 13, 14].

3.2 DLMS Algorithm

The LMS adaptation scheme suffers from a maximal longest path delay in the circuit and also has slower convergence speed [15]. The original structure is modified through minimizing the critical path delay by using the retiming technique. The gradient algorithm must be realized with delay coefficients to apply the retiming techniques. The modified algorithm is implemented by introducing the delay elements in an error feedback loop of LMS adaptive algorithm is known as delayed LMS (DLMS) [16] which is shown in Fig. 3 which is described as:

The feedback error signal

$$e_{k-m} = d_{k-m} - y_{k-m} \tag{9}$$

The filter taps coefficient expression

$$W_{k+1} = W_k + 2\mu e_{k-m} X_{k-m} \tag{10}$$

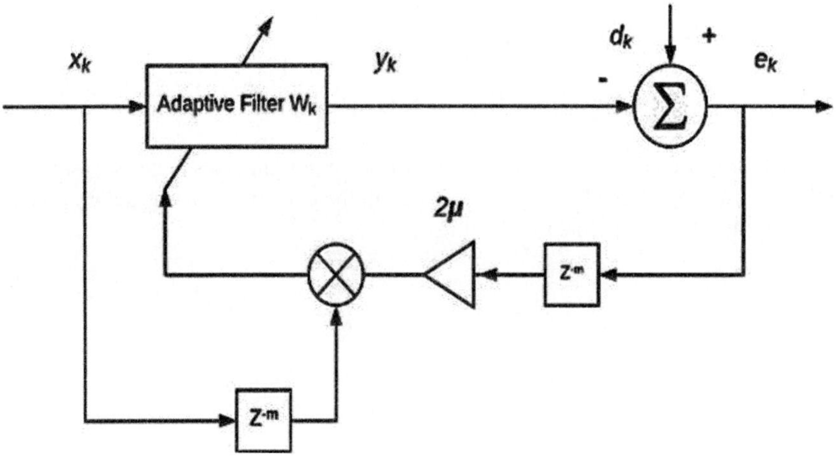

Fig. 3 Delayed LMS algorithm

where y_{k-m} is the delayed version of y_k and m is a number of the delay components [17]. Steady-state behavior of the LMS algorithm is slightly affected by the delay coefficients in DLMS structure, if μ is within specific bound.

4 Retiming Technique

Retiming is a convenient technique in the field of VLSI design. A circuit can be able to achieve a lower power utilization and lower critical path delay with high throughput rate using the retiming technique. It is a transformation technique in which the input/output attributes of the circuits does not change by relocating the delay elements in the circuit [13, 17, 18]. Thus, the critical path run through a combinational circuit in a feed-forward system can be reduced by the insertion of the delay coefficients or registers. In pipelining, further inclusion of additional register can alter the transfer functions of the system, whereas in retiming the registers are redistributed in a system to optimize the design. Filter structure is represented by using a data-flow graph (DFG) to apply a retiming. Retiming technique are categorized into two, namely cut-set retiming and feed-forward cut-sets retiming. In cut-set retiming, a data flow graph is retimed by applying a valid cut-set line. Set of edges known to be cut-set which is removed from the DFG to create disconnected subgraphs are G2 and G1. Then by adding m delay to each respective edge from G1 to G2 and removing of m delay from each edge from G2 to G1, a retimed DFG is created [18].

$$W_r(G_1 \to G_2) = W(G_1 \to G_2) + m \tag{11}$$

$$W_r(G_2 \to G_1) = W(G_2 \to G_1) - m \tag{12}$$

where the range of m lies between

$$-\min\{W(G_1 \to G_2)\} \le m \le \min\{W(G_2 \to G_1)\} \tag{13}$$

Pipelining or feed-forward cut-set is one of the important cases of cut-set retiming technique; it is applied when there is no edge in the cut-set during the subgraph G2 to subgraph G1, i.e. pipelined without loops.

5 Architecture

5.1 *Existing LMS Adaptive FIR Filter*

The 4-tapped delay line FIR adaptive filter is based upon the LMS adaptive algorithm which adjusts its tap coefficients [14, 11]. The direct form realization of LMS (DF-LMS) adaptive filter in Fig. 4 is described as:

$$y_k = \sum_{i=0}^{N-1} w_{i,k} x_{k-i} = w_{0,k} x_k + w_{1,k} x_{k-1} + \cdots + w_{N-1,k} x_{k-N+1} \quad (14)$$

$$w_{i,k+1} = w_{i,k} + 2\mu e_k x_{k-i} \quad (15)$$

The conventional LMS requires $2N$ adder, $2N$ multiplier, and a total of the $3N - 2$ delay elements in the structure. The critical path of the system is given as:

$$2\tau_m + (N+1)\tau_a \quad \text{for DF–LMS} \quad (16)$$

$$2\tau_m + \left(2 + \log_2 N\right)\tau_a \quad \text{for TDF–LMS} \quad (17)$$

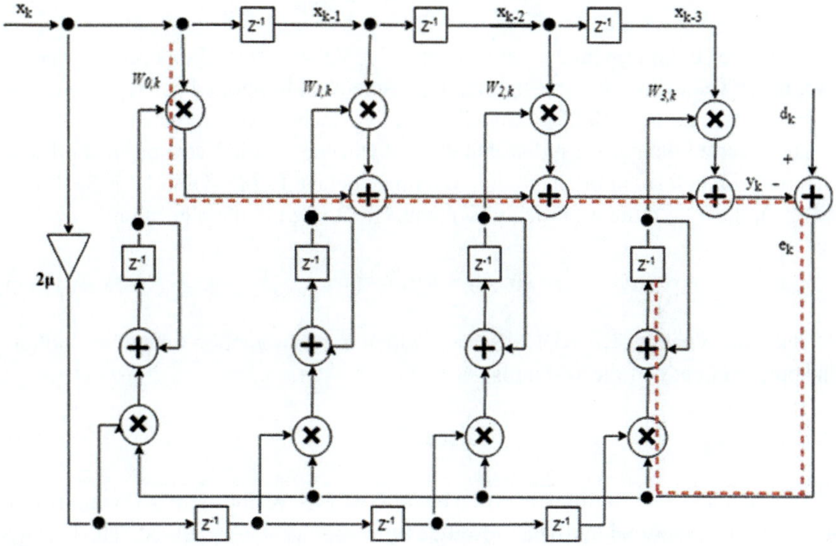

Fig. 4 Direct form realization of LMS (DF-LMS) FIR adaptive filter architecture

$$2\tau_m + 3\tau_a \quad \text{for TF–LMS} \tag{18}$$

where N is the filter length, τ_a is the computation time of an adder and τ_m is computation time of the multiplier.

5.2 Proposed De-noising Retimed LMS Adaptive FIR Filter

Retiming technique is used to increase the clock speed by minimizing the computation time of the critical path of the circuit. The conventional LMS adaptive FIR filter architecture is shown in Fig. 4. There are 2 multipliers and 5 adders in the critical path that limit the operational speed of the circuit. The critical path of filter structure is minimized by inserting the delay element between the combination of adder and multiplier. First delay element or additional register need to be included in the filter architecture and relocate the delay to reduce the clock period without changing the system functionality [18, 19]. The delay element for DF-LMS filter structure is given by $3N - 2$. Therefore, the number of delays is 10. The weight factor expression for the filter is given below:

$$w_{k+1} = w_k + 2\mu e_k x_k^T \tag{19}$$

The delayed LMS (DLMS) structure has 5N. Thus, for four-tap FIR filter, 20 delay elements are used. The weights factor expression for DLMS is given by:

$$w_{k+1} = w_k + 2\mu e_{k-m} x_{k-m}^T \tag{20}$$

There are six horizontal cut-set in 4-tap DF-LMS adaptive filter structure; therefore, the m value chosen to be six to compensate each horizontal cut-set, one delay for each edge. In the path of e_k as well as x_k, five delays are introduced.

The inserted delays are redistributed to obtain the retimed version of the LMS filter, i.e., direct form implementation of retimed delayed LMS (DF-RDLMS) shown in Fig. 5, the total number of delay elements are utilized in the algorithm:

$$N(N + 13)/2 = 34 \tag{21}$$

The critical path of DF-RDLMS is τ_m, which is a computation time of multiplier. The output latency of the system is:

$$\text{Latency} = N \tag{22}$$

2N adder and 2N multiplier are used in DF-RDLMS. As the increase in number of delay element, the total hardware resource utilization is also increased. The latency of un-retimed LMS adaptive filter is zero, and for retimed de-noising LMS adaptive filter latency is same as filter length N. From Fig. 5, the shaded delay element is the

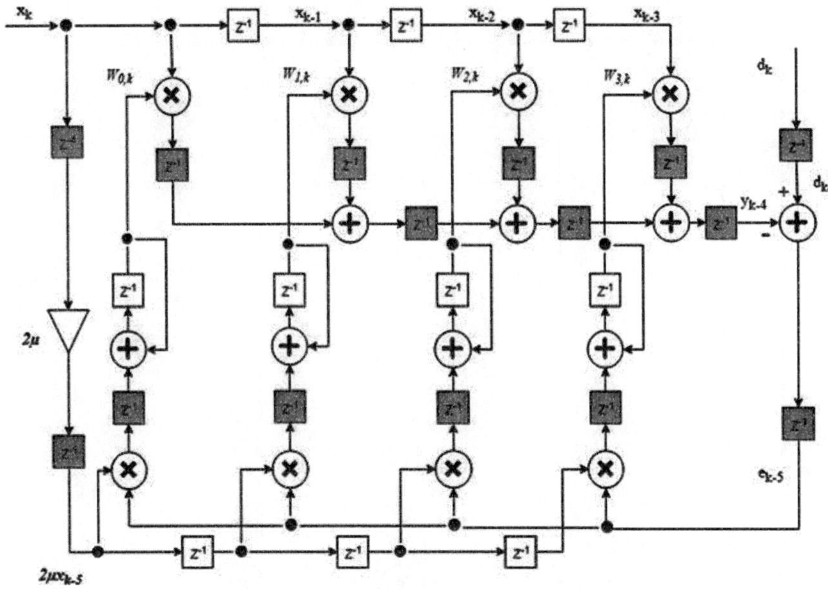

Fig. 5 Direct form realization of retimed delayed LMS (DF-RDLMS) adaptive FIR filter architecture

inserted delays in the filter architecture to increase the maximum operating frequency. For the 4-tap transposed direct form-retimed delayed LMS (TDF-RDLMS), adaptive filter architecture is shown in Fig. 6, and the *m* value is five because to compensate the five horizontal cut-set. The number of hardware element is 2*N* for both adder and multiplier same as un-retimed filter structure. Similarly, for the 4-tap transpose form-retimed delayed LMS (TF-RDLMS) filter architecture, as shown in Fig. 7, the m value is chosen to be four to compensate the four cut-sets.

The total delay elements calculation for the structures TF, TDF-RDLMS is as follows.

$$6N + 1 + 2\log_2 N \quad \text{for TDF–RDLMS} \tag{23}$$

$$5(N + 1) \quad \text{for TF–RDLMS} \tag{24}$$

$$\begin{aligned} \text{Latency} = 1 + \log_2 N & \quad \text{for TDF–RDLMS} \\ 2 \qquad & \quad \text{for TF–RDLMS} \end{aligned} \tag{25}$$

The critical path of a binary tree and transpose form filter architecture is to be τ_m (computation time of multiplier). As the order of filter increases, the operating frequency of filter reduces because of the increase in the maximum combinational path delay of the filter and thus has a tradeoff in resource area utilization. But the

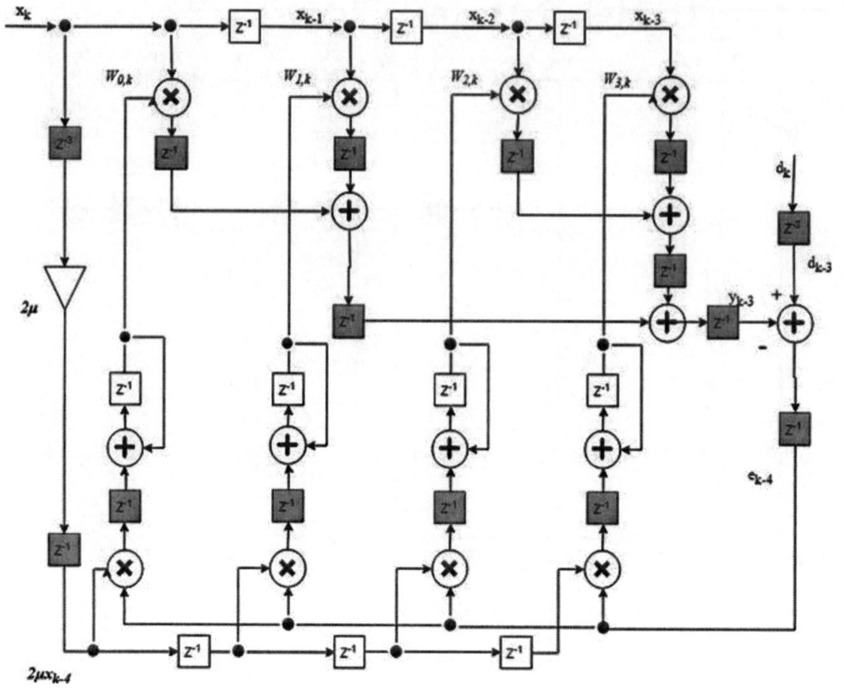

Fig. 6 Transposed direct form-retimed delayed LMS (TDF-RDLMS) adaptive FIR filter architecture

performance of the filter in means of signal-to-noise ratio (SNR) is increase based on the audio signal.

5.3 Proposed Fine-Grain Retimed LMS Adaptive FIR Filter

LMS algorithm has the wide spectrum of application in high-speed DSP because of its stable performance and considerable tolerance to error. An increase of a number of delay register in the filter leads to operating in high clock rate. In fine-grain pipelined processing, the pipelining approach is applied at the computation component level. Pipelining of the particular multiplier in certain clock cycles can able to break down the critical path in the architecture, thus improving the processing rate [17].

In the transpose form FIR filter, the output latency linearly dependent on the order of the filter. To achieve a finer-grain pipelining in transpose system, a significant number of pipeline register are require. The transpose form filter structure is constrained to single multiplier τ_{m} (computation time of multiplier) for better performance. In the direct form of an FIR filter, the output latency is not linearly dependent on the order of the filter length. Therefore, fine-grain pipelining in a direct form can

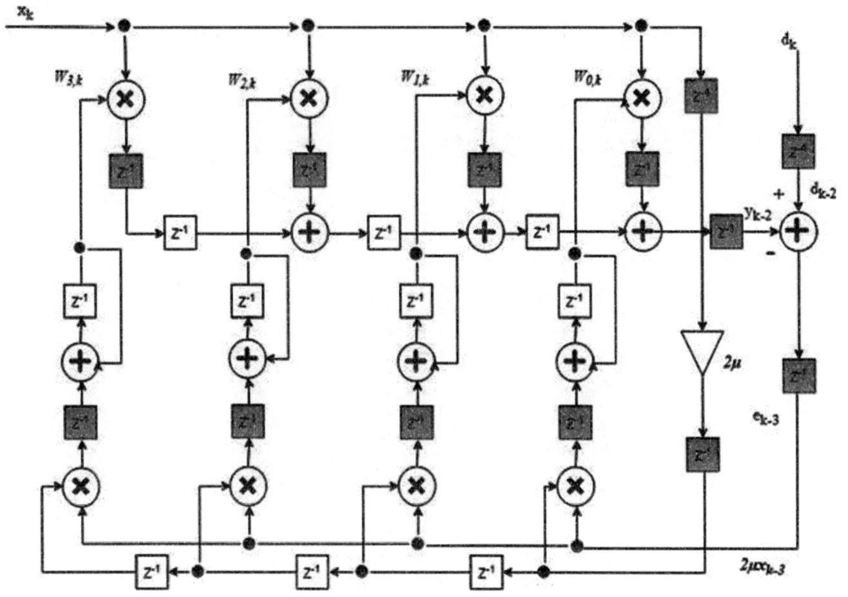

Fig. 7 Transpose form realization of retimed delayed LMS (TF-RDLMS) adaptive FIR filter architecture

be used to further improve the processing speed of the filter with a critical path of τ_a (computation time required by an adder).

6 Results

6.1 Synthesis Result

The synthesis analysis of adaptive FIR filter performs in accordance with maximum operating frequency, hardware resource, and power consumption, etc. The comparisons among the LMS, DLMS, and RDLMS algorithm have been made with various FPGA compilations, which are carried out in MATLAB Xilinx System Generator (XSG) and synthesized using Xilinx ISE tool.

Table 1 shows the result of the synthesis of different de-noising adaptive FIR filter structure implemented in the Virtex-5 FPGA series. From Table 1, the operational speed of retimed structure is better than conventional un-retimed structure, but the cost of the hardware is the tradeoff. The tradeoff is due to the additional delay element, which is used to minimize the critical path. The structure of DF-LMS has the large critical path of 24.28 ns than the DF-RDLMS filter of 9.104 ns. This shows the adaptive filter can be operated in higher clock rate or faster processing speed than the original LMS structure.

Table 1 Synthesis result of conventional, delayed, and retimed adaptive FIR filters structure on Xilinx Virtex-5 series

Adaptive FIR filters architecture	DF-LMS	TDF LMS	TF-LMS	DF-DLMS	TDF-DLMS	TF-DLMS	DF-RDLMS	TDF-RDLMS	TF-RDLMS
Slices	91	113	109	130	129	120	121	131	114
Flip-flops	160	160	161	193	193	193	401	401	369
4-LUTs	285	352	345	377	377	377	377	377	377
Bonded IOBs	49	49	49	49	49	49	49	49	49
DSP48s	8	8	8	8	8	8	8	8	8
Max frequency of operation (MHz)	41.181	42.066	53.141	51.446	55.106	53.158	109.842	108.096	111.383
Max comb path delay (ns)	24.283	23.772	18.818	19.438	18.147	18.812	9.104	9.251	8.978
Power utilization (mW)	448.11	448.12	448.31	448.15	448.19	448.18	448.35	448.34	498.74

Performance analyses of different filters architecture realization on different FPGA platforms are shown in Table 2 [13]. Figures 8 and 9 show the graphical analysis result on the basis of synthesis report. Figure 8 shows the comparison between different conventional, delayed, and proposed retimed de-noising adaptive FIR filter structure for maximum operational frequency in MHz. The TF-RDLMS gives the best result among all the filter structures in terms of operational speed at the same level of power consumption. Figure 9 presents the graphical analysis of critical path for distinct de-noise adaptive FIR filter in ns. The DF-LMS suffers larger critical path delay that minimizes the speed of filter architecture. Figure 10 represents the comparative analysis of transpose form-retimed DLMS architecture for three different FPGA implementation platforms. For instance, Spartan-6 FPGA provides minimum power consumption for the highest operating frequency, whereas the combinational path delay for the Virtex-5 FPGA platform is minimum. The pipelining at operator level is carried out in MATLAB Simulink block which optimize the speed of filter instead, of area constraint. Among the various structures, direct form fine-grain retimed adaptive algorithm gives the highest operating frequency of 119.66 MHz. Figure 11 shows the contrast between retimed structure and fine-grain retimed structure in the direct form in which fine-grained structure have the highest operating speed compare to the original structure with the same level of power consumption. But it consumes more hardware resource because of additional delay registers at the operator level.

6.2 Simulation Result

The proposed filter structures are designed in MATLAB (2010a). The proposed designs are synthesized using XILINX (ver. 13.2). The Simulink model for retimed delayed least mean square is designed in Xilinx System Generator (XSG) is shown in Fig. 12 which is a high-level software tool that gives a Simulink blocks libraries and cycle's accurate modeling of an arithmetic and logic operations as well as DSP operations. The original speech signal is evaluated through the filter structure. The actual speech signal is added with the white noise to generate the artificial corrupt signal. A primary input signal is given to the FIR filter followed by the comparator, which gives an error signal, i.e., difference between the corrupt signal and processed input signal. The error signal fed to the architecture to obtain an adaptive weight to remove the noise (unknown) from the speech signal. The simulation of different adaptive FIR filter is performed using Harvard sentence "THESE DAYS, A CHICKEN LEG IS A RARE DISH. THE HOGS WERE CHOPPED CORN AND GARBAGE" from SpEAR database and white noise from NOIZEUS database with filter tap of 4, a μ of $2^{-4,}$ i.e., step size with signal noise ratio (SNR) of 73 dB for this RDLMS de-noising adaptive filter.

The mean square error (MSE) is evaluated for conventional LMS filter and retimed LMS filter for the analysis of rate of convergence. The MSE expression for LMS adaptive FIR filter is as follows:

Table 2 Synthesis results for retimed adaptive FIR filter architectures

Adaptive FIR filters architecture	Xilinx Spartan-6 xc6slx16-2csg 324			Xilinx Virtex-4 4vsx35ff668-10			Xilinx Virtex 5 xc5vlx50t-1ff1136		
	DF-RDLMS	TDF-RDLMS	TF-RDLMS	DF-RDLMS	TDF-RDLMS	TF-RDLMS	DF-RDLMS	TDF-RDLMS	TF-RDLMS
Slices	110	117	116	306	292	276	121	131	114
Flip-flops	400	400	369	402	400	400	401	401	369
4-LUTs	392	388	373	287	287	287	377	377	377
Bonded IOBs	49	49	49	49	49	49	49	49	49
DSP48s	8	8	8	8	8	8	8	8	8
Max frequency of operation (MHz)	105.742	104.635	104.767	100.361	100.210	100.251	109.842	108.096	111.383
Max comb path delay (ns)	9.457	9.557	9.545	9.964	9.979	9.975	9.104	9.251	8.978
Power utilization (mW)	20.37	20.38	20.37	441.89	441.91	441.89	448.35	448.34	498.74

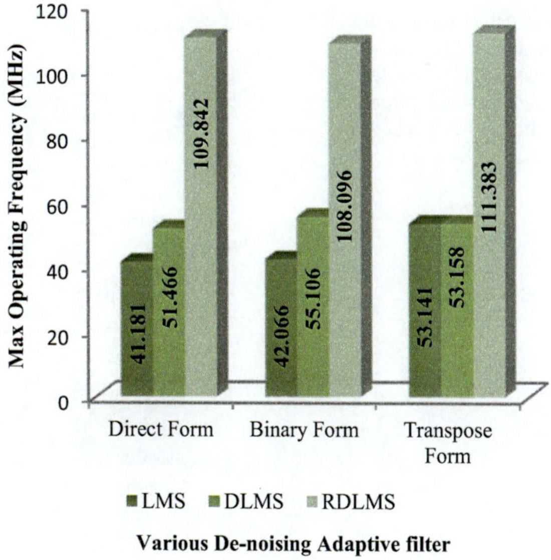

Fig. 8 Max operating frequency for various architectures realized on Xilinx Virtex-5 FPGA platform

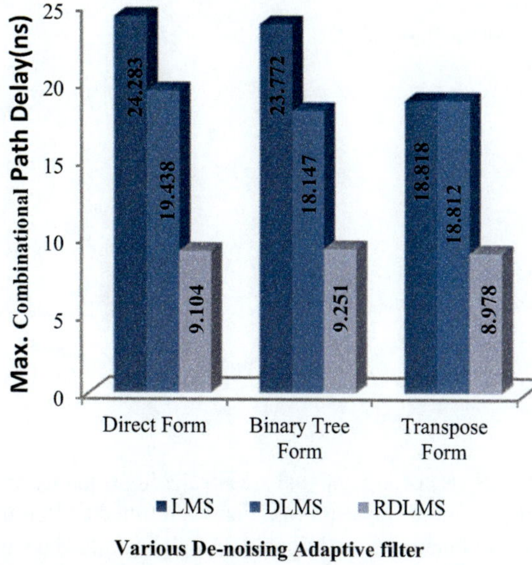

Fig. 9 Max combinational circuit path delay for various architectures realized on Xilinx Virtex-5 FPGA platform

$$\text{MSE} = \frac{1}{N}\sum_{i=0}^{N-1}(e_k - s_k)^2 \tag{26}$$

where N is the number of samples. Figure 13 shows the simulation results of mean square error for both LMS and RDLMS at $\mu = 2^{-4}$ with respect to number of

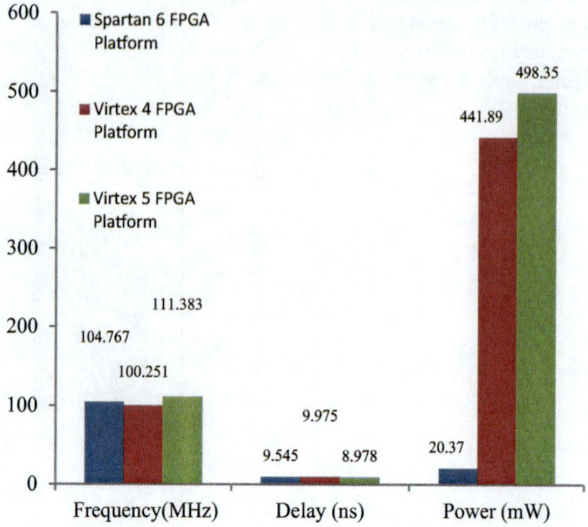

Fig. 10 Performance analysis of TF-RDLMS adaptive filter structure implemented for three different FPGA platforms

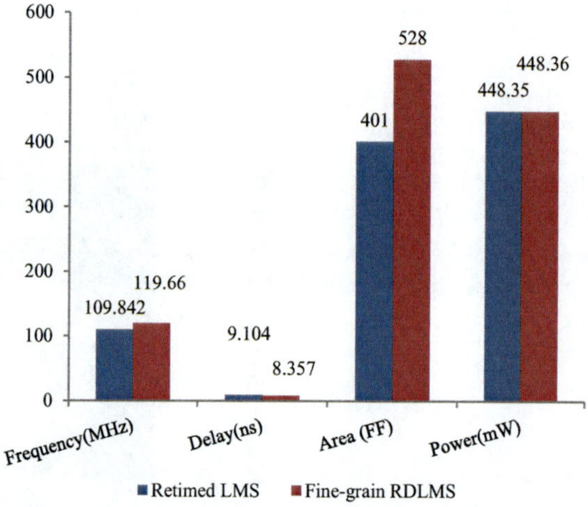

Fig. 11 Investigation on direct forms RDLMS and direct form fine-grain-RDLMS adaptive filter

samples (N) of input signal s_k. For the high-quality speech enhancement, the MSE value should be lowest. From Fig. 13, retimed LMS filter has a minimum MSE value when compared to traditional LMS algorithm. The simulation results are shown in Fig. 14. Simulink block is compiled in MATLAB and synthesized in Xilinx tool.

Table 3 shows the signal-to-noise ratio (SNR) value in decibel for three different de-noising adaptive FIR filter are direct form LMS, retimed DLMS and fine-grain RDLMS structures for 16,000 samples of input signals. This shows the fine-grain pipelining gives the highest performance enhancement with tradeoff in area utilization.

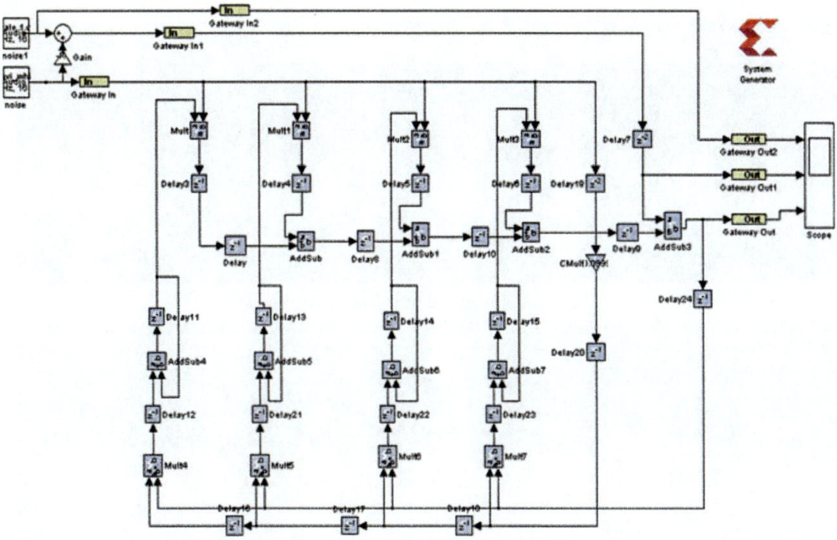

Fig. 12 Simulink model of transpose form-retimed delayed LMS (TF-RDLMS)

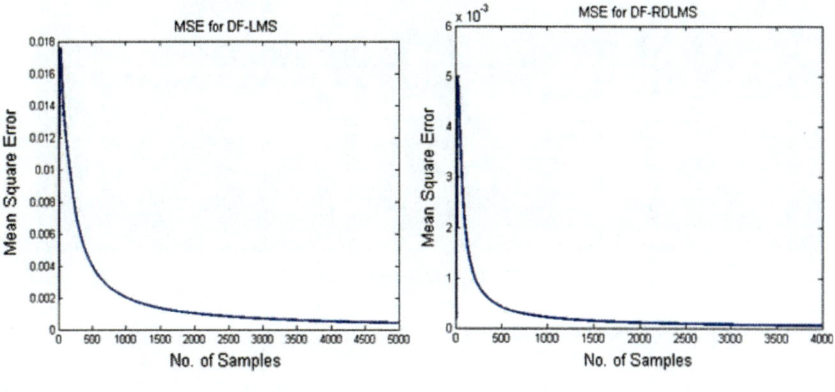

Fig. 13 Mean square error analysis with respect to number of samples of input signal for de-noising LMS and RDLMS

7 Conclusion

The proposed retimed high-speed adaptive filters, as well as fine-grained retimed adaptive FIR filter architecture, which is realized using different FPGA platform. For the retiming technique, RDLMS architecture of transpose form FIR filter implemented on Virtex-5 platform gives the finest performance among all the different structures implemented on Virtex-5 platform in provision of maximum operational speed or operating frequency of 111.383 MHz and critical path delay of 8.978 ns. For the fine-grained retimed adaptive filter architecture, the DF-RDLMS adaptive

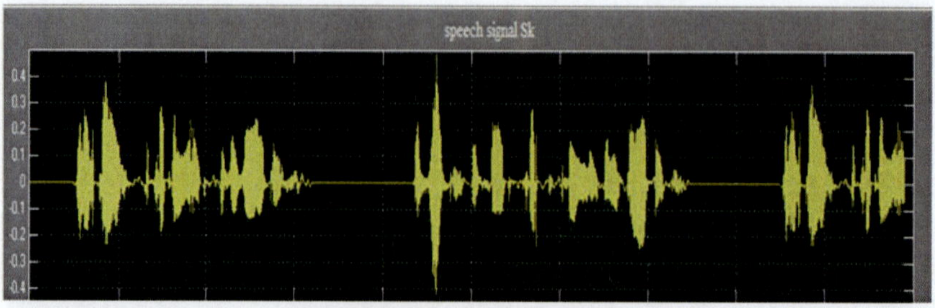

(a) The speech signal from a source

(b) Speech signal corrupted with white noise

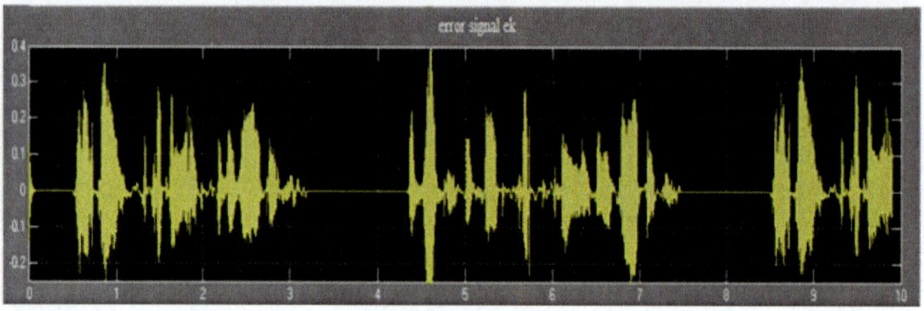

(c) Error signal

Fig. 14 Waveforms of a de-noising adaptive FIR filter

Table 3 SNR (dB) value of de-noising adaptive filter architecture for 16,000 samples

Architecture	LMS	RDLMS	FP-RDLMS
	70.98	73.25	84.54

FIR filter gives the finest performance in contrast with other forms of filters based on maximum frequency of operation and critical path delay with the same level of power consumption of 448.35 mW in Xilinx Virtex-5 FPGA platform. For various FPGA platforms, there is a significant difference in performance of a given de-noising filter. These de-noising VLSI structures are more convenient for small size filters with faster processing speed. The speech signal from SpEAR database and white noise from NOIZEUS database are being taken for an audio de-noising process. The minimum value of MSE for retimed structure compared to conventional LMS shows the faster convergence rate of RDLMS. A fine-grain retimed adaptive filter on Vitex-5 series FPGA platform achieves 8% improvement in clock frequency compared to a retimed structure because an insertion of register at operator minimize the critical path of operator which further reduces the critical path of overall filter.

References

1. Mula, S., Gogineni, V.C., Dhar, A.S.: Algorithm and VLSI architecture design of proportionate-type LMS adaptive filters for sparse system identification. In: IEEE Transaction on Very Large Scale Integration (VLSI) Systems, pp. 1750–1762 (2018)
2. Venkatesan, C., Kumar, P.K., Varatharajan, R.: FPGA implementation of modified error normalized LMS adaptive filter for ECG noise removal. In: Cluster Computing, pp. 1–9. Springer (2018)
3. Bahoura, M., Ezzaide, H.: FPGA implementation of parallel and sequential architectures for adaptive noise cancellation. In: Circuits Systems and Signal Processing, pp. 1521–1548. Springer (2011)
4. Dhal, M., Ghosh, M., Goel, P., Kar, A., Mohapatra, S., Chandra, M.: A unique noise canceller with advanced variable-step BLMS algorithm. In: IEEE International Conference on Advances in Computing Communications and Informatics (ICACCI), pp. 178–183, India (2015)
5. Behuria, A., Mishra, B.P., Goel, P., Kar, A., Mohapatra, S., Chandra, M.: A novel variable step-size feedback filtered-x LMS algorithm for acoustic noise removal. In: IEEE International Conference on Advances in Computing, Communications and Informatics (ICACCI), pp. 166–171, India (2015)
6. Liu, M., Wang, M., Liu, D., Zhao, B.: VLSI implementation of the modified sign-error LMS adaptive algorithm. In: J-Stage IEICE Electronics Express, pp. 2016–1001 (2016)
7. Son Vu, H., Chen, K.H.: A High performance feedback FxLMS active noise cancellation VLSI circuit design for in-ear headphones. In: Circuits System and Signal Process, pp. 2767–2785. Springer (2016)
8. Mohanty, B.K., Singh, G., Panda, G.: Hardware design for VLSI implementation of FxLMS- and FsLMS-based active noise controllers. In: Circuits System and Signal Process, pp. 447–473. Springer (2016)
9. Liu, M., Wang, M., Liu, D., Zhao, B.: Delay-optimized realization of 2-parallel delayed LMS adaptive FIR filter. In: J-Stage IEICE Electronics Express, pp. 2017–0225 (2017)
10. Mula, S., Gogineni, V.C., Dhar, A.S.: Algorithm and architecture design of adaptive filters with error nonlinearities. In: IEEE Transaction on Very Large Scale Integration (VLSI) Systems, pp. 2588–2601 (2017)
11. Haykin, S.S.: Adaptive Filter Theory, 3rd edn. Pearson Education India, Delhi (2008)
12. Tasleem, M., Ahmed, S.R.: Area and power efficient VLSI architecture of distributed arithmetic based LMS adaptive filter, In: IEEE International Conference on VLSI Design and International Conference on Embedded System, pp. 283–288 (2018)
13. Bujjibabu, P., Sirisha, K.: Design and implementation of efficient IIR LMS adaptive filter with improved performance, In: IEEE International Conference on Big Data Analytics and Computational Intelligence (ICBDACI), pp. 240–245 (2017)
14. Y Yi, R Woods, High speed FPGA-based implementations of delayed-LMS filters, Journal of VLSI signal processing systems for signal, image and video technology, pp. 113–131. Springer, (2005)
15. Goel, P., Chandra, M.: VLSI implementations of retimed high speed adaptive filter structures for speech enhancement, In: Micro- and Nanosystems Information Storage and Processing Systems, pp. 4799–4806. Springer (2018)

16. Zhao, B., Wang, M., Liu, M.: A high speed realization of the delayed dual sign LMS algorithm. J-Stage IEICE Electronics Express, p. 20180116 (2018)
17. Yagain, D., Krishna, A.V., Chennapnoor, S.: Design optimization platform for synthesizable high speed digital filters using retiming technique, In: IEEE International Conference on Semiconductor Electronics (ICSE), pp. 551–555. Malaysia (2012)
18. Parhi, K.K.: VLSI Digital Signal Processing Systems Design and Implementation, Student edn, Wiley, India (1999)
19. Elhossini, A., Areibi, S., Dony, R.: An FPGA implementation of the LMS adaptive filter for audio processing, In: IEEE International Conference on Reconfigurable Computing and FPGA's, pp. 1–8, Mexico (2006)

Recovery of Lost Information from Unclear Images Using MRF-Based Inpainting

Rebecca Fernando. I. M. S. B. and Dr. D. Jemi Florinabel

Abstract The work given in this paper explains the process of inpainting the images under global optimization technique. Initially, patch selection has been taken from an image by the clustering method to determine the portion of the unavailable information in an image. This takes place by the selection of the patch of lost portion in an image by grouping of the pixels in a region that gives a high quality of the image which has been followed by the refining of patches. The patch refinement takes place using high order singular value decomposition (HOSVD) that gets the cluster of patches and remove the unwanted pixels that are not needed for the inpainting. This collects the pattern from the patches that are clustered initially. At last, the weight value is obtained from the patches that has refined and the objective function has been included to get the clear image which gives a clear resemblance of the original image using Markov random field (MRF).

Keywords Image inpainting · Spectral clustering · HOSVD · Priority belief propagation

1 Introduction

Image inpainting [1] is the technique that is able to remove the portion of the image that avoids the view of information. This method takes the image as a source region and the hidden information in image as the target region. After that applying the inpainting algorithms [2] on the image, the refined image with the recovered information has been obtained. It helps to remove the damaged portion of the image that makes the image to be unclear.

R. Fernando. I. M. S. B. (✉) · Dr. D. J. Florinabel
Department of Computer Science and Engineering, Dr. Sivanthi Aditanar College of/Engineering, Tiruchendur, Thoothukudi, Tamil Nadu, India
e-mail: rebeccafdo777@gmail.com

Dr. D. J. Florinabel
e-mail: jemidelta@gmail.com

P. Suresh et al. (eds.), *Advances in Smart System Technologies*,
Advances in Intelligent Systems and Computing 1163,
https://doi.org/10.1007/978-981-15-5029-4_14

The image inpainting is used to get the information of the image that has been needed. The applications of image inpainting are audio recover, video restoration, infrared clearing, and red-eye removal and so on. The inpainting can be applied using structure-based inpainting [3] and texture-based inpainting [4]. In structural inpainting, the inpainting takes place using partial differential equations [5].

The structure layer in the image is taken as the input from which the tensor value is calculated using the luminous component of pixels that are extracted. Then, region of interest (ROI) [6] is calculated from the tensor value and the values applied to portion of image to be inpainted. Finally, the image is painted but result with a drawback of existing blurring effect in image.

In exemplar-based inpainting, the images are divided as the patch and the matches for the getting of lost information is derived. It obtained the multiple patches that are required to be inpainted using greedy algorithm. The patch to be inpainted is taken, and the priority value is calculated to get the way of inpainting the image. Then, the clear image is obtained, but it returns with the less error of inpaint the image, as the discontinuity of the pixels is difficult to obtain.

To improve the inpainting algorithm, the components like patch similarity and the patch priority measures are changed or updated using global based inpainting. In this method of inpainting takes the patches that has to be removed. The nodes of each patch are assigned with prior values such that they lead to find the target or missing region from the source region to be inpainted.

The MRF-based inpainting is done by taking nodes of the patch that covers the information in an image. The priority assigned to each node by the belief propagation (BP) algorithm. Under this, the nodes with priority are assigned to labels. The labels are needed to determine the values of patch that has to be inpainted.

2 Related Works

In anisotropic diffusion inpainting algorithm [7], the image is taken as the source space, from which the blur or the noise of the images can be removed. The anisotropic diffusion filter is applied at the noisy portion of image such that the images are smoothening by the gradient values got from the images using the edge stopping function. This algorithm has a drawback of not able to inpaint the image of high intensity, and the edges of the portion of images that are to be inpainted cannot be preserved when complete noise are removed.

In graph-cut exemplar-based inpainting algorithm [8], the virtual displacement map has been set by the extraction of patches from the portion of the image to be inpainted. After that, the energy function is computed from the virtual displacement map of the patches. The periphery of the patch portion to be inpainted is got out. Finally, graph-cut algorithm is applied to the boundary with energy function value and the image is inpainted. This returns with a drawback of the leftover of periphery that does not be in sequential for the hidden portion under the image.

In image melding process [9], the particles of the image-like color, texture, and region are the important characteristics through which the patch can be got for inpainting. The search space of patch is derived to make it easy method of inpainting. The image gradient value can be obtained from the image patch using a poison color solver. The gradient transaction of the patches are normalized and inpainting is made. But, the images still has got some blur on them.

In hierarchical super resolution inpainting algorithm [10], the patch of the image is calculated using the smooth term values. Further, patch sampling method has been applied to the extracted patches to get a low resolution image. The SR algorithm has been applied to the low resolution image such that the texture to be inpainted has been determined correctly. At last, the image is inpainted using belief propagation technique.

3 Proposed Work

The objective of the proposed work is to repair the data loss in an image and to create the new image which gives a clear resemblance of the original image. The proposed work takes place at three major steps: (a) selection of patches to get the patch that hides the image information, (b) refining the patch at applying HOSVD to obtain the portion of image to be inpainted, and (c) assigning patch using BP under MRF. The architecture of inpainting is in Fig. 1. This is given below:

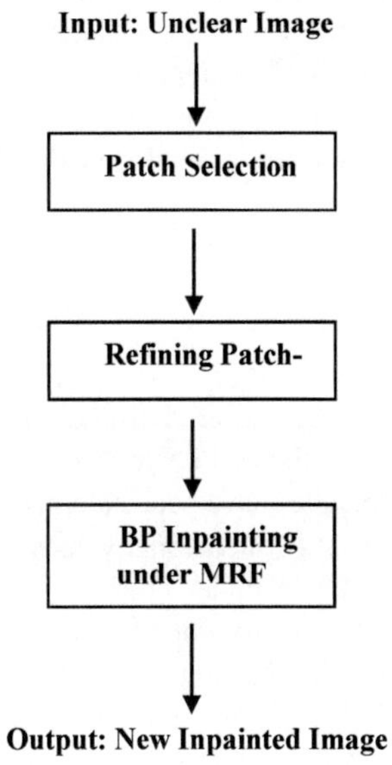

Fig. 1 Inpainting process

4 A. Selection of Patches

In image, patch is defined as the rectangular portion of the image. The patch can be split according to the size of image. The patch is to be taken from the unclear image is the portion that hides the information of the image. Initially, in exemplar-based inpainting the patches are selected using patch sparsity [11]. Initially, the image is partitioned into patches and the priority of the patches that needs to be inpainted obtained as smooth term values.

Then, the smooth term is given to the unclear portion of the images that contains the hidden information within it. The patches that are taken from the image to restore the image information are termed as candidate patch. The candidate patch is obtained from the images, and the patch matching algorithm [12] takes place. But, the images are left over with some errors or uncovered of image inpaint portions.

In the proposed algorithm, the first step is to take the unwanted portions that present on the images. For that, the clustering plays an important role of getting the part of images that does not give a full view of image. The spectral clustering [13] receives the patches as the input and group the unwanted portions that overlap on the original image. The spectral clustering is processed as the following architecture (Fig. 2).

The original image that contains information of not obvious to the viewers is inserted as input. The affinity matrix is measured from the image's patches has been computed to get the similarity of the whole portion (image) with the portion of the region to be inpainted from the image. The diagonal matrix has to be get from the affinity matrix. This is followed by the abstract of the eigenvalues which is used to get nearby information about the portion of the pixels in the images. Following eigenvalues, the number of clusters obtained using the difference between the similarity of the patch available on the patch hides the data and the patch of the original image. Finally, the cluster group obtained using k-means algorithm.

5 B. Refining Patches

Under the patch refining process, the clustered patches that are got from the clustering algorithm has to be refined to find the main patch from the portion of the image that has to be inpainted. Here, the patches have been calculated at two-dimensional stack since the image is taken as the two-dimensional values in the system. The group of patches that have been clustered is taken as input.

The tensor values computed in the two-dimensional matrix according to the size of patches. The tensor of the patches has to be rebuilt to the refining step under hard thresholding. The stack for the refining is calculated as the $k - 1$ patches where k is the number of clusters obtained for refining. The larger coefficients of the textures have been retrieved as the values from the stack that has to be refined.

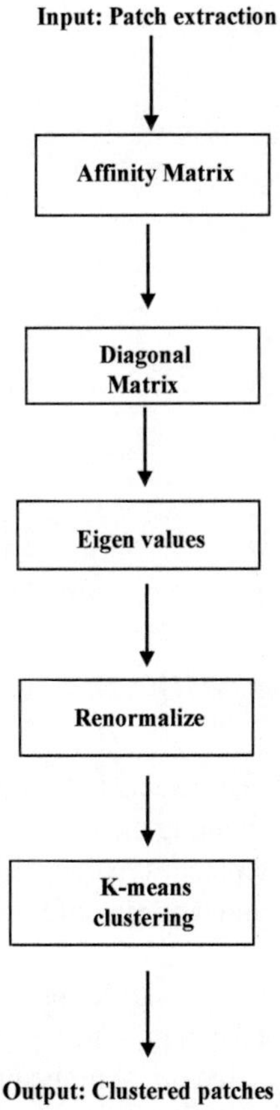

Fig. 2 Patch clustered under spectral clustering

Then, for the target patch of the source image, the patch to be applied for refining [14] has been taken as the stack of the patches. The decomposition of is done by calculation of the SVD to get the coefficient matrix and nullify the coefficients that comes under the upper thresholding. The tensor values are obtained by the inverse of the transform of the tensor matrix.

Tensor coefficients obtained are helpful to provide the details of the target patches that are to be used to inpaint the images. The tensor coefficients are needed since larger coefficients always have larger information about the images; the small coefficients may be either skipped or lost at the time of calculation of the weight average that carried out in exemplar-based inpaint algorithm.

The architecture of the patch refinement using HOSVD is given Fig. 3:

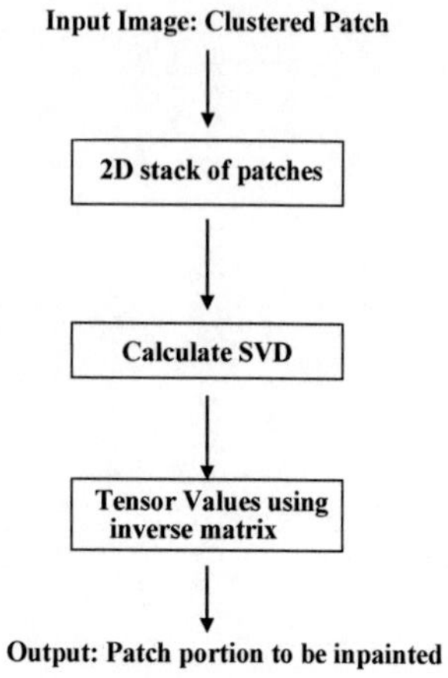

Fig. 3 Architecture of the patch refining process

6 Patch Inpainting Under MRF

As the patch portion to be inpainted is detected, the inpainting process takes place. In the patch-based method [15] of inpainting the images of lost images can be obtained by the filing of the regions of the patches extracted based on image segmentation [16] by the obtaining of the segmented portion of the object that has to be inpainted from the image.

In the proposed method of inpainting algorithms, the refined portion of image that has to be inpainted can be taken as the input. The weight of each pixels present in the patches are calculated based on the factors of refined patches that has been got from the HOSVD refinement. The weight value is given to the priority belief propagation algorithm [17]. The priority belief propagation was done as follows:

Step1 The refined patches of the HOSVD algorithm given as the input from which the group of pixels in the hidden portion can be taken as the group of nodes.
Step2 The node priority is given to the pixels to determine the number of pixels to be taken to make the image to be inpainted.
Step3 The priority is computed and the nodes where visited up to k nodes.
Step4 The label pruning is given to the visited node to know the important pixels that have to be inpainted.
Step5 The scheduling algorithm [18] of the patch pixels is done and the messages are updated to each pixel's node.
Step6 Repeat the above step until all nodes have been completed their visiting in the image pixels.

The architecture of the priority belief propagation is described in Fig. 4.

In Step1, the output got after the marking of target patch using HOSVD is made as input to the final stage of the inpainting process. The images are stored as the pixel collection from which the nodes have been assigned to the patch group that has to be removed. In Step2, the node priority is the value assigned as a key, value pair

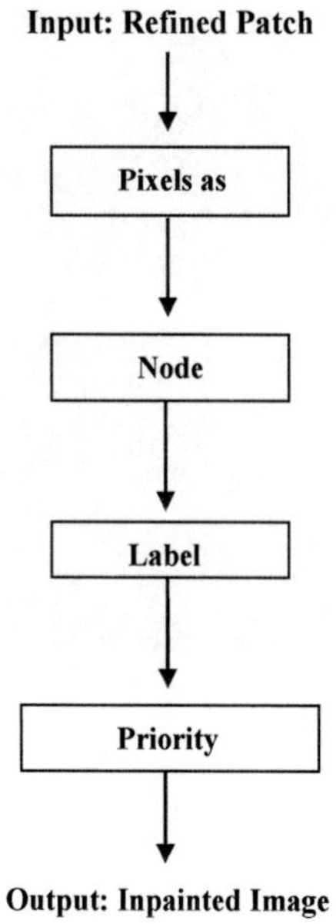

Fig. 4 Patches inpainted under MRF

such that the nodes are stored as the tree dataset. The processes of the determination of the number of the pixels that are friendly to the tree dataset are calculated as the priority of the node group so that the portion to be inpainted can be identified.

The nodes of the images are assigned and iterated by the priority function such that the node group can be prepared to be ready for application of inpainting. The priority function assigns the node group such that it reaches up to the k level as the images are stored in the way of $m * n$ with k number of clusters in the algorithm. Thus, the Step3 is completed and the label has been given to each node as given in Step4 in the group for the easy notify of the images' patches that are obtained for inpainting.

In Step5, the scheduling algorithm is the process of the creation of the schedule to the images such that the priority obtained from the node, the distance of each node, the time taken to reach the pixel portion of the hidden nodes are gathered. The schedule is developed using the tree dataset created by the system storage for the pixels and the values are generated. Step6 is repeated until the node groups have completed the task of the values of image portion that has to be inpainted.

7 Experimental Results

The algorithm is tested with the structure images, texture images and the real-world images. The input images can be with scratches or blur or with text. The image with lost information is given as input shown in Figs. 5 and 6.

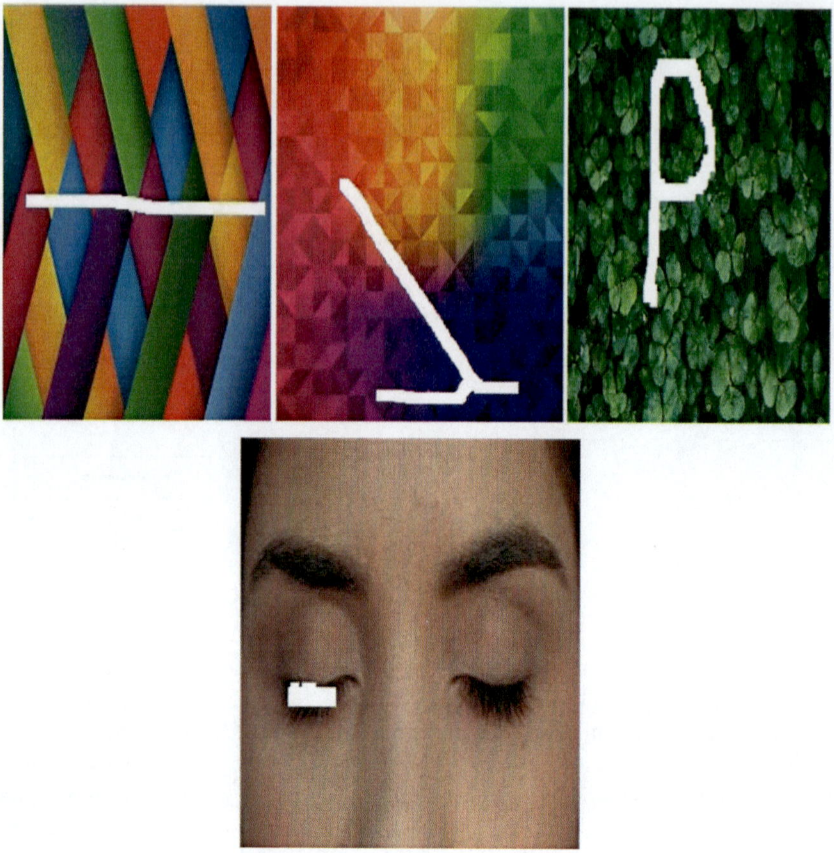

Fig. 5 Input images

First, the patch is been selected to know the portion that makes the information of images to be hide from the users. This is achieved by the detection of the position of target patch from the source patch under the clustering process that uses spectral clustering. The subgroup of the portions that are hidden within the images has been calculated such that the destination patches have been selected from the team of patches. The next gathered patches are compared with the similar patches of the image portions that are at near to the shown portions of the images. Finally, the patches that are to be inpainted are identified.

After the clustering of patches has been obtained, they are given as input to the refining of patches under HOSVD that returns the result as given in Fig. 7. The HOSVD is calculated from the pixels of clusters under the source region using the tensor values. The HOSVD can be got by the calculation of the coefficient matrix of the noise of the images that still available at the hidden portion of the images. Since the destination patch can be with large or small coefficients according to the image

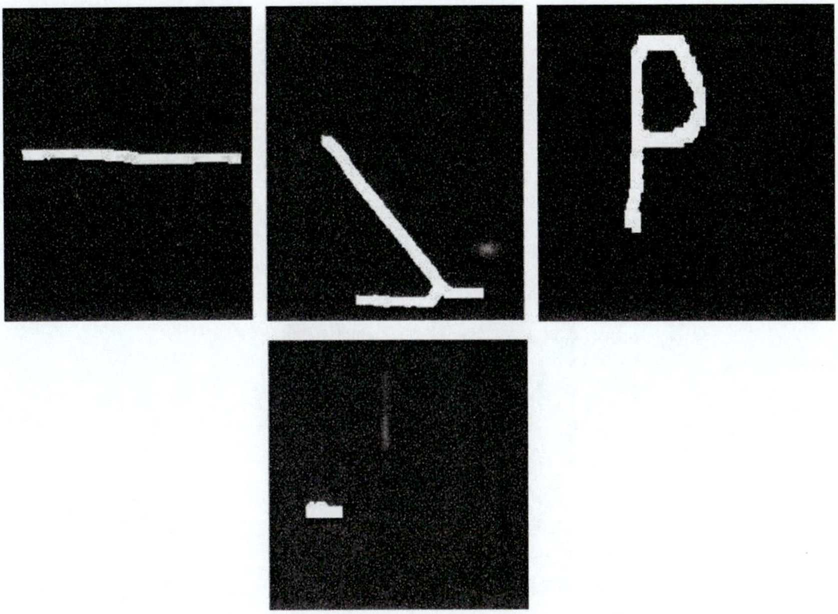

Fig. 6 Clustered patch

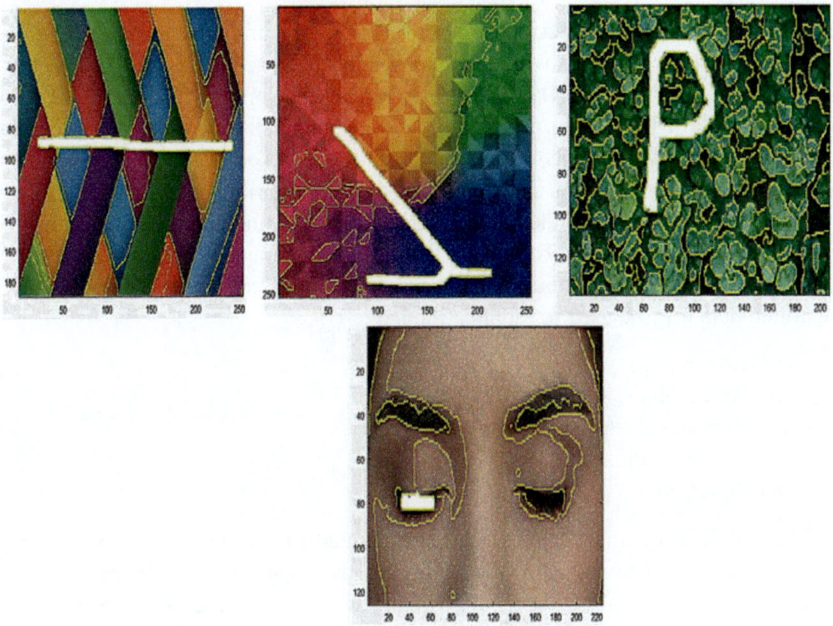

Fig. 7 Refined using HOSVD

Fig. 8 Inpainted images

size, the larger coefficient has been chosen and the obtaining the exact destination patch to be obtained for removing the damages has been acquired.

Finally, the output of HOSVD is delivered as the input to the final stage of inpainting. The MRF is the process of working under the undirected connected graph. Here, the MRF has been applied to each refined pixel of the image using the belief propagation algorithm. The belief propagation algorithm is used to determine the value of priorities from the patches selected for refining and the patches of the other portion of the images. The final result has been got as given in Fig. 8.

Thus, the inpainting of the images has been applied to the RGB images. The inpainted data may be used for application like

- signal processing,
- computer graphics,
- cyber forensics,
- logo removal in video and
- film restoration.

The PSNR has compared for different inpainted images given in Table 1. The PSNR values may be high or low depending on the image size. The work has been established for the images of the medium quality due to the time taking of the running time of the images in the work.

The set of images used for the working of inpainting using MRF are the

- flowers,
- faces,

Table 1 Comparison of PSNR values among different inpainted images

Image name	Elapsed time (in s)	PSNR values
Flower	6.128812	34.28
Face	25.668039	22.45
Bricks	25.438843	17.92
Pot	153.073731	25.47
Wall	82.086121	29.71

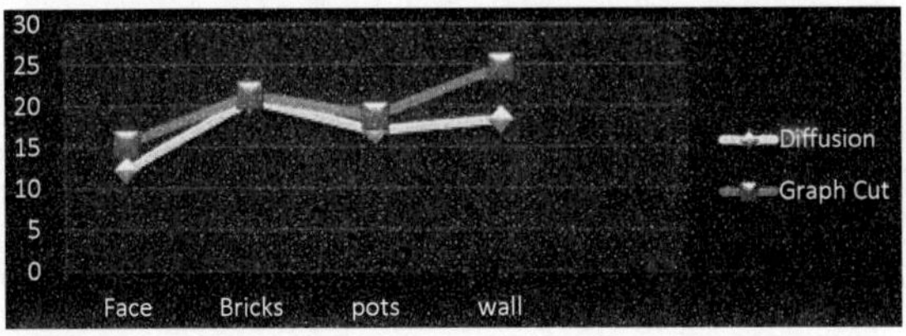

Fig. 9 PSNR versus images for inpainting

Table 2 Comparison with existing methods

Images	Diffusion	Graph cut	Image melding	Proposed work
Face	12.23	15.45	20.73	22.45
Bricks	20.76	21.34	24.15	27.56
pots	16.98	18.78	23.13	24.92
wall	18.33	24.69	28.87	29.71

- bricks,
- pots,
- wall.

The graph has been made using the above values (Fig. 9).

The comparison of the proposed work has been made with diffusion-based, graph cut, image melding using PSNR values as shown in Table 2.

The graph has been plotted using Table 2 to know the performance range (Fig. 10).

8 Conclusion

Thus, the image inpainting is taken under the MRF by three major steps. The patch selection is done based on the target region that has been retrieved as the portion to be inpainted with respect to the spectral clustering. This reduces the time of searching the patches that are needed to remove. This is followed by the refining of

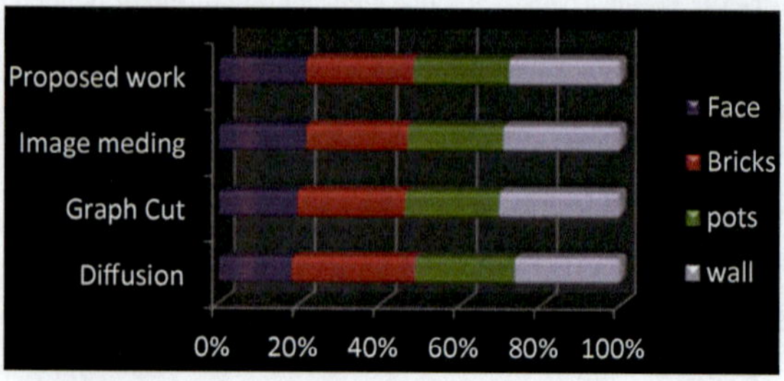

Fig. 10 Proposed inpainting with existence method

the patches from the source region. The HOSVD values calculated on the clustered patch. Finally, the Belief propagation algorithm has been applied under MRF for the complete inpainting of images.

References

1. Pushpalwar, R.T., Bhandari, S.H.: Image inpainting approaches—a review. In: IEEE 6th International Conference on Advance Computing (2016)
2. Patel, P., Prajapati, A., Mishra, S.: Review of different inpainting algorithms. IEEE Int. J. Comput. Appl. **59**(18), 30–34 (2012)
3. Li, H., Luo, W., Hwang, J.: Localization of diffusion-based inpainting in digital images. IEEE Trans. Inf. Forensics Secur. **12**(12) (2017)
4. Bhyssens, P., Daisy, M., Tschmperle, D., Lexoray, O.: Exemplar-based inpainting: technical review and new heuristics for better geometric reconstructions. IEEE Trans. Image Process. **24**(6), 1809–1824 (2015)
5. Alilou V.K., Yaghmaee, F.: Introducing a new fast exemplar-based inpainting algorithm. In: 22nd Iranian Conference on Electrical Engineering (ICEE), 20–22 May 2014
6. Sharma, N., Mehta, N.: Region filling and object removal by exempler-based image inpainting. Int. J. Inventive Eng. Sci. (IJIES) **1**(3) (2013)
7. Palma, C.A., Cappabianco, F.A.M., Ide, J.S., Miranda, P.A.V.: Anisotropic diffusion filtering operation and limitations—magnetic resonance imaging evaluation. Int. Fed. Autom. Control **47**(3), 3887–3892. 24–29 Aug (2014)
8. Lin, Y., Caselles, M.: Exemplar-based image inpainting using multiscale graph cuts. IEEE Trans. Image Process. **22**(5), 1699–1711 (2013)
9. Darabi, S., Shechtman, E., Barnes, C., Goldman, D.B., Sen, P.: Image melding: combining inconsistent images using patch-based synthesis. ACM Trans. Graph (2012)
10. Le Meur, O., Ebdelli, M., Guillemot, C.: Hierarchical super-resolution-based inpainting. J. Latex Class Files **22**(10) (2007)
11. Fan, Q., Zhang, L.: Improvement of patch selection in exemplar based image inpainting. In: 3rd IIAE International Conference on Intelligent Systems and Image Processing (2015)
12. Athishtalakshmi, M., Parimala, R., Pushpalatha, S., SenthamaraiSelvi, N.: A patch match algorithm for image completion using Fr based image inpainting. UISET (2015)
13. Ng, A.Y., Jordan, M.J. Weiss, Y.: On spectral algorithm: analysis and algorithm. In: 14th International Conference on Neural Information Processing System (2002)
14. Rajwade, A., Rangarajan, A., Rranerjee, A.: Image denoising using the higher order singular value decomposition. IEEE Trans. Pattern Mach. Intell. (2013)
15. Ruzic, T., Pizurica, A.: Context aware patch based image inpainting using Markov Random field modeling. IEEE Trans. Image Process (2013)

16. Hassan, E.T., Abbas, H.M., Mohamed, H.K.: Image inpainting based on image segmentation and segment classification. In: IEEE International Conference on Control System, Computing and Engineering, 29 Nov–1 Dec (2013)
17. Hsin, H.-F., Leou, J.J., Lin, C.S.: Image inpainting using structure-guided priority belief propagation and label transformation. In: International Conference on Pattern Recognition (2010)
18. Komodakis, N., Tziritas, G.: Image completion using efficient belief propagation via priority scheduling and dynamic pruning. IEEE Trans. Image Process. **16**(11), 2649–2661 (2007)

Understanding Textile Antenna by Reviewing and Simulating It for High Data Rates Applications

Asit Kittur and G. Vairavel

Abstract In the proposed study, different wearable antenna papers are studied, and their detailed analysis is done. The conclusion of the survey and demonstration of future work is analyzed. While going through the study, different substrate materials and conducing materials along with their permittivity are also studied. For beginners, design calculations are also given. Different antennas with different band and shape with different textile materials and different frequency are studied. Finally, for this study, fractal antenna for different shape and different frequency is simulated.

Keywords Conducting materials · Operating frequency · Substrate materials · Textile antenna

1 Introduction

Human efforts and his thinking are helping for speed of technology and its tremendous growth like wildfire. In the world of radio technology and antenna, the relation between man and machine has taken forward a big steps. The rectangular and circular micro-strip for wearable application can be seen designed.

Along with the wearable application of textile, it has capacity to conduct as well. If antenna is wearable, then the number of application increases in every field like sports, medical and army.

Though metallic antenna is highly conductive and is solid structured, it gives stable output. The major factor is output stability even though the antenna elongated or compressed.

As the rigid structure of antenna will be harmful and carrying such antenna onto body will cause discomfort, the antenna made of textile will have no point of any harm. This will be very useful for studying the area called "Body Area Network".

Asit Kittur · G. Vairavel (✉)
Department of ECE, Veltech Rangarajan Dr. Sakunthala R&D Institute of Science & Technology, Chennai, India
e-mail: vairavel@veltech.edu.in

P. Suresh et al. (eds.), *Advances in Smart System Technologies*,
Advances in Intelligent Systems and Computing 1163,
https://doi.org/10.1007/978-981-15-5029-4_15

For many purposes and interest, the body area network will be a great boost for being healthcare, sports, educational and emergency services like military applications, etc.

The space deformation technique is used for three-dimensional manipulation of wearable antenna deformation. The results are compared with published results for antennas bent over a cylindrical surface. The method is also applied to model crumpling and twisting of textile antenna [1].

The properties of E-textile for wearable application are assessed [2]. GPS application antenna is presented for 1.575 GHz. Jig holding fabric is developed to make sure stiffness for dielectric measurement process. It gives good return losses at resonant frequency. The radiation property of antenna is also studied. The good return loss of −37 db at 1.9 GHz resonant frequency is obtained.

The effect of antenna orientation towards the SAR values in other body parts must be considered. Effect of antenna on body and effect of body on antenna can be studied. The orientation of antenna horizontally gives 16% more SAR than it is placed vertically. Antenna resonant frequency is shifted by human body, and it can modify radiation pattern. Placement of antenna at different body location and different orientation is studied [3]. It is concluded that to place antenna on human body carefully, choose orientation and position.

The importance of artificial magnetic conductor is that it can absorb electromagnetic wave into human body and improve antenna performance. While doing so, the materials elasticity, wetness, bending effect, human body movement and moisture changes are taken into consideration [4]. It is concluded that radiation exposed to human body is less and directional pattern can be directional.

Magnetically coupled Antenna for the active circuitry can be implemented for suitable garment. For hybrid substrate, primary winding is implemented, and for monolithic substrate, secondary windings are implemented which reform heterogeneous transformer. Antenna and circuitary contacts are avoided by coupling technology [5]. The proposed technique is used for RFID tags. It is concluded that the system can be used for Wi-Fi communication. For future scope, instead of one turn primary and secondary winding, a two or more turn winding can be implemented.

Placing antenna at different distances from the human body and bending effect is considered. It is seen that human body shifts resonant frequency and radiation pattern is modified [6]. It is concluded that when curved textile is kept away from body for 1 mm and is bent, successive approximation rate is improved to 92.3%. For further research, effects using different antenna with different frequency and different body part can be investigated.

Back radiation is reduced by soft surface of patch antenna is proposed. The bent finite ground plane scenario and accurate model of electromagnetic property of the human body scenario are considered [7]. The resilience of textile to difficult conditions, such as moisture effect on electromagnetic property of the felt, also metallic thread to bending, crumbling, is performed.

Dual-band leaf-shaped textile antenna for ultra high frequency is used with denim as substrate [8]. The antenna frequency is applicable for long-term evolution (LTE) and WiMAX. Placing antenna perpendicular to the human body along with its future scope by placing it very close to the body is also studied.

A dual-band textile antenna for digital television and other application such as GSM 900, WLAN 2.45/5.8 GHz, Wimax 3.5 GHz, Hyperlan 5.2 GHz and UMTS-LTE is analyzed. Antenna is placed at different position with not much affected return loss. The return loss is further reduced by placing antenna at the back of human body. For receiving digital television and wireless communication, this soft antenna is good candidate in smart clothing environment [9].

A rectangular antenna having slots is anticipated. The antenna is of 200 MHz, X-band. The micro-strip antenna is excited by feeding with probe achieved. The slot gives appropriate bandwidth and high gain [10].

For body area network for wireless system, a textile micro-strip antenna with UWB range is proposed. Evaluation of antenna in free space is performed on body [11].

The investigation is done on tiny-shaped textile antenna on human, and the experiment is done in reverberation chamber. It is shown that material selection has major impact on body frequency detuning and efficiency levels. The investigation is done under bent condition, and it is shown that the performance of textile antenna is optimized [12]. It is concluded that for radiation efficiency, 20 mm is enough. For future measurement, multiband antennas with large-sized textile materials are envisaged.

For mobile router application, a folded textile antenna is proposed. Double feeding method on rectangular monopole is used for two antennas, and the isolation is increased by separating slit [13]. It is concluded that antenna array can increase the performance of router used for mobile, and also, the property of felt textile material is determined.

2 Wearable Textile Antenna

2.1 Various Substrate and Fabric Investigate

Different materials having different dielectric properties are mathematically designed on micro-strip patch antenna. Its effect on antenna parameter is also checked for designing textile antenna [14]. How different textile materials and different frequency with different band and shape for different antennas can be obtained is studied in Table 1.

Table 1 Analysis of fabric materials

Fabric Material	ε_r	Loss Tan σ	S_{11}	Gain (dBi)	Directivity (dBi)	Efficiency η (%)
Lycra	1.5	0.0093	−31	6.8	8.59	67
Cordura	1.9	0.0098	−29	5.9	7.93	64
Polyester	1.9	0.0098	−35	6.8	7.90	76
Cotton	1.5	0.0093	−31	6.8	6.59	67

The simulation of antennas with anisotropic fabric is done on IE3D. The antennas used have different loss tangent and permittivity. The cordura and lycra having same permittivity and loss tangent are slightly different are used to see its effect on comparative performance as shown in above table.

By examining the characteristics of four antennas, it is concluded that polyester is better than remaining three as its loss tangent is 0.0045 the lowest of all. Cordura and lycra are good for high directivity antenna.

Figure 1 shows return loss on x-axis and frequency on y-axis of cotton material. Having low gain, it is used for short range for lower data rates.

Figure 2 shows S_{11} graph plotted against frequency giving return loss for polyester a negative value, and also, frequency at resonance is 2.4 GHz which when compared with operational frequency, it found to be less.

It seems from Fig. 3 and Table 1, to increase efficiency, low substrate material must be selected, and to also minimize surface wave losses, low dielectric constant is useful. Inside the substrate, the propagation of wave is also helpful.

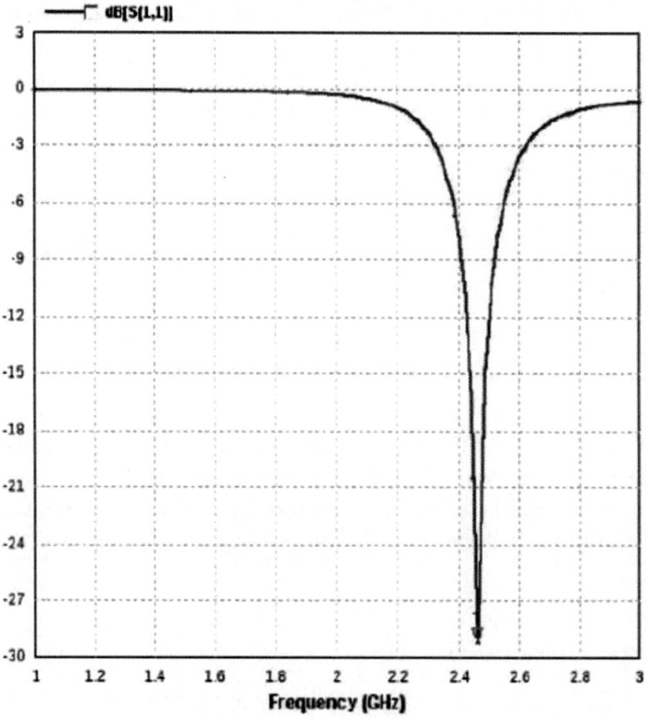

Fig. 1 Reflection coefficient of cotton

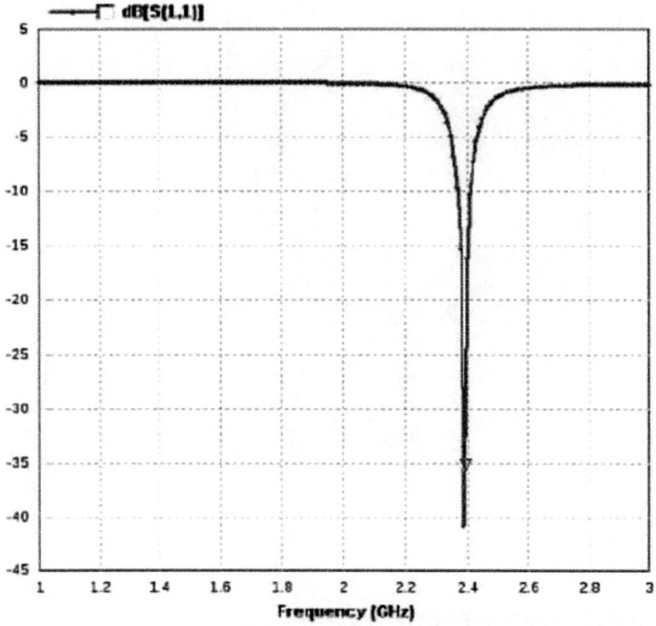

Fig. 2 Is reflection coefficient of lycra having same gain but higher directivity than cotton

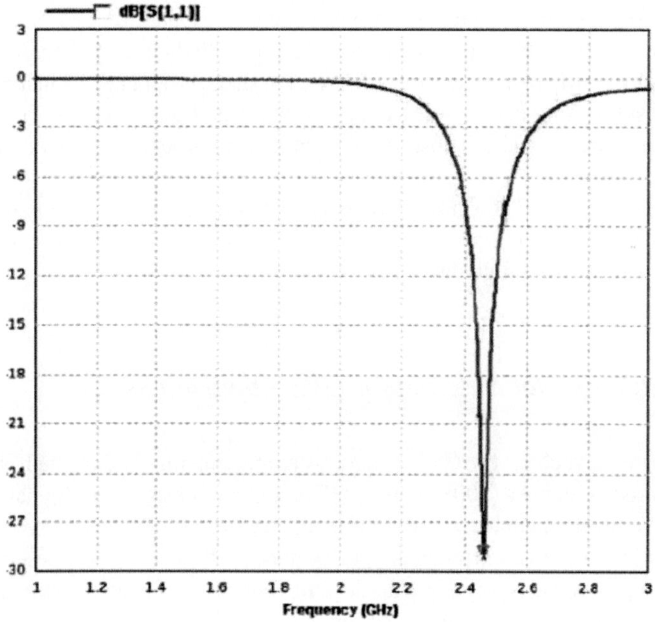

Fig. 3 Reflection coefficient of lycra

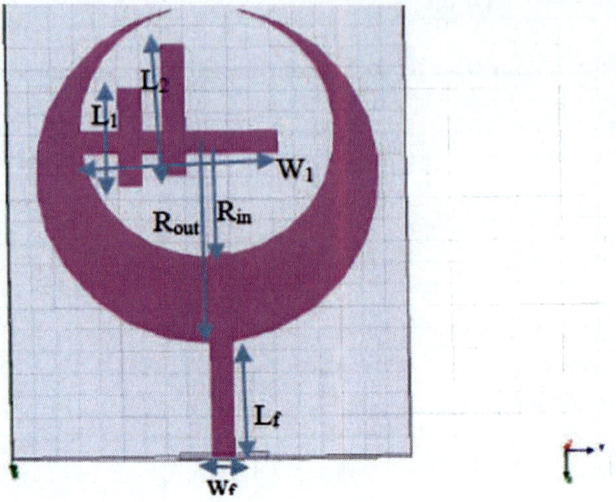

Fig. 4 Crescent-shaped antenna

2.2 *Crescent-Shaped Textile Antenna*

Two circular-shaped antennas with different radius are combined in such a way to achieve a crescent-shaped antenna so that obtained resonant frequency should be compatible with WLAN [15]. Using jeans substrate with thickness of 1 mm and area 22 × 18 mm, the strip line is used as feed. Crescent shape is obtained by including two circles and subtracting inner circle from outer circle as shown in Fig. 4. The antenna is placed on jeans dielectric substrate. It is placed on ground as perfect conductor sheet.

It seems from Fig. 5, for 5.8 GHz resonant frequency, bandwidth is 600 MHz from 5.5 to 6.1 GHz −15.05 dB of return loss. This concludes that crescent-shaped antenna can be used for high data rates.

2.3 *Textile Fractal Antenna for GPS Applications*

For applications such as L1 GPS, a new wearable antenna is designed [16]. The substrate used is flannel (100% cotton). The antenna designed is fractal antenna where caps of different material are integrated; cap integrated with wearable antenna is placed on human head, and antenna's performance is studied. The proposed antenna shows if Fig. 6 consists of identical slots in four arms. Ground textile substrate thickness is from 1 mm to 3 mm. The antenna is resonated at frequency of 1.602, 1.599, 1.577, 1.608 and 1.601 GHz at TF = 1, 1.5, 2, 2.5 and 3 mm, respectively.

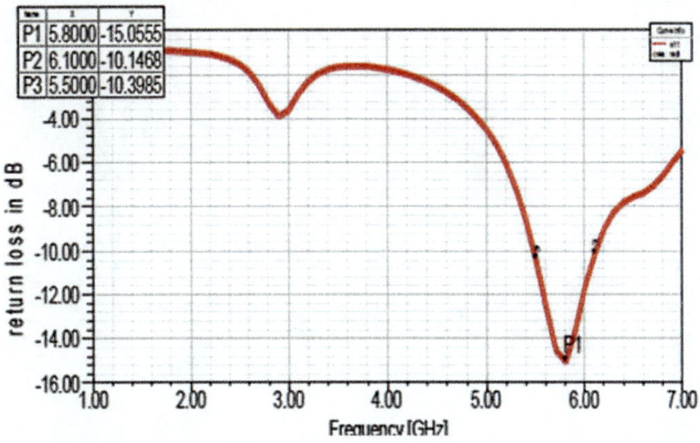

Fig. 5 Return loss against frequency

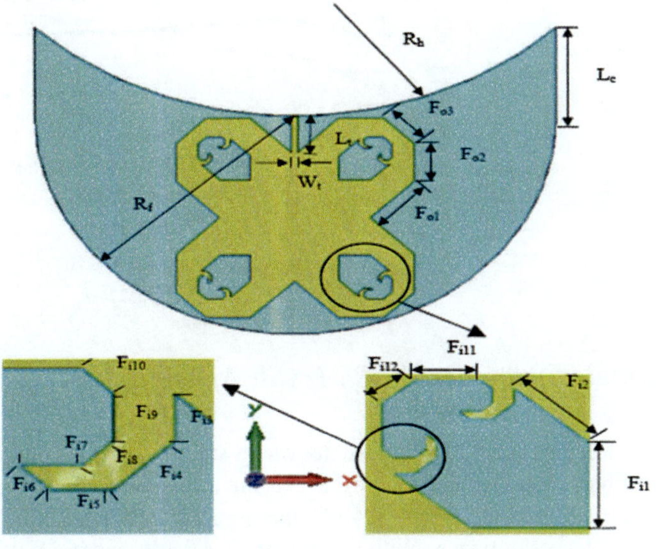

Fig. 6 Geometry of fractal antenna

Figure 7 shows reflection coefficient obtained for various substrate thickness. From Fig. 8, it seems that boost in antenna gain can be obtained by increasing substrate thickness.

The designed antenna is integrated on CAP, and its performance is investigated. The insertion of CAP increases resonant frequency which is compensated by the presence of human head which decreases resonant frequency. The designed antenna can be used for high data rates applications like GPS in L1 band.

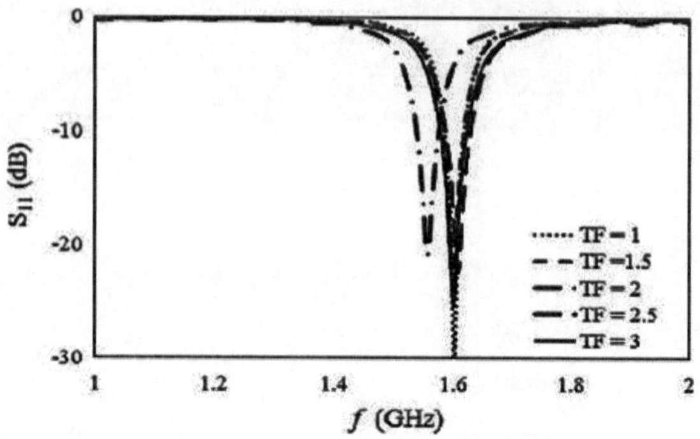

Fig. 7 Reflection coefficient of various substrate

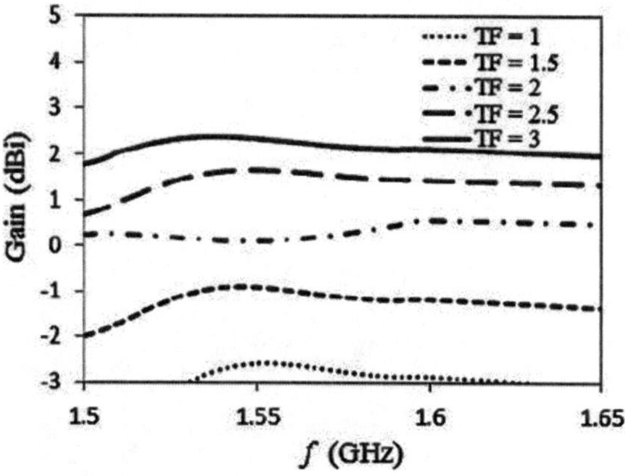

Fig. 8 Gain for different frequency

2.4 *Wearable Embroidered GPS Textile Antenna*

The work is designed for GPS applications where stitch pattern is contour filled. The complex human model and bending effects are also simulated. It gives return loss $S_{11} = -13$ dB and Gain = – 0.4 dBi operated for GPS L1 band (1575.42) giving reasonable performance. While simulating SAR, the margin obtained for electromagnetic radiation in the worst case is 24.3%.

For antenna details shown in Fig. 8, a $S_{11} = -36$ dB shown in Fig. 10.

For average human arm, the bending case is also considered in free space shown in Fig. 11, and the S_{11} is decreased to −29 dB which is still considered good return loss with minimum frequency shift shown in Fig. 12.

Stitching textile and filling contour are embroidered with *XL-550* embroidery machine. The stitching is denser as a number of stitching lines are constant [17]. The two layer cotton substrate textile is used as a patch antenna and ground plane of material WE-CF adhesive copper sheet is used as shown in Fig. 13.

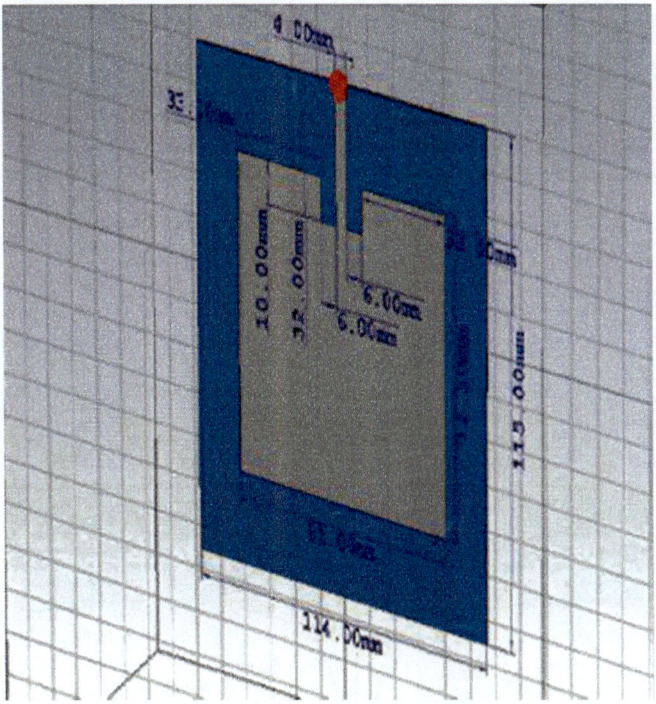

Fig. 9 Designed antenna

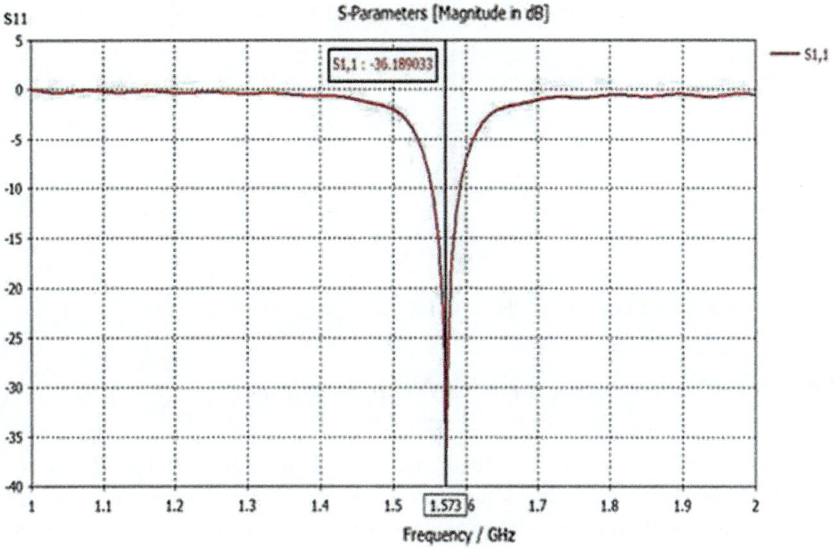

Fig. 10 Return loss of designed antenna

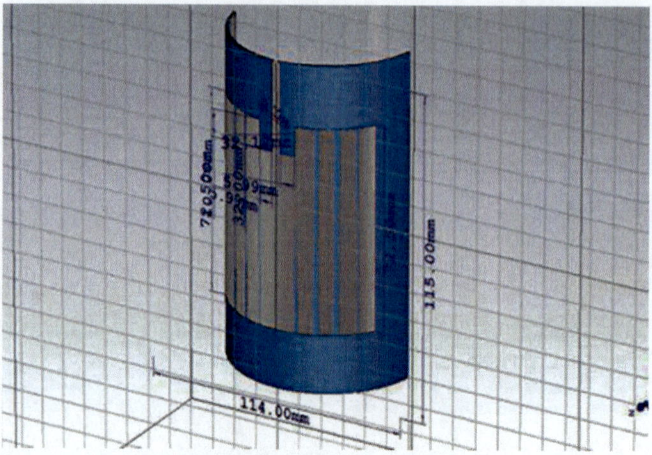

Fig. 11 Designed antennas for bending sample

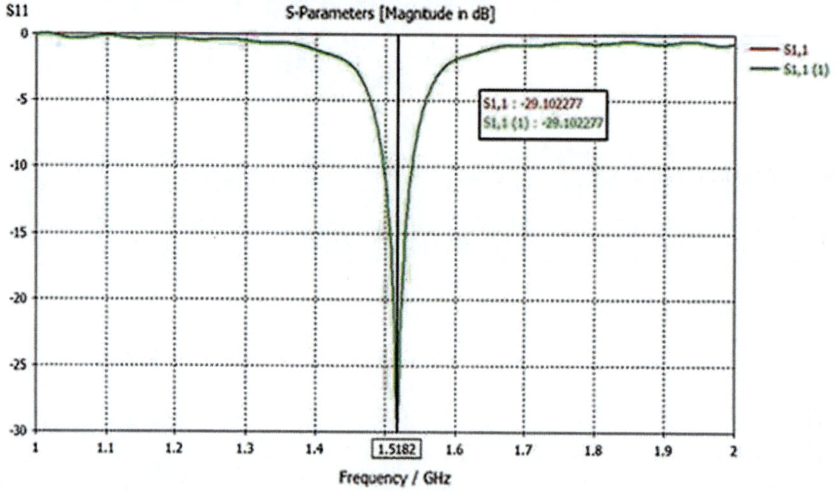

Fig. 12 Return loss for bending sample

2.5 A Textile Antenna for WLAN Applications

This study gives a wearable micro-strip circular disk antenna shown in Fig. 14 for WLAN has been tested. A conducting part is considered as a copper and polyester combination. Cotton fabric is used as a substrate [18]. The resonance method is used for accurate value for polyester and cotton combination (65:35). This technique is also used to measure permittivity of fabric which is 1.48.

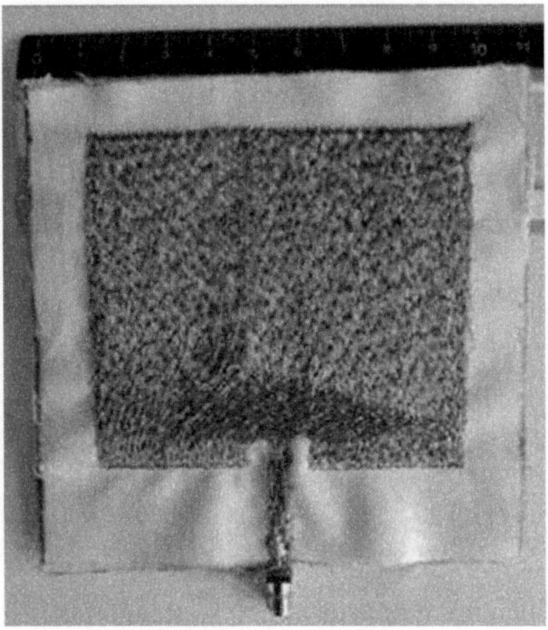

Fig. 13 GPS embroidered antenna patch on cotton fabric

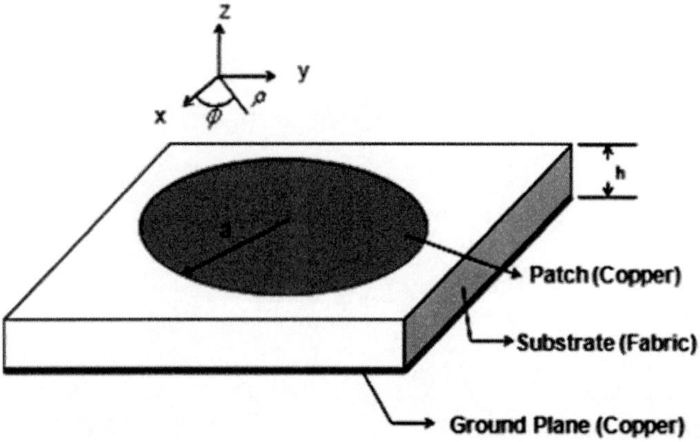

Fig. 14 Micro-strip patch antenna

From 2 to 3 GHz, simulations are carried out with each step comprised of 20 MHz as shown in Fig. 15. The frequency at which antenna resonates is 2.315 GHz having return loss up to −20.41 dB.

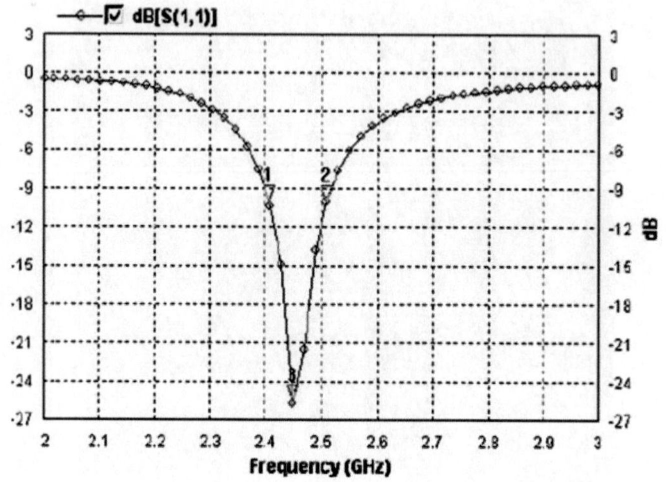

Fig. 15 Return loss plot

3 Design Calculation of Micro-Strip Antenna for Beginners

Design equation for micro-strip antenna requires the calculation of operating frequency f_0 height h and permittivity $\in_r$.

The wavelength is calculated by

$$\Gamma = \frac{c}{f} \tag{1}$$

$$W = \frac{c}{2f}\sqrt{\frac{2}{\in_r + 1}} \tag{2}$$

$$L = L_{\text{eff}} - 2\Delta L \tag{3}$$

where w = Patch width
C = Speed of light
F = Frequency of resonance
L = Patch length
$\in_r$ = Substrate dielectric constant
L_{eff} = Effective length and it is

$$L_{\text{eff}} = \frac{c}{2f\sqrt{\in_{\text{reff}}}} \tag{4}$$

The normalized extension in length can be calculated as

$$\Delta L = 0.412h\frac{(\in_{\text{reff}} + 0.3)\left(\frac{w}{h} + 0.264\right)}{(\in_{\text{reff}} - 0.258)\left(\frac{w}{h} + 0.8\right)} \tag{5}$$

where $\in_{\text{reff}}$= Effective dielectric constant and it is given by

$$\in_{\text{reff}} = \frac{\in_r + 1}{2} + \frac{\in_r + 1}{2}\left[1 + 12\frac{h}{w}\right]^{-1/2} \tag{6}$$

The substrate width and length are given by

$$L_g = L + 6h \tag{7}$$

$$W_g = W + 6h \tag{8}$$

where L_g and W_g are length of the substrate and width of substrate and h can be calculated as

$$h = \frac{0.0606\Gamma}{\sqrt{\in_r}} \tag{9}$$

Feed length $(L_F) = \frac{\Gamma_g}{4}$
Where Γ_g is guided wavelength and it is given by

$$\Gamma_g = \frac{\Gamma_g}{\sqrt{\in_{\text{reff}}}} \tag{10}$$

Antenna efficiency is given by

$$\eta = \frac{\text{gain}}{\text{directivity}} * 100\% \tag{11}$$

where η = Efficiency.

4 Properties of Textile Material

See Table 2.

Table 2 Properties of various insulators for textile antenna

Materials	Thickness (μm)	Permittivity $\in_r$	Conductivity in S/m	Loss tangent
Bi-adhesive tape	100	3	–	0
Denim	85	1.67	–	0.0019
Black foam	3940	1.575		–
Copper	18	–	5.8×10^7	–
SheildIT	17	–	1.8×10^5	–
Flectron	40	–		
Pyralux	50	3.2	–	0.001
Fleece	600	1.43	–	0.0024
Foam	300	1	–	0.020

5 FRACTAL Antenna Design

Tree-Shaped Fractal Antenna

For basic operation, antenna, as shown in Fig. 16, resonated for three different frequencies. For the first antenna, resonant frequency is 7.7 GHz with S_{11} is −20.29 dB and bandwidth is 298 MHz. For the second antenna, resonant frequency is 5.8 GHz, the S_{11} is −13.75 dB, and bandwidth is 298 MHz. The third antenna is designed for 8.4 GHz having S_{11} is −22.22 dB with bandwidth of 248 MHz, and its reflection coefficient plot is shown in Fig. 17.

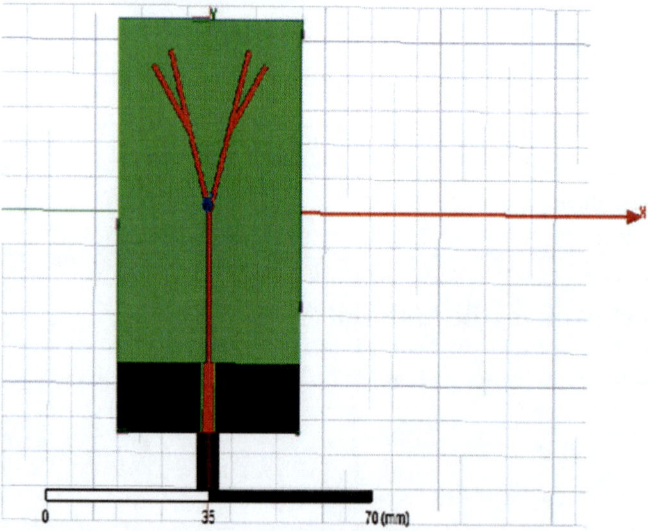

Fig. 16 Tree-shaped fractal antenna

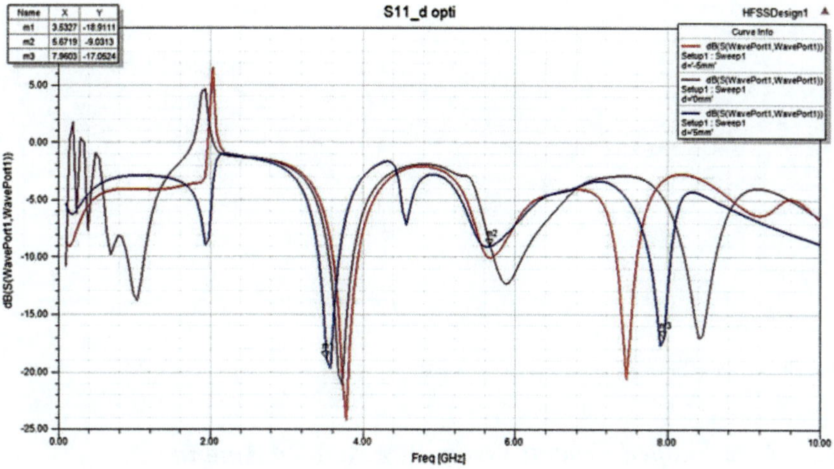

Fig. 17 S_{11} plot for tree-shaped fractal antenna

5.1 Optimized Tree-Shaped Fractal Antenna

Figure 18 shows the antenna is operated at three different frequencies. The first antenna is having frequency of 2.8 GHz giving return loss S_{11} of -16.43 dB and bandwidth of 99.4 MHz. The second antenna frequency is 16.68 GHz having S_{11} of -13 dB and bandwidth of 5.37 GHz. The frequency of the third antenna is 20.38 GHz, whose S_{11} is -12.10 dB and bandwidth is 2.74 GHz as shown in Fig. 19.

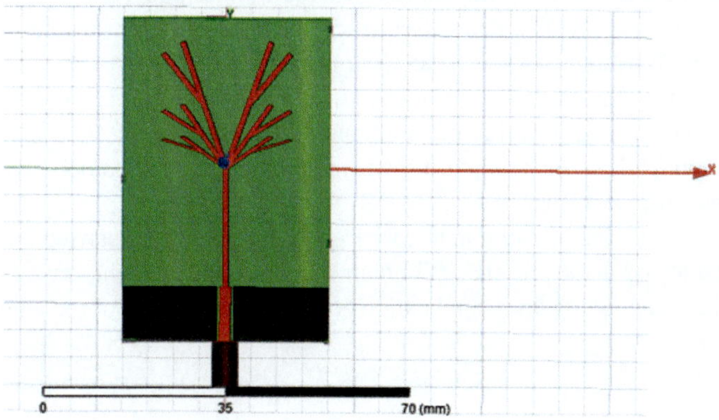

Fig. 18 Tree-shaped fractal optimization

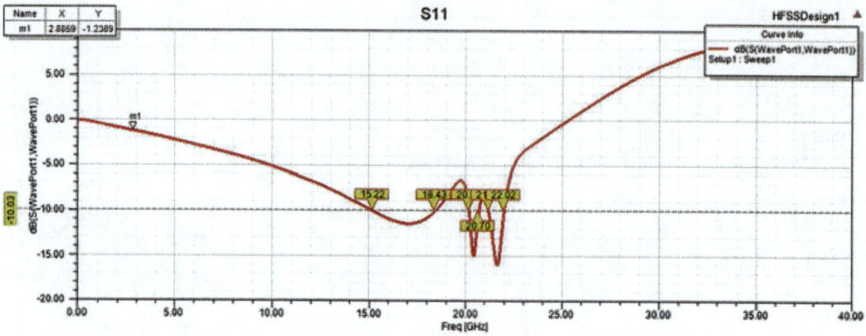

Fig. 19 S_{11} Plot for tree-shaped fractal optimization

5.2 *Tree-Shaped Fractal Vertically Extended Antenna*

For Fig. 20, the first frequency is 6.48 GHz whose S_{11} is -13.8 dB. The bandwidth for this is 2.34 GHz. The second frequency is 8.37 GHz having S_{11} is -9.93 dB and bandwidth is 598 MHz which can be seen in Fig. 21.

6 Constraints While Designing the Antenna

To continuously monitor the biometric data of human body, antennas need to be very close to human body. Considering the solid structure of antenna, it will be very harmful for human body. At the same time, the human body radiates frequency, and this may interfere with antenna frequency can cause different results. The back radiation caused by antenna will be also harmful for body.

As the textile has the property to bend, elongate and compress, its property changes, and hence, while doing measurement, different position should be considered. The applications such as putting wearable antenna on shoulders will help to calculate real-life problems.

Apart from antenna measurements, where measurements are carried out with no human present and in the presence of human, antenna measurements have to be done on body also. Depending upon applications of antenna, position of the antenna will also change.

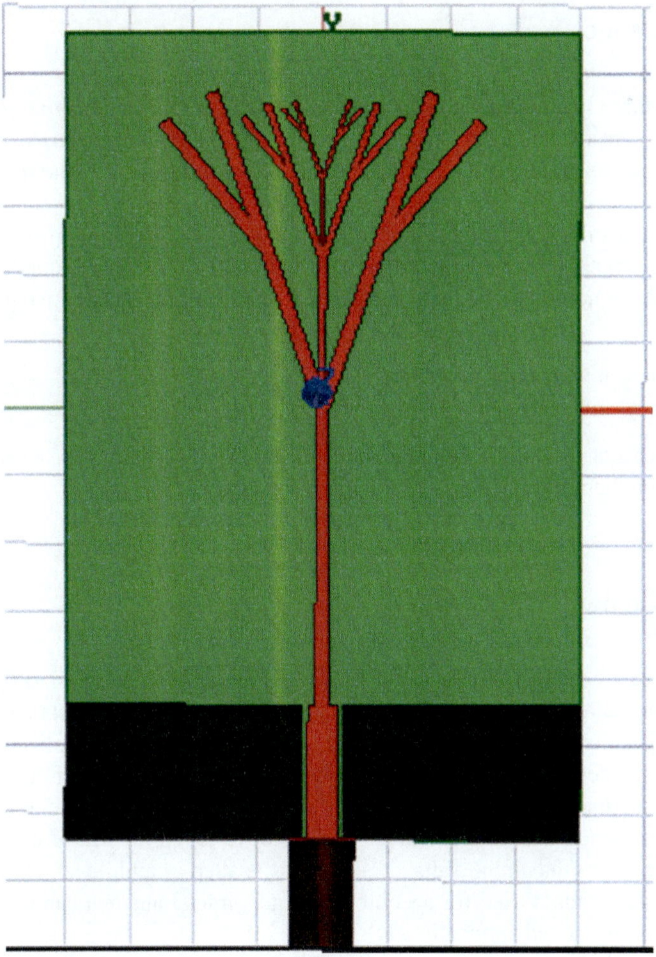

Fig. 20 Tree-shaped fractal vertically extended antenna

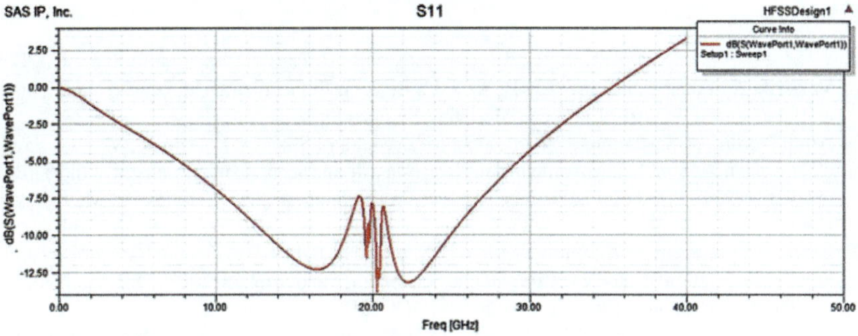

Fig. 21 S_{11} Plot for tree-shaped fractal vertically extended

7 Challenges

There are three main issues. Substrate material selection for antenna is first issue. In order to select suitable substrate, it is mandatory to know the electrical behavior of it. Second constraint is electrical characterization of different textile materials is required.

Second constraint is that the antenna performance under any kind of stress. Since antenna is integrated in textile, it will have different kind of bending, elongation and compression conditions. The main issue is to keep the all the antenna parameters at acceptable levels. Such parameters are

- Frequency at which it is operated
- Efficiency (human tissue to be prevented to degrade efficiency).

Third issue deals with the actual manufacturing. Antenna has to be robust enough against manufacturing tolerances is the main concern.

8 Conclusion

The purpose of this activity is to understand, simulate and synthesize the wearable (textile) antenna and its substrate materials for antenna elements and arrays. More specifically, some of the antennas are surveyed for different applications. For beginners, the design procedure of micro-strip antenna along with limitations while implementing the textile antenna is also briefly given. Finally, the fractal antenna's different version is also implemented. The effective length of fractal antenna can be increased to its maximum value. They are also called as multilevel. Unlike standard antenna which is used for specific frequency, fractal antenna can be used for wideband or multiband applications.

References

1. Vukovik, A., Zongyuah, S., Sewell, P., Aziz, L., Benson, T.: Geometric challenges of modeling textile antennas. In: 2017 European Modeling Symposium
2. Amin Nordin, M.S., Rahman, N.H.A., Ali, M.T., Sharatol Shah, A.A., Ahmad, V.: Full-wave electromagnetic simulation of antenna on electro-textile and accurate measurement of dielectric properties through precise adjustment jig. In: Proceedings of 2017 Asia Pacific Microwave Conference
3. Elias, N.A., Samsuri, N.A., Rahim, M.K.A., Othman, N., Jail, M.E.: Effect of human body and antenna orientation on dipole textile antenna performance and SAR. In: 2012 IEEE Asia-Pacific Conference on Applied Electromagnetics (APACE2012)
4. Abdullah, M.A.B., Rahim, M.K.A., Jalil, M.E.B., Samsuri, N.A., Murad, N.A.: Integrated two textile dipole antenna with dual-band textile artificial magnetic conductor. In: 2013 7th European Conference on Antenna and Propagation

5. Virili, M., Rogier, H., Alimenti, F., Mezzanotte, P., Roselli, L.: Wearable textile antenna magnetically coupled to flexible active electronic circuits. IEEE Antennas Wireless Propag. Lett. **13** (2014)
6. Elias, N.A., Samsuri, N.A., Rahim, M.K.A., Othman, N.: The effects of human body and bending on dipole textile antenna performance and SAR. In Proceedings of APMC 2012. Kaohsiung, Taiwan, 4–7 December 2012
7. Rajo-Iglesias, E., Gallego-Gallego, I., Inclan-Sanchez, L., Quevedo-Teruel, O.: Textile soft surface for back radiation reduction in bent wearable antennas. IEEE Trans. Antennas Propag. **62**(7), July 2014
8. Ramly, N., Rahim, M., Samsuri, N., Jalil, M., Majid, M., Fatin, N., Dewan, R.: Leaf-shaped dual band antenna textile performance for on-body application. In: 2014 IEEE APACE, 8–10 December 2014
9. Paul, D., Giddens, H., Paterson, M., Hilton, G., McGeehan, J.: Impact of body and clothing on a wearable textile dual band antenna at digital television and wireless communication bands. IEEE Trans. Antennas Propag. **61**(4), April 2013
10. Nawaz, H., Ali Abbasi, M.: On-body textile antenna design and development for body-centric wireless communication systems. In: IBCAST 2015, 13th–17th January 2015
11. Samal, P., Soh, P., Vandenbosch, G.A.E.: UWB all-textile antenna with full ground plane for off-body WBAN communications. IEEE Trans. Antenna Propag. **62**(1), January 2014
12. Byes, S., Soh, P., Huang, Y., Vandenbosch, G.A.E., Khiabani, N.: Measurement and performance of textile antenna efficiency on a human body in a reverberation chamber. IEEE Trans. Antenna Propag. **62**(2), February 2013
13. Choi, S., Lim, S.: 4*4n Textile MIMO Antenna Array for Mobile Router Applications
14. Potey, P.M., Tuckley, K.: Design of wearable textile antenna with various substrate and investigation on fabrics selection. In: 3rd International Conference on Microwave and Potonics (ICMAP 2018), 9–11 February 2018
15. Manikonda, R., Rajyalaxmi, V., Mallikharjuna Rao, V: Design and analysis of crescent shape textile antenna for WLAN application. In: 2017 IEEE 7th International Advance Computing Conference
16. Hassan, W.M., Attiya, A.M.: Textile antenna integrated with a cap for L1 GPS application. In: 2018, 35th National Radio Science Conference (NRSC 2018)
17. Gill, I., Fernandez-Gracia, R.: Wearable embroidered GPS textile antenna. In: 2017 Progress in Electromagnetic Research Symposium—Spring (PIERS)
18. Sankaralingam, S., Gupta, B.: A textile antenna for WLAN applications. In: 2009 International Conference on Emerging Trends in Electronic and Photonic Devices & Synthesis

Design and Analysis of Approximate Multiplier for Image Processing Application

S. Sathiyapriya and C. S. Manikandababu

Abstract Approximate computing is an emerging technology in digital circuits that improves the accuracy performance in error resilient application. Approximate compressors are the main key element, which is used for the design of approximate multiplier with the efficiency of power and area highly used for the image processing applications. The 4:2 approximate compressors are proposed to reduce the error which produces the output of exact 4:2 compressor. By using the improved compressor, two 4 × 4 multipliers are designed by three different stages of compressor with different accuracies which the errors are reduced by encoding its inputs using generate and propagate signals. The designed 4 × 4 multipliers are used as a building block for scaling up to 16 × 16 and 32 × 32 multipliers. The most accurate multiplier is identified by comparing the results of area, power, delay and power delay product (PDP) and analyzed by mean relative error distance (MRED). In image processing applications, the output image with higher quality and lower power consumption is achieved by the proposed effective multiplier.

Keywords Approximate computing · Multiplier · Error distance · Image processing applications

1 Introduction

Multiplier is one of the main blocks for different arithmetic blocks, which is widely used in different applications especially image processing and signal processing application [1, 2]. Power consumption and speed are the main critical parameters in the design of digital circuits. It is important to optimize these parameters of the

S. Sathiyapriya (✉)
VLSI Design, Sri Ramakrishna Engineering College, Coimbatore, India
e-mail: sathiyapriya238@gmail.com

C. S. Manikandababu
Department of Electronics and Communication Engineering, Sri Ramakrishna Engineering College, Coimbatore, India
e-mail: manikandababu.shelvaraju@srec.ac.in

P. Suresh et al. (eds.), *Advances in Smart System Technologies*,
Advances in Intelligent Systems and Computing 1163,
https://doi.org/10.1007/978-981-15-5029-4_16

multiplier. Normally, the optimization of one parameter is based on a constraint of the other parameter of the multiplier. Especially, the achievement of the desire speed performance with the limited low power constraint of system is a challenging one. Further, attaining a specified level of reliability is another barrier to achieve the system target performance [3]. The reduction of power and increase in speed are achieved by the different techniques at different levels of concepts have been recommended [4].

Approximate computing is an emerging trend in digital design which meets the power and speed specifications and relaxes the requisite of exact computation to gain the substantial performance improvement in terms of power, speed and area [5]. Approximate computing approaches are mainly focused on achieving the target specifications at the cost of reducing the computation accuracy. The approach is mainly used for error tolerant or resilient applications where the errors do not influence the desirable output for some applications [6]. These applications include multimedia processing, machine learning, signal processing and other error resilient computations. Approximate arithmetic units are mainly based on the removal of complexities from the arithmetic units circuits. Many of the previous works which are targeting on the approximate multiplier for the computational purposes are distinctly possible to have less accuracy when compared with the exact computational methods [7, 8].

These less accuracy results of the approximate multipliers lead to the error in applications. These inaccuracies are likely reduced by using the addition error recovery module in the design of multiplier. The error recovery module is to be added for the higher applications and larger circuits which are used in industry-oriented process [9]. The error in the larger industrial circuits may affect the whole process flow of the design. The approximate multiplier is widely used in signal processing, image processing and communication systems.

2 Related Works

Jiang et al. [10] paper deals with the comparative evaluation of approximate multipliers. The evaluation reveals that the precise results are not significant for the application of data processing. There is no obvious difference in the application of image processing by the presence of the precise approximate result due to the instinctual limitations of human beings. The current design of approximate multipliers and their comparative evaluation of their performances are reviewed. The approximate multipliers are used in the application of image processing such as image sharpening to evaluate their performances. The error distance (ED) and mean relative error distance (MRED) are also calculated for that approximate multiplier. Reviews are done only for the small-scale multiplier.

Moons et al. [11] paper deals with energy efficiency and accuracy of stochastic computing circuits in emerging technologies. The scaling size of the multiplier in integrated circuits technology may lead to the more inconstant and undetermined in the behavior of the circuits. The proposed technique is stochastic computing (SC) which is used to process the bit streams of the multiplier in the manner of long pseudo random to obtain the output with highly determinant and constant. This technique is also used to perform the analyses of multi-stage stochastic computing circuits. There is some error tolerance in the circuit which may lead to the high energy costs.

Chia Hao Lin et al. [12] paper deals with the high accuracy approximate multiplier with error correction. A system's performance is generally determined by the speed of the multiplier because the multiplier is typically the sluggish element in the real-time system. Multipliers need more hardware resources than the adder and subtractors. The accuracy of the multiplier is increased by using the error recovery module along with the multiplication circuit. Moreover, the proposed multiplier has high pass rate and high accuracy when compared with the existing multiplier.

Michael Orshansky et al. [13] paper deals with approximate computing: an emerging paradigm for energy-efficient design. The error distance is calculated to analyze the accuracy of the design which is relevant to the application. This proposed technique is widely used in image processing applications. The two test images are processed by using the proposed method and the parameters are observed. But there is a quality loss of image and some amount of error is tolerable to achieve the efficiency of power.

3 Exact Compressor

The exact 4:2 compressor is designed by using the two full adders. The exact 4:2 compressor is shown in Fig. 1.

The Boolean expression for the 4:2 compressors is given by

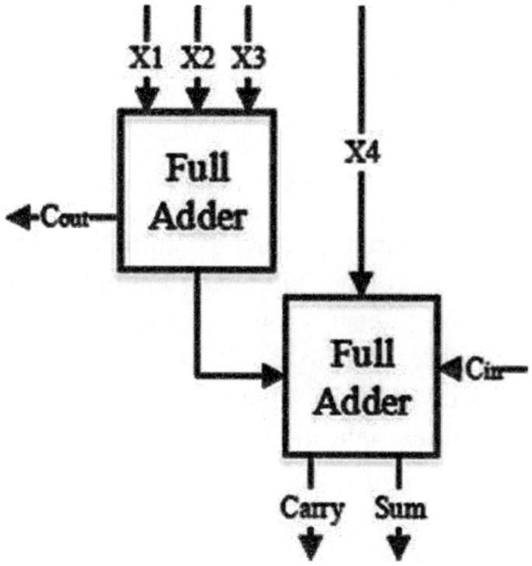

Fig. 1 Exact 4:2 compressor

$$\text{Sum} = x1 \oplus x2 \oplus x3 \oplus x4 \oplus C_{\text{in}} \tag{1}$$

$$\text{Cout} = (x1 \oplus x2).x3 + \overline{(x1 \oplus x2)}.x1 \tag{2}$$

$$\text{Carry} = (x1 \oplus x2 \oplus x3 \oplus x4).C_{\text{in}} + \overline{(x1 \oplus x2 \oplus x3 \oplus x4)}.4 \tag{3}$$

The exact 4:2 compressor is implemented to decrease the partial product of the multiplication for 4 × 4 multiplier [14]. These are introduced in normal multiplier for the reduction of original partial product accumulation of the multiplier. The following table gives the partial product of the four-bit multiplier and its respective product bits. The half adders used in the multiplication are a simpler design. Hence, the approximation of multiplier is not needed.

4 Proposed Multiplier Design

4.1 Approximate 4:2 Compressors

The exact compressor is mainly implemented in multiplier to reduce its hardware cost. In the above-mentioned exact 4:2 compressor, C_{out} does not have a high impact on the output and its performance. Hence, the C_{out} is neglected from the compressor. By neglecting the C_{out}, there are some incorrect values in the approximate outputs that create the error in the output. To manage the source of inaccuracy, the inputs to the approximate compressors are encoded into propagate and generate signals. The propagate and generate signals to compressors are given by

$$P_{i,j} = pp_{i,j} + pp_{j,i} \tag{4}$$

$$G_{i,j} = pp_{i,j} + pp_{j,i} \tag{5}$$

The illustration of general partial product of multiplication is illustrated below (Fig. 2). The one of the operands of the multiplier is multiplied with the other operand called multiplicand which forms the first row of multiplication. The other rows of the multiplication are arranged according to the bit position of the multiplier commonly called as partial product array. The partial product in the same rows is summed up altogether to obtain the final multiplication result (Fig. 2).

Stage 7	Stage 6	Stage 5	Stage 4	Stage 3	Stage 2	Stage 1	Stage 0
	$pp_{3,3}$	$pp_{3,2}$	$pp_{3,1}$	$pp_{3,0}$	$pp_{2,0}$	$pp_{1,0}$	$pp_{0,0}$
		$pp_{2,3}$	$pp_{2,2}$	$pp_{2,1}$	$pp_{1,1}$	$pp_{0,1}$	
			$pp_{1,3}$	$pp_{1,2}$	$pp_{0,2}$		
				$pp_{0,3}$			
Y_7	Y_6	Y_5	Y_4	Y_3	Y_2	Y_1	Y_0

Fig. 2 Partial product of multiplication

4.2 *Two Approximate 4 × 4 Multipliers*

M1 and M2 are the representation of two approximate multipliers which are designed by using the actual partial product of the multiplication. The main difference between these two multipliers is that the multiplier M1 considers the previous stage carry C_4 generated by the partial product terms pp which uses the exact full adder.

Figure 3 represents the block diagram of the multiplier M1. In multiplier M2, the longest path C_4 gets neglected. By neglecting this path, there is no existence of higher amount of error to the output. The latency of the circuit is reduced by neglecting the longest path, and it is the common technique.

Figure 4 represents the block diagram of the multiplier M2. There are three stages of compressors are used to reduce the partial product generation of the multiplier. To obtain the output Y_3, there are five signals. Hence, there is a need of 5:2 compressors. These five signals are merged to form four signals because the proposed design is the 4:2 approximate compressor designs. There are three stages of approximate multiplier are employed to obtain the output Y_2, Y_3 and Y_4. C_4, the longest path, is neglected. The outputs Y_1, Y_5 and Y_7 are obtained by using exact full adder and half adder. The output Y_7 is obtained by the carry C_6 from the previous stage of the multiplier.

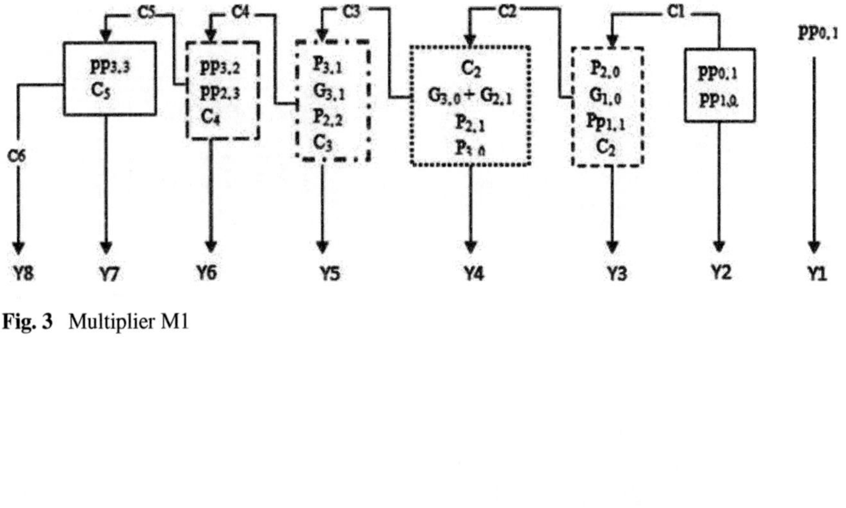

Fig. 3 Multiplier M1

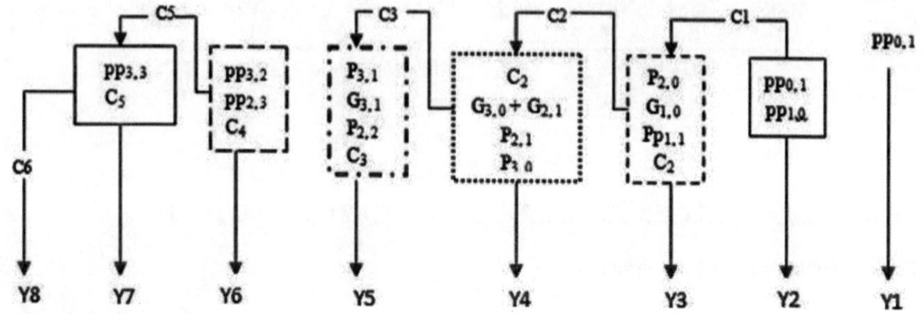

Fig. 4 Multiplier M2

4.3 *Scaling to Large Multiplier*

The larger multipliers are constructed by using the designed 4 × 4 multiplier M1 and M2 by combining the multipliers in an array structure (Fig. 5) [15]. For example, to construct 8 × 8 multiplier by using 4 × 4 multipliers, the two eight-bit operands *A* and *B* are partitioned into two four-bit nibbles. α_H and α_L are the two nibbles for *A*. For *B* partial product, β_H and β_L are the two nibbles. From these four nibbles, two nibbles are multiplied using 4 × 4 multipliers, and the partial products are shifted based on the importance of the nibble and added together to obtain the final multiplication result by using the Wallace Tree's structure (Fig. 5).

Table 1 represents the building of 2*n* × 2*n* multiplier by using an *n* × *n* multiplier. The combination of multiplier to build the multiplier to the larger scale is given by.

Fig. 5 Building 2n × 2n multipliers using n × n multiplier

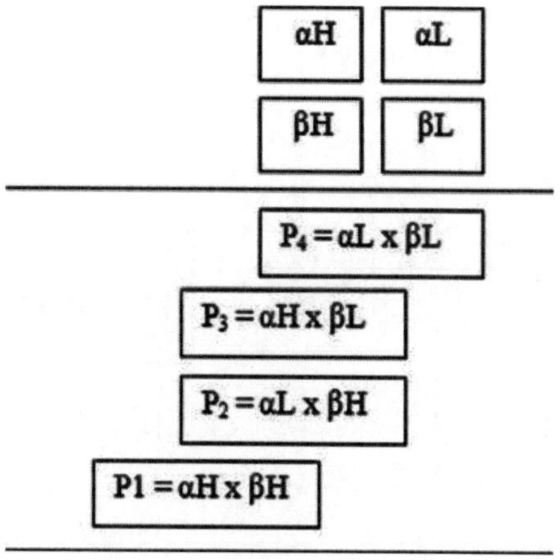

Table 1 Construction of 8 × 8, 16 × 16 and 32 × 32 multiplier

Size	Design	P1	P2	P3	P4
8 × 8	M8_1	M1	M1	M1	M1
	M8_2	M2	M2	M2	M2
	M8_3	Exact	M1	M1	M1
	M8_4	Exact	M2	M2	M2
	M8_5	Exact	Exact	Exact	M1
	M8_6	Exact	Exact	Exact	M2
16 × 16	M16_1	M8_1	M8_1	M8_1	M8_1
	M16_2	M8_2	M8_2	M8_2	M8_2
	M16_3	M8_3	M8_3	M8_3	M8_3
	M16_4	M8_4	M8_4	M8_4	M8_4
	M16_5	M8_5	M8_5	M8_5	M8_5
	M16_6	M8_6	M8_6	M8_6	M8_6
32 × 32	M32_5	M16_5	M16_5	M16_5	M16_5
	M32_6	M16_6	M16_6	M16_6	M16_6

5 Accuracy Analyses

The output accuracy is the important metric for the approximate multiplier design which is analyzed by using the exact output of the multiplier. Mean relative error distance (MRED) is the metric which is used to analyze the accuracy of the output. The error distance (ED) is analyzed by using the formula,

$$\mathrm{ED} = \left|M' - M\right| \tag{6}$$

M' represents the approximate output, and M represents the exact output. The relative error distance (RED) is the important metric to calculate mean relative error distance (MRED).

$$\mathrm{RED} = \mathrm{ED}/M \tag{7}$$

MRED is calculated by the average of relative error distance of the output for all the input cases.

6 Results and Discussion

The proposed multiplier designs are implemented and analyzed using Xilinx ISE design suite 13.2, and the simulation is taken by using Isim simulator.

6.1 Power

Power is estimated on the basis of number and types of compressors. From the comparison table, multiplier 8_2, multiplier 16_2 and multiplier 32_6 have the low power delay product (PDP).

6.2 Transistor Count

Transistor count is defined as the number of cells, LUT, slices which are taken for the design of multiplier. Based on the number of transistor count, the area of the multiplier is determined.

6.3 Delay

The delay of the multiplier is determined by the number of reductions in partial product reduction in the multiplier. The reduction of the delay in multiplier will increase the efficiency and speed of the proposed multiplier (Table 2).

Table 2 Comparison result of proposed multipliers

MUL size	MUL type	LUT	Delay (ns)	Power (mW)	PDP (nJ)
8 × 8	M 8_1	66	16.4	97	1.599
	M8_2	**52**	**13.6**	**93**	**1.265**
	M8_3	82	21.09	120	2.531
	M8_4	76	21.07	125	2.634
	M8_5	128	21.2	149	3.159
	M8_6	124	21.2	142	3.011
16 × 16	M16_1	305	23.952	**142**	3.401
	M16_2	**233**	**20.877**	158	**3.298**
	M16_3	363	28.463	209	5.948
	M16_4	329	28.367	204	5.786
	M16_5	539	28.521	254	7.244
	M16_6	529	28.521	255	7.272
32 × 32	M32_5	2245	36.56	479	17.53
	M32_6	**2181**	**36.51**	**477**	**17.00**

The bold characters denotes the best multiplier in means of LUT, delay, Power and PDP in their respective size of multiplier

From Fig. 6, Mul8_2 has the low power consumption of 93 mW, requirement of lower area of 52 LUT (lookup table) and high speed performance with less delay of 13.60 ns.

From the above graph (Fig. 7), Mul16_2 has the low power consumption of 158 mW, requirement of lower area of 233 LUT (lookup table) and high speed performance with less delay of 28.463 ns.

From Fig. 8, Mul32_6 has the low power consumption 477 mW, requirement of lower area of 2181 LUT (lookup table) and high speed performance with less delay of 30 ns.

Table 3 represents the accuracy analysis of the different designs of multiplier by using the proposed multiplier. Decrease in the mean relative error distance increases in the accuracy of the multiplier.

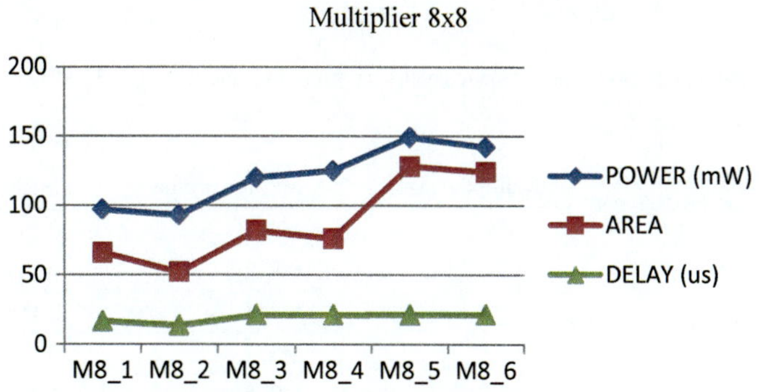

Fig. 6 Comparison graph for performance metric (power, area and delay) of 8 × 8 multiplier

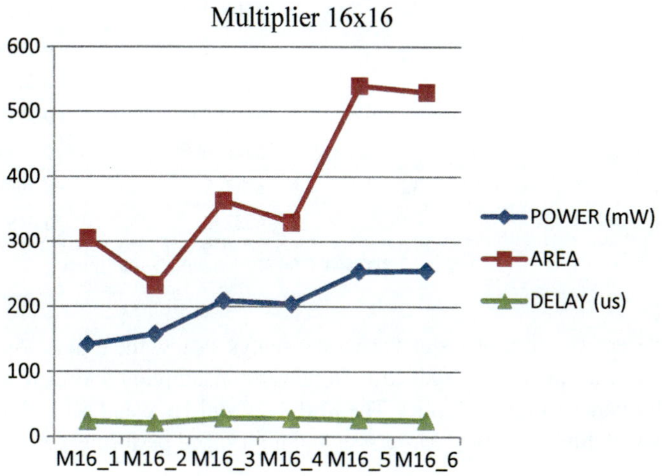

Fig. 7 Comparison graph for performance metric (power, area and delay) of 16 × 16 multiplier

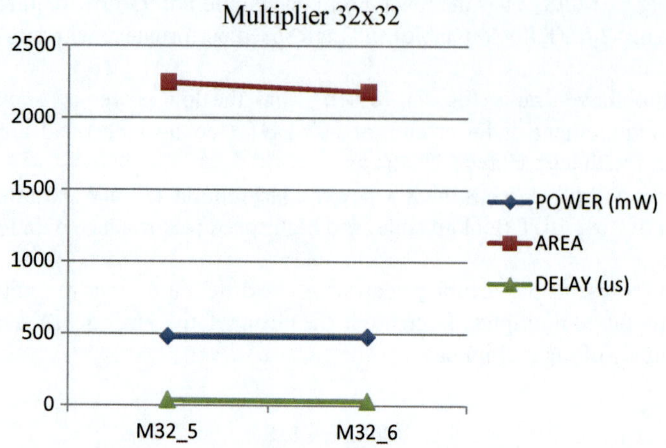

Fig. 8 Comparison graph for performance metric (power, area and delay) of 32 × 32 multiplier

Table 3 Accuracy comparison for the designed multiplier

Multiplier size	Multiplier type	MRED
8 × 8	M8_1	0.7420
	M8_2	0.2971
	M8_3	0.1558
	M8_4	0.186
	M8_5	1.9119
	M8_6	**0.1595**
16 × 16	M16_1	0.3872
	M16_2	0.3897
	M16_3	0.2364
	M16_4	0.2239
	M16_5	0.1717
	M16_6	**0.0733**
32 × 32	M32_5	0.2971
	M32_6	**0.018**

The bold represents the most accuracy multiplier

The different metric analysis and accuracy analysis show the efficient multiplier with high accuracy, less area, reduction in power consumption and increase in the speed performance of the multiplier. The Mul 8_2, Mul 16_2 and Mul 32_6 are the multipliers with high accuracy, less power and high speed performance.

7 Applications

The efficacy of the proposed multiplier designs is estimated on the basis of application where the proposed designs are implemented. The main application of the proposed design is image processing application such as image sharpening, image smoothening and image multiplication [16]. Here, the designs are used for image sharpening. In image processing applications, the sharpness of the image is improved by using image sharpening algorithms. Here, the algorithm is implemented by using the special 5 × 5 spatial filter that processes the 5 × 5 blocks of pixels on image [17–19].

$$G = \begin{bmatrix} 16 & 64 & 112 & 64 & 16 \\ 64 & 256 & 416 & 256 & 64 \\ 112 & 416 & 656 & 416 & 112 \\ 64 & 256 & 416 & 256 & 64 \\ 16 & 64 & 112 & 64 & 16 \end{bmatrix} \tag{5}$$

The above G matrix represents the impulse 5 × 5 spatial filter. The filter is implemented in image processing, and the quality is measured on the basis of peak signal-to-noise ratio (PSNR) and mean squared error (MSE).

The image processing is done by using MATLAB tool. The multiplier 16_2, multiplier 16_5 and multiplier 16_6 have the better quality of image with high resolution by means of peak signal-to-noise ratio (PSNR) and mean squared error (MSE).

$$\mathrm{MSE} = \frac{1}{mn} X \sum_{i=0}^{m-1} \sum_{j=0}^{n-1} \left[\widehat{S}(i,j) - S(i,j) \right]^2 \tag{6}$$

$$\mathrm{PSNR} = 10X \log_{10}\left(\frac{\mathrm{MAX}^2}{\mathrm{MSE}} \right) \tag{7}$$

Equations (6) and (7) are used to evaluate PSNR and MSE [20].

The image $x(i)$ is taken as the reference image to obtain the sharpened image (Fig. 9). The image $x(j)$ is obtained by using the 5 × 5 spatial filter from the image $x(i)$. The efficacy of the sharpened image is analyzed by using the calculation of PSNR of the sharpened image (Fig. 9).

The efficacy of the sharpened image is analyzed by using the calculation of PSNR of the sharpened image (Fig. 10). The PSNR value of the sharpened image is 30.8614.

8 Conclusion

The efficient approximate multipliers are designed using approximate compressors with less area, low power consumption and high speed performance. The accuracy

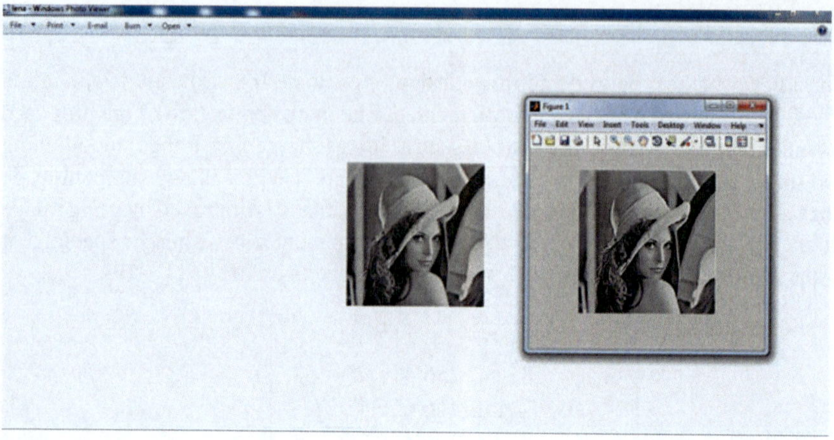

Fig. 9 Sharpening of image

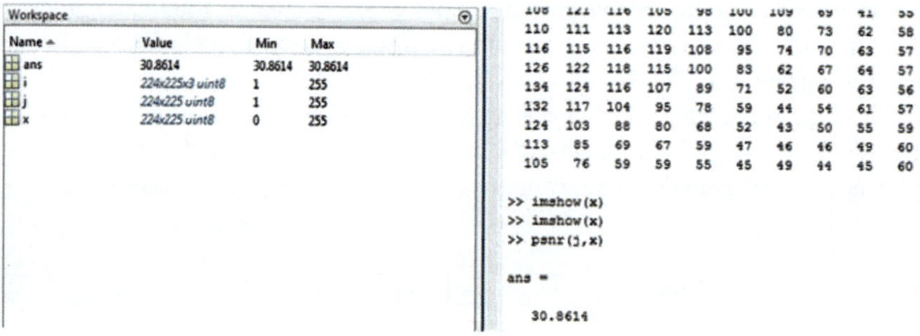

Fig. 10 PSNR of the sharpened image

of these multipliers is analyzed by calculating the error distance, mean relative error distance. By using the analysis of accuracy, the multipliers with good performance metrics are identified. The proposed multipliers are designed by using the Xilinx ISE design suite. The multipliers are scaled up to larger multipliers, and the accuracies are evaluated by means of mean relative error distance. The multiplier with high performance rate is used for the image processing application such as image sharpening using MATLAB.

References

1. Schlachter, J., Camus, V., Palem, K.V., Enz, C.: Design and applications of approximate circuits by gate-level pruning. IEEE Trans. Very Large Scale Integr. Syst. **25**(5), 1694–1702 (2017)
2. Narayanamoorthy, S., Moghaddam, H.A., Liu, Z., Park, T., Kim, N.S.: Energy- efficient approximate multiplication for digital signal processing and classification applications. IEEE Trans.

Very Large Scale Integr. Syst. **23**(6), 1180–1184 (2014)
3. Zhong, H., Xu, W., Xie, N., Zhang, T.: Area-efficient min-sum decoder design for high-rate quasi-cyclic low-density parity-codes in magnetic recording. IEEE Trans. Magn. **43**(12), 4117–4122 (2007)
4. Venkataramani, S., Ranjan, A., Roy, K., Raghunathan, A.: AxNN: energy-efficient neuromorphic systems using approximate computing. In: International Symposium on Low Power Electronics and Design, pp. 27–32 (2014)
5. Roy, K., Raghunathan, A.: Approximate computing: an energy-efficient computing technique for error resilient applications. In: IEEE Computer Society Annual Symposium on VLSI, pp. 473–475 (2015)
6. Sanchez-Clemente, J., Entrena, L., Hrbacek, R., Sekanina, L.: Error mitigation using approximate logic circuits: a comparison of probabilistic and evolutionary approaches. IEEE Trans. Reliab. **65**(4), 1871–1883 (2016)
7. Lau, M.S.K., Ling, K.V., Chu, Y.C.: Energy-aware probabilistic multiplier: design and analysis. In: International Conference on Compilers, Architecture, and Synthesis for Embedded Systems, pp. 281–290 (2009)
8. Wang, Z., Bovik, A.C., Sheikh, H.R., Simoncelli, E.P.: Image quality assessment: from error visibility to structural similarity. IEEE Trans. Image Process. **13**(4), 600–612 (2004)
9. Liu, C., Han, J., Lombardi, F.: A low-power, high-performance approximate multiplier with configurable partial error recovery. Des., Autom. Test Eur. **2014**(1), 1–4 (2014)
10. Jiang, H., Liu, C., Liu, L., Lombardi F., Han, J.: A review, classification and comparative evaluation of approximate arithmetic circuits. ACM J. Emer. Technol. Comput. Syst. **13**(4) (2017). Article No. 60
11. Moons, B., Verhelst, M.: Energy-efficiency and accuracy of stochastic computing circuits in emerging technologies. IEEE J. Emerg. Sel. Top. Circ. Syst. **4**(4), 475–486 (2014)
12. Lin, C.H., Lin, I.C.: High accuracy approximate multiplier with error correction. In: IEEE International Conference on Computer Design, pp. 33–38 (2013)
13. Han, J., Orshansky, M.: Approximate computing: an emerging paradigm for energy-efficient design. In: IEEE European Test Symposium, pp. 1–6 (2013)
14. Adde, P., Le Bidan, R.: A low-complexity soft-decision decoding architecture for the binary extended Golay code. In: IEEE International Conference on Electronics, Circuits and Systems (ICECS), pp. 705–708 (2012)
15. Liu, C.: Design and analysis of approximate adders and multipliers. Master's Thesis, University of Alberta, Canada (2014)
16. Shah, M.: Future of JPEG XT: privacy and security. Ph.D. Thesis, University of Texas Arlington, Texas, USA (2016)
17. Yang, J., Zhu, G., Shi, Y.Q.: Analyzing the effect of JPEG compression on local variance of image intensity. IEEE Trans. Image Process. **25**(6), 2647–2656 (2016)
18. Mahdiani, H.R., Ahmadi, A., Fakhraie, S.M., Lucas, C.: Bio-inspired imprecise computational blocks for efficient VLSI implementation of soft-computing applications. IEEE Trans. Circ. Syst. I Regul. Pap. **57**(4), 850–862 (2010)
19. Pham, C.C., Jeon, J.W.: Efficient image sharpening and denoising using adaptive guided image filtering. IET Image Proc. **9**(1), 71–79 (2015)
20. Babar, Z., Ng, S.X., Hanzo, L.: EXIT-chart-aided near-capacity quantum turbo code design. IEEE Trans. Veh. Technol. **64**(3), 866–875 (2015)

Miniaturized and Highly Randomized True Random Bit Generator

N. Kamlesh Raj and C. S. Manikandababu

Abstract This paper introduces true random bit generator (TRBG) with a higher quality of randomness based on chaos circuit and unpredictable linear feedback shift register (LFSR). This miniaturized true random bit generator made of two stages of randomness. The first stage produces the continuous random variable, and the second stage provides the Metastability to the circuit. The chaos circuit gives immunity from the side channel attack, and the circuit has Metastability for every events of bit generation and also ensures a 1 Mbps bit rate data output. This generator can be used as the seed generator for any encryption design (like image encryption), the true random generator is designed for the cheaper devices like complex programmable logic devices (CPLDs) and field programmable gate arrays (FPGAs) for any lightweight encryption design. The National Institute of Standard and Technology (NIST) tests of randomness are applied for testing the randomness of TRBG.

Keywords LFSR · Chaos circuit · Phase detector · Sequential chaos circuit based ring oscillator (SCRO's) · NIST

1 Introduction

The random bit generators become a very crucial component in the digital world, the digital world contains many sensitive and personal information which are needed to be protected from attackers/hackers, in this condition, we need a cryptography system which has to be highly efficient in encrypting and decrypting data using cryptographic algorithm, and this type of system uses a key component called random bit generator. There are two types of random bit generators, which are pseudo random

N. Kamlesh Raj (✉)
VLSI Design, Sri Ramakrishna Engineering College, Coimbatore, India
e-mail: nkamleshraj2617@gmail.com

C. S. Manikandababu
Department of Electronics and Communication Engineering, Sri Ramakrishna Engineering College, Coimbatore, India
e-mail: manikandababu.shelvaraju@srec.ac.in

P. Suresh et al. (eds.), *Advances in Smart System Technologies*,
Advances in Intelligent Systems and Computing 1163,
https://doi.org/10.1007/978-981-15-5029-4_17

bit generator (PRBG) and true random bit generator (TRBG). The pseudo random bit generator generates a sequence of bit which almost looks random bits, but they are generated with algorithms, which makes PRBG deterministic. However, the true random bit generators are truly random, and every bit generated by this generator is unpredictable. The TRBG is unpredictable because they use the randomness in the nature and capture it to produce the random bits. The random signals in nature are sensed using the sensors, and these signals are captured and processed to get digital output from the continuous random input signal using sampling and quantization techniques. However, true randomness can also be generated in the hardware of FPGA using the principle of Metastability and introducing chaos in design. This technique reduces the use of extra hardware like sensor and ADC, etc. So, this type of design can be used in lightweight applications. The random bit generators used in many applications such as they are used in cryptography systems, password/PIN generator, seed/key generator for encryption and decryption system, authentication system, asymmetric encryption, digital signature algorithm also like lottery generator and games.

Most of the software functions are processed in the processor of system and which in turn uses the hardware generated data, and this data are generated by the cryptographic system which has its own random bit generator. The cryptographic coprocessors are used in the advanced processor to aid the ease of encryption. Most of the software-based system with limited hardware interventions needs the encryption. In such cases, the cryptographic coprocessors are used. The Internet of things (IoTs) are one of such lightweight system which uses the cryptographic coprocessors, which provides the protection from side channel attack (SCA) [1] and also provides high randomness to the system. Most of the devices are found in a form of SOC nowadays, which helps to work in discrete form as well as it helps in performing most sophisticated operations very easily, so the random bit generator also found in the SOCs, which is very helpful in the areas of the lightweight system. The increase in the popularity of the use SOC devices embedded with the sophisticated microcontroller and the programmable circuits like Zynq, Vertex and Cyclone devices. We use specialized and complex circuits for the extraction of randomness, but there is another way to extract the randomness by using the reconfigurable devices [2], [3], well these devices also use some complex circuits like digital clock manager (DCM), programmable delay line (PDL) or phase lock loop (PLL). [4, 5]. The physical processes used for harvesting the random numbers usually done by amplifying the noise in circuits [6], Metastability of the circuit [7] and also the resolve time and uncertainty of the either logic states. The non-deterministic chaos circuits are used in this to produce the true randomness.

The paper is organized in following order: In Sect. 2, theoretical background is explained for all the subcircuit used for the true random number generator, in Sect. 3, proposed design and the operation are explained, and in Sect. 4, the results are given.

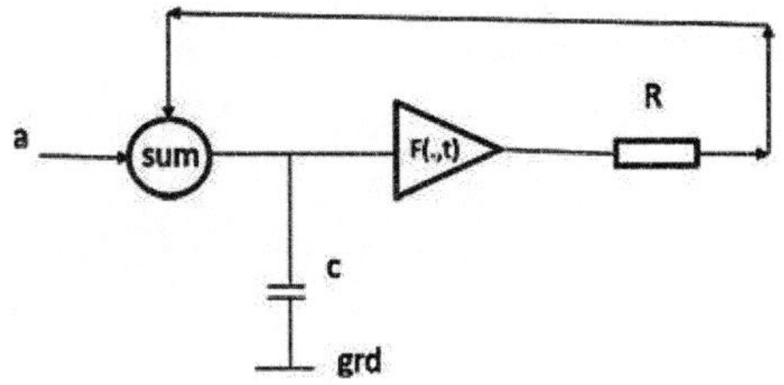

Fig. 1 Typical electrical model of ring oscillator

2 Theoretical Background

In the theoretical background, the following five concepts will be explained, and it will be used in design of TRBG, which are as follows:

A. Ring oscillator,
B. Chaos circuit,
C. Phase detector,
D. Metastable circuit and
E. LFSR.

2.1 Ring Oscillators

Typically, ring oscillators are made of repeated sequence of not gate, the last not gate output is connected to the input of the ring oscillator, and it acts as feedback [8]. The lumped low pass filter is needed to be added with the ring oscillator for analysis in the case of electrical model. The low pass filter is made up of resister (R) and the capacitor (C) which are connected in parallel. The electrical model for the ring oscillator is shown in Fig. 1. There resistor and capacitor values determine the active region behavior of the inverter. The two switchable ring oscillators are used in the proposed design.

2.2 Chaos Circuit

The design of deterministic circuit is simple, and by using those deterministic circuits, we can easily produce chaos in the circuit [9]. The time series produced by the chaos generator (shown in Fig. 2) is hard to predict by the observer. There chaos generated distortion helps to make the circuit output unpredictable. The time domain behavior becomes complex, but it is not the result of chaotic circuit intervention. The chaotic circuit equations produce the very complex circuit behavior in terms of timing analysis, and also, entropy and probability density function of circuit is in acceptable range. The 'unpredictability' comes from sensitivity of the circuit's internal condition, which affects the circuits state in terms of time.

Fig. 2 Chaotic circuit

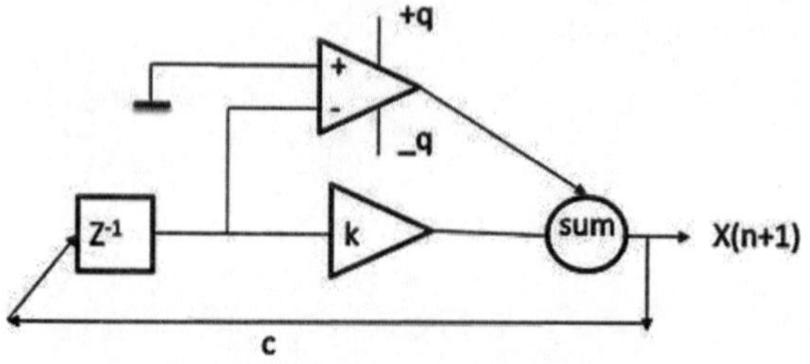

The chaotic map is used to define deterministic discrete time system. Where *fm*: *S* tends to *S* in the state-space.

$$\mathrm{x}[n+1] = fm(\mathrm{x}[n]) \tag{1}$$

2.3 *Phase Detector*

The phase detector is defined as the device used to detect the difference in the phase of two input voltages. In this project, the D flip flop (DFF) with XORed input is used for the phase detection which is a simple circuit compared to other circuits given below and is more stable and can be controlled based on time and its good for timing analysis. There are following type of phase detector is present and few are mentioned below:

1. XOR gate
2. Dual Port D flip flop
3. Master–slave D Flip flop
4. PLL.

2.4 *Metastable Circuit*

The metastable circuit is needed to stabilize the random generator circuit; the metastable circuit consists of the SR latch followed by the LFSR which is controlled by the clock generator at the speed of 10 us. The Metastability is involved in each and every events of the bit generation. So, the circuit becomes stable and less affected by the temperature changes as well as frequency [10]. The Metastability is mainly introduced in the second stage because to generate the output, we need the control in the output appearance with the same frequency of clock with the ring oscillator generated frequency of operation.

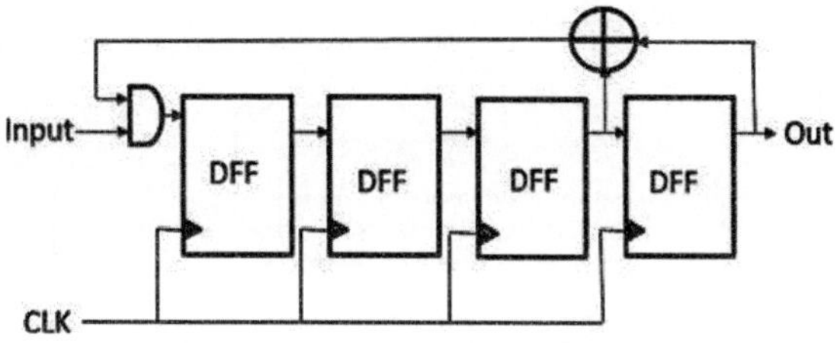

Fig. 3 LFSR

2.5 *LFSR*

The LFSR is a pseudo random bit generators, and the random generator used in this project is four-bit unpredictable linear feedback shift register [11] (shown in Fig. 3). It is placed at the place of CB4 (four-bit counter) [1] in the metastable circuit given by the Krzysztof golofit [1, 12]. The bits can be limited at the output based on the number of flip flop used by the LFSR. Here, in this design, it is limited to the four-bit. The LFSR with flip flop and few XOR gate is better than counters (CB4) [1] because LFSR has higher clock frequency as well as design requires lower number of logic gates compared to counter of same bit. There are many complex LFSR are available, but in this design, primitive LFSR is used to keep the design simple and effective.

The primitive polynomial equation of four-bit LFSR is given in Eq. 2:

$$f(x) = x^4 + x + 1 \tag{2}$$

where the highest power of the 'x' shows the number of flip flop used and maximum length of sequence generated by LFSR is $2^n - 1$.

3 Design of TRBG (Proposed)

The design is consisting of ring oscillator of five phase walk (not gates), chaos circuit made of NAND gate, the phase detector made of XOR gate followed by the D flip flop, SR latch and the LFSR. Figure 4 given below shows a proposed design of true random bit generator.

The TRNG is split into two half or into two stages of randomness, the first stage is composed of sequential chaos circuit based of ring oscillator generator and phase detector, and the second stage of randomness is generated by the latch and LFSR. The two switchable ring oscillators are used which produce two different phases of output, and phase detector receives it as input and produces a response which acts as a feedback for input of ring oscillator. This feedback passes through chaos circuit before it goes to input section of design. The whole first part is controlled by a single clock. The second stage is also controlled by same clock. The LFSR acts as both

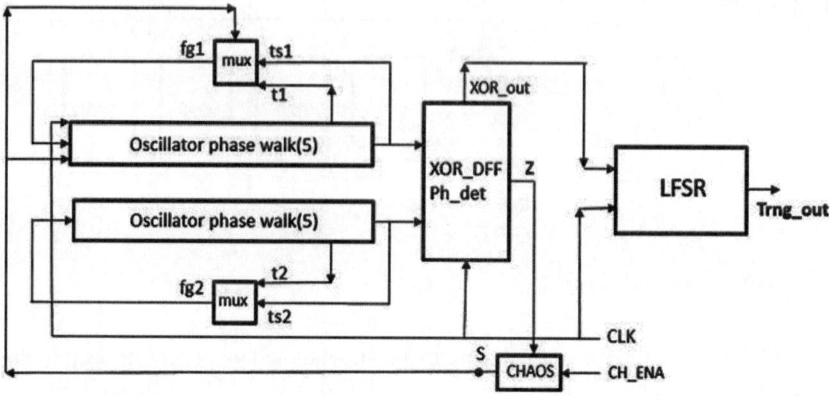

Fig. 4 Block diagram of proposed TRBG

serial in parallel out/serial in serial out and also a random bit generator. The LFSR used in design is a four-bit random bit generator.

The Metastability of circuit is the main concept of design, and the first stage produces a chaotic random output which is fed to the input of LFSR. The LFSR is deterministic as well as it generates pseudo random bits. However, the input of LFSR is random and chaotic which makes the truly LFSR unpredictable in sense. Where the first stage has the chaos circuit which introduces the delay in the circuit which creates a Metastability at the input of phase detector. So, the phase detector used is D flip flop, it compares the output of XOR with the clock, due to the introduction of delay by chaos circuit, the output of DFF becomes uncertain (shown in Fig. 5).

4 Results

The random bit generated by the proposed true random bit generator is given below:

010101011101101000111101001011011110000101011110100101111001001000100101000100101111110011010000011...

The waveform generated for the TRNG is given below shown in Fig. 6 for a single instance.

The design is implemented in the Spartan 6 FPGA (XC6SLX25 and XC6SLX45) and also in the CPLD (XC2C256 and XC2C64). The design is tested by NIST Design suit and passed all the test of randomness. For the clock pulse of 1 micro second, the random output data of 1 Mbps obtained without any requirement of additional post processing.

The comparison between proposed and existing true random bit generator is given below, in this, the macro cells, *p*-terms and clock to pad (output pad) delay values are given for each of the RBG. Three different phase walks are compared in above

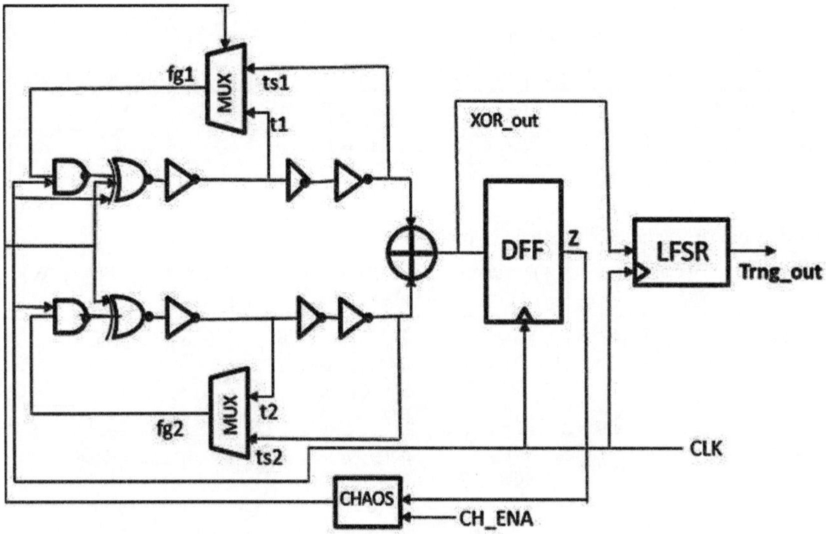

Fig. 5 Proposed design of TRBG

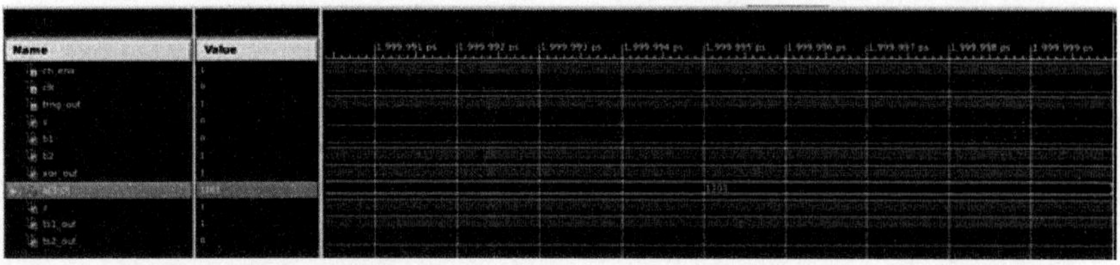

Fig. 6 Output waveform of *n*-sequence

Table 1, the area utilization is obtained which is not changed much, but there is time delay obtained between input to output pad which makes the proposed design faster, and also, the gate count is reduced for ring oscillator.

In Table 2, the two different phase detectors are compared for optimized selection of circuit for design. Here, they are compared in terms of area and speed. Where, Table 2 shows that XOR-DFF are faster as well as area efficient compared to Master–slave DFF. Table 3 gives the comparison of proposed and existing true random bit

Table 1 Comparison of phase walk of SCRO's

Phase walk	Area (slice LUT's, I/OB's)	Timing (s-ts1_out, clk-ts1_out) (ns)	Gates
7	2, 3	6.979, 6.145	7
5	2, 3	6.899, 6.264	5
3	2, 3	6.325, 6.785	3

Table 2 Comparison of phase detector

Phase detector	Slice LUT's	Slice register	Source timing	Destination timing (ns)
Master slave DFF	1	4	0.972	6.333
XOR-DFF	1	0	0.687	5.187

Table 3 Comparison table of existing and proposed

Reference	Category	Device	Scale/block used	Clock pad to o/p delay	Power (W)
Proposed	Meta.-chaotic	Xilinx XC6SLX25	7/(1%) slice FF and registers, 5/1(%) slice LUT's	7.500 ns	0.036
Proposed	Meta.-chaotic	Xilinx XC2C256	35/(14%) (macrocells), 54/(6%) (*P*-terms)	7.500 ns	0.036
Piotr Zbigniew [1]	Meta.-chaotic	Xilinx XC2C256	51/(19%) (macrocells), 90/(10%) (*P*-terms)	8.400 ns	0.09

generators. Table 4 shows the randomness tests done for the sample bits generated by proposed TRBG. The pval is obtained for all the tests of randomness, and this test result clearly shows that the proposed design passes all the tests of randomness.

5 Conclusion

The TRBG design based on the chaotic circuit and Metastability concept. The unpredictability of the true random number generator is achieved in deterministic LFSR, which is a Pseudo Random Bit Generator. This chaos circuit utilizes the time variance occurs in the circuit due to delayed feedback together with the oscillatory Metastability circuit. This circuit is very small and simple in structure. So, it can be easily implemented in any programmable devices. This design is developed for implementing in FPGAs because most of random generators are developed for CPLD and higher memory programmable devices. The design is very small so that it takes only few LUTs and Slices, and also, it produces the high quality of randomness. The randomness is verified using the NIST design Suit. Another advantages of this design are that it gives Metastability in each event of bit generation. This TRBG can be used in cheaper programmable devices for lightweight operation like IoTs, lottery generators and can be used in encryption for seed generation.

Table 4 NIST statistical test for checking the randomness

TEST	TRBG (sample o/p bit) pval
Freq._test_(Monobit)	0.59
Freq._test_within_a_block	0.10
Run_test	0.90
Longest_run_of_ones_in_a_block_test	0.31
Binary_matrix_rank_test	0.59
Spectral_test	0.51
Non_overlapping_template_matching test	0.89
Overlapping_template_matching_test	0.28
Maurer's_universal_statistical_test	0.80
Linear_complexity_test	0.74
Serial_test	0.88
Approximate_entropy_test	0.55
Cumulative_sums_test	0.05
Random_excursions_test	0.30
Random_excursions_variant_test	0.38

References

1. Wieczoreck P.Z., Golofit, K.: True random number generator based on flip flop resolve time instability boosted by random chaotic source. IEEE Trans. Circuits Syst. 1279–1291 (2018)
2. Li, D., Lu, Z., Zou, X., Liu, Z.: PUFKEY: a high-security and high throughput hardware true random number generator for sensor networks. Sensors 26251–26266 (2015)
3. Wieczorek, P.Z., Golofit, K.: Metastability occurance based physical unclonable functions for FPGAs. Electron. Lett. **50**(4), 281–283 (2014)
4. Majzoobi, M., Koushanfar, F., Devada, S.: FPGA-based true random number generation using circuit Metastability with adaptive feedback control. In: 13th International Workshop on Cryptographic Hardware and Embedded Systems (CHES), pp. 17–32. Springer, Heidelberg (2011)
5. Wieczorek, P.Z.: An FPGA implementation of the resolve time based true random number generator with high quality control. IEEE Trans. Circuits Syst. 3450–3459 (2014)
6. Jun B., Kocher, P.: The Intel random number generator. White Paper, Intel corporation. Cryptography, Research, Inc., Santa Clara, CA, USA (1999)
7. Robson, S., Leung, B., Gong, G.: Truly random number generator based on a ring oscillator utilizing last passaging time. IEEE Trans. Circuits Syst. 937–941 (2014)
8. Wieczorek P.Z., Golofit, K.: Dual-Metastability time-competitive true random number generator. IEEE Trans. Circuits Syst. 134–145 (2014)
9. Addabbo, T., Alioto, M., Fort, A., Rocchi, S., Vignoli, V.: A feedback strategy to improve the entropy of a chaos-based random bit generator. IEEE Trans. Circuit Syst. 326–337 (2006)
10. Hata, H., Ichikawa, S.: FPGA implementation of Metastability-based true random number generator. IEICE Trans. Inf. Syst. 426–436 (2012)

11. Han, M., Kim, Y.: Unpredictable 16 bits LFSR-based true random number generator. In: IEEE International SoC Design Conference (ISOCC), pp. 284–285, South Korea (2017)
12. Drutarovsky M., Varchola, M.: New high entropy element for FPGA based true random number generators. In: 12th International Workshop on Cryptographic Hardware and Embedded Systems (CHES), pp. 351–365. Springer, Heidelberg (2010)

DeepPhish: Automated Phishing Detection Using Recurrent Neural Network

M. Arivukarasi and A. Antonidoss

Abstract Phishing attacks are one among the foremost common and least defended safety threats these days. We have got an inclination to gift associate technique that uses tongue method techniques to analyze text and see tangential statements that are indicative of phishing assaults. Our approach is novel compared to the previous paintings as a result of it makes a specialty of the seasoning language matter content contained among the assault, taking part in linguistics analysis of the matter content to come back across malicious reason. a novel strategy dependent on RNN. This implies it as a versatile and quick acting proactive discovery framework that does not require full substance investigation.

Keywords Phishing attacks · Recurrent neural network · Linguistics analysis · E-mails

1 Introduction

Phishing recognition systems do endure low location exactness and high false caution particularly when our phishing approaches are presented. In addition, the most widely recognized procedure utilized boycott-based strategy which is wasteful in reacting to exuding phishing assaults since enrolling new space has turned out to be simpler; no far-reaching boycott can guarantee an ideal state-of-the-art database. Moreover, page content examination has been utilized by a few systems to beat the false negative issues and supplement the vulnerabilities of the stale records. Besides, each page content review calculation has distinctive way to deal with phishing site identification with changing degrees of exactness. Along these lines, ensemble can be believed to be

M. Arivukarasi (✉) · A. Antonidoss
Hindustan Institute of Technology and Science, Chennai, India
e-mail: arivu.prem@gmail.com

A. Antonidoss
e-mail: aantonidoss@hindustanuniv.ac.in

P. Suresh et al. (eds.), *Advances in Smart System Technologies*,
Advances in Intelligent Systems and Computing 1163,
https://doi.org/10.1007/978-981-15-5029-4_18

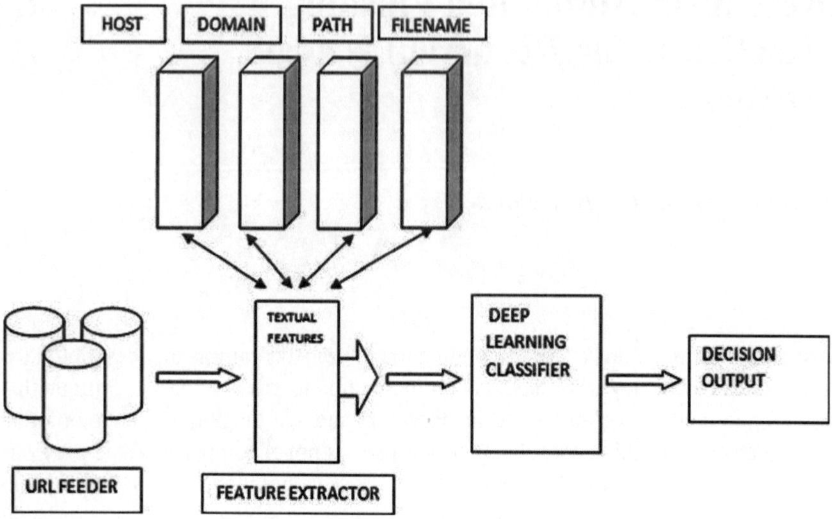

Fig. 1 RNN model

a superior arrangement as it can join the similitude in exactness and diverse blunder discovery rate properties in chosen algorithms (Figs. 1 and 2).

2 Related Work

2.1 Phishing Detection

Phishing URL discovery should be possible by means of proactive or responsive methods. On the responsive end, we discover administrations, for example, Google Safe Browsing API3. This kind of administrations uncovered a boycott of vindictive URLs to be questioned. Boycotts are developed by utilizing diverse strategies, reporting honey pots or by creeping the Web looking for known phishing qualities [1, 2]. The Phishing URL suggests web clients which stays in danger until the URL is submitted and the boycott is refreshed. In addition, since most of phishing destinations are dynamic for not exactly multi day [2, 3], their main goal is finished when they are added to the boycott. Proactive strategies alleviate this issue by breaking down the qualities of a site page progressively so as to survey the potential danger of a site page. Hazard evaluation is done through an arrangement demonstrate [4]. Boosting [5, 6], arbitrary woods [7], inert Dirichlet assignment [8], online gradual learning [9], and neural systems [10]. A few of these techniques utilize a variety of site attributes, which imply that so as to assess a site, first it must be rendered before the calculation can be utilized. This includes a lot of time to the assessment procedure [9, 10]. Utilizing URLs, rather than substance investigation, lessens the

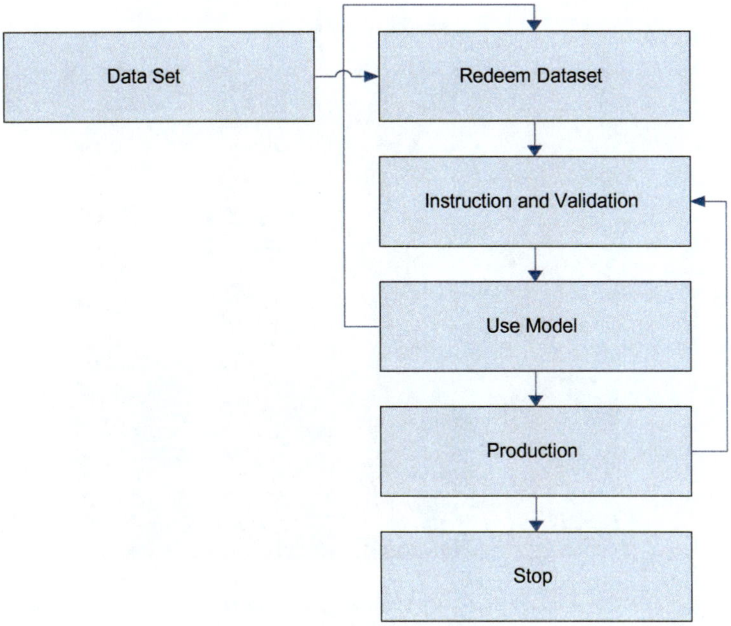

Fig. 2 Execution of the data preprocessing module

assessment time in light of the fact that just a restricted bit of content is breaking down (Figs. 3 and 4).

Recently, the use of machine learning strategies for URL order has been picking up consideration. A few examinations proposing the utilization of order calculations to identify phishing URLs have gone to the light as of late [9, 5, 7]. These examinations are chiefly centered around making highlights through master learning and lexical investigation of the URL. At that point, the phishing site's trademark is utilized as

Fig. 3 Dataset processing technique

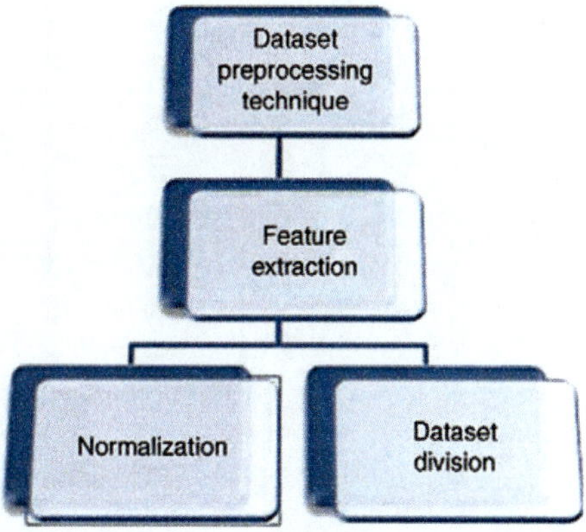

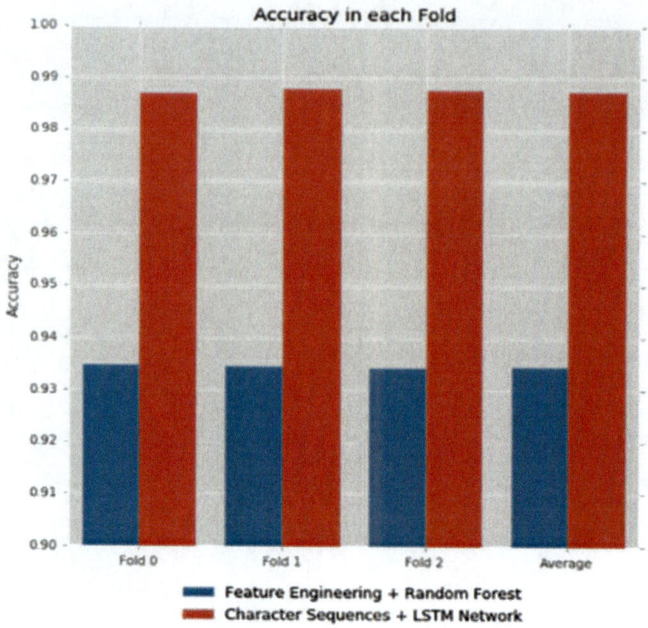

Fig. 4 Result for LSTM and RNN

quantitative contribution to the model. The model thusly figures out how to perceive examples and affiliations; the information sources must follow so as to name a site as genuine or vindictive (Table 1).

Table 1 Feature dataset—long URL

ip_address	ssl_connection	long_url	dots	at_symbo	hexadecimal	frame	redirect	submit
0	0	72	3	0	0	0	0	1
0	0	232	5	0	0	0	0	1
0	0	56	3	0	0	0	0	1
0	0	173	5	0	0	0	0	1
0	0	35	2	0	0	0	0	1
0	0	220	2	0	0	0	0	1
0	0	45	2	0	0	0	0	1
0	0	135	3	0	0	0	0	1
0	0	218	4	0	0	0	0	1

3 Proposed Model

3.1 RNN

To comprehend the idea of back-propagation through time, you unquestionably need to comprehend the ideas of forward and back-propagation first. I will not delve into the subtleties here in light of the fact that would be way out of the farthest point of this blog entry (Table 2).

So, I will endeavor to give you a meaning of these ideas that is as straightforward as could be allowed yet enables you to comprehend the general idea of back-propagation through time. In neural systems, you fundamentally do forward propagation to get the yield of your model and check if this yield is right or off base, to get the blunder. Presently, you do backward propagation, which is only going in reverse through your neural system to locate the halfway subordinates of the blunder as for the loads, which empowers you to subtract this incentive from the loads. Those subordinates are then utilized by gradient descent, a calculation that is utilized to iteratively limit a given capacity. At that point, it changes the loads up or down, contingent upon which diminishes the mistake. This represents the neural network gets the hang of amid the preparation procedure. So, with back-propagation, you essentially attempt to change the loads of your model while preparing. The picture beneath outlines the idea of forward propagation and backward propagation flawlessly at the case of a feedforward neural network (Table 3).

Table 2 Feature dataset

ip_address	ssl_connection	long_url	dots	at_symbol	hexadecimal	frame	redirect	submit
0	0	0.229437	0.076923	0	0	0	0	1
0	0	0.922078	0.153846	0	0	0	0	1
0	0	0.160173	0.076923	0	0	0	0	1
0	0	0.666667	0.153846	0	0	0	0	1
0	0	0.069264	0.038462	0	0	0	0	1
0	0	0.87013	0.038462	0	0	0	0	1
0	0	0.112554	0.038462	0	0	0	0	1
0	0	0.502165	0.076923	0	0	0	0	1
0	0	0.861472	0.115385	0	0	0	0	1

Table 3 Algorithm comparison

Methods	Time for training	Evolution time	Memory consume
RF	2.95+0.12	942.12+95.03	288.0
LSTM	238.7+0.78	280.90+64.43	0.580

3.2 LSTM

Long short-term memory (LSTM) systems are an augmentation for intermittent neural systems, which fundamentally expands their memory. Hence, it is appropriate to gain from imperative encounters that have long time slacks in the middle. The units of a LSTM are utilized as building units for the layers of an RNN, which is then regularly called a LSTM arrange. LSTMs empower RNNs to recollect their contributions over an extensive stretch of time. This is on the grounds that LSTMs contain their data in a memory that is much similar to the memory of a PC on the grounds that the LSTM can peruse, compose and erase data from its memory. This memory can be viewed as a gated cell, where gated implies that the cell chooses whether or not to store or erase data (e.g., in the event that it opens the entryways or not), founded on the significance it doles out to the data. The doling out of significance occurs through loads, which are additionally learned by the calculation. This basically implies that it learns after some time which data is vital and which is not. In a LSTM, you have three doors: input, overlook and yield entryway. These doors decide to give new info access (input entryway), erase the data since it is not imperative (overlook entryway) or to give it a chance to affect the yield at the present time step (yield entryway). You can see a representation of an RNN with its three entryways underneath.

4 Implementation

There is such a significant number of machine learning calculations, and every calculation has its own working component. In this article, we have clarified decision tree algorithm, since I think; this calculation is a straightforward and amazing one. At first, as we referenced above, phishing area is one of the arrangement issues. In this way, this implies that we require marked occasions to manufacture location component. In this issue, we have two classes: (1) phishing and (2) real. When we compute the highlights that we have chosen our necessities and purposes, our dataset looks like in figure underneath. In our precedents, we chose 12 highlights, and we determined them. Consequently, we created a dataset which will be utilized in preparing period of machine learning calculation. Onto the part you have most likely been hanging tight for this time: preparing machine learning calculations. To have the capacity to test the execution of our calculations, I initially played out an 80/20 train-test split, part our decent informational collection into two pieces. To abstain from overfitting, I utilized the exact regular resampling system of k-overlay cross-approval. This implies that you separate your preparation information into k parts (folds) and after that fit your model on k-1 overlap before making expectations for the kth holdout crease. You at that point rehash this procedure for each and every overlay and normal the subsequent forecasts.

To show signs of improvement feeling of which calculation would perform best on our information, how about we rapidly spot-check probably the most mainstream grouping calculations:

- Classification trees
- Support vector classifier
- Random forest classifier
- XGBoost Classifier a decision tree can be considered as an enhanced settled if-else structure. Every component will be checked one by one. A precedent tree show is given underneath. Creating a tree is the fundamental structure of identification component. Yellow and circular formed ones speak to highlights, and these are called hubs. Green and precise ones speak to classes, and these are called leaves. The length is checked when a precedent arrives, and afterward, alternate highlights are checked by the outcome. At the point when the voyage of the examples is finished, the class that an example has a place will turn out to be clear.

4.1 Preprocessing

To understand a dataset reasonable with the end goal of this examination, the phish information gathered from PhishTank was revamped, and some inferred highlights were included. Likewise, the dataset position that was downloaded from PhishTank archive must be changed from dab csv group (.csv) to SQL database design (.sql) so as to make it worthy for use with PHP. In light of the open-source nature of PhishTank, the majority of the highlights required for this task were excluded and thusly most highlights were separated physically utilizing PHP code. In the dataset gathered from PhishTank, a portion of the highlights, for example, "phish_detail_url," "submission time," "confirmed," "verification time" and "on the web" were prohibited, and new highlights separated and added to the dataset. Here, both the phishing and non-phishing sites have been tried for alive properties ensuring that the information to be utilized must be right and accessible for further advance which include extraction. Then,

4.2 Feature Engineering

This part covers the dataset method of gathering and planning including the extraction of the majority of the highlights proposed for use in this investigation. PhishTank vault is utilized as the main wellspring of phishing dataset, while the non-phish dataset is gathered physically utilizing Google Web crawler. Essentially, the yield from highlight extraction will be utilized as contribution to assessing the individual classifiers. The information handling methods including the dataset insights. the element extraction process, information check, data normalization; technique and criteria utilized for standardization. the dataset division; as far as dataset gathering and the percentage of phishing and non-phishing dataset utilized with support so as to expand the execution of the classifier preparing procedure to more readily enhance the precision of the outcome. What's more, the likewise talk about the achievement of this section in agreement to the goals of this undertaking. As to understand a dataset reasonable with the end goal of this task, the phish information gathered from PhishTank was revamped, and some inferred highlights were included. Additionally, the dataset organize that was downloaded from PhishTank vault must be changed from spot csv arrange (.csv) to SQL database design (.sql) so as to make it adequate for use with PHP. On account of the open-source nature of PhishTank, the vast majority of the features required for this task were excluded and all things considered most highlights were separated physically utilizing PHP code.

URL: Long URLs can be utilized to shroud the suspicious piece in the location bar. Albeit experimentally demonstrated the solid technique for foreseeing the scope of length that legitimizes a site as phishing or non-phishing, however then it is a criteria utilized with different highlights in identifying suspicious destinations.

Dots: A protected Web site page connect contains at most five spots. On the off chance that perhaps, there are excess of five dabs in a page; at that point, it might be perceived as a phishing join.

IP address: A few sites are facilitated with IP address rather than a fully qualified space name. This is a suspicious demonstration since the greater part of the genuine site never again utilizes this strategy in light of security reasons. Likewise, since most phishing sites remain online temporarily, this component can be considered as one of the extremely important phishing identification highlights.

SSL connection: It is essential for an installment or e-commerce payment Web page to be verified in with the end goal that the information transmitted from and to the site is encoded. Additionally, it very well may be utilized to affirm the personality of a site by utilizing SSL declaration which incorporates explicit data with respect to the site.

"@": The phishing URL may incorporate the "@" symbol somewhere inside the location in light of the fact that the Internet browser, when perusing a Web address, overlooks everything to one side of the @ image, consequently, the location ebay.com

Hexadecimal: Specific to phishing is hex-encoded URLs. In light of a legitimate concern for similarity, most mail client operators, Internet browsers and HTTP servers all comprehend essential hex-encoded character.

4.3 Classification Approach

(a) ***Data Verification***

Information gathered physically should be checked so as to find out the alive status particularly on account of phishing as it is realized that phishing site for the most part keeps going for a constrained timeframe. Consequently, every URL must be confirmed before preparing.

(b) ***Data Normalization***

There are numerous strategies for information standardization that incorporate min–max standardization (go change), z-score standardization and normalization by decimal scaling. Min–max standardization plays out a direct change on the first information. Assume that min(a) and max(a) are the base and the most extreme qualities for trait A. Min–max normalization maps an esteem v(A) -> u in the range (temp-mina, temp-maxa) by processing as appeared in condition (1). To modify the standardization output to wanted scale, extend change strategy was chosen.

(c) ***Dataset Division***

After information preparing, the dataset is partitioned into three sets for training and testing reason and to research the precision of result. Two stages of information division are utilized; the initial step is to isolate the information into three unique gatherings and after that pick diverse level of phishing and non-phishing for each gathering,

5 Result and Discussion

The dataset for phishing site is downloaded from PhishTank tried to affirm it is on the Web, and after that, the highlights are extricated from every site. For non-phishing site, a Web crawler is utilized to extricate the dataset. First, we assessed the execution of the conventional component building in addition to the classification calculation technique. We made 14 highlights dependent on the URL's lexical and factual examination. At that point, we prepared an irregular timberland classifier with 100 choice trees.

6 Conclusions

To assess the methodologies, we utilized a database that included one million real URLs from the common crawl database and one million phishing URLs from PhishTank. This is urgent as there are memory-limited applications, for example, versatile applications. For this situation, the RF display is eccentric and LSTM ought to be picked. In our investigation of the approaches, we discovered advantages and disadvantages for the two techniques. The LSTM demonstrates a generally higher expectation execution without the need of master information to make the highlights. The drawback is that internal operations cannot be deciphered effectively.

References

1. Sahil Dhankhad, A., Mohammed, E.A.: Supervised machine learning algorithms for credit card fraudulent transaction detection: a comparative study. In: IEEE International Conference on Information Reuse and Integration for Data Science (2018)
2. Kalai Selvi, N., Rajalakshmi, S., Padmavathi.: Credit card fraud Detection using learning to rank approach. In: IEEE International Conference on Computation of Power, Energy, Information and Communication (ICCPEIC) (2018)
3. Sahingoz, O.K., Buber, E., Demir, O., Diri, B.: Machine Learning Based Phishing Detection from URLs. Elsevier, pp. 345–357, September (2018)
4. Xuan, S., Liu, G., Li, Z.: Random forest for credit card fraud detection. IEEE Trans. Neural Netw. Learn. Syst. (2018)
5. Boutaba, R., Salahuddin, M.A.: A comprehensive survey on machine learning for networking: evolution, applications and research opportunities. J. Internet Serv. Appl., June (2018)
6. Moskovitch, R., Shabtai, A., Edri, M., BarAd, O., Elovici, Y.: Keeping pace with the creation of new malicious PDF files using an active-learning based detection framework. In: Security Informatics, February (2018)
7. Khanzadeh, M., Akula, R., Zhang, F., Zhang, S., Medal, H.: Botnet detection using graph-based feature clustering. J. Big Data, 12 May (2017)
8. Guigou, F., Deruyver, A., Collet, P.: Foundations and applications of artificial Intelligence for zero-day and multi-step attack detection. EURASIP J. Inf. Secur., April (2018)
9. Yang, P., Zhao. G.„ Zeng, P.: Phishing website detection based on multidimensional features driven by deep learning. IEEE Access, vol. 7 January (2019)
10. Yuan, X.: Deep learning-based real-time malware detection with multi-stage analysis. In: 2017 IEEE International Conference on Smart Computing (SMARTCOMP) May (2017)

Hybrid Approach Using Machine Learning Techniques in Credit Card Fraud Detection

S. Sivanantham, S. R. Dhinagar, P. Kawin, and J. Amarnath

Abstract Crime financial services have a serious problem by credit card fraud. Every year billions of dollars are lost by credit card fraud. Lack of research and studies on analyzing real-world credit card data is there. In this paper, we apply machine learning algorithms to detect the credit card fraud using credit card datasets and compare their performance. We use hybrid algorithm model involving voting classifier to detect the fraud. The result positively indicates that the voting classifier method gives good accuracy, precision, recall, and F1 score in detecting fraud in credit card transactions. More than that, the work identifies the important variables in the dataset that may lead to higher accuracy in credit card fraud detection, and on the whole, we discuss about various types of classifiers and its performance in detecting fraudulent transactions.

Keywords Credit card fraud · Machine learning · Voting classifier · Random forest · Gradient boosting

1 Introduction

Fraud detection may gain profit to fraud person, but it is a criminal deception. To avoid fraud, there are two mechanisms: One is prevention and other one is detection. Proactive method is used in prevention mechanism, and similarly, detection mechanism is used for find fraudster attempts in transaction. While purchasing and

S. Sivanantham (✉) · S. R. Dhinagar · P. Kawin · J. Amarnath
Department of IT, Adhiyamaan College of Engineering, Hosur, India
e-mail: sivanantham17@gmail.com

S. R. Dhinagar
e-mail: s.r.dhinagar@gmail.com

P. Kawin
e-mail: Kawinpadmanaban@gmail.com

J. Amarnath
e-mail: amaranth.j@gmail.com

P. Suresh et al. (eds.), *Advances in Smart System Technologies*,
Advances in Intelligent Systems and Computing 1163,
https://doi.org/10.1007/978-981-15-5029-4_19

making payment by credit card, illegal fraud occurs. Transaction may occur in physical or digital manner in credit card [1]. Nowadays, the person with credit card stores cookies in Web site like credit card number, expiry date with verification number in Web site, etc.

In e-commerce Web sites, the payment method in the form of credit card system has been increased, and cash on delivery (COD) is rarely used. In Malaysia, credit card transaction is 320 million in the year of 2011, and it is increased to 360 million in the year of 2015. Due to increased transaction, fraud also has amplified. These fraudsters hide their location and ID through Internet. In financial industry, there is a big threat by fraudsters.

The merchants have to bear the loss of customers including charges, fees, and so on, by the effect of fraudster. Eliminating the fraud cases is effective for reducing credit card fraud. There have been many data mining and machine learning methods used to detect fraud from credit card transactions [2]. In this paper, machine learning techniques like logistic regression, random forest, gradient boosting, and voting classifier algorithms are used in detection of credit card fraud. The performance of these techniques is evaluated by a real-world credit card dataset and with some benchmark. The predictions made by machine learning methods like random forest, logistic regression, and gradient boosting are used by voting classifier in order to get an accurate result for fraud detection [2]. In this hybrid model, voting in majority (voting classifier) seems to be reliable, and over three months of credit card transaction is available in the dataset for research.

1.1 Logistic Regression

Logistic regression is one of the statistical methods for analyzing a dataset with one or more independent variables which determine an outcome. The outcomes are measured with a dichotomous variable (in which there are only two possible outcomes). The outcome estimated for each independent variable is always binary (1/0, yes/no, true/false). Dummy variable is used for maintaining those binary/categorical outcomes. It is implicit that logistic regression shall be a special case of linear regression. Logistic regression works by estimating probability of existence of an event by fitting data to a logistic function [3].

1.2 Gradient Boosting

A simple ensemble for building many independent predictions and combining them with the use some model averaging techniques is called bagging.

Boosting is an ensemble technique in which the predictors have been made sequentially but not independently.

Gradient boosting is a machine learning technique normally used for regression and classification problems. Gradient boosting helps in producing a strong predictive model as an ensemble of weak predictors like decision tree, Bayesian classifier, random forest, etc. The predictive model is built in a stage-wise fashion as other boosting method do, and this generalizes them by allowing optimization of a differentiable loss function in each stage [4].

1.3 Random Forest

Random forest is a one of the data mining cum machine learning algorithms like before, as the name implies. The forest of data which is built by this algorithm is an ensemble of decision tree which has been trained by the "bagging" method maximum of the time. The hyperparameters of the decision tree are approximately the same as that of the random forest; there is no necessity for mingling a decision tree with a bagging classifier.

Random forest will add additional randomness to the model when the trees are grown-up morphologically. Despite of searching for the most important feature when splitting a node, this algorithm searches for the best features among a random subset of features. This will result in a wide range of diversity which will generally result in a better prediction model [5].

Hence, in the random forest, only a random subset of the features will be taken into consideration while splitting a node by using this algorithm. We can also make random trees by using random thresholds additionally for each feature instead of searching for the best possible thresholds as the normal decision tree do.

1.4 Voting Classifier

The voting classifier is a meta-classifier used for linking analogous or conceptually different machine learning classifiers for classification via majority or plurality voting concept [6]. Figure 1 shows that a very interesting ensemble solution is offered by the voting classifier, which is not an actual classifier but a wrapper for a set of different classifiers that are trained and evaluated in parallel in order to exploit the different peculiarities of each technique used.

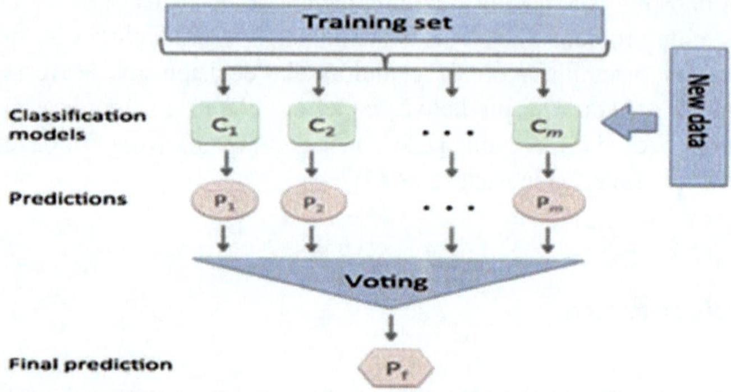

Fig. 1 Voting classifier

2 Related Work

In these days, many types of research have been undertaken by researchers in the development of credit card fraud detection systems (CFDS) using machine learning algorithm. Let us review some worthy contributions toward CFDS and machine learning algorithm approaches in this section.

Natekin and Knol [4] have discussed about gradient boost model (GBM); they told that gradient boosting model is the best model compared to available bagging and boosting concepts.

Randhawa et al. [7] have discussed about AdaBoost and majority voting; they suggest majority voting gives optimal detection rate with AdaBoost classifier [7].

Biau [5] has worked the random forest and logistic regression and suggest that logistic regression gives the best result with random forest boost [5].

Ravisankar et al. [8] have discussed about support vector machine, logistic regression, MLLF, and GMDH; they suggest that the logistic regression gives good accuracy compared to other methods [8].

Chen et al. [9] used an integration of multiple models like random forest, decision tree, rough set theory, and back-propagation neural network to build a fraud detection model for corporate financial statements. They suggest that the hybrid model gave the highest classification accuracy compared to other combinations [9].

Li et al. [10] have developed a hybrid model using principal component analysis and random forest for insurance fraud detection, and their inception is that the hybrid prediction model is time and cost efficient [10].

Many of the existing works focus on improving the accuracy rate for fraud detection. The review shows clearly that the hybrid machine learning approaches provide better performance compared to single model.

3 Experimental Setup

In this proposed work, we have selected logistic regression (LR), gradient boosting (GB), and random forest (RF) to work with voting classifier.

The voting classifier takes the best parameters from these three algorithms to give the best accuracy, precision, recall, and F1 score. The voting classifier is mostly used for improving the accuracy.

The entire experiment is set up in Anaconda framework at Jupyter notebook and evaluated in Kaggle credit card dataset.

3.1 About Kaggle Credit Card Dataset

Kaggle dataset contains transaction of payment using credit card containing fraudulent transactions. The attributes start from v1, v2 in sequels manner up to v28. This dataset is strictly avoided by using alphabets and other components except members. Background will not provide larger data or information, and the main features include time and amount. Time is recorded every second, where the amount gets transacted within a period that is taken into account. The dataset contains 20% of fraud transaction and has 80% of normal transaction [11].

3.2 About Anaconda

The two open-source languages used often for data science in Anaconda are:

- Python.
- R programming language.

To maintain the lack of absence of packages, here we use predictive analytics, and majorly some of them are helpful to build the entire application such as scale processing and data science.

Various dataset formats, namely ARFF, CSV, C4.5, XRFF, BSI, Json instance files, MATLAB ASCII files, etc., are supported in Anaconda Jupyter notebook. To name a few, the data analytics methods like preprocessing, clustering, classification, regression, association, etc., could be employed in Anaconda Navigator [12].

Figure 2 shows the implementation of the proposed work. The raw data is loaded into the Jupyter Notebook. Then, the data has gone to the preprocessing steps. After preprocessing, the model selection technique is applied. Then, we selected the best model for prediction. Once the model selection is over, next logistic regression, random forest, gradient boosting, and voting classifier are applied in order. To improve the accuracy, precision, recall, and F1 score, the k-fold cross-validation is

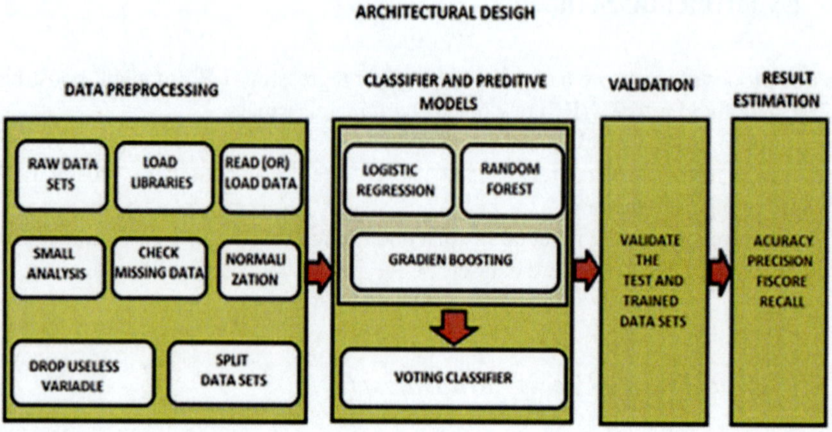

Fig. 2 Steps to implement the proposed work

used. From the result of confusion matrix, TP, TN, FP, and FN are obtained and used for calculation of accuracy, precision, recall, and F1.

4 Result and Discussion

The results of the proposed work include the calculation of accuracy, detection rate (DR), false alarm rate (FAR), precision, recall, and F1 score. The above said measures are calculated from confusion matrix considering true positive (TP), true negative (TN), false positive (FP), and false negative (FN) [13].

Accuracy, detection rate (DR), and false alarm rate (FAR), precision, F1 score, and recall are calculated as,

Accuracy = Number of correct prediction/Total number of prediction.
Recall = TP/TP + FN.
Precision = TP/TP + FP.
Detection rate = TP/TP + FN * 100.
False Alarm rate = FP/TN + FP.
F1 Score = 2 * Precision * Recall/Precision + Recall.

In the first stage of the experiment, the four proposed schemes are implemented without validation, and the results are recorded.

In the second stage of the experiment, the four proposed schemes are implemented with k-fold validation.

Table 1 Performance of machine learning algorithm without validation

Approach used	Precision	Recall	F1 score
Logistic regression	0.825	0.584	0.684
Gradient boosting	0.912	0.348	0.504
Random forest	0.932	0.764	0.840
Voting classifier	0.923	0.809	0.862

Table 2 Performance of machine learning algorithm with validation

Approach used	Precision	Recall	F1 score
Logistic regression	0.065	0.921	0.121
Gradient boosting	0.912	0.348	0.504
Random forest	0.945	0.775	0.852
Voting classifier	0.923	0.783	0.847

At each stage, the voting classifier gives good result with LR, GB, and RF combination. Next, we discuss the accuracy, detection rate, and false alarm rate for all machine learning model.

The false alarm rate is decreasing in order for all models. From Tables 1, 2, and 3, it is clear the voting classifier gives best results of accuracy 99.98%, detection rate 99.98%, and false alarm rate 1.910 compared to logistic regression, gradient boosting, and random forest classifier.

In Table 3, the accuracy increases in all models in order. Voting classifier provides maximum accuracy and false alarm rate (Figs. 3 and 4).

Figure 5 shows that with no doubt, voting classifier gives good result for this dataset problem. It is very evident that voting classifier, when combined with LR, GB, and RF outperforms the individual methods.

Table 3 Result estimation for the proposed method

Approach used	Accuracy (%)	Detection rate (%)	False alarm rate
Logistic regression	99.90	99.97	4.157
Gradient boosting	99.88	99.99	6.516
Random forest	99.94	99.99	2.359
Voting classifier	**99.98**	99.98	**1.910**

Fig. 3 Performance of machine learning algorithm without validation

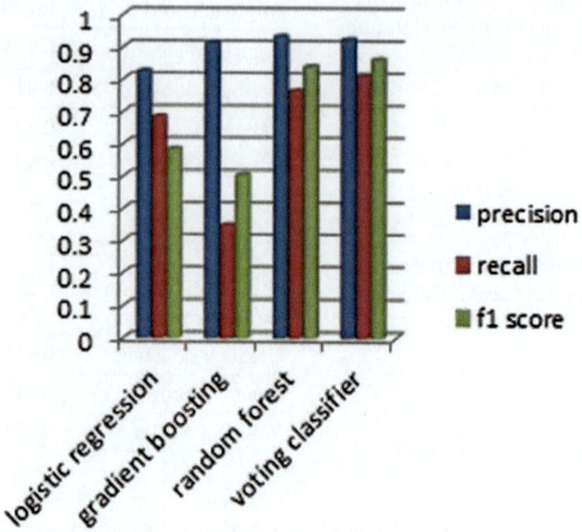

Fig. 4 Performance of machine learning algorithm with validation

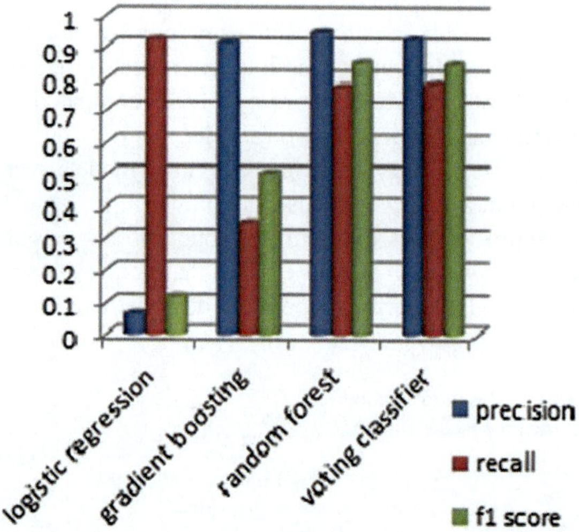

5 Conclusion

Credit card fraud detection model proposed is a hybrid model involving LR, GB, RF, and voting classifier. The hybrid ones always outperform the individual techniques. On result analysis, it is clear that voting classifier gives maximum accuracy 99.98%, random forest and gradient boosting give maximum detection rate 99.99%, and voting classifier gives maximum false alarm rate 1.910.

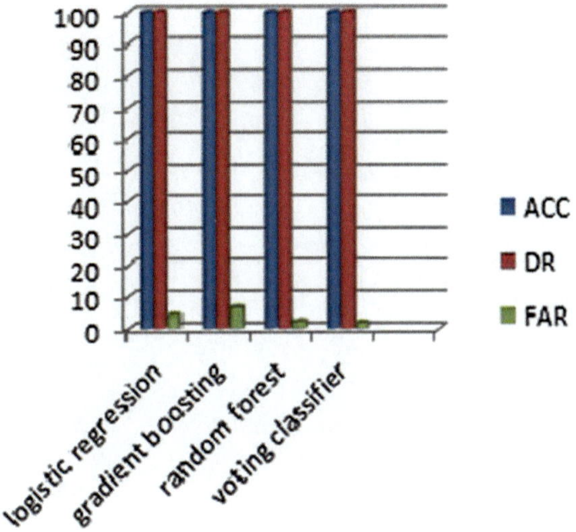

Fig. 5 Result estimation for this method

References

1. Delamaire, L., Abdou, H., Pointon, J.: Credit card fraud and detection techniques: a review. J. Banks Bank Syst. **4**(2), 57–68 (2009)
2. Adewumi, A.O., Akinyelu, A.A.: A survey of machine-learning and nature-inspired based credit card fraud detection techniques. Int. J. Syst. Assur. Eng. Manag. **8**, 937–953 (2017)
3. Peng, C.Y.J., Lee, K.L., Ingersoll, G.M.: An introduction to logistic regression analysis and reporting. J. Educ. Res. **96**(1), 22–31 (2010)
4. Natekin, A., Knol, A.: Gradient boosting machines, a tutorial. Front. Neuro Robot. Acad. Art. **7**(21), 1–21 (2013)
5. Biau, G.: Analysis of a random forests model. J. Mach. Learn. Res. **13**(7), 1063–1095 (2012)
6. Chen, H., Lin, Y., Tian, Q., Xu, K.: A comparison of multiple classifier combinations using different voting-weights for remote sensing image classification. Int. J. Remote Sens. **39**(11), 208–219 (2018)
7. Randhawa, K., Loo, C.K., Seera, M., Lim, C.P., Nandi, A.K.: Credit card fraud detection using AdaBoost and majority voting. IEEE Trans. **6**, 14277–14284 (2017)
8. Ravishankar, R., Ravi, V., Raghava Rao, G., Bose, I.: Detection of financial statement fraud and feature selection using data mining techniques. J. Decis. Support Syst. **50**(2), 491–500 (2011)
9. Chen, F.H., Chi, D.J., Zhu, J.Y.: Application of random forest, rough set theory, decision tree and neural network to detect financial statement fraud–taking corporate governance into consideration. In: Proceeding of International Conference on Intelligent Computing, pp. 221–234. Springer, Berlin (2014)
10. Li, Y., Yan, C., Liu, W., Li, M.: A principle component analysis based random forest with the potential nearest neighbor method for automobile insurance fraud identification. J. Appl. Soft Comput. **70**, 1000–1009 (2017).
11. Credit card fraud detection dataset for researchers. Available online: https://www.kaggle.com/creditcardfraud
12. Anaconda complete manual. Available online: https://www.anaconda.org
13. Powers, M.W.: Evaluation: from precision, recall and F-measure to ROC, informedness, markedness & correlation. J. Mach. Learn. Technol. **2**(1), 37–63 (2011)

An Experimental Study and Analysis of Impact on Mobile Sink in Wireless Sensor Networks

V. C. Diniesh, G. Murugesan, M. Joseph Auxilius Jude, E. M. Jayanth, N. Rishikesh, and K. Nanthini

Abstract Targeting a growing number of promising applications sink mobility has been measured as a good strategy to prolong the lifetime of mobile wireless sensor network MWSNs. Mobile sink helps indirectly achieving uniform energy—consumption and provides load balancing to enhance data collection from the sensor node. Generally routing in MWSN is more complicated as it operates in low power, battery constraint with limited resources. A recent development in mobile sink leads to researcher to design routing algorithms for MWSN but still many unanswered gaps like shortest path, node mobility prediction, route reconstruction, and maintaining network coverage under movement and multi-sink approach for data collection towards critical events from sensor nodes. This paper address taxonomy of various data collection approaches under the routing protocol schemes sensor field. Based on this we classify routing schemes as location-based and query-driven approaches. Finally, discussion summarizes the detection of some unanswered issues and also gives a path to researchers to design and enhancement to improve existing routing schemes for future direction.

Keywords Mobile wireless sensor network · Mobile sink · Mobility routing · Query-driven

1 Introduction

The Internet of Thing (IoT) has recently given attention to researchers and industry and contemplates to support in emerging applications in smart world, habitat monitoring, health care, tracking object, and Intelligent Transport System (ITS). Mobile IoT is a network consists of protocol stack, Application-layer (COAP, HTTP)

V. C. Diniesh (✉) · G. Murugesan · M. Joseph Auxilius Jude · N. Rishikesh · K. Nanthini
Kongu Engineering College, Perundurai, Erode, Tamil Nadu, India
e-mail: vcdiniesh@gmail.com

E. M. Jayanth
Ubisoft, Pune, India

P. Suresh et al. (eds.), *Advances in Smart System Technologies*,
Advances in Intelligent Systems and Computing 1163,
https://doi.org/10.1007/978-981-15-5029-4_20

Network layer (routing), Medium Access Control (MAC), and Radio Duty Cycle (RDC) layer [1, 2]. In traditional networks, every layer plays a role associated with a metrics is usually designed for a particular application environment here nodes reveal M-to-1 network in collecting sensed data to Pan coordinator or gateway node. On the other end, multiple mobile sinks in IoT network are considered as widespread new technology from smaller to finer devices that assist balanced energy in a rigorous environment [3, 4]. In addition to static sink, node equipped with mobile sink plays a vital role in several environments to alleviate energy-hole problem, lengthen network lifetime, forecast of topology changes; brings a challenging role for data transmission in WSN [5,6].

Early data advisement mechanism in WSN is one PAN coordinator (base station) and multiple coordinators deployed in an area with line of sight. Finally, sensor nodes sense the information and send data towards the base station in multihop fashion with the development of emerging mobile application many mechanisms aspire at multi-level mobile sink has been designed and proposed in sensor field. Due to mobility sink, topology in WSN environment becomes dynamic nature, and the existing routing protocol mechanism deals with static and this gives a path to researchers for the design of improved routing schemes for mobile sink based WSN.

The remaining of this paper is covered as follows: Sect. 3 presents the related works encircle the several existing schemes and architecture towards mobile sinks in WSN. To analyze multiple sink Sect. 4 describes Experimentation and hardware testbed setup in Vinton Network Lab and results are discussed in detail and the paper ends up with a conclusion and future direction in Sect. 5.

2 MWSN Architecture

The mobile sink node architecture as shown in Fig. 1 differs from static WSN because of the sink node travel in and around the sensor environment for data collection. They are different types of nodes available in MWSN namely sensor node, cell head, and Mobile sink. First, sensor node is a normal node deployed in a sensor environment and sense the information. Upon sensing the information, these nodes disseminate the data forward to the cell head or cluster head. Second cell head node in WSN acts as a gateway node, coordinately collects packets received from sensor nodes, and forward to sink. Lastly, mobile sink is the same as sensor node but depends on applications involved in human personal digital assistant (PDA), Internet of Vehicles (IoV), Animal monitoring for data collection.

At present, they are many number of data collection schemes developed on sink mobility for MWSN. Based on which we classify location-based and query-driven schemes as shown in Fig. 2. In this location-based schemes, sink node continuously advertises sensor network environment in order to keep an update of the location to all sensors for data. On the other end, they are several applications which require reinforced data of neighboring node in the particulate network with other for data delivery to mobile sink and this can be further achieved by query-driven approach.

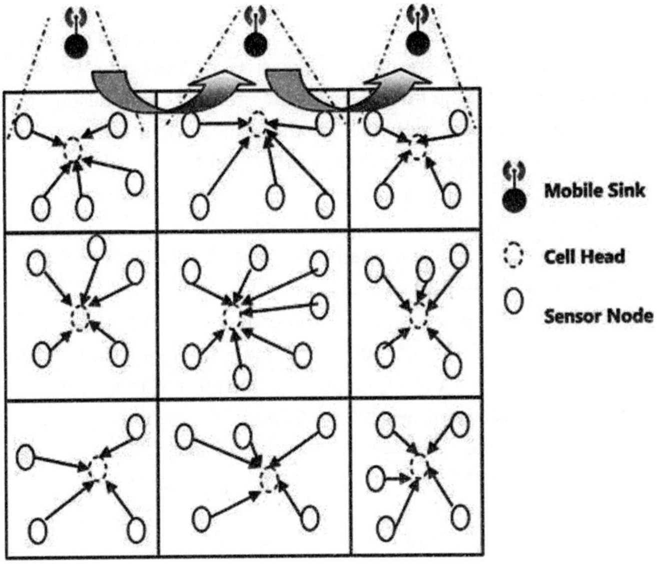

Fig. 1 Mobile sink node architecture

Fig. 2 Sink mobility schemes

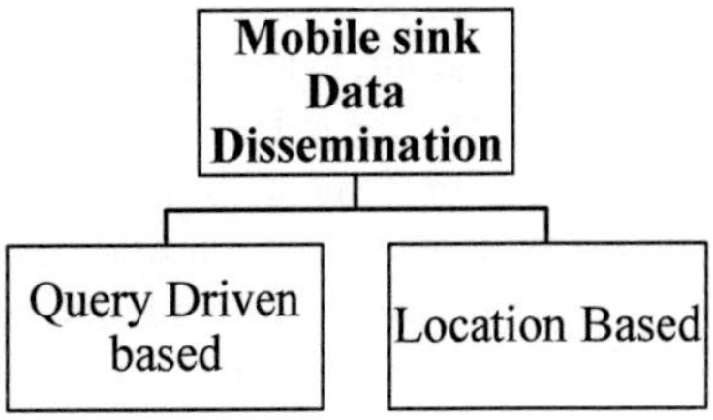

3 Related Works

To prolong the lifetime of sensor networks while maintaining optimum route adjustment and balance the energy node for mobile sink, virtual grid-based dynamic routes adjustment scheme (VGDRA) is proposed in [7]. VGDRA scheme constructs the network region based on virtual grid technique. VGDRA schemes build the sensor region into several partitions uniformly according to the number of sensors. Here each region appoints cell head closest to the connectivity of neighboring sensors. According to the sink node movement cell head dynamically adjust the routs and inform position and alerts to send the data. The proposed scheme update the mobile sink location aims to govern the routes reconstruction and maintain the reliable route reconstruction to enhance the lifetime of sensor node. Here cell head is continuously forwarding the data of the entire network and leads to sink-hole or hot spot problems.

Sharma et al. [8] have proposed rendezvous-based routing protocol (RRP) for mobile sink WSN. RRP address the latency of end to end data transmission, so that formation of network divides into vertical and horizontal strip across the network. In RRP nodes are deployed randomly creates a virtual cross at centre called as rendezvous region, construct the nodes across the middle region based on tree structure act as a powerful backbone network. In this paper, focus on two different data transmission methods: (1) Tree is constructed directed to the sink and source nodes transmitted data to sink. (2) Sink broadcast its position to the tree and source nodes receive update location from the tree and send data forward to sink. A good feature of this RPP is achieved by adopting methods 1 and 2 makes a reliable end to end data transmission, increase in packet reception, network lifetime.

Predictive routing for mobile sinks WSN is proposed in the literature [9], which adopts two lightweight routing protocol schemes namely Predictive and Predictive_PB. The author believes the concept of a milestone approach plays a vital role by enhancing (1) spreading the sink location to the neighboring nodes and by means of received information that can reroute the data packets towards the mobile sink. In addition to that predictive routing maintains periodic broadcast to the entire network to grant localization of sink node in order to attain high packet reception and reduce energy. However periodic broadcast makes high attention in network leads to enlarging listening time in sensor.

Two-Tier Data Dissemination protocol (TTDD) in [10] intends to bring new challenges for a mobile sink in large-scale WSNs. Here TTDD protocol divide network into many virtual grid structures so that propagation of message is avoided, data transmission attain in mobile sink effectively. On the other hand, forming grid structure by mobile node guide to energy-hole problem, unwanted topology formation and queries from multiple mobile sinks avoids excess energy consumption and collision in network. In critical condition, TTDD not agile to transmit immediately because it depends on query-driven approach that makes high energy consumption.

A multiple mobile sink data dissemination called MSDD protocol has been proposed by Xie et al. [11]. The main idea behind MSDD is to construct two-tier grid formation by the mobile sink and by global agent and hierarchical monitoring mechanism gives additional benefits to track sink locations to make sinks quickly respond during emergency stipulation. In the MSDD approach instead of the formation of grid structure, it constructs sole two-tier structure in network environment. Based on this, the protocol extends the agent mechanism for the sink tracking approach to transmit the data immediately. Furthermore, it optimizes path from source to destination node by adopting grid cell head which will be the shortest distance to send data to sink and neighboring nodes make reliable data forwarding under multihop to all connection in sensor environment.

Khan et al. [12] have proposed query-driven virtual grid-based data dissemination (QDVGDD) for WSN using mobile sink. It is an extended version of VGDRA, and most of the mechanism based on an event or periodic sensing of nodes. There are many applications in WSN that require query-based data collection using mobile sink in the network. However, QDVGDD is based on query-driven mechanism that aims to enhance data delivery rate to mobile sink with high quality of service (QoS). This protocol minimizes the control overhead by adopting dynamic virtual infrastructure according to network deployment. On the other hand, unpredictable network changes due to mobility make to transmit more control messages and lead to unwanted wakeup, collision, packet loss and high energy consumption (Table 1).

4 Experimental Evaluation

In this section, we explain the experimental network setup and core module parameters setting, the scenarios and topological configuration are described.

4.1 Network Setup

At this juncture, network setup environment carried out at Vinton network lab in ECE department with 37″ × 43″ dimensions. Here Core module consists of Jennic JN5168 mote embedded with IEEE 802.15.4 WSN network stack [13]. Here lab consists of four regions as shown in Fig. 3a which has two faculty cabin and two lab areas. Each region comprises of cluster head (CH) at center region and sensor nodes (SN) deployed randomly sense the data/information transmits to CH. Sensor node (SN) consists of core module, battery holder, and temperature and light sensor placed above the stack, and to transmit information it has integrated antenna range up to 70 m.

4.2 Result and Discussion

In order to execute realistic experiments, we attached a mobile sink as shown in Fig. 3b to human body attached with the hand and connected with PC for power and collect information and sent to cloud for storing and retrieving data. Here mobile sink (MS) will start a journey from region 1 to region 4 and collecting data from cluster head (CH) located at each region and comes back to origin point (START/END). As can be observed from the graphs, that each sensor nodes (SN) periodically sense temperature and light information and is collectively sending the data to CH. In the meantime when MS moves to a particular region it collects data from CH and sends to cloud. Here we use open source cloud platform (think speak) to collect the data [14]. Once MS start travels to all regions come to original point and make either replacement or recharging of sensor nodes. As shown in Fig. 4a, b, it shows a collection of data both light and temperature from SN to CH of a particular region and CH to MS. It observed packet reception using mobile sink gives more efficient data collection at some period of time and repeated packet is discarded and makes less energy utilization leads to enhancement of senor network lifetime.

Table 1 Summary of mobile sink routing protocol based on data collection schemes

Mobility schemes	Protocols	Focus	Mobility coverage	Remark
Location-based	VGDRA [7]	(a) Mobility management (b) Minimize the route construction	Extensive and limited	Optimum routes and only mobile sink with next hop
	Rendezvous [8]	(a) Reducing end to end latency (b) Mobility management cost	Tree construction and extensive	No consideration on scalability and additional overhead
	Predicative based [9]	(a) Achieve data packet reception and reduce energy consumption (b) Flooding of packets to determine the predictive/estimated sink location	Extensive and limited	Broadcast scheme leads to unwanted listening time in sensor nodes
Query-based	TTDD [10]	(a) Two-tier data dissemination (b) Query forwarding process leads to successive delivery to the sink	Extensive	Grid structure leads to energy-hole problem and queries from multiple sink leads to excess energy and collision
	MSDD [11]	(a) Targeting the position of sink location towards emergencies situation (b) Hierarchical monitoring reliable data transfer	Extensive and based on hierarchy	Additional overhead for constructing master sink and periodic wakeup in environment leads to computation and high energy
	QDVGDD [12]	(a) QOS to mobile sink (b) Query and response packet	Extensive	Query on multiple cells leads to collision and leads to packet drop and periodic wakeup time gives to high energy consumption

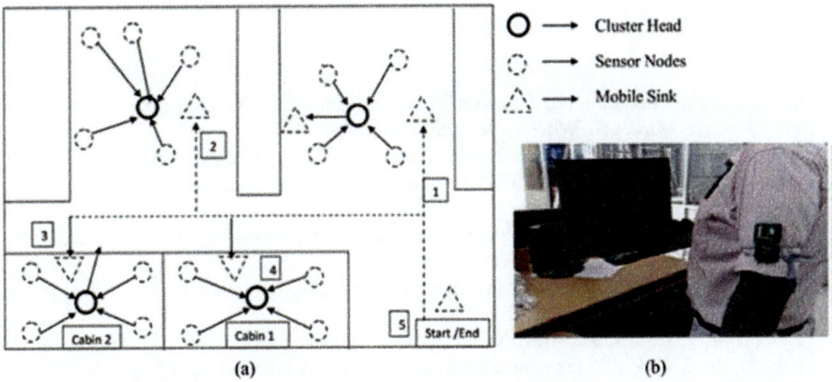

(a) (b)

Fig. 3 Experimental evaluation. **a** Experimental setup with CH and sensors distributed across four regions at Vinton network lab. **b** Mobile sink (MS) attached with shoulder

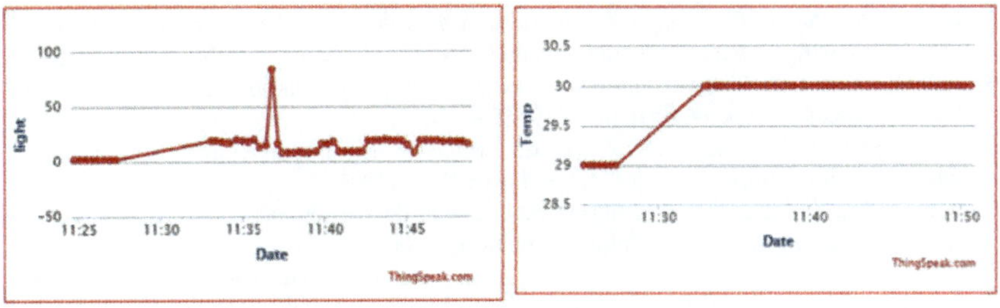

(a) (b)

Fig. 4 Mobilesink (MS) collecting data from CH in all regions and transmit to cloud platform (think speak) in terms of **a** light, **b** temperature

5 Conclusion and Future Work

Routing is one of the current challenging tasks for recent years in mobile sensor networks (MWSN). In this paper, we have presented an extensive literature review of various existing routing schemes that utilize sink mobile and have categorized them based on their location and query-driven approach. Based on our observation, the mobile sink with a hierarchical routing based approach and cross-layer approach will provide an efficient routing solution in all topology both static and dynamic (mobility), control overhead, and processing delay.

The future scope of this paper is the inclusion of priority, optimization technique, different mobility speed, and different packet generation rate. This makes a path for researchers for developing to address this weakness for designing a routing algorithm for mobile sink more effectively. We believe that optimization makes mobile sink travel across the network based on the shortest path, traveling sales problem (TSP), and priority to collect the data for improving and enrichment in the coming years.

References

1. Qutqut, M.H., Al-Sakran, A., Almasalha, F., Hassanein, H.S.: Comprehensive survey of the IoT opensource OSs. IET Wireless Sens. Syst. 1–17 (2018)
2. Wang, Z.M., Basagni, S., Melachrinoudis, E., Petrioli, C.: Exploiting sink mobility for maximizing sensor network lifetime. In: Proceedings of the 38th Hawaii International Conference on System Sciences (2005)
3. Chang, J.-H., Tassiulas, L.: Maximum lifetime routing in wireless sensor networks. IEEE/ACM Trans. Netw. **12**(4), 609–619 (2004)
4. Papadimitriou, I., Georgiadis, L.: Maximum lifetime routing to mobile sink in wireless sensor networks. In: IEEE SoftCom (2015)
5. Keskin, M.E., Altinel, K., Aras, N., Ersoy, C.: Lifetime, maximization in wireless sensor networks using a mobile sink with nonzero traveling time. Comput. J. **54**(12), 871–884 (2011)
6. Yun, Y., Xia, Y.: Maximizing the lifetime of wireless sensor networks with mobile sink in delay-tolerant applications. IEEE Trans. Mob. Comput. **9**(9), 1308–1318 (2010)
7. Khan, A.W., Abdullah, A.H., Razzaque, M.A., Bangash, J.I.: VGDRA: a virtual grid-based dynamic routes adjustment scheme for mobile sink-based wireless sensor networks. IEEE Sens. **15**(1) (2015)
8. Sharma, S., Puthal, D., Jena, S.K., Zomaya, A.Y., Ranjan, R.: Rendezvous based routing protocol for wireless sensor networks with mobile sink. J. Supercomput. **73**, 1168–1188 (2016)
9. Shin, K., Kim, S.: Predictive routing for mobile sinks in wireless sensor networks: a milestone-based approach. J. Supercomput. **62**, 1519–1536 (2012)
10. Luo, H., Ye, F., Cheng, J., Lu, S., Zhang, L.: TTDD: two-tier data dissemination in large-scale wireless sensor networks. Wireless Netw. **11**, 161–175 (2005)
11. Xie, D., Wu, X., Li, D., Sun, J.: Multiple mobile sinks data dissemination mechanism for large scale wireless sensor network. China Commun. **1**, 1–8 (2014)
12. Khan, A.W., Bangash, J.I., Ahmed, A., Abdullah, A.H.: QDVGDD: query-driven virtual grid based data dissemination for wireless sensor networks using single mobile sink. Wireless Netw. **25**(1), 241–253 (2017)
13. Jennic data sheet (Online). https://www.nxp.com/docs/en/datasheet/JN5168-001-MXX.pdf
14. Think speak (Online). https://thingspeak.com/channels/700259/private_show

Quick Search Optimization Algorithm-Based Implementation of Virtual Power Plant for Distribution Network

K. Lakshmi, T. Kesavan, R. Kavin, S. Sheebarani Gnanamalar, M. Senthilkumar, and V. Gomathy

Abstract Virtual power plant (VPP) is one of the developing concepts for integrating of renewable energy source (RES) photovoltaic (PV), air turbines (WT), or integrated heat and power generators as a single energy plant, coordinated and constructed. This paper proposes quick search optimization algorithm based monitoring and control of the virtual power station in the distribution network. The proposed algorithm is used to manage electrical power in the distribution network to reduce the purchased power of the network. This objective is achieved through optimal selection of renewable-based distributed generators, control of load and optimization of energy storage components. In these proposes, two main renewable energy sources of wind power and solar power are integrated with grid to manage the energy in VPP. In this, quick search algorithm is used for forecasting the generate power from windmill and solar cell based on wind circulation and earth temperature, respectively, and also calculating power demand which depends on the load condition. The MATLAB software is used to model the VPP and the performance analysis of generating power of sources and power demand of load.

K. Lakshmi · T. Kesavan (✉) · R. Kavin · S. Sheebarani Gnanamalar · M. Senthilkumar · V. Gomathy
Department of Electrical and Electronics Engineering, Sri Krishna College of Engineering and Technology, Coimbatore, Tamilnadu, India
e-mail: t.kesavan87@gmail.com

K. Lakshmi
e-mail: klakshmi01@gamil.com

R. Kavin
e-mail: kavin882@gmail.com

S. Sheebarani Gnanamalar
e-mail: sheebaranis@skcet.ac.in

M. Senthilkumar
e-mail: senthilkumarm@skcet.ac.in

V. Gomathy
e-mail: gomathyv@skcet.ac.in

P. Suresh et al. (eds.), *Advances in Smart System Technologies*, Advances in Intelligent Systems and Computing 1163,
https://doi.org/10.1007/978-981-15-5029-4_21

Keywords Virtual power plant · Renewable energy sources · Wind power generation system · Distributed energy resources

1 Introduction

Electrical energy is essential for all home appliances, commercial and industrial loads. Electrical power is generated from various conventional energy sources and renewable energy sources. Generated power [1] from various energy sources is not fully transferred and distributed to load side. Conventional Electrical Management System (CEMS) is used to combine some of the renewable energy sources with grid for distributing the power to load side without wastage of electrical energy for continuous power supply [2]. But conventional EMSs faces the challenge of continuously screening, scheduling and balancing the load in distribution side. In conventional energy management system is used to integrate (RES) renewable energy sources with energy storage elements and connected to load. Because all RES are not generating power at all time. But power is required at all time in load side. In this time, grid power supply is connected to load side to reach the power demand [3]. Controlling of generate power from RES is one of the major issues in CEMS. Conventional energy management is needed in continuously monitoring and controlling is required distribution side depends on load condition. Generating power from RES may be connected to energy storage element when load demand is reached level [4]. In this paper, virtual power plant (VPP) is proposed to overcome the drawback of CEMS. VPP is mainly used for continuously monitoring, controlling and scheduling of DER in load side [5].

The main objective of this research is to reduce the usage of grid power which is in load side and decrease the amount of purchasing the power from grid supply. This operation is achieved by correctly selecting the RES to load based on the demand. In this paper, local search algorithm is proposed in VPP for forecasting and distributing the generate power to load side without losses. Generating power from RES is connected to energy storage element, when the power is more than load demand.

In this paper, Sect. 2 consists of general bock diagram and operation of virtual power plant. Section 3 describes scheduling and monitoring concept of LSO. Section 4 explains simulation and results of LSO-based VPP for 24 h. Section 5 has the conclusion of the paper.

2 Virtual Power Plant

It is a new idea for energy management system in distribution side. VPP is used to integrate the various energy sources of wind, solar, grid power and energy storage element. VPP is acting as single power plant to the load [6]. The general block

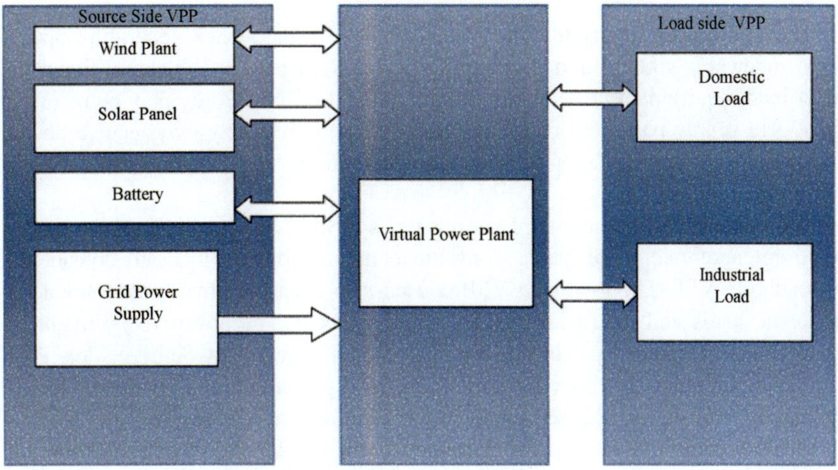

Fig. 1 General block diagram of VPP

diagram of VPP is shown in Fig. 1. The proposed VPP can be classified into source side VPP and load side VPP based on the operation.

Source side VPP consists of some of renewable energy sources like solar PV panel, wind power plant, battery storage and grid power supply. Load side VPP consists of various load VPPs and load side VPP based on the operation [7]. In this paper, domestic loads and industrial loads are considered in load side VPP. The source side VPP is used for monitoring the all RES, battery and grid power supply, and generating power of all RES can be forecasted separately based on different environmental conditions [8].

The wind plant is not generating constant power at all time. Generating power from wind is varied depending on the wind circulation on the place. In wind plant, source side VPP is used to forecast the generating power for the next 24 h based on the previously observed values of wind circulation in area, height of the tower, number and dimensions of blades. Generating power is continuously monitored and distributed to load side based on the load requirement. The wind power may be connected to battery when generating power from wind is more than load requirement. The source side VPP is also used to forecast the generating power from solar panel based on the measured values of inclination angle of solar panel, climatic condition and number of panels. The solar power is monitored by VPP and distributed to load side. Solar power may be connected to energy storage element when generating power is more than load demand. Battery is automatically charging and discharging based on the operation renewable sources and load requirement. The battery is selected based on the renewable sources and load requirement. The battery is charging when renewable sources generate more power compared to power requirement from load. The battery is discharging when load requirement is more than generating power from renewable source [9].

The grid power supply may be connected to load side when generating power from renewable sources and discharging power from battery are not sufficient to meet load requirement [10]. The minimum amount of power only may be received from grid supply based on load requirements. The grid supply may receive power from renewable sources by VPP when RES generates more power compared to load requirement power [11].

The load side VPP consists of various types of load, and it calculates the power requirement of load. In this paper, domestic loads and industrial loads are considered in load side VPP. The load side VPP is used to predict the power requirement of domestic loads and industrial loads for coming 24 h. The power requirement of loads can be predicted by load side VPP based on the previously observed details of load. Various renewable sources, battery and grid power supply may be connected to load by load side VPP based on the power requirement of load [12].

The load side VPP is used for continuously monitoring the power consumption and demand of load. If the power demand is high, the load side VPP is used to connect more number of RES to load based on the power requirements. If the power demand is less, the load side VPP is used to connect minimum number of RES to load, and other renewable sources may be connected to battery and grid power supply [13].

3 Quick Search Algorithm

Quick search (LSO) algorithm is proposed in VPP for forecasting the generating power from various RES and predicting the power requirement based on the load variation. The main objective of LSO is to optimally select the RES to place the load. The working of LSO algorithm is drawn as flowchart as shown in Fig. 2. In flowchart, P_l is power required for load side, P_w is the generating power from windmill, P_s the generating power from solar panel, P_{g1} is the generating power in the first stage.

In this paper, domestic loads and industrial loads are considered in VPP, and the power requirement of both loads can be predicted based on the previous details.

$$P_{\text{Total}} = \sum_{g=1}^{24} P_i + \sum_{g=1}^{24} P_d \tag{1}$$

The total power demand of industrial load and domestic load is expressed in Eq. 1, where P_{Total} is the total amount of power required in distribution side, P_i is the power required for industrial load and P_d is the power required for domestic load. In this procedure, windmill is considered as the nearest power plant to both loads. If generating power from windmill meets the power demand of load, there is no need to search another RES and grid supply.

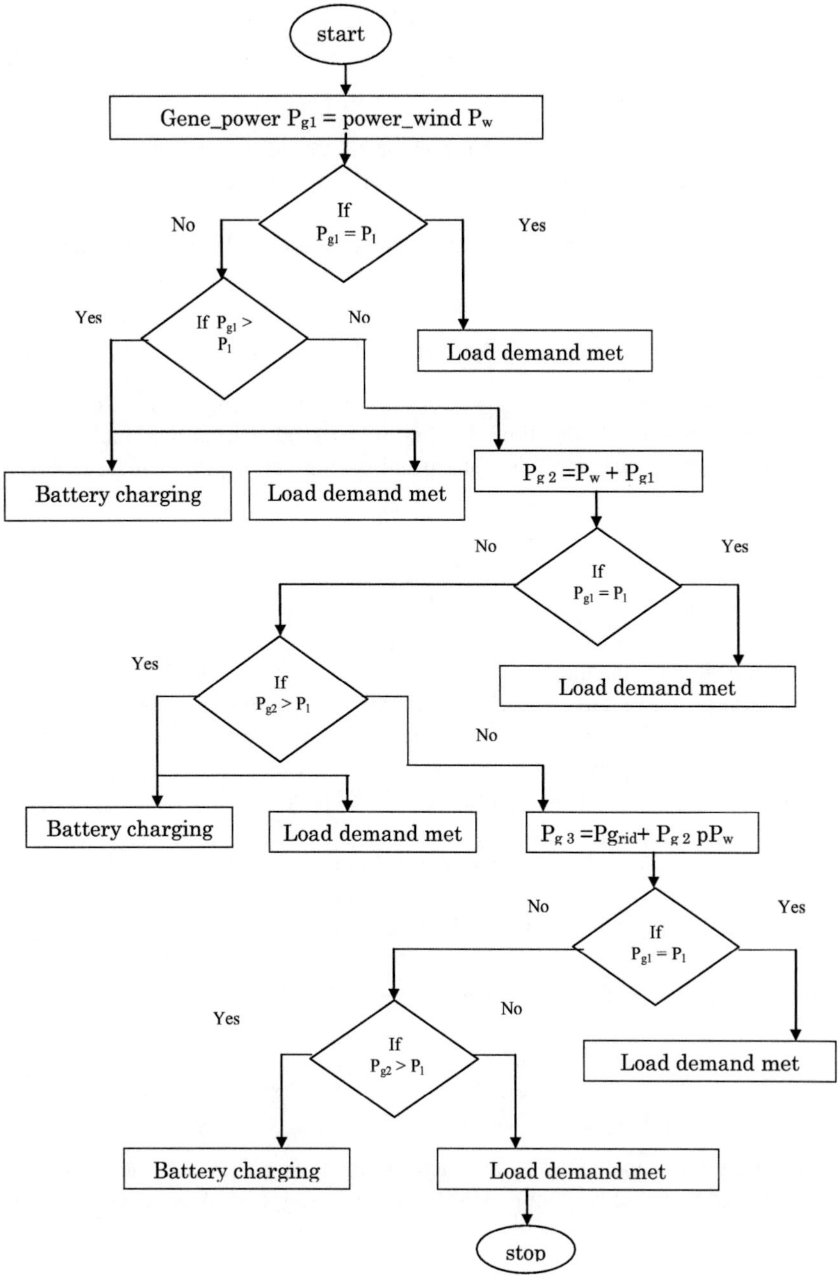

Fig. 2 Flow chart of LSO algorithm

$$P_{g1} = \sum_{gw=1}^{24} P_{gw} \tag{2}$$

If Eq. 2 is satisfied, wind power meets the requirement of both load demands. Other renewable sources of solar power are directly connected to energy storage element. If $P_{g1} > P_{Total}$, and Eq. 3 shows power saved in energy storage element, and it is possible only when the generating power from wind is more than the load demand.

$$P_{SE} = P_{g1} - P_{Total} \tag{3}$$

If Eq. 2 is not satisfied, the power demand of load is higher than the power generating from wind. In this time, LSO searches another local renewable source which is nearest to load side. The solar panel is considered another local RES for both loads.

$$P_{g2} = \sum_{gs=1}^{24} P_{gs} + P_{g1} \tag{4}$$

From Eq. 4, the generating power from solar panel and windmill is connected to load side for power requirement. If Eq. 4 is satisfied, combined renewable sources meet the demand of load power, and grid power supply is not required to load side. Equation 5 shows the power given to the energy storage element; when $P_{g2} > P_{Total}$, generating power is more than requirement of load power.

$$P_{SE} = P_{g2} - P_{Total} \tag{5}$$

The battery supply may be connected to load, if generating power of any renewable source is suddenly varied or reduced due to some environmental condition. The Eq. 4 is shows the combined power of windmill, solar panel and storage element.

$$P_{g3} = \sum_{gs=1}^{24} P_{gs} + \sum_{gw=1}^{24} P_{gw} + P_{SE} \tag{6}$$

The grid power supply may be considered to the load when Eq. 2 is not satisfied, and combined power of windmill and solar panel does not fulfill the power demand of load.

$$P_{g4} = \sum_{gs=1}^{24} P_{gs} + \sum_{gw=1}^{24} P_{gw} + P_{grid} \tag{7}$$

where P_{gw} is the generating power from windmill, P_{gs} is the generation power from solar panel, P_{SE} is the energy stored in energy storage element and P_{grid} is the power getting from grid supply. In Eq. 4, combined power from windmill, solar panel and grid power supply may be connected to load. In LSO algorithm, power demand of load can be solved by renewable sources. But grid power supply may be connected to load side when power requirement of load is more than generating power from renewable sources. In this time, based on the load demand, minimum power only may be connected from grid supply.

4 Simulation and Results

In this paper, load flow study has completed for domestic loads and industrial load for 24 h by VPP. The distribution network of 33 nodes is shown in Fig. 3. General electric board source is connected to all nodes. There are three loads are considered in this network. Load 1 is located in node 10, and it is considered as domestic load.

Load 2 and Load 3 are located in node 22 and 30, respectively, and both are considered as industrial load. Power requirement of Load 1 is predetermined, and 1.25 MW power is averagely needed in Load 1 for 24 h. Load 1 required maximum power of 2.6 MW at time of 17 hour and minimum power of 1.1 MW from the time of 0–1 and 4–5 hours as shown in Fig. 4. Power demand of Load 2 is calculated, and 1.4 MW power is averagely needed for 24 h. Load 2 required maximum power of 2.5 MW at time of 17 and required minimum power 1.1 MW at time of 5. Power demand of Load 3 is predetermined as shown in Fig. 4, and 1.7 MW power is averagely needed for 24 h. The maximum power of load 3 is 2.7 MW and minimum power is 1.4 MW. These three loads are predetermined for 24 h by load side VPP.

The generating power of wind power plant and solar PV panel is forecasted for 24 h by source side VPP. The wind power plant is generating average power of 1 MW for 24 h. Figure 5 shows power output of wind power plant from 0 to 24 h. In morning time, the wind power plant generates power of above 1 MW from 0 to

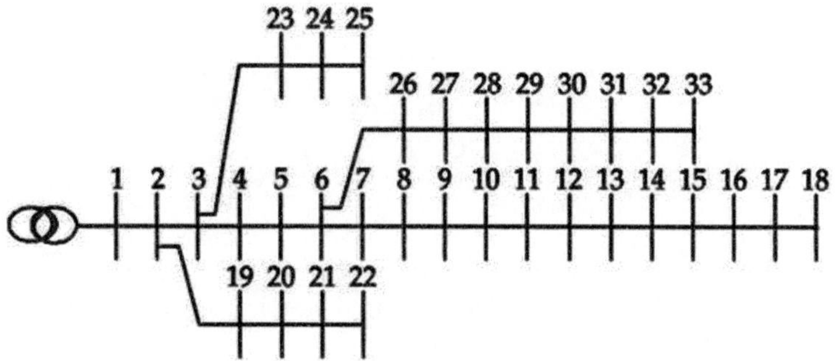

Fig. 3 The distribution network of 33 nodes

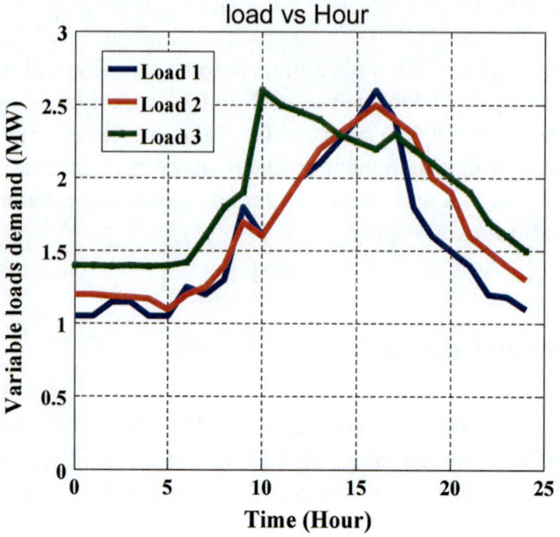

Fig. 4 Power demand curve of various loads

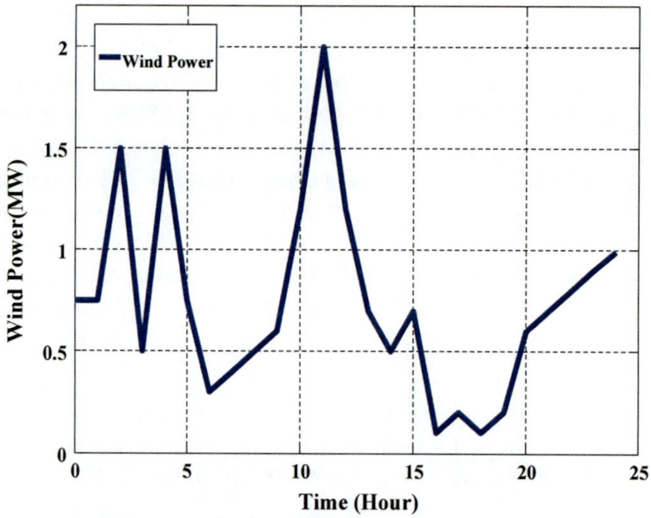

Fig. 5 Characteristics curve of wind turbine unit

5 a.m. and below 0.5 MW from 6 to 8 a.m. and the maximum power of 2 MW has been generated from 10 a.m. to 12 p.m. The wind power plant has generated below 0.2 MW at evening and above 0.5 MW at night time.

The characteristic curve of solar panel in a summer day is shown in Fig. 6. The solar PV panel has generated average of 0.5 MW power for 24 h. In summer day,

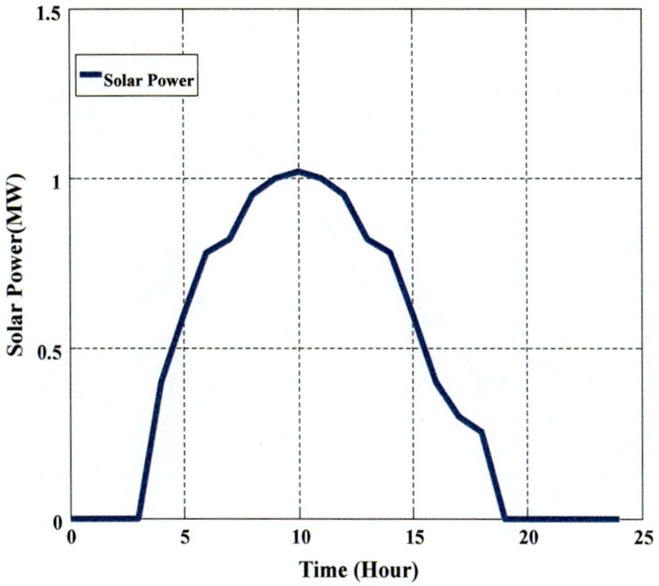

Fig. 6 Characteristic curve of solar panel in a summer day

the solar panel has generated the power from the time of 5–19.30, and it produced maximum power of 1 MW from the time of 10–12.

The power demand of three loads has compensated by load side VPP as shown in figures. Figure 7 shows that required power of Load 1 has satisfied load side virtual power plant. In Load 1, LSO is used to check every minute of load demand and compare with source side VPP. The power demand of Load 1 is 1.2 MW at 12 a.m., and wind plant has generated the power of 0.75 MW at same time.

Power demand of Load 1 is not met by wind power plant and 0.45 MW of extra power is get by grid supply and same procedure has followed to other load requirements. The wind plant and solar panel have generated the power of 2 MW and 1 MW respectively at 10 a.m. and the power requirement of load is 1 MW for same time. Energy storage element gets power of 1 MW from load side VPP. Load 1 required power demand of 2.2 MW at 12 p.m., and wind plant generates 0.5 MW and solar panel produced 0.7 MW for same time. In this instant, insufficient power of 1 MW has received from Energy storage element and the same procedure has followed for 24 h.

Figure 8 shows that required power of Load 2 has satisfied load side virtual power plant. In Load 2, LSO algorithm is used to check every minute of load demand and compare with source side VPP. The power demand of Load 2 is 1.25 MW at 12 a.m., and wind plant has generated the power of 0.75 MW at same time. Load 2 is not fulfilled by wind power plant, and 0.5 MW power is required more from wind power plant, and solar panel is also not may be connected grid power supply for 0.5 MW and same procedure has followed to other load requirements. The wind plant and solar

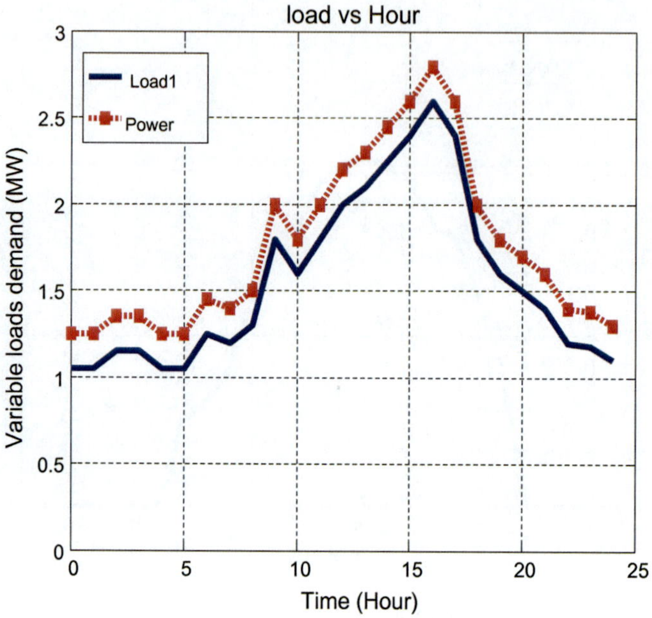

Fig. 7 Required and generated power of Load 1

Fig. 8 Required and generated power of Load 2

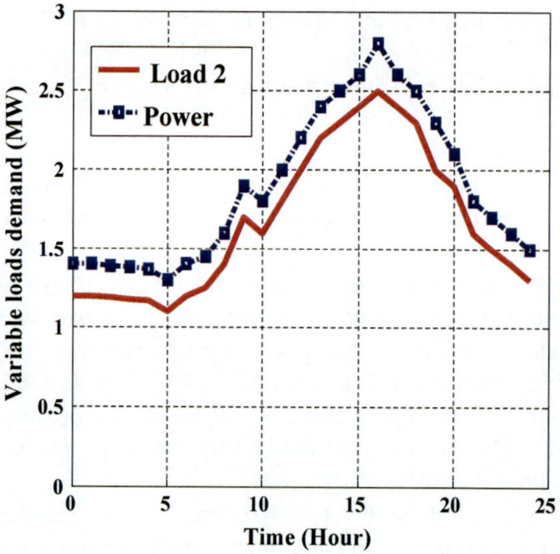

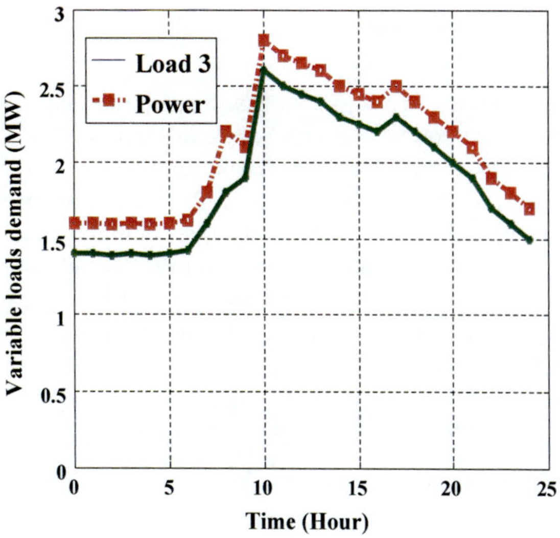

Fig. 9 Required and generated power of Load 3

panel have generated the power of 2 MW and 1 MW, respectively, at 10 a.m., and the power requirement of load is 1.7 MW for same time. Energy storage element gets power of 1.3 MW from load side VPP. Load 2 required power demand of 2.2 MW at 12 p.m., and wind plant generates 0.5 MW, and solar panel produced 0.7 MW for same time. In this instant, insufficient power of 1 MW has received from energy storage element, and the same procedure has followed for 24 h.

Figure 9 shows that required power of Load 3 has satisfied by load side virtual power plant. In Load 3, LSO algorithm is used to check every minute of load demand and compare with source side VPP. The power demand of Load 3 is 1.4 MW at 12 a.m. and wind plant has generated the power of 0.75 MW at same time. Load 3 is not fulfilled by wind power plant and 0.65 MW power is required more from wind power plant and Solar panel is also not may be connected grid power supply for 0.65 MW and same procedure has followed to other load requirements. The wind plant and solar panel have generated the power of 2 MW and 1 MW, respectively, at 10 a.m., and the power requirement of load is 2.6 MW for same time. Energy storage element gets power of 0.4 MW from load side VPP. Load 3 required power demand of 2.4 MW at 12 p.m., and wind plant generates 0.5 MW, and solar panel produced 0.7 MW for same time. In this instant, insufficient power of 1 MW has received from energy storage element and grid power supply, and the same procedure has followed for 24 h. The requirement power of all three loads can be fulfilled without losses and very fast response. LSO algorithm is programmed in load side VPP to retain the generating power of above 0.25 MW compared to required power of load side.

5 Conclusion

The quick search algorithm-based monitoring and controlling of virtual power plant in distribution network is presented in this paper. The electrical energy of distribution side has managed, and purchasing energy from grid supply has minimized. In this paper, wind power plant and solar PV panel have integrated with grid to manage the energy in VPP. Renewable sources are optimally selected and placed in distribution side based on the required power of load. The selection procedure of LSO algorithm has discussed with flow chart. The performance curve of generation power from wind and solar panel is monitored, and power demand of various loads is satisfied using load side VPP.

References

1. Yamuna, S., Kesavan, T.: Harmonic compensation in residential distribution system with Mppt. Int. J. Appl. Eng. Res. **10**(20), 15737–15741 (2015)
2. Othmana, M.M., Hegazyb, Y.G., Abdelaziza, A.Y.: Electrical energy management in unbalanced distribution networks using virtual power plant concept. Electr. Power Syst. Res. **145**, 157–165
3. Liang, Z., et al.: Risk-constrained optimal energy management for virtual power plants considering correlated demand response. IEEE Trans. Smart Grid **10**(2), 1577–1587 (2017)
4. Kesavan, T., Lakshmi, K., Sheeba Rani, S., Kavin, R., Senthilkumar, M.: Design and study of interleaved step up DC converter with high level gain for the application of solar photovoltaic module. Int. J. Recent Technol. Eng. (IJRTE) **7**(6), 1426–1431 (2019). ISSN: 2277–3878
5. Kavin, R., Kesavan, T., Sheebarani Gnanamalar, S., Rameshkumar, K.: Optimal charging and discharging planning for electric vehicles in energy saving system. IEEE Conference Proceedings, 978-1-5386-9533-3ccv, 976–978 (2019)
6. Lakshmi, K., Vasantharathna, S.: Hybrid artificial immune system approach for profit based unit commitment problem. J. Electr. Eng. Technol. **8**, 959–968 (2013)
7. Kavin, R., Kesavan, T., Dr. Nandagopal, V., Malini, T.: Fuzzy based Ev charging with reduced power fluctuation under renewable power consumption constraint. Int. J. Pure Appl. Math. **119**(18), 1691–1706 (2018)
8. Sheeba Rani, S., Kesavan, T., Gomathy, V., Anie Selva Jothi, A.: Effectiveness of pitch control scheme in load balance of WECS. J. Adv. Res. Dynam. Control Syst. **10**(11), 517–523 (2018)
9. Lombardi, P., Powalko, M., Rudion, K.: Optimal operation of a virtual power plant. In: Power and Energy Society General Meeting (2009)
10. Farret, F.A., Simoes, M.G.: Integration of Alternative Sources of Energy. Wiley, Hoboken. ISBN-13: 978-0-471-71232-9 (2006)
11. Nikonowicz, Ł.B., Milewski, J.: Virtual power plants—general review: structure, application and optimization. J. Power Technol. **92**(3), 135–149 (2012)
12. Kavin, R., Kesavan, T., Anbumani, S.: A smart monitoring of faults in power transformers and maintenance based on Wi-Fi. Int. J. Eng. Res. **6**(8), 382 (2017)
13. Bansal, K., Meliss, M.: Renewable Energy Sources and Conversion Technology. Tata Mc Graw Hill

Building a Kernel Image of RTEMS on Host Operating System

A. Susheel and S. Selvendran

Abstract In this paper, we built a kernel image of a real-time operating system (RTOS) using the host operating system. Ubuntu 18.04 LTS is chosen as the host operating system as its performance and support features are better as compared with other operating systems. RTEMS is chosen as real-time operating system because, it is open-sourced, and its performance is comparable with the proprietary real-time operating systems. Building a kernel image of RTEMS can be achieved by the following steps. By source building, kernel image of RTEMS is generated which is compatible with the generic architectures. As the applications depend on the hardware used, we need the kernel image which is compatible with the desired hardware. It is achieved through cross-compilation. The executable files of RTEMS kernel image are generated by make the file process. These files can be ported to the desired hardware and applications can be achieved.

Keywords Real-time operating system · RTEMS · Source building · Cross-compilation · Make file

1 Introduction

There are different kinds of operating systems that serve for a different purpose [1]. In this paper, it is planned to work with time-constraint based applications, so the real-time operating system (RTOS) is preferred [2]. Adhoc techniques were widely used for real-time applications which include US Army missile systems, submarine systems, and robotics [3]. But, because of limitations in the hardware resources like supporting single-processor environment, license policies aren't reliable [4]. To overcome this, hardware resources must be upgraded, so that license policies do not limit the users. This can be achieved by using multiprocessor environment or distributed systems for real-time applications instead of single-processor environment [5]. Working on real-time applications using distributed systems increase

A. Susheel (✉) · S. Selvendran
CVR College of Engineering, R.R District, Telangana, India
e-mail: susheelankem115@gmail.com

P. Suresh et al. (eds.), *Advances in Smart System Technologies*,
Advances in Intelligent Systems and Computing 1163,
https://doi.org/10.1007/978-981-15-5029-4_22

the processing capability and efficiency of the system [6]. To build this system is a complex and complicated process. Using multiprocessor environment for running real-time applications, it is implementable. RTEMS has been chosen as RTOS [7] since it supports multiprocessor environment with rate monotonic real-time scheduler. Further, it is open-sourced, and its performance is comparable with the proprietary RTOS. It supports high-end applications which include missile systems, satellite applications, Rover which was sent by NASA to Mars [8–10]. Applications using RTOS are based on the hardware used. Hence, Raspberry Pi is used as the target hardware [9]. It has quad-core processor and processing capability which is preferred for real-time applications.

Kernel is the lowest level, so RTEMS kernel image file is not available for embedded board. Hence, kernel image for the required embedded board is configured. Further, source building, cross-compilation, and Make File are done to generate the kernel image files for RTEMS for the required board.

2 Implementation

Building a kernel image for RTEMS involves the following three stages:

- Source Building
- Cross Compilation
- Make File.

2.1 Source Building

For different host environments, RTEMS provides pre-built compilers for all RTEMS architectures. In addition to pre-built compilers, by using the tool RTEMS source builder, patch is downloaded, built, and installed in the compiler.

For source building, the following different packages are to be installed:

1. Build Essential
2. Python development
3. Binutils
4. Git.

Build Essential
This package is needed for the building of debian packages. It contains essential information list of packages for building.

Python Development
This package is a dependency package. It depends on debian default python version.

Binutils

These are a group of program design tools that are used for generating and managing binary programs, profile data, object files, libraries and assembly source code.

Git

Git is a directory content manager that was premeditated to knob huge projects such as Linux kernel with rapidity and efficiency. Cloning is done to download files from the git repository. Gnu is an integrated distribution of compilers for several major programming languages. They include C, C++, Ada, Java, etc. Gdb enables to see another program inside while it executes.

Steps Involved in Source Building Installation of Ubuntu prerequisites:

```
$ sudo apt-get install build-essential
$ sudo apt-get install git
$ sudo apt-get install python-dev
$ sudo apt-get build-dep binutilsgcc g++ gdb unzip git//The place where RTEMS tools and projects reside is decided
$ cd $HOME
$ mkdir development
$ cd development
$ mkdirrtems
$ cd rtems
$ mkdir compiler
```

Build the RTEMS ARM cross-compiler

1. Check out the RTEMS source builder tool

```
$ cd $HOME/development/rtems
$ git clone git://git.rtems.org/chrisj/rtems-source-builder.git
```

2. Check all the dependencies of source builder are present

```
$ cd rtems-source-builder//Checking RTEMS Source Builder environment
$ source-builder/sb-check 5.0/rtems-arm
```

3. Add the toolchain bin directory to the path and it is ready to download and build RTEMS

```
$ cd $HOME PATH=$HOME/development/rtems/compiler/5.0/bin:$PATH
```

4. Now, logout and log back in. Open a terminal and type: $ arm-rtems5.0-gcc-v (Fig. 1).

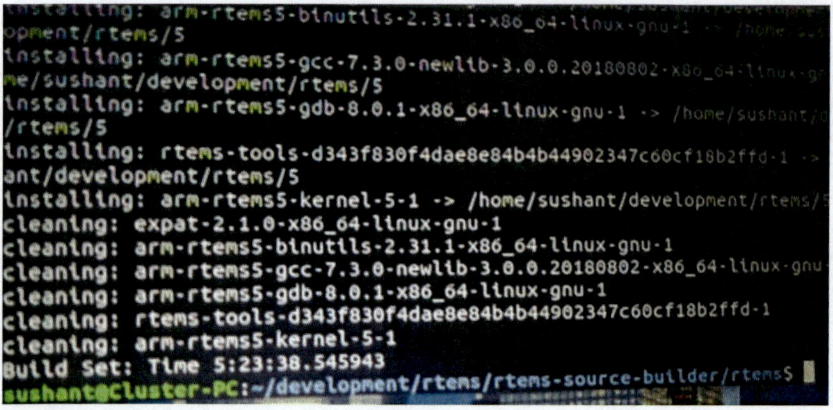

Fig. 1 Source built with build set: time 5 h 23 min 38 s

2.2 *Cross Compilation*

A cross compiler is a compiler skilled in generating executable code for a platform other than the one on which the compiler is running. In this process, the board support packages are generated for the desired hardware. The steps involved for cross-compilation are as follows:

Step 1: To prepare and download RTEMS:

$ cd $HOME/development/rtems
$ git clone git://git.rtems.org/rtems.git rtems-git
$ cd rtems-git
$ ./bootstrap

Step 2: To Configure RTEMS:

$ cd $HOME/development/rtems
$ mkdir build-rtems-rpi
$ cd build-rtems-rpi
$ ../rtems-git/configure–target=arm-rtems5.0\
–enable-rtemsbsp=raspberrypi \
–enable-tests=samples \
–enable-networking \
–enable-posix \
–prefix=$HOME/development/rtems/bsps/5.0 (Figs. 2, 3 and 4).

Fig. 2 Board support packages of generic architectures

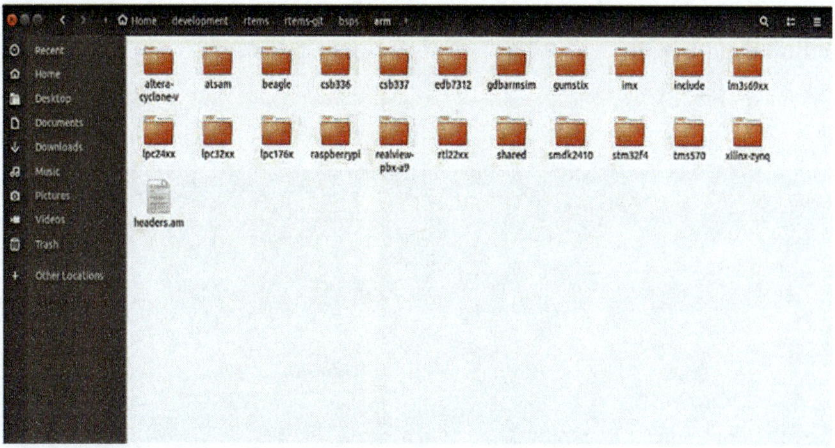

Fig. 3 Board support packages of arm architecture

2.3 *Make File*

In this process, executable kernel files are generated which can be ported into the target hardware. The steps involved to generate the kernel executable files are

target architecture: arm.
available BSPs: Raspberry Pi.
'make all' will build the following BSPs: Raspberry Pi.

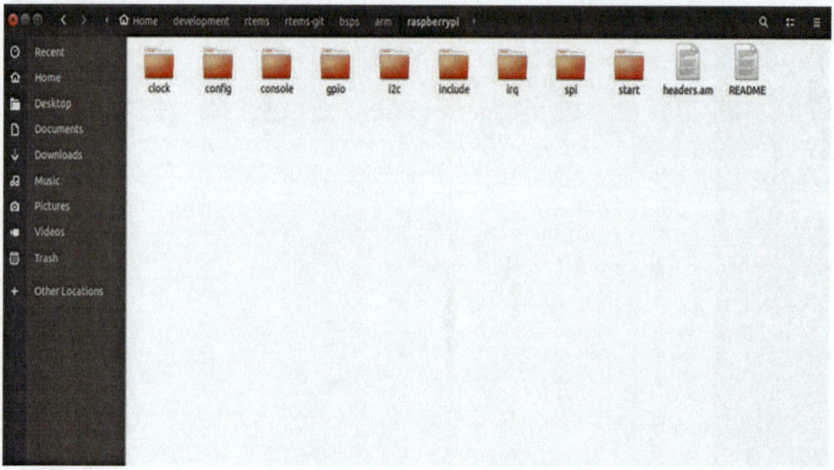

Fig. 4 Board support packages of Raspberry Pi generated after cross-compilation

other BSPs can be built with 'make RTEMS_BSP="bsp1 bsp2 ..."'

To Build and Install

$ make (Fig. 5).

Fig. 5 Executable files generated after make file process

3 Conclusion

Kernel image of real-time operating system is built. RTEMS is chosen as the real-time operating system. It is open-sourced. The features of RTEMS are comparable with the proprietary based real-time operating systems. They support high-end applications. Source building is the first step to generate kernel image for RTEMS. In this process, build tools are generated which is a collection of packages like gcc or gdb. In the second step, cross-compilation is performed to generate the board support packages for the target device. Finally, executable kernel files of RTEMS are generated. These kernel image executable files can be ported into target hardware. Soft and Hard real-time applications are possible after porting these executable files into the Raspberry Pi. Hard real-time applications are ballistic missile systems, rover, aviation simulators, and surgical applications. Soft real-time applications are automatic door opening systems in trains and traffic lights controlling systems. Our future plan is to port these kernel images into Raspberry Pi and run applications on scheduling, to realize the key feature offered by RTOS.

References

1. Stallings, W.: Operating Systems, Internals and Design Principles, 7th edn., pp. 47–101. Pearson, Prentice Hall, Upper Saddle River, NJ (2005)
2. Stankovic, J.A.: Real Time Operating Systems, pp. 237–253. Kluwer Academic Publishers, The Netherlands (2004)
3. Li, Q., Yao, C.: Real-Time Concepts for Embedded Systems, pp. 1–16. CRC Press, Boca Raton (2003)
4. Army Missile Command Redstone Arsenal Al Guidance and Control: DTIC ADA276687: Real-Time Executive for Military Systems C Applications User's Guide, pp. 1–3. Directorate Defense Technical Information Centre (1993)
5. Coulouris, G., Dollimore, J., Kindberg, T., Blair, G.: Distributed Systems: Concepts and Design, 5th edn., pp. 1–3. Pearson Edu. Ltd., Edinburg (2011)
6. Magnoni, L.: Modern messaging for distributed systems. J. Phys. Conf. Ser. **608**(1), 012038 (2015)
7. Straumann, T.: Open source real time operating overview—journal archive. In: 8th International Conference on Accelerator & Large Experimental Physics Control Systems, San Jose, California (2001)
8. Upton, E., Halfacree, G.: Raspberry Pi User Guide. Wiley, Hoboken (2012)
9. Sukvichai, K., Wongsuwan, K., Kaewnark, N., Wisanuvej, P.: Implementation of visual odometry estimation for underwater robot on ROS by using Raspberry Pi 2. In: International Conference on Electronics, Information and Communications (ICEIC), Da Nang, Vietnam (2016)
10. Chakravorty, D., Islam, S., Rana, T.K.: IoT based patient guidance system using Raspberry Pi. In: 2nd International Conference on Electronics, Materials Engineering & Nano-Technology (IMENTech), Kolkata, India (2018)

Compact Planar Monopole UWB MIMO Antenna for Diversity Applications

S. Kolangiammal and G. Vairavel

Abstract A compact four elements planar Monopole UWB MIMO antenna for diversity applications is presented in this paper. The unit cell monopole antenna is designed on 1.6 mm thick FR4 substrate with a compact size of 20 × 20 × 1.6 mm^3 and is fed by a 50 Ω microstrip line. The antenna made to operate between 3.1 and 10.6 GHz by making necessary modifications in the ground plane and the radiator thereby covering the entire UWB range. The unit cell is replicated three times and is placed orthogonally to achieve reduced mutual coupling. The designed antenna achieves mutual coupling less than −20 dB. A four-port diversity antenna is designed by replicating the unit cell and placing it orthogonally to attain high isolation. The overall size $W \times L \times h_s$ of the four-port antennas is 40 × 40 × 1.6 mm^3, where h_s is the height of the substrate.

Keywords Ultra-wideband (UWB) antenna · Diversity · Mutual coupling · MIMO

1 Introduction

The evolution of wireless communication has always been towards a more reliable, greater throughout, and capable of loss resistive for providing the best quality of service. In view of this, Ultra-Wideband (UWB) technology provides a lucrative option especially due to its provision of higher bandwidth and low power spectral density. UWB has been specified within the frequency ranges of 3.1–10.6 GHz having the bandwidth of 7.5 GHz, holding the potential for a more efficient, high data rate communication system. The proposed antenna will be of compact size and will be

S. Kolangiammal (✉)
Department of Electronics and Communication Engineering, SRM Institute of Science and Technology, Chennai, India
e-mail: kolangis@srmist.edu.in

G. Vairavel
Department of Electronics and Communication Engineering, Vel Tech Rangarajan Dr. Sakunthala R&D Institute of Science and Technology, Chennai, India
e-mail: vairavel@veltech.edu.in

P. Suresh et al. (eds.), *Advances in Smart System Technologies*,
Advances in Intelligent Systems and Computing 1163,
https://doi.org/10.1007/978-981-15-5029-4_23

fabricated on the FR4 substrate which can have a thickness of about 1.6 mm low cost and easy availability makes FR4 ideal for using it as substrate. The ε_r value of FR4 substrate is 4.3. One side of the substrate contains the radiating element and the other side will be partially defected ground plane. The antenna's feed line is helped to excite the antenna element. The feed can be micro strip line or coaxial line. The important constraint for antenna designers is the reduction of antenna size. When the antennas are placed close at vicinity, mutual coupling occurs. The main concern in the MIMO antenna design is to decrease the effect of mutual coupling and to increase gain. In [1], upturned L shaped strips employed in the ground plane to reduce the mutual coupling. A tree-like structure is inserted on the ground plane to attain wideband isolation as in [2]. High isolation can be accomplished by inserting parasitic stubs between antenna elements [3], by etching T-shaped slots in the ground plane [4] or by the orthogonal placement of antenna elements [5]. Spacing between the antennas plays a crucial role in increasing the isolation between the diversity antennas if spacing between the antenna is increased the compactness of the antenna is lost. The radiation pattern produced by the antenna is characterized based on S parameters, Envelope Correlation Coefficient (ECC), and Diversity gain (DG). By altering the structure of the ground and the radiating element the inductive and capacitive loading of the antenna can be achieved which helps in the radiation characteristics. In [6], pattern and polarization diversities can be achieved by adding a monopole with pair of ILAs and each ILAs are fed in anti-phase excitations. The UWB antenna in [7], a ground stub was introduced to achieve high impedance bandwidth without using decoupling structures. In [8], metamaterial-inspired isolators can be used to achieve high isolation and minimization. The two elements MIMO antenna in [9], an orthogonal arrangement of antenna elements were employed to obtain polarization diversity and high isolation. In [10], the high isolation can be achieved through a perpendicular arrangement of two printed monopole antenna elements with 3 mm spacing. In [11], the designed array accomplished a reduced mutual coupling and a small envelope correlation coefficient through a folded patch with a supporting wall fed by a folded planar monopole with a 3D U-slot cut. In [12], two antenna elements are excited at different modes in the ground plane and achieves desired radiation patterns and reduced mutual coupling without any additional decoupling structures. In [13], VSWR ≤ 2 and improved impedance performance can be achieved by etching central metal of the patch and inserting a curve in the ground plane just beneath the radiator. The S_{11} of the antenna should be maintained less than -10 dB. Isolation between the antennas should be maintained greater than 15 dB.

2 Proposed Antenna Design

2.1 Unit Cell Design

The top view and back view of the unit cell are shown in Figs. 1 and 2, respectively and the measurements of different parameters are shown in Table 1. The width of the feed is designed based on the analytical line impedance calculation such that 50 Ω impedance is achieved. A 50 Ω impedance is suitable for connecting antennas to feed. One half of the substrate is used for designing the ground plane for proper radiation in higher frequencies. The proposed microstrip fed unit cell antenna is fabricated on FR4 substrate. The ε_r value of FR4 substrate is 4.3 and the loss tangent value of FR4 substrate is 0.02. A copper layer (PEC) of 0.035 mm thickness is deposited on the substrate to form the ground and radiator. The radiating part of the antenna is simple circle with four sub circle slots and ground plane is defected ground structure and it covers half of the substrate. In order to reduce the capacitance, the gap between the ground and the radiator is increased by making semicircle cut in the ground plane. A primary circle of radius 5 mm is the main radiator. A secondary four sub circle slots etched in the radiator. Two sub circle slots at the top of the radiator which have a radius of 1.5 mm and two sub circle slots at the bottom of the radiator which have a radius of 1 mm. The semicircle cut which has a radius of 1.15 mm is inserted in the ground to achieve increased isolation and the enhanced bandwidth. The overall size of the unit cell antenna is $20 \times 20 \times 1.6\ \text{mm}^3$.

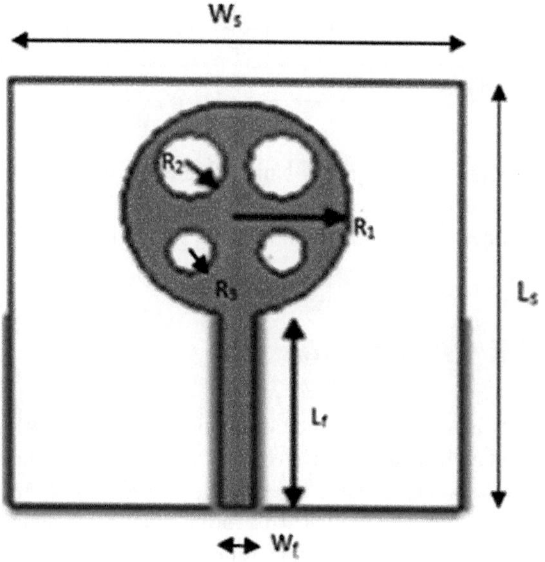

Fig. 1 Top view of unit cell antenna

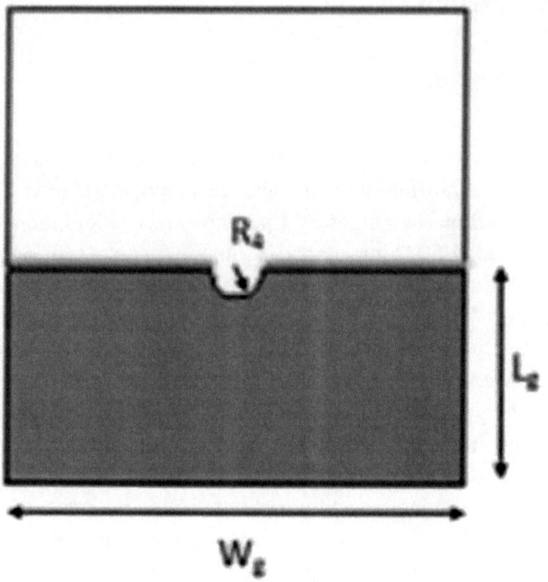

Fig. 2 Back view of unit cell antenna

Table 1 Geometricalparameters of the proposed monopole UWB unit cell antenna

Parameters	W_s	L_s	W_f	L_f	W_g
Units (mm)	20	20	1.5	10	20
Parameters	L_g	R_1	R_2	R_3	R_4
Units (mm)	9	5	1.5	1	1.15

The scattering parameter (S_{11}) of the Unit cell antenna is below −10 dB over the desired Bandwidth and is shown in Fig. 3. The VSWR of the Unit cell antenna is less than 2 throughout the entire bandwidth and is shown in Fig. 4.

2.2 *Monopole UWB MIMO Antenna*

A four-port diversity antenna is designed by replicating the unit cell and placing it orthogonally to attain high isolation. To have good isolation, the scattering parameters of the MIMO antenna should be less than −15 dB. All radiating elements are placed in the same plane and are accomplished by orthogonal arrangement of radiating elements. The overall size $W \times L \times h_s$ of the four-port antenna is $40 \times 40 \times 1.6\ \text{mm}^3$, where h_s is the height of the substrate. This produces the layout for multiple input and multiple output antenna. Different geometries have been proposed in the final antenna design. The top view and back view are shown in Figs. 5 and 6, respectively.

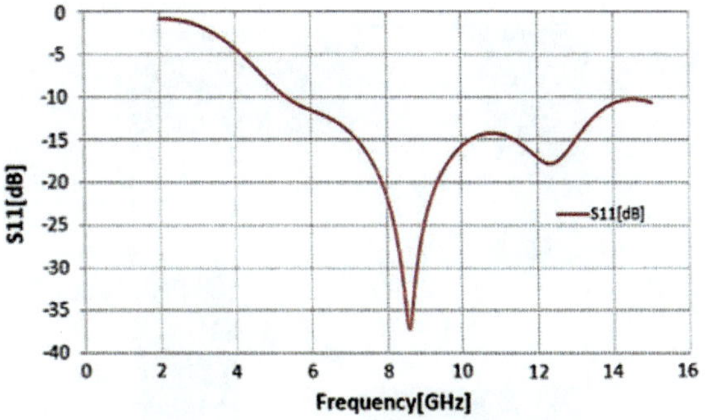

Fig. 3 S_{11} parameters of the monopole UWB unit cell antenna

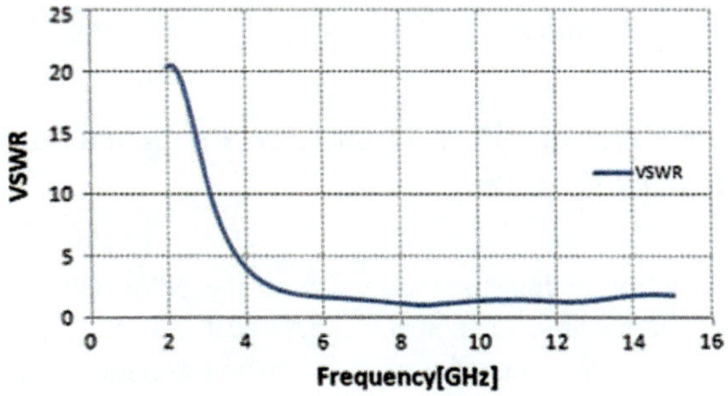

Fig. 4 VSWR of the monopole UWB unit cell antenna

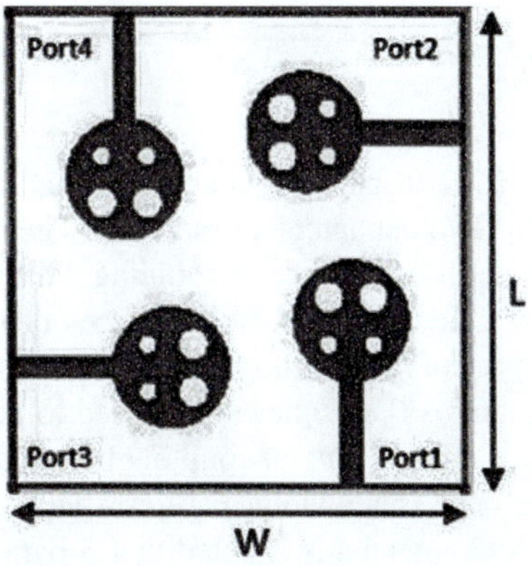

Fig. 5 Top view of MIMO antenna

Fig. 6 Back view of MIMO antenna

3 Analysis of Diversity Performance of Proposed Monopole UWB MIMO Antenna

The proposed diversity antenna is analyzed with the help of their *S* parameters, VSWR, ECC, Diversity Gain, Efficiency, Realized Gain, and Far-field radiation pattern. Simulation and analysis of proposed antenna radiation characteristics are done by using CST STUDIO SUITE.

3.1 Mutual Coupling

The monopole UWB MIMO antenna has four antennas that are close to each antenna will suffer interference due to other antenna's radiation. The measure of this interference in terms of *S* parameter is called mutual coupling. When two or more antennas are kept adjacent to each other there are more chances of mutual coupling which can affect the radiation pattern. The orthogonal arrangement of antenna elements and modified ground structures (DGS) have been used to reduce mutual coupling. The antenna designed here offers a mutual coupling between two antenna elements less than −20 dB for the entire bandwidth. Figures 7 and 8 show the *S* parameter for monopole UWB MIMO antenna, it is noted that *S* parameter for the designed monopole UWB MIMO antenna is obtained well below −20 dB and Fig. 9 shows VSWR for the proposed monopole UWB MIMO antenna. The VSWR is plotted against the frequency for the monopole UWB MIMO antenna which should be ideally less than 2 is obtained from the proposed antenna throughout the entire bandwidth. Thus, the etching of the ground plane with the slot leads to better impedance matching performance and high isolation. Hence the desired result is obtained.

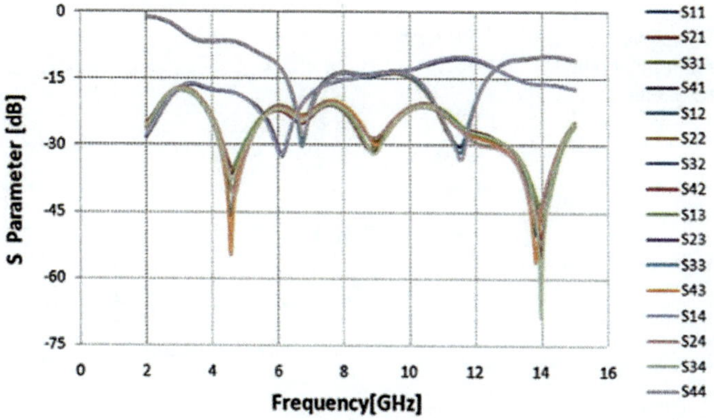

Fig. 7 *S* parameters for monopole UWB MIMO antenna

Fig. 8 *S* parameters for monopole UWB MIMO antenna

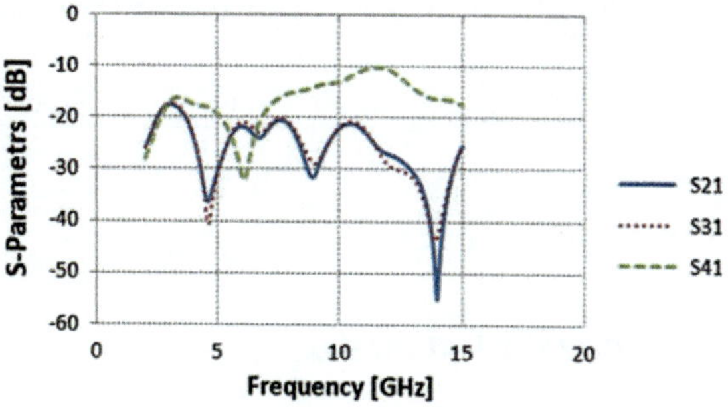

Fig. 9 VSWR of monopole UWB MIMO antenna

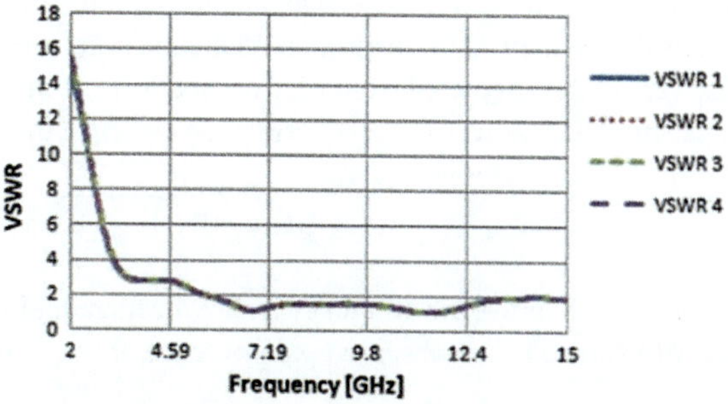

3.2 Envelope Correlation Coefficient (ECC)

The correlation between the radiating elements can be measured from 3D radiation patterns which is the important diversity parameter of MIMO antenna. Since the computation of ECC is complex it can be calculated from *S* parameters. The ECC explains how independent two antenna's radiation pattern is. If the radiation patterns

Fig. 10 ECC versus diversity gain of monopole UWB MIMO antenna

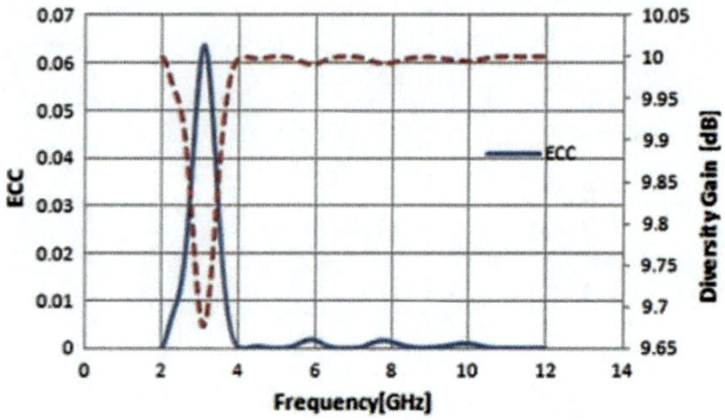

of two antennas are completely independent of each other. The Envelope Correlation Coefficient obtained would be zero but its practical limits are less than 0.5. The ECC can be calculated using *S* parameter formula given in Eq. (1) [14]. It is observed from Fig. 10 that ECC is much less than 0.001 which can be perceived as a good ECC.

$$\mathrm{ECC} = \frac{\left|S_{11}^{*}S_{12} + S_{21}^{*}S_{22}\right|^{2}}{\left(1 - |S_{11}|^{2} - |S_{21}|^{2}\right)\left(1 - |S_{22}|^{2} - |S_{12}|^{2}\right)} \tag{1}$$

3.3 Diversity Gain (DG)

To maintain the reliability and quality of the wireless communication system, the DG of the monopole UWB MIMO antenna should be high and its ideal value is 10 dB. The value of the DG of the proposed antenna is calculated using the formula given in Eq. (2) [1]. It is depicted from Fig. 10 that the maximum value of DG is 9.99999 dB. The simulated value of DG is very much closer to the ideal value.

$$\mathrm{DG} = 10\sqrt{1 - \mathrm{ECC}^{2}} \tag{2}$$

Figure 11 shows the efficiency and realized gain of the proposed monopole UWB MIMO antenna. The observed gain over the entire frequency band from 0 to 4 dB.

Fig. 11 Efficiency and gain of monopole UWB MIMO antenna

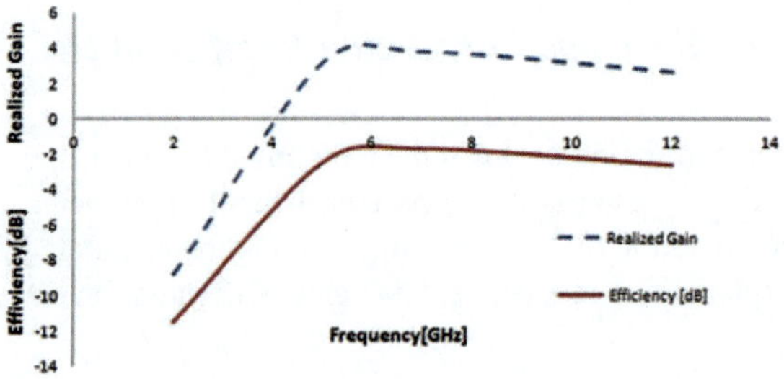

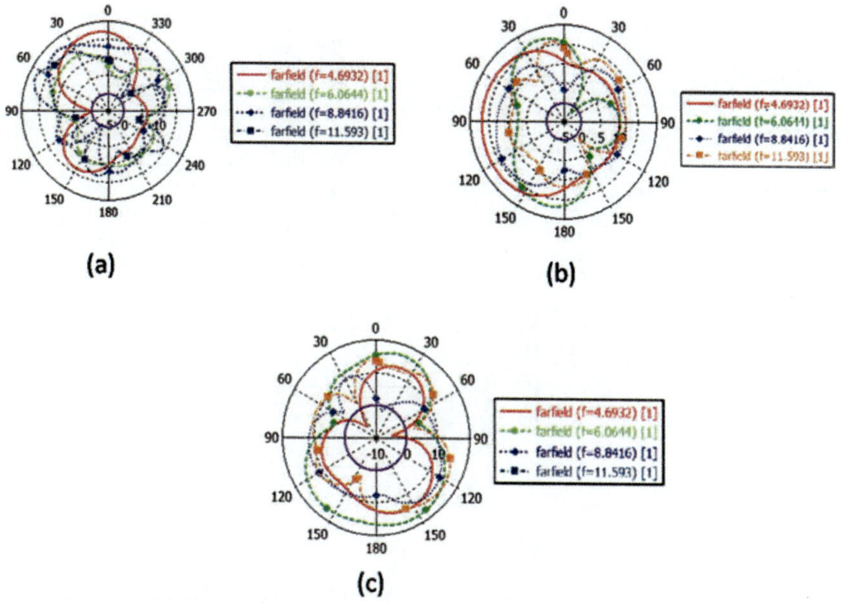

Fig. 12 Simulated radiation pattern when port 1 excited. **a** *XY* plane, **b** *XZ* plane, **c** *YZ* plane

The simulated radiation patterns in *XZ*, *XY*, and *YZ* planes at 4.6932 GHz, 6.0644 GHz, 8.8416 GHz, 11.593 GHz frequencies for the designed monopole UWB MIMO antenna at port 1 and port 2 are depicted in Figs. 12 and 13, respectively. When one port is excited all other ports are terminated by the matched load. Orthogonal arrangement of antenna elements, the radiation patterns in *XY* plane of antenna excited at port 1 and port 2 are orthogonal to each other.

4 Conclusion

An efficient solution to reduce mutual coupling between ultra-wideband antenna is proposed through the orthogonal arrangement of antenna elements. The designed antenna achieves mutual coupling less than −20 dB. A compact radiating antenna optimized for ultra-wideband applications is prepared with a substrate thickness of 1.6 mm and design dimensions of 40 × 40 mm^2. Employing diversity techniques through the orthogonal placement of antenna elements and compact antenna array is attained.

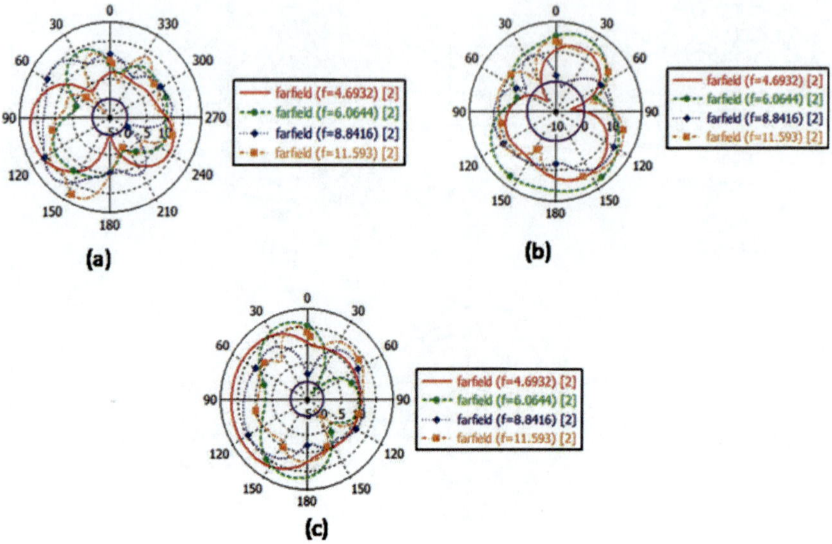

Fig. 13 Simulated radiation pattern when port 2 excited. **a** *XY* plane, **b** *XZ* plane, **c** *YZ* plane

References

1. Chandel, R., Gautam, A.K., Rambabu, K.: Tapered fed compact UWB MIMO-diversity antenna with dual band-notched characteristics. IEEE Trans. Antennas Propag. **66**(4), 1677–1684 (2018)
2. Zhang, S., Ying, Z., Xiong, J., He, S.: Ultrawideband MIMO/diversity antennas with a tree-like structure to enhance wideband isolation. IEEE Antennas Wirel. Propag. Lett. **8**, 1279–1282 (2009)
3. Liu, L., Cheung, S.W., Yuk, T.I.: Compact MIMO antenna for portable UWB applications with band-notched characteristic. IEEE Trans. Antennas Propag. **63**(5), 1917–1924 (2015)
4. Mao, C.-X., Chu, Q.-X.: Compact coradiator UWB MIMO antenna with dual polarization. IEEE Trans. Antennas Propag. **62**(9), 4474–4480 (2014)
5. Zhang, S., Lau, B.K., Sunesson, A., He, S.: Closely-packed UWB MIMO/diversity antenna with different patterns and polarizations for USB dongle applications. IEEE Trans. Antennas Propag. **60**(9), 4372–4380 (2012)
6. Wang, X., Feng, Z., Luk, K.-M.: Pattern and polarization diversity antenna with high isolation for portable wireless devices. IEEE Antennas Wirel. Propag. Lett. **8**, 209–211 (2009)
7. Wu, Y., Ding, K., Zhang, B., Li, J., Wu, D., Wang, K.: Design of a compact UWB MIMO antenna without decoupling structure. Int. J. Antennas Propag. **2018** (2018)
8. Wang, F., Duan, Z., Li, S., Wang, Z.-L., Gong, Y.-B.: Compact UWB MIMO antenna with metamaterial-inspired isolator. Prog. Electromagnet. Res. **84**, 61–74 (2018)
9. Zhu, J., Li, S., Feng, B., Deng, L., Yin, S.: Compact dual polarized UWB quasi-self-complementary MIMO/diversity antenna with band-rejection capability. IEEE Antennas Wirel. Propag. Lett. **15**, 905–908 (2016)
10. Koohestani, M., Moreira, A.A., Skrivervik, A.K.: A novel compact CPW-fed polarization diversity ultra wideband antenna. IEEE Antennas Wirel. Propag. Lett. **13**, 563–566 (2014)
11. Jiang, Z.H., Zhang, L., Zhang, Y., Yu, C., Cai, L., Zheng, S., Hong, W.: A compact triple band antenna with a notched ultra wideband and its MIMO array. IEEE Trans. Antennas Propag. **66**(12), 7021–7031 (2018)

12. Zhao, X., Yeo, S.P., Ong, L.C.: Planar UWB MIMO antenna with pattern diversity and isolation improvement for mobile platform based on the theory of characteristic modes. IEEE Trans. Antennas Propag. **66**(1), 420–425 (2018)
13. Alsath, M.G.N., Kanagasabai, M.: Compact UWB monopole antenna for automotive communications. IEEE Trans. Antennas Propag. **63**(9), 4204–4208 (2015)
14. Tian, R., Lau, B.K., Ying, Z.: Multiplexing efficiency of MIMO antennas. IEEE Antennas Wirel. Propag. Lett. **10**, 183–186 (2011)

A Tree-Shaped Wearable Conductive Fabric Patch Antenna for ISM Band Applications

T. Annalakshmi, S. Ramesh, and M. Swetha Lakshmi

Abstract In today's world, women safety has become a great concern in most societies. In this paper, we present an innovative conductive fabric antenna that can be used for tracking purpose to ensure safety of women traveling alone. Precise location of the traveling woman can be tracked by their friends or family, and immediate help could be dispatched in case of any emergency. The novelty of the designed antenna is that it is made in attractive shape and can be easily assimilated into clothes. The designed antenna operates in ISM band with center frequency of 2.4 GHz. It provides a reasonable return loss of −17.8 dB with maximum gain of 5.4 dB at 2.4 GHz. Due to its appealing shape and flexibility, the antenna can be worn without being obtrusive and can blend with the design of dresses.

Keywords Tree shape · Conductive fabric · ISM band · Return loss · Bending performance

1 Introduction

Currently, various compact gadgets are consistently structured because of scaling down. One such electronic improvement is wearable that can create, transmit, tweak, and identify electrons while continually remaining joined on an individual. Wearable antenna has developed from different progressions in innovation. Wearable antenna encourages keen correspondence without being prominent [1]. They can be entwined with regular garments which make them simple to wear [2]. They should comprise great sturdiness independent of mileage and bowing of the textures. The littler size, lightweight, and ideal expense of wearable antenna make them appealing and most

T. Annalakshmi (✉) · M. Swetha Lakshmi
New Prince Shri Bhavani College of Engineering and Technology, Chennai, India
e-mail: lakshmishanmu15@gmail.com

S. Ramesh
SRM Valliammai Engineering College, Chennai, India

P. Suresh et al. (eds.), *Advances in Smart System Technologies*,
Advances in Intelligent Systems and Computing 1163,
https://doi.org/10.1007/978-981-15-5029-4_24

looked for after innovation in the correspondence advertise [3]. Wearable antenna can be actualized in a wide scope of ventures for an assortment of purposes.

2 Antenna Design and Implementation

In this paper, a flexible conductive fabric antenna is designed. The antenna uses conductive fabric as the patch material. Conductive fabrics help in the development of colorful antennas which are lightweight and robust and provide equally good performance when compared with classic antennas. The conductivity of the fabric is 2.5×10^5 S/m, and the thickness is about 0.08 mm. Denim jean materials with a relative permeability of 1.7 and thickness of 1 mm are used as a substrate [4]. Lower the relative permeability of substrate, higher the performance of antenna [5]; hence, the denim jean material has been chosen. Conductive fabric used for patch is also used to construct the ground. Tree antenna dimensions are calculated from the Eqs. (1) and (2). The desired center frequency for the antenna is chosen as f_r. The thickness of chosen substrate is denoted by *h*, and ε_r is the relative permeability of the substrate.

$$a = \frac{F}{\left\{1 + \frac{2h}{\pi \varepsilon_r F}\left[\ln\left(\frac{\pi F}{2h}\right) + 1.7726\right]\right\}^{1/2}} \tag{1}$$

$$F = \frac{8.791 \times 10^9}{f_r \sqrt{\varepsilon_r}} \tag{2}$$

The antenna consists of rectangular-shaped substrate on which tree-shaped patch is designed. The tree patch is made up of four concentric circles each with a radius of 22.8 mm. To increase gain and bandwidth, a defective ground structure made of a hollow rectangular slot is constructed. The wider feed strip lowers characteristic impedance [6], so the designed antenna is fed with a feed of 4 mm width. Table 1

Table 1 Measurements of conductive fabric tree antenna

Parameters	Size in mm
Circular patch radius (a)	22.8
Patch thickness	0.08
Substrate length	130
Substrate width	120
Substrate thickness	1
Ground thickness	0.08
Ground slot outer square length × width	80 × 80
Ground slot inner square length × width	68 × 68

describes the complete measurements of tree-shaped antenna. The ground is made as defective ground by making the square slot with a thickness of 6 mm. The defective ground improves the return loss.

The designed structure of the conductive fabric tree antenna is shown in Fig. 1. Figure 1a, b depicts the patch and defective ground structures of simulated antennas. Figure 1c, d shows the patch and ground configurations of fabricated antenna.

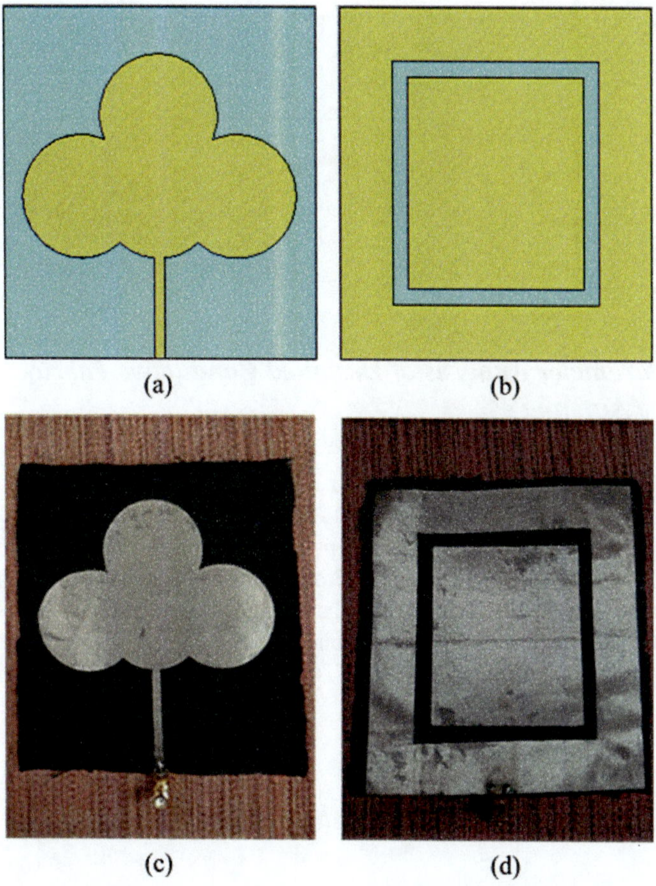

Fig. 1 Structure of conductive fabric tree antenna: **a** patch of simulated antenna, **b** defective ground of simulated antenna, **c** patch of fabricated antenna, and **d** defective ground of fabricated antenna

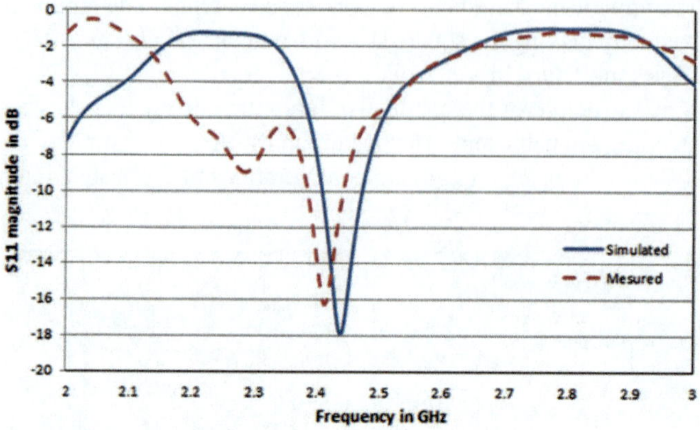

Fig. 2 Return loss of simulated and measured tree antenna

3 Antenna Performance

3.1 S-Parameter Analysis of Designed Conductive Fabric Tree Antenna

The return loss of designed tree antenna is depicted in Fig. 2. At 2.4 GHz, the tree antenna produces a return loss of −17.8 dB for simulation and −16.2 dB for measurement of prototype antenna. The predicted graph in Fig. 2 shows that the simulated frequency and S11 magnitude are almost matched with the measured prototype antenna. The antenna covers the bandwidth of the ISM band which supports low-power wireless transmission. The ground is made as defective ground by making the square slot with a thickness of 6 mm. The defective ground improves the return loss and bandwidth [7].

3.2 Radiation Pattern of Tree Antenna

Figure 3 represents the 3D radiation pattern of the tree antenna. At 2.4 GHz, the tree antenna produces an optimum gain of 5.2 dB.

The E-plane and H-plane patterns of the tree antenna for 2.4 GHz frequency are shown in Fig. 4. The designed tree antenna radiates in a bi-directional pattern with high gain in the selected application band 2.4 GHz.

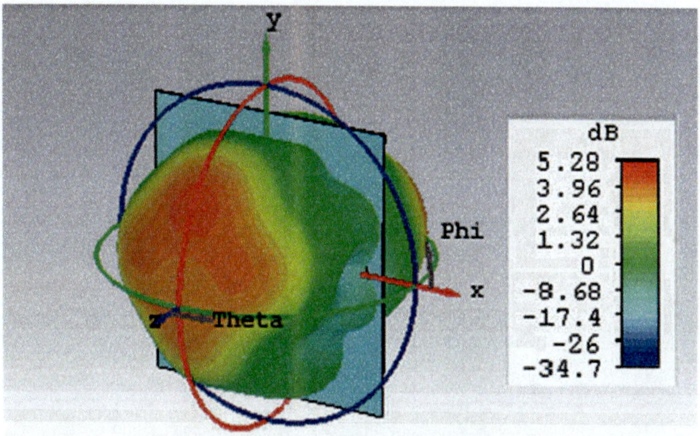

Fig. 3 Radiation pattern of tree antenna

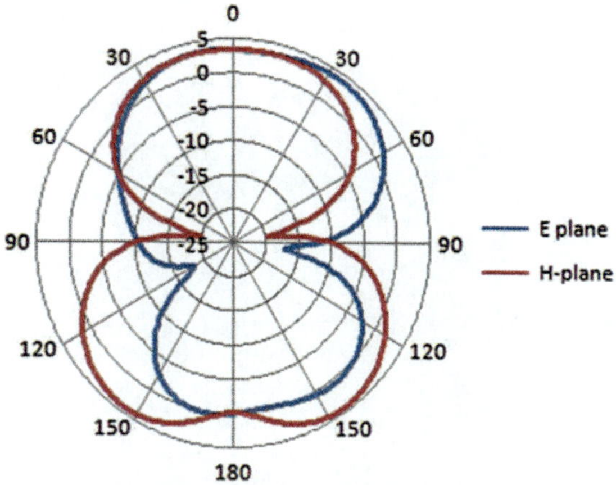

Fig. 4 E-plane and H-plane patterns of tree antenna

3.3 Bending of Designed Conductive Fabric Antenna

The bending of conductive fabric antenna is studied at different angles of 40°, 60°, and 90°. Figures 5, 6, 7, and 8 depict the antenna structures for corresponding bending angles. The designed conductive fabric antenna is flexible, and the structure of the antenna remains intact irrespective of the bending effects.

Fig. 5 0° bending

Fig. 6 40° bending

3.4 Performance Analysis of Designed Conductive Fabric Antenna Under Different Angles of Bending

S-parameter comparison of the tree antenna for various angles of bending is represented in Fig. 9. At lower bending angles, the antenna covers greater spectrum bandwidth than at higher bending angles. For the bending angles of 40° and 60°, the designed antenna covers the desired ISM band, but for higher bending angle of 90°,

Fig. 7 60° bending

Fig. 8 90° bending

the resonant frequency gets a bit shifted to 2.3 GHz. However, for both higher and lower bending angles, the designed antenna provides good impedance matching with return loss lower than −10 dB.

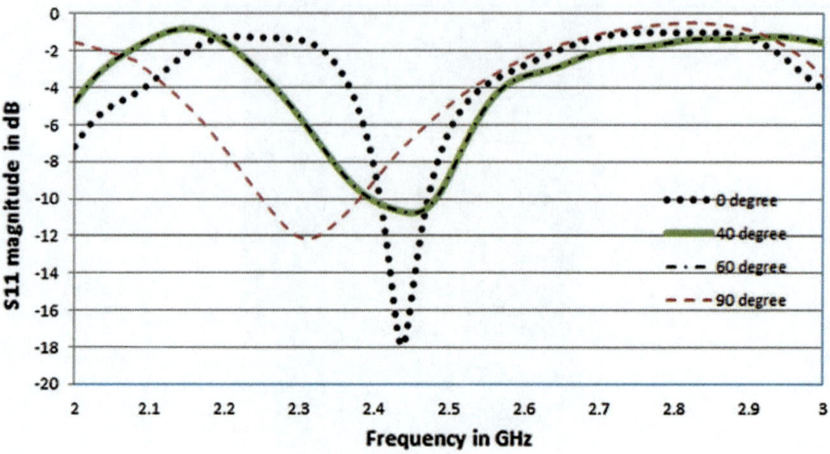

Fig. 9 Evaluation of return loss of tree antenna for various bending angles

3.5 *Gain Result of Designed Conductive Fabric Antenna*

The designed conductive fabric tree antenna gives a good gain over a wide range of frequencies. The antenna provides an optimal gain of 5 dB over the entire bandwidth, and at 2.4 GHz, it produces a gain of 5.4 dB. The simulated gain of the designed conductive fabric tree antenna is shown in Fig. 10.

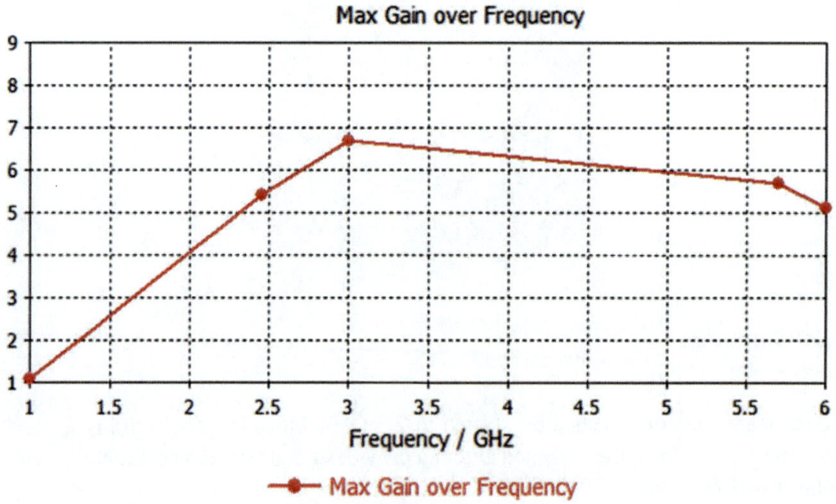

Fig. 10 Simulated gain of conductive fabric tree antenna

4 Conclusions

A novel, conductive fabric antenna in the attractive shape of tree was presented in this paper especially for babies and women's tracking purpose. The tree-shaped antenna resonates at 2.4 GHz frequency corresponding to ISM band and produced a low return loss of −16 dB. It provides good impedance matching irrespective of bending effects. The proposed tree antenna works well even in the bending for different angles with slight deviation in the resonated frequency. The tree antenna radiates in omni-direction with an optimum gain of 5.4 dB at designed frequency. The applications of ISM band antenna widely covered with near-field communication (NFC) devices, monitoring purpose for women's safety, wireless computer networks (Wi-Fi), and ext. The overall performance and structure of the antenna make it suitable for integrating it in clothes for everyday use.

References

1. Wang, F., Arslan, T.: Body coupled monopole UWB antenna for wearable medical microwave imaging application. In: IEEE-APS Topical Conference on Antennas and Propagation in Wireless Communications (APWC), pp. 146–149 (2017)
2. Mendes, C., Peixeiro, C.: On-body transmission performance of a novel dual-mode wearable micro strip antenna. IEEE Trans. Antennas Propag. **66**, 4872–4877 (2018)
3. Zhong, J., Lee, C.W., Papantonis, D., Kiourti, A., Volakis, J.L.: Body-worn 30:1 bandwidth tightly coupled dipole array on conductive textiles. IEEE Antennas Wirel. Propag. Lett. **17**(5), 723–726 (2018)
4. Yahya, R., Kamarudin, M.R., Seman, N.: Effect of rainwater and seawater on the permittivity of denim jean substrate and performance of UWB eye-shaped antenna. IEEE Antennas Wirel. Propag. Lett. **13**, 806–809 (2014)
5. Balanis, C.A.: Antenna Theory: Analysis and Design, 3rd edn. Wiley, New York (2009)
6. Kiourti, A., Volakis, J.L.: High-geometrical-accuracy embroidery process for textile antennas with fine details. IEEE Antennas Wirel. Propag. Lett. **14**, 1474–1477 (2015)
7. Poffelie, L.A.Y., Soh, P.J., Yen, S., Vandenbosch, G.A.E.: A high-fidelity all textile UWB antenna with low back radiation for off-body WBAN applications. IEEE Trans. Antennas Propag. **64**(2), 757–760 (2016)

Experimental Validation of Magnetic Field for Three-Phase Cables After Conductor Splitting and Phase Mixing

P. Damodaren, R. Rengaraj, D. Rohit, G. R. Venkatakrishnan, and S. Santhoosh Aravind

Abstract Electricity plays a vital role in our day-to-day life. When electricity flows, close to the cables, both electric and magnetic fields exist close to the appliances. In electric cables, the extremely low-frequency magnetic fields (ELF-MF) are highly pronounced as they are inevitable in a power system. ELF magnetic fields' exposure has increased public concern on potential health risks. The biological effects on exposure to ELF magnetic fields which include leukemia, brain tumor, neurological effects, and cardiovascular effects have unavoidably led to imposing legal limits. Such appalling side effects of ELF magnetic fields have made it an exigent issue. The solutions found in literature are active shielding, conductor splitting, twisting, and phase mixing. However, to achieve the objective magnetic field (<1 μT), this problem has to be tackled from the design methodology. Unlike the design methods extended in the literature, evolutionary algorithms are used to determine the optimal amount of conductor splitting and their best possible arrangement in a linear manner. The resulting optimized conductor configuration producing the least magnetic field is experimentally validated using three-axis magnetometer. This idea can be further utilized to develop extremely low-frequency magnetic field (ELF-MF) cables of different shapes and sizes with better aesthetic.

Keywords Extremely low-frequency magnetic field (ELF-MF) · Conductor splitting · Phase mixing · 3-axis magnetometer · Modified genetic algorithm (MGA)

P. Damodaren (✉) · D. Rohit
Siechem Wires and Cables, Chennai, India
e-mail: pd@siechem.com

R. Rengaraj · G. R. Venkatakrishnan
SSN College of Engineering, Chennai, Tamil Nadu, India
e-mail: rengarajr@ssn.edu.in

S. Santhoosh Aravind
Department of EEE, SSN College of Engineering, Chennai, Tamil Nadu, India
e-mail: santhooshsundar@gmail.com

P. Suresh et al. (eds.), *Advances in Smart System Technologies*,
Advances in Intelligent Systems and Computing 1163,
https://doi.org/10.1007/978-981-15-5029-4_25

1 Introduction

At the moment, electricity is undoubtedly the most significant and indispensable commodity. Researchers and engineers work relentlessly to obtain the most efficient methods of producing, transmitting, and distributing the electricity [1–3]. The most intriguing problem with the power cables is the production of extremely low-frequency magnetic field (ELF-MF). In the electromagnetic spectrum, ELF-MF occupies the lower part that is 0–3000 Hz [4–7].

At home, environment and workplace peoples are exposed to ELF-MF due to the widespread use of electricity. Continuous exposure to ELF-MF has proved to cause a plethora of health effects [8]. The harmful effects of ELF-MF have proven to cause childhood leukemia and neurological diseases as per the epidemiologic studies. In addition, magnetic field exposure causes electromagnetic hypersensitivity which in turn leads to physiological distress [9]. Recent studies show that ELF-MF can alter the sleeping habits. Change in Melatonin levels caused due to ELF-MF is supposed to be carcinogenic. ELF-MF produces adverse biological effects on humans like bone marrow stromal cells differentiation, neurological, and cardiovascular disorders [10]. In order to ensure safety from exposures to magnetic fields, the results obtained from International Commission for Non-ionizing Radiation protection (ICNIRP) and IEEE guidelines are adopted by most of the countries. The appalling side effects of ELF-MF discussed above make it an exigent issue.

ICNIRP has imposed exposure limits on these electric and magnetic fields for both workers and members of the general public [11–15]. Most countries adopt these limits from the international guidelines as their own regulatory framework. Biological studies reveal that prolonged exposure to ELF-MF leads to acute effects on the central nervous system tissues. The detailed survey results obtained from several studies show that there was an increased prevalence in childhood leukemia associated with time-averaged magnetic fields greater than 0.4 μT (Table 1).

The magnetic fields from 50 Hz power cables must be under the ICINIRP exposure limit. The attention value limit has to be employed in children's playgrounds, residential buildings, and schools where people are staying for 4 h or more per day. The objective limit has to be achieved in modern power cables, so that they are called as extremely low magnetic field "Green Cables."

The Green Cables promise to uphold the conceptions of sustainability as they are harmless and environmentally friendly. In a densely populated country like India, it is saddening to know for the fact that the Green Cable technology is not present yet [16, 17]. The main challenge in establishing this idea is to produce them in a cost effective manner.

Table 1 ICINIRP regulatory guidelines

ICINIRP regulatory guidelines	Magnetic fields (μT)
Exposure limit	100
Attention value limit	10
Objective limit	3

Humans are being exposed to electric and magnetic fields produced by both natural and man-made sources. The magnitude of the fields from man-made sources has a greater magnitude (several orders) when compared to the fields from natural sources; it requires immediate attention in order to reduce them. It has now become important to assess the intensity and control the magnetic field under power lines especially at the vicinity of the factories, schools, and houses which should keep the emissions under the permissible values. In addition to these, to deploy communication equipment such as telephone trans-receivers, the power towers are increasingly used. The people who work there, installing and maintaining such equipment, experience higher field exposures which are extremely dangerous to the health. The most intriguing problem with power cables is the production of extremely low-frequency magnetic field (ELF-MF).

The operating frequency of electric power cables is around 50 or 60 cycles per second, or hertz (Hz). In certain appliances, the magnetic field values can be in the order of a few hundred microtesla (μT) [18–20]. The magnetic fields of underneath power lines can be about 20 μT, and for electric fields, it can be several thousand volts per meter. The residential power-frequency magnetic fields in an average at homes are much lower of about 0.07 μT in Europe and 0.11 μT in North America. Mean values of the electric field in the home are up to several tens of volts per meter. Electric cables in the forthcoming years are expected to achieve a very low magnetic field at their vicinity and are set with an objective limit of 3 μT [21]. However, to achieve the objective magnetic field (<1 μT), this problem has to be tackled from the design methodology. The solutions found in literature are passive and active shielding, conductor splitting, twisting, and phase mixing. Passive shielding uses ferromagnetic shields and thick eddy current shields to shield static and ELF-MF. However, the materials required for ferromagnetic shield are expensive and very heavy. The alternate method of active shielding uses magnetic fields to superpose the disturbing magnetic field having the same magnitude but in opposite direction. Despite its efficiency, high current requirement at power frequency appears as a major drawback for this method. The other popular method used is cable twisting. When the cable conductors are bulky, twisting is not practically possible, and it was also observed that twisting tends to reduce magnetic field at some regions, whereas it seemed to have increased at other regions.

Conductor splitting and phase mixing are cost-effective and efficient solutions. Their efficiency is supposed to be higher than the stated methods because they have to be implemented from the initial design. The design has to be done carefully so as to determine the correct position for the split conductors in the arrangement without violating the ampacity and thermal limit of the cable.

Determining Configuration Using Modified Genetic Algorithm

Due to the process of conductor splitting, the number of subconductors increases and the permutation of the conductor arrangement increases. Thus, it is difficult to calculate the magnetic field for all arrangement of conductors. MGA promises to minimize this convolution and provides the optimal arrangements with minimum magnetic field.

Table 2 Total configuration for various values of N_s (in numbers)

N_s (subconductors used)	Total permutations
2	90
3	1680
4	34,650
5	756,756

$$\text{Total number of permutations} = \frac{N_t!}{(N_s!)^n}$$

where n represents the number of phases. Each phase conductor is split into "N_s" subconductors.

Thus, $N_t = n \times N_s$ represents the total number of conductors in the arrangement (Table 2).

2 Experimental Setup

Experiment is conducted to measure the magnetic field for each one of the cable configuration obtained from MGA, with the help of 3-axis magnetometer. Lamp load is connected to the three-phase cables which draw 4.2 A current per phase, and the magnetic field is calculated by keeping the 3-axis magnetometer at a distance of 5 cm from the center of cable configuration. Similarly, the magnetic field for different cable configuration after conductor splitting is obtained using the same setup. The equipment used in the experiment are:

3×4 mm^2 cables for no splitting configurations (1×4 mm^2 per phase);
6×2.5 mm^2 cables for two splitting configurations (2×2.5 mm^2 per phase);
9×1.5 mm^2 cables for three splitting configurations (3×1.5 mm^2 per phase);
12×1 mm^2 cables for four splitting configurations (4×1 mm^2 per phase);
SENSYS GmbH 3-axis magnetic field sensor, FGM3D/100;
SENSYS FGM3D TD spectrum analyzer;
three-phase lamp load.

The voltage was set to 230–240 V, and the load was set to 2400 W such that approximately 4.2 A current flows through each phase. The disturbance caused due to minute fluctuations in the voltage level is minimal, and thus, they are neglected. While the magnetometer is placed near the cable configuration, the magnetic field is noted down from the peak value of graph shown in spectrum analyzer, and it is found that as the distance between the center of the cable and the magnetometer increases, the magnitude of magnetic field decreases (Figs. 1, 2 and 3).

Fig. 1 Test bench

Fig. 2 Lamp load

Fig. 3 Three-axis magnetometer

3 Results and Discussion

3.1 No Splitting

1 × 4 mm^2 cable is used for each phase carrying a current of 4.2 A. Magnetic field due to the three-phase cable consisting of 4 mm^2 conductor per phase is 1.22 μT (Fig. 4).

3.2 Two Splitting

2 × 2.5 mm^2 cable is used for each phase carrying a current of 4.2 A. Table 3 shows the magnetic field values for the various configurations obtained through two conductor splitting and phase mixing (Fig. 5).

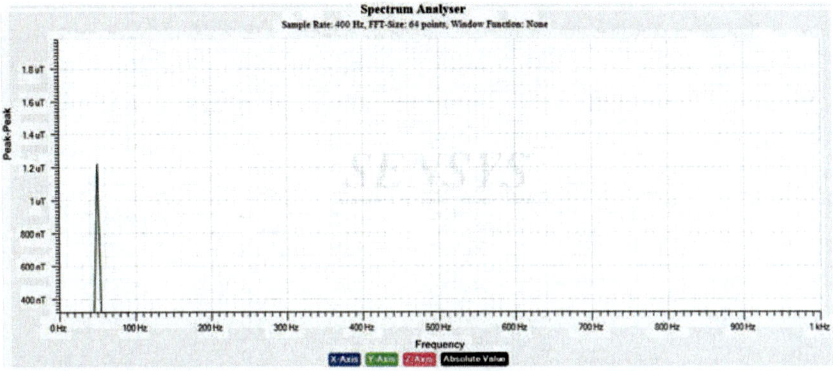

Fig. 4 Absolute magnetic field for best configuration of 1 × 4 mm^2 per phase conductor configuration

Table 3 Experimental values for conductor configuration for $N_s = 2$

Configuration		B_x	B_y	B_z	Absolute value
RYBBYR	123321	0.12	0.44	0.065	0.44
RBYYBR	132231	0.077	0.475	0.134	0.48
RYYBBR	122331	0.16	1	0.2	1
RYBRYB	123123	0.192	1.15	0.184	1.12
RRYYBB	112233	0.845	2	0.62	2
RRBBYY	113322	0.068	2.18	0.096	2.12

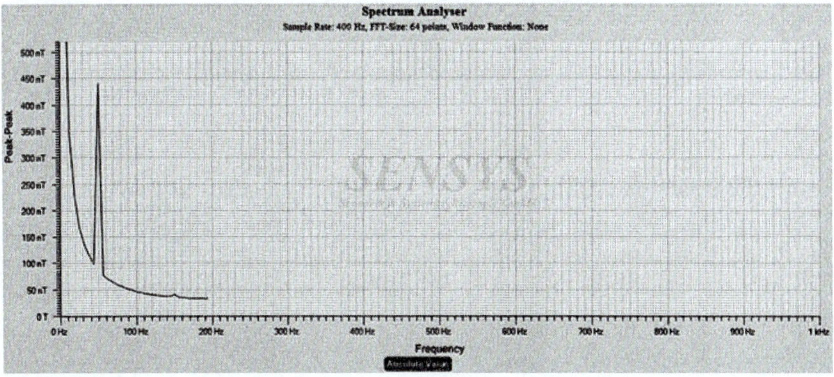

Fig. 5 Absolute magnetic field for best configuration of 2 × 2.5 mm^2 per phase conductor configuration

Table 4 Experimental values for conductor configuration for $N_s = 3$

Configuration		B_x	B_y	B_z	Absolute
YBRBRYRYB	231312123	0.08	0.336	0.078	0.34
RYBYBBRRY	123233112	0.235	0.425	0.13	0.445
RYBBYRBRY	123321312	0.432	0.456	0.168	0.528
RYBYBRYRB	123231213	0.46	0.64	0.18	0.68
RRYBBYBRY	112332312	0.68	1.08	0.26	1.12
RYRRBBYBY	121133232	0.96	1.72	0.26	1.76
YYYRRBBBR	222113331	0.76	2.36	0.41	2.36
BBBRRYRYY	333112122	1.47	3	0.17	3.06
RRRYYYBBB	111222333	1.6	3.6	0.3	3.6

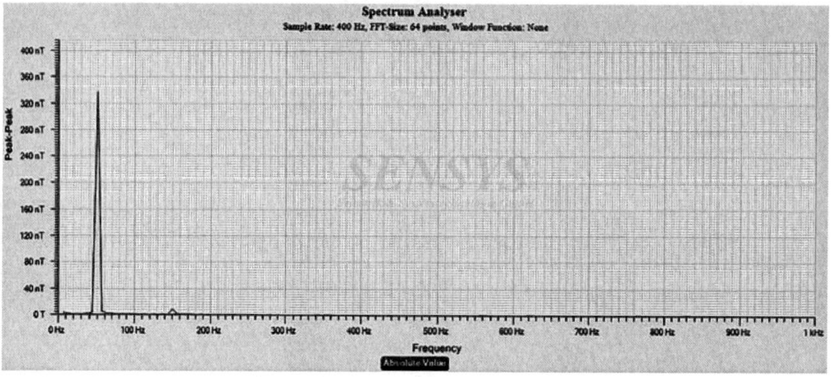

Fig. 6 Absolute magnetic field for best configuration of 3 × 1.5 mm^2 per phase conductor configuration

3.3 Three Splitting

3 × 1.5 mm^2 cable is used for each phase carrying a current of 4.2 A. Table 4 shows the magnetic field values for the various configurations obtained through three conductor splitting and phase mixing (Fig. 6).

3.4 Four Splitting

4 × 1 mm^2 cable is used for each phase carrying a current of 4.2 A. Table 5 shows the magnetic field values for the various configurations obtained through four conductor splitting and phase mixing (Fig. 7).

Table 5 Experimental values for conductor configuration for $N_s = 4$

Configuration		B_x	B_y	B_z	Absolute value
YRBBRYBRYYRB	213312312213	0.078	0.126	0.057	0.132
BRYBRYYRYBRB	312312212313	0.118	0.14	0.078	0.156
RYBYBRBYRRBY	123231321132	0.159	0.168	0.147	0.21
BRRYYBYBYRRB	311223232113	0.232	0.36	0.156	0.374
RYBBRYRBYBYR	123312132321	0.417	0.475	0.207	0.528
RYYRBBRBYBRY	122133132312	0.46	1.18	0.83	1.27
RRYBBBBYRYYR	112333321221	0.86	1.78	0.97	1.86
RRYBRBBBYYRY	112313332212	1.32	2.7	0.64	2.72
YYBRBYYBBRRR	223132233111	1.48	3.04	0.62	3.08
RRYRYRYBBBBY	112121233332	1.7	3.64	0.56	3.72
RRRRYYBYYBBB	111122322333	1.8	3.92	0.58	3.96
YYYRYRRRBBBB	222121113333	1.84	3.92	0.6	4
YYYYRRRBRBBB	222211131333	1.92	4.06	0.54	4.12
BBBBRRRRYYYY	333311112222	2.16	4.5	0.54	4.56

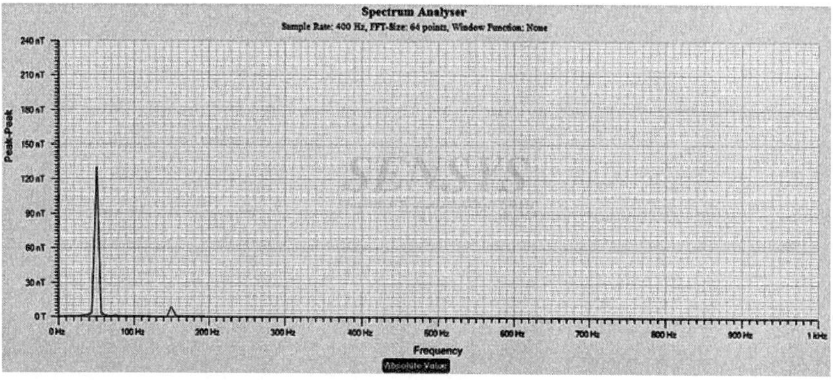

Fig. 7 Absolute magnetic field for best configuration of 4 × 1 mm^2 per phase conductor configuration

Results for Best Combination in Four Conductor Splitting with Cobalt Glass Tape

Best configuration obtained from MGA of four conductor splitting (4 × 1 mm^2 cable) is taped using cobalt glass tape. This cable carries a current of 4.2 A per phase, and it is shown in Fig. 8. The net magnetic field observed was 0.09 μT (Fig. 9).

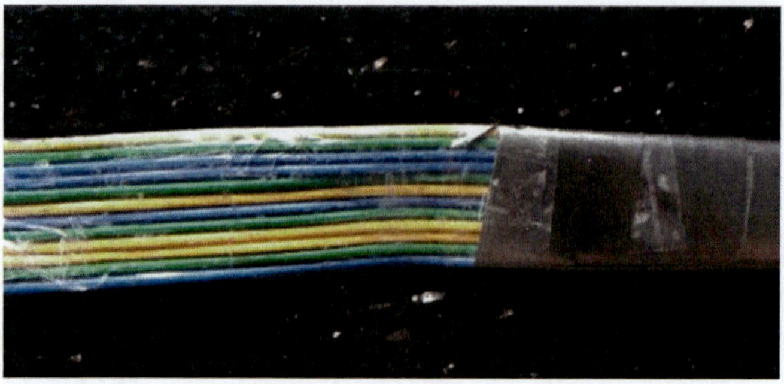

Fig. 8 Best configuration in four conductor splitting with cobalt glass tape

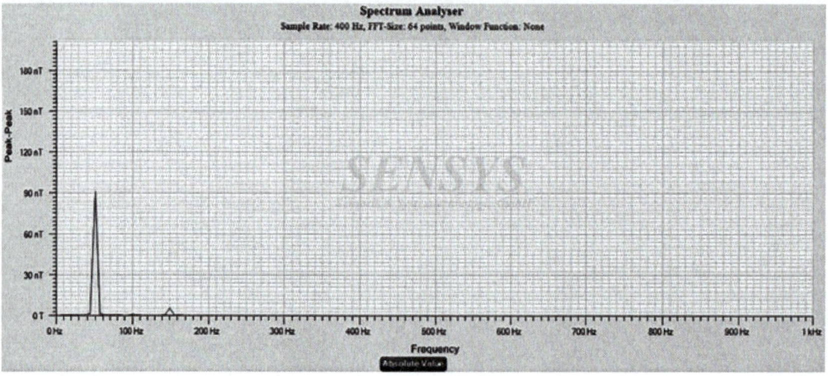

Fig. 9 Absolute magnetic field for best configuration of 4 × 1 mm^2 per phase conductor configuration with cobalt glass tape

4 Conclusion and Future Scope

The results and tabulations presented in this project report show the ELF magnetic field at the vicinity of the cable changes when the configuration of the conductors inside the cable is changed. A decreasing trend in the magnitude of magnetic field at the point of observation is observed when conductor splitting and phase mixing were implemented in the cable design. These solutions bring about an effective change in reducing the magnetic to be under attention limit of exposure. The theoretical simulation and practical results show that the magnitude of magnetic field can be kept under the objective limit by implementing these methods. Hence, conductor splitting and phase mixing have been proved to be a low cost and an easy solution in order to reduce the magnetic field from single-phase electric cables. Also the scope for development in this solution lies in the different combinations of conductor splitting and mixing. The study of magnetic field which was carried out theoretically and practically in this project has been confined only for linear configuration. The design strategy can be formulated in future by analyzing the various other possibilities of conductor splitting and phase mixing to obtain the Green Cable in other shapes with better aesthetics.

References

1. Vander-Hyde, D., Fiori, I., Paoletti, F.: Testing the magnetic field produced from twisted wire. California State University, Fullerton
2. Bauer Biotec [Online]. Available: http://www.bauer-biotec.com/downloads/Living-with-EMF-Radiation.pdf
3. Ministry of Health [Online]. Available: http://www.health.govt.nz/publication/electric-and-magnetic-fields-and-your-health
4. ICNIRP guidelines for limiting exposure to time-varying electric and magnetic fields (1 Hz–100 kHz). Health Phys. **99**(6), 818–836 (2010)
5. Hardell, L., Sage, C.: Biological effects from electromagnetic field exposure and public exposure standards, 0753-3322/$ ©2007 Elsevier Masson SAS. All rights reserved. https://doi.org/10.1016/j.biopha.2007.12.004
6. WHO: Electromagnetic fields and public health exposure to extremely low frequency fields [Online]. Available: https://www.who.int/pehemf/publications/facts/fs322/en/¼
7. National Center for Biotechnology Information [Online]. Available: https://www.ncbi.nlm.nih.gov/pubmed/12570073
8. Celozzi, S., Garzia, F.: Active shielding for power-frequency magnetic field reduction using genetic algorithms optimization. IEE Proc. Sci. Meas. Technol. **151**(1) (2004)
9. Ehrich, M., Fichte, L.O., Luer, M.: Magnetic field reduction of power cables by multi splitting of conductors and compensation effects. In: Asia-Pacific Conference on Environmental Electromagnetics CEEM* 2000, Shanghai, China, 3–7 May 2000
10. Ehrich, M., Kuhlmann, J., Netzler, D.: Models of a cable bunch formed by twisted three-phase cables. Department of Allgemeine Theoretische Elektrotechnik, Universitat der Bundeswehr Hamburg, Germany, ISBN: 0-7803-3608-9
11. Giaccone, L.: Optimal layout of parallel power cables to minimize the stray magnetic field. Electr. Power Syst. Res. **134**, 152–157 (2016)
12. Man, K.F., Tang, K.S., Kwong, S.: Genetic algorithms: concepts and applications. IEEE Trans. Ind. Electron. **43**(5) (1996)
13. Li, L., Zhan, Z.-H., Gong, Y.-J., Chen, W.-N., Zhang, J., Li, Y.: Differential evolution with an evolution path: a DEEP evolutionary algorithm. IEEE Trans. Cybern. **45**(9) (2015)
14. Liao, S., Dourmashkin, P., Belcher, J.W.: MIT, Sources of magnetic field [Online]. Available: http://web.mit.edu/viz/EM/visualizations/coursenotes/modules/guide09.pdf
15. Ross, C.L., Siriwardane, M., Almeida-Porada, G., Porada, C.D., Brink, P., Christ, G.J., Harrison, B.S.: The effect of low-frequency electromagnetic field on human bone marrow stem/progenitor cell differentiation. Stem Cell Res. **15**, 96–108 (2015)
16. Kaune, W.T., Zaffanella, L.E.: Analysis of magnetic fields produced far from electric power lines. IEEE Trans. Power Deliv. **7**(4), 2082–2091 (1992)
17. Zaffanella, L.E.: Magnetic field management for overhead transmission lines: a primer. General Electric Company EPRI High Voltage Transmission Research Center, EPRI ATR1103328 (1994)
18. Gouramanis, K., Demoulias, C., Labridis, D.P., Dokopoulos, P.: Distribution of non-sinusoidal currents in parallel conductors used in three-phase four-wire networks. Electric Power Syst. Res. **79**, 766–780 (2009)
19. Du, P.Y., Wang, X.H.: Electrical and thermal analyses of parallel single-conductor cable installations. IEEE Trans. Ind. Appl. **46**(4), 1534–1540 (2010)

20. Kovac, N., Sarajcev, I., Poliak, D.: Nonlinear-coupled electric-thermal modeling of underground cable systems. IEEE Trans. Power Deliv. **21**, 4–14 (2006)
21. Karady, G.G., Murray, J.C.: Optimization of the flat power cables electrical insulation system. In: Conference Record of the 1992 IEEE International Symposium on Electrical Insulation, Baltimore, MD, USA, 7–10 June 1992

Achieving More Page Views Through Search Engine Optimization

T. Anuradha, T. Lakshmi Surekha, N. Praveena, and B. Swapna

Abstract A user who wants to search for any data online uses a search engine which displays a list of relevant websites based on their ranking in the search engine result pages. In general, the user will go through only the sites which are displayed in the top positions of the results page. So, the owners of the websites should see that their site is listed in the initial positions to get more views. Especially a commercial or business website should get more viewers to their website compared to that of their competitors. Search engine optimization is an art of designing a website by using different techniques to get a good ranking by the search engine. This paper proposes some new SEO techniques. The experimentation is done with Google search engine, and the results proved that the new techniques, if followed along with the existing techniques, can result more traffic to a website.

Keywords Optimization · Page views · SERP · Technical SEO · Word press

1 Introduction

According to Internet live stats [1], there are more than 1.5 billion web sites in the Internet, providing different kinds of information. At present, a user who wants to know any information is basically depending on the Internet to get the information he/she wants to get. For that, they need to use some search engine. Almost all search

T. Anuradha (✉) · T. Lakshmi Surekha · N. Praveena
VR Siddhartha Engineering College, Vijayawada, AP, India
e-mail: atadiparty@gmail.com

T. Lakshmi Surekha
e-mail: lakshmi.surekha2412@gmail.com

N. Praveena
e-mail: praveena.4u@gmail.com

B. Swapna
TCS, Chennai, India
e-mail: swapna9282@gmail.com

P. Suresh et al. (eds.), *Advances in Smart System Technologies*,
Advances in Intelligent Systems and Computing 1163,
https://doi.org/10.1007/978-981-15-5029-4_26

engines work in the similar way to provide most relevant and best information. When the user types the keywords related to the information, the search engines search the Internet based on the keywords and display a list of related web sites in the form of search engine results page (SERP) [2]. The setting of a website in SERP is depending on the ranking given to the site by the search engine [3–5]. The main aim of the owners of the websites always will be to get good ranking to their sites by the search engines because the user will go through only the websites which are in the top positions in the first page of SERPs [6]. The sites which have a lot of good information also may sometimes never be visited by the users if the sites are at the last pages of SERPs [7]. It is very essential for any site especially a commercial site to be in the top position of the first page of SERPs for getting more user traffic. Search engine optimization is a collection of approaches to be followed by a website to get good page ranking by a search engine [8, 9]. This paper proposes some new SEO techniques and describes various traditional SEO techniques that can be applied at different levels. The techniques were applied on a product comparison website and analyzed with Google Analytics, and the results proved that the application of these techniques leads to improved user count and also the number of views by the users.

2 SEO Process

Search engine optimization (SEO) is the way of applying some techniques to get more number of users and more page views to the site [10, 11]. It is also called organic or free or editorial search. SEO process has different phases. Websites which follow these phases will get a good rank by search engines among all available related websites. Search engine optimization can be done at three levels, namely 1. technical SEO, 2. on page SEO, and 3. off page SEO.

2.1 Technical SEO

This can be considered as the first stage of SEO. It deals with the technical aspects of the site like loading time, simple templates, reducing the number of page redirects, visual optimization, etc., which make the search engine crawlers to easily access the website. The problems in crawling and indexing can negatively impact page ranking [12].

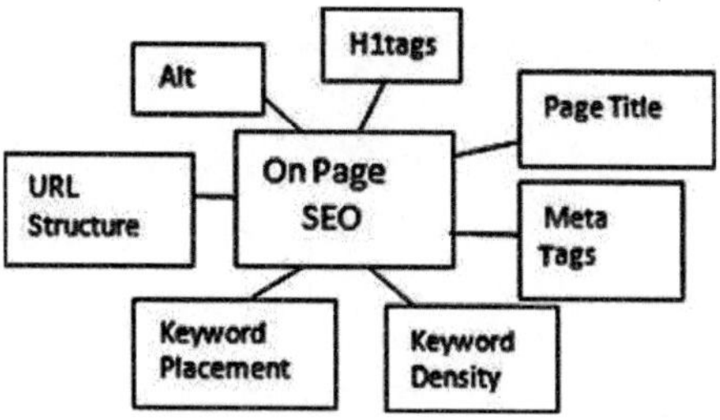

Fig. 1 On page SEO techniques

2.2 *On Page SEO*

On page SEO deals with a set of approaches incorporated within the web page itself to improve the site's ranking by the search engine [13, 14]. Figure 1 shows the main techniques of on page SEO, and the following points briefly explain them.

- Separate title for each page: If the website has multiple pages, every page should have a different title, and it should consist of focus terms of the page.
- Description tags in head section: The page head section should contain the page's focus term description.
- Short URL: The page URL should be simple and should contain focus terms. URL location also affects the SEO.
- Content descriptors as heading tags: Page content should be in readable format and should be maintained in smaller parts with HTML heading tags [15].
- Minimize focus term repetition: The focus terms should not be repeated more number of times. More usage of the word within a page may not be acceptable by search engine [16].
- Image optimization with title and alternate text: Generally, an image draws more attention compared to text. If there are images in the page, they should have titles similar to that page or sometimes any relevant term. Alternative text should be given for an image.
- Proper linking within the site: Links given within the website are also important in optimization.

2.3 *Off Page SEO*

These techniques are applied external to the webpage. Once the webpage is placed in the search engine's indexing and has on page SEO techniques, then these techniques need to be applied to further improve webpage rank and user traffic. Figure 2 represents some of the off page search engine optimization techniques, and the following points briefly explain them.

Fig. 2 Off page SEO techniques

- Register with social networking sites: Registering in famous networking sites and sharing the website details in those sites make more people know about the website.
- Promotion in blog directories: A special blog with respect to the site and its promotion in blog directories may help to get more user views.
- Register with a search engine: A search engine takes some time to locate a new website. If any new site wants quicker recognition, it should register with famous search engines.
- Photo sharing: Pictures help to understand any information more clearly. If there are pictures on the website or blog, they should be uploaded on Flickr, Picasa, etc.
- Video marketing: If there are any videos about the website, they should be uploaded to the video sharing sites like YouTube for more user views.
- Article submission: Writing about the website in article submissions can help to increase user traffic.
- Advertising in shopping network: Advertising the products on online shopping sites can increase the website ranking for ecommerce sites.
- Answering questions: Answering the questions related to the site on popular question and answer sites like Yahoo Answers can increase the website's ranking.

2.4 New Techniques in SEO

Along with the traditional techniques, the website owners should device new methods to get better ranking to their site from any search engine by carefully following the process of how search engines like Google are giving rank to the sites. Some of the new techniques proposed in this research are given below:

- Focus on all header tags: The web designers focus mostly will be on H1, H2, H3, and H4 tags as they look bigger, but other tags like H5 and H6 also need to be taken care.
- PPT inclusion: Attractive PowerPoint presentations having the slides with information presented in simple and direct manner with keywords can be used to get more user attention.

- Long-tail bold keywords: These are the sets of combined keywords having more than two words in the set. Sometimes customers search for a product by giving specific keywords. They may be the probable customers who are going to buy the product. To attract them, long-tail keywords having specific words can correctly describe the product need to be used, and they should be made bold.
- Conversion from PPT to PDF format: PDF formats are easily downloadable and readable compared to PPT. So, the websites which are in PPT can be converted to PDF, and thereby, they can have a PDF link.

3 Experimental Work

The traditional and new SEO techniques were applied in the design of a website called compareegg.in. This is a product price comparison site which shows the details of various products like mobiles, computers, other electronic products and home appliances, their prices in all major cities in India, the product prices on different online shopping sites, and which are the best deals. The experimentation was done by placing the site in Google search engine. The website was created with Word Press tool [17], a content management system that supports various plug-ins which satisfy SEO methods. Then, site was made to be indexed in Google through Google Search Console. The content in the web pages was updated by using the keywords [18] after checking their position in free keyword position checker tool called SERPLAB which gives the position of various keywords in Google search engine. The experimentation was done in step-by-step manner. First on page SEO techniques were applied, number of users and website traffic were observed through Google Analytics, then off page and new techniques were applied, and again Google Analytics were taken.

4 Experimental Results

The results show the process from title and keyword selection to getting the page views and traffic changes in Google Analytics to one of the pages related to Moto M 64 GB gold mobile in compareegg.in web site. Figure 3 shows the process of setting title, slug, and meta-description for the webpage by using Word Press tool.

Figure 4 shows the compareegg.in website Google Search Console information. It is necessary to add the website in the Google Search Console to make the Google analyze individual webpage's traffic in the website.

Figure 5 shows keyword position checker using SERPLAB tool. Here, based on how the users are going to search for the product, different long-tail keywords related to the product were selected and their position was identified. It shows among the keywords used like 'moto m64gb price', 'moto m 64gb price', 'moto m price', 'moto m 64 gb price in india', 'moto m64gb price' got position 3, 'moto m 64gb price' was in position 21, and the others were in the positions more than 21. It was observed

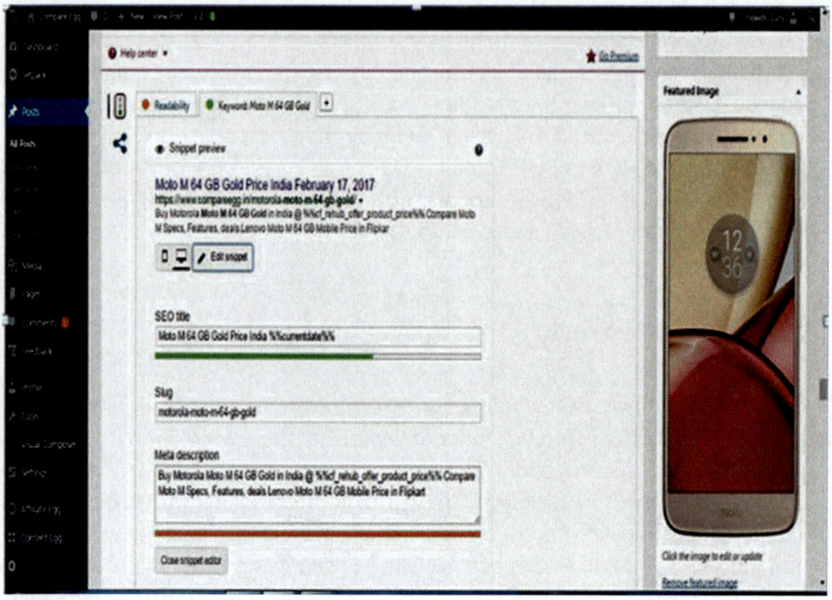

Fig. 3 SEO title, meta description, focus keyword

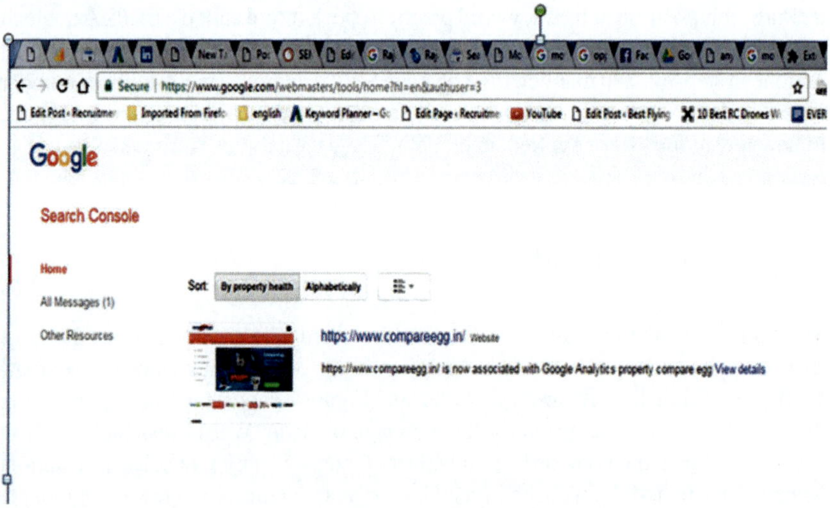

Fig. 4 Compareegg.in website Google search console

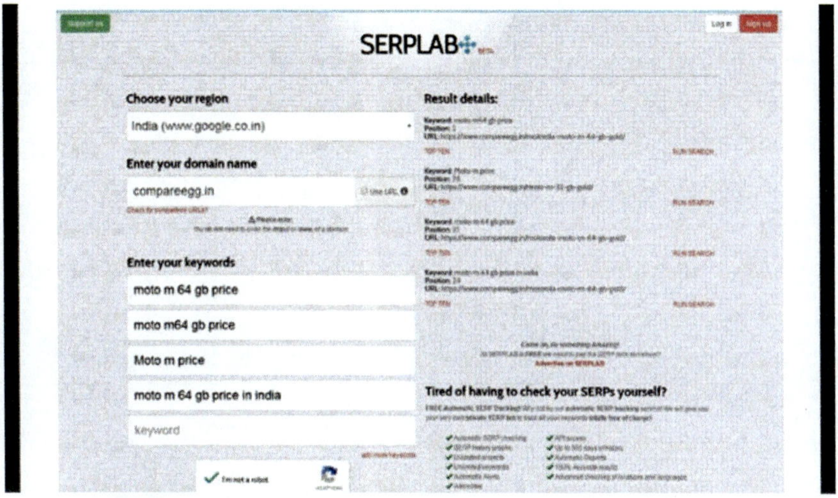

Fig. 5 SERPLAB for keyword position checker

that even with a space change from ‘m64’ to ‘m 64’, the position of the keywords changed from 3 to 21.

Then, On page SEO techniques were applied to the website, and the traffic analysis was observed over a month’s period [19]. Figure 6 shows the traffic analysis results of the site by Google Analytics after on page SEO. It was observed that during a month’s

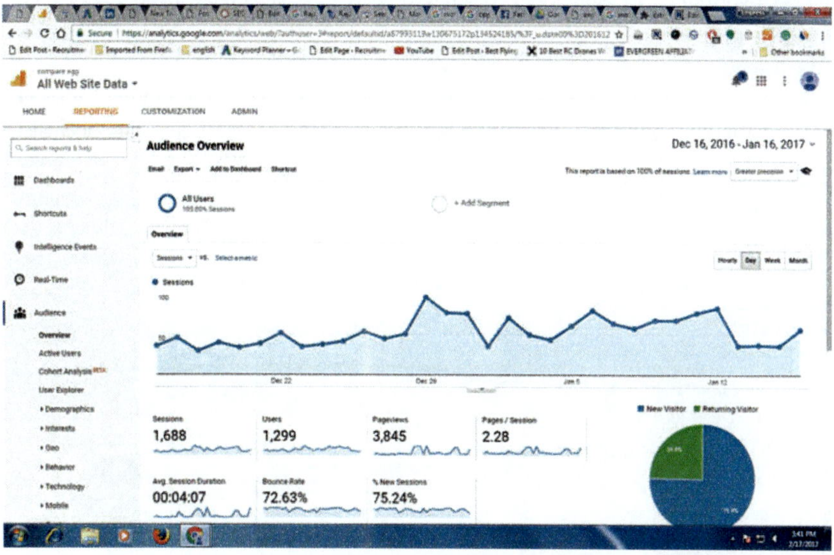

Fig. 6 Google analytics for compareegg.in website after on page SEO

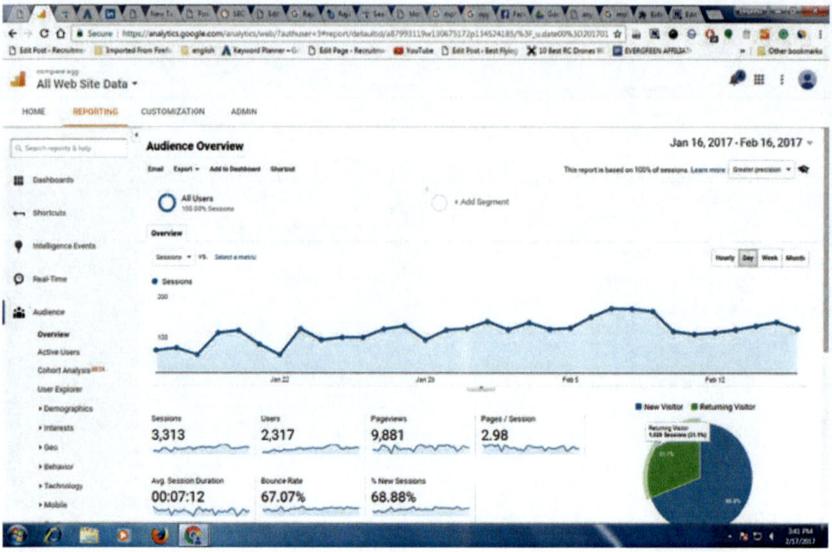

Fig. 7 Google analytics for compareegg.in website after off page SEO

period after on page SEO, the number of users visited the page was 1299 and the total number of views for the page was 3845. After that off page SEO techniques were applied and traffic analysis was observed. Figure 7 shows the compareegg.in website traffic analysis by Google Analytics over a month after off page SEO. It is observed that after applying Off Page SEO, number of users increased from 1299 to 2317, i.e., almost two times and page views increased from 3845 to 9881, approximately by three times.

Then, new SEO techniques were applied and traffic was observed continuously for two months. Figure 8 shows the result after the first month of applying new SEO

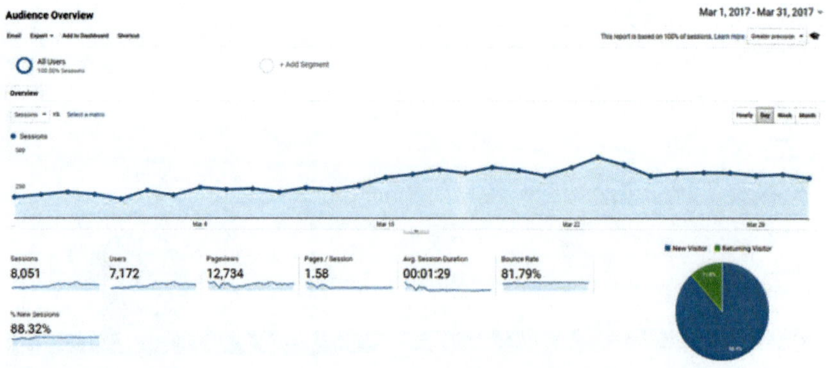

Fig. 8 Google analytics for compareegg.in website after first month of new SEO

Fig. 9 Google analytics for compareegg.in website after second month of new SEO

techniques. It was observed that the number of users increased from 2317 to 7172, by almost three times and page views increased from 9881 to 12,734, by 1.25 times approximately. Figure 9 shows the result after the second month of applying new SEO techniques. It was observed that the number of users increased by almost 1.2 times, i.e., from 7172 to 8628, and there was a slight decrease in page views also from 12,734 to 12,666.

4.1 Discussion

Previous researchers mostly concentrated on page ranking after application of SEO. In [4], the authors applied on page and off page optimization and observed their web page ranking improved from 1 to 4. In [14], with the application of traditional and their proposed methods, authors achieved top position to their web page. The current research focuses on traffic analysis. Altogether, comparison of Figs. 6 and 9 shows that application of traditional and new SEO techniques proposed in this paper has increased the number users by almost eight times and number of page views by almost four times.

5 Conclusions

The traditional and new SEO techniques were applied on a product price comparison website, and the experimentation proved that on page SEO is compulsory to get user traffic, off page SEO techniques further improved the traffic, and the proposed new techniques proved that they can be used to get even more traffic.

Acknowledgements Authors are thankful to the web development team of Compareegg.in for allowing one of the authors to work with them.

References

1. Internet livestats. http://www.internetlivestats.com/total-number-of-websites/. Last accessed 2018/12/18
2. Article #1.3 How do search engine rank web pages. https://www.rankwatch.com/learning/content/13-how-do-search-engine-rank-web-pages. Last accessed 2018/12/20
3. D'souza, R., Kulkarni, A., Mirza, I.A.: Automatic link generation for search engine optimization. Int. J. Inf. Educ. Technol. **2**(4), 401–409 (2012)
4. Gunjan, V.K., Kumari, M., Kumar, A., Rao, A.: A search engine optimization with Google. Int. J. Comput. Sci. Issues **9**(1), 206–215 (2012)
5. Cui, M., Hu, S.: Search engine optimization research for website promotion. In: International Conference of Information Technology, Computer Engineering and Management Sciences 2011, pp. 100–103. IEEE (2011)
6. Hatab, R.: Improve website rank using search engine optimization (SEO). MS thesis. https://www.researchgate.net/publication/275637141_Improve_Website_Rank_Using_Search_Engine_OptimizationSEO. Last accessed 2019/01/10
7. Jacobson, M.: How far down the search engine results page will most people go? https://www.theleverageway.com/blog/how-far-down-the-search-engine-results-page-will-most-people-go/. Last accessed 2019/01/12
8. Kumar, M.S., Arpana, M.: Survey on search engine optimization techniques to achieve high page rank. Imperial J. Interdisc. Res. **2**(8) (2016)
9. Ur Rehman, K., Khan, M.N.: The foremost guidelines for achieving higher ranking in search results through search engine optimization. Int. J. Adv. Sci. Technol. **2**(5), 101–110 (2013)
10. Hussien, A.S.: Factors affect search engine optimization. Int. J. Comput. Sci. Netw. Secur. **14**(9), 28–33 (2014)
11. Nath, C., Ahuja, L.: Search engine optimization (SEO): improving website ranking. Int. J. Eng. Res. Technol. **3**(4) (2014)
12. Perzyńska, K.: What is technical SEO? A complete and simple guide. https://unamo.com/blog/seo/technical-seo. Last accessed 2018/12/26
13. Su, A.J., Hu, Y.C., Kuzmanovic, A., Koh, C.K.: How to improve your search engine ranking: myths and reality. ACM Trans. Web (TWEB) **8**(2) (2014)
14. Swapna, B., Anuradha, T.: Achieving higher ranking to web pages through search engine optimization. In: Proceedings of International Conference on Computational Intelligence and Data Engineering, pp. 105–112. Springer, Singapore (2018)
15. Aashusukhija 231: What is Google crawling and indexing? https://www.shoutmeloud.com/google-crawling-and-indexing.html. Last accessed 2018/02/12
16. Patel, N.: The ultimate guide to off-page SEO. https://neilpatel.com/blog/everything-you-need-to-know-about-off-page-seo/. Last accessed 2018/02/18
17. Jensen, K.: Reasons to use wordpress in your website design. https://www.bopdesign.com/bop-blog/2015/09/4-reasons-to-use-wordpress-in-website-design/. Last accessed 2018/12/18
18. Free keyword position tracker tools every SEO needs. https://marketingkeytech.com/free-keyword-position-tracker-tools/. Last accessed 2018/12/18
19. Esquivel, M.: How to compare periods of time in Google analytics. https://www.wholewhale.com/tips/how-to-compare-periods-of-time-in-google-analytics/. Last accessed 2018/12/28

Industrial Informatics for Quality Assurance and Real-Time Defect Detection Through Computer Vision

Gonçalo Marques and Rui Pitarma

Abstract Quality assurance can be assumed as the absence of defects in the manufacturing process to deliver products, solutions or services to customers with high quality. On the one hand, due to the availability of cost-effective components such microcontrollers, cameras, and communication interfaces, the deployment of real-time defect detection systems using computer vision technologies for several applications has been increasing in the recent past. On the other hand, regarding the requirements of several industrial areas, there is an effective need for industrial applications to detect defects in real-time leading to optimization and efficiency on the manufacturing process. This paper proposes a low-cost and easy-to-install automatic real-time detection system based on the Raspberry Pi platform. This system shall be referenced as ***Coil Spring Control Box***. This solution aims to guarantee the placement of the correct coil spring on the respective vehicle in real-time. This system incorporates a Raspberry Pi 2 as processing unit, a low-cost full HD camera, a Wi-Fi adaptor, a touch screen for visual output, and a radio module for communication with the automation network. This architecture allows the real-time detection of wrong coil spring references, avoiding the need to correct defects at the end of the assembly line. The results are promising, and this system can be assumed as a low-cost and reliable approach for defect detection through computer vision.

Keywords Computer vision · Industrial informatics · Raspberry Pi · Pattern recognition

G. Marques (✉)
Instituto de Telecomunicações, Universidade da Beira Interior, 6201-001 Covilhã, Portugal
e-mail: goncalosantosmarques@gmail.com

G. Marques · R. Pitarma
Polytechnic Institute of Guarda, Av. Dr. Francisco Sá Carneiro, nº 50, 6300–559 Guarda, Portugal
e-mail: rpitarma@ipg.pt

P. Suresh et al. (eds.), *Advances in Smart System Technologies*,
Advances in Intelligent Systems and Computing 1163,
https://doi.org/10.1007/978-981-15-5029-4_27

1 Introduction

The emergence of Industry 4.0 has changed the scope of operator responsibilities in the manufacturing context as the operator adopts an extensive variety of tasks. Thus, technological support is necessary to guarantee product quality [1].

Cyber-physical systems (CPS) are informatic devices that allow communication between the physical world and software infrastructures [2].

Industry 4.0 merges manufacturing digitalization with CPS. People and machines cooperate and share data using Big Data and cloud computing technologies [3]. Furthermore, 82% of organizations that implement smart manufacturing increase productivity, with a 49% increase in product quality and a 45% increase in client satisfaction [4].

Industry 4.0's most significant challenge is the interoperability, and an integrated and universal technology is needed to support the purposes, levels and steps of the manufacturing process [5]. An industrial Internet-of-Things hub (IIHub) which incorporates a customized access module (CA-Module), access hub (A-Hub) and local service pool (LSP) for heterogeneous equipment interoperability is proposed by the authors of [6].

Innovation is the main key for economic value optimization which leads to faster product transitions for both local and external markets [7].

Cloud computing changed the industrial process, with cloud adoption, the access to the manufacturing critical data can be done in real-time for enhanced manufacturing. Cloud manufacturing leads to efficient manufacturing systems and becomes to be progressively globalized and distributed [8].

A technique to classify the phases of manufacturing processes that must be followed in order to answer to strategic objectives variations for smart manufacturing is proposed by the authors of [9].

Due to the accessibility of cost-effective and high-quality cameras, the study of computer vision approaches for different scenarios has been increasing [10]. Several solutions for real-time defect detection in multiple industry areas can be found in the literature. An automatic vision-based system for quality assurance of web textile goods is proposed by Cho et al. [11], and a real-time vision-based defect inspection for high-speed steel products is proposed by the authors of [12]. An automated inspection system for automobile parts which can be applied in several cases that uses laser scanned point cloud data is proposed by Hong-Seok and Mani [13]. Xiaolong Bai et al. present an effective and efficient method to detect defects in images using computer vision-based inspection of electronic chips or dies in semiconductor production lines [14]. A defect detection method for wheel bearings with entropy, time-spectral kurtosis (TSK) and support vector machine (SVM) where the possible outliers in the short-time Fourier transform (STFT) amplitude series are initially estimated and preprocessed with information entropy is proposed by the authors of [15].

This paper presents an industrial informatics system for quality assurance through real-time defect detection named ***Coil Spring Control Box***. To design a cost-effective solution, the Raspberry Pi 2 and a low-cost HD camera was chosen. The Raspberry Pi is a cost-effective approach for a set of a miniaturized microcontroller to make the technology accessible to a wide audience [16]. This solution incorporates a hardware prototype for ambient data collection and analysis, and a PHP web server is used to handle the data for the verification process. The ***Coil Spring Control Box*** is an exclusively Wi-Fi-sourced solution based on open-source technologies, with numerous advantages compared to existing systems, such as its easy installation, scalability and modularity.

The rest of the manuscript is organized as follows: Sect. 2 describes the methods and materials used in the development of the solution; Sect. 3 explains the solution operation and experimental results, and the conclusion is presented in Sect. 4.

2 Materials and Methods

The ***Coil Spring Control Box*** solution has been developed as a cost-effective, reliable solution which supports easy and quick configuration and installation. Figure 1 represents the developed system.

A succinct presentation of all components used is described below:

- **Raspberry Pi 2 Model B** is the second generation of computer board; it incorporates a Broadcom BCM2837 Quad-core (ARM Cortex-A53) SoC processor, a 1 GB LPDDR2 memory, a 10/100 BaseT socket Ethernet, 4$\times$ USB 2.0 ports, HDMI and a 40-pin GPIO Header.
- **Raspberry Pi Touch Display** is an 800 $\times$ 480 touch screen monitor module for Raspberry Pi that can be attached using a ribbon cable that connects to the DSI port. This display supports 10 touch points.
- **150 Mbps Wireless N Nano USB Adapter** is a miniature Wi-Fi dongle with wireless transmission capabilities up to 150 Mbps. It supports advanced security protocols such as 64/128 WEP, WPA, PA2/WPA-PSK/WPA2-PSK (TKIP/AES).
- **APC802 Radio Communication Module** is an economic solution for wireless data communications. This module supports UART/TTL, RS232 and RS485 interface and removes the requirement for packetizing and data encoding. It is an enhanced version of APC220 module which offers up to 2800 m of data transmission range.
- **C920 HD PRO** is a USB webcam which can make stream 1080p videos with a frame rate of 30 fps with no frame rate reduction. This camera incorporates built-in two stereo microphones with automatic noise reduction and light adjustment.

Table 1 presents the system cost. The prices were obtained from Amazon (accessed on 20 February 2019). The total system cost is around 340$.

Figure 2 represents the system architecture and connection interfaces used. The C920 camera and the wireless adaptor are connected to the Raspberry using USB interface.

Fig. 1 Coil spring control box system prototype

Table 1 Cost of the *coil spring control box* system

Component	Cost (USD) ($)
Raspberry Pi 2 model B (kit)	48.99
Raspberry Pi touch display	74.00
150 Mbps wireless N nano USB adapter	14.99
APC802 radio communication module	83.96
Box and cables	30.00
C920 HD PRO	89.00
Total	340.94

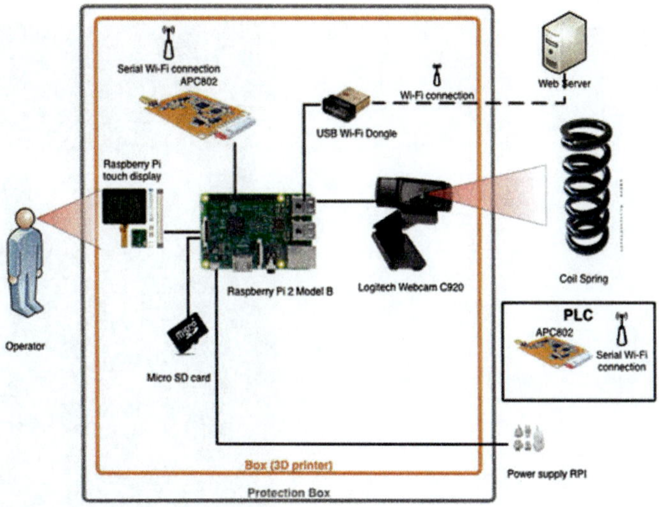

Fig. 2 *Coil spring control box* connection diagram

The ***Coil Spring Control Box*** operating system is the Raspbian Jessie, and OpenCV 3 is used for image analysis [17]. This solution is an automatic system that uses computer vision to ensure correct placement of the rear coil springs in the respective vehicle in real-time. In addition, the system also indicates the operator the respective coil spring reference to be placed in the vehicle. Figure 3 shows an example of a rear coil spring used in the tests. This system is powered using a 5.1 V 2.5 A power supply.

3 Discussion and Results

The proposed system was tested in a laboratory, and no false-positive situations were reported. However, there are some situations in which the coil spring is not detected. This issue is related to deficiencies in the marking performed by the supplier, in the majority of the times. On the one hand, by providing real-time supervision the system avoids costs related to repair time in the quality tests. On the other hand, this system can improve quality indicators.

Eight different references of rear coil springs are tested. Table 2 presents the references and the correspondent supply marker colours.

The system functionality can be divided into three steps, as shown in Fig. 4. The ***Coil Spring Control Box*** is connected to the Intranet using Wi-Fi communication and is linked to the automation network by using an APC 802 radio module.

The system is synchronized with the assembly line using the automation network. First, the PLC sends the reference of the vehicle to the ***Coil Spring Control Box*** using

Fig. 3 Rear coil spring

Table 2 Coil spring references and marker colours used

Ref.	Colour sequence
9685775180	blue/blue/orange
9685775080	blue/violet/violet
9685774980	blue/red/green
9685775380	red/green/orange
9685775280	red/green/violet
9686058080	white/orange/violet

the APC radio module. The system is wirelessly connected to the industrial server and makes an HTTP request to the PHP web server to get the reference of the coil spring. Then, the system uses computer vision to detect the three circles and then get the colour of each circle. Furthermore, the detected colours are assigned to a specific reference. When the detected rear coil spring is not correct, an alarm will be raised to the automation network.

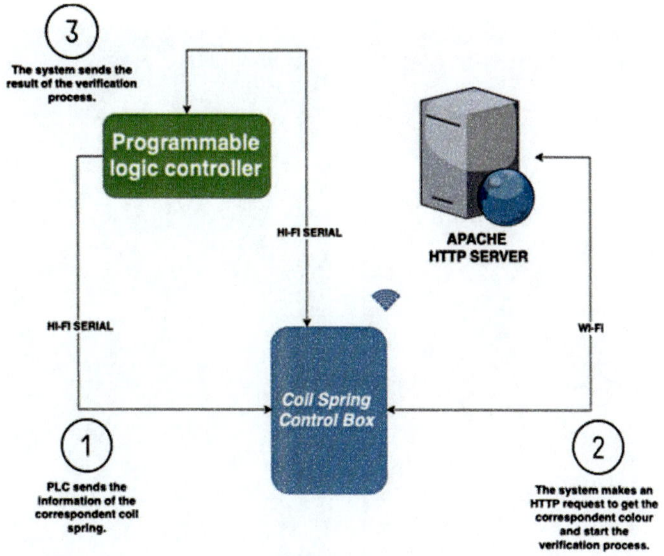

Fig. 4 *Coil spring control box* system architecture

This solution uses a Raspberry Pi 2 as processing that incorporates an ARM Cortex A7 CPU and 1 GB of SDRAM, in order to increase the frames per second (FPS) and decrease the latency the authors use multithreading.

The results show that the Raspberry Pi and OpenCV are suitable low-cost platforms for computer vision. On the one hand, the OpenCV-Python has several advantages such as the ease of use, visualization, debugging and interoperability. On the other hand, using Python programming language is possible to create wrappers that can be used as Python modules extending the C/C++ code. Furthermore, the code is faster than the original C/C++ scripts.

The system uses computer vision to provide real-time image analysis to check if the coil spring is correct or not, and display the colour sequence of the correct coil spring in the screen to inform the operator. If the operation is correct, the screen turns to green and informs "Correct" (Fig. 5b). On the contrary, if the wrong coil reference is detected, the production line stops and an audio alarm is triggered. In the case that coil spring reference is not detected, the screen stays with a red background and continues to do the image analysis until the automation network sends another coil spring reference (Fig. 5a).

The graphical user interface (GUI) was developed using Tkinter that is the Python's standard GUI library.

This system uses OpenCV and performs colour detection using HSV colour space to ensure the correct coil spring that will be placed in the vehicle with the colours defined by the reference.

The coil spring detection time is less than 1 second using a video resolution of 1920 × 1080 pixels. Therefore, the application broadly meets the requisites of the process.

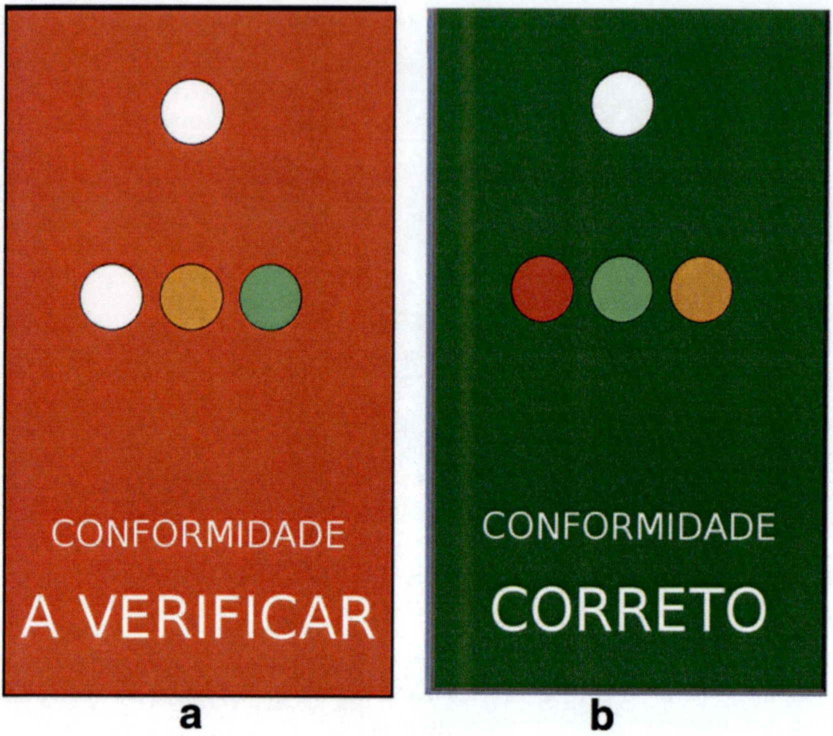

Fig. 5 System output, **a** in the process of verification; **b** confirmation has completed

For colour real-time detection, we use cv2.inRange OpenCV method that returns a binary mask with the image that falls into the upper and lower limit ranges. Figure 6 shows the real-time image colour analysis process done in a laboratory using OpenCV-Python using contour analysis in a binary image using the algorithm proposed by Suzuki and Abe [18].

The camera selection was conducted with the aim to create a cost-effective solution since the main objective was to test the functional architecture of the proposed system. Considering that the system is projected to be used in indoor environments where energy is easily accessible, there was no great concern regarding the selection of energy-efficient components. Python programming language provides an important platform for testing with new computer vision algorithms and is recommended for computer vision development.

The system is a wireless solution which leads to several advantages in terms of installation. The installation process only requires a 230 V power source.

The results achieved are favourable, indicating an important contribution to real-time quality assurance through defect detection systems. Nonetheless, the proposed solution has some limitations. The ***Coil Spring Control Box*** needs further experimental validation to improve system calibration and accuracy.

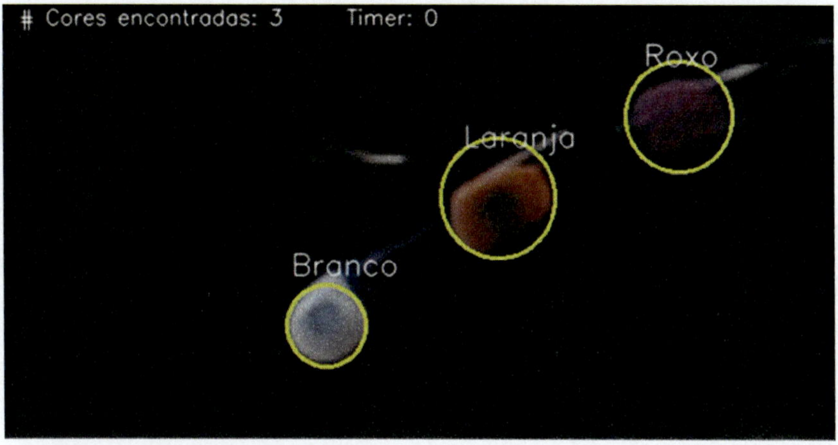

Fig. 6 OpenCV-python template matching

As a future work, this study will be focused on the enhancement of the presented method more in depth to increase accuracy and to decrease the inspection time. Efforts have been being made to empower the system by widening the application area of the solution. In addition to the free-form surface defects inspection module, a flatness measuring module is being developed to carry out a variety of tasks by a single system. The authors also plan to develop a web Portal for data analytics and visualization for the plant manager. This portal should incorporate dashboards for quality assurance analysis.

4 Conclusion

This manuscript presented a wireless system composed for enhanced default detection in real-time. This low-cost solution is easy to install and was developed based on open-source technologies. In the development of this solution, a Raspberry Pi 2 was used as processing unit, a low-cost full HD camera, Wi-Fi, a touch screen for communication for information output and a radio module for communication with the automation network.

This architecture allows the real-time detection of wrong coil spring references, thus avoiding the need to correct defects at the end of the assembly line and also indicates to the operator that the coil spring reference to place the vehicle in real time. By detecting defaults in real-time, we can reduce the production costs and improve product quality.

The system can be easily installed by the average user as it uses wireless communication technologies and just requires a 230 V power supply. The ***Coil Spring Control Box*** is a Wi-Fi solution based on open-source technologies, with numerous advantages compared to existing systems, such as its easy installation, scalability and modularity.

The results obtained are promising, representing a significant approach for enhanced quality assurance system based on open-source technologies and computer vision at low cost. Nevertheless, the proposed solution has some limitations. The system needs further experimental validation to improve system accuracy.

In the future, hardware and software improvements have also been planned to introduce artificial intelligence algorithms in the system to improve the detection of springs with marking problems.

References

1. Gorecky, D., Schmitt, M., Loskyll, M., Zuhlke, D.: Human-machine-interaction in the industry 4.0 era, pp. 289–294 (2014)
2. Jazdi, N.: Cyber physical systems in the context of Industry 4.0, pp. 1–4 (2014)
3. Lu, Y.: Industry 4.0: a survey on technologies, applications and open research issues. J. Ind. Inf. Integr. **6**, 1–10 (2017)
4. Shrouf, F., Ordieres, J., Miragliotta, G.: Smart factories in Industry 4.0: a review of the concept and of energy management approached in production based on the Internet of Things paradigm, pp. 697–701 (2014)
5. Kang, H.S., et al.: Smart manufacturing: past research, present findings, and future directions. Int. J. Precis. Eng. Manuf. Green Technol. **3**(1), 111–128 (2016)
6. Tao, F., Cheng, J., Qi, Q.: IIHub: an industrial internet-of-things hub toward smart manufacturing based on cyber-physical system. IEEE Trans. Ind. Inform. **14**(5), 2271–2280 (2018)
7. Davis, J., Edgar, T., Porter, J., Bernaden, J., Sarli, M.: Smart manufacturing, manufacturing intelligence and demand-dynamic performance. Comput. Chem. Eng. **47**, 145–156 (2012)
8. Xu, X.: From cloud computing to cloud manufacturing. Robot. Comput. Integr. Manuf. **28**(1), 75–86 (2012)
9. Jung, K., Morris, K.C., Lyons, K.W., Leong, S., Cho, H.: Mapping strategic goals and operational performance metrics for smart manufacturing systems. Procedia Comput. Sci. **44**, 184–193 (2015)
10. Koch, C., Georgieva, K., Kasireddy, V., Akinci, B., Fieguth, P.: A review on computer vision based defect detection and condition assessment of concrete and asphalt civil infrastructure. Adv. Eng. Inform. **29**(2), 196–210 (2015)
11. Cho, C.-S., Chung, B.-M., Park, M.-J.: Development of real-time vision-based fabric inspection system. IEEE Trans. Ind. Electron. **52**(4), 1073–1079 (2005)
12. Yun, J.P.: Real-time vision-based defect inspection for high-speed steel products. Opt. Eng. **47**(7), 077204 (2008)
13. Hong-Seok, P., Mani, T.U.: Development of an inspection system for defect detection in pressed parts using laser scanned data. Procedia Eng. **69**, 931–936 (2014)
14. Bai, X., Fang, Y., Lin, W., Wang, L., Ju, B.-F.: Saliency-based defect detection in industrial images by using phase spectrum. IEEE Trans. Ind. Inform. **10**(4), 2135–2145 (2014)
15. Chen, B., Yan, Z., Chen, W.: Defect detection for wheel-bearings with time-spectral kurtosis and entropy. Entropy **16**(1), 607–626 (2014)

16. Kirby, J., Chapman, L., Chapman, V.: Assessing the Raspberry Pi as a low-cost alternative for acquisition of near infrared hemispherical digital imagery. Agric. For. Meteorol. **259**, 232–239 (2018)
17. Brahmbhatt, S.: Embedded computer vision: running OpenCV programs on the Raspberry Pi. In: Practical OpenCV, pp. 201–218. Apress, Berkeley, CA (2013)
18. Suzuki, S., Abe, K.: Topological structural analysis of digitized binary images by border following. Comput. Vis. Graph. Image Process. **30**(1), 32–46 (1985)

A New 75-Level Inverter Topology with Reduced Number of Switch Count

V. Thiyagarajan

Abstract In modern era, many researchers attempt to develop different multilevel inverter (MLI) topologies with reduced number of circuit components. This paper aims to develop a new 75-level asymmetric-type inverter topology based on the series–parallel connection of several DC sources with reduced number of circuit components. The proposed 75-level inverter structure can create positive, negative, and zero output levels without any additional circuit components. As compared with other recently presented improved topologies, the proposed inverter reduces the number of DC sources, switches, and the number of conducting switches. At last, to confirm the performance, simulation results for 75-level asymmetrical operation of the inverter are presented in this paper.

Keywords Multilevel inverter · Asymmetric · Reduced switch · 75-level · THD

1 Introduction

Multilevel inverters (MLI) attract more attention as a new approach to power conversion to create an improved power quality voltage waveform in the field of renewable energy resources. It creates an AC staircase output voltage waveform with high power quality with the help of an array of power semiconductor switches and several input DC sources. The important merits of MLI include reduction in the total harmonic distortion (THD), low dv/dt stress, less switching losses, and minimum electromagnetic interference [1–3]. The conventional MLI includes flying capacitor inverters, cascaded H bridge (CHB) inverters, and diode-clamped inverters [4, 5]. The MLI can be either a symmetric-type or asymmetric-type inverters. In a symmetric type inverter, all input DC voltage sources have same magnitude, whereas in an asymmetric-type inverter, each input DC voltage sources have different magnitude [6].

A new basic structure for MLI with H bridge inverter is presented [1, 2]. This basic structure can generate only positive output levels, and thus, to create negative

V. Thiyagarajan (✉)
SSN College of Engineering, Chennai, Tamil Nadu 603110, India
e-mail: thiyagarajanv@ssn.edu.in

P. Suresh et al. (eds.), *Advances in Smart System Technologies*,
Advances in Intelligent Systems and Computing 1163,
https://doi.org/10.1007/978-981-15-5029-4_28

output levels, an additional H bridge inverter is added across the output terminals of the basic unit. In [3], a basic module with the combination of H and half-bridge units is presented. This inverter module requires more number of sources to create larger output levels. A combination of CHB inverter with a double-level unit is presented in [4]. This combination increases the output voltage level to nearly twice that of a CHB inverter. However, increase in the number of switches is the main drawback of this topology. The topology presented in [5–7] requires more switches, and it increases the cost and size of the inverter. The topologies presented in [3–9] have inherent ability to create negative output levels without any additional unit. However, these topologies require and use high variety of power electronic switches and DC sources and thus, lost its modularity.

In this paper, a new asymmetric-type 75-level inverter topology with reduced switch count is presented. This paper is organized as follows: Sect. 2 presents a new 75-level inverter topology. The comparison of the proposed MLI with other recently developed topologies is presented in Sect. 3. Section 4 presents the simulation results of 75-level inverter topology, and the conclusion is made in Sect. 5.

2 Proposed Inverter Structure

The proposed 75-level inverter topology is shown in Fig. 1. The proposed inverter consists of 6 DC sources and 13 switches which create 75-level output voltage during asymmetric condition where the magnitude of DC sources is different.

The switching conditions for different output levels obtained during both positive and negative mode of operations are given in Table 1, and the operation of the proposed MLI is given in Fig. 2. The main advantage of the proposed inverter is the inherent ability to create negative output levels, without any additional circuit components. The switches $S_8, S_9, S_{10}, S_{11}, S_{13}$ and $S_2, S_3, S_4, S_5, S_{13}$ are turned ON to

Fig. 1 Proposed inverter

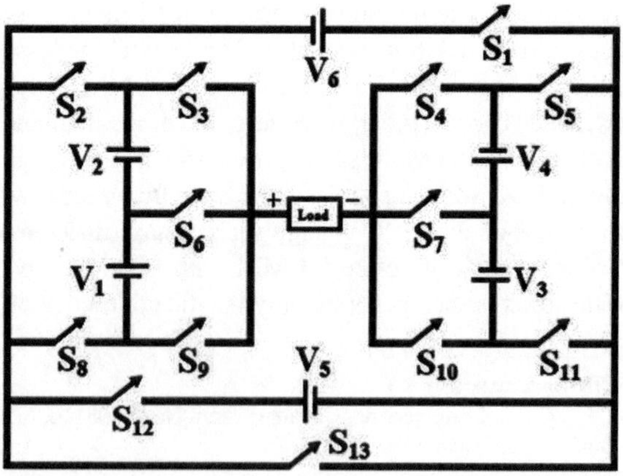

Table 1 Switching conditions for different output levels

Output level	Conducting switches	
	Positive cycle	Negative cycle
0	$S_8, S_9, S_{10}, S_{11}, S_{13}$	$S_2, S_3, S_4, S_5, S_{13}$
1	$S_6, S_8, S_{10}, S_{11}, S_{13}$	$S_2, S_4, S_5, S_6, S_{13}$
2	$S_3, S_8, S_{10}, S_{11}, S_{13}$	$S_2, S_4, S_5, S_9, S_{13}$
3	$S_2, S_9, S_{10}, S_{11}, S_{12}$	S_1, S_3, S_4, S_5, S_8
4	$S_2, S_6, S_{10}, S_{11}, S_{12}$	S_1, S_4, S_5, S_6, S_8
5	$S_8, S_9, S_{10}, S_{11}, S_{12}$	S_1, S_2, S_3, S_4, S_5
6	$S_6, S_8, S_{10}, S_{11}, S_{12}$	S_1, S_2, S_4, S_5, S_6
7	$S_3, S_8, S_{10}, S_{11}, S_{12}$	S_1, S_2, S_4, S_5, S_9
8	$S_1, S_2, S_7, S_9, S_{11}$	$S_3, S_5, S_7, S_8, S_{12}$
9	$S_1, S_2, S_6, S_7, S_{11}$	S_5, S_6, S, S_8, S_{12}
10	$S_1, S_7, S_8, S_9, S_{11}$	$S_2, S_3, S_5, S_7, S_{12}$
11	$S_1, S_6, S_7, S_8, S_{11}$	$S_2, S_5, S_6, S_7, S_{12}$
12	$S_1, S_3, S_7, S_8, S_{11}$	$S_2, S_5, S_7, S_9, S_{12}$
13	$S_2, S_7, S_9, S_{11}, S_{13}$	$S_3, S_5, S_7, S_8, S_{13}$
14	$S_2, S_6, S_7, S_{11}, S_{13}$	$S_5, S_6, S_7, S_8, S_{13}$
15	$S_7, S_8, S_9, S_{11}, S_{13}$	$S_2, S_3, S_5, S_7, S_{13}$
16	$S_6, S_7, S_8, S_{11}, S_{13}$	$S_2, S_5, S_6, S_7, S_{13}$
17	$S_3, S_7, S_8, S_{11}, S_{13}$	$S_2, S_5, S_7, S_9, S_{13}$
18	$S_2, S_7, S_9, S_{11}, S_{12}$	S_1, S_3, S_5, S_7, S_8
19	$S_2, S_6, S_7, S_{11}, S_{12}$	S_1, S_5, S_6, S_7, S_8
20	$S_7, S_8, S_9, S_{11}, S_{12}$	S_1, S_2, S_3, S_5, S_7
21	$S_6, S_7, S_8, S_{11}, S_{12}$	S_1, S_2, S_6, S_5, S_7
22	$S_3, S_7, S_7, S_{11}, S_{12}$	S_1, S_2, S_5, S_7, S_9
23	$S_1, S_2, S_4, S_9, S_{11}$	$S_3, S_5, S_8, S_{10}, S_{12}$
24	$S_1, S_2, S_4, S_6, S_{11}$	$S_5, S_6, S_8, S_{10}, S_{12}$
25	$S_1, S_4, S_8, S_9, S_{11}$	$S_2, S_3, S_5, S_{10}, S_{12}$
26	$S_1, S_4, S_6, S_8, S_{11}$	$S_2, S_5, S_6, S_{10}, S_{12}$
27	$S_1, S_3, S_4, S_8, S_{11}$	$S_2, S_5, S_9, S_{10}, S_{12}$
28	$S_2, S_4, S_9, S_{11}, S_{13}$	$S_3, S_5, S_8, S_{10}, S_{13}$
29	$S_2, S_4, S_6, S_{11}, S_{13}$	$S_5, S_6, S_8, S_{10}, S_{13}$
30	$S_4, S_8, S_9, S_{11}, S_{13}$	$S_2, S_3, S_5, S_{10}, S_{13}$
31	$S_4, S_6, S_8, S_{11}, S_{13}$	$S_2, S_5, S_6, S_{10}, S_{13}$
32	$S_3, S_4, S_8, S_{11}, S_{13}$	$S_2, S_5, S_9, S_{10}, S_{13}$
33	$S_2, S_4, S_9, S_{11}, S_{12}$	$S_1, S_3, S_5, S_8, S_{10}$
34	$S_2, S_4, S_6, S_{11}, S_{12}$	$S_1, S_5, S_8, S_9, S_{10}$
35	$S_4, S_8, S_9, S_{11}, S_{12}$	$S_1, S_2, S_3, S_5, S_{10}$
36	$S_4, S_6, S_8, S_{11}, S_{12}$	$S_1, S_2, S_5, S_6, S_{10}$

(continued)

Table 1 (continued)

Output level	Conducting switches	
	Positive cycle	Negative cycle
37	$S_3, S_4, S_8, S_{11}, S_{12}$	$S_1, S_2, S_5, S_9, S_{10}$

create zero voltage across the load during positive and negative cycles, respectively. Another important advantage is that it requires only five conducting switches to create any output level. The maximum voltage obtained during positive cycle is ($V_1 + V_2 + V_3 + V_4 + V_5$), and during negative cycle is $-(V_1 + V_2 + V_3 + V_4 + V_6)$. It is very important that the switching combinations (S_3, S_6, S_9), (S_4, S_7, S_{10}), (S_2, S_8), and (S_5, S_{11}) should not be turned ON simultaneously as it creates short circuit across the voltage sources.

3 Comparison Study

In this section, the proposed MLI is compared with recently presented existing MLI topologies during asymmetrical mode of operation. The various comparison indices used are number of DC sources, switches, number of ON-state switches, and the ability to create negative level. The comparison results are presented in Table 2 for a given 75-level inverter. It is shown that there are 78.5 and 60% reductions in number of sources for the proposed 75-level inverter as compared with topologies presented in [3, 5–7]. Similarly, there are 82 and 69% reductions in number of switches as compared with topologies presented in [4, 7]. It is also shown that the proposed topology requires minimum number of conducting switches as compared with the aforementioned topologies.

4 Simulation Results

The simulation results of proposed 75-level asymmetric inverter are presented in this section. The magnitude of the voltage sources is taken as $V_1 = V_2 = V_{dc} = 5$ V, $V_3 = V_4 = 15\ V_{dc} = 75$ V, and $V_5 = V_6 = 5\ V_{dc} = 25$ V, and thus, the obtained maximum voltage is 175 V. A nearest level control method is used to control the proposed MLI in which the nearest level to the reference voltage is selected at any instant of time. Therefore, the output voltage is nearly sinusoidal with minimum error with respect to the reference voltage. The total standing voltage across the switches in the proposed inverter is equal to 169 V_{dc} (i.e., equal to 845 V). The 75-level output voltage waveform and its THD are shown in Fig. 3a, b, respectively. It is observed that THD of the 75-level output voltage waveform is obtained as 1.06%. The simulation results of load voltage and load current waveform for various load power factor are shown in Fig. 4, and the corresponding results are given in Table 3. It is inferred that

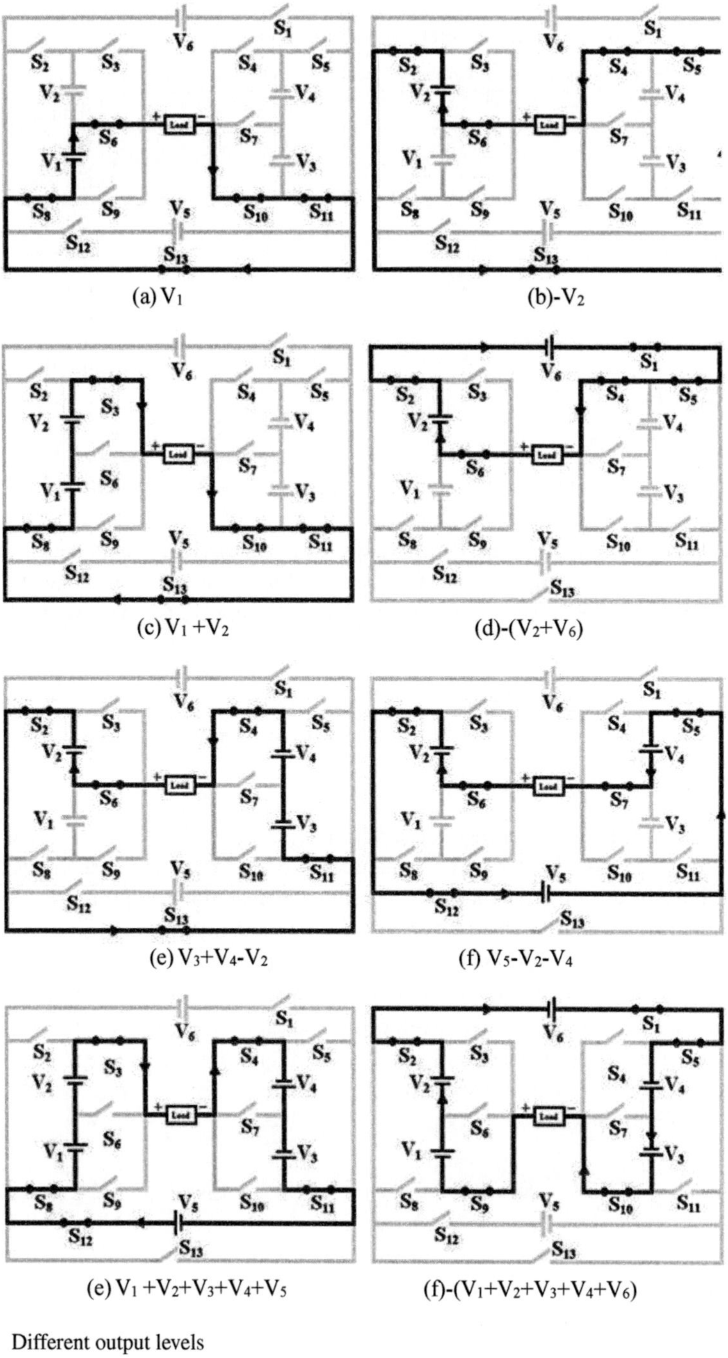

(a) V_1 (b)-V_2

(c) $V_1 + V_2$ (d)-(V_2+V_6)

(e) V_3+V_4-V_2 (f) V_5-V_2-V_4

(e) V_1 +V_2+V_3+V_4+V_5 (f)-(V_1+V_2+V_3+V_4+V_6)

Fig. 2 Different output levels

Table 2 Comparison analysis for 75-level inverter

Topology	Number of sources	Number of switches	ON-state switches	Negative level
Ref. [1]	11	23	6	With H-bridge
Ref. [2]	6	15	5	With H-bridge
Ref. [3]	28	18	6	Inherent
Ref. [4]	19	72	11	Inherent
Ref. [5]	15	16	9	Inherent
Ref. [6]	15	18	9	Inherent
Ref. [7]	15	42	8	Inherent
Ref. [8]	9	26	9	Inherent
Proposed	6	13	5	Inherent

due to the low-pass filtering characteristic of the series RL load, the output waveform contains less THD and is nearly sinusoidal.

The comparison of the THD of the output voltage waveform obtained from the different proposed inverter topologies is shown in Table 4. It is observed that the THD of the output voltage waveform is reduced considerably by the proposed 75-level inverter.

5 Conclusion

In this paper, a new 75-level inverter structure with reduced number of switches is proposed. The proposed inverter can generate both positive and negative output levels and facilitate the asymmetric operation regarding the magnitude of input DC sources. The comparison study between the proposed 75-level inverter and other recently developed topologies is presented in this paper. The study shows that the proposed 75-level inverter has superior features in view of number of sources, switches, and ON-state switches over other recently developed topologies. To validate the performance of the suggested 75-level inverter topology, the simulation results were presented. The result shows that the proposed inverter produces high output voltage containing low value of harmonics.

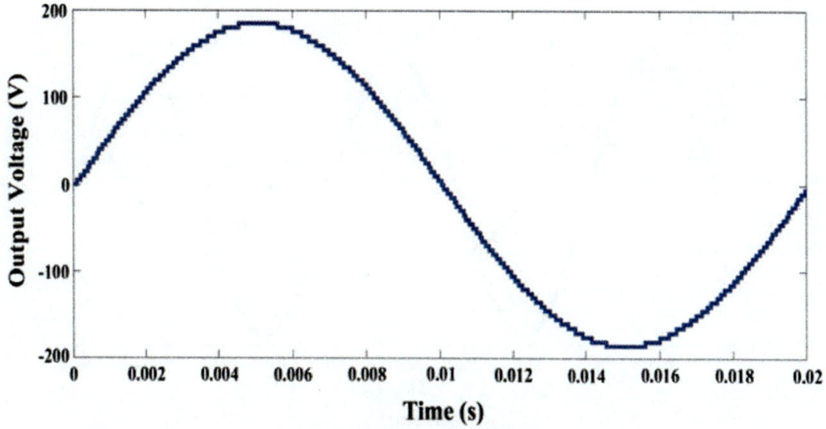

(a) Output Voltage

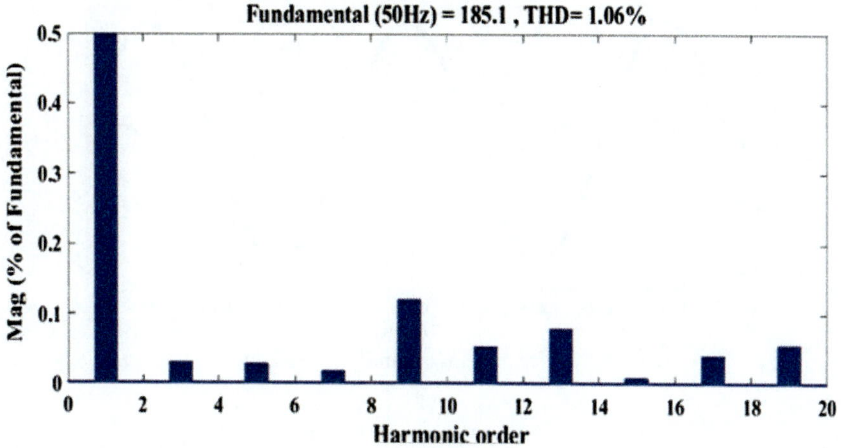

(b) THD

Fig. 3 Simulation results—75 level

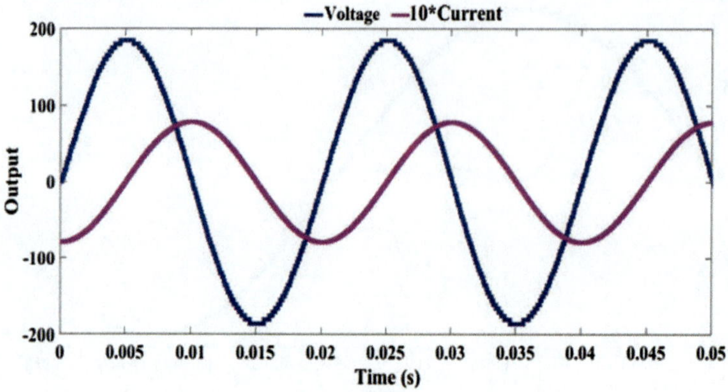

(a)75mH

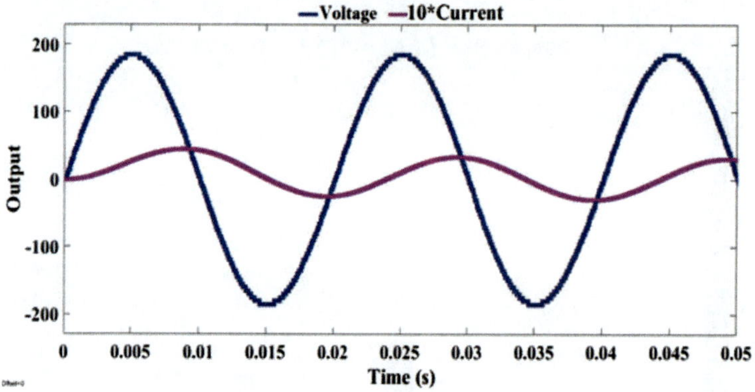

(b) 15Ω,185mH

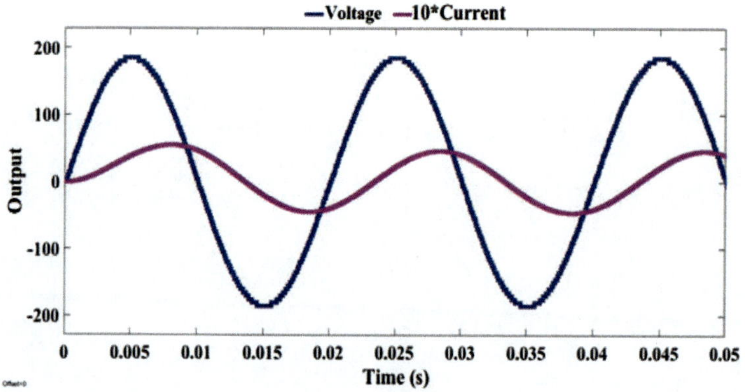

(c) 20Ω,110mH

Fig. 4 Simulation results—75 level for various load p.f.

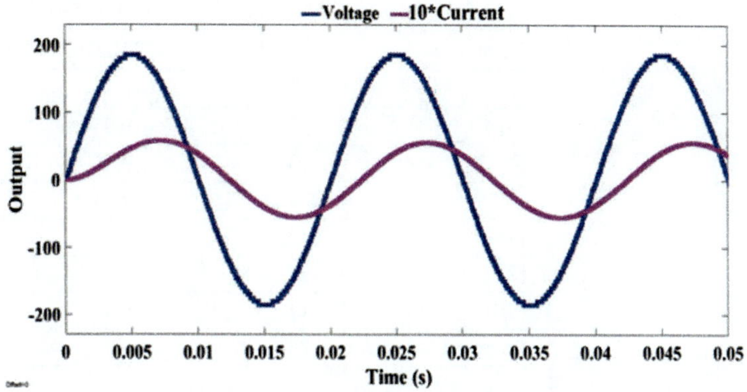

(d) 25Ω,70mH

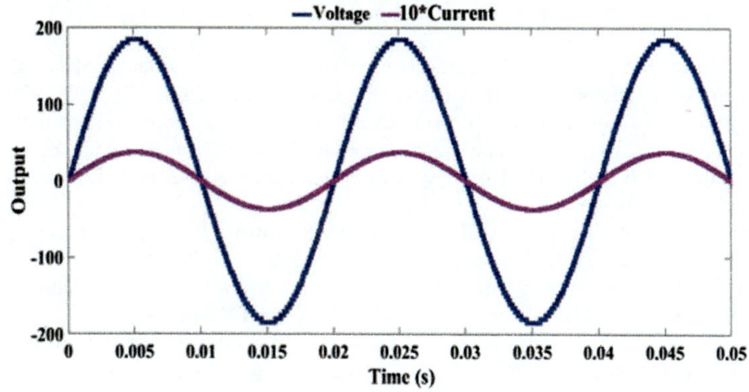

(e) 50Ω

Fig. 4 (continued)

Table 3 Details of load variation study for 75-level inverter

S. No.	Load	Power factor	Load voltage THD (%)	Load current THD (%)
1	75 mH	0	1.06	0.03
2	15 Ω, 185 mH	0.3	1.06	0.04
3	20 Ω, 110 mH	0.5	1.06	0.06
4	25 Ω, 70 mH	0.75	1.06	0.09
5	50 Ω	1	1.06	1.06

Table 4 Comparison of THD

References	Level	THD (%)
[2]	31	2.7
[3]	17	10.95
[4]	13	9.46
[6]	9	14.23
[7]	15	7.36
[8]	29	4.62
[9]	17	4.87
Proposed	75	1.06

References

1. Babaei, E., Laali, S., Bayat, Z.: A single-phase cascaded multilevel inverter based on a new basic unit with reduced number of power switches. IEEE Trans. Indus. Electron. **62**(2), 922–929 (2015)
2. Thiyagarajan, V., Somasundaram, P.: Multilevel inverter topology with modified pulse width modulation and reduced switch count. Acta Polytech. Hung. **15**(2), 141–167 (2018)
3. Odeh, C.I., Obe, E.S., Ojo, O.: Topology for cascaded multilevel inverter. IET Power Electron. **9**(5), 921–929 (2016)
4. Prabaharan, N., Palanisamy, K.: Analysis of cascaded H-bridge multilevel inverter configuration with double level circuit. IET Power Electron. **10**(9), 1023–1033 (2017)
5. Babaei, E., Laali, S., Alilu, S.: Cascaded multilevel inverter with series connection of novel H-bridge units. IEEE Trans. Industr. Electron. **61**(12), 6664–6671 (2014)
6. Prayag, A., Bodkhe, S.: Novel basic block of multilevel inverter using reduced number of on-state switches and cascaded circuit topology. Adv. Electr. Eng. **2017**, 1–16 (2017)
7. Thamizharasan, S., Baskaran, J., Ramkumar, S., Jeevananthan, S.: Cross-switched multilevel inverter using auxiliary reverse connected voltage sources. IET Power Electron. **7**(6), 1519–1526 (2014)
8. Ajami, A., Mokhberdoran, A., Oskuee, M.R.J.: A new topology of multilevel voltage source inverter to minimize the number of circuit devices and maximize the number of output voltage levels. J. Electr. Eng. Technol. **8**(6), 1328–1336 (2013)
9. Thiyagarajan, V., Somasundaram, P.: Modeling and analysis of novel multilevel inverter topology with minimum number of switching components. CMES Comput. Model. Eng. Sci. **113**(4), 461–473 (2017)

Stud App—Student Data Management Application

Robin P B, Pavan Kurian, and Justin Mathew

Abstract Nowadays, in most of the engineering colleges, students are using the conventional methods for applying for duty leaves, medical leaves, other permissions, etc. Also, the students' data regarding their personal information, academic performance, etc., are kept in computer databases or in special record books. The students submit the hard copy of the certificates of their activities to their faculty advisors. The activity points earned by each student have to be calculated and uploaded to the university websites by the faculty staff. As the number of students is high, it is a tedious task. It will be easier for the students and faculties if there is a facility which enables them to apply for leaves, submit certificates, store and access academic details of students, etc., via an application. This paper explains a web application that will enable the students and faculties to quickly process the leave applications and other requests, update and access students' data, upload certificates for activity points, etc.

Keywords Duty leave · Activity points · Permissions · Web application

1 Introduction

1.1 General Background

For student data management, activity points' calculation and permissions' granting purposes, most colleges are using the conventional methods [1]. This involves a lot

R. P. B (✉) · P. Kurian · J. Mathew
Department of Computer Science & Engineering, Saintgits College of Engineering, Kottayam, India
e-mail: robinpbabu55@gmail.com

P. Kurian
e-mail: pavankurian12345@gmail.com

J. Mathew
e-mail: justin.mathew@saintgits.org

P. Suresh et al. (eds.), *Advances in Smart System Technologies*,
Advances in Intelligent Systems and Computing 1163,
https://doi.org/10.1007/978-981-15-5029-4_29

of paper works. This paper explains a web application that will enable the students and faculties to quickly process the leave applications and other requests, update and access students' data, upload certificates for activity points, etc. [2]. The application provides two separate views, one for students and the other for the faculties. Each of them can login using their registered email address or university register number and password [3]. The biggest merits of the application are that it makes processing easier and faster and also eliminates the overhead of keeping separate hard copy of records for each student.

The students submit their requests to their staff advisors. As soon as a student applies for some permission or upload a certificate for activity points, the concerned faculties will get a notification regarding the same [4]. The faculty will then approve it or forward it to the higher authority as per the need. As soon as the request is approved or rejected, the student will get notified about the same. The application enables the faculties to access students' data easily via this app. Also, the activity points calculation will be easier as the points are awarded to the students within a short time after the submission of certificates. Hence, the student will be able to know the current status of activity points and can plan his activities accordingly.

1.2 Relevance

In the present world, time is one of the most valuable things. This web application helps us to save a lot of time by reducing the amount of time required for completing various functions by providing a user-friendly, simple and easy interface which can be used by the faculties as well as students. Also, this will reduce the amount of paper work needed. The overhead of keeping separate hard copy of different data of each student is eliminated.

1.3 Socio-economic Importance

As there are large number of students, using paper for keeping all their information will increase the cost. By using this application, we can reduce the use of paper for record-keeping purposes. This avoids the use of paper for keeping academic, personal and other information of students. The expenses for buying papers and record books for storing this information can be avoided, and it will be a step towards the Go Paperless Initiative. Thus, this application has a positive impact on social and economic aspects.

2 Existing System

2.1 Overview of Existing System

Most of the student management systems that are currently available are focusing on attendance management, faculty contact information sharing and exam result publishing. Also, most of these systems are not very user friendly. The features like activity points calculation, permissions managements are not available in them. These features, if present, will be very helpful in managing these types of requests easily and quickly. So in the existing systems, apart from the attendance management, and some other limited features, most of the other types of functions are done by paper work. Time required for completing the paper works is high, and it also requires a lot of paper. Storing these hard copy records is another overhead. As the number of students is quite high, the number of hard copy records to be stored will also be high.

2.2 Limitations of Existing Systems

The existing systems have a lot of limitations. These limitations have to be overcome in order to develop a system which will minimize the use of paper to the maximum extent. Some of the major limitations of the existing systems are:

- No permissions management functionality
- No activity points management functionality
- Not very user friendly
- Slow processing.

3 Objective and Proposed Innovation

3.1 General Objective

The general objectives of this application are to eliminate the overhead of keeping separate hard copy records of students' data and to simplify the procedures for availing duty leaves, medical leaves, etc. Thus, a lot of time can be saved, and the amount of paper work can be considerably reduced. The system also ensures location independent access to data quickly and easily.

3.2 Proposed System

More focus is given to features such as activity points management and permissions requesting. As soon as a student submits the certificates to staff advisor via this app, he/she can check it and assign activity points. Thus, the student will be able to know how much more points are required and plan his activities accordingly. Requesting permissions from head of the department and staff advisors will become a lot more easier. Once a student submits a request, the concerned authority will get a notification about the same immediately. He can approve or reject the request via this app after examining the request. The proposed system eliminates the use of paper completely. Student data management, activity points management and permissions requesting can be done using this app. It enables the users to access the data independent of their location. As the data are stored in a server, it can be accessed by the users even if they are not present in the institution. By using this app, unexpected holidays will no more affect obtaining permissions and accessing important data. Also, the overhead of keeping separate hard copy records of different kinds of data can be eliminated by using this application.

4 Design

There are two separate views, one for students and the other for faculties. The students and faculties can login using the 'Student Login' and 'Faculty Login', respectively (Fig. 1).

4.1 Student View

In the student's home page, there are four buttons. On clicking the 'General Info' button, the personal information and academic information of the student can be viewed. Personal information like name, age, date of birth, blood group, parent's details, etc., are available. Academic information like admission number, university register number, roll no, department, etc., are available.

On clicking the 'Academics' button, the academic data of semester exams including course code, course name, grade and number of attempts are available. The academic data for all the eight semesters will be displayed here. On clicking the 'Permissions' button, then student can see three buttons: one for submitting new requests, one for viewing the pending requests and one for viewing the requests that are approved. Submitting new requests will be an easy process as the user only has to fill in a few details and then just click on the Submit button.

In the 'Certificates' tab, the user can upload new certificates and submit them to the staff advisor for evaluation. The staff advisor verifies the certificate and gives

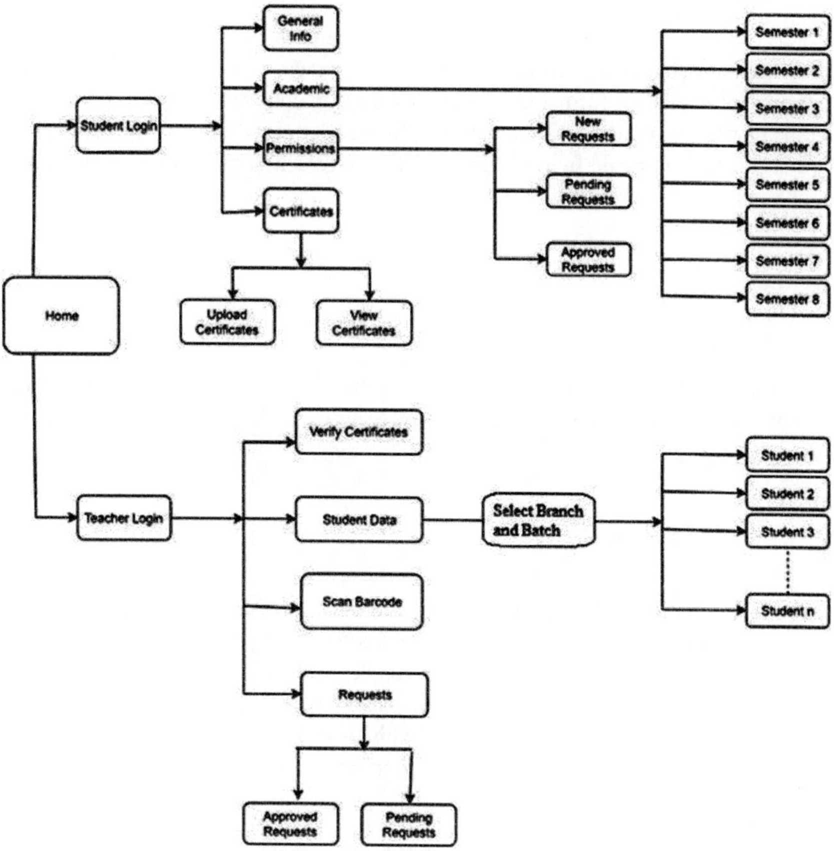

Fig. 1 Activity diagram

activity point to it. The student can view the list of certificates that is already verified and activity points awarded by the staff advisor by clicking on the 'View Certificates' button.

4.2 *Faculty View*

In the faculty home page, there are four buttons. On clicking the 'Verify Certificates' button, the certificates submitted by the students under the advisory of that faculty can be seen. The faculty can verify the certificates and give activity points to the students. In the 'Student Data' tab, the faculty can view the details of the students by selecting the branch and batch of the student. On clicking the 'Scan Barcode' button, the faculty can scan the barcode on the identity card of a student. If a faculty finds a student indulging in inappropriate activities inside the campus, he/she may ask the student his identity card and scan the barcode in the identity card. The details of that student including his/her name, branch, semester, mobile number, staff advisor, contact information of staff advisor and contact information of parents will be displayed on scanning the barcode. In the 'Requests' tab, the faculty can see the list of requests made by the students. He can reject, approve or forward it to a higher authority accordingly.

5 Results

There are two different views, one for the students and the other for the faculties. Here are some snapshots of the student view (Figs. 2, 3, 4, 5, 6, 7, 8 and 9).

6 Conclusion and Future Scope

The paper explains Stud App, a web application that will enable the students and faculties to quickly process the leave applications and other requests, update and access students' data, upload certificates for activity points, etc. The application provides two separate views, one for students and the other for the faculties. Each of them can login using their registered email address and password. The biggest merits of the application are that it makes processing easier and faster and also eliminates

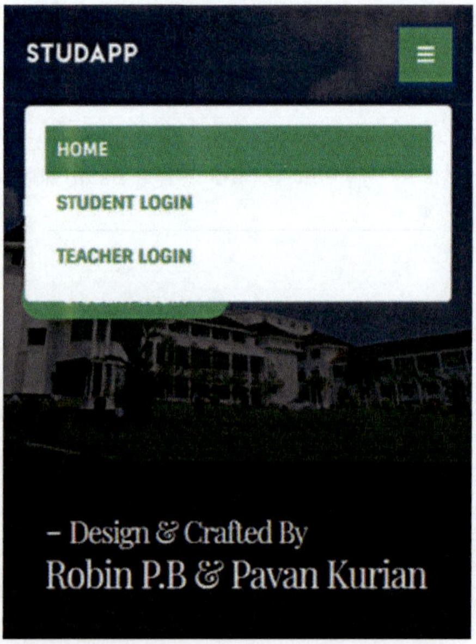

Fig. 2 Login page

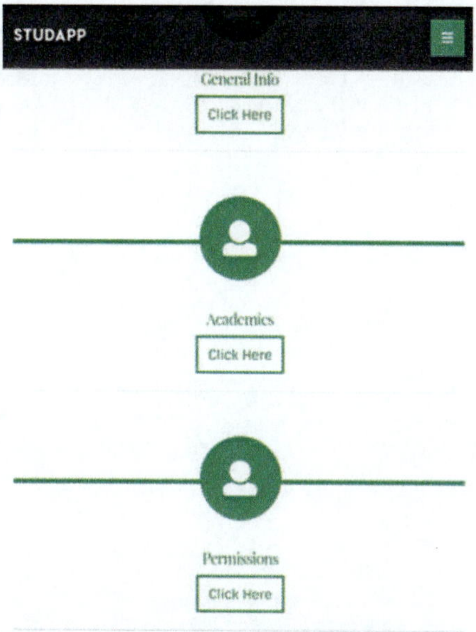

Fig. 3 Student dashboard

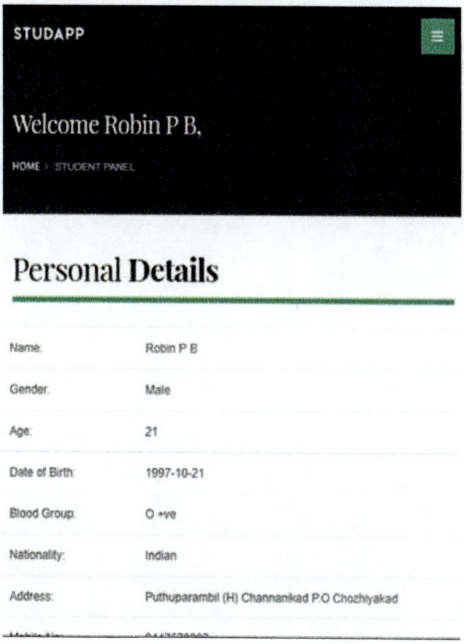

Fig. 4 Student data

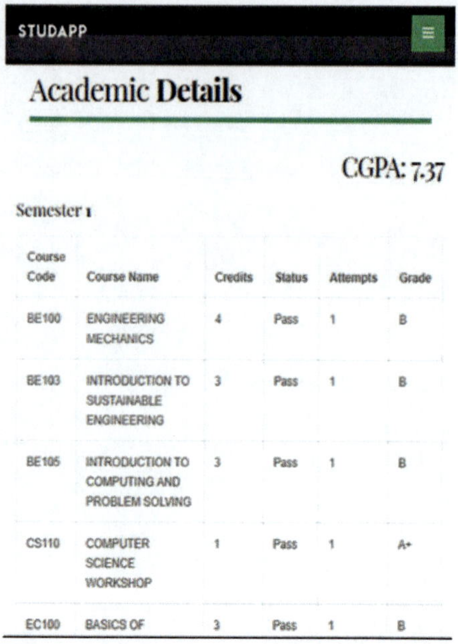

Fig. 5 Academic details

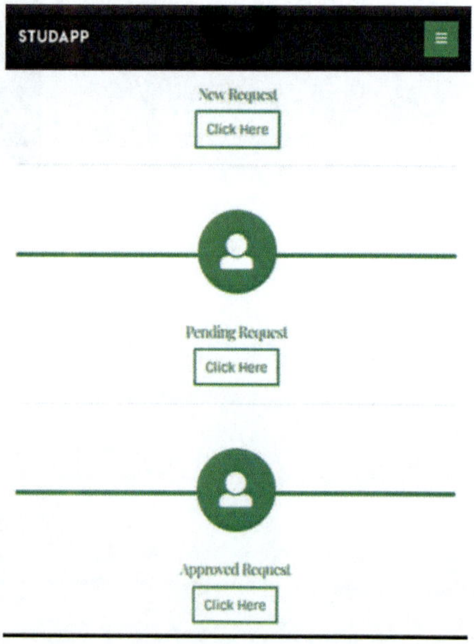

Fig. 6 Permissions tab

Fig. 7 New request

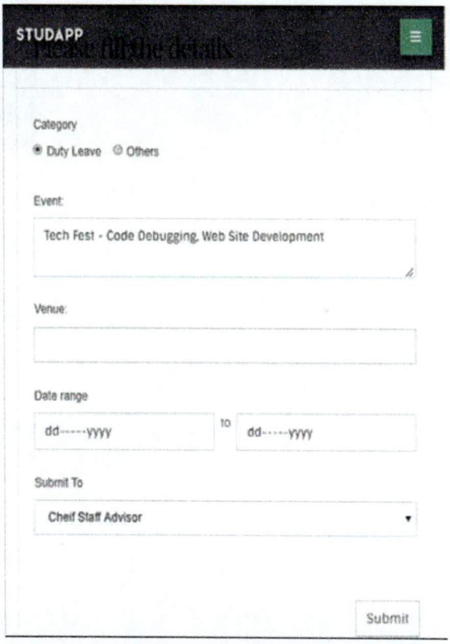

Fig. 8 Pending requests

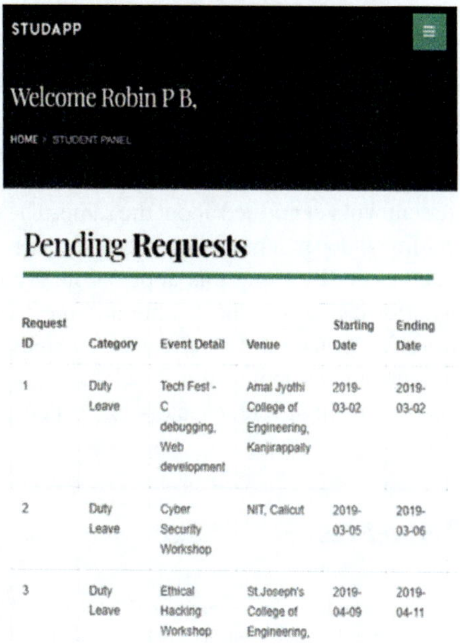

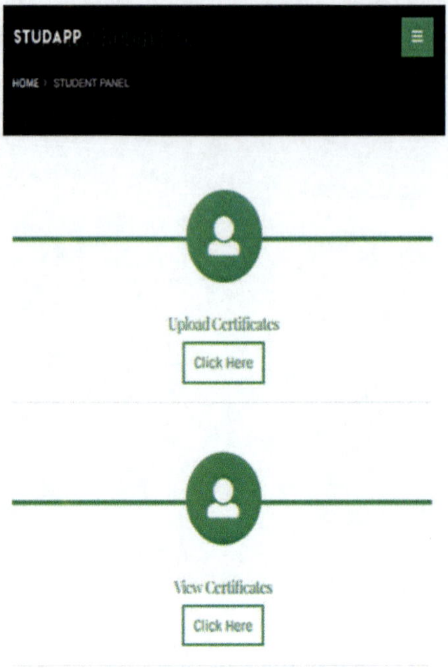

Fig. 9 Upload certificates

the overhead of keeping separate hard copy of records for each student. As soon as a student apply for some permission or upload a certificate, the concerned faculties will get a notification regarding the same. The faculty will then approve it or forward it to the higher authority as per the need. As soon as the request is approved or rejected, the student will get notified about the same. The student data can be accessed by faculties anytime independent of their location. Activity points calculation of students will be much easier by using this application. By using this application, the students and faculties can save a lot of time and the paper usage can be considerably reduced. More functionalities like attendance management, exam result publication, etc., can be added to this application. The leave-applying functionality discussed in this paper can be used in different sectors like offices, schools, industries, etc.

References

1. Bharamagoudar, S.R., Geeta, R.B., Totad, S.G.: Web based student information management system. Int. J. Adv. Res. Comput. Commun. Eng. **2**(6) (2013)
2. Jewel, M.R., Rakib Uddin, M., Mohenul Islam, M., Asadushjaman, M., Ahmed, K.T., Repon Hossen, M., Saha, S.: An android app for university management system. IOSR J. Mob. Comput. Appl. (IOSR-JMCA) **4**(I) (2017). e-ISSN: 2394-0050, P-ISSN: 2394-0042

3. Mohitha, H.V., Mohan, A., Reji, J., Babu, B.S., Tintumol, M.G.: Design and implementation of improved system for college automation and management. Int. J. Innov. Res. Comput. Commun. Eng. **5**(5) (2017)
4. Kalmegh, P.A., Gavfale, R.R., Shelke, R.R.: Student information system with working. Int. J. Res. Appl. Sci. Eng. Technol. (IJRASET) **4**(III) (2016)

Automatic Wheat Grain Grading System Using Physical and Chemical Characteristics

S. Rajalakshmi, G. Dineshraj, R. Brindha Priyadharshini, and R. Divya Brindha

Abstract Agriculture is the backbone of our country. Automation in agriculture is a part of Digital India. The grade classification of grain/fruit/vegetable/spices, etc., can be automated. Most of the available agricultural product classification systems analyze only certain defining characteristics of the sample studied. Increasing demand for wheat and its products calls for a quick and reliable classification of the wheat. Our aim is to develop a wheat grading system by incorporating physical and chemical characteristics of wheat. Physical features like the shape, size, and color gradient of wheat samples are obtained from scanned grain images using image-processing techniques. Chemical features like moisture and protein contents are measured using equipments. These features are trained in a neural network for classification of wheat into five standard grades. Results show that classification accuracy is improved by combining physical and chemical features of wheat grains. The analysis of the performance of neural networks with other machine learning techniques like support vector machine, Naive Bayes, and decision tree shows that neural networks give better accuracy when compared to others.

Keywords Wheat grading · Agriculture automation · Physical and chemical features · Embedded system · Image processing · Neural networks

S. Rajalakshmi (✉) · G. Dineshraj · R. Brindha Priyadharshini · R. Divya Brindha
Department of Computer Science and Engineering, SSN College of Engineering, Chennai, Tamil Nadu 603110, India
e-mail: rajalakshmis@ssn.edu.in

G. Dineshraj
e-mail: dinesh12028@cse.ssn.edu.in

R. Brindha Priyadharshini
e-mail: brindha12024@cse.ssn.edu.in

R. Divya Brindha
e-mail: divya12029@cse.ssn.edu.in

P. Suresh et al. (eds.), *Advances in Smart System Technologies*,
Advances in Intelligent Systems and Computing 1163,
https://doi.org/10.1007/978-981-15-5029-4_30

1 Introduction

Agriculture is the important part in deciding the Indian economy. Due to the advancement of technology in agriculture, the need for the accurate prediction of quality of agricultural products is increased [1]. Conventional wheat grading system might be detrimental to the agricultural market as they are time-consuming. Existing wheat systems (Indian or otherwise) involve many manual procedures which include manual weight determination, separating shrunken and broken kernels, and finally determination of the class of wheat itself [2, 3]. If we go for automation, most of the methods involve computer vision techniques only [4, 5] and do not take the chemical properties (moisture, protein) into consideration.

Rice [6] and wheat grading [7] have some similarity in the technologies required for grading. Rice grading technology is quite well researched, with SVM techniques achieving up to 86% accurate classification [8]. Integrated testing of multiple grain types (rice, wheat, corn) [9] has achieved 85% accuracy, but is limited to only one or two varieties of the grain. So, wheat grading research would be of good use in the current scenario.

In already experimented technologies in grain testing [10], a neural network is a popular option to identify the grade. Rice cultivar identification using back propagation with two hidden layers and 36-6-5-5 and 36-9-6-5 topologies were extremely successful, with between 98.8 and 100% accuracy [11]. Wheat classification using artificial neural networks (ANN) obtained 85% accuracy with one hidden layer, using only morphological features [12].

Based on the examination of these previous researches, we aim to achieve good accuracy using image processing and ANN, by also including chemical features along with the physical properties in the analysis. So far, chemical features have not been used in most of the grain grading attempt. However, chemical features play a key role in the grade classification. Indian Agmark standards (Manual on Standards of Wheat) specify chemical standards (like protein content, moisture content, maximum acceptable pesticide limits, maximum acceptable insect feces, etc.) as a crucial part in grade specification. This paper takes a look at moisture information collected from the wheat and its importance in the final grading. Our proposed approach also examines the importance of color gradient and protein analysis, all as crucial chemical features of the grade.

2 Materials and Methodology

2.1 Wheat Grain Samples

Based on the agro-climatic and cultural conditions, at present there are about 293 varieties in wheat grains given by the Central Sub-committee on Crop Standards, Notification and Release of Varieties for Agricultural Crops, Central Variety Release

Table 1 Wheat varieties released in India during 1965–2018 [13]

Species	No. of varieties released by CVRC	No. of varieties released by SVRC	Total
Bread wheat (triticum aestivum)	246	132	378
Durum wheat (triticum durum)	37	22	59
Dicoccum wheat (triticum dicoccum)	06	01	07
Triticale	04	–	04
Total	293	155	448

Committee (CVRC), and 155 varieties released by State Varietal Release Committee (SVRC) of the states. Out of these, 378 were bread wheat, 59 durum wheat, and 7 dicoccum wheat, besides 4 triticale [13]. Table 1 shows the number of wheat releases according to the wheat varieties.

This work on grade classification primarily focuses on samples of different grades of durum, dicoccum, and organic wheat samples. Different samples of wheat are collected from various regions of Tamil Nadu in India. Around 500 samples are collected and used as dataset for the proposed work.

2.2 *Methodology—Proposed System*

The process can be divided into two major sections. First, the focus is on extracting the physical characteristics of the wheat grain, i.e., the physical appearance of the grains. Grain quality is characterized by factors like color, composition, bulk density, aroma, size, and shape [14]. Secondly, the focus is to extract the chemical characteristics of wheat grains. Chemical features include the moisture and protein content of grains. Wheat is normally harvested with a moisture content of 10–12% [15]. In most countries, moisture content is not a part of the grading system, though it affects quality of wheat grain, as it is inversely related to dry matter loss. The moisture content is measured with the help of grain moisture meter (GMM). The protein content plays a major role in deciding the grade and quality of wheat. Wheat contains five different classes of protein: (i) albumin (soluble in water), (ii) globulin (soluble in salt solution), (iii) gliadin (soluble in 70% aqueous ethanol), (iv) proteose (soluble in water), and (v) glutenin (soluble in dilute acid or alkali). The protein content of the wheat can be obtained by using infrared spectroscopy method.

Figure 1 illustrates the workflow of the proposed system. The physical characteristics are extracted using the image processing techniques applied on the wheat grain sample images. The chemical characteristics are extracted using grain moisture meter and near infrared spectroscopy method. All the information will be processed using

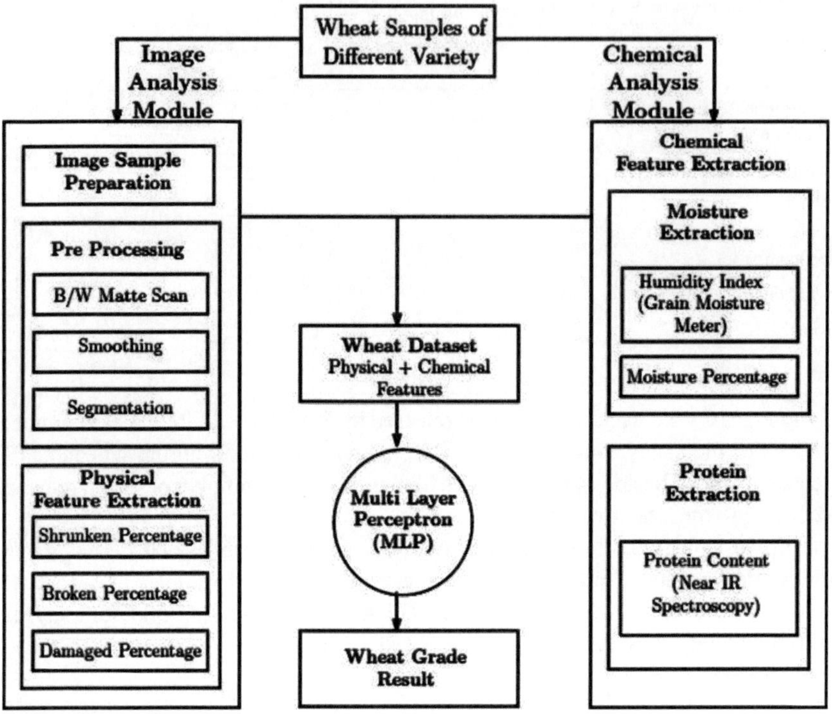

Fig. 1 Architecture for wheat grain grading system

multilayer feed-forward neural network or multilayer perceptron (MLP) methodology to obtain the grade of the wheat grains instantly. The proposed algorithm is aimed to obtain maximum accuracy in the grading of wheat. The proposed system consists of three modules:

1. Physical features—image analysis module.
2. Chemical features—chemical analysis module.
3. Wheat grade classification—machine learning techniques.

3 Image Analysis Module—Physical Feature Extraction

3.1 *Shape, Size, and Broken Kernel Detection*

Wheat samples are scanned using CIS scanner with resolution set to 200dpi. Different grades of samples are randomly arranged on surface of flatbed scanner, and scanned with black background (Fig. 2) and also with white background (Fig. 3). There are few specifications to satisfy and ensure that kernels are not overlapping or hiding each other. There are approximately 50–80 kernels per image.

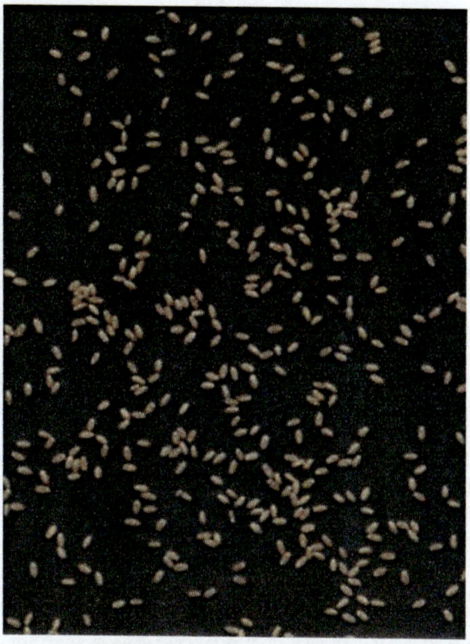

Fig. 2 Wheat scan against black background

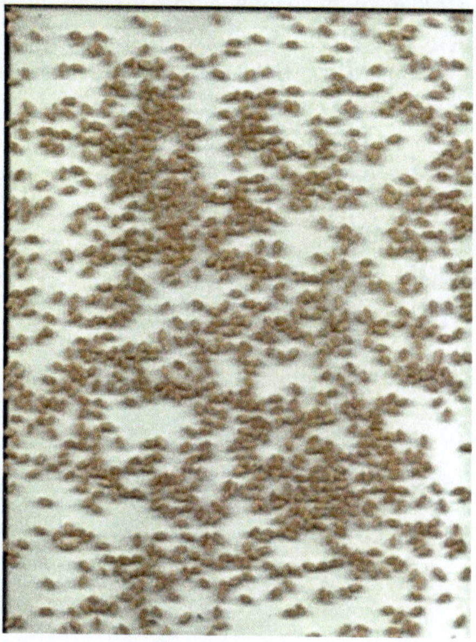

Fig. 3 Wheat scan against white background

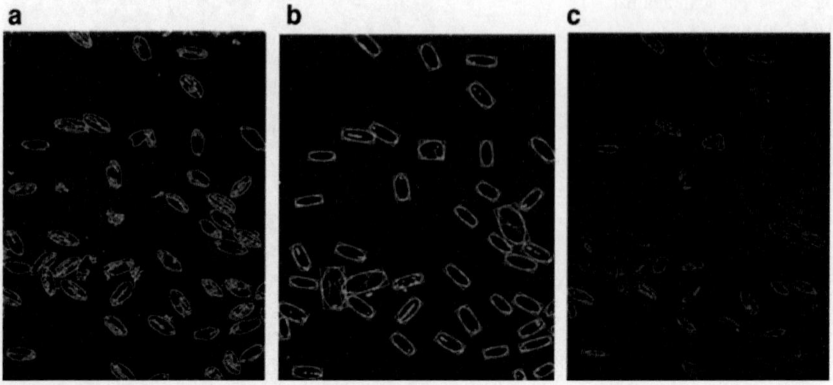

Fig. 4 **a** Outline using Canny edges, **b** ellipse contour identification, and **c** concave/convex hulls (broken kernel)

The images are processed using OpenCV 3.0 [16], by first extracting the outlines of the wheat kernels using Canny edge detection [17], and the resulting image is shown in Fig. 4a. As a part, the pre-processing is performed on the images, which included slight shadow elimination on the images with white background. Areas of interest are identified by eliminating white areas surrounding the wheat kernels [14], so that there is a rough approximation of whether a pixel belonged inside a kernel or outside.

Depending upon the nature of scan and resolution, the resolution specifications need to be modified. But, dynamic thresholds are avoided due to similar nature of the images. Firstly, the pre-processed image is examined for elliptical shapes aimed to identify the shape of the kernels and worked under the assumption that all the kernels are whole. The resultant image is analyzed for contours detection using the steps mentioned in Algorithm 1.

Algorithm 1 Ellipse Contour Identification

Input: Input image
Output: Output image with marked contours
contour_image ← Apply Canny Edge detection on input image to mark the edge of the wheat kernels
foreach identified contour ∈ contour_image **do**
  if pixel size(contour) ≤ 6 **then**
    Continue to next contour
  end
  Fix ellipse points on contour
  Obtain the containing rectangle box of points for contour
  if box_width ≥ box_height ×3 or box_height ≥ box_width ×3 then
    Continue to next contour
  end
  Draw the ellipse and rectangle of contour in contour image
  Display centres, dimensions, rotation angle of rectangles
end
return contour_image

The collections of contour arrays are then searched for shapes that would be elliptical or roughly elliptical. After experimenting with the threshold to identify ellipses and their minimum sizes, the threshold is adjusted to the images of the samples being used. Identification of ellipses is followed by finding the containing rectangles of the ellipses [18], as shown in Fig. 4b. This would give the size of the object, the centers of the rectangle, the length and breadth of the rectangle corresponded to the kernel parameters.

The percentage of amount of shrunken kernel is calculated as summation of size of all the kernels (size_i) and whole divided by total number of kernels (N) and size of the biggest kernel (S) identified by ellipse contour as shown in Eq. 1.

$$\text{percentage}_{\text{shrunken}} = \frac{\sum_{i=1}^{N} \text{size}_i}{N * S} \quad (1)$$

Extraction of these features are tried with many images to ensure that no non-kernel shapes are identified as ellipses and to reduce pixels detected falsely as inside (false positives) and vice versa (false negatives). Accuracy is crucial as these ellipses and edges in combination would test pixels in later stages to see if they are kernel or non-kernel pixels.

The next stage involves detection of broken and insect-bitten kernels. For broken kernels, a low threshold is used while searching for ellipses to include background space in ellipse formation, to identify incomplete kernels, and then test against the originally detected canny edges. The broken kernels are identified from Canny edges. Thus, percentage of broken kernels is calculated using Eq. 2.

$$\text{percentage}_{\text{broken}} = \frac{n * 100}{N} \quad (2)$$

where n is the number of broken kernels based on the shrunken percentage and N is the total number of kernels in the given image.

The original images are reprocessed for finding concave shapes (insect bites), as shown in Fig. 4c. Such kernels are compared again to eliminate duplicates among broken and bitten kernels. By analyzing the scans in this manner, the physical features of interest like kernel shape, size, and brokenness, insect-bitten are extracted.

3.2 Color and Damaged Kernel Detection

The color of an object is determined by wavelength of light reflected from the surface of the wheat granules. Hence, it depends on the wavelength. These spectral variations provide a unique key to machine vision and image analysis [19]. Color histogram is graphical representation of the RGB values across the wheat kernels shown in Fig. 5. Using the RGB values, one can determine whether the wheat kernels are damaged or not. And hence, one can obtain the damaged kernel.

A histogram is a graphical representation of data. It divides the plane into bins and evaluates or plots for each bin. The image is divided into 256 bins. The y-axis represents the number of pixels. Figure 5 shows the values plotted for different color variables (*R*, *G*, and *B*). Each color can be represented using a combination of these three values.

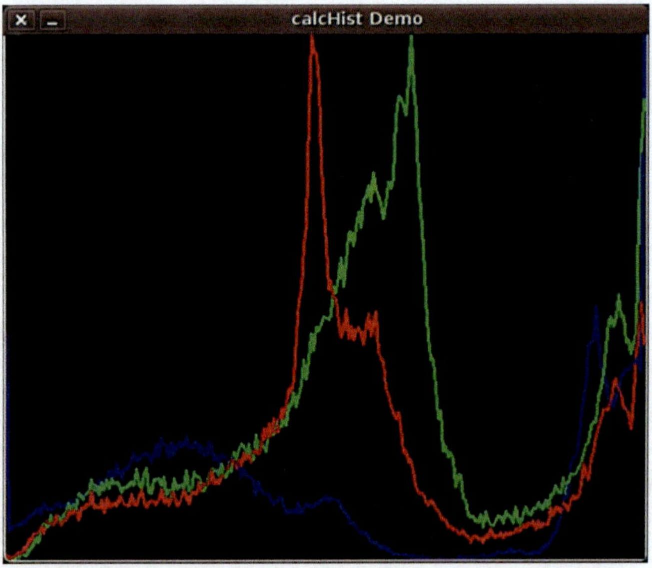

Fig. 5 Color histogram representation

Fig. 6 Damaged kernel circled in red

A separate line is plotted in graph for each color by traversing through each pixel in the image. Using the color histogram, the distribution of the each component *g*(*r*, *g*, *b*) is obtained which is further used to detect the deviation (*d*) from the normal color distribution *h*(*r*, *g*, *b*) for finding the damaged kernels [20]. Equation 3 is used to determine the deviation of damaged kernel color distribution from the normal color distribution. Figure 6 shows the damaged kernels in the input image marked with red-colored circle.

$$d^2(h, g) = \sum_R \sum_G \sum_B (h(r, g, b) - g(r, g, b))^2 \tag{3}$$

4 Chemical Analysis—Chemical Feature Extraction

Apart from morphological features, chemical features are also obtained. Moisture is an important determinant in grade decisions. Agmark standards specify moisture content as one of the nine parameters to differentiate grade [21]. Moisture is important as it determines what happens to wheat during storage. Studies prove that low moisture affects flour protein, flour extraction, dough-mixing characteristics, and alkaline noodle color [22], while heavy moisture favors quick storage deterioration, by allowing mold, insect infestation, and reducing protein [23].

A handheld device grain moisture meter shown in Fig. 7 is used to calculate the moisture in the wheat samples. Sample wheat is scooped into the devices holding area, and after calibrating to wheat grains, the moisture reading is noted [24].

Protein is the most important chemical feature in the wheat grading, yet it is the most neglected among the many automated grading methods. Near-infrared spectroscopy seems to have been established as the most trustworthy method in determining protein content and the effect of protein on grade, and their modeling accuracy

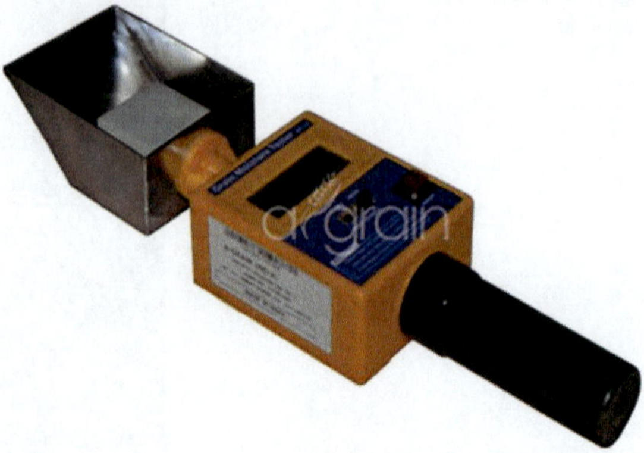

Fig. 7 Grain moisture meter [24]

has been studied [25]. The wheat samples are also tested for protein using the same method, and protein value is also used together with other gathered features to decide the grade.

5 Wheat Grade Classification—Machine Learning Techniques

5.1 Multilayer Perceptron

Multilayer perceptron (MLP) is a supervised learning multilayer feed-forward neural network [26]. It learns by back propagating the error and corrects the learning by modifying the bias and weights applied to the neurons in each layer. The neural network (Z) is fully connected, and each neuron is connected with certain weights (w), to every other node in the following layer. Bias (b) is an extra input to neurons, and it is always 1 to make sure that even when all the inputs are none (all 0 s), there is going to be activation in the neuron. Equation 4 is used to calculate the forward propagation value from N neurons (x) in previous layer to neuron in next layer.

$$Z = b + \sum_{i=1}^{N} (x_i)(w_i) \tag{4}$$

Figure 8 shows the trained neural network model for wheat grade classification. MLP network consists of five input neurons, three neuron hidden layer, and five output neurons to obtain the highest accuracy after multiple cross-validation with same dataset.

Each input neuron corresponds to the extracted features' shape, size, color, moisture, and protein of the wheat kernels. The output neurons (y grades) correspond to probability of five grades of wheat given by Eq. 5. The neuron class with highest

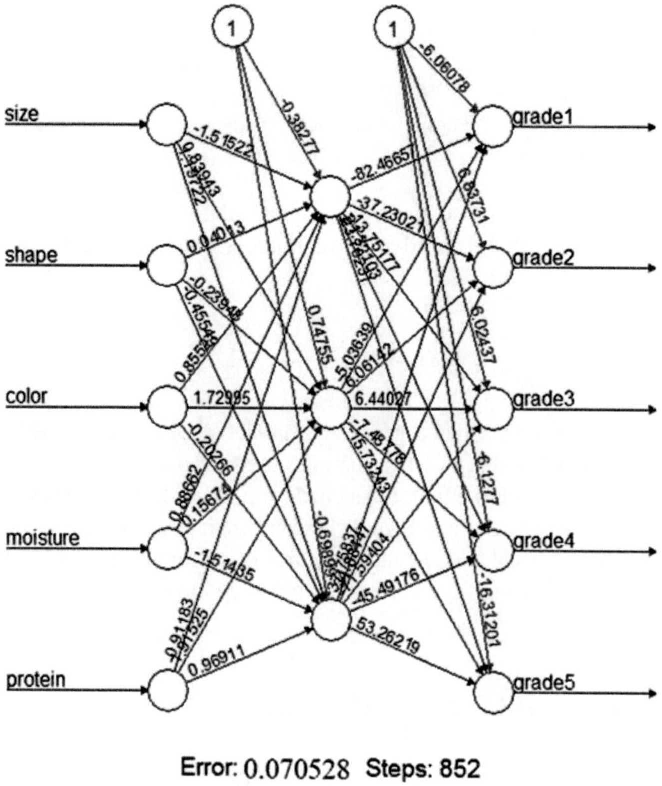

Fig. 8 MLP network and weight parameters

probability is taken as the grade of the given wheat sample.

$$y_{\text{grades}} = P(Z) \tag{5}$$

R programming language is used to implement the machine learning techniques. The neural network visualization done using plot () function is shown in Fig. 8. For the neural network to be nonlinear, activation functions are used. The activation function is the actual function that maps the weighted input neuron values to the output wheat grades. Tanh activation function shown in Eq. 6 is applied for our network.

$$f(x_i) = \tanh(x_i) = \frac{2}{1 + e^{-2x_i}} - 1 \tag{6}$$

Now the test data is fed to the trained neural network for finding the grade. The error is calculated for the model trained with only morphological properties of wheat, i.e., shape, size, and color, and the model trained with both morphological

Table 2 Error measures for the system with various features

Features	MSE
Physical features (alone)	0.568963
Physical + chemical features	**0.070528**

and chemical features. The loss function computes the error for each test example. Mean squared error (MSE) is used as a loss function. The formula to calculate the MSE is shown in Eq. 7.

$$\text{MSE} = \frac{\sum_{i=1}^{n} (y_i - \widehat{y_i})^2}{n} \tag{7}$$

where $\widehat{y_i}$ and y_i are the predicted grade and the actual grade of ith wheat sample input, and n the total number of wheat samples used for training.

Table 2 shows the error rate for the system developed with physical characteristics alone and system with both physical and chemical characteristics. Results show that on combining the physical and chemical features of wheat grain samples, less error rate is achieved and in turn more accuracy is obtained in grade classification.

5.2 *Importance of Chemical Attributes*

Most of the existing wheat classification systems consider only the morphological features. Chemical features are equally important as they play a major role in deciding the rate of decay. This is very crucial in deciding the quality of the wheat grains. The gwplot() from the model created from MLP plots the relationship between chemical and morphological features with respect to grade of wheat (GW). From Fig. 9, it is clearly determined that protein and moisture features are equally important to morphological features like shape and size. The vertical alignment of the plots says the dependency of the attributes like protein, moisture, size, and shape to the grade. More the vertical alignment, more the attribute affinity toward grading; thus, Fig. 9 shows the vertical alignment of protein and moisture plots, which depicts protein and moisture are equally important along with physical features.

5.3 *Comparison with Other Classification Models*

Since both morphological and chemical features are taken into consideration for the classification network, the accuracy of the system will be also increased. Different machine learning algorithms like support vector machine (SVM) [27], Naive Bayes, and decision tree are also tested for the same wheat dataset. Figure 10 shows the

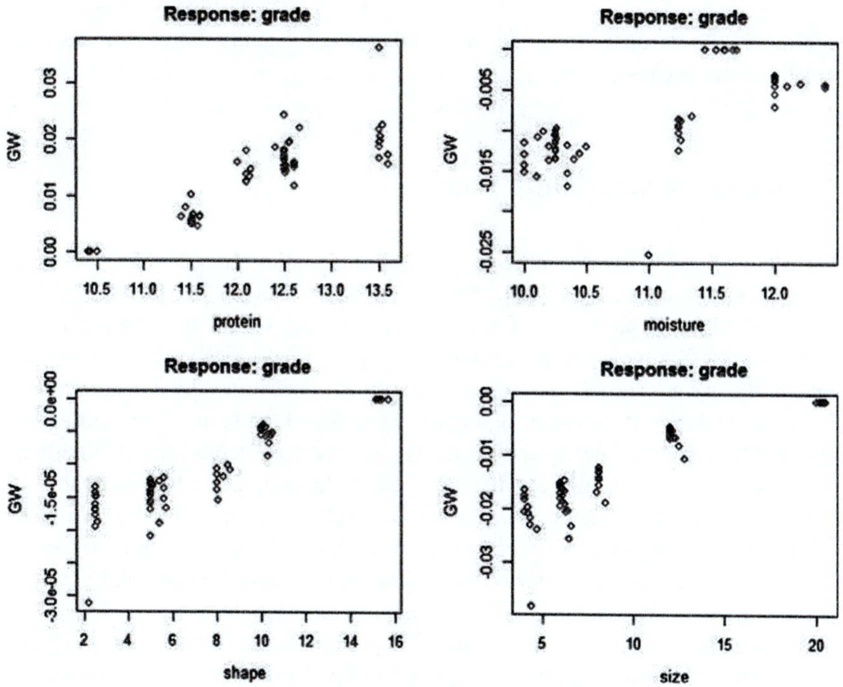

Fig. 9 Attribute importance with respect to wheat grade

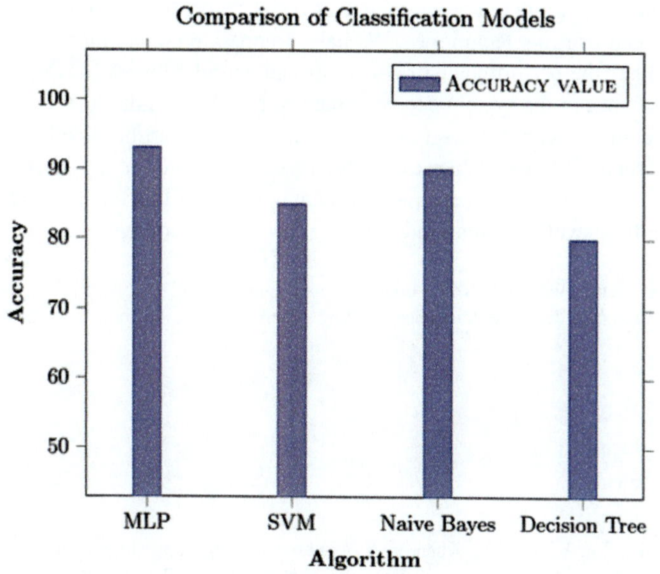

Fig. 10 Algorithm comparison for wheat grade classification

accuracy of the various systems that are implemented. Results show that multilayer perceptron has highest accuracy rate of 93% when compared to other algorithms.

6 Conclusion and Future Scope

Wheat cultivation being a significant source of income in India, an effective solution to identify grades to sell in international market is necessary. It is observed that along with physical properties of wheat such as shape, size, and color, the chemical properties like moisture and protein content are also important in deciding the grade of the grains. The health of the wheat is nearly determined by moisture. Thus, the cross-validation of attribute reputation identifies that chemical properties are equally vital when calculating the grade. Though there are a plethora of classification models can be considered, MLP provides better accuracy when compared to SVM, Naive Bayes, and decision tree. MLP is found to be scalable and highly robust with increase in dataset over time. Image-processing techniques are used for identifying the shape, size, and color of wheat kernels (physical features) and physical equipments like GMM and near IR spectroscopy are used to extract moisture and protein content of wheat kernels (chemical features), respectively. Finally, a cheap and cost-effective hardware solution is created using Raspberry Pi which incorporates image-processing techniques and neural network pre-programmed with international standard to aid the farmers.

Since there are limited amount of image data, we could not analyze the performance of deep learning techniques. We have planned to collect more data samples and investigate the performance of convolutional neural network (CNN) technique. The performance of the system can be enhanced by adding more features like artificial stain, common bunt (disease infected), earth pellets, fertilizer pellets, and hard vitreous kernels. We would like to further investigate on combing similar kind of grains like corn, oats, barley, rice, and millets with wheat that can classify the type of grain on first level and then grade of the grain at the second level.

Acknowledgements We would like to thank the management of SSN College of Engineering for funding this project to do research in wheat grading system which will be helpful to the agriculture people.

References

1. Patel, K.K., Kar, A., Jha, S.N., Khan, M.A.: Machine vision system: a tool for quality inspection of food and agricultural products. J. Food Sci. Technol. **49**(2), 123–141 (2012)
2. MacDonal, A.A.: Classification and grading. In: Wheat: Production, Properties and Quality, vol. 4, pp. 37–46 (1994). Springer, Berlin. https://doi.org/10.1007/978-1-4615-2672-8

3. Alfatni, M.S.M., Rashid, A., Shariff, M., Abdullah, M.Z., Marhaban, M.H.B., Saaed, O.M.B.: The application of internal grading system technologies for agricultural products review. J. Food Eng. **116**, 703–725 (2013)
4. Raj, M.P., Swaminarayan, P.R.: Applications of image processing for grading agriculture products. Int. J. Recent Innov. Trends Comput. Commun. **3**(3), 1194–1201 (2015)
5. Zapotoczny, P.: Discrimination of wheat grain varieties using image analysis: morphological features. In: European Food Research and Technology. Springer, Berlin (2011). https://doi.org/10.1007/s00217-011-1573-y; ISSN: 1438-2385
6. Mandal, D.: Adaptive neuro-fuzzy inference system based grading of basmati rice grains using image processing technique. Appl. Syst. Innov. **1**(2), 19 (2018)
7. Raj, M.P., Swaminarayan, P.R., Saini, J.R.: Machine learning paradigm for grading new released and old wheat cultivars (Triticum Durum & Triticum Aestvium) of Gujarat Region of India. In: Proceedings of SmartCom-2016, Springer CCIS Series. ISSN: 1865-0929 (2016)
8. Kaur, H.., Singh, B.: Classification and grading rice using multi-class SVM. Int. J. Sci. Res. Publ. **3**(4), 624– 630 (2013)
9. Sidnal, N., Uttam, V., Patil, P.P.: Grading and quality testing of foodgrains using neural network. Int. J. Res. Eng. Technol. **02**(11), 545–549 (2013). ISSN: 2321-7308
10. Raj, M.P., Swaminarayan, P.R., Saini, J.R., Parmar, D.K.: Applications of pattern recognition algorithms in agriculture: a review: Int. J. Adv. Netw. Appl. **6**(5), 2495–2502 (2015). ISSN: 0975-0290
11. Golpour, I., Parian, J.A., Chayjan, R.A.: Identification and classification of bulk paddy, brown, and white rice cultivars with colour features extraction using image analysis and neural network. Czech J. Food Sci. **32**, 280–287 (2014)
12. Khoshroo, A., Arefi, A., Masoumiasl, A., Jowkar, G.H.: Classification of wheat cultivars using image processing and artificial neural networks. Agric. Commun. **2**(1), 17–22 (2014)
13. Gupta, A., Singh, C., Kumar, V., Tyagi, B.S., TIWARI, V., Chatrath, R., Singh, G.P.: Wheat varieties notified in India since 1965. ICAR Indian Institute of Wheat & Barley Research, Karnal, Haryana (2018). www.iiwbr.org/wp-content/uploads/2018/08/wheat-varieties-notified-in-india.pdf
14. Sharma, D., Sawant, S.D.: Grain quality detection by using image processing for public distribution. In: International Conference on Intelligent Computing and Control Systems (ICICCS), 2017, pp. 1118–1122 (2017)
15. Basati, Z., Rasekh, M., Abbaspour-Gilandeh, Y.: Using different classification models in wheat grading utilizing visual features. Int. Agrophys. **32**(2), 225–235 (2018)
16. Opencv tutorials. http://docs.opencv.org/2.4/doc/tutorials/tutorials.html
17. Canny Edge Detector OpenCV 2.4.13.0 Documentation. http://docs.opencv.org/2.4/doc/tutorials/imgproc/imgtrans/cannydetector/cannydetector.html
18. Alemayehu, H.D.: Development of automatic sesame grain grading system using image processing techniques. J. Food: Microbiol. Saf. Hygiene 82 (2017)
19. Saini, M., Singh, J., Prakash, N.R.: Analysis of wheat grain varieties using image processing—a review. Int. J. Sci. Res. **3**(6), 490–495 (2014)
20. Jeong, S.: Histogram-Based Color Image Retrieval. Psych221/EE362-Reports (2001)
21. Manual on Standards of Wheat. http://agmarknet.nic.in/Wheatmanual.html
22. Guttieri, M.J., Stark, J.C., O'Brien, K., Souza, E.: Relative sensitivity of spring wheat grain yield and quality parameters to moisture deficit. Crop Sci. **41**(2), 327–335 (1999)
23. Butt, M.S., Nasir, M., Akhtar, S., Sharif, K.: Effect of moisture and packaging on the shelf life of wheat flour. Int. J. Food Saf. **4**, 1–6 (2004)
24. Grain Moisture Meter—https://www.a-grain.in/digital-grain-moisture-tester-meter.html
25. Delwiche, S.R.: Protein content of single kernels of wheat by near-infrared spectroscopy. J. Cereal Sci. **27**(3), 241–254 (1998)
26. Günther, F., Fritsch, S.: Neural net: training of neural networks. R J. **2**(1), 30–38 (2010). ISSN 2073-4859
27. Meyer, D.: Working of support vector machines. https://cran.r-project.org/web/packages/e1071/vignettes/svmdoc.pdf

28. Yimyam, P., Clark, A.F.: Agricultural produce grading by computer vision using genetic programming. In: IEEE International Conference on Robotics and Biomimetics (ROBIO), pp. 458–463 (2012). https://doi.org/10.1109/robio.2012.6491009

Safety Device for Children Using IoT and Deep Learning Techniques

S. Rajalakshmi, S. Angel Deborah, G. Soundarya, V. Varshitha, and K. Shyam Sundhar

Abstract The safety and security of children is a major problem in the current era. The children are too young to take care of themselves. We cannot monitor the children at all times in school, play area, and outside place. In this paper, we discuss the concept of child safety device based on Internet of things. The aim of this device is to provide safety to the child by allowing the parent to locate the child and view their surroundings. This device can be used to monitor the temperature and motion of the child. If any problem persists, the GSM mobile communication module automatically sends a text message to the parent as SMS. The other features of the device are emergency light and alarm buzzer which are activated when the button is pressed by the child in a distress situation to seek the attention of the bystanders. The accelerometer and vibration sensors are used to detect the motion of the child. The camera is used to capture the environment of the child. The image taken is processed using convolutional neural network (CNN) which predicts the background like play area, railway station, beach, road, or classroom. The GPS module is used to record current location of the device which is used to track the device if the child is missing. Hence, this device provides a security cover to the child in today's time.

S. Rajalakshmi (✉) · S. Angel Deborah · G. Soundarya · V. Varshitha · K. Shyam Sundhar
Department of Computer Science and Engineering, SSN College of Engineering, Kalavakkam, Chennai, Tamil Nadu 603110, India
e-mail: rajalakshmis@ssn.edu.in

S. Angel Deborah
e-mail: angeldeborahs@ssn.edu.in

G. Soundarya
e-mail: soundarya15105@cse.ssn.edu.in

V. Varshitha
e-mail: varshitha15121@cse.ssn.edu.in

K. Shyam Sundhar
e-mail: shyamsundhar15312@cse.ssn.edu.in

P. Suresh et al. (eds.), *Advances in Smart System Technologies*,
Advances in Intelligent Systems and Computing 1163,
https://doi.org/10.1007/978-981-15-5029-4_31

Keywords IoT · Child safety · Deep learning · Raspberry pi · GSM module · GPS module · Camera · SOS light · Alarm buzzer · Distress · Accelerometer · Vibration sensor · Temperature sensor

1 Introduction

The Internet of things (IoT) is a collection of Web-enabled devices like embedded sensors and communication hardware that collect data, send, and act on those data. It is also called as Internet of everything (IoE) [1]. Most of the automation systems like wearable devices, healthcare monitoring, smart cars, home automation systems, smart phones, and lighting controls are based on IoT systems. The aspects of IoT are the sensor devices and the server side architecture that supports them. There are many cases like kidnapping children, abusement, and children getting lost in crowded places that urge the need for a safety device to protect and safeguard children [2, 3]. Nowadays, many safety devices are available for tracking the location. They are mainly based on Bluetooth [4, 5] or Wi-Fi [6] where the range of the network is very small. In proposed device, GSM technology is used that is present everywhere and has a wider range of coverage. Location of the child will be sent as SMS automatically using GSM technology.

The platform used for running this device is the Raspberry Pi and the GSM module SIM900A is used to provide the functions of sending and receiving SMS, making calls, and connection to the Internet. Additional measures like SOS light, alarm buzzer, vibrator, accelerometer, temperature sensor, and camera are added to enhance the safety measure. The SOS light and alarm buzzer can be activated by the parent when the child is lost in a crowd so that the child can attract the attention of the bystanders and they can help them until the child is reunited with their parents. This can also be activated from the child side by pressing a button. The accelerometer and the vibration sensor are used to detect whether the child is motionless or the child/device falls down. It then makes an alert to the parent along with the location of the child. The temperature sensor is used to check for any abnormal temperature rise in the child's body temperature. The camera is used to get the images of the surrounding environment of the child and the environment is predicted using image processing techniques. The image along with the prediction is sent to the parent.

A mobile application is also developed that connects the device and the parent mobile. The app is used to activate the SOS light and buzzer and also used to request the required information (like location, temperature, and image) from the device and the information is sent via SMS. The device automatically sends the current location of the child as latitude and longitude positions and the surrounding image to parent through message/mail if the child is in unsafe situation.

2 Literature Survey

This paper [7] discusses about the wearable embedded system for detecting accidents while running. Fatal accidents are the cause of unintentional death after road traffic injuries. The people present in the surrounding can be called for help or to provide first aid will distinguish the fatal and not fatal accidents. An accident can happen even during group activities like dance or team sport. If the person is alone or out of sight, the device can be used to detect an accident and raise an alarm to alert other people. This paper focuses on a wearable device that is used to detect accidents occurred during the running practice. A model is trained on normal activities using SVM algorithm and used to classify the unknown situation. In order to avoid false detection, the consciousness of the person is observed and an alarm is raised if the person is unconscious after the abnormal event.

This paper [8] focuses on the protection and safety of girls and women by developing a wearable device. The objective is to analyze the physiological signals in conjunction with the body and raw accelerometer data from the triple axis accelerometer is used to determine the body position. Raw data is collected and activity recognition is performed using a specialized machine learning algorithm. Sensor data are sent wirelessly to an open-source cloud platform for real-time monitoring, and analyzing is done on MATLAB simultaneously. The device is programmed to monitor the parameters continuously and take action on any dangerous situation. This can be implemented by detecting the change in the signals and send alerts/notifications to the desired person.

This paper [9] discusses about Bluetooth low energy (BLE). BLE is a form of wireless communication designed especially for short-range communication that connects billions of devices in the next few years. This paper focuses mainly on Bluetooth low energy and explores various applications. Within a short range, the user can receive message and calls by pressing the switch in the wearable device. Device notification is performed by LED and vibrator motor. Liquid crystal display is used for the incoming calls and messages. Device and mobile are connected via Bluetooth.

This paper [10] discusses about the smart wearable device for little children. It uses SMS text-based communication as an interface between the wearable device and the parent. The parent can access the information about the child by sending text keywords like 'LOCATION', 'UV', 'TEMPERATURE', 'BUZZ', 'SOS'. The device will reply back with the UV radiation index and surrounding temperature so that parent can keep track whether the temperate or UV radiation is not suitable for the child. It also provides the accurate location of the child which can be viewed in Google maps by clicking the message. SOS light and alarm buzzer are implemented as a secondary measure which can be activated by the parent via SMS to grab the attention of bystanders if the child is in distress. In this work, they have not considered the motion of the child (conscious/unconscious).

This paper [11] discusses about the parent's perception on the child safety device. Parents are concerned about the safety of their child, especially those who are in the age group between six to eleven years. If both the parents working, then it is difficult to monitor the child wherever they go. Nowadays there are many tracking devices available in the market to ensure kids safety. Therefore, a survey was done on 110 parents of South Bangalore who bought a smart wearable device for their kids. And it is found that most of the parents were happy with it. This shows the real need for a device that ensures the safety of a child. The importance of the wearable device is discussed in [12, 13]. The papers [14, 15] discuss the use of deep learning techniques in embedded systems where as papers [16, 17] discuss the use of cloud computing and IoT techniques in safety wearable devices. A survey on the research challenges in wearable devices are discussed in [18].

3 Proposed System

The backbone of the device is the Raspberry Pi board. The GSM module is used to send and receive SMS and calls are made through the Internet. The current location of the child can be identified using the GPS module. The other features like SOS light, alarm buzzer, temperature sensor, vibration sensor, camera, and the accelerometer are interfaced to the Pi board. By pressing the button which is interfaced with the device, SOS light, alarm buzzer gets activated to grab the attention of the bystanders and also the location and the surrounding image of the child is captured and sent to the parent. Also, the device sends an alert to the parent, if the child/device falls down or if the child is motionless for a long time. This can be identified by observing the values of vibrator and accelerometer. If the temperature of the child gets increased, an alert is sent to the desired person. The camera is used to get the surrounding place images of the child and the environment is predicted using deep learning techniques. The image along with the prediction is sent to the parent. A mobile application is developed as an interface between the mobile and the device. This mobile application is installed in the parent mobile that contains few buttons like Get Location, Get Temperature, Get Image, Activate Alarm and LED. If any of these buttons is pressed, it activates the device which triggers the desired sensor to fetch the data and the parent receives the information as a text message.

4 Architecture Diagram

The architecture diagram is shown in Fig. 1. Raspberry Pi board is interconnected with temperature sensor, vibration sensor, accelerometer, GPS module, alarm buzzer, SOS light, camera, and GSM module. The data from the sensors are analyzed and alert message is sent via GSM module if child is unsafe.

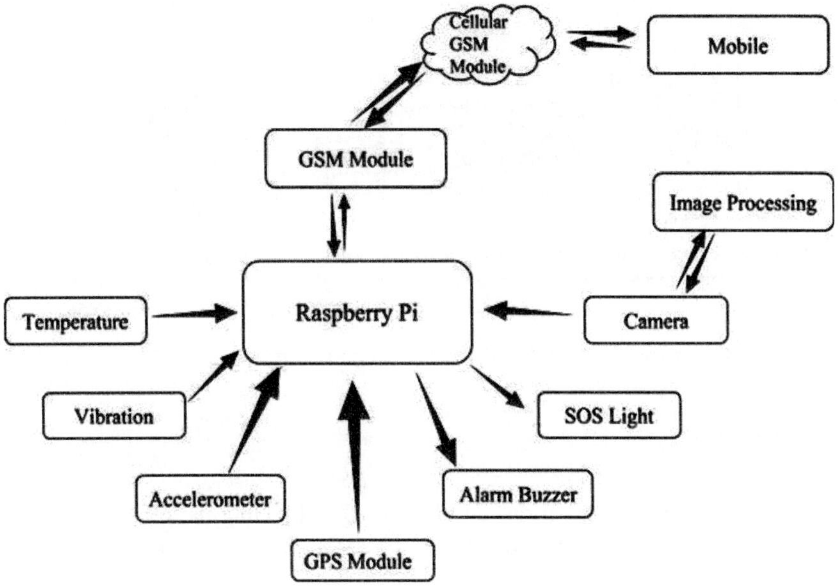

Fig. 1 Architecture diagram of the proposed system

5 Image Processing Using CNN Technique

A convolutional neural network (CNN) is a deep learning technique for supervised learning. CNNs are applied to natural language processing, image processing, speech processing, and other kinds of cognitive tasks. Neural networks consist of individual units called neurons. Each neuron in the one layer is connected to every other neuron in the next layer. Data is passed from the input layer to the output layer through these neurons. Each individual node performs a mathematical calculation and an activation function. Then, it transmits the output data to all the nodes in the next layer. In between the input and output layer, there are one or more hidden layers. In CNN, the in between layers are convolutional layer. Each layer performs the convolution operation. The overall process of CNN [19] is given in Fig. 2.

5.1 Tools Used

The various tools used for the development of CNN are as follows:

1. Python
2. Keras
3. Tensorflow
4. Numpy
5. Matplotlib.

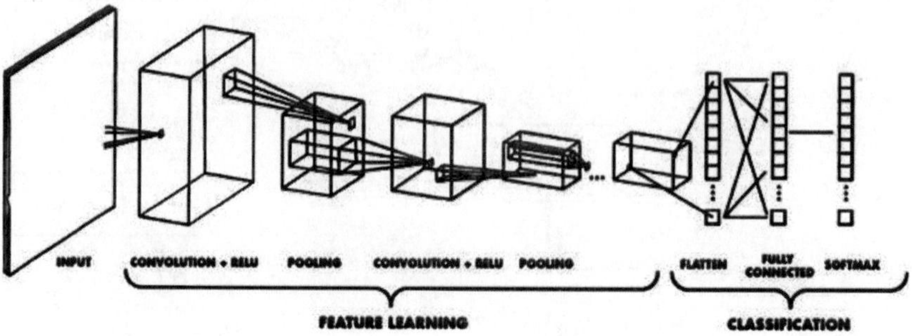

Fig. 2 Convolutional neural network [19]

5.2 Process

This section describes the various tasks involved in building the CNN model for identifying the background environment in which the child is present.

(a) Dataset collection
(b) Data augmentation
(c) Construct the CNN structure
(d) Build the model–Training
(e) Testing.

(a) *Dataset Collection*:

In order to build the deep learning model, huge amount of data is needed. The model can learn certain relations and common features related to the objects from the data. We have collected the images of various background environments like beach, classroom, park, road, and railway station, and used them for training and testing.

(b) *Data Augmentation*:

Data augmentation is a process used to reduce overfitting and class imbalance problem, where we increase the amount of training data using information available from the training data itself. Some of the data augmentation techniques are flip, rotate, scale, translate, crop, zoom, and interpolation techniques are constant, edge, reflect, symmetric, and wrap. We have used rotate, shear, zoom, shift, and flip techniques for augmentation and reflect technique for interpolation. After augmentation, there are 16,383 images. Training set contains 13,521 images and test set has 2862 images.

(c) *Construct the Network*:

Next, the network structure for the deep neural network (CNN) is constructed. It consists of four layers.

(i) Input layer
(ii) Convolution layer
(iii) Nonlinear layer
(iv) Pooling layer
(v) Fully connected layer
(vi) Output layer.

(i) *Input layer*:

The pixel values of the background image are given to the input layer. We have taken the two-dimensional image for constructing the model.

(ii) *Convolution layer*:

Next layer in the network is the convolution layer. The image (matrix with pixel values) from the input layer is given to this layer. The filter matrix (kernel or core) of size 2×2 or 3×3 is applied on the input image. The filter performs the convolution operation on the input image. The filter multiplies its values with the original pixel values of input image. The resultant values are summed up to a single value. This convolution operation is performed fully by sliding the filter on the entire input image. The resultant convoluted matrix is given as input to the next layer.

(iii) *Nonlinear layer*:

After each convolution layer, the nonlinear layer is added. This is also called as activation layer. Activation function is used to make the convoluted output nonlinear. The network should be sufficiently intense to model the output response variable that depends upon the input variables. Rectified linear units (ReLU) function is used for activation. The formula for ReLU is shown in Eq. 1.

$$f(x) = \max(x, 0) \tag{1}$$

where x is the input to the activation function and it returns x if it is positive else 0 is returned.

(iv) *Pooling layer*:

The pooling layer follows the nonlinear layer. Pooling layer is used to reduce the dimensionality and the parameters of the network. The main aim of pooling layer is to reduce the size of the image and hence the image volume by performing down sampling. Since each convolutional layer extracts some important features, the whole detailed image is not needed further. Maxpooling is used in our network. This combination of convolution, activation, and maxpooling layers act as the hidden layer and there are one or more hidden layers in the network.

(v) *Fully connected layer*:

A flatten layer is attached after completing the series of convolutional, nonlinear, and pooling layers. Flatten layer is used to convert the high dimensional data into single dimensional data that is given as input to the fully connected layers. Fully connected layers are like normal neural network layers in which every neuron in one layer is connected to every neuron in the next layer. ReLU function is used for activation in this layer also. Dropout layer can be added to avoid overfitting problem.

(vi) *Output layer*:

The output layer consists of 5 neurons which depend upon the number of classes in the given data. Softmax function is used for activation in the output layer. Softmax function produces the output in the range of 0–1. It is equivalent to the categorical probability distribution which tells the probability of the given sample belongs to each class. The formula to calculate the output σ using softmax function is shown in Eq. 2.

$$\sigma(z)_j = \frac{e^{\beta z_j}}{\sum_{i=1}^{k} e^{\beta z_j}} \tag{2}$$

where z is the input vector to the output layer, k denotes the number of classes, and j denotes the index of a particular output unit.

(d) *Build the Model*:

Once the deep neural network is constructed, training is done using the input dataset. To enhance the accuracy, number of epochs can be increased. Adam gradient descent technique is used for optimization.

(e) *Testing*:

The model can be tested using random unique images that are not used for training. The model predicts the class of the test image. Accuracy metric is used to measure the quality of the network.

6 Modules Split-Up

The proposed approach is divided into the following modules.

6.1 Alert Measures

When the child is in unsafe mode the *Piezo Buzzer* and *LED* are activated to seek the attention of bystanders who are in a certain distance. Message and mail will be sent to the parent during this situation. This alert will be automatically activated if the device detects an unsafe mode, or externally pressed by the child on critical conditions.

6.2 Motion Detection

Vibration sensor *WYPHSW-420* is used to detect the removal of the device from the child. The device can be snatched away or would have fallen down. This *no motion* detection triggers the Raspberry Pi to send the message to the parent with the appropriate information such as location, surrounding images, and temperature. There will be a situation in which, the child may be in an unconscious state. The accelerometer *ADXL345* is used to check whether the child is motionless for a long time and alerts the parent with appropriate information.

6.3 Location Identification

Parent can requests the device for location through mobile app. The GSM receives the request and creates a connection medium between the mobile and device. It obtains the location from GPS module via Raspberry Pi. This GSM module sends back the message with accurate coordinates (latitude, longitude) to the mobile that can be located via Google maps.

6.4 Capturing Surroundings

The camera is interfaced with the device to take the pictures of surroundings. This can be activated, if the parent requests the device for pictures or the button is externally pressed by the child. These pictures are sent to the parent by e-mail.

6.5 *Temperature Detection*

Temperature sensor *LM35DZ* interfaced with the device is used to detect the temperature of the child. When the device is removed, temperature of the body gets changed and this activates the device to send an alert to the parent.

6.6 *Mobile Application*

An android application is developed and installed in the parent's mobile. The app contains the options like *TEMPERATURE, CAMERA, SOS, LOCATION, and ALARM*. The parent can get required information about the child by activating the device using these keywords.

6.7 *Analyzing Data*

The data from the sensors like temperature, accelerometer, and vibration are analyzed using rule-based approaches and alert measure is taken if the child is in unsafe situation. We have identified nine rules that lead to unsafe state; else the child is in safe state.

6.8 *Image Processing*

The image taken by the camera is analyzed using the deep learning technique (CNN). It is used to identify the background environment of the child. The various categories are school, play area, beach, railway station, or road. The details about the model building are discussed below.

6.8.1 Dataset

Table 1 contains the information about the total number of images in the train and test dataset for each class. The five categories are beach, classroom, park, road, and railway station. The model is built using 13,521 images and tested with 2862 images. The CNN model provides 97.5% accuracy for training data and 83.15% for test data.

Table 1 Dataset for CNN

Category	Train data	Test data	Total no of images
Beach	2701	530	3231
Classroom	2690	581	3271
Park	2658	564	3222
Road	2762	594	3356
Railway station	2710	593	3303
Accuracy	97.5	83.15	

The accuracy of the system is calculated using Eq. 3.

$$\text{Accuracy} = \frac{\sum_{i=1}^{n} \text{TP}_i}{\sum_{i=1}^{n} N_i} \tag{3}$$

where n is the number of classes in the dataset, TP_i is the true positive value of ith class and N_i is the total number of samples in ith class. TP_i is calculated as the total number of images in *ith* class that are correctly classified into the same class.

6.8.2 Construction of the Model

The dimensions of all the input images are 480 × 360 pixels. The numbers of epochs are set to 100 and batch size is set to 32 in order to improve the accuracy. There are six hidden layers in the network. Each hidden layer is a combination of convolutional, activation, and maxpooling layers. ReLu is used as an activation function in hidden layers. The output from hidden layer is given to the flatten layer that converts to a single dimensional array. The next layer is dense layer also known as fully connected layer with 512 neurons. To prevent overfitting, the dropout layer is added to drop some connections of neurons from the dense layer. The last layer is the output layer with 5 neurons. Softmax function is used for activation in the output layer. Adam optimization function is the combination of properties of AdaGrad and RMSProp optimization techniques. It is used to handle the noise in data and to optimize the model.

7 Block Diagram

The circuit connection for the safety device is shown in Fig. 3. It shows the interconnection between Raspberry Pi, camera, GSM, GPS, accelerometer, temperature, and vibration sensor.

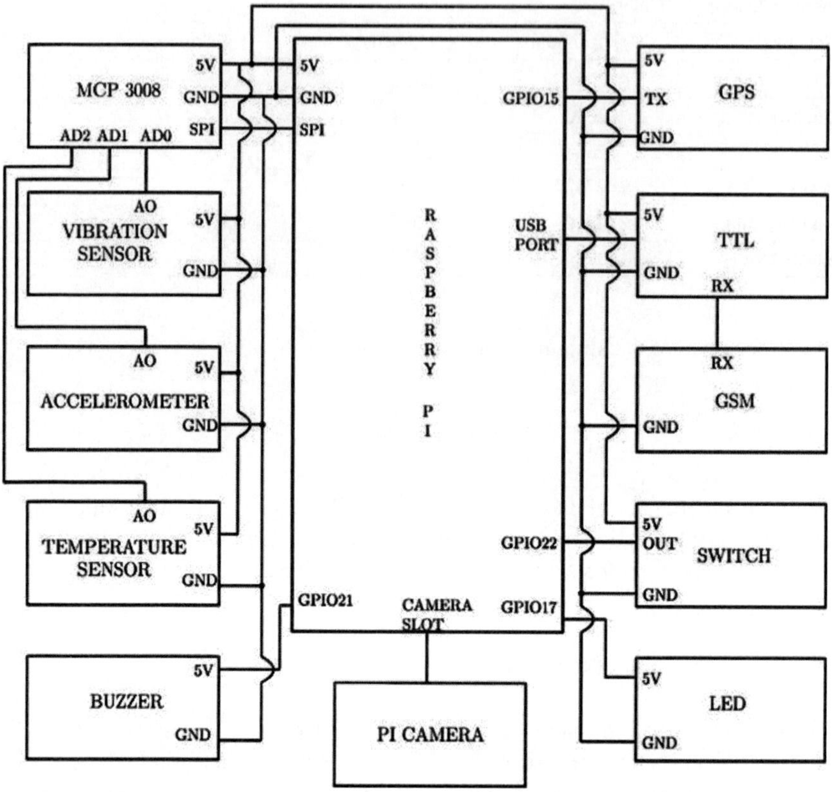

Fig. 3 Block diagram of the device

8 Flow Diagram

Initially, the connection has been established between the mobile and the device. The Raspberry Pi is activated if the parent sends request or if the button is pressed. The data from the sensors like temperature, accelerometer, and vibration sensor are extracted by the device and analyzed. The output of the system is either safe or unsafe. If it is unsafe, the GSM module sends the location, surrounding pictures, and temperature of the body to the parent through message and e-mail. CNN model is used to identify the surrounding environment as school, play area, beach, railway station, or road. The flow diagram for the overall system is shown in Fig. 4. The procedure for checking the safety of the child is discussed in Algorithm 1.

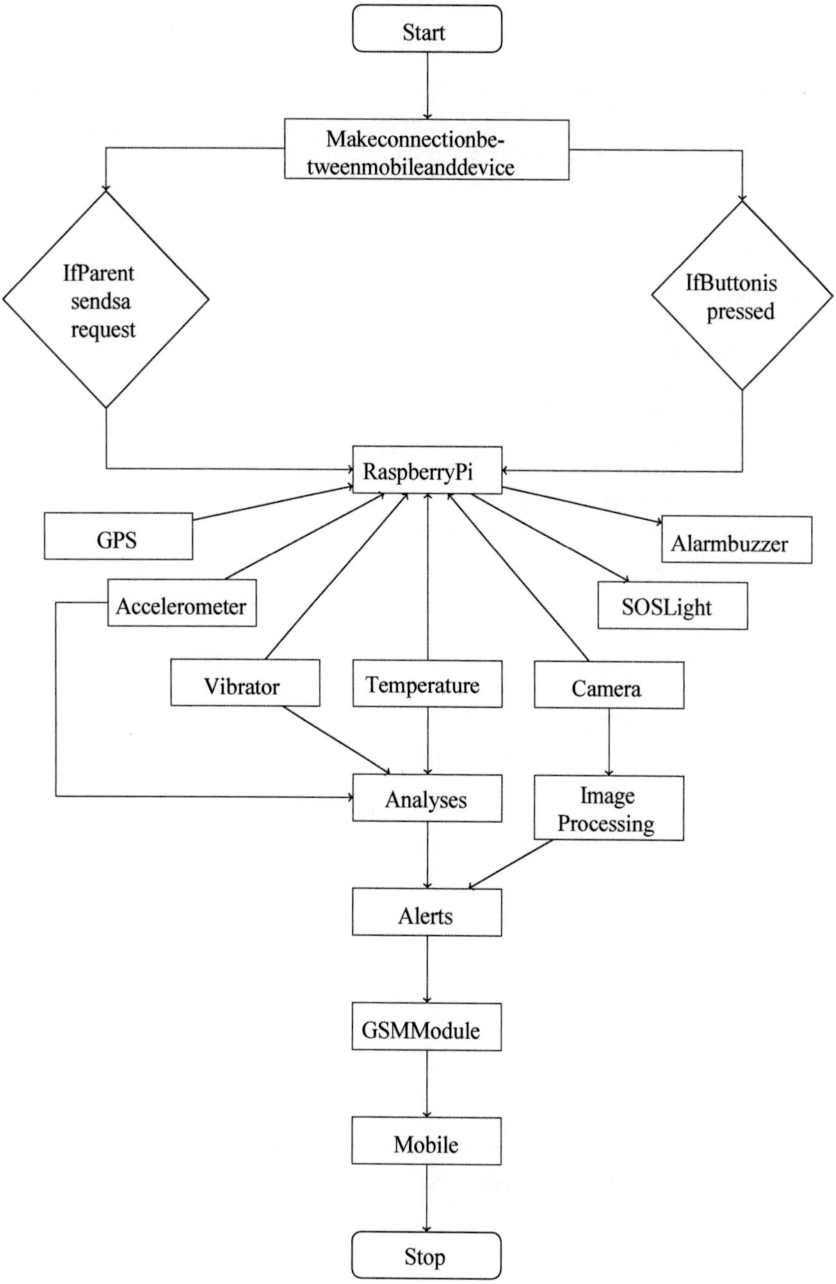

Fig. 4 Flow of the proposed system

Algorithm 1 Safety checking

Algorithm1: Safety checking

Require: Temperature Sensor, Vibration Sensor, Accelerometer, Led, Buzzer, Button, Camera, Raspberry Pi, GSM, GPS
Input: Device is activated and monitoring continuously
Output: Alert send to parent if child is unsafe

```
Make connection between mobile & device
if Parent sends request through mobile App then
        if request == GPS then
            send Location
        else if request == TEMPERATURE then
            send Body Temperature
        else if request == IMAGE then
            send Surrounding Image
        else if request == LED & BUZZER then
            activate Led
            activate Buzzer
        end if
end if
if Button pressed by child then
        L1: activate Led
        activate Buzzer
        send Location to parent as message and mail
        send Image to parent as mail
end if
if No Movement || High Movement || High Vibration || Temperature Variation then
        goto L1
end if
```

9 Test Cases

The various situations for safety checking are discussed in this section. Alert is sent to the parent, if the child is in unsafe state.

1. If vibration is low, no movement is detected and child may be unconscious.
2. If the device is snatched or thrown away, vibration and accelerations are high.
3. If the device falls down, low acceleration is detected.
4. If the device is hit by something, then vibration will be high.
5. If the child is lost in the crowd, parent can activate the SOS light and buzzer in the device from mobile application to find the child.
6. If there is a temperature variation, the device is removed from the child.
7. If button is pressed by the child on critical situations pictures, temperature, and location are sent to the parent and buzzer and SOS light are activated to seek attention of neighbors.
8. The parent can also access the device to know the status of the child via a mobile application.
9. Sensor data are being updated in a server for every 5 min of interval. When the server has not received any data from the device for a particular time period, then the device is either powered off or disconnected from the network. Alert message will be sent to the parent in this situation.

10 Conclusion

Safety of the children is a critical problem in current world. We cannot watch the activity of the child throughout the day. So to ease the process and to ensure the safety of the children, we have developed a child safety device. The device is developed with features like finding the location, body temperature, surrounding pictures, and motion of a child. Parent can activate the device through mobile application to know the status of the child. Child can activate the device on critical situations to seek the attention of nearby people. The image captured by the camera is processed using deep learning technique. Convolutional neural network model is used to identify the surrounding environment as beach, railway station, play area, classroom, or road. The data from the sensors (temperature, accelerometer, and vibration) are monitored continuously and alert is sent to the parent if any unusual activity is detected.

We would like to further analyzed the persons in the image and classify them as known/unknown person. The image can be processed further by identifying the stranger or the unknown person who follows the child for long time (who appears in most of the frames). We can train the model with images of relatives/friends/teacher for this purpose, so that the anomalies can be identified easily.

The device could not recognize the drop of the signal in GSM module. At that time, alert to the parent may not be sent correctly. This causes a serious problem. To overcome this, network bandwidth can be monitored and if there is any sign of lower range of the network, prior alert is sent to the parent.

Acknowledgements We would like to thank the management of SSN College of Engineering for funding this project to develop a safety device which will be helpful to the society.

References

1. Dorsemaine, B., Gaulier, J.P., Wary, J.P., Kheir, N., Urien, P.: Internet of things: a definition & taxonomy. In: 9th International Conference on Next Generation Mobile Applications, Services and Technologies, pp. 72–77. IEEE (2015)
2. Huang, Z., Gao, Z., Lu, H., Zhang, J., Feng, Z., Xia, H.: An mobile safety monitoring system for children. In: 10th International Conference on Mobile Ad-hoc and Sensor Networks, pp. 323–328. IEEE (2014)
3. Raflesia, S.P., Lestarini, D., et al.: An integrated child safety using geo-fencing information on mobile devices. In: 2018 International Conference on Electrical Engineering and Computer Science (ICECOS), pp. 379–384. IEEE (2018)
4. Bhagwat, P.: Bluetooth: technology for short-range wireless apps. IEEE Internet Comput. **5**(3), 96–103 (2001)
5. Braam, K., Huang, T.C., Chen, C.H., Montgomery, E., Vo, S., Beausoleil, R.: Wristband vital: a wearable multi-sensor microsystem for real-time assistance via low-power bluetooth link. In: 2015 IEEE 2nd World Forum on Internet of Things (WF-IoT), pp. 87–91. IEEE (2015)
6. Datar, RV.: Wifi and wimax: break through in wireless access technologies. In: IET Conference on Wireless, Mobile and Multimedia Networks, pp. 141–145 (2008)
7. Carletti, V., Greco, A., Saggese, A., Vento, M., Vigilante, V.: A wearable embedded system for detecting accidents while running. In: 13th International Joint Conference on Computer Vision, Imaging and Computer Graphics Theory and Applications (VISIGRAPP), pp. 541–548 (2018)
8. Jatti, A., Kannan, M., Alisha, R.M., Vijayalakshmi, P., Sinha, S.: Design and development of an IoT based wearable device for the safety and security of women and girl children. In: IEEE International Conference on Recent Trends in Electronics, Information & Communication Technology (RTEICT), pp. 1108-1112. IEEE (2016)
9. Mandanka, K., Mistry, S., Tank, B.: Design and development of wearable device using bluetooth low energy. In: 2017 International Conference on Computing Methodologies and Communication (ICCMC), pp. 762–766. IEEE (2017)
10. Moodbidri, A., Shahnasser, H.: Child safety wearable device. In: 2017 International Conference on Information Networking (ICOIN), pp. 438–444. IEEE (2017)
11. Vaz, P., Aarthy, C., Advithya, C.: Parents perception on child safety wearable device—a survey in South Bangalore. Int. J. Manage. Soc. Sci. Res. (IJMSSR) **6**(8), 7–10 (2017). ISSN: 2319-4421
12. Haripriya, R., Hemashree, S., Indhirani, S., Kamala Jothi, S.: Child security wearable gadget. Int. J. Pure Appl. Math. **118**(20), 313–318 (2018)
13. Moustafa, H., Kenn, H., Sayrafian, K., Scanlon, W., Zhang, Y.: Mobile wearable communications. IEEE Wireless Commun. **22**(1), 10–11 (2015)
14. Lane, N.D., Bhattacharya, S., Georgiev, P., Forlivesi, C., Kawsar, F.: An early resource characterization of deep learning on wearables, smartphones and internet-of-things devices. In: International Workshop on Internet of Things Towards Applications, pp. 7–12. ACM (2015)
15. Lane, N.D., Bhattacharya, S., Georgiev, P., Forlivesi, C., Kawsar, F.: Squeezing deep learning into mobile and embedded devices. IEEE Pervasive Comput. **16**(3), 82–88 (2017)
16. Mu, Y., Wang, T., Jia, Y.: Research on framework of underground wearable devices framework based on cloud computing. In: 8th International Symposium on Computational Intelligence and Design (ISCID), vol. 1, pp. 458–461. IEEE (2015)
17. Santhi, V., Ramya, K., Tarana, A.P.J., Vinitha, G.: Iot based wearable health monitoring system for pregnant ladies using cc3200. Int. J. Adv. Res. Methodol. Eng. Technol. 56–60 (2017)
18. Seneviratne, S., Hu, Y., Nguyen, T., Lan, G., Khalifa, S., Thilakarathna, K., Hassan, M., Seneviratne, A.: A survey of wearable devices and challenges. IEEE Commun. Surv. Tutor. **19**(4), 2573–2620 (2017)
19. Cornelisse, D.: An intuitive guide to convolutional neural networks, https://medium.freecodecamp.org (2018)

Enlargement of Ubiquitous Security Organism for Cloud Computing

R. Dhaya, R. Kanthavel, M. Devi, Fahad Algarni, and Pooja Dixikha

Abstract Cloud storage empowers clients to tenuously pile their information and appreciate the on-request excellent cloud requests deprived of the weight of neighborhood hardware and software the executives. But there is some security hazards headed for the precision of files or documents or information in the cloud in added with the loss of storage in the existing which utilize the homomorphic token. In order to discourse this delinquent, a new effective security mechanism is proposed called data aware routing (DAR). This scheme reduces the data contention by partitioning the data storage, and using the concept of logical IP, it makes the data secured in the cloud. It also preserves data integrity and reduces the transmission time by partitioning the data in the cloud. DAR proved to be an efficient scheme in terms of reducing the transmission time during data request

Keywords Cloud storage · Security · Data aware routing · Contention · Data integrity

---

R. Dhaya (✉) · M. Devi
Department of Computer Science, King Khalid University, Sarat Abidha Campus, Abha, Kingdom of Saudi Arabia
e-mail: dhayavel2005@gmail.com

M. Devi
e-mail: devinov6@gmail.com

R. Kanthavel
Department of Computer Engineering, King Khalid University, Abha, Kingdom of Saudi Arabia
e-mail: kanthavel2005@gmail.com

F. Algarni
Department of Computing and Information Technology, University of Bisha, Bisha, Kingdom of Saudi Arabia
e-mail: fahad.a.algarni@gmail.com

P. Dixikha
Department of ECE, Sona College of Technology, Salem, Tamil Nadu, India

P. Suresh et al. (eds.), *Advances in Smart System Technologies*,
Advances in Intelligent Systems and Computing 1163,
https://doi.org/10.1007/978-981-15-5029-4_32

1 Introduction

Cloud computing implies Internet-based improvement, where 'cloud' here, is a metaphor for 'Internet' and 'computing', which means the utilization of 'PC innovation.' It is an administration that encourages clients to play out the assignments over the Internet [1]. Cloud stockpiling empowers customers to remotely store their information and welcome the on-ask for incredible cloud applications without the heaviness of neighborhood equipment and programming the board [2]. Despite the fact that the advantages are clear, such an administration is additionally giving up clients' physical ownership of their re-appropriated data, which definitely presents novel security hazards in the direction of the correctness of the data in cloud [3]. So as to discourse this anxiety, we have proposed a new efficient security mechanism called data aware routing (DAR) which makes the cloud computing more securable [4]. Initially, by using data partition algorithm, the information is distributed to each and every node. Then, by using our proposed scheme, with the concept of logical IP, we ensure the data confidence in cloud data storage [5].

2 Existing System

In the existing system, a versatile appropriated stockpiling reliability assessing part, utilizing the homomorphic token and spread destruction coded information is used for secure cloud storage [6]. The proposed arrangement empowers customers to audit the cloud stockpiling with very lightweight correspondence and count cost [7]. The assessing result ensures strong cloud stockpiling rightness guarantee, just as in the meantime achieves fast information botch confinement, i.e., the distinctive confirmation of getting raucous server [8]. Normally, cloud stockpiling enables customers to remotely store their information and welcome the on-ask for sublime cloud applications without the heaviness of close-by equipment and programming the board [9]. Disregarding the way that the advertisement vantages are clear, such an organization is furthermore surrendering customers' physical responsibility for re-appropriated information [10], which unquestionably exhibits new security threats toward the rightness of the information in cloud. So considering the cloud information are dynamic in nature, the proposed structure further support secure and capable amazing exercises on redistributed information, including square change, crossing out, and append. The arrangement ensures cloud information storage [11]. At first, the homomorphic token is displayed. The token figuring work has a spot with a gathering of comprehensive hash work, spared the homomorphic properties, which can be immaculately consolidated with the check of destruction coded information. Along these lines, it is showed up at decide a test response tradition for checking the capacity exactness and recognizing getting rambunctious servers [12]. The framework for record recuperation and screw up recovery reliant on annihilation helping code is in like manner plot. If the adversary changes the information deters among any of the information stockpiling servers, this checking plan can viably recognize the strike with high probability. For whatever time allotment that the information change is gotten, the customer will moreover make sense of which server is coming up short [13]. This can be cultivated by taking a gander at the response regards [14]. The as of late recovered squares would then have the capacity to be redistributed to the motivating boisterous servers to keep up the precision of capacity. Cancelation helping code [15] in the record transport course of action is to give redundancy fairness vectors and affirmation the information unwavering quality [16]. By utilizing the homomorphic token with flowed affirmation of annihilation coded information, the arrangement achieves information botch confinement [17].

Major Problems in Existing System: The existing distributed storage integrity auditing mechanism has the following problems,

1. The major concern is that the increasing security issues in the existing scheme.
2. During the retrieval of information or during data transfer, it causes loss of data storage.
3. Scalability problems with sequential processing. When several number of nodes are constantly requesting for data, it causes scalability problems.
4. During data transfer, it is difficult to migrate massive amounts of data.

To overcome these problems, an efficient security mechanism is proposed with the reference of various papers related to this paper, which is described below in three modules.

3 Proposed System

To make the cloud computing more securable and also to achieve efficient cloud storage, a new efficient security mechanism is proposed which is called as data aware routing (DAR). Objectives to implement this new scheme,

- To reduce the transmission time by increasing the response time during files transfer.
- Ensuring more security in cloud computing by using the data aware routing (DAR), i.e., to convert normal cloud computing into more securable and to preserve data integrity.
- Effective data storage by reducing the latency and to reduce data contention.
- To increase the data transfer rate, to improve connection scalability, and to reduce data corruption by data partitioning.

In data aware routing, by comparing the original IP and logical IP, the data sharing is permitted. Therefore, it makes the computing more securable. By partitioning the data storage, the transmission time is reduced. The overall proposed system structure which is the combination of three modules is shown in Fig. 1. In Fig. 1, initially, cloud network is formed, in which data storage is partitioned and the information is

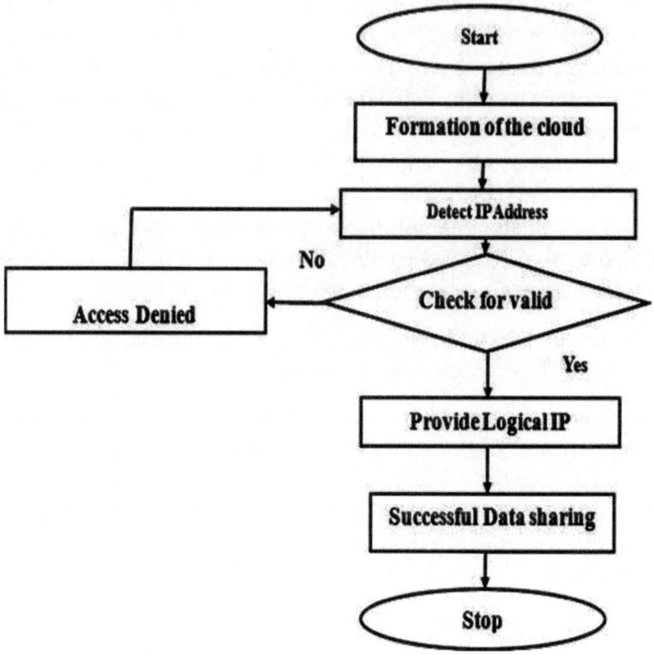

Fig. 1 Overall flowchart of the proposed system

distributed among the nodes. During the data request, it detects the IP address of the node. Then, IP address is checked for validity. If the IP is valid, then it is provided with logical IP. If not the access is denied. The logical IP is given for avoiding IP conflict. By comparing the logical and original IP, it is permitted for sharing. If the data is not confidential, it is accessed normally and has a successful data sharing. If the data is confidential, then it will be accessed with some authentication.

The information is efficiently distributed to the nodes and data is shared in an easy and efficient manner which is requested. This paper work is divided into the following three modules that are (1) formation of cloud, (2) data partition algorithm, and (3) security algorithm. In the first module, cluster head maintains all the node information. Each node is interconnected with each other node within the cloud network. Each node knows the details of the information that are stored in other nodes. In the second module, data is distributed to each node, i.e., data storage is partitioned depending on the memory of each node accordingly by data partition algorithm which reduces the data contention [18]. In the last module, to make the computing more securable data aware routing is implemented. With the concept of logical IP, data storage is made secured.

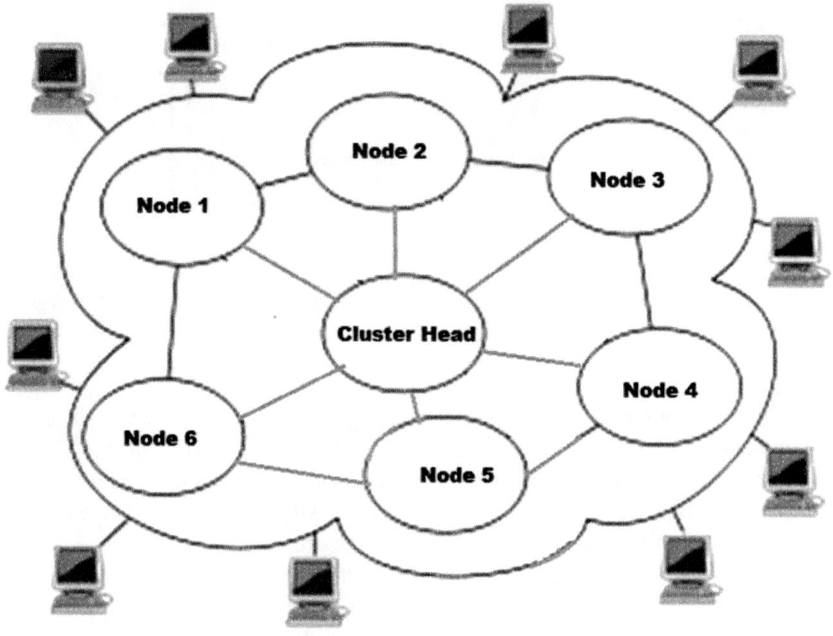

Fig. 2 Formation of cloud network

4 Formation of Cloud Systems

Cluster head maintains all the node information. Each node is interconnected with each other node within the cloud network. Each node knows the details of the information stored in other nodes [19]. It is proposed that when a data is requested to another node, it checks for the same and provide the information. But if the information is not available, it guides it to request to the particular node which has that information. In Fig. 2, it is shown that each node is connected to each other nodes and also the information stored in each node is known by other nodes. In addition, the cluster head helps maintains all the node information [20].

5 Data Partition Algorithm

Functional Operations: To reduce data contention, the data is partitioned and distributed to the nodes efficiently. Also each and every node knows the other nodes information, thus it guides the data request sent by any node to direct which node is having the particular information. In Fig. 3, when the client node demands for the information, master node which maintains all the nodes retrieves the information to the client node with the help of data storage.

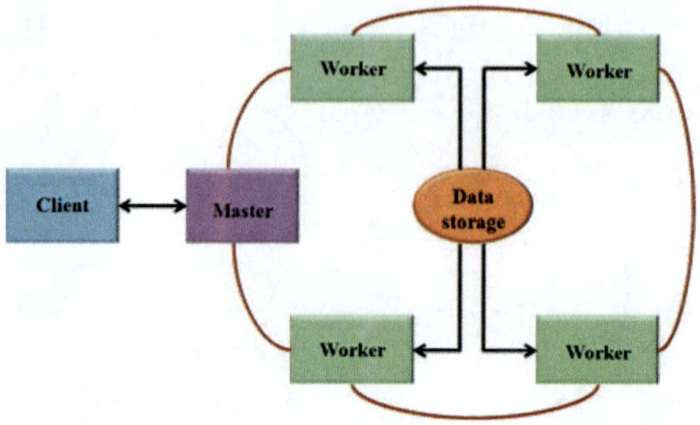

Fig. 3 Block diagram of overall structure

All the nodes are connected to the data storage, at the same time, to reduce the data contention as it is partitioned and distributed to all the nodes. Therefore, the transmission time is reduced efficiently with this data partition method. Thus, whenever there is a data request, the data is transferred easily. The requested data can be shared only under the condition that it is not confidential information or information that can be retrieved with legal authentication. Otherwise, normal information can be shared when it is requested. The client sends the request to the master node that governs all the worker nodes. It knows about the information and other details about them. There is a central data storage that is partitioned among the worker nodes based on the memory space. Then, the data transmission is done accordingly.

Algorithm

Step 1: The data is partitioned and distributed to the nodes.
Step 2: Each node knows the details of each and every other node.
Step 3: Each node consults every other node for any data request send by any node and guide for the same.
Step 4: During data request check the information can be shared or not.

- If the data can be shared, go to step 6.
- If not, Access is denied.

Step 5: Check the requested information is confidential,

- If the demanded data is trustworthy and can be pooled to the authenticated user, data sharing is permitted.
- If the demanded data is trustworthy and cannot be pooled, Access is denied.

Step 6: Ensure right information is provided for the right node and permit data sharing.

In the algorithm steps, it is shown that initially the data storage is partitioned and distributed to the nodes. When there is a data request, it retrieves the data quickly. As the data storage is partitioned, and each and every node knows other nodes storage information, the requested data is retrieve shortly. This reduces the transmission time during the data transfer. During data retrieval, there are some conditions to be checked. Firstly, the information is to be checked, whether it is a confidential information or not. If the information is confidential and can be shared with legal authentication, then it is shared with some authentication. Before data transfer ensure that it is sharing right data with right node. Then, the data sharing is permitted.

6 Security Algorithm

Functional Operations: During the process of data request, there is a fortuitous of damage of information or data integrity is not maintained due to security problems. Therefore, to make it more secure, a security algorithm is proposed which preserves the data integrity and also ensure storage correctness. In this algorithm, initially, it detects the IP address of the node which request for some information to other node. If the IP is valid and also under the cloud, it is provided with logical IP with the check whether the logical IP does not already exist. While checking for valid IP, it checks not only for the IP under the cloud, but also it checks the format of the IP addresses. If the IP address for a node have the null value, it is not at all recognized as a node and the node will be dropped with the concept of logical IP it makes the computing secure and also is secured as only authenticated node is retrieving the information. Further, checking for IP conflicts, if there is an IP conflict, by the use of logical IP, the node can be identified easy and guide for the data retrieval. This can be done by comparing the original and logical IP of the particular node. Thus, it ensures more security. If the any of the node's IP addresses conflicts with some other node's IP address, there is problem or confusion in data transfer. To avoid this problem, it is proposed to provide logical IP, for the node which gives a data request. Then, data transfer is done in a secure manner. Transmission time for data transfer is noted for each and every data transfer to any of the nodes. Therefore, when the different nodes request constantly, it avoids scalability problems as the cloud data storage is partitioned efficiently. The transmission time is noted for when the first data transfer is started processing. Then, the same is noted for the second transfer, and the process continues. The transmission time is noted for all the data transfers. The data transfer is shown by viewing the size of all the nodes. This security algorithm efficiently makes the data transfer secure and avoids the loss of data storage by means of partitioning the same using data partition algorithm. Mainly with the concept of logical IP the data transfer is secured. Thus, by implementing this algorithm, un-securable data transfer is avoided and loss of data storage is prevented. This algorithm mainly concentrates on the security by providing access only to authenticated nodes. It also reduces the data contention because only the confidential and secured information are being shared. There are many steps to implement this. That is shown in the security algorithm explained below.

Algorithm Steps

Step 1: Cluster head detects the IP address.

Step 2: Check whether the IP is valid and also under the cloud,

(1) If the IP is valid, check whether Logical IP exist or not.
(2) If the IP is not valid, Access denied.

Step 3: Check for IP conflict.
Step 4: If Logical IP is not already existing,

- If two same IP, provide new unique Logical IP.
- If not, provide logical IP normally.

If Logical IP is already existing,

- Compare Original IP and Logical IP.

Step 5:
- If both the IP addresses match, it is permitted for data sharing.
- If both the IP addresses do not match, access denied and goes to Step 1.

7 Results and Discussion

Experimental Results of the Proposed System: In the proposed system, there is a connection between five nodes. Each node is of varying memory space containing text files like word document, PDF file, and also image files. The central data storage is partitioned among the nodes as required. As the data is partitioned efficiently, the transmission time is reduced and speed of transmission is increased considerably. The response time is increased because each node in the cloud knows about the details of all other nodes. So that the time for searching the required file is decreased. These are the two main parameters that are implemented using the data partition and security algorithm which is named as data aware routing. Scalability is also achieved so that any number of nodes can be added at any time. Data contention is reduced because there is a partition of data storage according to the memory space. Eavesdropping cannot be done as each node is authenticated. All the nodes in the cloud are provided with an alternative logical IP so that even if there is an IP conflict successful data transfer takes place. The logical IP is checked with the original IP whenever data transmission takes place. It avoids unauthenticated users to access. When a node wrongly requests a file that is not in the destination node, it guides the requesting node by providing the details about the node containing that file. So that it helps the node in finding the destination node shortly. Ultimately, these concepts help in reducing the transmission time and increasing the response time. The outputs representing these implementations are given below.

Cloud Formation of the Proposed System: The cloud network is designed with five nodes. Each node is connected to each other node. Also, each node knows the details of every other node. Figure 4 clearly shows when any one of the nodes in the cloud is clicked it provides the node information that contains the details about the text files like: Word document, PDF file, and image file that are stored in that node. It also contains information about other nodes and details about the node size.

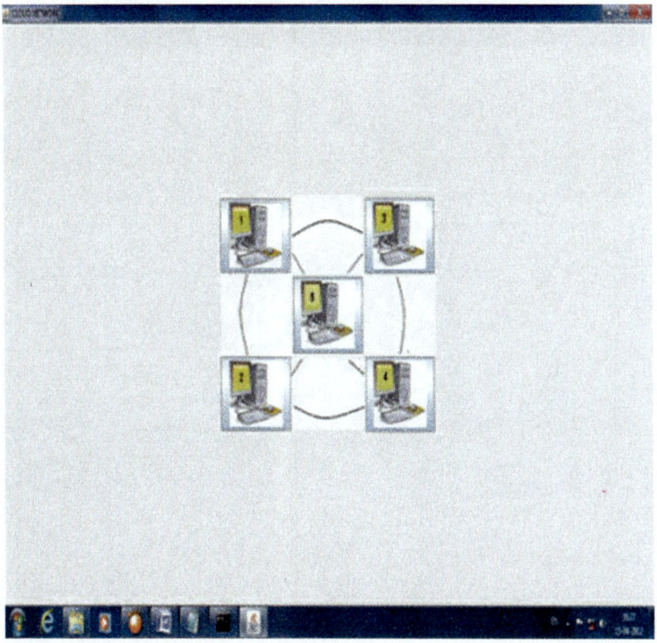

Fig. 4 Formation of cloud network

It provides a provision that allows the user to request for the data. That shows the various destinations for the node to make data request. The data request by a node 1 to other nodes is shown in Fig. 5. This condition is same for all the other nodes.

Data Request From the Node 1: Fig. 5 shows the data request from node 1 to other nodes. It can be done by requesting any of other nodes that are under the cloud formed in the paper. The IP address of source node is checked for the validity. The validity checks for the standard format of the IP address of the node and also checks whether the node is under the cloud. Once, we choose the destination node the IP conflict is checked. IP conflict exists when a node makes request to the same node again, so that a logical IP is provided to that node. If there is no such conflict, then also a logical IP is provided and allowed to share the data without any further procedures. Initially, each node has some memory space before transmission. Once a data is transmitted their node size is increased that is shown in Fig. 6.

Variation of Node Size During Data Transfer: Fig. 6 shows the size of the node before data transfer. It mentions the original space of the node containing its own data and files before data transmission. Figure 7 shows the size of the node after data transfer. That is, it mentions the space of the node after the data and files are transmitted from other node.

Checking IP Conflict Between the Nodes: Fig. 8 shows that there is no IP conflict between those nodes during data request. Since there is no IP conflict they are allowed

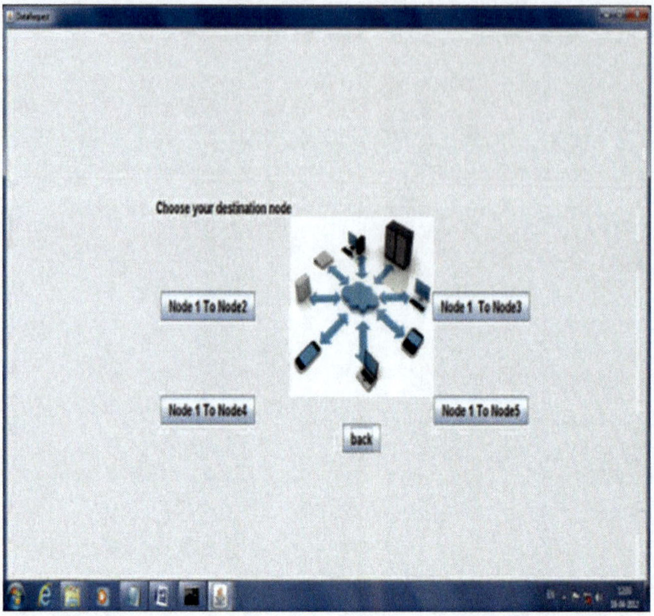

Fig. 5 Data request from node 1

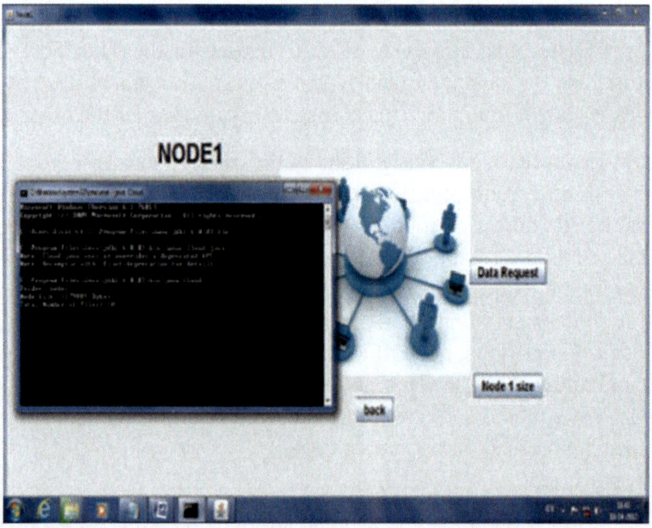

Fig. 6 Node size before data transfer

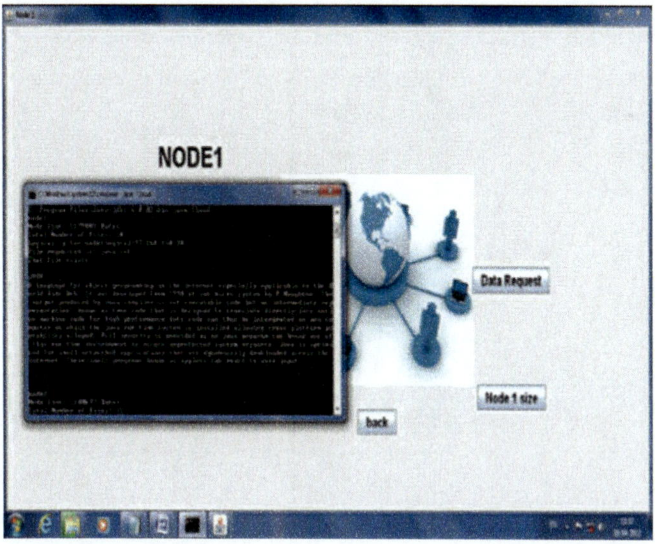

Fig. 7 Node size after data transfer

Fig. 8 No IP conflict

Fig. 9 IP conflict between node 1 and node 4

to transmit data as requested. Figure 9 shows a conflict between node 1 and node 4. Since they have same IP, new logical IP is provided. Then, sharing takes place using this new logical IP.

Providing Logical IP: For the security concern, here, the logical IP is provided to node 1 in Fig. 10. When again the request is sent by the same node the existence of the logical IP is checked. This Fig. 11 shows the existence of the logical IP when the node again requests for the data to the same node to which the data request is already sent. The above figures show the details about the data security through authentication by providing logical IP. Now the following figures explain the file retrieval by using the data partition algorithm. It also shows that there is no loss of data storage and also ensures the correctness of the data.

File Retrieval: Fig. 12 shows there is no IP conflict so there is a sharing of information. Here, a text document is requested that is confidential to that particular node. So when request is sent a security check is done, where the username and password are to be provided in order to access the information. If correct username and password are provided the information is retrieved. Image files can also be retrieved. Figure 13 shows the retrieval of the particular image which is requested. As the image is available in the node, it is retrieved. If the image is not available, it guides the node to send the request to the node which has the requested image as shown in Fig. 14.

Graph for the Proposed System: After implementing, the proposed efficient scheme, time taken for the transmission is noted and graph is created. Figure 15

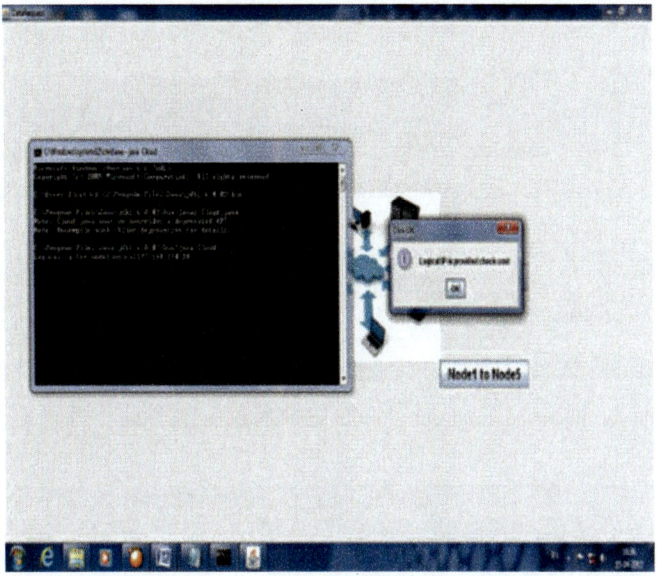

Fig. 10 Providing logical IP for node 1

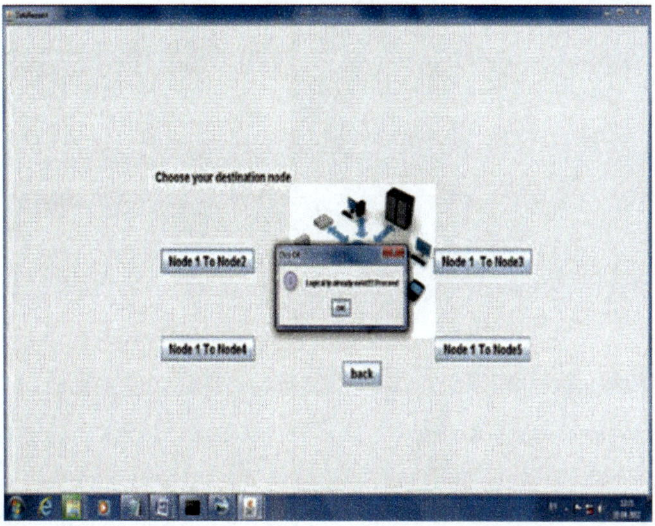

Fig. 11 Existence of logical IP for node 1

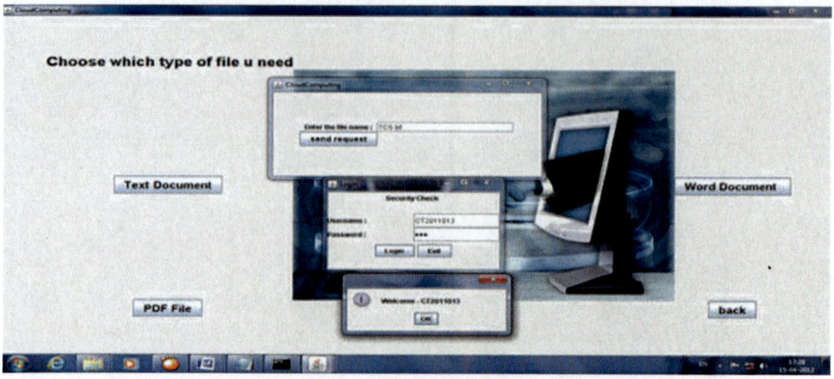

Fig. 12 Authentication to get confidential information which can be shared

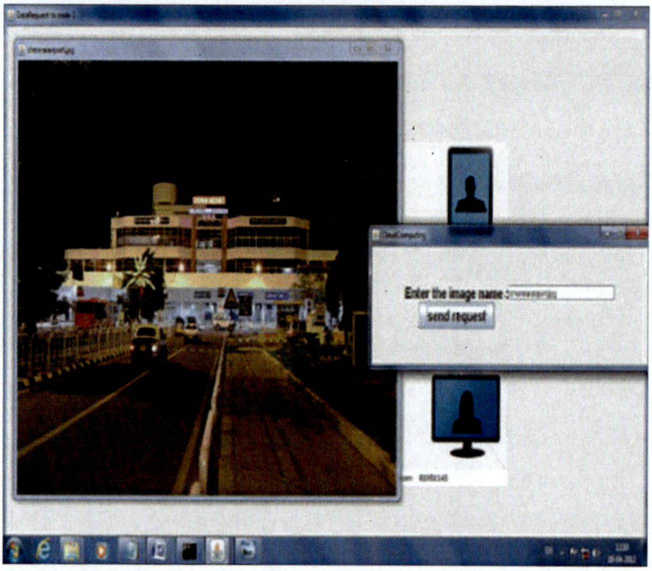

Fig. 13 Retrieving the requested image

shows the graph which is for the number of nodes and the data transmission time of the proposed system.

Figure 16 demonstrates the graph which shows the values of the existing system and the values of the projected system. From the figure, it is understood that the data transmission time is considerably reduced when compared to the existing system. Thus, the data aware routing is efficiently implemented.

Graph: Compared With Existing System: Fig. 17 shows the percentage ratio of the constraints proposed in this paper. Thus, while comparing with the existing system

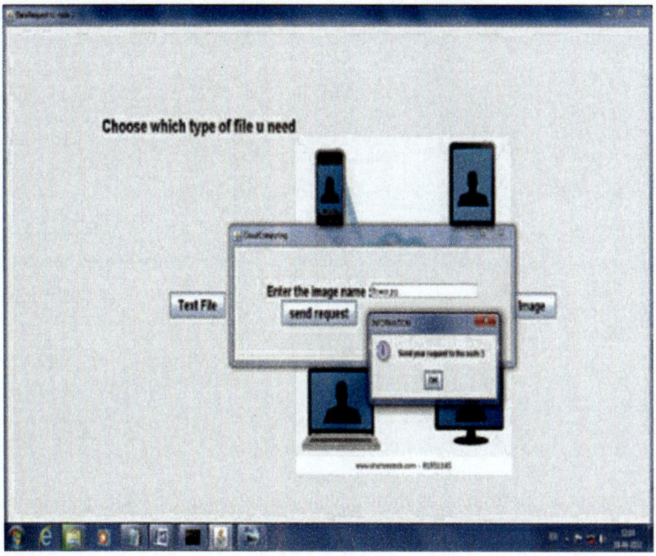

Fig. 14 Requesting image

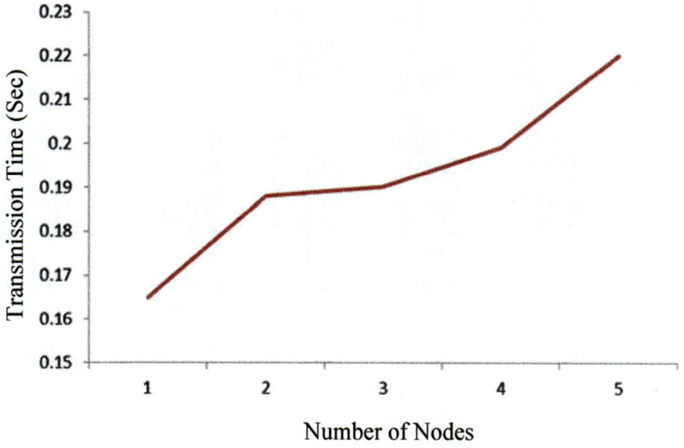

Fig. 15 Graph for the proposed system

the reliability, scalability, and transmission time is reduced, thereby the response time is increased. Therefore, the data transfer is easily done and the loss of data storage is avoided.

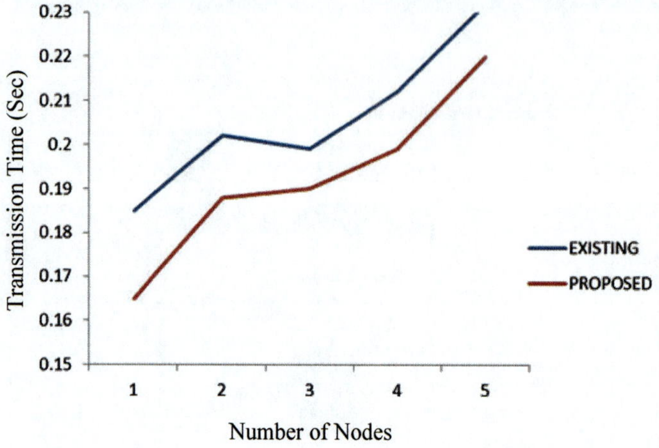

Fig. 16 Node versus data transmission time reduction matched to the existing system

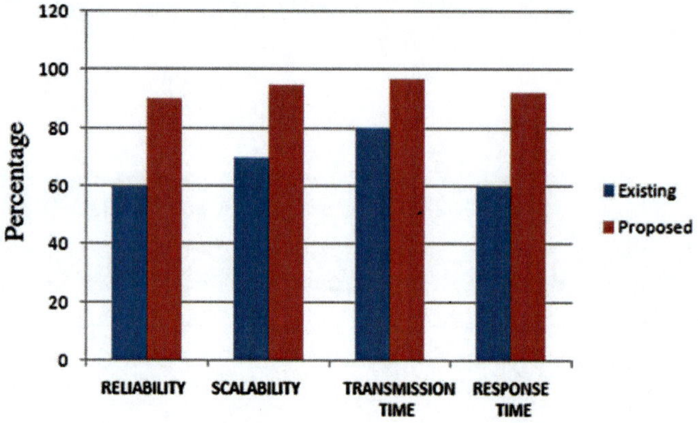

Fig. 17 Bar chart showing node versus data transmission time reduction compared to the existing system

8 Conclusion

The data aware routing is implemented in this paper which initially partitions the data in the cloud storage. Further by providing logical IP, the data integrity is preserved. Thus, the transmission time is reduced by the data partition and also security risks are avoided by comparing the original and logical IP. This scheme ensures the storage correctness, and preserves high security during file transfer and attain the cloud data truthfulness and obtainability to implement the excellence of reliable cloud storage provision for consumers. The reliability is strongly achieved by the security algorithm which is implemented here. Thus, whenever the nodes are constantly giving request, it reduces that scalability problem because of data partition algorithm. Thus, this scheme efficiently made the cloud computing more securable and also made the secure cloud data storage.

Future Work: The proposed scheme efficiently reduces the transmission time during file transfer and also preserves high security. Even though the proposed scheme provides these things, it needs further investigation on the data storage. It is hard for this scheme to identify the server which is misbehaving. So it is need to be checked in efficient manner. Thus, some additional concentration should be given on the error localization, only then it can classify the disobedient server and accurate the information stowing consequently.

References

1. Erway, C., Kupcu, A., Papamanthou, C., Tamassia, R.: Dynamic provable data possession. In: Proceedings of CCS'09, 2009, pp. 213–222
2. Wang, C., Ren, K., Lou, W., Li, J.: Towards publicly auditable secure cloud data storage services. IEEE Netw. Mag. **24**(4), 19–24 (2010)
3. Bowers, K.D., Juels, A., Oprea, A.: Proofs of retrievability: theory and implementation. In: Proceedings of ACM Workshop on Cloud Computing Security (CCSW'09), 2009, pp. 43–54
4. Ateniese, G., Burns, R., Curtmola, R., Herring, J., Kissner, L., Peterson, Z., Song, D.: Provable data possession at untrusted stores. In: Proceedings of CCS '07, pp. 598–609, 2007
5. Wang, C., Wang, Q., Ren, K., Lou, W.: Privacy-preserving public auditing for storage security in cloud computing. In: Proceedings of IEEE INFOCOM'10, San Diego, CA, USA, Mar 2010
6. Bowers, K.D., Juels, A., Oprea, A.: Hail: a high-availability and integrity layer for cloud storage. In: Proceedings of CCS'09, 2009, pp. 187–198
7. Shah, M.A., Baker, M., Mogul, J.C., Swaminathan, R.: Auditing to keep online storage services honest. In: Proceedings of HotOS'07. Berkeley, CA, USA: USENIX Association, 2007, pp. 1–6
8. Wang, C., Wang, Q., Ren, K., Lou, W.: Ensuring data storage security in cloud computing. In: Proceedings of IWQoS'09, July 2009, pp. 1–9
9. Curtmola, R., Khan, O., Burns, R., Ateniese, G.: Mr-pdp: Multiple-replica provable data possession. In: Proceedings of ICDCS'08. IEEE Computer Society, 2008, pp. 411–420
10. Ateniese, G., Pietro, R.D., Mancini, L.V., Tsudik, G.: Scalable and efficient provable data possession. In: Proceedings of SecureComm'08, 2008, pp. 1–10
11. Nareshvurukonda B., Rao, T.: A study on data storage security issues in cloud computing. Procedia Comput. Sci. **92**, 128–135 (2016)
12. Wang, Q., Wang, C., Li, J., Ren, K., Lou, W.: Enabling public verifiability and data dynamics for storage security in cloud computing. In: Proceedings of ESORICS'09, vol. 5789 of LNCS, Sept 2009, pp. 355–370. Springer, Berlin
13. Dodis, Y., Vadhan, S., Wichs, D.: Proofs of retrievability via hardness amplification. In: Proceedings of the 6th Theory of Cryptography Conference (TCC'09), San Francisco, CA, USA, Mar 2009
14. Wang, C., Wang, Q., Ren, K., Lou, W.: Towards secure and dependable storage services in cloud computing. In: Proceedings of IWQoS '11, May 2011, pp. 1–14
15. Wang, Q., Ren, K., Lou, W., Zhang, Y.: Dependable and secure sensor data storage with dynamic integrity assurance. In: Proceedings of IEEE INFOCOM'09, Rio de Janeiro, Brazil, Apr 2009
16. Juels, A., Burton, J., Kaliski, S.: PORs: proofs of retrievability for large files. In: Proceedings of CCS '07, pp. 584–597, 2007
17. Abo-alian,A., Badr, N.L., Tolba, M.F.: Data storage security service in cloud computing: challenges and solutions. In: Multimedia Forensics and Security, pp 25–57 (2016)

18. Schwarz, T.S.J., Miller, E.L.: Store, forget, and check: using algebraic signatures to check remotely administered storage. In: Proceedings of ICDCS '06, pp. 12–12, 2006
19. Ismail, M., Yusuf, B.: Ensuring data storage security in cloud computing with advanced encryption standard (AES) and authentication scheme (AS). Int. J. Inf. Syst. Eng. **4**(1), 18–39 (2016)
20. Lillibridge, M., Elnikety, S., Birrell, A., Burrows, M., Isard, M.: A cooperative internet backup scheme. In: Proceedings of the 2003 USENIX Annual Technical Conference (General Track), pp. 29–41, 2003

A New 4-2 Compressor for VLSI Circuits and Systems

Inamul Hussain and Saurabh Chaudhury

Abstract The compressor is a useful element that is widely used in VLSI circuits and systems. It is generally used as a processing element. In this work, a 4-2 compressor has been designed with XOR-XNOR module and multiplexer module. The XOR-XNOR module is consists of six transistors and the multiplexer is consists of transmission gate. The compressor has been designed and simulated in Synopsys tool by using 90 nm CMOS technology. Average power dissipation, worst case delay and power delay product are computed. The same has been observed by varying the supply voltage. A study of the performance comparison has been carried out with existing compressors. It is found that the proposed compressor has low power consumption and the best PDP.

Keywords Compressor · Delay · MUX · Power · PDP · XOR-XNOR · VLSI

1 Introduction

Low power consumption and high speed of operations are the main demands of electronics consumers [1]. Thus, an abrupt growth of fast and efficient electronics is observed to fill up the recent needs. Hence, to design and develop such systems are become the premiere goal of VLSI designers [2–6]. The compressor is a useful element that is widely used in VLSI circuits and systems [7]. It is generally used as a processing element. It is used to perform an addition operation [8]. One of the main applications of compressors is in the multiplier circuit for faster operation [9–13], where partial products are added by using the compressor to reduce complexity. The use of multiplier circuits is in DSP processor, image processing, digital filters, communication systems, etc. [8, 9]. So, the performance of these applications depends on the performance of the compressor in terms of speed and power consumption. So it is important to design compressor circuits for VLSI applications.

I. Hussain (✉) · S. Chaudhury
Department of Electrical Engineering, NIT Silchar, Silchar, Assam 788010, India
e-mail: ihinamul07@gmail.com

P. Suresh et al. (eds.), *Advances in Smart System Technologies*,
Advances in Intelligent Systems and Computing 1163,
https://doi.org/10.1007/978-981-15-5029-4_33

A compressor circuit is denoted by m-n compressor, where m and n represent inputs and outputs bits, respectively [6]. It is also denoted as m:n. The simplest compressor is 3-2, which is nothing but a full adder. Different types of compressors such as 3-2, 4-2, 4-3, 5-2, 5-3, 6-2, 6-3, etc., are designed for different purposes [10, 11]. The 4-2 compressor is one of the popular compressors that are widely used in VLSI circuits and systems. Thus, in this work, a 4-2 compressor has been designed.

In the literature, different 4-2 circuits have been studied which are designed by different approaches [14–18]. A 4-2 compressor has been designed based on CMOS logic and transmission gate, but it has to compromise with power consumption and delay [15]. Another 4-2 compressor has been designed by double pass-transistor logic (DPL) [17]. Though it has a good speed of operation, it has worse power dissipation. The number of the transistor is also high in this design. The transistor count has been reduced by a new 4-2 compressor, which has been designed by 10T XOR-XNOR module with a demerit of high power consumption [18]. Thus, in this work, a 4-2 compressor has been designed with 6T XOR-XNOR module to reduce transistor count and improve the performance. The design has been implemented based on bulk-Si based MOSFET technology [19].

The rest of the paper has been presented as, in Sect. 2, a brief overview of the 4-2 compressor is given, in Sects. 3 and 4, the detail of proposed compressor and performance analysis have reported, respectively. Section 5 concludes the paper.

2 Overview of 4-2 Compressor

A compressor circuit is denoted by the m-n compressor, where m and n represent inputs and outputs bits, respectively [7]. It is also denoted as m:n. A 4-2 compressor is one of the popular compressors that are widely used in VLSI circuits and systems. It is used multiplier circuit to reduce partial products. Thus, it can reduce four partial products into two [14]. Along with the 4 inputs and 2 outputs, 4-2 compressor has a carry input (Cin) from the previous operation and a carry output (Cout) to propagate. The block diagram is shown in Fig. 1. The inputs are A, B, C, D and Cin. The outputs are Sum, Carry and Cout.

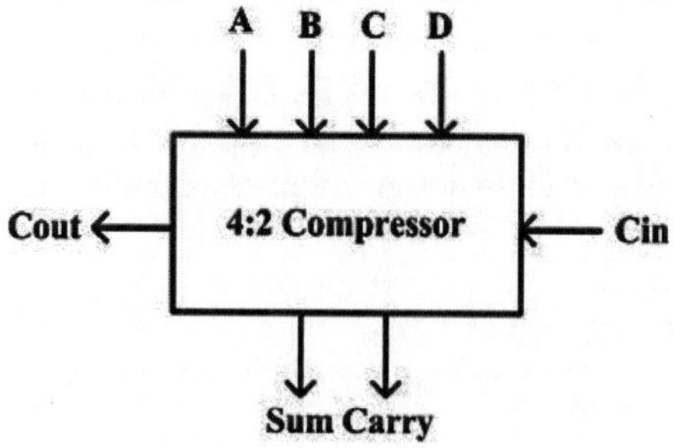

Fig. 1 Block diagram of 4-2 compressor

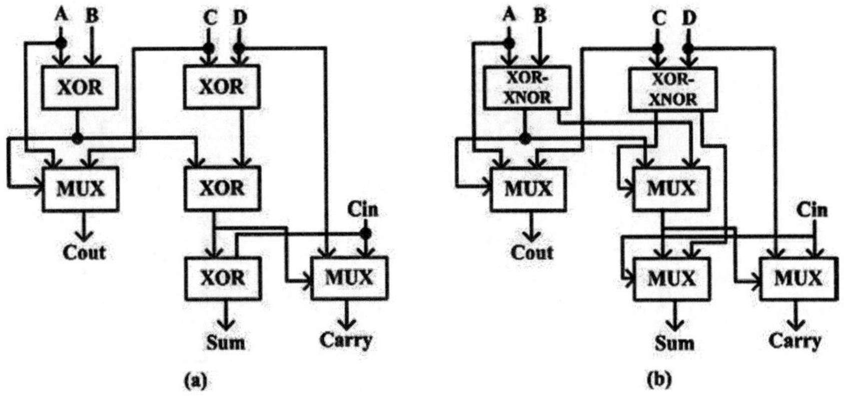

Fig. 2 Architectures of 4-2 compressor, **a** conventional, **b** modified

The most two popular architectures [17, 18] have been shown in Fig. 2. Figure 2a shows a conventional architecture of the 4-2 compressor by using XOR gates and MUX. It has more number of XOR gates that increase delay and power consumption. To address the delay and power consumption, a modified architecture has been designed as shown in Fig. 2b based on XOR-XNOR module and MUX [16].

3 The Detail of the Proposed 4-2 Compressor

The proposed compressor has been designed based on the modified architecture as shown in Fig. 2b. The XOR-XNOR module has been designed by six transistors [5] and MUX is by transmission gate logic [1]. XOR-XNOR module consumes very high power, thus, the module used here is a low power hybrid XOR-XNOR module. It is called hybrid because it is designed by two logic styles pass-transistor logic and CMOS logic. PTL-based gives a fast output, whereas CMOS gives a full swing output In this module, initially, XNOR output is obtained followed by XOR output by inverting the outputs of former. The MUX is designed by transmission gate logic to get full swing outputs. Figure 3a, b show XOR-XNOR module and MUX module, respectively.

4 Simulation Results of the Proposed 4-2 Compressor

The compressor has been designed and simulated in Synopsys tool by using 90 nm CMOS technology. Average power dissipation, worst case delay and power delay product are computed. The same has been observed by varying the supply voltage. The results have been reported in Table 1.

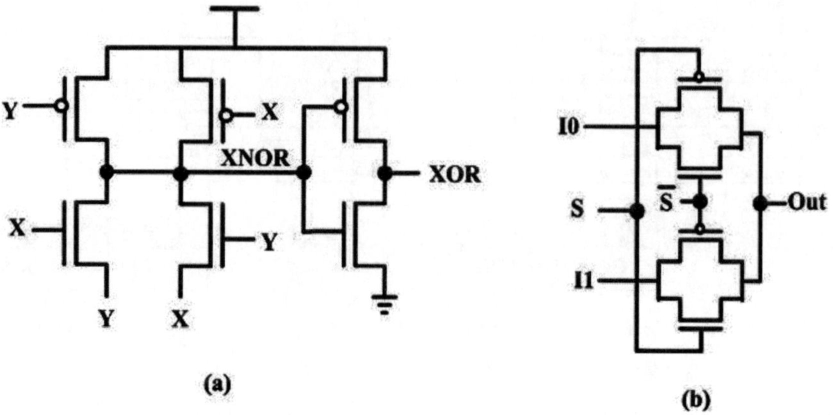

Fig. 3 **a** XOR-XNOR module and **b** transmission gate MUX

Table 1 Simulation results at 1.2 V power supply

Compressor	Delay (ps)	Power(μW)	PDP (fJ)	Transistor count	References
CMOS	250.1	60.0	15.0	72	[15]
DPL	170.0	35.6	6.1	62	[17]
10T	220.7	38.5	8.5	60	[18]
Proposed	178.5	15.5	2.8	32	–

Table 1 also shows the comparison of the performance of the designed compressor with the few existing compressors at 1.2 V power supply. A comparison graph of delay, power and PDP has been shown in Fig. 4a–c, respectively. It is observed that the designed compressor has satisfactory performances. The proposed compressor has better performance because it has been designed by the improved architecture. The XOR-XNOR module used is energy efficient. Moreover, the number of transistor count in the proposed compressor has been reduced to 32 that shrink the complexity of the circuit.

5 Conclusion

In this work, a 4-2 compressor has been designed for VLSI applications. It is designed by 6T XOR-XNOR module and MUX module with modified architecture. The design and simulation have been done in the Synopsys tool by using 90 nm CMOS technology. The performance of the compressor is observed by varying the power supply. A comparison of the performance with existing circuits is also done. Though in some cases, the compressor does not show full swing outputs, it has the best power consumption and PDP. Thus, it can be used for VLSI circuits and systems.

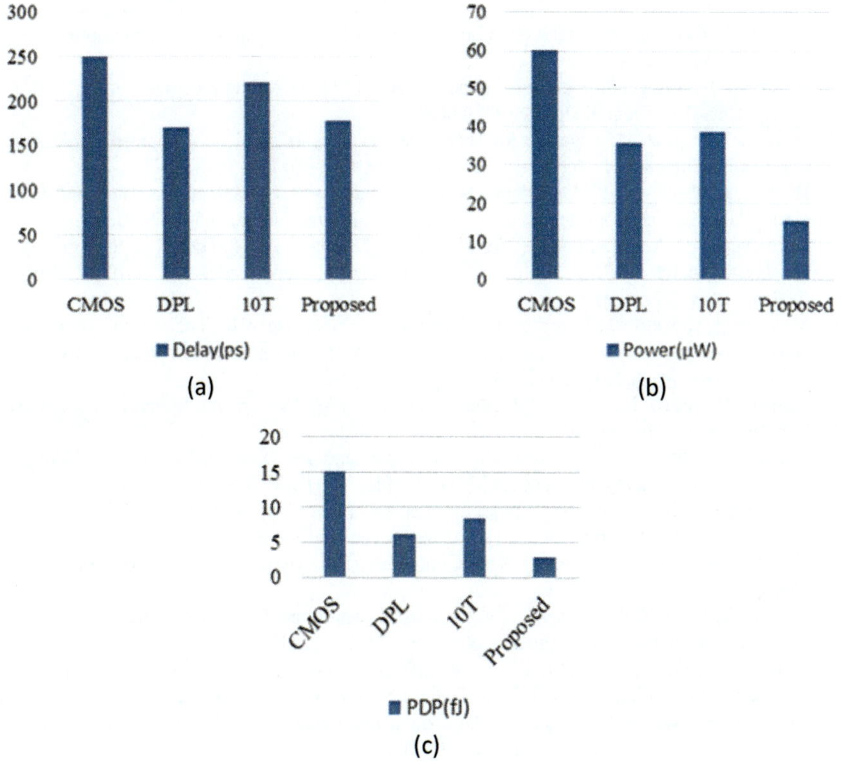

Fig. 4 Comparisons of different parameters, **a** delay, **b** power dissipation, **c** PDP

In the future work, the compressor may be used in fast multiplier for partial product reduction.

Acknowledgements The authors would like to acknowledge TEQIP cell, NIT Silchar and MHRD, India for giving necessary funding and assistantship.

References

1. Hussain, I., Chaudhury, S.: A comparative study on the effects of technology nodes and logic styles for low power high-speed VLSI applications. Int. J. Nanopart. **12(1/2)**, 122 (2020)
2. Hussain, I., Chaudhury, S.: Performance comparison of 1-bit conventional and hybrid full adder circuits. In: Bera, R., Sarkar, S., Chakraborty, S. (eds.) Advances in Communication, Devices and Networking. Lecture Notes in Electrical Engineering, vol. 462 (2018)
3. Rabaey, J., Chandrakasan, A., Nikolic, B.: Digital Integrated Circuits a Design Perspective, 2nd edn. Prentice Hall Inc., New Jersey (2003)
4. Weste, N.H.E., Harris, D., Banerjee, A.: CMOS VLSI Design—A Circuit and Systems Perspective. Pearson (2006)

5. Hussain, I., Chaudhury, S.: CNFET based low power full adder circuit for VLSI applications. Nanosci. Nanotechnol.-Asia **9**, 1 (2019)
6. Hiremath, Y.: A novel 8-bit carry select adder using 180 nm CMOS process technology. Int. J. Emerg. Eng. Res. Technol. **2**(6), 187–194 (2014)
7. Hussain, I., Kumar, M.: Design and performance analysis of a 3-2 compressor by using improved architecture. J. Act. Passiv. Electron. Dev. **12**(3–4), 173–181 (2017)
8. Hussain, I., Pandey, C. K., Chaudhury, S.: Design and analysis of high performance multiplier circuit. 2019 Devices for integrated circuit (DevIC), Kalyani, India, 245–247 (2019)
9. Dorojevets, M., Kasperek, A.K., Yoshikawa, N., Fujimaki, A.: 20-GHz 8 × 8-bit parallel carry-save pipelined RSFQ multiplier. IEEE Trans. Appl. Supercond. **23**, 1300104–1300108 (2013)
10. Mokrian, P., Howard, G.M., Jullien, G., Ahmadi, M.: On the use of 4: 2 compressors for partial product reduction. In: IEEE Canadian Conference on Electrical and Computer Engineering (CCECE), vol. 1, pp. 121–124 (2003))
11. Hussain, I., Kumar, M.: A fast and reduced complexity Wallace tree multiplier. J. Act. Passiv. Electron. Dev. **12**, 63–71 (2017)
12. Khan, S., Kakde, S., Suryawanshi, Y.: VLSI implementation of reduced complexity Wallace multiplier using energy efficient CMOS full adder. In: IEEE International Conference on Computational Intelligence and Computing Research, Enathi, pp. 1–4 (2013)
13. Asif, S., Kong, Y.: Performance analysis of Wallace and radix-4 Booth-Wallace multipliers. In: 2015 Electronic System Level Synthesis Conference (ESLsyn), San Francisco, CA, pp. 17–22 (2015)
14. Prasad, K., Parhi, K.K.: Low-power 4-2 and 5-2 compressors. In: 35th Asilomar Conference on Signals, Systems and Computers, vol. 1, pp. 129–133 (2001)
15. Chang, C.-H., Gu, J., Zhang, M.: Ultra low-voltage low-power CMOS 4-2 and 5-2 compressors for fast arithmetic circuits. IEEE Trans. Circuits Syst. I **51**, 1985–1997 (2004)
16. Nirlakalla, R., Rao, T.S., Prasad, T.J.: Performance evaluation of high speed compressors for high speed multipliers. Serbian J. Electr. Eng. **8**, 293–306 (2011)
17. Veeramachaneni, S., Krishna, M.K., Avinash, L., Puppala, S.R., Srinivas, M.B.: Novel architectures for high-speed and low-power 3-2, 4-2 and 5-2 compressors. In: 20th International Conference on VLSI Design. Held jointly with 6th International Conference on Embedded Systems, pp. 324–329 (2007)
18. Tonfat, J., Reis, R.: Low power 3-2 and 4-2 adder compressors implemented using ASTRAN. In: 3rd IEEE Latin American Symposium on Circuit and Systems (LASCAS), pp. 1–4 (2012)
19. Hussain, I., Singh, A., Chaudhury, S.: A review on the effects of technology on CMOS and CPL logic style on performance, speed and power dissipation. 2018 IEEE Electron Devices Kolkata Conference (EDKCON), Kolkata, India, 332–336 (2018)

A Novel High Gain Modified SEPIC Converter for Renewable Energy Sources

R. Thulasi Lakshmi and B. V. Manikandan

Abstract This paper presents an performance enhanced SEPIC converter topology without and with including a magnetic coupling in order to attain an elevated voltage gain. This proposed converter can be used in renewable energy applications such as in automobiles and DC grid networks. The proposed converter topology delivers a low source voltage and reduced switch voltage, therefore, giving growth in efficiency for increased gain. The power electronic converter with coupled inductor and without magnetic coupling is simulated through MATLAB. The efficiency increase is mainly important, primarily for battery functioned power systems and extraordinary cost power systems.

Keywords DC-DC enhanced SEPIC converter · Static gain · Solar · PI controller · ANFIS controller

1 Introduction

Large gain DC-DC converters are presently vital study space as needed to the large demand of this technology for renewable energy source usage such as fuel cells, transportable electronic equipment's and constant power source [1]. High step up gain is needed for loads with DC or AC highest voltage advanced than 10 intervals the supply voltage.

When an enlarged gain [2] is wanted the expected thing here is the use of isolated power electronic converter. The proposed converters used with solar photovoltaic gives us the increased production due to the higher cost with non-renewable source of energy. The avoidance of the transformer can surge the efficiency and power focus of the power electronic converter with an output voltage about ten and twenty times the input source with a D value equal to 0.82. A gain around five [3] is a partial value for the conventional SEPIC converter in this work and this case is measured as a normal gain. A converter with gain higher than twenty is taken into

R. Thulasi Lakshmi (✉) · B. V. Manikandan
Mepco Schlenk Engineering College, Virudhunagar District, Sivakasi, India
e-mail: thulasilakshmi.r@gmail.com

P. Suresh et al. (eds.), *Advances in Smart System Technologies*,
Advances in Intelligent Systems and Computing 1163,
https://doi.org/10.1007/978-981-15-5029-4_34

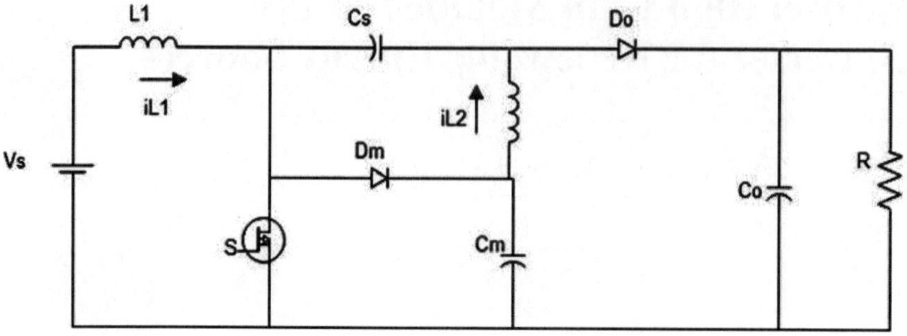

Fig. 1 DC-DC enhanced SEPIC converter less magnetic coupling

account an awfully high gain to this proposed work. Energy shortage and ecological waste product issues have increased the employment of healthful energy renewable resources like wind, electric cell and solar energy system. A PV system could be a grid thought of to provide real-world solar energy. The reduced voltage stress across the switch permits the operaticlosed of MOSFET with reduced closed resistance. This device will accomplish a high change of boost relaticlosed lacking of using a high switching frequency electrical device with as well as magnetic coupling.

2 Proposed Converters—DC-DC Enhanced SEPIC Converter Without Magnetic Coupling

2.1 Power Circuit

CCM [4] mode implies that the converter current never becomes zero. Closedce the converter operates in above zero current mode, the ripples through L_1 are often expected negligible [5]. The input supply is taken from the solar panel. Figure 1 shows DC-DC enhanced SEPIC converter less magnetic coupling.

2.2 Modes of Operation for DC-DC Enhanced SEPIC Converter Without Magnetic Coupling

The modes of operaticlosed of proposed DC-DC enhanced SEPIC converter while not including magnetic coupling has two modes of operaticlosed, that is, explained below.

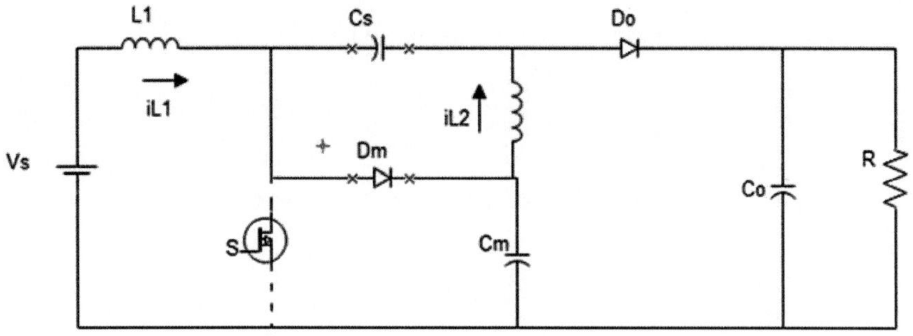

Fig. 2 DC-DC enhanced SEPIC converter under switch opened and diode closed topology

2.2.1 Mode1

When MOSFET is opened and also the energy in inductance L_1 is transferred to the output through Cs, Do, Cm from starting to end of the diode D_M.

Therefore, the switch voltage is closede and the same to the C_M voltage. The energy in L_2 is made to flow to the output via diode $D_{o.}$ The circuit for this mode is given in Fig. 2.

2.2.2 Mode2

When the power switch S is closed, the diodes D_M and D_o are reverse biased and the energy storage device L_1 and L_2 stores the energy. The input voltage is useful to the inductance L_1, and therefore the voltage $V_{CS} - V_{CM}$ is applied to the energy storage device L_2. The circuit for this mode2 is given in Fig. 3.

The C_M capacitor voltage is calculated as follows.

$$\frac{V_{CM}}{V_i} = \frac{1}{1 - D} \tag{1}$$

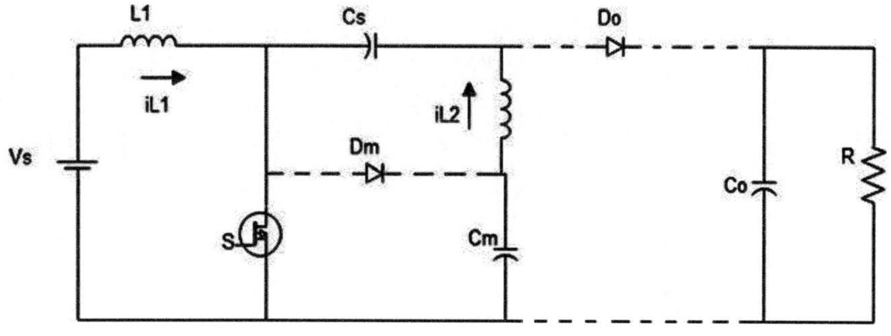

Fig. 3 DC-DC enhanced SEPIC converter under switch closed and diode opened topology

The energy storage device values are calculated by using the below formula.

$$L_1 = L_2 = \frac{V_i D}{\Delta i_L f} \tag{2}$$

The output current, diode current can be found easily by Eq. (3).

$$\mathrm{ID_0} = \mathrm{ID_M} = I_0 = \frac{P_0}{V_0} \tag{3}$$

The output voltage ripple can be calculated by the following equation.

$$\Delta V_C = \frac{V_i}{1 - D} \tag{4}$$

Duty cycle of the proposed converter is used to close and open the power switch. Here, the MOSFET is the switch used. The value of duty cycle can be calculated as follows.

$$D = \frac{V_0 - V_i}{V_0 + V_i} \tag{5}$$

2.3 Power Circuit with Magnetic Coupling

Functional operating characteristics of classical SEPIC converters are less improved compared with the proposed DC-DC enhanced SEPIC converter with including a magnetic coupling structure. The static gain increases [6] in higher amount than in the proposed DC-DC enhanced SEPIC converter without including a magnetic coupling. The transformer leakage inductance prevents the leakage current. Figure 4

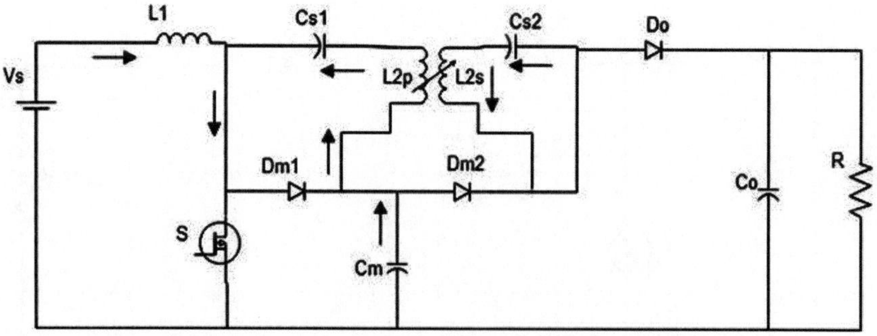

Fig. 4 DC-DC enhanced SEPIC converter with coupled inductor

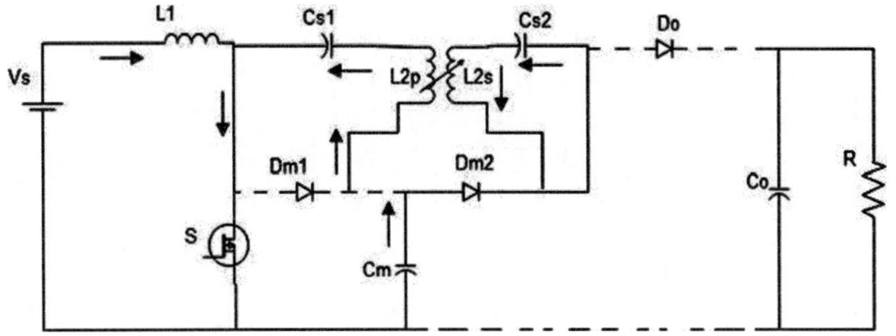

Fig. 5 DC-DC enhanced SEPIC converter with coupled inductor under switch closed and diode opened topology

represents the power circuit of a proposed DC-DC enhanced SEPIC converter with magnetic coupling.

2.4 *Operating Stages*

The modes of operation of proposed DC-DC enhanced SEPIC converter with magnetic coupling are divided into five modes.

2.4.1 Mode1

When the power switch S is closed, the energy storage device L_1 is storing the charges. The C_{S2} is excited by inductance L_{2S} and diode D_{m2}. The inductance outflow does not allow sudden change in current. Unidirectional device D_{m2} is under forward biased state and the diode D_{m1} and D_0 are under reverse biased state. The circuit for this mode is given in Fig. 5.

2.4.2 Mode2

At the end of mode1, when the diode D_{M2} is reverse biased, when the power switch is opened, the passive elements L_1 and L_2 accumulates a charge in it, and their current increases rapidly. The diode D_{m2} is under forward biased condition. The circuit for this mode is shown in Fig. 6.

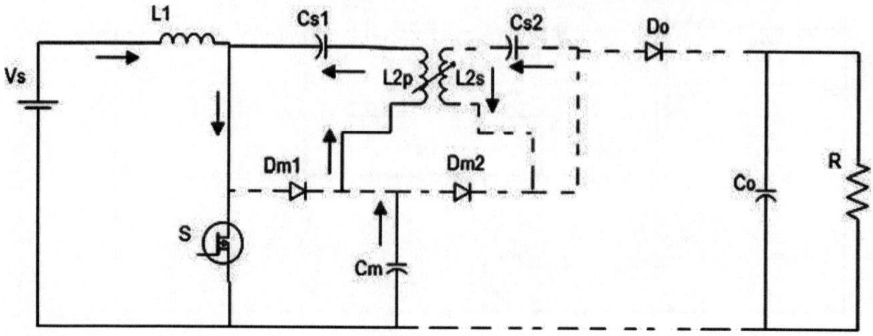

Fig. 6 DC-DC enhanced SEPIC converter with coupled inductor under switch closed and diode opened topology

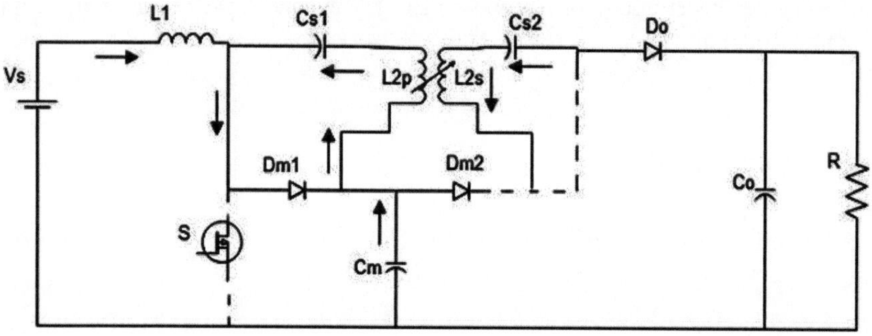

Fig. 7 DC-DC enhanced SEPIC converter with coupled inductor under switch opened and diode D_{m2} opened topology

2.4.3 Mode3

At the end of mode2, when the MOSFET is opened, the charges in L_1 is fed to C_M. Now there occurs power travel to output via C_{S1}, L_2 and D_o. The diode D_{m1} and D_0 are in forward biased state. The circuit for this mode is presented in Fig. 7.

2.4.4 Mode4

At the end of mode3, the power transmitted to C_M is completed, D_{m1} is reverse biased. The power travel to the load is continued up to the time, once the MOSFET is under conducting mode. The diode D_{m1} and D_{m2} are under forward biased state. The circuit for this mode is presented in Fig. 8.

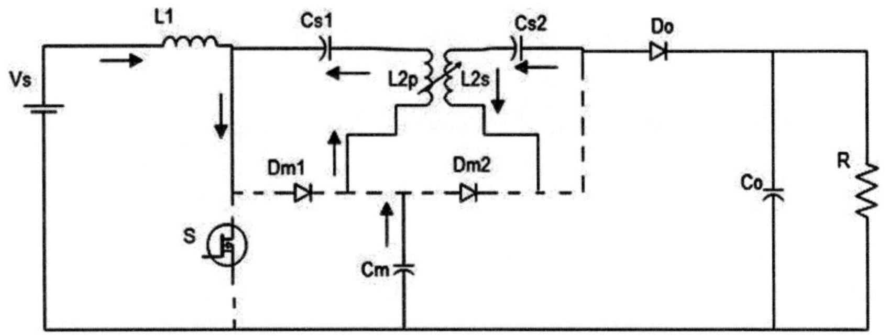

Fig. 8 DC-DC enhanced SEPIC converter with coupled inductor under switch opened topology and diode D_0 closed topology

2.4.5 Mode5

When MOSFET is in closed state, the current in D_o reduces rapidly. Instantaneous change in current is controlled by the inductance L_{2S}, thus lowering the unidirectional device opposite recovery current complexity. While D_o is not under forward biased condition, the converter proceeds to the initial operating mode. The diode D_o is under forward biased state. Figure 9 shows the circuit of

The number of turns (n) can be calculated as follows.

$$\frac{V_0}{V_i} = \frac{1}{1-D}(1+n) \quad (6)$$

$$n = \frac{N_{L2P}}{N_{L2S}} \quad (7)$$

The voltage across the diode can be found by the below equation,

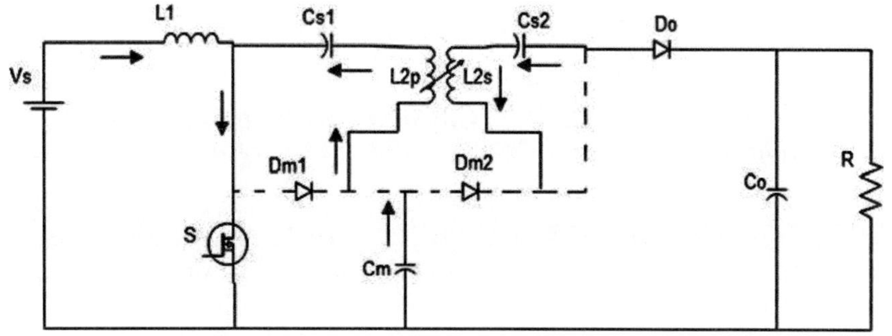

Fig. 9 DC-DC enhanced SEPIC converter with coupled inductor under switch closed and diode opened topology

$$\mathrm{VD}_0 = \mathrm{VD}_{\mathrm{M2}} = V_0 - V_{\mathrm{CM}} = \frac{nV_i}{1-D} \tag{8}$$

The secondary winding of an energy storage device is given by

$$L_{2P} = n^2 L_{2S} \tag{9}$$

Primary winding and input side inductance can be calculated from the formula of a DC-DC enhanced SEPIC converter without including a magnetic coupling.

The source side capacitance is found from the below equaticlosed.

$$C_{S1} = C_{\mathrm{M}} = \frac{I_0 R}{\Delta V_C f} \tag{10}$$

3 PI Controller

PI regulator will remove [7] obligatory oscillations and stable state fault resulting in act of on-off controller and P controller, respectively. However, introducing basic mode has a depressing effect on speed of the reaction and largely stability of the system. Thus, PI regulator will not enlarge the speed of the response. The PI controller takes into calculation the desired output of the converter and its definite output and thus finds the error. Based on this inaccuracy, controller works to provide the input needed to achieve the preferred output. PI controller provides enhanced regulation. It has two terms, proportional and integral term. The PI controller is used for simulation in both the proposed DC-DC enhanced SEPIC converter with and without magnetic coupling.

4 Simulation Results

4.1 DC-DC Enhanced SEPIC Converter Without Magnetic Coupling

MATLAB–Simulink model of enhanced SEPIC converter less magnetic coupling in closed loop with PI controller and including PV modeling is shown in Fig. 10.

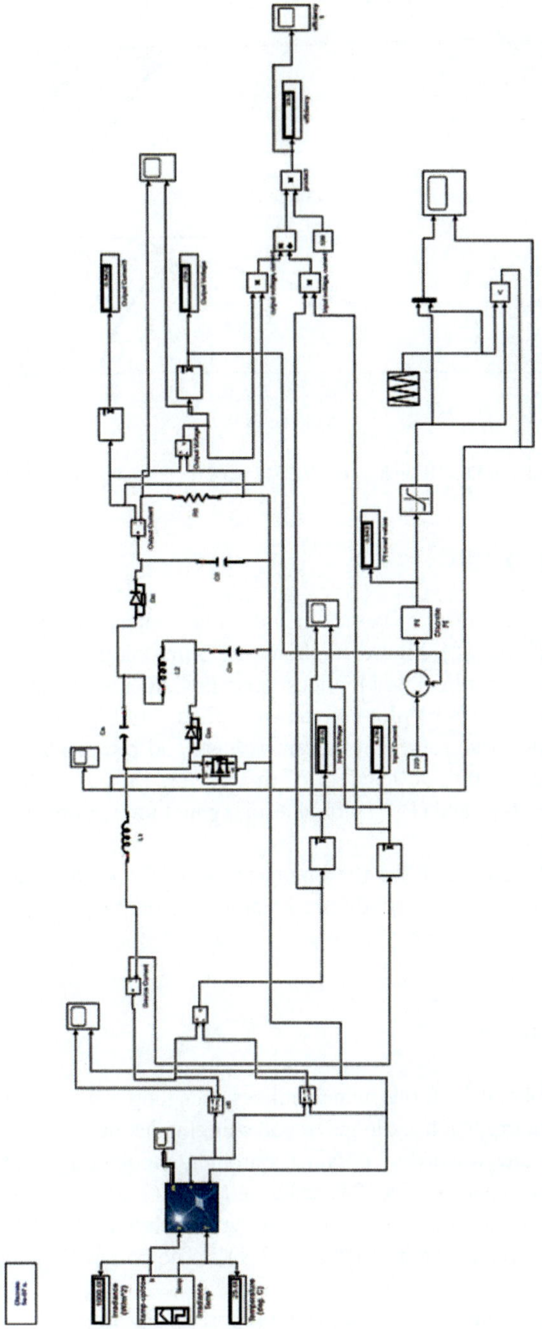

Fig. 10 MATLAB–Simulink model of enhanced SEPIC converter less magnetic coupling in closed loop (PI controller) and including PV modeling

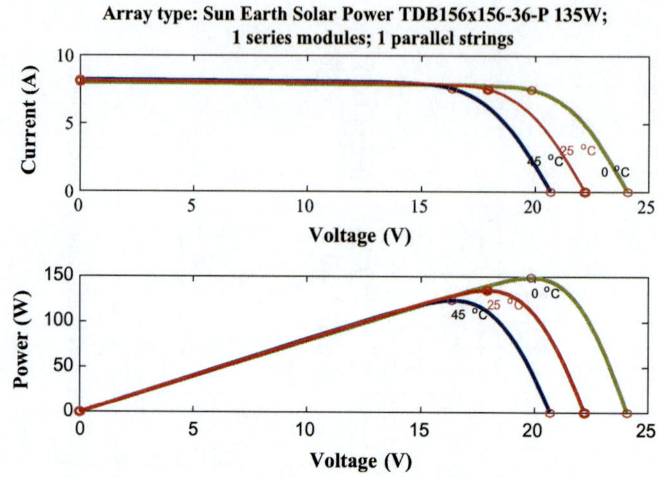

Fig. 11 *I-V* and *P-V* plot of DC-DC enhanced SEPIC converter less coupled inductor

4.1.1 Open Loop System

The DC-DC enhanced SEPIC converter includes one switch, two inductors, three capacitor and two diodes [8]. Figure 10 shows the Simulink model of the DC-DC enhanced SEPIC converter in closed loop less coupled inductor. The current-voltage (*I-V*) and power-voltage (*P-V*) plot are shown in Figs. 11, 13, and 17. The green color and blue color curve represent current-voltage and power-voltage plot of a solar panel for a temperature of 0 and 45 °C in Figs. 11, 13, and 17. The red color curve represents the (*I-V*) and (*P-V*) plot of a solar panel for a temperature of 25 °C in Figs. 11, 13, and 17 (Fig. 12).

With no magnetic coupling, for the input voltage of 22 V DC, output voltage obtained is 200 V DC in open loop. Figure 12 shows the output current and voltage waveform

4.1.2 Closed Loop System

A closed loop control scheme is one, which utilizes open loop system as its frontward path but has one or more feedback loops or paths linking its input and its output.

The output voltage waveform obtained through simulation in Fig. 10 the MATLAB–Simulink model of DC-DC enhanced SEPIC converter less coupled inductor. In Fig. 14 assuming that, the DC output of solar is 22 V DC, the input voltage given to the proposed converter is 22 V DC. The output obtained is 220 V DC.

Table 1 shows the designed values used in proposed converter.

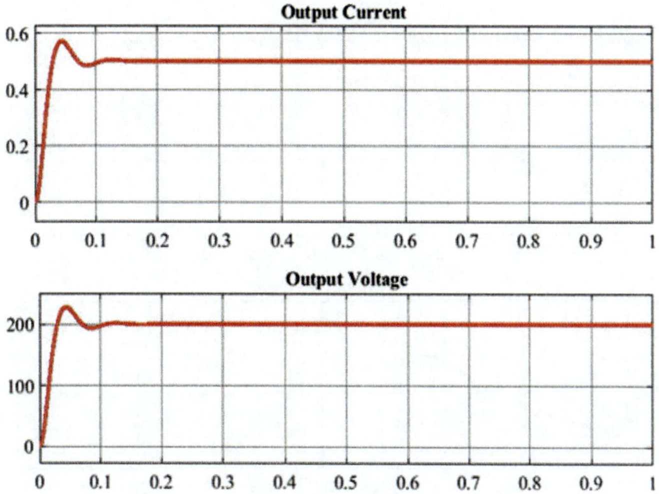

Fig. 12 Output current and output voltage waveform DC-DC enhanced SEPIC converter less coupled inductor

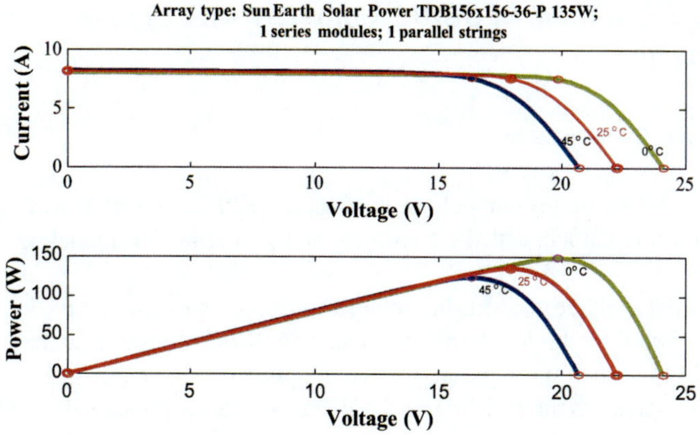

Fig. 13 *I-V* and *P-V* plot of DC-DC enhanced SEPIC converter less coupled inductor

Table 2 shows the load regulation of a proposed DC-DC modified SEPIC converter without including a magnetic coupling. Table 3 shows the simulation result obtained through MATLAB–Simulink for a proposed converter.

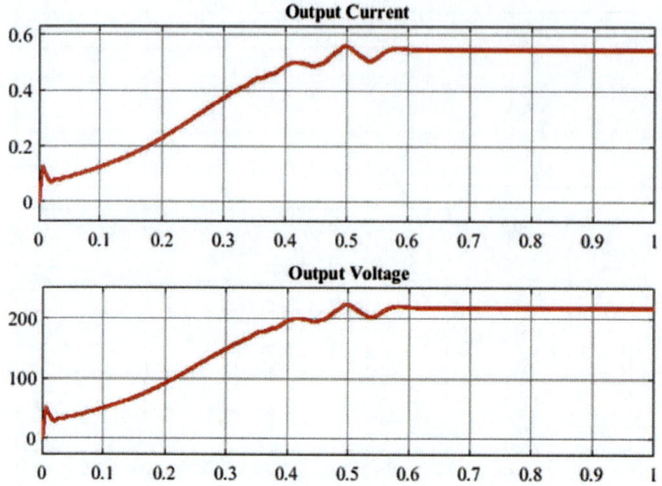

Fig. 14 Output current and voltage waveform of DC-DC enhanced SEPIC converter less coupled inductor

4.2 DC-DC Enhanced SEPIC Converter with Coupled Inductor

4.2.1 Open Loop System

The MATLAB–Simulink model of enhanced SEPIC converter with magnetic coupling in closed loop with PI controller and including PV modeling is shown in Fig. 15.

The output voltage waveform obtained through simulation in Fig. 16, the MATLAB–Simulink model of DC-DC enhanced SEPIC converter without coupled inductor. In Fig. 15 assuming that, the DC output from PV panel is 22 V DC, the input voltage given to the converter is 22 V DC. The output obtained is 434 V DC.

4.2.2 Closed Loop System

PI Controller

The output voltage waveform obtained through simulation in Fig. 16, the MATLAB–Simulink model of DC-DC enhanced SEPIC converter with coupled inductor. In Fig. 17 assuming that, the DC output from solar is 22 V DC, the input voltage given to the power circuit is 22 V DC. The output obtained is 440 V DC.

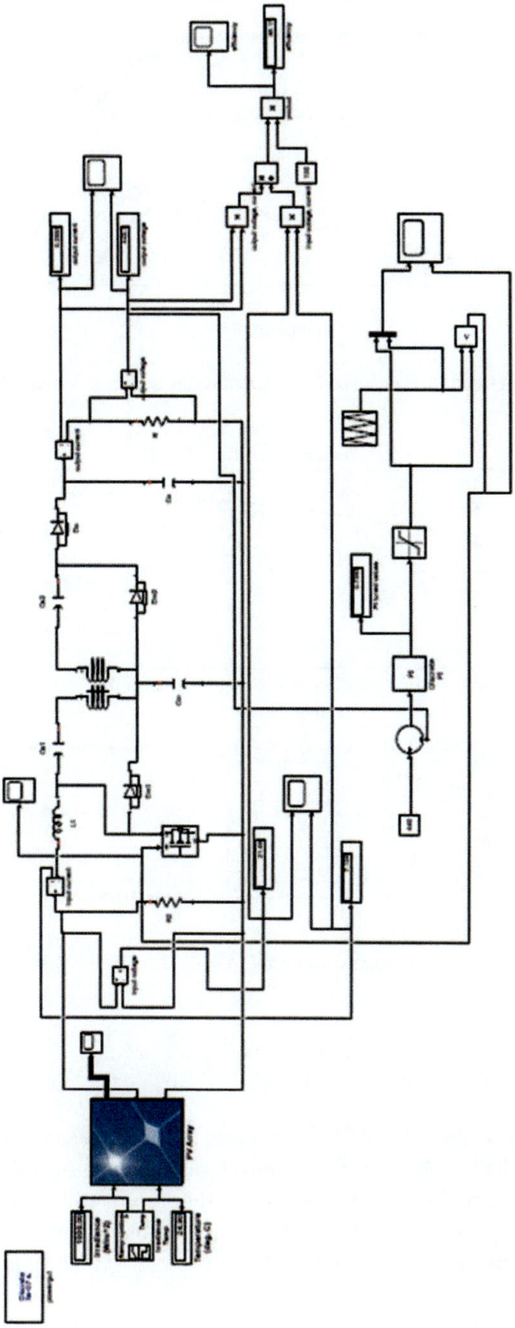

Fig. 15 MATLAB–Simulink model of enhanced SEPIC converter with magnetic coupling in closed loop (PI controller) and including PV modeling

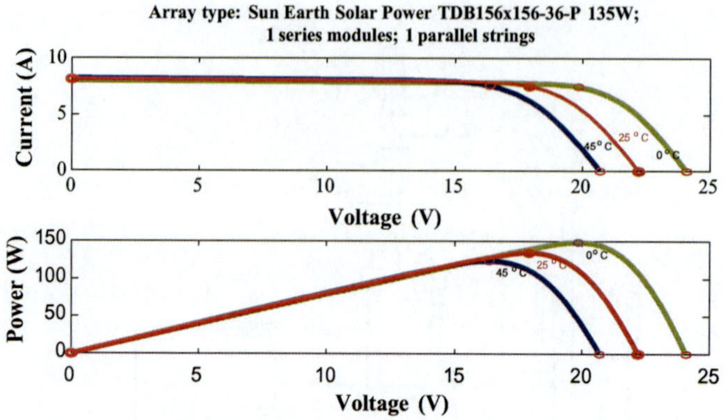

Fig. 16 *I-V* and *P-V* plot of DC-DC enhanced SEPIC converter with coupled inductor

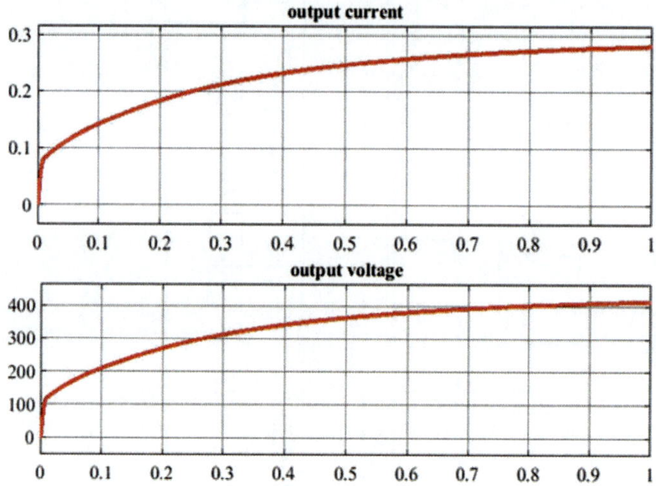

Fig. 17 Output current and output voltage waveform of DC-DC enhanced SEPIC converter with coupled inductor

ANFIS Controller

The output voltage waveform of DC-DC enhanced SEPIC converter with coupled inductor obtained with the help of ANFIS controller is shown in Fig. 18. It provides line and load regulation, peak overshoot, peak undershoot and settling time are getting still reduced than using conventional PI controller. The time indices parameters are good in proposed converter with coupled inductor topology. The output voltage waveform is shown in Fig. 19.

Table 4 shows the MATLAB–Simulink result of the proposed DC-DC enhanced SEPIC converter with including magnetic coupling.

Table 1 Ratings of proposed converter

Parameters	Enhanced SEPIC converter without magnetic coupling	Enhanced SEPIC converter with magnetic coupling
Input voltage (V_i)	22 V	22 V
Input current (I_i)	7.45 A	7.45 A
switching frequency (f)	24 kHz	24 kHz
output voltage (V_o)	220 V	440 V
output current (I_o)	0.6 A	0.3 A
Duty cycle	0.82	0.82
output power (P_o)	135 W	135 W
Static gain	10	20

Table 2 Load regulation

System	Type of controller	Input voltage (V)	Output voltage (V)	Rise time in (s)	Settling time in (s)
Open loop	Without controller	22	200	–	–
Closed loop	With PI controller	21	220	0.1186	0.6000
		18	220	0.1180	0.3667
		17	220	0.1173	0.1366

Table 3 Simulation result

S. No.	Parameters	Simulated values
1	Input voltage (V)	19.99
2	Input current (A)	6.775
3	Output voltage (V)	219.2
4	Output current (A)	0.578
5	Efficiency (%)	93.22
6	K_p	0.0006
7	K_i	0.016

Table 5 shows the load regulation and improved time indices parameter including an ANFIS controller of a proposed DC-DC enhanced SEPIC converter without including a magnetic coupling.

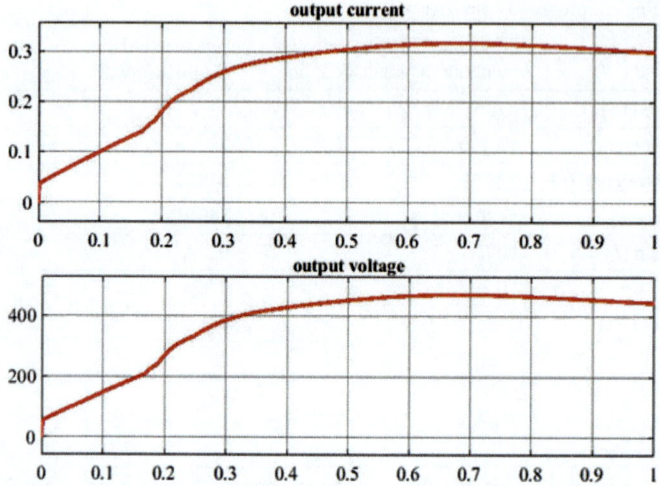

Fig. 18 Output current and output voltage waveform of DC-DC enhanced SEPIC converter with coupled inductor

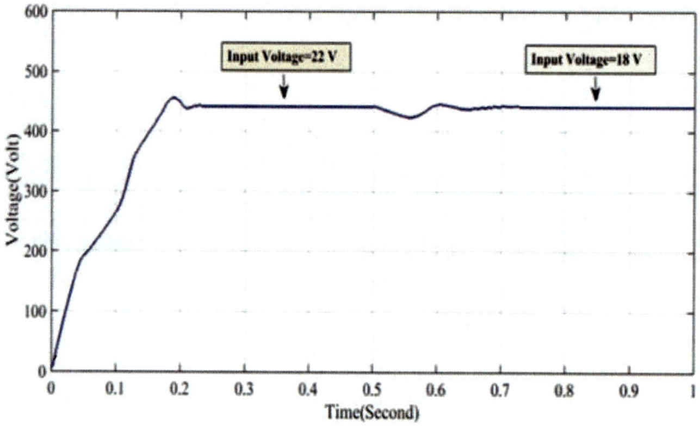

Fig. 19 Output voltage waveform of ANFIS closedtroller

Table 4 Simulation result

S. No.	Parameters	Simulated values
1	Input voltage (V)	21.81
2	Input current (A)	7.812
3	Output voltage (V)	442.7
4	Output current (A)	0.298
5	Efficiency (%)	96.84
6	K_p	0.001
7	K_i	0.011

Table 5 Load regulation

System	Type of closedtroller	Input voltage (V)	Output voltage (V)	Rise time in (s)	Settling time in (s)
Open loop	Without controller	22	434	–	–
Closed loop	With PI controller	22	440	0.2100	0.3471
		18	440	0.2160	0.1312
		17	440	0.2165	0.3565
	With ANFIS controller	22	440	0.1820	0.2210
		18	440	0.1822	0.2211

5 Conclusion

Dual topologies of large gain including as well as not including coupled inductor are simulated through MATLAB simulation. The proposed topology of DC-DC enhanced SEPIC converter lacking coupled inductor produces the static gain of 10 and the proposed DC-DC enhanced SEPIC converter with magnetic coupling provides gain as 20. The efficiency of power electronic converter less magnetic coupling is equivalent to 93.22% with an input voltage of 22 V and an output voltage of 220 V and output power of135 W.

The efficiency of a power electronic converter with magnetic coupling is equal to 96.2% with input voltage of 22 V and an output voltage to 440 V and output power of 135 W.

References

1. Gules, R., Dos Santos, W.M., Dos Reis, F.A., Romaneli, E.F.R., Badin, A.A., Member IEEE: A modified sepic converter with high static gain for renewable applications. IEEE Trans. Power Electron. **12**(11) (2013)
2. Meneses, D., Blaabjerg, F., Garcia, O., Cobos, J.A.: Review and comparison of step-up transformer less topologies for photovoltaic AC-module application. IEEE Trans. Power Electron. **28**(6), 2649–2663 (2013)
3. Hu, X., Gong, C.: A high gain input-parallel output-series DC/DC converter with dual coupled inductors. IEEE Trans. Power Electron. **30**(3), 1306–1317 (2015)
4. Silveira, G.C., Tofoli, F.L., Bezerra, L.D.S., Torrico-Bascopé, R.P.: A nonisolated DC–DC boost converter with high voltage gain and balanced output voltage. IEEE Trans. Ind. Electron. **61**(12), 6739–6746 (2014)
5. Park, K., Moon, G., Youn, M.-J.: Nonisolated high step-up boost converter integrated with SEPIC converter. IEEE Trans. Power Electron. **25**(9), 2266–2275 (2010)
6. Henn, G., Silva, R., Praça, P., Barreto, L., Oliveira, D.: Interleaved boost converter with high voltage gain. IEEE Trans. Power Electron. **25**(11), 2753–2761 (2010)

7. Seong, H.-W., Kim, H.-S., Park, K.-B., Moon, G.-W., Youn, M.-J.: High step-up DC-DC converters using zero-voltage switching boost integration technique and light-load frequency modulation control. IEEE Trans. Power Electron. **27**(3), 1383–1400 (2012)
8. Li, W., Zhao, Y., Deng, Y., He, X.: Interleaved converter with voltage multiplier cell for high step-up and high-efficiency conversion. IEEE Trans. Power Electron. **25**(9), 2397–2408 (2010)

Marine Navigation and Ocean Crisis Prevention System

N. Pradheep, S. T. Aarthy, C. Sabarinathan, A. Seetharam, and U. S. Venkateshwara

Abstract In order to catch fishes such as halibut, sole and shrimps, fishermen deploy a technique known as bottom trawling. While this technique helps to obtain such a variety of fishes, it also destroys the sea bed as the whole net is dragged along the seafloor. Also, these fishermen at times cross the maritime boundaries leading to international disputes. Our embedded system helps to overcome the above-specified problems by notifying the government about the activity and the boat identity number.

Keywords Bottom trawling · Marine navigation · Project loon

1 Introduction

1.1 Synopsis

There have been disputes between Tamil Nadu and Sri Lanka for the past years for crossing the international maritime boundary line (IMBL) and exploiting the neighboring country's resources [1, 2]. While some fishermen cross unknowingly, few of them do it intentionally and are being arrested across the border. In the past

N. Pradheep (✉) · S. T. Aarthy · C. Sabarinathan · A. Seetharam · U. S. Venkateshwara
SRM Institute of Science and Technology, Kattankulathur, India
e-mail: pradheepn_nna@srmuniv.edu.in

S. T. Aarthy
e-mail: aarthy.s@ktr.srmuniv.ac.in

C. Sabarinathan
e-mail: sabarinathan_chandramohan@srmuniv.edu.in

A. Seetharam
e-mail: seetharama_va@srmuniv.edu.in

U. S. Venkateshwara
e-mail: venkateshwarau_sak@srmuniv.edu.in

P. Suresh et al. (eds.), *Advances in Smart System Technologies*,
Advances in Intelligent Systems and Computing 1163,
https://doi.org/10.1007/978-981-15-5029-4_35

25 years, more than 750 Indian fishermen were killed by the Sri Lankan government for crossing the international border [3]. One of the other important issues that are happening in the sea is environment endangering which is bottom trawling. It destroys the sea bed which is the home for various aquatic species [4]. The Sri Lankan government has issued a ban on bottom trawling. In order to prevent these issues, we have devised an embedded system that sends a message to the control section at the shore so that the fishermen who crossed the border can be rescued before they are arrested [5]. Also, if a fisherman uses bottom trawling technique, it can be identified and informed to the control section and necessary action can be taken to prevent the same in the future.

1.2 Bottom Trawling

The act of dragging the fishing net along the seafloor is called bottom trawling. It is also known as "dragging". Bottom trawling is divided into two types. Dragging a net at the bottom most part of the sea is called benthic trawling. Dragging a net slightly above the benthic zone is called as demersal trawling. It is undiscriminating and it causes severe damages the seafloor. This method catches all living organisms and inanimate objects it runs into. Due to this, the unwanted organisms are thrown dead, endangered species included, which can live for several years. Up to 90% of a drag's total catch is by catch, which is the collateral damage caused by catching such endangered species. The food and shelter to various aquatic species given by enormous areas of deep sea habitats are destroyed by wide and heavy nets. The marine ecosystem is permanently damaged by such habitat destructions of greater magnitude (Fig. 1).

Fig. 1 Demonstration of bottom trawling. A boat dragging the fishing net along the sea floor

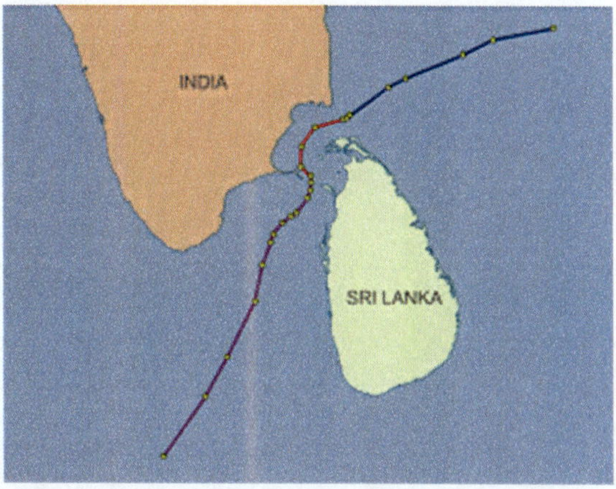

Fig. 2 International maritime boundary line between India and Sri Lanka

1.3 International Maritime Boundary Line (IMBL)

The marine region of the Earth is partitioned just like the ground surface. This imaginary separation line is called a maritime boundary. Crossing the IMBL may provoke international disputes between the concerned nations. It is equivalent to illegally entering a nation and snatching its resources. Indian fishermen crossing Sri Lanka's maritime boundary are arrested and mostly never returned. If the coast guards are intimated immediately of a boat crossing the international border, the Indian patrol boats can rescue them before getting arrested (Fig. 2).

1.4 Communication

Communication from the deep sea to the coast guards can be difficult. RF communication over sea buoys cannot be implemented practically because the salinity of the seawater evaporates to the atmosphere and makes the area above the sea level a superconductor. So, only noise will be received if this method is implemented. For effective and reliable communication, underwater optical communication can be used if the systems are already laid. If underwater communication is to be set up from scratch, it will not be cost-efficient. Hence, Google's Project Loon helps us in this situation. It was originally designed to provide Internet access to rural areas using balloons (loon) that will be flying in the stratosphere (Fig. 3).

Fig. 3 Deep sea communication using Google's Project Loon

2 System Design

2.1 Boat Section

The boat section is the embedded system that is to be installed on the boats. It contains the components present in the diagram. The location of the border is predefined before installation. The GPS module continuously fetches the current location of the boat and feeds it to the microcontroller. The microcontroller, in turn, checks the fed values with the preset values to determine if the boat crosses the border. A threshold is set. After which, the alarm buzzer beeps to indicate that the fisherman is near the maritime boundary. Once the GPS coordinates match with the preset value, the boat number, along with its present location is sent to the control section. Similarly, an altimeter is attached to the fishing net. The depth of the sea bed is predefined in the embedded system. If the difference between the predefined value and the depth found by the altimeter is 0, the information about the boat using destructive fishing method is sent to the control section (Fig. 4).

2.2 Control Section

The control section is the receiver part of this system which is at the shore. This section receives the information from the boat section and displays it on an LCD for ease of access. If a boat crosses the IMBL, the buzzer goes off in the control section and the live location of the boat can be seen in the display. Also, if bottom trawling is done, the buzzer buzzes and the information of the boat are shown in the display (Fig. 5 and Table 1).

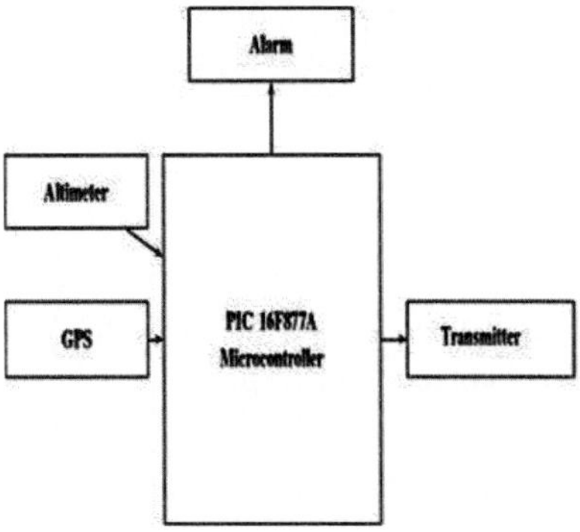

Fig. 4 Design of the boat section on the left and the control section on the right

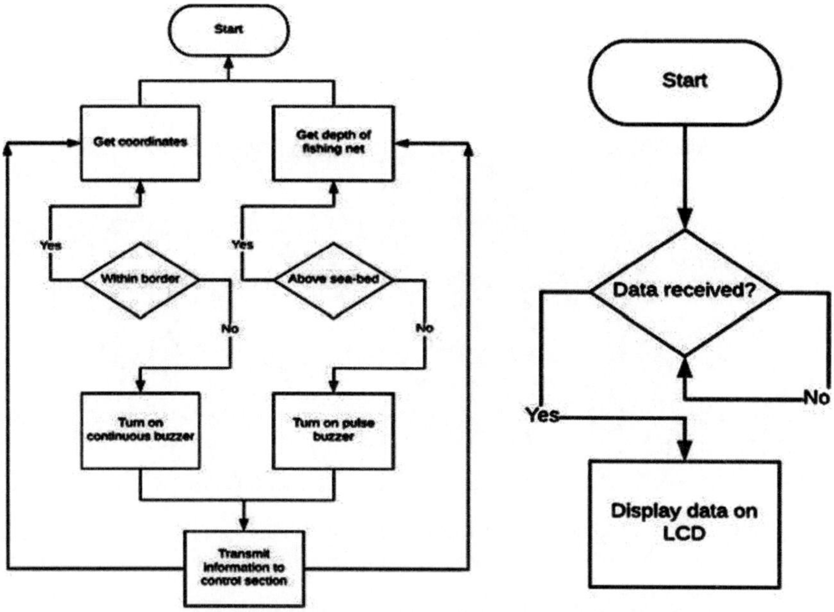

Fig. 5 Process flow of the boat section on the left and the control section on the right

Table 1 Sample coordinates tested against preset border coordinates

Latitude	Longitude	Border crossed
9.190159	79.503520	No
9.431886	79.476459	No
9.141595	79.874415	Yes
9.508802	79.862033	Yes

Preset IMBL coordinates: 9.127301, 79.561430

The distance between the preset IMBL coordinates and the GPS fetched coordinates can be found using the following mathematical formula.

$$\sqrt[2]{(x_2 - x_1)^2 + (y_2 - y_1)^2}$$

where,

x_1, y_1 is the latitude and longitude of the preset IMBL.

x_2, y_2 is the latitude and longitude of the current location fetched by the GPS (Fig. 6 and Table 2).

Fig. 6 Circuit of the boat section on the left and the control section on the right

Table 2 Sample heights tested against the preset depth of sea bed

Depth of net	Difference	Destructive fishing
3000	900	No
3500	400	No
3890	10	Yes
3900	0	Yes

Preset depth of sea bed: 3900 m

3 Conclusion

This system serves as a device for safer navigation and maintaining a healthy marine environment. Since there is prevailing heat between India and Sri Lanka due to maritime boundary crossing issues, this system is designed to minimize the unfortunate incidences that occur due to it. This cost-efficient and compact system also helps to conserve the marine environment by helping to reduce bottom trawling.

4 Future Work

Even though this system is compact, it can be made much more portable by redesigning the circuit using new methods such as printed electronics. The modules used can be optimized to improve accuracy and efficiency. The system can be integrated into smartphones for increased portability. With 5G just around the corner, network slicing can be implemented to create a private network for communication in the deep sea for which the coverage can be provided using project loon's data hopping technology.

References

1. Wikipedia: Bottom Trawling
2. Wikipedia: Maritime boundary
3. Kim, D.: A survey of balloon networking applications and technologies (OnlinePaper), 20 Dec 2013
4. Project Loon: Innovating the connectivity worldwide. In: 2017 2nd IEEE International Conference on Recent Trends in Electronics, Information & Communication Technology (RTEICT)
5. Underwater wireless optical communication using a directly modulated semiconductor laser. In: IEEE Conference, May 2015

High-Gain and High-Efficiency Solid-State Converter for Fuel Cell Applications

Muhaidheen, Tamilarasi, and Muralidharan

Abstract Usage of fuel cell is now in increasing trend as it becomes an invariable component of electric vehicle technology. This paper deals with high efficient and elevated gain DC/DC converter for fuel cell applications toward future renewable energy system. High efficiency is obtained by employing single switch in the converter. Extended voltage gain with reduced voltage stress is obtained by proposed a hybrid converter which combines self-lift Cuk-boost topology. The proposed topology is inherently suitable for half-bridge inverter topology which requires galvanic isolation and three-wire supply. Simulation has been carried out to verify the proposed converter with variable load using MATLAB–Simulink environment. Comparison has been done with ideal DC supply and fuel cell supply.

Keywords DC/DC converter · Static voltage gain · Self-lift Cuk converter · Voltage source inverter · Half-bridge inverter and Grid-connected load

1 Introduction

Increasing demand of electrical energy, without any environmental degradation, will be only meet out by utilizing renewable energy sources like solar photovoltaic generation system, windmill power generation system, fuel cell-based power generation system, etc., utilizing fuel cell for electrical power generation that requires fuel handling system. PEMFC stack modeling is performed and numerical equations are derived to express three polarizations with the function of pressure and flow rate of hydrogen and air [1, 2].

Muhaidheen · Tamilarasi · Muralidharan (✉)
Mepco Schlenk Engineering College, Viruthunagar District, Sivakasi, India
e-mail: smurali@mepcoeng.ac.in

Muhaidheen
e-mail: muhai@mepcoeng.ac.in

Tamilarasi
e-mail: tamil260696@yahoo.com

P. Suresh et al. (eds.), *Advances in Smart System Technologies*,
Advances in Intelligent Systems and Computing 1163,
https://doi.org/10.1007/978-981-15-5029-4_36

Internal combustion engine of vehicles converts the hydrogen into mechanical energy either by burning the hydrogen or by chemical reaction of hydrogen with oxygen in fuel cell to operate the electrical vehicle. Research manuscript [3] proposes the integration fuel cell system with electrical vehicles.

The output of the fuel cell produced is not enough for operating all the loads. That means fuel cell provides the low amount of power, current, and voltage. So, it requires the power electronic subsystem to produce the desired output. Boost converter just boosts the input according to the duty cycle of semiconductor switch. Research manuscript [4] analyzes the characteristics of conventional boost power converter and provides a simulation of it. The boost converter produces output with a lot of harmonics and disturbances. Generally, the output voltage is controlled by sensing the digital form of output voltage and computed error voltage is given to the semiconductor switch [5]. Closed loop of boost converter using controller should eliminate that disturbance. The controller is designed using small-signal model in [6].

The Cuk converter with self-lift provision is analyzed and designed as per [7]. Here, Cuk converter with self-lift provision is tested in both different conduction modes of operation with simulation model. This Cuk converter with self-lift provision is suitable for the renewable energy resources [8]. It shows the design of Cuk converter with self-lift provision with MPPT enabled solar PV panel. The advantages of Cuk converter with self-lift provision are explained as in [9] with transformerless topology, using single switch converter and common ground. It also produces high efficiency than other classical boost converters with small ripples. The voltage boosting ability is higher than classical Cuk converter. Another method of voltage lift is by adding more number of components that double output DC/DC converter [10]. Reasonable voltage gain and reduced stress can be obtained by combining both classical boost converter with Cuk with self-lift enabled power converter [11].

In this research manuscript, AC supply is obtained by using conventional half-bridge configuration-based inverter in order to supply the AC loads. Hsieh et al. [12] presents the structure, operation, and analysis of the conventional half-bridge configuration-based inverter.

Menaka et al. [13] outlines the design of symmetric multilevel inverter using evolutionary algorithms for harmonic profile optimization. Fuel cell can be utilized for different applications such as power conditioning [14], DGS environment [15] and etc.

2 Proposed Converter

2.1 Features

It is a hybrid version of single switch-based DC power electronic converter which combines both DC boost power converter and Cuk converter with self-lift features.

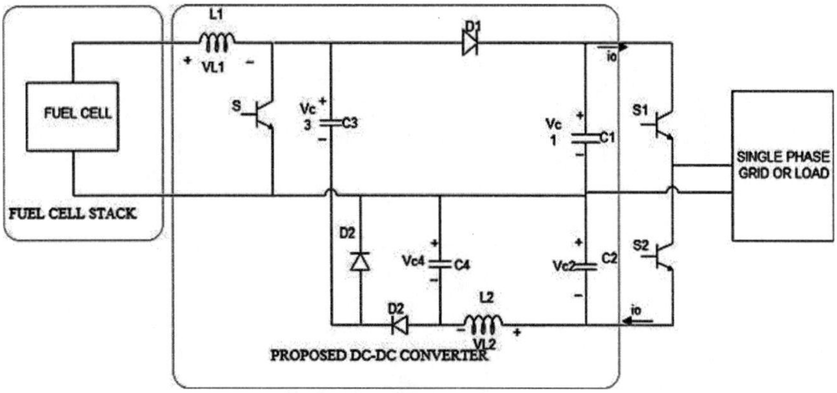

Fig. 1 Proposed hybrid converter supplying grid

Since it uses a single switch, the extended efficiency can be obtained through the proposed DC power converter than their counterparts owing to the low switching losses.

Boosting operation and ripple reduction can be achieved through the source side inductor and output side capacitor, respectively, of the hybrid DC/DC converter. Moreover, natural galvanic isolation is inherently available in the converter which may be suitable for AC supply generation through half-bridge configuration-based power inverter as shown in Fig. 1.

2.2 Modes of Operation

This operation of single-switch converter will be explained in terms of two modes of operation based on the turn-on and turn-off state. These modes of operations are classified by using the gating pulse given to the switch. The mode 1 is shown in Fig. 2.

When the switch S is on (δT), the passive inductor L_1 gets charge through supply voltage V_i. At the same time, capacitors C_1, C_2, C_3 discharge their charge into the circuit and inductor L_2 gets charged. The diode D_1 and D_3 in off state. And the remaining diode D_2 is in on state to deliver the charge in the circuit.

The mode 2 operation is shown in Fig. 3. It is inverse of the mode 1 operation. When the switch S is off $(1-\delta)T$, at the time gate pulse is not given to the switch. Here, input supply and charge of inductor L_1 pass into circuit, so the input is boosted. The capacitor C_3 is getting charge. The diodes D_1 and D_3 are in on state. The diode D_2 is in off state. The on-state diodes are helping to deliver charge to the load.

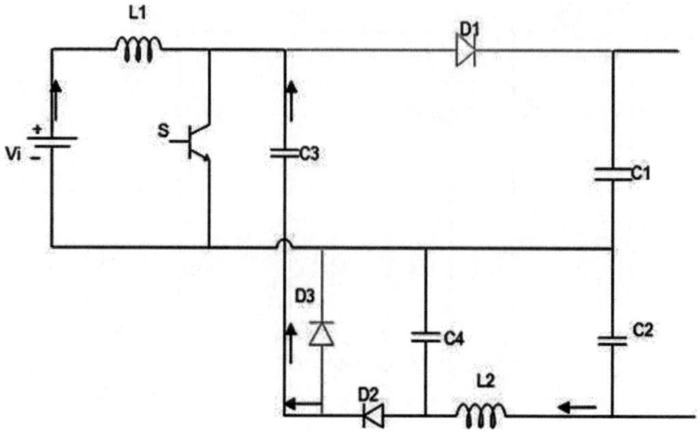

Fig. 2 Mode 1 operation

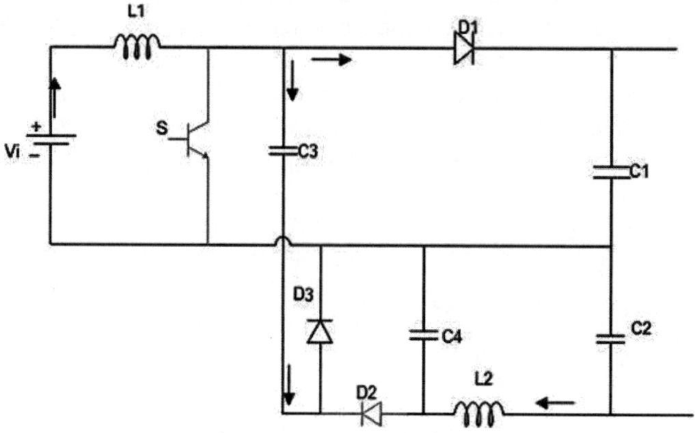

Fig. 3 Mode 2 operation

2.3 Mathematical Modeling of Hybrid DC Power Converter

The following section covers the detailed analysis of converter by assuming all passive elements and semiconductor switch are ideal for generalization and converter operates at steady state in CCM.

2.3.1 Static Voltage High Gain

By considering CCM in the DC power converter, the voltage across the both passive element inductor L_1 and inductor L_2 is

$$V_{L1} = \delta V_i + (1-\delta)(V_i - V_{C1}) = 0 \tag{1}$$

$$V_{L2} = \delta(V_{C3} - V_{C2}) + (1-\delta)(V_{C4} - V_{C2}) \tag{2}$$

$$V_{C1} = V_{C3}; \quad V_{C2} = V_{C4}$$

where t_1 is on period and t_2 is off period,

At switch ON time:

$$V_i = \frac{L_1(I_2 - I_1)}{t_1} = \frac{L_1 \Delta I}{t_1} \tag{3}$$

From the above equation,

$$\Delta I = \frac{t_1 V_i}{L_1} \tag{4}$$

At switch OFF time,

$$V_i - V_{C1} = \frac{L_1(I_1 - I_2)}{t_2} = \frac{-L_1 \Delta I}{t_2} \tag{5}$$

From above Eq.

$$\Delta I = \frac{t_2(V_i - V_{C2})}{L_1} \tag{6}$$

Compare the ΔI Eqs.

$$\frac{V_i t_1}{L_1} = \frac{t_2(V_i - V_{C1})}{L_1}; \quad V_i t_1 = t_2 V_{C2} - V_i t_2$$

$$V_i(t_1 - t_2) = t_2 V_{C1}; \quad V_{C1} = \frac{V_i(t_1 + t_2)}{t_2} \tag{7}$$

Here, t_1 = on period = δT; t_1 = off period = $(1-\delta)T$

$$\therefore V_{C1} = \frac{V_i(\delta T + (1-\delta)T)}{(1-\delta)T}; \quad V_{C1} = \frac{V_i}{(1-\delta)} \tag{8}$$

Similarly voltage across $C_2(V_{C2})$ is,

$$V_{C2} = \frac{1}{(1-\delta)} V_i \tag{9}$$

The obtained output voltage is the aggregate of voltage across two output passive element capacitor V_{C1} and capacitor V_{C2},

$$V_O = V_{C1} + V_{C2}; \quad V_O = \frac{2}{(1-\delta)} V_i \tag{10}$$

From Eq. 10, it is proved that the proposed converter yields double times the output voltage of the classical boost converter.

2.3.2 Capacitor and Inductor Design

The passive element capacitors C_1 and C_2 are designed to assume the small ripple voltages according to the average voltage of capacitors. The input current I_{L1} is derived from voltage gain of the projected converter,

$$V_1 = V_O I_O; \quad I_{L1} = \frac{2}{(1-\delta)} \tag{11}$$

Assume input current I_{L1} is divided equally into the capacitors C_1 and C_3
At switch ON time,

$$\Delta V_{C3} = \frac{1}{C_3}\int_0^{t_1} I_{C3}\mathrm{d}t; \quad \Delta V_{C3} = \frac{I_{L1}(1-\delta)T}{2C_3} \tag{12}$$

And we know that,

$$C_3 = \frac{\Delta Q}{\Delta V_{C3}} = \frac{\Delta I_{L2}}{8\Delta V_{C2} f} \tag{13}$$

$$C_1 = \frac{\Delta Q}{\Delta V_{C1}} = \frac{I_{L1}(1-\delta)}{2\Delta V_{C1} f} = \frac{I_O}{\Delta V_{C1} f} \tag{14}$$

$$C_2 = \frac{\Delta Q}{\Delta V_{C2}} = \frac{\Delta I_{L2}}{8\Delta V_{C2} f} \tag{15}$$

C_1 and C_2

$$C_4 = \frac{I_{L1}(1-\delta)T}{\Delta V_{C4}} = \frac{I_O(1-\delta)}{\Delta V_{V4}} \tag{16}$$

The projected converter inductances are carefully designed to limit the current ripple ΔI_{L1}.

$$L_1 = \frac{\delta V_i}{\Delta I L_1 f} \tag{17}$$

And the inductor L_2 is obtained from the current ripple ΔI_{l2},

$$L_2 = \frac{I_0(1 - \delta)}{8 \Delta I L_2 f^2 C_4} \tag{18}$$

2.3.3 Voltage Stress and Current Stress on the Switch

The voltage stress at the power switch can be determined by using output capacitor C_1, ripple voltage ΔV_{C1} and voltage across capacitor C_1 (V_{c1}) at the time of switch is in open.

$$V_S = \frac{V_{C1} + \frac{\Delta V_{C1}}{2}}{1 + \delta} \tag{19}$$

When the switch is in open, the capacitors C_4 and C_3 considerably large to eliminate the voltage ripple ΔV_{C4} and ΔV_{C3}. Then, the voltage emphasis across the two diodes is given below,

$$V_{D1} = V_{C1} + \frac{\Delta V_{C1}}{2} \tag{20}$$

$$V_{D2,3} = V_{C2} + \frac{\Delta V_{C2}}{2} \tag{21}$$

Equations (19)–(21) show reduction of voltage stress at the switch than the proposed converter. The steady-state switch current I_S is,

$$I_S = \frac{2 I_0 \delta}{1 - \delta} \tag{22}$$

The current across the diode in the steady-state operation is almost equal to output current $I_0 . I_{D1} = I_{D2} = I_{D3} = I_0$.

It is evident from Eq. (22) that the average current of the switch is much lower than the output current.

3 Simulation Results with Conventional DC Supply

The simulation of the projected DC power converter with inverter configuration can be implemented in the MATLAB–Simulink whose values are depicted in Table 1.

Table 1 System parameters

Component	Parameter	Values
Proposed DC/DC converter	Duty cycle	70%
	Input DC voltage	24 V
	Output DC voltage	160 V
	Input power	447.72
	Output DC power	437.75 W
	Device switching frequency	10 kHz
	C_1 and C_2	1000 μF
	C_3 and C_4	470 μF and 10 μF
	L_1 and L_2	1.9 mH and 1 mH
	Load	49 Ω
	Efficiency	98.21%
	Voltage gain Lecture Notes	6.6
DC proposed converter (PI controller)	K_p	0.001
	K_i	0.12
Fuel cell (300 W)	Nominal operating voltage	36 V
	Nominal operating current	8.3 A
Fuel cell fed proposed converter (PI controller) and with an inverter	K_p	0.0015
	K_i	0.085
Single-phase half-bridge inverter	DC input voltage	160 V
	AC output voltage	80 V
	Sine wave frequency	50 Hz
	Switching frequency	10 kHz
	Load	49 Ω, 15 mH
DC source fed converter with inverter (PI controller)	K_p	0.1
	K_i	0.2

The proposed DC power converter produces the output as 160 V with the gain of 6.6 for the input of 24 V (Fig. 4).

The proposed converter is designed with voltage ripples of capacitors are $\Delta V_{C1} \approx 0.05\%$, $\Delta V_{C2} \approx 0.1\%$, $\Delta V_{C3} \approx 0.05\%$ and $\Delta V_{C4} \approx 0.2\%$. And current ripples are low in this proposed converter $\Delta I_{L1} \approx 20\%$ and $\Delta I_{L2} \approx 2\%$.

3.1 Open Loop

To verify the improvement of the proposed DC power converter, a Simulink model (Fig. 5) has been simulated under open-loop control. The switching pulse for the

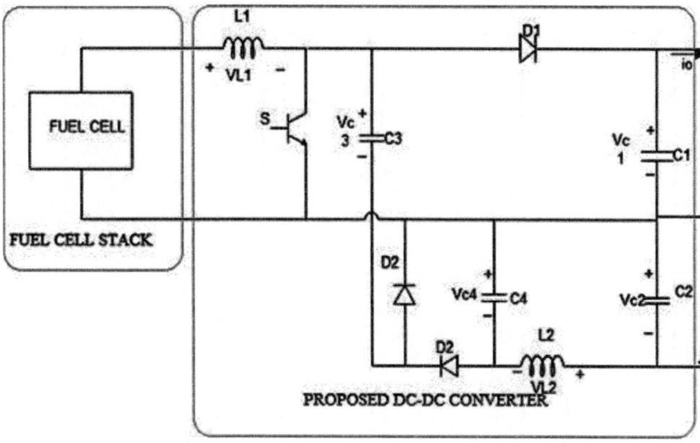

Fig. 4 Overall DC/DC converter with inverter configuration

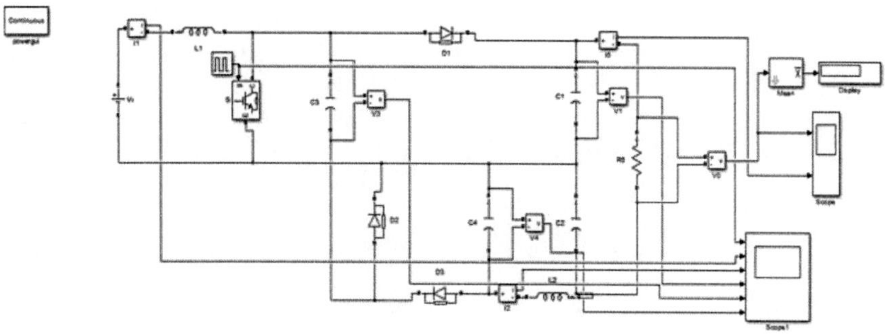

Fig. 5 Open-loop proposed DC power converter

converter is given in Fig. 6. The simulation of the proposed DC power converter is designed with the power rating is around 400 W. In open-loop simulation, 98.21% of efficiency is obtained with the proposed DC power converter.

The current through the inductors L_1 and L_2 is shown in Fig. 7, respectively.

The voltage stress at all capacitors C_1, C_2, C_3 and C_4 is shown in Fig. 8, respectively.

The voltage and current of the proposed DC power converter are shown in Figs. 9 and 10, respectively.

From the above observation, more than 70% overshoot has been noted which may be not suitable for passive components.

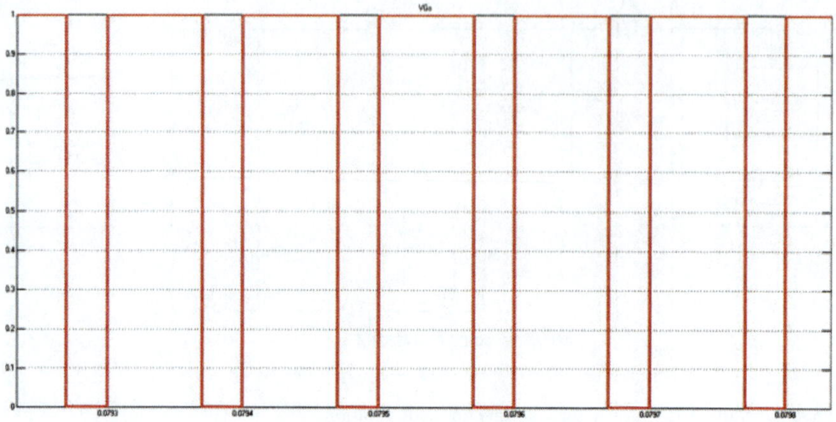

Fig. 6 Gate pulse of open-loop DC power converter

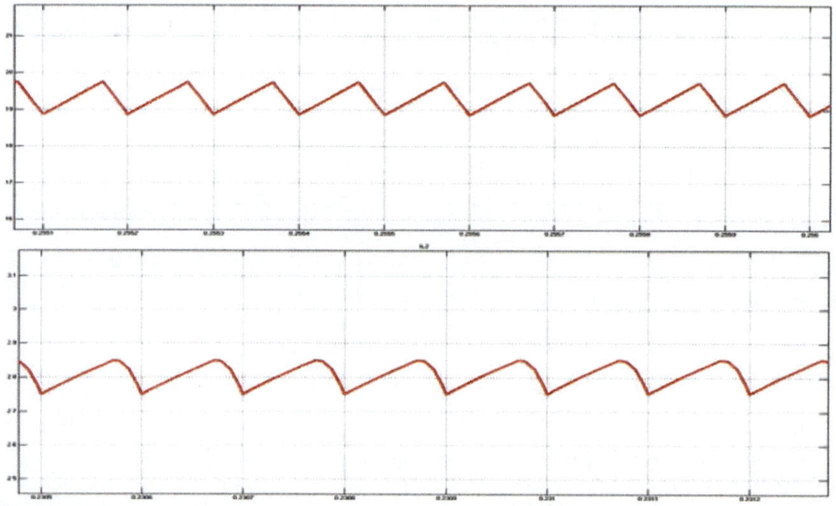

Fig. 7 Inductors L_1 and L_2 current profile of proposed DC power converter

3.2 *Closed Loop*

Closed-loop control may be excercised in order to reduce the peak overshoot and obtain better voltage regulation from DC power converter, inevitable. The PI control has been tried in the model shown in Fig. 11 to reduce the overshoot by using trial and error method. The switching pulse for the closed-loop DC power converter is shown in Fig. 12.

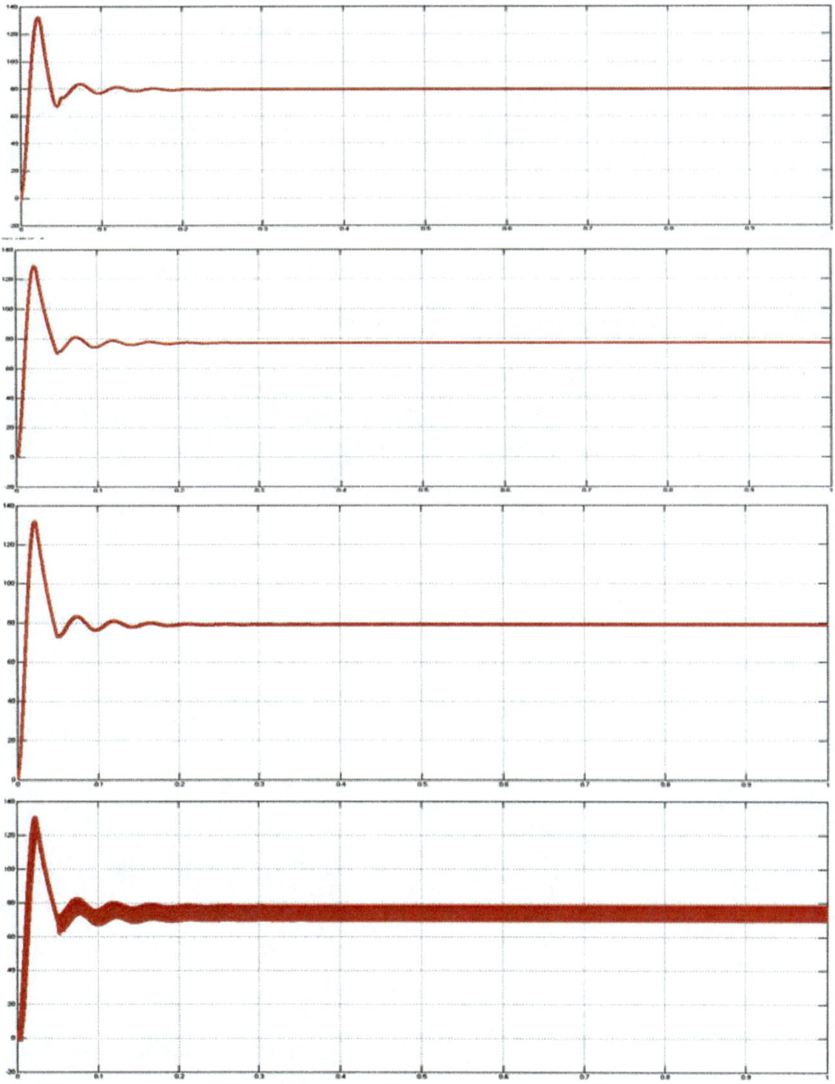

Fig. 8 Capacitor C_1, C_2, C_3 and C_4 voltage of the proposed DC power converter

By observing above output voltage and current waveforms (Figs. 13 and 14), overshoot has been completely eliminated and the response time will be around 0.1 s which is less than the open-loop control scheme.

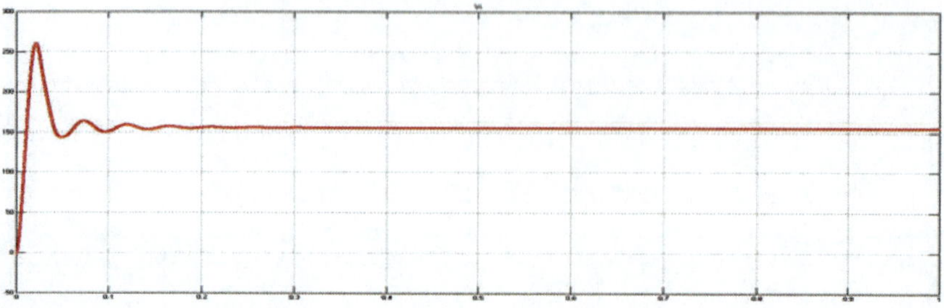

Fig. 9 Output voltage of the proposed DC power converter

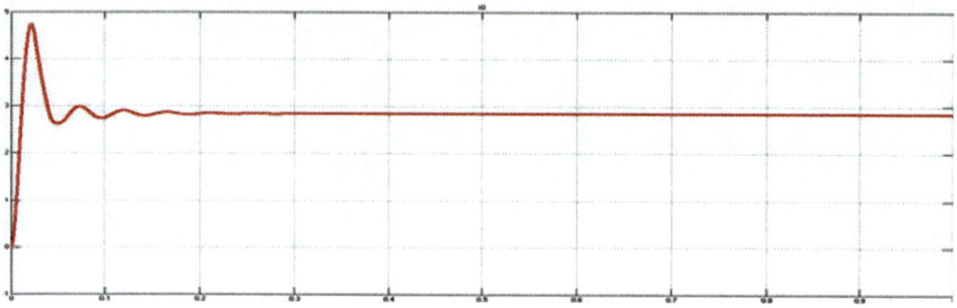

Fig. 10 Output current of the proposed DC power converter

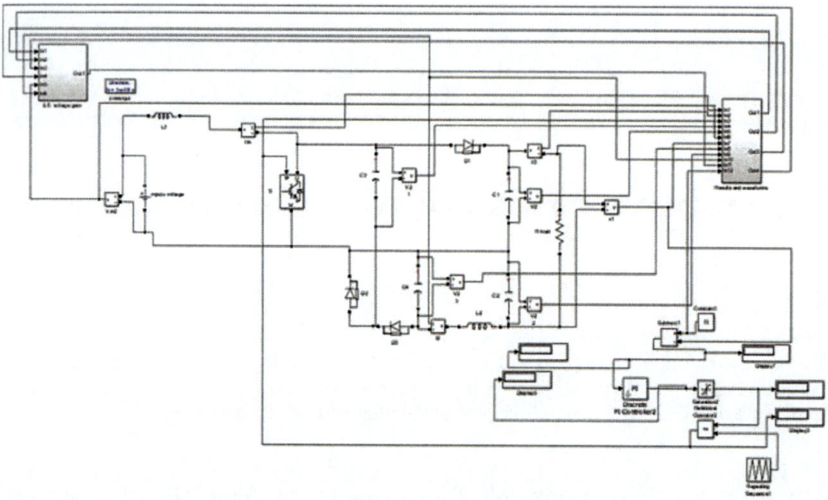

Fig. 11 Closed loop of single switch hybrid DC/DC converter

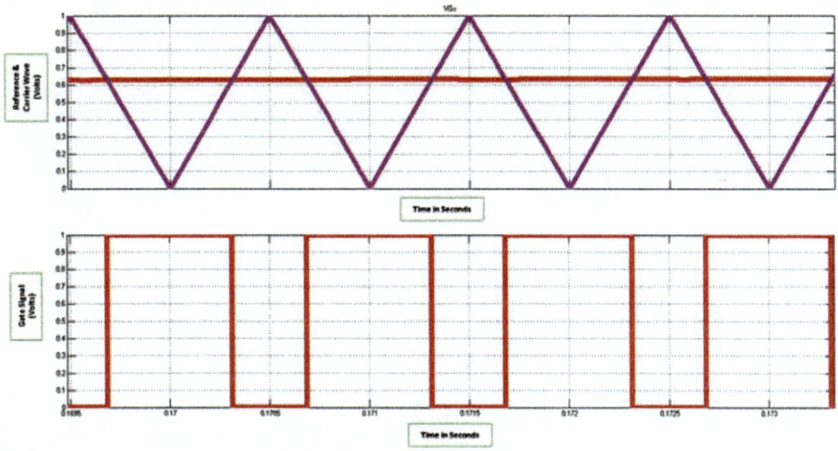

Fig. 12 Gate pulse of closed-loop proposed DC/DC converter

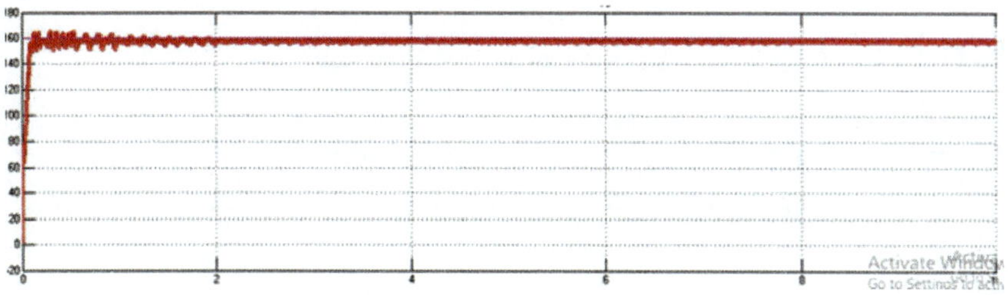

Fig. 13 Voltage of closed-loop proposed DC/DC converter

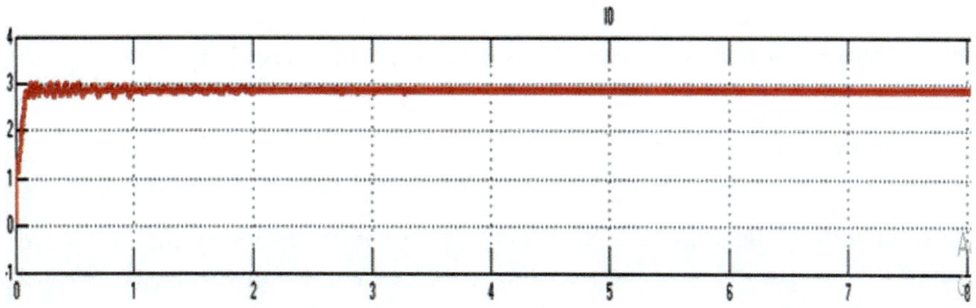

Fig. 14 Current of closed-loop proposed DC/DC converter

4 Simulation Results with Fuel Cell (300 W)

The simulation of the proposed DC power converter with inverter configuration with PEM fuel cell is done in the MATLAB–Simulink as given in Fig. 15 whose design values are depicted in Table 1.

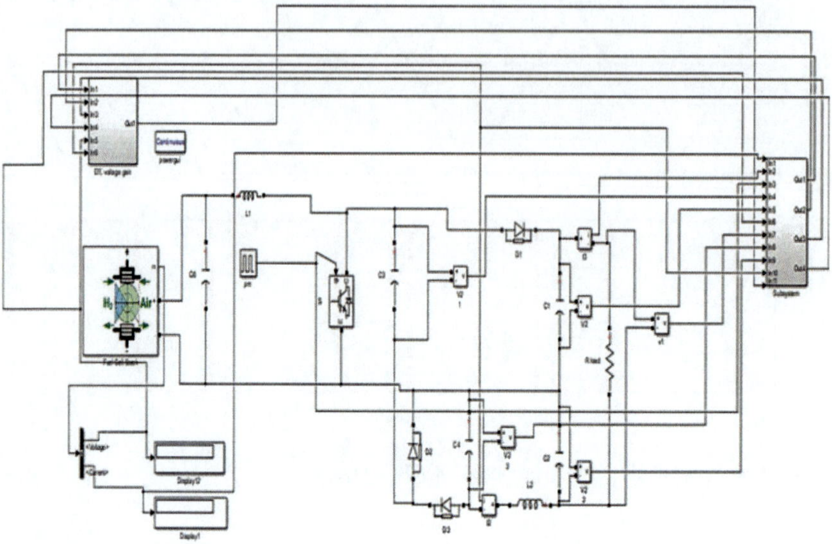

Fig. 15 Fuel cell fed proposed DC power converter with open-loop control

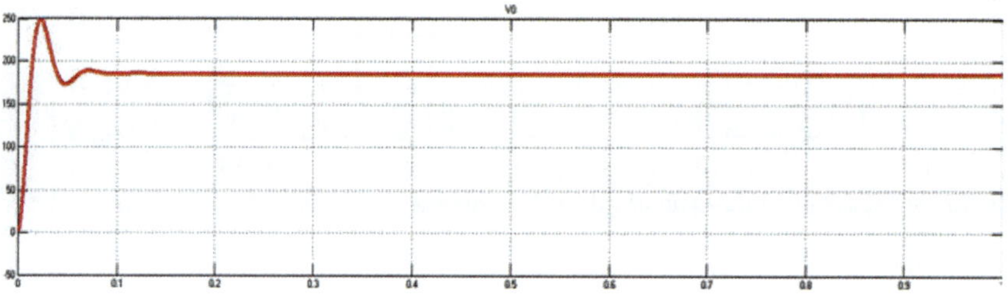

Fig. 16 Voltage of fuel cell fed proposed DC power converter under open loop

4.1 Open Loop

The simulation of the proposed DC power converter with the 300 W fuel cell power rating is done. Under open-loop condition, the efficiency was found as 98.21% (Figs. 16 and 17).

More than 70% overshoot has been noted in the voltage profile as in Sect. 3.

4.2 Closed Loop

To enhance the performance of the fuel cell-based DC power converter, closed-loop control is implemented. The PI control was tried as shown in Fig. 18 to reduce

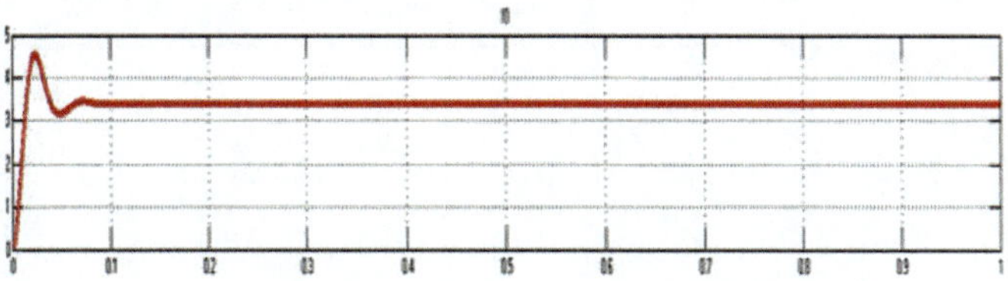

Fig. 17 Current of fuel cell fed proposed DC power converter under open loop

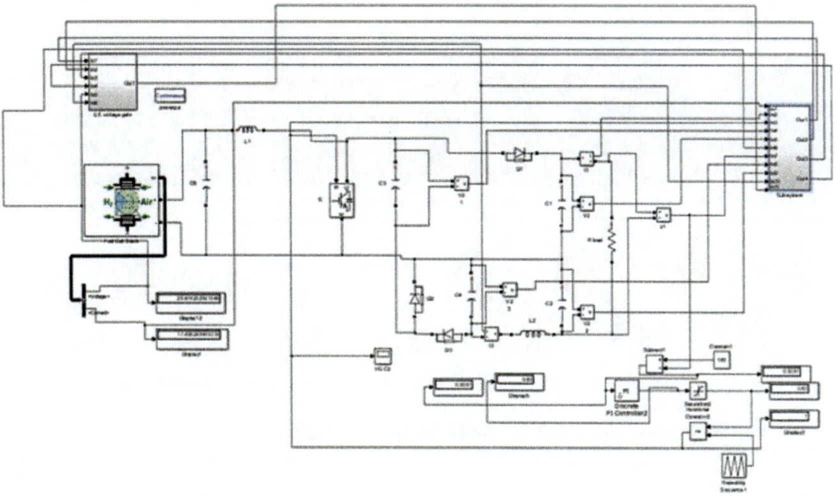

Fig. 18 Fuel cell fed proposed DC power converter with closed-loop control

the overshoot by using trial and error method whose values are shown in Table 1. Figures 21 and 22 show voltage and current of the fuel cell power source fed DC converter, respectively (Figs. 19 and 20).

The rise time of the voltage and current will be around more than 0.15 s which is mainly due to initial dynamics of PEM fuel cell power source. The output voltage and current of an inverter and given in Fig. 21, respectively.

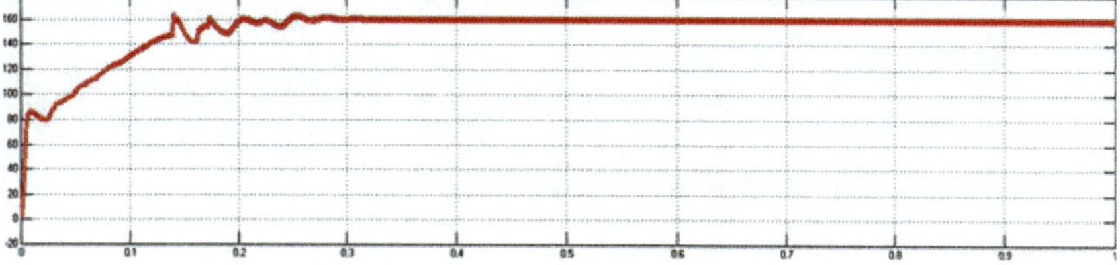

Fig. 19 Voltage profile of fuel cell fed DC power converter under closed-loop condition

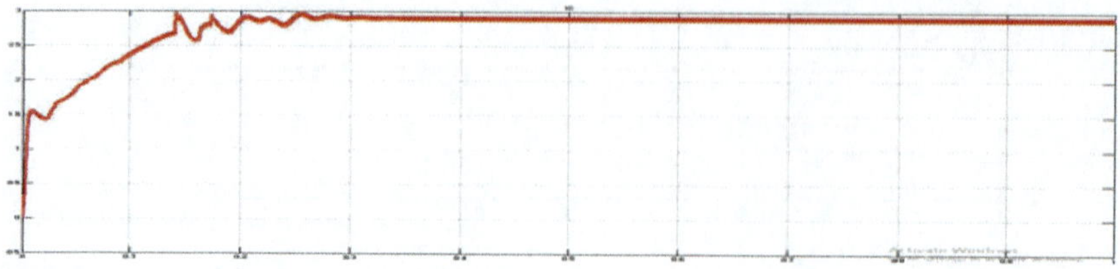

Fig. 20 Output current of fuel cell fed proposed DC power converter under closed-loop condition

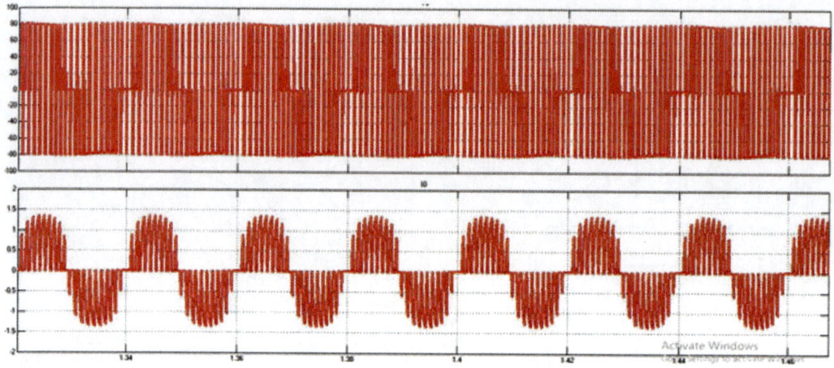

Fig. 21 Inverter output voltage and current of the DC converter

5 Performance Analysis

The performance of the proposed converter in terms of voltage gain versus duty cycle against which the conventional converter is depicted in Fig. 22.

It is evident from the above statics that desired output voltage can be obtained by using a small duty cycle with the proposed converter. The stress of both voltage and current at the converter component has reduced to half of the output voltage and it is shown in passive component voltage/current waveforms as depicted in Sects. 3 and 4 (Fig. 23).

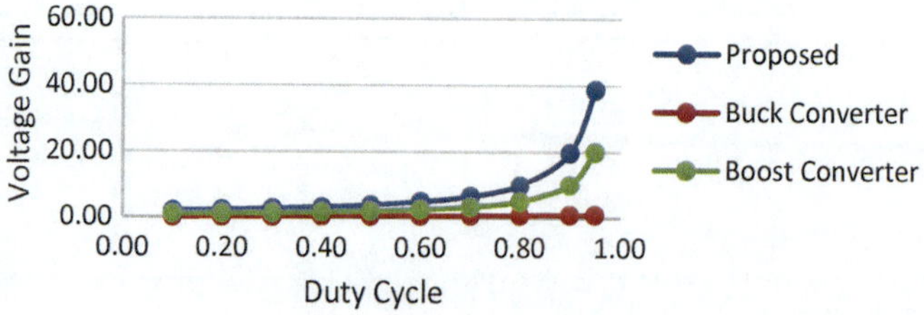

Fig. 22 Duty cycle versus voltage gain comparison

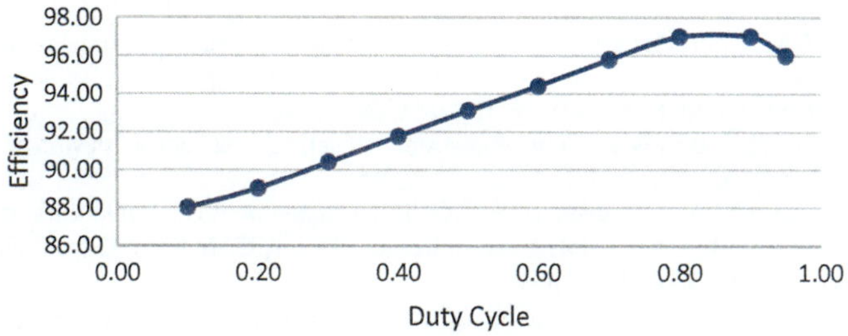

Fig. 23 Duty cycle versus efficiency of the proposed converter

6 Conclusion

This paper discussed the performance of single-switch hybrid DC power converter with extended voltage gain supplied with PEM fuel cell power source under closed-loop operation. The scope of the proposed DC power converter is it provides the neutral point in the output side which makes easier to combine its operation with single-phase half-bridge inverter. This will enable to deliver AC grid. Also, the proposed topology consists of less number of semiconductor devices leading to reduced switching loss. Analysis of the proposed DC power converter proves with reduced stress at power electronic switches resulting in longer life of the devices. Moreover, the transformerless topology of the proposed converter has made this topology most suitable for the fuel cell applications.

References

1. Choe, G.-Y., Kim, J.-S., Kang, H.-S., Lee, B.-K., Lee, W.-Y.: Proton exchange membrane fuel cell for high efficiency fuel cell balance of plant (BOP). In: International Conference on Electrical Machines and System (ICEMS), Nov 2007
2. Luckose, L., Urlaub, N.J., Wiedeback, N.J., Hess, H.L., Johnson, B.K.: Proton exchange membrane fuel cell (PEMFC) modeling in SCADA/EMTDC. In: IEEE Electrical Power and Energy Conference, pp. 11–16, Nov 2011
3. Wee, J.-H.: Application of proton exchange membrane fuel cell system. Renew. Sustain. Energy Rev. 1720–1738 (2007)
4. Attanasio, R., Cacciato, M., Gennaro, F., Consoli, A.: An innovative boost converter for fuel cell stationary generation system. In: The 30th Annual Conference of the IEEE Industrial Electronics Society, pp. 2831–2836, Dec 2004
5. Bryant, B., Kazimierczuk, M.K.: Modelling the closed-current loop of PWM boost DC-DC converter operation in CCM with peak current-mode control. IEEE Trans. Circuits Syst.—I: Regul Pap **52**, 2404–2412 (2005)
6. Bryant, B., Kazimierczuk, M.K.: Voltage-loop power-stage transfer function with MOSFET delay for boost PWM converter operation in CCM. IEEE Trans. Ind. Electron. **54**(1), 347–353 (2007)
7. Zhu, M., Luo, F.L.: Enhanced self-lift cuk converter for negative-to-positive voltage conversion. IEEE Trans. Power Electron. **25**(9), 2227–2233 (2010)

8. Latha Devi, M., Chilambarasan, M.: Design and simulation of incremental conductance MPPT using self lift cuk converter. In: 2013 International Conference on Renewable Energy and Sustainable Energy (ICRESE'13), pp. 105–111, Oct 2014
9. Ashokan, A., Paul, B., Joy, J.: Comparative study of self lift cuk converters with high voltage gain. Int. J. Eng. Res. Gener. Sci. **3**, 127–133 (2015)
10. Zhu, M., Luo, F.L.: Development of voltage lift technique on double-output transformerless DC/DC converter. In: The 33rd Annual Conference of the IEEE Industrial Electronics Society (IECON), pp. 1983–1988, Mar 2008
11. Fernão Pires, V., Foito, D., Baptista, F.R.B., Fernando Silva, J.: A photovoltaic generator system with a DC/DC converter based an on integrated boost-cuk topology. Solar Energy **136**, 1–9 (2016)
12. Hsieh, F.-H., Wang, H.-K., Chang, P.-L., Wu, H.-C.: Nonlinear dynamic behaviors in voltage-mode controlled single-phase half-bridge inverters via varying input voltage. In: 2010 First International Conference on Pervasive Computing, Signal Processing and Applications, pp. 230–234, Nov 2010
13. Menaka, S., Muralidharan, S.: Application of evolutionary algorithms for harmonic profile optimization in symmetric multilevel inverter used in medical electronic equipments. Curr. Signal Transduct. Ther. **13**, 12–20 (2018)
14. Muhaideen, M., Prabakaran, K., Muralidharan, S.: Design and development of power conditioning circuit for fuel cell. Int. J. Appl. Eng. Res. **10**(9), 7507–7511 (2015)
15. Muhaideen, M., Satheesh Prabu, B., Muralidharan, S.: Design and development of Z source inverter in DGS environment. Int. J. Appl. Eng. Res. **10**(9), 7463–7468 (2015)

Detection of Byzantine Attack in Cognitive Radio Network

A. Saraswathi Pooja, Rampa Theresa, Satya Govind, and Anuja Aloy Mary

Abstract The latest technology is cognitive radio networks which are used widely in present days. The cognitive was introduced by Mitola and Maguire in 1999. The main concept of network of cognitive radio network is one of which unauthorized secondary end user could utilize the vacant channel in the band of spectrum of the authorized primary end users. Main reason for spectrum scarcity could be cognitive radio network. Due to the acceptance of the CRN, technologies have initiated several attacks in cognitive radio. One of the major attacks is the byzantine attack. The concept of SSDF attack is the unauthorized user can change sensing information to disturb other secondary users' decision on the primary user state. This attack is called as SSDF or byzantine attack. This attack can be recognized by density-based SSDF detection scheme which will rectify SSDF attack. Mainly this scheme is used because it will omit the false sensing reports from attackers, so that the PU activity can be detected by the secondary user.

Keywords Cognitive network radio · Data falsification · Licensed user

1 Introduction

Cognitive radio network is the most important and widely used wireless network technology in the recent time to solve the issues dealing depleting spectrum and improvising the utilization of spectrum. Cognitive radio which can adapt itself to the radio

A. S. Pooja · R. Theresa · S. Govind (✉) · A. A. Mary
Vel Tech Rangarajan Dr. Sangunthala R &D Institute of Science and Technology, Chennai, India
e-mail: vtu6443@veltechuniv.edu.in

A. S. Pooja
e-mail: pooja.durai19@gmail.com

R. Theresa
e-mail: vtu6589@veltechuniv.edu.in

A. A. Mary
e-mail: Draloyanujamary@veltechuniv.edu.in

P. Suresh et al. (eds.), *Advances in Smart System Technologies*,
Advances in Intelligent Systems and Computing 1163,
https://doi.org/10.1007/978-981-15-5029-4_37

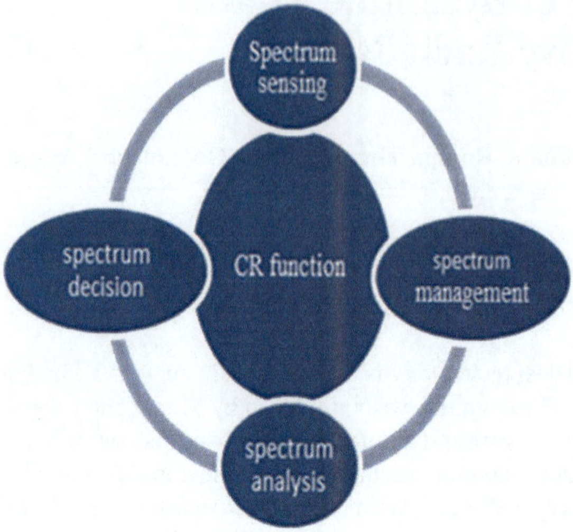

Fig. 1 Cognitive cycle

frequency environment is required by the users for communicating. To communicate from sender to receiver, certain frequency band authorization is required. Secondary users are unlicensed users who use those channels which are not used by licensed primary users at the same time. Because of the access method, some unlicensed user can access the primary users' channel which will introduce attack, which causes wide damage to CRN. Cognitive radio is configurable radio device, and the main tasks of the cognitive radio are identifying unused bandwidth spectrum in the licensed band without interfacing the licensed user, transmit power control, channel identification, and dynamic spectrum management. There are two main characteristics of cognitive radio:

- Cognitive capability
- Configurability (Fig. 1).

The cognitive cycle has four important phases:

- Sensing of the spectrum
- Management of the spectrum
- Sharing of the spectrum
- Mobility of the spectrum.

The above four tasks are done by cognitive radio for secured communication. The cognitive cycle which has tasks is implemented in the cognitive engine.

Spectrum Sensing: The main function of this phase is to monitor the interface temperature over the spectrum to identify the unused channel so that the utilization of spectrum can be improved which is one of the most major problems in the wireless network technology.

Spectrum Management: The second phase's main function is to select the unused channel which will have the different unique property according to the time-varying environment. CR has to monitor and decide the best channel to satisfy the QoS requirement.

Spectrum Sharing: The third phase in the cognitive cycle is spectrum sharing. This spectrum sharing technique is useful for saving energy, which can be done by utilizing lower frequency bands that will control the transmit power. Dynamic radio change is one of the properties of spectrum sharing, and this property is due to the behavior of cognitive radio (operating frequency range will be changing continuously). In this technique, the channels are the basic spectrum unit. Channels are also utilized by the licensed primary user, so secondary user must vacate the channel whenever the authorized user reaches. This spectrum sharing is the main challenge that takes place in cognitive radio network.

Spectrum Mobility: The last phase of cognitive cycle is spectrum mobility. In this phase, handover will take place which means the unlicensed secondary user will occupy the channels which are not used by the original user but secondary user surely has to monitor the activity of the origin user. In case the primary user returns to use the channel which is using by the unlicensed user, then the secondary user has to leave the occupied channel and hop to another channel, this process called handover.

2 Architecture

This chapter discusses the architecture of cognitive radio network. CNR is composed of various types of network. The main aim of network infrastructure is to improve the whole network usage. The architecture of the network designed in such a way that can provide more packets per unit bandwidth. Cognitive radio network is classified into two:

- Centralized approach
- Distributed approach.

2.1 Centralized Approach

In this approach, the centralized entity (e.g., base station) is used to control spectrum allocation, medium access, and secondary use (Fig. 2).

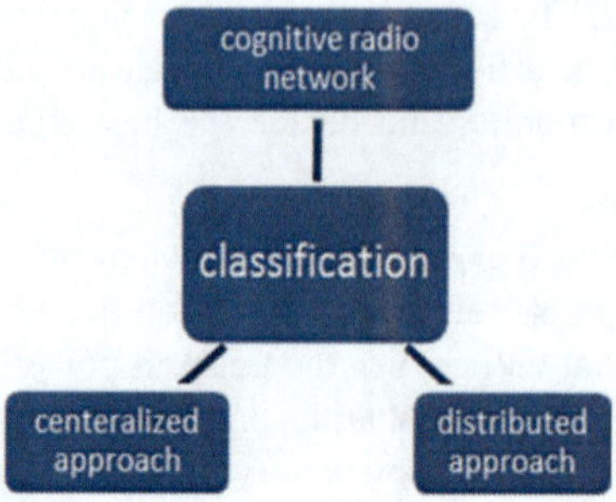

Fig. 2 Classification of CRN

2.2 *Distributed Approach*

In this approach, CR users are used to manage spectrum allocation and access.

Cognitive radio network system is classified into two types:

- Primary system
- Cognitive radio system (Fig. 3).

Primary system: This system operates in one or many fixed or constant spectrum bands, and primary systems are classified into two kinds:

Primary system licensed band: This system has more priority to use the spectrum band because it is used in the licensed bands.

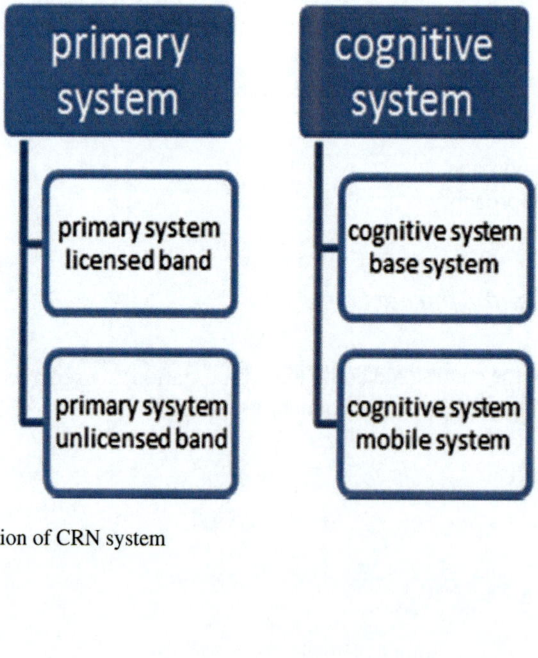

Fig. 3 Classification of CRN system

Primary system unlicensed band: In this, system will operate in one unauthorized band which will interfere with another one.

System of cognitive radio: It includes a constant frequency band which is in operation mode, and this is unauthorized for the band usage. The modus operandi for communication within the system dynamically done through licensed frequency band popularly known as spectrum hole. Base station and mobile station are the major parts of the network.

Cognitive radio base station: It is a constant component present in the base system, and it has capacities of cognitive radio. It will show the framework of the cognitive radio system and provide licensed frequency band which is also known as spectrum hole, mobility management, and security management to the mobile system.

Cognitive radio mobile station: It is a movable device with cognitive radio capabilities. It recognizes authorized frequency band and used to meet up with mobile station or base station.

3 Byzantine Attack

In cognitive radio network, technologies have introduced several attacks and one of the main attacks is the byzantine attack. This attack is also called as SSDF attack. There is two main aim of byzantine attack, 1. detection probability of the primary user system is reduced because the operation of the primary user system is distracted, 2. the probability of false alarm is increased.

The main reason for the presence of SSDF attack is due to the less radio spectrum. Byzantine attack makes the situation critical and uncontrollable, which is the great obstacle to the cognitive radio system and this is because unlicensed user will change the sensing data or information in order to distract the secondary user's decision on primary user state. To rectify issues, spectrum sensing is widely used. The main function of spectrum sensing is to stop complicate interference with authorized primary users and detect the spectrum which is accessible for raising the spectrum usage. Shadowing and multipath fading are some issues which occur in the cognitive radio network; to rectify these issues, cooperative spectrum sensing technique is widely used and CSS will increase the detecting performance by exploiting spatial diversity (Fig. 4).

3.1 Components of Cooperative Spectrum Sensor

Signal detection technique: Method is utilized to accumulate primary users' signals on authorized channel. The information basis is given to the spectrum sensing.

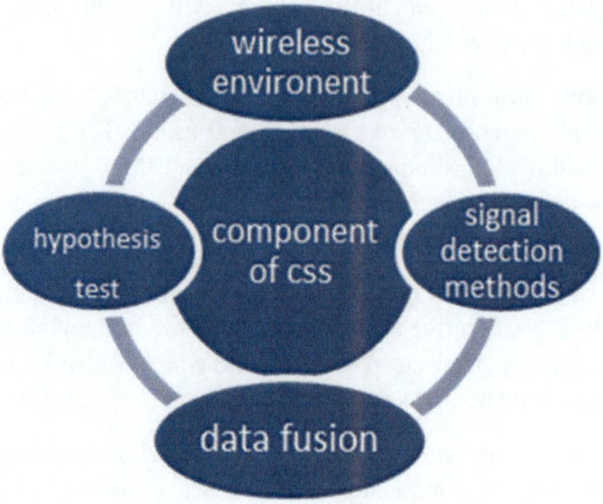

Fig. 4 Components of cooperative spectrum sensing

Hypothesis testing: This testing is called statistic technique because it will give the confirmation of present and absent of primary users which is actually based on the sensing report.

Data fusion: Data fusion will detect licensed frequency band by initiating the combine decision of multiple secondary users.

3.2 Parameter of Byzantine

Attack outline: Cooperative spectrum are of two types (Fig. 5).

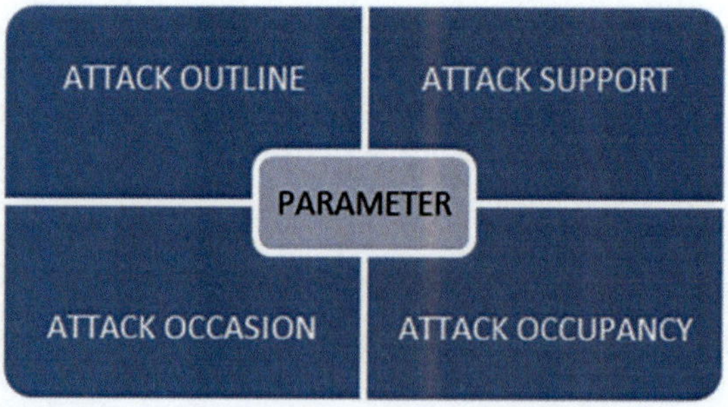

Fig. 5 Parament of byzantine attack

3.2.1 Centralized

Decentralized

It is clubbed format of the node receivers while clubbing the repost of sensor, thereby affecting the system by introducing attack.

Attack support: Gives information based on malicious user falsify measurement and solves how to introduce attack.

Attack occasion: This says about introducing attack at a particular slot. False users might conclude whether to attack or not to attack at particular probability by including the sensing report.

Attack occupancy: It explains about which one to attack and same to percentage of a false user in all secondary users which are critically attacked by attackers.

4 Attack Model Based on Byzatine

There are several models in byzantine attack and some are mentioned below:

- Centralized Independent probabilistic small-scale model
- Decentralized Independent probabilistic small-scale model
- Centralized dependent probabilistic small-scale model
- Decentralized dependent non-probabilistic small-scale model
- Centralized independent probabilistic massive model.

Centralized Independent probabilistic small-scale attack (CIPS): The most routine attack is CIPS. In CIPS, false users are less in number and each hostile user contributes independently by introducing the band attack which is a basis of its own observation for the present slot as much is not known about the actual state of channel; apart from false user, several other users sense the output, and the pervious slot is the result of the fusion of final output sensing.

Centralized dependent probabilistic small scale (CDPS): In this attack model, there will be good coordination between malicious users to falsify data. This attack will avoid collision between attackers and cause serious damage to global sensing performance which will lead to an abnormal high correlation of attackers.

Centralized dependent non-probabilistic small-scale attack model(CDNS): Non-probability opportunity of attack is considered by users which are unauthorized completely exploiting that attack basis timeless thereby reaching more secret attack considering the expectancy of attack meaning calculating the impact of attack prior modifying the false data to improve attack predictancy.

Decentralized independent probabilistic small-scale (DIPS) attack: This attack considers clubbed structure formed with a basis of attack and opportunity of attack.

It will use self-observation as well as its neighbors report as attack outline, which will give a clear idea of the users monitoring report and actual slot channel.

5 Density-Based SSDF Detection

A distributed method to counteractive the byzantine attack in cooperative spectrum sensing called density-based detection. The main concept of DBSD is it will omit all false sensing report first and then detect unauthorized user. DBSD performs well in CSS which is shown at the end of the simulation. The probability density of the variable is estimated by a method called kernel density estimator. The probability of reached authorized user signals is estimated by kernel density estimator by considering random sensing report. Following are the simulation performance based on DBSD (Fig. 6).

Figure 7 indicates that success probability for DBSB to detect the primary user activity versus beta with total 10, 30, and 50 number of secondary users which are considered as malicious users simulated to launch byzantine attack. When beta decreases, PU detection success probability increases.

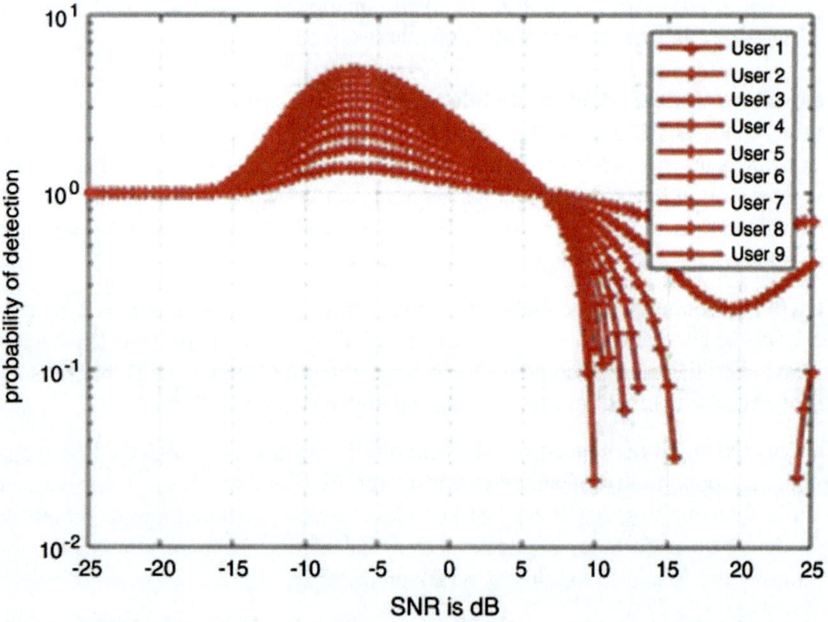

Fig. 6 Probability of detection versus SNR probability of detection versus SNR will show the number of user and it will indicate probability of detection in *Y*-axis and SNR in *X*-axis

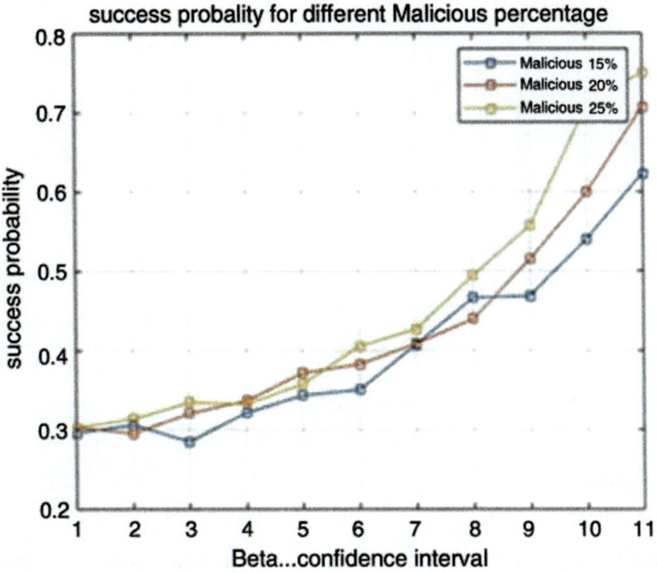

Fig. 7 Beta versus success probability

Figures 8 and 9 Function of a percentage malicious user is PU detection success probability with values $N = 1$, beta = 0.25, receptively, and PU detection success probability reduces gradually when the percentage of false users are more in number.

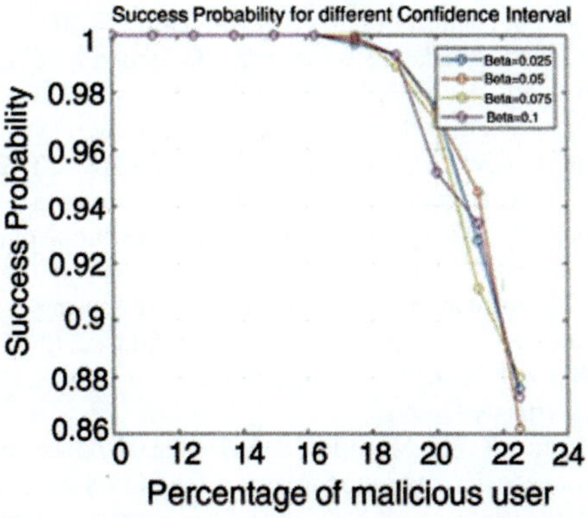

Fig. 8 Success probability vs Percentage of malicious user

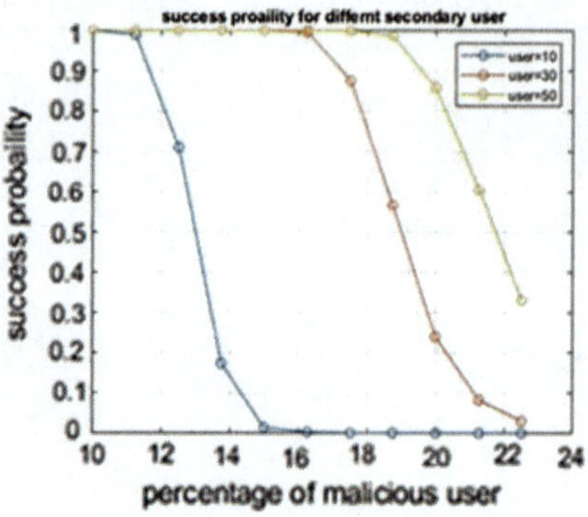

Fig. 9 Success probability vs Percentage of malicious user

6 Conclusion

This paper will explain about the detection of byzantine attack by using density-based method because it will omit all the false users report in the cognitive radio network and gather all secondary users report and find the status of the primary user.

References

1. Chen, C., Song, M., Xin, C.S.: A density based scheme to countermeasure spectrum sensing data falsification attacks in cognitive radio networks. IEEE (2013)
2. Farmani, F., Abbasi Jannat-Abad, M., Berangi, R.: Detection of SSDF attack using SVDD algorithm in cognitive radio networks. IEEE (2011)
3. Khare, A., Saxena, M., Thakur, R.S., Chourasia, K.: Attacks & preventions of cognitive radio network-A survey. Int. J. Adv. Res. Comp. Eng. Technol. (IJARCET) **2**, (2013)
4. Wang, H., Lightfoot, L., Li, T.: On PHY-layer security of cognitive radio: collaborative sensing under malicious attacks. In: 44th Annual Conference on Information Sciences and Systems (CISS). Princeton, NJ (2010)
5. Magdalene, A.H.S., Thulasimani, L.: Analysis of spectrum sensing data falsification (SSDF) attack in cognitive radio networks: a survey. J. Sci. Eng. Educ. **2**, (2017). (ISSN-2455-5061)
6. Zhao, T., Zhao, Y.: A new cooperative detection technique with malicious use suppression. In: ICC 2009. Dresden, Germany (2009)
7. Sodagari, S., Attar, A., Leung, V.C.M., Bilen, S.G.: Denial of service attacks in cognitive radio networks through channel eviction triggering. In: GLOBECOM 2010, Miami, FL (2010)
8. Li, H.: Cooperative spectrum sensing via belief propagation in spectrum-heterogeneous cognitive radio systems. In: IEEE Wireless Communication and Networking Conference (2010)
9. Oikonomou-Filandras, P.A., Wong, K.K., Zhang, Y.: Informed scheduling by stochastic residual belief propagation in distributed wireless networks. IEEE Wirel. Commun. Lett. (2015)
10. Chen, Y., Yang, L., Ma, S., Yuan, X.: Detecting primary user emulation attacks based on PDF-BP algorithm in cognitive radio networks. IEEE (2016)

Lopels—The AI Based Multi-vendor Loyalty Platform

Christy Maria James, Darin Manoj, Greety Margret Pious, Jaise P. Jose, and V. Jeyakrishnan

Abstract A customer loyalty program is a reward program offered by the shopkeepers to customer who frequently make purchases. Loyalty programs are one of the advanced marketing strategies that merchants adopt to influence their customers to visit back their shop again and again. But the present system is not founded to be effective in many areas such that it is not economical for small scale merchants to implement expensive loyalty system in their shops and customers need to carry a bunch of loyalty cards with them. Also, customer retention is low with the present system such that the loyalty cards do not understands customer behavior or their shopping patterns. The proposed system is an AI powered multi-vendor loyalty mobile application, where merchants can implement an economically viable loyalty system in their shops and it will be a hassle-free shopping experience for shoppers. The system uses AI technology to influence customers more to visit again to the shops by collecting their shopping data and analyzing it. So, this system will be an economically viable loyalty platform and an effective marketing tool.

Keywords Business intelligence · Loyalty system · Artificial intelligence · Multi-vendor loyalty program

---

C. M. James · D. Manoj · G. M. Pious · J. P. Jose (✉) · V. Jeyakrishnan
Computer Science and Engineering Department, Saintgits College of Engineering, Kottukulam Hills, Pathamuttom, Kottayam District, Kerala, India
e-mail: jaise97@gmail.com

C. M. James
e-mail: christymerry97@gmail.com

D. Manoj
e-mail: darinmanoj005@gmail.com

G. M. Pious
e-mail: greetypious@gmail.com

V. Jeyakrishnan
e-mail: jeyan.krishnan@gmail.com

P. Suresh et al. (eds.), *Advances in Smart System Technologies*, Advances in Intelligent Systems and Computing 1163,
https://doi.org/10.1007/978-981-15-5029-4_38

1 Introduction

The Indian Economy is highly dependent on the income from small scale shopkeepers. Through the emergence of various e-commerce sites, the business at various small-scale shops get reduced. In order to compete with e-commerce markets, merchants are trying various marketing strategies like giving discounts, offers, gifts etc. One of the advanced methods of these various approach is "Loyalty Program". Statistics shows us that almost 61% retailers cite customer retention as their biggest obstacle and their average profit per customer can increase by 25–100% by a 5% increase in customer loyalty. The ultimate aim of these loyalty programs is to influence the buying prospects of the customers.

The proposed system is an AI based loyalty platform which influence the customers to shop again at the same outlet and also to create new loyal customers for the shops. Also, the project is a multi-vendor platform, where multiple categories of shops meet together and cooperate together to bring a chain of loyalty.

2 Existing Systems

2.1 Loyalty Cards

One of the mostly used loyalty system is card-based loyalty programs, where the shopkeeper will give a card to the customers after their first purchase. So, if the customer came back to the shop with this card, the shopkeeper will give some rewards to the customer either as a gift, discount or as cashback. So, shopkeeper need to print cards and they need a software to manage this loyalty program. The limitations of the existing card system are as follows:

- Customer need to carry one card for each shop, in detail, if he/she visits 40 shops, they need to carry 40 cards with them. It is difficult to handle a bunch of cards.
- Customers can't check their collected points, its validity and other offers just by looking the card. Either they need to visit the shop or need to contact the merchant.
- Customers can't redeem the rewards collected from a shop in any other shops.
- Shopkeepers need to spend a huge amount of money for cards, card printing machine, scanning/swiping machine and for software to maintain loyalty program. So, a rough figure of Rs. 70 is to be spent by the shopkeeper for each customer. This makes small shopkeepers not to try this program in their shops.
- Another limitation of this system is that, there is no enough security. A stranger who get this card can redeem the points without any permission from card owner.

2.2 Loyalty Apps

Some shops are now using mobile application-based loyalty programs. But, among these shops, most of the shops uses a dedicated mobile application for their shop. So, the shoppers need to have separate mobile applications for each shop. This will cause some difficulty for customers as they need to download different apps and handle it.

And some apps are there in the market where it lists various loyalty programs together in a single application. In such an application, customers only need to handle a single mobile application for their various loyalty programs.

3 Proposed System

As every country is moving towards the goal of digitalization in each area, all types of offline transactions are switched to online modes. This project also changes the traditional offline loyalty card system to QR code based mobile application. Here, both merchants and shoppers are using mobile application to handle loyalty program. The system is a multi-vendor loyalty platform where shoppers can use this one stop application for various loyalty programs they are participating.

Both merchants and shoppers are identified using their own unique QR code. Figure 1 shows the home page of customers application, where his/her QR code will

Fig. 1 Customers app home page

be displayed. They can scan these QR codes to transfer and redeem their rewards. One of the issues that faced by the customers with the present loyalty system was that they can't redeem their collected rewards from a shop in any other shops. The proposed system solves this issue such that, it introduces a chain of shops where the shoppers can redeem their reward collected from one shop to any other shop in this chain. Merchants are transferring actual money as rewards instead of transferring points. So that, when shoppers redeem their rewards at any other shops, there won't be any financial issue. Since this system is a chain of loyalty, there will be different categories of shops in this chain. So, there occurs a great probability of customer retention as shoppers can redeem the rewards only at this chain.

3.1 Integrating AI with Lopels

The ultimate aim of all loyalty program is to increase customer retention and thereby improving businesses. But, when considering the present system, customer retention is very much low. This product makes use of Artificial Intelligence to increase customer retention and also to bring new customers to the shops. The system captures various shopping data of customers and analyze it using marketing patterns. Using those results, the system influences those customers to visit shops again and again. So, this project will be a perfect marketing tool and an economical loyalty platform for merchants.

Whenever a shopper visits a shop and use this product, the product collects the various data's like customer name, shop name, location, category of shop, bill amount. Also, the product continuously captures the location of the customer, records which all shops and category of shop he/she visits. So, after some period of time, the system will learn the shopping behavior of the customer and identifies or recommend merchant about customer behavior. Hence, merchants can identify whether the customer is a potential customer for him or not. Also, the merchant can recommend his shop to the customer, whoever is nearby in his shop. The application goes countless with AI in loyalty system.

3.2 Location Based Clustering

One of the applications that the system proposes is that, whenever a shopper reaches a particular location where our chain of shops exist, the system displays the nearby shops to the shopper. So, each shop is in a particular cluster or group based on location. Whenever a shopper reaches a location, the system identifies the exact and appropriate cluster for the shopper and displays the shops with respect to the cluster. For this, it uses unsupervised machine learning algorithm, where clustering is made through k-means clustering algorithm. This algorithm uses iterative refinement to produce final result.

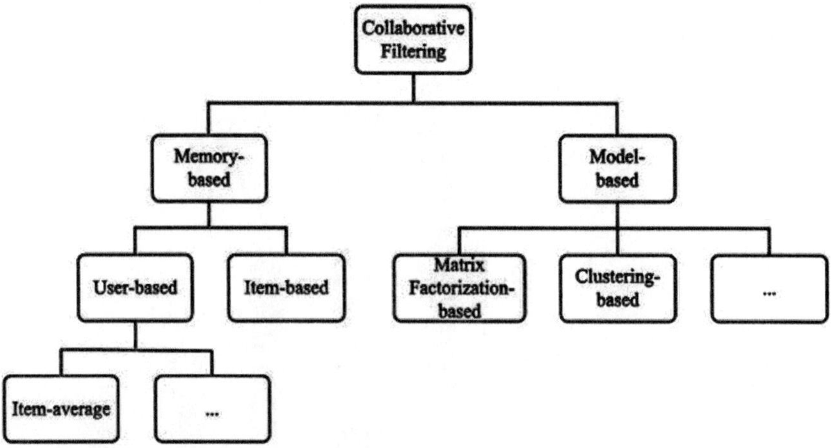

Fig. 2 Collaborative filtering in recommender systems

3.3 Recommendation Based on Behavior

Recommendation of shops based on customer shopping behavior is the next application that the system introduces. Firstly, the system learns about the shopping behavior of each customer, where do they go, what do they buy etc. So, using this knowledge, the system can recommend shoppers about various shops that he/she might be interested to visit at that time.

For this, Collaborative filtering (CF) recommendation algorithm is used. It is one of the most commonly used recommendation algorithms. The basic concept in this algorithm is that, to recommend someone about a thing, the system should learn about what are the interests of other similar people, their behavior and then recommend those things to the user. The two approaches in this algorithm are user-based filtering and item-based filtering.

As shown in Fig. 2. [1], the system uses memory-based approach and it uses user rating data to find the similarity between users or items. Both user-based and item-based approaches are needed in the proposed solution. User based approach recommends on the basis of rating given by similar users.

3.4 Business Intelligence with Lopels

Knowing the potential customers is a big challenge for businesses. The proposed system makes this task easier for business managers such that, the various shopping data of every customer is analyzed and presented intelligently. The following features are available in manager's dashboard.

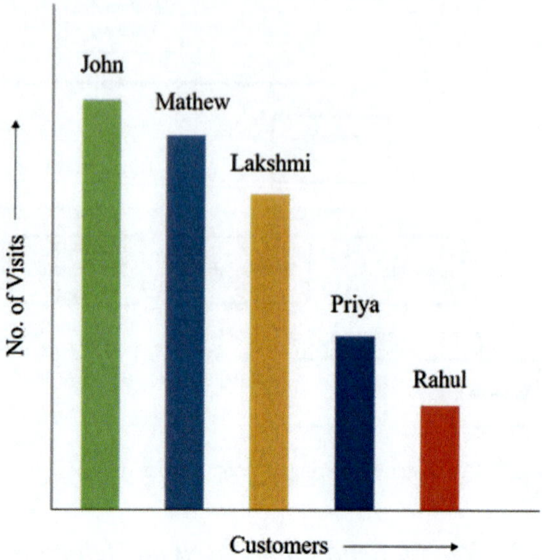

Fig. 3 Best customer analysis report

Finding the Best Customers. This will display the best customer of that shop. This can be based on the number of visits or based on the total bill amount. Figure 3 shows an analyzed report of best customer of a particular shop. With this report, the managers can offer some special prizes or discounts for the best customers.

Recommended Customers. The system predicts and recommends a list of new customers to the managers based on their shopping behavior. For example, a customer who visits a textile shop may visit a footwear shop also. So, such customers are potential customers for the footwear shop. Location of the customer will be also one another parameter for recommendation. Therefore, by getting the recommended customers, the managers can send some special offers to attract the customers.

Send Offers and Notifications. Managers can use of the data to set a new marketing strategy and send custom offers and notification for a selected group of customers.

Competitor Analysis. This feature will report about the customer acquisition by each competitor in same category. The system will analyze the complete data of each category and generates result as shown in Fig. 4.

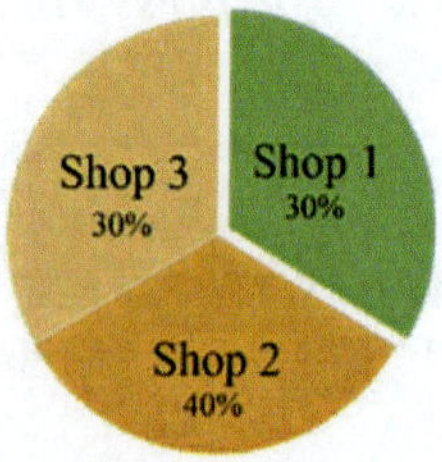

Fig. 4 Customer acquisition report

4 Advantages

Both shoppers and merchants are the beneficiaries of this proposed solution. Shoppers gets a hassle-free shopping experience as they don't need to carry a bunch of loyalty cards with them or don't need to download various loyalty apps in their phone. The proposed solution is a one stop application for all loyalty programs where customers can also redeem the reward collected from one shop in any other shop. The system is an economically viable loyalty platform such that even small-scale shopkeepers can try loyalty programs in their shop without investing huge money.

Also, with the integration of artificial intelligence in the system, the platform came to be a best marketing tool also for merchants as they can target their potential customers easily with this system. Shoppers also receive personalized loyalty offers and they will be more attracted to visit the shop again and again thereby increasing the customer retention rate of each shops.

5 Applications

The system has a lot of applications as it collects various shopping data of each customer in different shops. The data's can be properly analyzed and use it depending on the need. The merchants get the best marketing platform through this system as they can target their potential customers without any effort. They can create customized offers for their customers based on the shopping behavior of each customers.

6 Conclusion

The proposed system will be a perfect one stop multi-vendor loyalty program and also the best marketing tool for merchants. The AI integration made it possible for numerous applications in future. The shoppers receive a good shopping experience through this platform as they receive customized loyalty rewards based on their previous shopping behavior. The system improves the customer retention much more than the existing systems. Also, merchants don't need to spend a huge amount to start loyalty program in their shop. So. Small scale shopkeepers too can try loyalty-based marketing strategy in their outlets.

References

1. Anon: [online] Available at: https://www.researchgate.net/figure/Classification-of-collaborative-filtering-algorithms_fig4_303556519. Accessed 20 Mar 2019
2. Chakraborty, I.: Hierarchical Bayesian modeling for clustering sparse sequences in the context of user profiling in customer loyalty program. In: 2018 IEEE International Conference on Big Data (Big Data), pp. 4678–4683. IEEE
3. Khan, I.: Impact of customer satisfaction and retention on customer loyalty. Int. J. Sci. Technol. Res. **1**(2), 106–110 (2012)
4. Gronholdt, L., Martensen, A., Kristensen, K.: The relationship between customer satisfaction and loyalty: cross-industry differences. Total Qual. Manag. **11**(4–6), 509–514 (2000)
5. Ning, F., Ming-li, Z., Sai-li, T.: Impact of loyalty programs on relationship benefits and customer loyalty: a customer perspective. In: 2010 International Conference on Management Science & Engineering 17th Annual Conference Proceedings, pp. 533–538. IEEE
6. Zakaria, I., Rahman, B.A., Othman, A.K.: The relationship between loyalty program and customer loyalty in retail industry: a case study. In: 2012 International Conference on Innovation Management and Technology Research, pp. 33–38. IEEE
7. Lu, G., Lu, P., Qi, X.: Customer loyalty prediction and implementation using data mining. In: 2011 International Conference on Computer Science and Service System (CSSS), pp. 3120–3123. IEEE
8. Uncles, M.D., Dowling, G.R., Hammond, K.: Customer loyalty and customer loyalty programs. J. Consum. Mark. **20**(4), 294–316 (2003)
9. Sharp, B., Sharp, A.: Loyalty programs and their impact on repeat-purchase loyalty patterns. Int. J. Res. Mark. **14**(5), 473–486 (1997)
10. Lewis, M.: The influence of loyalty programs and short-term promotions on customer retention. J. Mark. Res. **41**(3), 281–292 (2004)
11. García Gómez, B., Gutiérrez Arranz, A., Gutiérrez Cillán, J.: The role of loyalty programs in behavioral and affective loyalty. J. Consum. Mark. **23**(7), 387–396 (2006)
12. Berman, B.: Developing an effective customer loyalty program. Calif. Manag. Rev. **49**(1), 123–148 (2006)
13. Otto, J., Van Luchene, A.: Retail DNA LLC, 2012. Method and system to manage multiple party rewards using a single account and artificial intelligence. U.S. Patent 8,306,937
14. Forbes.com.: Council post: It's Time For An AI Transformation To Revitalize Brand Loyalty Programs. [online] Available at: https://www.forbes.com/sites/forbesagencycouncil/2018/06/27/its-time-for-an-ai-transformation-to-revitalize-brand-loyalty-programs/. Accessed 20 Mar 2019
15. Trevino, A.: Introduction to K-means Clustering. [online] Datascience.com. Available at: https://www.datascience.com/blog/k-means-clustering. Accessed 20 Mar 2019
16. Stats and Bots. (2019). Recommendation System Algorithms. [online] Available at: https://blog.statsbot.co/recommendation-system-algorithms-ba67f39ac9a3. Accessed 20 Mar 2019

Finding Missing Child in Shopping Mall Using Deep Learning

Nevil Susan Abraham, Rithu Ann Rajan, Riya Elizabeth George, Salini Gopinath, and V. Jeyakrishnan

Abstract Recent reports show that there is a multifarious number of children missing each year. This is higher mostly in the case of India. Most of the cases involve missing children from the shopping mall. This paper takes a deep learning perspective in solving this serious issue with the help of face recognition due to its accuracy when a huge amount of data is under consideration. The main focus is to design an efficient system that can be used in shopping malls to find the missing child using deep learning. The system makes use of real-time analysis of video captured by the surveillance camera to detect the most recent presence of missing child. When the child goes missing the parent approaches the security who accesses the already uploaded photo by the parent of the missing child using the parent's phone number. The facial features are extracted from the photo and compared with the video being captured by the camera for face recognition. When a match is found an alert is sent to the security within the area in which the child was found.

Keywords Face recognition · Deep learning · Real-time analysis

1 Introduction

Children are a gift from God. They are a great asset to the parents. Their smile creates happiness in our lives. If a child goes missing it creates panic to the parents. Nowadays we all go to the shopping malls for purchasing. In some cases, if the parents are busy shopping, the children may let go off their parent's hands and goes missing. Deep learning is a part of machine learning which is based on learning the data representations. Learning are of various types such as supervised, unsupervised and semi-supervised. Deep learning contains a variety of algorithms. The proposed

---

N. S. Abraham · R. A. Rajan (✉) · R. E. George · S. Gopinath · V. Jeyakrishnan
Saintgits College of Engineering, Kottayam, India
e-mail: rithuannrajan@gmail.com

S. Gopinath
e-mail: salinigk97@gmail.com

P. Suresh et al. (eds.), *Advances in Smart System Technologies*,
Advances in Intelligent Systems and Computing 1163,
https://doi.org/10.1007/978-981-15-5029-4_39

system is a way to find the missing child in a shopping mall. This system not only helps to find the child but also provides an alert to find the missing child. This system uses machine learning for comparing the face of the child with the original photo and the images captured from the surveillance camera. Two separate websites are provided for the parent as well as the security in the shopping mall. The parent can register into the website provided for them and can upload a recent photograph of their children. In case if their child goes missing, they can approach the nearby security and inform that their child has been lost. The security then logs into the security's website and gives the phone number of the parent. When the phone number is entered the photographs of their children will be shown. From there the security can search for the missing child. Then it shows where the child has recently appeared and in front of which camera and the corresponding floor. When it shows the location of the child, the security sends an alert message to the security on the corresponding floor in which the child was seen. The security on that floor hands over the child to the parents [1].

2 Literature Survey

Many researchers have been trying to develop effective methods or ways to find out the lost child in a shopping mall or a market. The use of simpler microcontrollers to very complex designs can be seen in various design and implementation models. Some of the contributions of the research are summarized below.

Liu and Li [2] proposed a novel paradigm that uses sensors and sensor-rich smartphones to locate the lost child in a crowd. It uses Mobile Crowd Sourcing/Sensing (MCS) which has the potential for locating and searching in a better efficient means at a good rate. It tells the novel innovative idea in the application for searching and finding the lost child in a crowd with the help of MCS. This approach uses an anchor network for references which is fixed or points as fingerprinting points. This proposed system is not effective for knowing the location of the open area where the child is present. For obtaining sufficient measurements, it takes several assistants based on the hop as an aid to access the participators for MCS-based collaborative localization. Optimization approaches on Semi-definite Programming (SDP) were proposed to indulge the location and measurements of range in a highly efficient manner.

Ito et al. [3] proposed a system of finding the lost child using BLE beacons. They conducted a trial observation in a building with five floors. The children were given with BLE beacon. Parents carried smartphones with them for detecting the BLE beacon which is present with the child. Also, they can be used for detecting the location by fixed beacon. The preparation based on nineteen patterns for testing the possible location combination of the phone used, foreground application status or background application status, floors analysis, lost children count and witnesses count to give back the position accuracy and floor accuracy where the child is present. The final result attained was with low accuracy, that is, the floor detection possibility resulted was 75.9.1%, and the location detection possibility on that corresponding floor was 77.9%. Thus the disadvantage is that the result of the system produced was with low accuracy.

The International Journal of Soft Computing Applications [4] explains the Multi-purpose Tracking System which provides the detection and tracking of the lost child that could happen in a mall or any other supermarket. This system marks the disadvantages of existing systems. The two modules present in the system design are parent module and child module. There is a system of child safety present and when this child safe violation is encountered, an allotted sensor present in the child module will generate a signal. This signal is sent to the controller from those sensors and follows the transmitter to the parent module leads to take the proper decision so that processes towards violation handling are able to start. The parent will be able to set and reset the system to work outdoor or indoor and can calculate the distance between both that is child and parent at any interval of time. For outdoor distance calculation, Global Positioning System (GPS) is used, while for calculation of indoor distance the change in amplitude of RF signal is used.

Eden et al. [5] proposed a system named as 'Finding lost children' which explains the finding of the child. It uses a system that is based on the Content-Based Image Retrieval and attribute search. Also, it explains and shows different series of evaluations that may or may not include a set realistic disaster drill with a method of reunification of children to their parents. It will be very difficult to find the public health infrastructure that accelerates reunification in an effective manner.

3 Proposed System

In this proposed system, there is a server-side website and a client-side website. In the server-side website, there is a login for an admin and the parent. The parent can create a login by first registering. While registering the parent has to give the basic information like parent name, phone number, email id, username, password and can then register. Using this username and password the parent can login and upload his child's photo. This uploading of the child's photo can be done either while entering the shopping mall or when the child goes missing. The admin at the server-side manages the security registration and assigns the area for each security. At the client-side, there is a login for an admin and each security. The client-side admin manages the camera and assigns each camera to each area. When the parent finds out that his/her child has gone missing, all that the parent has to do is go and contact the nearby security [6].

The security then login and enters the parent's phone number and clicks on a find button. Then the system compares the photo uploaded by the parent with the real-time video captured by the surveillance camera using machine learning techniques. Once the child's face is detected in the camera, it notifies the security within that area and further actions are taken. Then the child is handed over to parents [7].

4 Methodology

Consider a parent entering the shopping mall with his/her child. While the parent is engrossed in shopping items, there is a high chance of the child wandering away. This creates worry and tension among the parents and relatives. This project helps in finding the missing child and handing over to the parent [8].

4.1 Functional Modules

Area Module
In this paper, each floor in the shopping mall is considered as an area. The server-side admin can manage each area. He can add and view the area.

Camera Module
A surveillance camera is assigned in each area. These cameras are monitored by the client-side admin. He assigns these cameras to each area.

Parent Module
Parent registers and creates an account using his phone number, email id, username and password. He then uploads a recent image of his child.

Security Module
Security is the client of this system. He accesses the image of \missing child by entering the parent's phone number. Then the search is initiated.

System Training Module
When a search is initiated, the image of the child is compared with the real-time video captured by the surveillance camera. The image of the child given as input is classified and the photo which is the best emulation will be detected from the captured video. This is done by first training a deep learning model so that it identifies the missing child flawlessly from the video. The Convolutional Neural Network (CNN) is espoused here for face recognition. This model is then pretrained for attaining facial features [9, 10].

The child recognition is done by the comparer and the features are extracted by convolution network. The comparer performs a match between two images, uploaded image and the image captured by video streaming. Haar cascade detectors are used for extracting and comparing the facial features from the video [11] (Fig. 1).

Found Lost Child Module
After a match of the child is found, an alert is sent to the security on the floor in which the missing child was found [12].

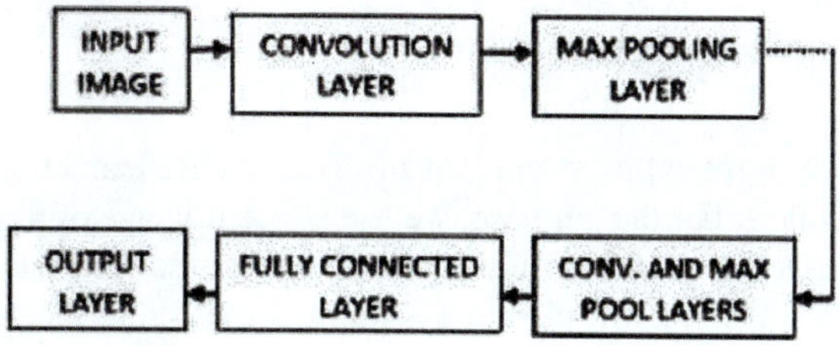

Fig. 1 Image classification using CNN

5 Results and Discussion

After the search is initiated if the child is found, status with the message child found, child's image, the camera number in which the match was found and the floor in which the child was spotted will be displayed on the website.

When the child is found an alert message will be sent to the security on the floor in which the child was found.

If a match of the child is not found, a message written matching not found will pop up on the screen.

The system using deep learning shows more accuracy than that of the sensor-based system. The accuracy keeps on increasing with successive epochs is shown in Fig. 2.

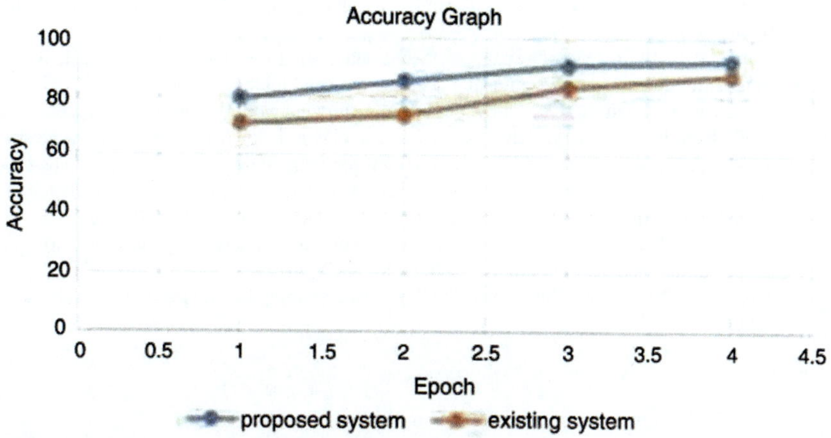

Fig. 2 Accuracy graph of existing and proposed system

6 Conclusion and Future Scope

In this paper, we proposed a system that involves feature extraction, comparison, verification and alert. For this purpose, we use a recent image of the missing child and real-time image captured from the camera. Our proposed system is more effective than existing systems. We will plan to extend this research to the cases of a child being kidnapped or any other unusual activity. Monitoring the child's current location using GPS can be added for future applications.

References

1. Saranya, J., Selvakumar, J.: Implementation of children tracking system on android mobile terminals. In: 2013 International Conference on Communication and Signal Processing (2013)
2. Liu, K., Li, X.: Finding nemo: finding your lost child in crowds via mobile crowd sensing. In: 2014 IEEE 11th International Conference on Mobile Ad Hoc and Sensor Systems (2014)
3. Ito, A., Xiang, F., Hatano, H., Uchiyama, H.: A study on observing a lost child using BLE Beacons. In: 2018 Sixth International Symposium on Computing and Networking Workshops (CANDARW)
4. In. J. Soft Comput. Appl. (4), 57–68 ©EuroJournalsPublishing,Inc.2009 http://www.eurojournals.com/IJSCA.htm. ISSN: 1453-2277
5. Eden, A., Christoudias, C.M., Darrell, T.: Finding lost children. In: 2011 IEEE Workshop on Person-Oriented Vision (2011). https://doi.org/10.1109/pov.2011.5712362
6. Dandekar, A.R., Nimbarte, M.S.: A survey: verification of family relationship from parents and child facial images. In: 2014 IEEE Students' Conference on Electrical, Electronics and Computer Science
7. Chandran, P.S., Byju, N.B., Deepak, R.U., Nishakumari, K.N., Devanand, P., Sasi, P.M.: Missing child identification system using deep learning and multiclass SVM
8. Ye, X., Chen, X., Chen, H., Gu, Y., Lv, Q.: Deep learning network for face detection
9. Vo, A.T., Tran, H.S., Le, T.H.: Advertisement image classification using convolutional neural network. In: 2017 9th International Conference on Knowledge and Systems Engineering (KSE)
10. Coskun, M., Ucar, A., Yildirim, O., Demir, Y.: Face recognition based on convolutional neural network. In: 2017 International Conference on Modern Electrical and Energy Systems (MEES)
11. Kim, S., An, G. H., Kang, S.-J.: Facial expression recognition system using machine learning. In: 2017 International SoC Design Conference (ISOCC) (2017)
12. Akbulut, Y., ùENGÜR, A., Budak, Ü., Ekici, S.: Deep learning based face liveness detection in videos

Pentadecagonal Fuzzy Number and Its Arithmetic Operations

Edithstine Rani Mathew and Sunny Joseph Kalayathankal

Abstract We are introducing a novel form of fuzzy number called pentadecagonal fuzzy number. Till now no one has introduced pentadecagonal fuzzy number. Here, we discuss pentadecagonal fuzzy number and its arithmetic properties using alpha-cut. AMS Classification: 03E72.

Keywords Fuzzy number (FN) · Dodecagonal fuzzy number

1 Introduction

Dutta [1] did a comparative study on arithmetic operations using alpha-cut and the same without the use of alpha-cut. Arithmetic operation of nonagonal fuzzy number was studied by Felix and Christopher [2]. Victor and Dhivya [3] proposed properties of hendecagonal fuzzy number. Operations on fuzzy number by Dubois and Prade [4], study on decagonal fuzzy number by Felix and Victor [5], fuzzy arithmetic by Kauffman [6] and study on Triskaidecagonal fuzzy number using alpha cut by Rajkumar and Helen [7] gave insight to our proposed work. Incertitude is a problem which comes when we operate in pentadecagonal number. Obsecurity can be rectified by the use of modified forms of FN. We are introducing a new type of FN. Pentadecagonal FN is studied upon a linguistic environment which is uncertain.

E. R. Mathew (✉)
Department of Mathematics, Catholicate College, Pathanamthitta, Kerala, India
e-mail: edithstine@gmail.com

S. J. Kalayathankal
Department of Mathematics, Kuriakose Elias College, Mannanam, Kerala, India
e-mail: sunnyjoseph2014@yahoo.com

P. Suresh et al. (eds.), *Advances in Smart System Technologies*,
Advances in Intelligent Systems and Computing 1163,
https://doi.org/10.1007/978-981-15-5029-4_40

2 Motivation

Modeling with uncertainty can be solved in the best way with fuzzy sets. How to define a membership function of a pentadecagonal fuzzy number (PDFN) is the challenge before the decisionmaker. Here, we attempt to characterize PDFN which can be a superior decision of a leader in various circumstances.

3 Preliminaries

Definition: Fuzzy number is a normalized set of real line, which is convex in nature. Membership value of each element lies in the interval [0,1].

Definition: Arithmetic FNs are represented as $AA_{\overline{PD}}$and $AB_{\overline{PD}}$. Linguistic variable (LV) is converted to pentadecagonal FNs $\left[t_i, t_j\right]$ and $[t_k, t_l]$ is denoted by $LA_{\overline{PD}}$and $LB_{\overline{PD}}$.

Definition: If $T = t_0, t_1 \ldots, t_g$ is finite, ordered set and if it is with odd linguistic terms (LT) where s_i implies ith linguistic term (LT), $i \in 0,\ 1,\ 2 \ldots g$ then T can be considered as LT set. $(g + 1)$ will be the cardinality of T. The set T should have the properties which are as follows

- Ordered set: $t_i \geq t_j$ if $i > j$
- Negation operator: $\neg(t_i) = t_j$ where $j = g - 1$
- There is a maximization operator: max $\left(t_i, t_j\right) = t_i$ if $t_i \geq t_j$
- Minimization operator: min $\left(t_i, t_j\right) = t_i$ if $t_i \geq t_j$

Let $\overline{T} = t_l, t_{l+1}, \ldots, t_u$ where $t_l, t_{l+1}, \ldots., t_u \in T, t_l \geq t_u, t_l$ is the lower limit, t_u is the upper limit, l, $u \in 0$, 1, …,g, respectively. Then, $\overline{T}$ can be considered as a linguistic term which is uncertain.

4 Pentadecagonal Fuzzy Number (PDFN)

Definition

A PDFN $A_{\overline{PD}} = (a_{*1},\ a_{*2},\ a_{*3},\ a_{*4},\ a_{*5},\ a_{*6},\ a_{*7},\ a_{*8},\ a_{*9},\ a_{*10},\ a_{*11},\ a_{*12},\ a_{*13},\ a_{*14},\ a_{*15})$ should satisfy the following conditions

(1) $\mu_{A_{\overline{PD}}}(x)$ is a continuous in [0, 1]
(2) $\mu_{A_{\overline{PD}}}(x)$ is continuous and strictly increasing on $[a_{*1}, a_{*2}]$, $[a_{*2}, a_{*3}]$, $[a_{*3}, a_{*4}]$, $[a_{*4}, a_{*5}]$,$[a_{*5}, a_{*6}]$, $[a_{*6}, a_{*7}]$, $[a_{*7}, a_{*8}]$.
(3) $\mu_{A_{\overline{PD}}}(x)$ is continuous and strictly decreasing on $[a_{*8}, a_{*9}]$, $[a_{*9}, a_{*10}]$, $[a_{*10}, a_{*11}]$, $[a_{*11}, a_{*12}]$, $[a_{*12}, a_{*13}]$, $[a_{*13}, a_{*14}]$, $[a_{*14}, a_{*15}]$.

The PDFN can be denoted as

$A_{\overline{PD}} = (a_{*1}, a_{*2}, a_{*3}, a_{*4}, a_{*5}, a_{*6}, a_{*7}, a_{*8}, a_{*9}, a_{*10}, a_{*11}, a_{*12}, a_{*13}, a_{*14}, a_{*15})$ where

$a_{*1}, a_{*2}, a_{*3}, a_{*4}, a_{*5}, a_{*6}, a_{*7}, a_{*8}, a_{*9}, a_{*10}, a_{*11}, a_{*12}, a_{*13}, a_{*14}, a_{*15}$ are real numbers. The membership function is as follows.

$$\mu_{A_{\overline{PD}}}(x) = \begin{cases} \frac{1}{7}\left(\frac{x-a_{11}}{a_{*2}-a_{*1}}\right) & a_{*1} \le x \le a_{*2}. \\ \frac{1}{7}+\frac{1}{7}\left(\frac{x-a_{*2}}{a_{*3}-a_{*2}}\right) & a_{*2} \le x \le a_{*3}. \\ \frac{2}{7}+\frac{1}{7}\left(\frac{x-a_{*3}}{a_{*4}-a_{*3}}\right) & a_{*3} \le x \le a_{*4}. \\ \frac{3}{7}+\frac{1}{7}\left(\frac{x-a_{*4}}{a_{*5}-a_{*4}}\right) & a_{*4} \le x \le a_{*5}. \\ \frac{4}{7}+\frac{1}{7}\left(\frac{x-a_{*5}}{a_{*6}-a_{*5}}\right) & a_{*5} \le x \le a_{*6}. \\ \frac{5}{7}+\frac{1}{7}\left(\frac{x-a_{*6}}{a_{*7}-a_{*6}}\right) & a_{*6} \le x \le a_{*7}. \\ \frac{6}{7}+\frac{1}{7}\left(\frac{x-a_{*7}}{a_{*8}-a_{*7}}\right) & a_{*7} \le x \le a_{*8}. \\ 1-\frac{1}{7}\left(\frac{x-a_{*8}}{a_{*9}-a_{*8}}\right) & a_{*8} \le x \le a_{*9}. \\ \frac{6}{7}-\frac{1}{7}\left(\frac{x-a_{*9}}{a_{*10}-a_{*9}}\right) & a_{*9} \le x \le a_{*10}. \\ \frac{5}{7}-\frac{1}{7}\left(\frac{x-a_{*10}}{a_{*11}-a_{*10}}\right) & a_{*10} \le x \le a_{*11}. \\ \frac{4}{7}-\frac{1}{7}\left(\frac{x-a_{*11}}{a_{*12}-a_{*11}}\right) & a_{*11} \le x \le a_{*12}. \\ \frac{3}{7}-\frac{1}{7}\left(\frac{x-a_{*12}}{a_{*13}-a_{*12}}\right) & a_{*12} \le x \le a_{*13}. \\ \frac{2}{7}-\frac{1}{7}\left(\frac{x-a_{*13}}{a_{*14}-a_{*13}}\right) & a_{*13} \le x \le a_{*14}. \\ \frac{1}{7}-\frac{1}{7}\left(\frac{x-a_{*15}}{a_{*15}-a_{*14}}\right) & a_{*14} \le x \le a_{*15}. \\ 0 & x > a_{*15}. \end{cases} \tag{1}$$

where $0 < k_1 < k_2 < 1$(Table 1).

Table 1 Linguistic table of pendecagonal fuzzy number

LT	LV
No progress	(0, 0, 0, 0, 0, 0, 0, 0.03, 0.06, 0.09, 0.12, 0.15, 0.18, 0.21, 0.24)
Very low progress	(0.20, 0.22, 0.24, 0.26, 0.28, 0.30, 0.32, 0.34, 0.36, 0.38, 0.40, 0.42, 0.44, 0.46, 0.48)
Low progress	(0.31, 0.33, 0.36, 0.39, 0.42, 0.45, 0.48, 0.51, 0.54, 0.57, 0.60, 0.63, 0.66, 0.69, 0.70)
Medium	(0.40, 0.43, 0.46, 0.49, 0.52, 0.55, 0.58, 0.61, 0.64, 0.67, 0.70, 0.73, 0.76, 0.79, 0.80)
High progress	(0.60, 0.62, 0.64, 0.66, 0.68, 0.70, 0.72, 0.74, 0.76, 0.78, 0.80, 0.82, 0.84, 0.86, 0.88)
Very high progress	(0.60, 0.64, 0.68, 0.72, 0.76, 0.80, 0.84, 0.88, 0.92, 0.96, 1, 1, 1, 1, 1)

Definition: Equality of two PDFN

Two PDFN $A_{\overline{PD}} = (a_{*1},\ a_{*2},\ a_{*3},\ a_{*4},\ a_{*5},\ a_{*6},\ a_{*7},\ a_{*8},\ a_{*9},\ a_{*10},\ a_{*11},\ a_{*12},\ a_{*13},\ a_{*14},\ a_{*15})$

$B_{\overline{\text{PD}}} = (b_{*1}, b_{*2}, b_{*3}, b_{*4}, b_{*5}, b_{*6}, b_{*7}, b_{*8}, b_{*9}, b_{*10}, b_{*11}, b_{*12}, b_{*13}, b_{*14}, b_{*15})$ are equal if $a_{*1} = b_{*1}, a_{*2} = b_{*2}, a_{*3} = b_{*3}, a_{*4} = b_{*4}, a_{*5} = b_{*5}, a_{*6} = b_{*6}, a_{*7} = b_{*7}, a_{*8} = b_{*8}, a_{*9} = b_{*9}, a_{*10} = b_{*10}, a_{*11} = b_{*11}, a_{*12} = b_{*12}, a_{*13} = b_{*13}, a_{*14} = b_{*14}, a_{*15} = b_{*15}$

5 Operations on PDFNs

Consider $G_{\overline{\text{PD}}}$ and $H_{\overline{\text{PD}}}$ be two linguistic terms. Let them be arbitrary and uncertain in nature.

$G_{\overline{\text{PD}}} = (a_{*1}, a_{*2}, a_{*3}, a_{*4}, a_{*5}, a_{*6}, a_{*7}, a_{*8}, a_{*9}, a_{*10}, a_{*11}, a_{*12}, a_{*13}, a_{*14}, a_{*15})$ and $H_{\overline{\text{PD}}} = (b_{*1}, b_{*2}, b_{*3}, b_{*4}, b_{*5}, b_{*6}, b_{*7}, b_{*8}, b_{*9}, b_{*10}, b_{*11}, b_{*12}, b_{*13}, b_{*14}, b_{*15})$ be their corresponding PDFNs. The various operations on PDFNs are as follows

Addition

$$\begin{aligned} G_{\overline{\text{PD}}} + H_{\overline{\text{PD}}} = (&a_{*1} + b_{*1}, a_{*2} + b_{*2}, a_{*3} + b_{*3}, a_{*4} + b_{*4}, a_{*5} + b_{*5}, a_{*6} + b_{*6},\\ &a_{*7} + b_{*7}, a_{*8} + b_{*8}, a_{*9} + b_{*9}, a_{*10} + b_{*10}, a_{*11} + b_{*11}, a_{*12} + b_{*12},\\ &a_{*13} + b_{*13}, a_{*14} + b_{*14}, a_{*15} + b_{*15}) \end{aligned}$$

Subtraction

$$\begin{aligned} G_{\overline{\text{PD}}} - H_{\overline{\text{PD}}} = (&a_{*1} - b_{*15}, a_{*2} - b_{*14}, a_{*3} - b_{*13}, a_{*4} - b_{*12}, a_{*5} - b_{*11}, a_{*6} - b_{*10},\\ &a_{*7} - b_{*9}, a_{*8} - b_{*8}, a_{*9} - b_{*7}, a_{*10} - b_{*6}, a_{*11} - b_{*5}, a_{*12} - b_{*4},\\ &a_{*13} - b_{*3}, a_{*14} - b_{*2}, a_{*15} - b_{*1}) \end{aligned}$$

Multiplication

$$\begin{aligned} G_{\overline{\text{PD}}} \times H_{\overline{\text{PD}}} = (&a_{*1} \times b_{*1}, a_{*2} \times b_{*2}, a_{*3} \times b_{*3}, a_{*4} \times b_{*4}, a_{*5} \times b_{*5}, a_{*6} \times b_{*6}, a_{*7}\\ &\times b_{*7}, a_{*8} \times b_{*8}, a_{*9} \times b_{*9}, a_{*10} \times b_{*10}, a_{*11} \times b_{*11}, a_{*12}\\ &\times b_{*12}, a_{*13} \times b_{*13}, a_{*14} \times b_{*14}, a_{*15} \times b_{*15}) \end{aligned}$$

Division

$$\frac{G_{\overline{\text{PD}}}}{H_{\overline{\text{PD}}}} = \left(\frac{a_{*1}}{b_{*1}}, \frac{a_{*2}}{b_{*2}}, \frac{a_{*3}}{b_{*3}}, \frac{a_{*4}}{b_{*4}}, \frac{a_{*5}}{b_{*5}}, \frac{a_{*6}}{b_{*6}}, \frac{a_{*7}}{b_{*7}}, \frac{a_{*8}}{b_{*8}}, \frac{a_{*9}}{b_{*9}}, \frac{a_{*10}}{b_{*10}},\right.$$

$$\frac{a_{*11}}{b_{*11}}, \frac{a_{*12}}{b_{*12}}, \frac{a_{*13}}{b_{*13}}, \frac{a_{*14}}{b_{*14}}, \frac{a_{*15}}{b_{*15}}\Bigg)$$

6 Numerical Example

Let $\begin{array}{l} LG_{\overline{PD}} = (0.20,\ 0.22,\ 0.24,\ 0.26,\ 0.28,\ 0.30,\ 0.32, \\ \quad 0.34,\ 0.36,\ 0.38, 0.0.40,\ 0.42,\ 0.44,\ 0.46,\ 0.48) \end{array}$ and $\begin{array}{l} LH_{\overline{PD}} = (0.40, 0.43, 0.46, 0.49, 0.52, 0.55, 0.58, 0.61, 0.64, \\ \quad 0.67, 0.70, 0.73, 0.76, 0.79, 0.82) \end{array}$ be two LV transform into PDFN then various arithmetic operations are as follows.

Addition

$$LG_{\overline{PD}} + LH_{\overline{PD}} = (0.6, 0.65, 0.7, 0.75, 0.8, 0.85, 0.9, 0.95, \\ 1, 1.05, 1.1, 1.15, 1.2, 1.25, 1.3)$$

Subtraction

$$LG_{\overline{PD}} - LH_{\overline{PD}} = (-0.62, -0.57, -0.52, -0.47, -0.42, -0.37, -0.32, \\ -0.27, -0.22, -0.17, -0.12, -0.07, -0.02, 0.03, 0.08)$$

Multiplication

$$LG_{\overline{PD}} \times LH_{\overline{PD}} = (0.08,\ 0.0946,\ 0.1104,\ 0.1274,\ 0.1456,\ 0.165,\ 0.1856,\ 0.2074, \\ 0.2304,\ 0.2546,\ 0.28,\ 0.3066,\ 0.334,\ 0.363,\ 0.393)$$

Division

$$\frac{LG_{\overline{PD}}}{LH_{\overline{PD}}} = (0.5, 0.511628, 0.521739, 0.530612, 0.538462, 0.545455, \\ 0.551724, 0.557377, 0.5625, 0.567164, 0.571429, \\ 0.575342, 0.0.578947, 0.582278, 0.585365)$$

Definition

A PDFN $\overline{PD}$ can be defined in another way also.

$P_1(a), Q_1(b), R_1(c), S_1(d), T_1(e), U_1(f), V_1(g), P_2(a), Q_2(b), R_2(c), S_2(d), T_2(e), U_2(f), V_2(g)\ a \in [0, 0.14285714],$
$b \in [0.14285714, 0.28571429],\ c \in [0.28571429, 0.42857143],$ and $g \in$
$d \in [0.42857143, 0.57142857],\ e \in [0.57142857, 0.71428571],$
$f \in [0.71428571, 0.85714286]$
$[0.85714286, 1]$ where

$$\begin{array}{ll}
P_1(a) = \frac{1}{7}\left(\frac{x-a_{*1}}{a_{*2}-a_{*1}}\right) & P_2(a) = \frac{1}{7}\left(\frac{x-a_{*15}}{a_{*15}-a_{*14}}\right) \\
Q_1(b) = \frac{1}{7} + \frac{1}{7}\left(\frac{x-a_{*2}}{a_{*3}-a_{*2}}\right) & Q_2(b) = \frac{2}{7} - \frac{1}{7}\left(\frac{x-a_{*13}}{a_{*14}-a_{*13}}\right) \\
R_1(c) = \frac{2}{7} + \frac{1}{7}\left(\frac{x-a_{*3}}{a_{*4}-a_{*3}}\right) & R_2(c) = \frac{3}{7} - \frac{1}{7}\left(\frac{x-a_{*12}}{a_{*13}-a_{*12}}\right) \\
S_1(d) = \frac{3}{7} + \frac{1}{7}\left(\frac{x-a_{*4}}{a_{*5}-a_{*4}}\right) & S_2(d) = \frac{4}{7} - \frac{1}{7}\left(\frac{x-a_{*11}}{a_{*12}-a_{*11}}\right) \\
T_1(e) = \frac{4}{7} + \frac{1}{7}\left(\frac{x-a_{*5}}{a_{*6}-a_{*5}}\right) & T_2(e) = \frac{5}{7} - \frac{1}{7}\left(\frac{x-a_{*10}}{a_{*11}-a_{*10}}\right) \\
U_1(f) = \frac{5}{7} + \frac{1}{7}\left(\frac{x-a_{*6}}{a_{*7}-a_{*6}}\right) & U_2(f) = \frac{6}{7} - \frac{1}{7}\left(\frac{x-a_{*9}}{a_{*10}-a_{*9}}\right) \\
V_1(g) = \frac{6}{7} + \frac{1}{7}\left(\frac{x-a_{*7}}{a_{*8}-a_{*7}}\right) & V_2(g) = 1 - \frac{1}{7}\left(\frac{x-a_{*8}}{a_{*9}-a_{*8}}\right)
\end{array}$$

Definition of PDFN with α cut

The α-cut is stated as $\overline{\mathrm{PD}} = \left\{x \in X/\mu_{A_{\overline{\mathrm{PD}}}}(x) \geq \alpha\right\}$ where $\alpha \in [0, 1]$, $P_1(a), Q_1(b), R_1(c), S_1(d), T_1(e), U_1(f), V_1(g)$ is finite and confined. It is increasing on [0,0.14285714], [0.14285714,0.28571429], [0.28571429,0.42857143], [0.42857143,0.5714285], [0.57142857,0.71428571], [0.71428571,0.85714286], and [0.85714286,1]. In this interval, the function is continuous also $P_2(a), Q_2(b), R_2(c), S_2(d), T_2(e), U_2(f), V_2(g)$ is finite and confined. It is decreasing on [0,0.14285714], [0.14285714,0.28571429], [0.28571429,0.42857143], [0.42857143,0.57142857], [0.57142857,0.71428571], [0.71428571,0.85714286], and [0.85714286,1] .Here, it is continuous also.

$$\overline{\mathrm{PD}}_\alpha = \begin{cases} [P_1(\alpha), P_2(\alpha)], & \alpha \in [0, 0.14285714) \\ [Q_1(\alpha), Q_2(\alpha)], & \alpha \in [0.14285714, 0.28571429) \\ [R_1(\alpha), R_2(\alpha)], & \alpha \in [0.28571429, 0.42857143) \\ [S_1(\alpha), s_2(\alpha)], & \alpha \in [0.42857143, 0.57142857) \\ [T_1(\alpha), T_2(\alpha)], & \alpha \in [0.57142857, 0.71428571) \\ [U_1(\alpha), u_2(\alpha)], & \alpha \in [0.71428571, 0.85714286) \\ [V_1(\alpha), v_2(\alpha)], & \alpha \in [0.85714286, 1) \end{cases} \tag{2}$$

Definition: If $P_1(x) = \alpha$ and $P_2(x) = \alpha$, then operations on α-cut can be obtained as $\begin{array}{l}[7\alpha(a_{*2} - a_{*1}) + a_{*1}, 7\alpha(a_{*15} - a_{*14}) + a_{*14}]+ \\ [7\alpha(b_{*2} - b_{*1}) + b_{*1}, 7\alpha(b_{*15} - b_{*14}) + b_{*14}]\end{array}$.

In the same way, we can obtain the operations on α-cut for $[Q_1(\alpha), Q_2(\alpha)]$, $[R_1(\alpha), R_2(\alpha)]$, $[S_1(\alpha), S_2(\alpha)]$, $[T_1(\alpha), T_2(\alpha)]$, $[U_1(\alpha), U_2(\alpha)]$, $[V_1(\alpha), v_2(\alpha)]$.

$$\begin{aligned}\left[Q_1(\alpha), Q_2(\alpha)\right] = {} & [7\alpha(a_{*3} - a_{*2}) + 2a_{*2} - a_{*3}, -7\alpha(a_{*14} - a_{*13}) + 2a_{*14} - a_{*13}] \\ & + [7\alpha(b_{*3} - b_{*2}) + 2b_{*2} - b_{*3}, -7\alpha(b_{*14} - b_{*13})) + 2b_{*14} - a_{*13}], \\ & \alpha \in [0.14285714, 0.28571429)\end{aligned}$$

$$\begin{aligned}[R_1(\alpha), R_2(\alpha)] = {} & [7\alpha(a_{*4} - a_{*3}) - 2a_{*4} + 3a_{*3}, -7\alpha(a_{*13} - a_{*12}) + 3a_{*13} - 2a_{*12}] \\ & + [7\alpha(b_{*4} - b_{*3}) - 2b_{*4} + 3b_{*3}, -7\alpha(b_{*13} - b_{*12}) + 3b_{*13} \\ & - 2b_{*12}], \ \alpha \in [0.28571429, 0.42857143)\end{aligned}$$

$$\begin{aligned}[S1(\alpha), S2(\alpha)] = {} & [7\alpha a * 3 - a * 4 + 4a * 4 - 3a * 5, -7\alpha a * 12 - a * 11 - 3a \\ & * 11 + 4a * 12] + [7\alpha(b * 5 - b * 4) + 4b * 4 - 3b * 5, -7\alpha \\ & (b * 12 - b * 11) - 3b * 11 + 4b * 12], \\ & \alpha \in [0.42857143, 0.57142857)\end{aligned}$$

$$\begin{aligned}[T_1(\alpha), T_2(\alpha)] = {} & 7\alpha(a_{*6} - a_{*5}) - 4a_{*6} + 5a_{*5}, -7\alpha(a_{*11} - a_{*10}) + 5a_{*11} - 4a_{*10}] \\ & + [7\alpha(b_{*6} - b_{*5}) - 4b_{*6} + 5b_{*5} - 7\alpha(b_{*11} - b_{*10}) + 5b_{*11} - 4b_{*10}, \\ & \alpha \in [0.57142857, 0.71428571)\end{aligned}$$

$$\begin{aligned}[U_1(\alpha), U_2(\alpha)] = {} & 7\alpha(a_{*7} - a_{*6}) - 5a_{*7} + 6a_{*6}, -7\alpha(a_{*10} - a_{*9}) + 6a_{*10} - 5a_{*9}] \\ & + [7\alpha(b_{*7} - b_{*6}) - 5b_{*7} + 6b_{*6}, -7\alpha(b_{*10} - b_{*9}) + 6b_{*10} - 5b_{*9}, \\ & \alpha \in [0.71428571, 0.85714286)\end{aligned}$$

$$\begin{aligned}[V_1(\alpha), v_2(\alpha)] = {} & [7\alpha(a_{*8} - a_{*7}) - 6a_{*8} + 7a_{*7}, -7\alpha(a_{*9} - a_{*8}) - 6a_{*8} + 7a_{*9}] \\ & + [7\alpha(b_{*8} - b_{*7}) - 6b_{*8} + 7b_{*7}, -7\alpha(b_{*9} - b_{*8}) - 6b_{*8} + 7b_{*9}], \\ & \alpha \in [0.85714286, 1)\end{aligned}$$

Hence, PDFN in terms of α-cut is as follows

$$\overline{PD}_\alpha = \begin{cases} [7\alpha(a_{*2}-a_{*1})+a_{*1}, 7\alpha(a_{*15}-a_{*14})+a_{*14}]+[7\alpha(b_{*2}-b_{*1})+b_{*1}, \\ 7\alpha(b_{*15}-b_{*14})+b_{*14}], \alpha\in[0, 0.14285714) \\ [7\alpha(a_{*3}-a_{*2})+2a_{*2}-a_{*3}, -7\alpha(a_{*14}-a_{*13})+2a_{*14}-a_{*13}] \\ +[7\alpha(b_{*3}-b_{*2})+2b_{*2}-b_{*3}, \\ -7\alpha(b_{*14}-b_{*13})+2b_{*14}-a_{*13}], \alpha\in[0.14285714, 0.28571429) \\ [7\alpha(a_{*4}-a_{*3})-2a_{*4}+3a_{*3}, -7\alpha(a_{*13}-a_{*12})+3a_{*13}-2a_{*12}] \\ +[7\alpha(b_{*4}-b_{*3})-2b_{*4}+3b_{*3}, \\ -7\alpha(b_{*13}-b_{*12})+3b_{*13}-2b_{*12}], \alpha\in[0.28571429, 0.42857143) \\ [7\alpha(a_{*3}-a_{*4})+4a_{*4}-3a_{*5}, -7\alpha(a_{*12}-a_{*11})-3a_{*11}+4a_{*12}] \\ +[7\alpha(b_{*5}-b_{*4})+4b_{*4}-3b_{*5}, \\ -7\alpha(b_{*12}-b_{*11})-3b_{*11}+4b_{*12}], \alpha\in[0.42857143, 0.57142857) \\ [7\alpha(a_{*6}-a_{*5})-4a_{*6}+5a_{*5}, -7\alpha(a_{*11}-a_{*10})+5a_{*11}-4a_{*10}] \\ +[7\alpha(b_{*6}-b_{*5})-4b_{*6}+5b_{*5}, \\ -7\alpha(b_{*11}-b_{*10})+5b_{*11}-4b_{*10}], \alpha\in[0.57142857, 0.71428571) \\ [7\alpha(a_{*7}-a_{*6})-5a_{*7}+6a_{*6}, -7\alpha(a_{*10}-a_{*9})+6a_{*10}-5a_{*9}] \\ +[7\alpha(b_{*7}-b_{*6})-5b_{*7}+6b_{*6}, \\ -7\alpha(b_{*10}-b_{*9})+6b_{*10}-5b_{*9}], \alpha\in[0.71428571, 0.85714286) \\ \left[7\alpha\left(a_{(*8}-a_{*7}\right)-6a_{*8}+7a_{*7}, -7\alpha(a_{*9}-a_{*8})-6a_{*8}+7a_{*9}\right] \\ +[7\alpha(b_{*8}-b_{*7})-6b_{*8}+7b_{*7}, \\ -7\alpha(b_{*9}-b_{*8})-6b_{*8}+7b_{*9}], \alpha\in[0.85714286, 1) \end{cases} \tag{3}$$

7 New Operations on Pentadecagonal Fuzzy Number

Let $\mathrm{LG}_{\overline{\mathrm{PD}}} = (b_{*1}, b_{*2}, b_{*3}, b_{*4}, b_{*5}, b_{*6}, b_{*7}, b_{*8}, b_{*9}, b_{*10}, b_{*11}, b_{*12}, b_{*13}, b_{*14}, b_{*15})$ and

$LH_{\overline{PD}} = (c_{*1}, c_{*2}, c_{*3}, c_{*4}, c_{*5}, c_{*6}, c_{*7}, c_{*8}, c_{*9}, c_{*10}, c_{*11}, c_{*12}, c_{*13}, c_{*14}, c_{*15})$ be the corresponding PDFN for all $\alpha \in [0, 1]$.

Addition:

We can denote the addition operation of α-cut as $\mathrm{LG}_{\overline{\mathrm{PD}}} + \mathrm{LH}_{\overline{\mathrm{PD}}}$.

$$\mathrm{LG}_{\overline{\mathrm{PD}}} = (0.20,\ 0.22,\ 0.24,\ 0.26,\ 0.28,\ 0.30,\ 0.32,\ 0.34,\ 0.36,\ 0.38,\ 0.0.40,\ 0.42,\ 0.44,\ 0.46,\ 0.48)$$

$$\mathrm{LH}_{\overline{\mathrm{PD}}} = (0.40,\ 0.43,\ 0.46,\ 0.49,\ 0.52,\ 0.55,\ 0.58,\ 0.61,\ 0.64,\ 0.67,\ 0.70,\ 0.73,\ 0.76,\ 0.79,\ 0.82)$$

Here, interval remains the same for $\alpha \in$ [0,0.14285714),$\alpha \in$ [0.14285714,0.28571429), $\alpha \in$ [0.28571429,0.42857143), $\alpha \in$ [0.42857143,0.57142857), $\alpha \in$ [0.57142857,0.71428571), $\alpha \in$ [0.71428571,0.85714286), and $\alpha \in$ [0.85714286,1) .

$$\mathrm{LG}_\alpha + \mathrm{LH}_\alpha = (0.35\alpha + 0.6, -0.35\alpha + 1.28) \text{for all} \alpha \in [0, 1].$$

when $\alpha = 0, \mathrm{LG}_\alpha + \mathrm{LH}_\alpha = [0.6, 1.3]$
when $\alpha = 0.14285714, \mathrm{LG}_\alpha + \mathrm{LH}_\alpha = [0.65, 1.25]$
when $\alpha = 0.28571429, \mathrm{LG}_\alpha + \mathrm{LH}_\alpha = [0.7, 1.2]$
when $\alpha = 0.42857143, \mathrm{LG}_\alpha + \mathrm{LH}_\alpha = [0.75, 1.15]$
when $\alpha = 0.57142857, \mathrm{LG}_\alpha + \mathrm{LH}_\alpha = [0.8, 1.1]$
when $\alpha = 0.71428571, \mathrm{LG}_\alpha + \mathrm{LH}_\alpha = [0.85, 1.05]$
when $\alpha = 0.85714286, \mathrm{LG}_\alpha + \mathrm{LH}_\alpha = [0.9, 1]$
when $\alpha = 1, \mathrm{LG}_\alpha + \mathrm{LH}_\alpha = [0.95, 0.95]$

In the above calculations, we can see that the values lie in the specified limit. Therefore, we can conclude that pentadecagonal fuzzy number satisfies the property of addition.

Subtraction:

$\mathrm{LG}_{\overline{\mathrm{PD}}} = (b_{*1}, b_{*2}, b_{*3}, b_{*4}, b_{*5}, b_{*6}, b_{*7}, b_{*8}, b_{*9}, b_{*10}, b_{*11}, b_{*12}, b_{*13}, b_{*14}, b_{*15})$ and

$\mathrm{LH}_{\overline{\mathrm{PD}}} = (c_{*1}, c_{*2}, c_{*3}, c_{*4}, c_{*5}, c_{*6}, c_{*7}, c_{*8}, c_{*9}, c_{*10}, c_{*11}, c_{*12}, c_{*13}, c_{*14}, c_{*15})$ be the corresponding pentadecagonal fuzzy number for all $\alpha \in [0, 1]$. The subtraction operation of α-cut denoted as $\mathrm{LG}_{\overline{\mathrm{PD}}} - \mathrm{LH}_{\overline{\mathrm{PD}}}$.

$$\mathrm{LG}_{\overline{\mathrm{PD}}} = (0.20, 0.22, 0.24, 0.26, 0.28, 0.30, 0.32, 0.34, \\ 0.36, 0.38, 0.0.40, 0.42, 0.44, 0.46, 0.48)$$

$$\mathrm{LH}_{\overline{\mathrm{PD}}} = (0.40, 0.43, 0.46, 0.49, 0.52, 0.55, 0.58, 0.61, \\ 0.64, 0.67, 0.70, 0.73, 0.76, 0.79, 0.82)$$

For $\alpha \in$ [0,0.14285714), $\alpha \in$ [0.14285714,0.28571429), $\alpha \in$ [0.28571429,0.42857143), $\alpha \in$ [0.42857143,0.57142857), $\alpha \in$ [0.57142857,0.71428571), $\alpha \in$ [0.71428571,0.85714286), and $\alpha \in$ [0.85714286,1) arithmetic intervals are same.

$$\mathrm{LG}_\alpha - \mathrm{LH}_\alpha = (0.35\alpha - 0.62, -0.35\alpha + 0.08) \text{for all} \alpha \in [0, 1].$$

when $\alpha = 0, \mathrm{LG}_\alpha - \mathrm{LH}_\alpha = [-0.62, 0.08]$
when $\alpha = 0.14285714, \mathrm{LG}_\alpha - \mathrm{LH}_\alpha = [-0.57, 0.03]$
when $\alpha = 0.28571429, \mathrm{LG}_\alpha - \mathrm{LH}_\alpha = [-0.52, -0.02]$
when $\alpha = 0.42857143, \mathrm{LG}_\alpha - \mathrm{LH}_\alpha = [-0.47, -0.07]$
when $\alpha = 0.57142857, \mathrm{LG}_\alpha - \mathrm{LH}_\alpha = [-0.42, -0.12]$

when $\alpha = 0.71428571$, $\text{LG}_\alpha - \text{LH}_\alpha = [-0.37, -0.17]$
when $\alpha = 0.85714286$, $\text{LG}_\alpha - \text{LH}_\alpha = [-0.32, -0.22]$
when $\alpha = 1$, $\text{LG}_\alpha - \text{LH}_\alpha = [-0.27, -0.27]$

In the above calculations, we can see that the values lies in the specified limit. Therefore, we can conclude that PDFN satisfies the property of subraction.

Multiplication

$\text{LG}_{\overline{\text{PD}}} = (b_{*1}, b_{*2}, b_{*3}, b_{*4}, b_{*5}, b_{*6}, b_{*7}, b_{*8}, b_{*9}, b_{*10}, b_{*11}, b_{*12}, b_{*13}, b_{*14}, b_{*15})$ and

$\text{LH}_{\overline{\text{PD}}} = (c_{*1}, c_{*2}, c_{*3}, c_{*4}, c_{*5}, c_{*6}, c_{*7}, c_{*8}, c_{*9}, c_{*10}, c_{*11}, c_{*12}, c_{*13}, c_{*14}, c_{*15})$ be the corresponding PDFN for all α ∈ [0, 1]. The multiplication operation of α-cut denoted as $\text{LG}_{\overline{\text{PD}}} \times \text{LH}_{\overline{\text{PD}}}$.

$$\text{LG}_{\overline{\text{PD}}} = (0.20, 0.22, 0.24, 0.26, 0.28, 0.30, 0.32, 0.34, 0.36, 0.38, 0.0.40, 0.42, 0.44, 0.46, 0.48)$$

$$\text{LH}_{\overline{\text{PD}}} = (0.40, 0.43, 0.46, 0.49, 0.52, 0.55, 0.58, 0.61, 0.64, 0.67, 0.70, 0.73, 0.76, 0.79, 0.82)$$

For $\alpha\in$ [0,0.14285714),$\alpha\in$ [0.14285714,0.28571429),$\alpha\in$ [0.28571429,0.42857143), $\alpha\in$ [0.42857143,0.57142857),$\alpha\in$ [0.57142857,0.71428571),$\alpha\in$ [0.71428571,0.85714286), and α∈ [0.85714286,1) arithmetic intervals are same.

$$\text{LG}_\alpha \times \text{LH}_\alpha = \left(0.0294\alpha^2 + 0.098\alpha + 0.08, 0.0294\alpha^2 - 0.2156\alpha + 0.3936\right) \text{ for all } \alpha \in [0, 1].$$

when $\alpha = 0$, $\text{LG}_\alpha \times \text{LH}_\alpha = [0.08, 0.3936]$
when $\alpha = 0.14285714$, $\text{LG}_\alpha \times \text{LH}_\alpha = [0.0946, 0.3634]$
when $\alpha = 0.28571429$, $\text{LG}_\alpha \times \text{LH}_\alpha = [0.1104, 0.3344]$
when $\alpha = 0.42857143$, $\text{LG}_\alpha \times \text{LH}_\alpha = [0.1274, 0.3066]$
when $\alpha = 0.57142857$, $\text{LG}_\alpha \times \text{LH}_\alpha = [0.1456, 0.28]$
when $\alpha = 0.71428571$, $\text{LG}_\alpha \times \text{LH}_\alpha = [0.165, 0.2546]$
when $\alpha = 0.85714286$, $\text{LG}_\alpha \times \text{LH}_\alpha = [0.1856, 0.2304]$
when $\alpha = 1$, $\text{LG}_\alpha \times \text{LH}_\alpha = [0.2074, 0.2074]$

In the above calculations, we can see that the values lies in the specified limit. Therefore, we can conclude that PDFN satisfies the property of multiplication.

Division

$$\text{LG}_{\overline{\text{PD}}} = (b_{*1}, b_{*2}, b_{*3}, b_{*4}, b_{*5}, b_{*6}, b_{*7}, b_{*8}, b_{*9}, b_{*10}, b_{*11}, b_{*12}, b_{*13}, b_{*14}, b_{*15}) \text{ and}$$

$\mathrm{LH}_{\overline{\mathrm{PD}}} = (c_{*1}, c_{*2}, c_{*3}, c_{*4}, c_{*5}, c_{*6}, c_{*7}, c_{*8}, c_{*9}, c_{*10}, c_{*11}, c_{*12}, c_{*13}, c_{*14}, c_{*15})$ be the corresponding pentadecagonal fuzzy number for all $\alpha \in [0, 1]$. The division operation of α-cut denoted as $\frac{\mathrm{LG}_{\overline{\mathrm{PD}}}}{\mathrm{LH}_{\overline{\mathrm{PD}}}}$.

$$\mathrm{LG}_{\overline{\mathrm{PD}}} = (0.20,\ 0.22,\ 0.24,\ 0.26,\ 0.28,\ 0.30,\ 0.32,\ 0.34,\ 0.36,\ 0.38,\ 0.0.40,\ 0.42,\ 0.44,\ 0.46,\ 0.48)$$

$$\mathrm{LH}_{\overline{\mathrm{PD}}} = (0.40,\ 0.43,\ 0.46,\ 0.49,\ 0.52,\ 0.55,\ 0.58,\ 0.61,\ 0.64,\ 0.67,\ 0.70,\ 0.73,\ 0.76,\ 0.79,\ 0.82)$$

For $\alpha\in$ [0,0.14285714), $\alpha\in$ [0.14285714,0.28571429), $\alpha\in$ [0.28571429,0.42857143), $\alpha\in$ [0.42857143,0.57142857), $\alpha\in$ [0.57142857,0.71428571), $\alpha\in$ [0.71428571,0.85714286), and $\alpha\in$ [0.85714286,1) arithmetic intervals are same.

$$\frac{\mathrm{LG}_{\overline{\mathrm{PD}}}}{\mathrm{LH}_{\overline{\mathrm{PD}}}} = \left(\frac{0.14\alpha + 0.20}{0.21\alpha + 0.4}, \quad \frac{-0.14\alpha + 0.48}{-0.21\alpha + 0.82}\right) \text{for all } \alpha \in [0, 1].$$

when $\alpha = 0$, $\frac{\mathrm{LG}_{\overline{\mathrm{PD}}}}{\mathrm{LH}_{\overline{\mathrm{PD}}}} = [0.5, 0.5853]$

when $\alpha = 0.14285714$, $\frac{\mathrm{LG}_{\overline{\mathrm{PD}}}}{\mathrm{LH}_{\overline{\mathrm{PD}}}} = [0.5116, 0.5822]$

when $\alpha = 0.28571429$, $\frac{\mathrm{LG}_{\overline{\mathrm{PD}}}}{\mathrm{LH}_{\overline{\mathrm{PD}}}} = [0.5217, 0.5789]$

when $\alpha = 0.42857143$, $\frac{\mathrm{LG}_{\overline{\mathrm{PD}}}}{\mathrm{LH}_{\overline{\mathrm{PD}}}} = [0.5306, 0.5753]$

when $\alpha = 0.57142857$, $\frac{\mathrm{LG}_{\overline{\mathrm{PD}}}}{\mathrm{LH}_{\overline{\mathrm{PD}}}} = [0.5384, 0.5714]$

when $\alpha = 0.71428571$, $\frac{\mathrm{LG}_{\overline{\mathrm{PD}}}}{\mathrm{LH}_{\overline{\mathrm{PD}}}} = [0.5454, 0.5671]$

when $\alpha = 0.85714286$, $\frac{\mathrm{LG}_{\overline{\mathrm{PD}}}}{\mathrm{LH}_{\overline{\mathrm{PD}}}} = [0.5517, 0.5625]$

when $\alpha = 1$, $\frac{\mathrm{LG}_{\overline{\mathrm{PD}}}}{\mathrm{LH}_{\overline{\mathrm{PD}}}} = [0.5573, 0.5573]$

In the above calculations, we can see that the values lie in the specified limit. Therefore, we can conclude that pentadecagonal fuzzy number satisfies the property of division.

8 Applications

In real-life applications, few parameters will be imprecise. For example, time, value, distance, supply, and so on. These parameters are represented using fuzzy numbers to get the optimum solution. Fuzzy model provides exactness, similarity, and adaptability which are vital in real-life application due to recent advancement in neural network, artificial intelligence, supply chain management, etc.

9 Conclusion

We have introduced pentadecagonal fuzzy number. Here, we present the arithmetic operations on pentadecagonal fuzzy number with example. Also, we have discussed the properties of pentadecagonal fuzzy number using alpha-cut.

References

1. Dutta, P., Boruah, H., Ali, T.: Fuzzy arithmetic with and without using α-cut method: A comparative study. Int. J. Latest Trends Comput. **2**(1) (2011) (E-ISSN: 2045-5364)
2. Felix, A., Christopher, S., Victor Devadoss, A.: A nonagonal fuzzy number and its arithmetic operation. Int. J. Math. Appl. **3**(2), 185–195 (2015)
3. Victor Devadoss, A., Dhivya, A.D., Felix, A.: A hendecagonal fuzzy number and its vertex method. Int. J. Math. Appl. **4**(1), 87–98 (2016)
4. Dubois, D., Prade, H.: Operations on fuzzy numbers. Int. J. Syst. Sci. **9**(6), 613–626 (1978)
5. Felix, A., Victor Devadoss, A.: A new decagonal fuzzy number under uncertain linguistic environment. Int. J. Math. Appl. **3**(1), 89–97 (2015)
6. Kauffman, A.: Introduction to fuzzy arithmetic: Theory and application. Van Nostrand reinhold, New york (1980)
7. Rajkumar, A., Helen, D.: A new arithmetic operations of Triskaidecagonal fuzzy number using alpha cut. Adv. Intell. Syst. Comput. Soft comput. Theories Appl. **583** (2016). ISBN: 978-981-10-5686- 4

Weed Identification in Agriculture Field Through IoT

P. Karthikeyan, M. Manikandakumar, D. K. Sri Subarnaa, and P. Priyadharshini

Abstract IoT is an evolving technology in recent days, and it will represent the future of communication and computing. Every people will be conditional on agriculture. Due to this, the dependency of smart farming in agriculture with the use of IT technologies is needed to improve our traditional agriculture process. By using this smart technology in agriculture field, it is able to control the pesticides and maintain a precision agriculture by this monitoring performance. Raspberry Pi and its camera module play a vital role in IoT-based smartweed identification in modern agriculture. This automatic weed identification system allows detecting the minute weedy plants in an agriculture field. This monitoring is being done under precession using real sensors, and also, it provides the segmentation of plant. For this purpose, we use convolutional neural networks (CNNs) and precision method and are used to perform crop and weed identification which is used to classify the weeds among grass and broadleaf, so here we aim to identify the weed and apply the specific herbicide to weed detected plants.

Keywords Raspberry pi · Pi camera · K-means clustering · Support vector mechanism · CNN algorithm

1 Introduction

By using this precision agriculture, it provides a higher impact on agriculture with a low-cost input and leads to depletion in environmental pollution. In this modern technology, the precision agriculture and food production are anticipated to increase the utilization of the latest technologies and sensors in field. According to this, support vector mechanism is used to identify the weed plant from the agriculture

P. Karthikeyan (✉) · M. Manikandakumar · D. K. Sri Subarnaa · P. Priyadharshini
Department of Information Technology, (CS-IS), Thiagarajar College of Engineering, Madurai 625010, India
e-mail: pkarthikeyantce@gmail.com

D. K. Sri Subarnaa
e-mail: subarnaa1995@gmail.com

P. Suresh et al. (eds.), *Advances in Smart System Technologies*,
Advances in Intelligent Systems and Computing 1163,
https://doi.org/10.1007/978-981-15-5029-4_41

field through machine learning algorithm and these are trained previously with the expert knowledge which is gained from farmer's data report from their agricultural management. It is designed in such a way which helps the farmers to take right decision on right time for their agriculture field in order to prevent from pesticide in their agriculture. It is used with sustainable system for monitoring the basic activities such as a soil moisture, plant growth, pests, and measuring the diseases and software tools that are used to analyze these data which is gathered from the farmers, and it will provide alerts to the farmers. Based on the alert message, the farmers to take decision based on the information provided by the services.

Many researchers have been undergone in the past few years in sequence to detect this crop and weed plant from the agriculture field through image processing with IoT systems. Martin et al. [1] proposed a system which is used to identify the weed species using their texture and by shape feature by machine learning algorithm. By using some of wireless sensor networks (WSNs) in field, it will increase the productivity, efficiency, and profit to their agricultural production [2]. By using this wireless network, we can able to know the real-time information about the agriculture field and this will gather all the information and sent to the end user from this information; the end user can take the decision based on the information required. WSNs are the main component for Internet of things (IoT) where different species information is collected and gathered to provide the best result through this form anywhere we can get to know the current work of our field through the Internet. By integrating the wireless sensor network with IoT, we can have the precision agriculture.

Area-specific Herbicides spraying has been developed in various systems in order to identify the infected part. By using these systems, the optical sensors named as photodiodes which is able to recognize the particular plants and soil by their texture and reflectance. The most famous ones are detected spray, weed seeker, and spray vision. These systems cannot differentiate between crop and weeds. Jeremy et al. [3] proposed a machine vision system for exactness sprayer. It differentiates between weed and plant using blob detection and Gabor filter. But, this system cannot detect the weed and pant in intra-row. There are some research work on potato crop and wavelet sub image techniques in weed detection [9–10].

2 Four Layers IoT Agriculture Architecture

A hypothetical model for precision agriculture is to classify the weed plant and is proposed by analyzing the machine learning algorithm through image processing technique associated with IoT. The comprehensive structure of IoT will compromise many material devices, and this will have three-layer structure.

The top layer is to integrate with the administration layer in agriculture, and it is connected to IoT application because it is connected with the user interface layer. By connecting with user interface layer, it makes it easier for farmer's where they get the information related to their agriculture field. According to this principle, it

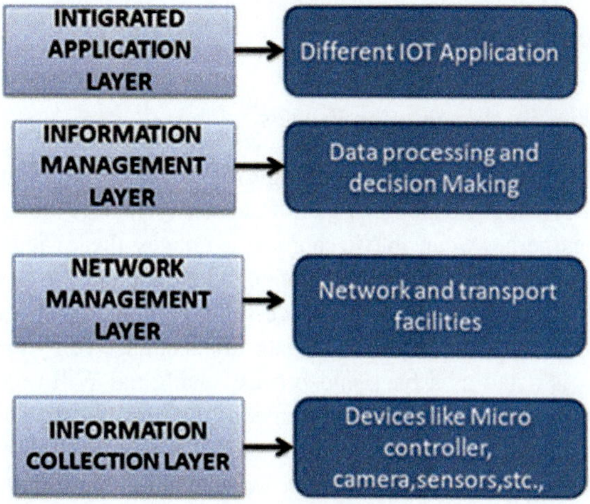

Fig. 1 Layers of IoT architecture

makes the farmers can take right decision at right time in order to protect their crop from without affecting any crop and make it as healthy for better food production. The second layer is information management layer where it will gain information about the current status of field; it is maintained by continuous scanning, and it will classify data and take decision. The third layer is network management layer which is used for transmission purpose here the technologies like Wi-Fi, GSM, RFID, and Bluetooth, Zigbee can be used. The fourth layer is information collection layer which will collect all the information from the camera and sensor and transmit to the storage area, and this collected information is used to classification purpose by the MATLAB GUI where the data is gained from agriculture area. Figure 1 shows the four-layer IoT structure [4].

Each device in the network will assign a separate IP address for identification purpose. For instance, suppose if we use a temperature sensor to identify the temperature of our surrounding, then in the network the address will be created with this the device ID T1 will use for the communication network. For this addressing, IPV6 and IPv4 will be useful. By using these identification methods, we can gain a clear result for each item within the network. IoT sensors will have been used as smart sensors for the accurate prediction of the environment by wireless sensing devices. These observed data's from agriculture field are sent to a cloud storage through gateway process where this gateway is connected to the Internet. From the cloud, the data are again sending to farmer's computer or smartphone. This received data is viewed in analyzed manner, and also, it provides some decision to the farmers. By the gained data, the farmers will take right decision [4].

3 Implementation

1. **IoT Hardware Implementation Process**

The prototype of this proposed methodology for IoT agriculture applications has been performed by using Raspberry Pi and Pi camera which is illustrated in Fig. 2. By using this Pi camera, we can able to capture the images; these captured images are stored in a cloud storage, and the storage image is done for analysis process; it is sent to the user side for the service. After this process, it is computed with a single gateway using nRF24L01 wireless interfaces. The gateway is connected by the Internet, and the cloud server is acted as a back-end process, using the Wi-Fi technology. The gateway collects data from the field and transmits the sensitive data to the cloud for performing data analytics. The back-end server will receive the data from Pi, stores the data, performs data analytics, and creates visualization representation for easier data elucidation.

The back-end is accountable for ease use of the end-user ability of accessing the stored data. This is achieved by implementing several resources like data visualization, data analytics, data storage, and in addition to providing a software tools and appropriate application program interface (API) through which it can reach the other end so that they can access the data. In our proposed architecture, we implement the back-end layer via the cloud-based server which is shown in Fig. 3. Here in this cloud storage, it is used as Dropbox cloud storage. Once it is stored in cloud database, we can view our captured image in the database which is shown in Fig. 4

2. **Binary Classification**

The images between the crops and weed are developed in MATLAB GUI [5]. Here carrot image dataset is used to classify the image in order to find the weed plant. The digital images is computerized image, and these images are stored in a type of 24-bit color pixel format which will provide a high resolutions of 1920 × 1080 pixels which is standardized computer format saved in RGB color space in the JPEG or in PNG

Fig. 2 Pi setup with camera module

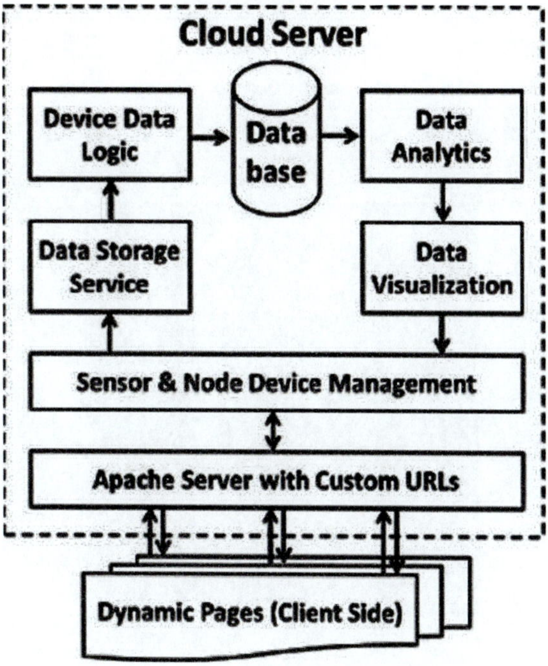

Fig. 3 Cloud server architecture

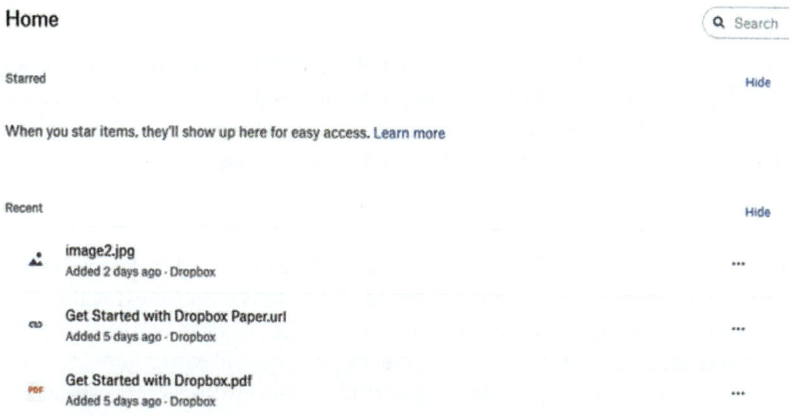

Fig. 4 Uploaded image in dropbox

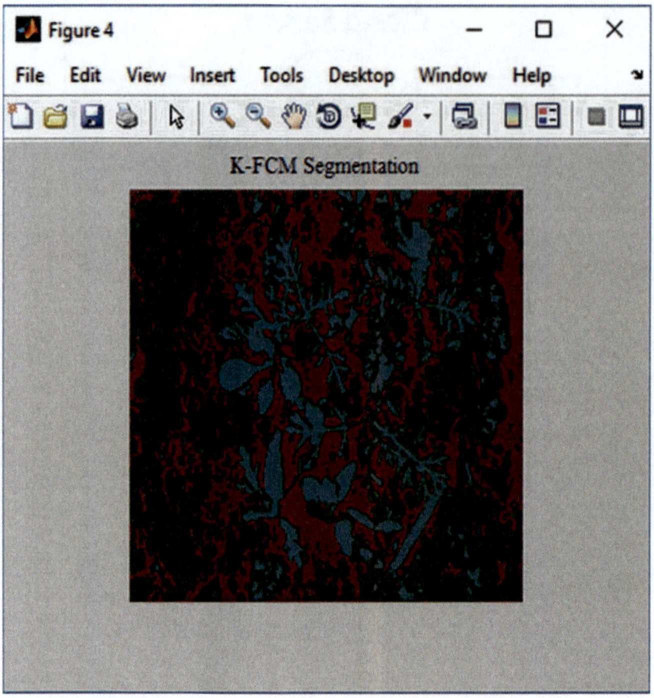

Fig. 5 Segmented image

format. The image data pixel and location of the ROI were specified considering the following: (1) the number of crop rows to be detected and (2) the imaged ROI that contains enough resolution (in pixels) to identify unambiguously green plants (crop and weed) from soil. The other step is explained in the subsections.

1. **Preprocessing**

Preprocessing is a method where the images will be displayed in the lowest level of abstraction. Preprocessing is a type of process which is used for extracting the data which has been given in high-quality input data. The objective of image acquisition, i.e., image preprocessing is to improve the image dataset like to remove the noise. So that it will allow enhancing the image quality [5]. By using this preprocessing method, we can resize the image into fixed size. There is some image which is important for showing the high quality of image for additionally processing, although the geometric transformations of images (e.g., rotation, scaling, and translation) are classified based on the machine learning algorithm, and then these image preprocessing methods will use some similar or same techniques are used. This preprocessing process will include data whitening methodology [6].

2. **Data Whitening**

By this powerful connection between pixels in the image, we can able to view many clustering centers and will be generated; this process is known as clustering. This process is used to delete the pixel correlation between the images. This includes PCA process and ZCA process for whitening method, and they both can ensure that the use of PCA could eliminate the correlation among the data under premise of the zero-mean. PCA whitening is a process of collecting the data from after gained from the external source of PCA transformation. Each pixel from the image dataset is set to have a fixed dimension; after some analysis, there will be a reduction. In some case, the data whitening process would lead to feature extraction, where it leads to zero, and there is some regularization which leads to reuse of feature extra value to realize the image scale, the use of constant to the feature value before taking the square root and the reciprocal will lead to gain the whiting intensity value [6]. Choosing the accurate E would lead to greenness composition so we can achieve the effect of low-pass filter. The whiting process is given by [6]

$$W_{\text{ZCA}} - \text{whiten}, i = \frac{UX}{\sqrt{\gamma + \mu}}$$

3. **Image Segmentation**

Discrimination is the main process which is used for gaining the maximum accuracy for the given input image as possible in wide crop rows, between crops and weeds with similar signatures, where it is used for crucial specific treatments in precision agriculture methodology. In this segmentation, the use of *K*-means algorithm is one of the easiest unsupervised learning algorithms that allow solving the known clustering problem. By the use of *k*-means classification algorithm, we can view the greenness detected area from the given data set based on their certain number of clusters and are assigned (assume *k* clusters) fixed a priori, which is fixed prior. The main intension is to define k centers, one for each cluster. These centers should be placed based on different locations of cluster that causes different results. The best option is to place them as much as possible far away from each other. And in Fig. 6, we can see the clustering centers after PCA whitening, i.e., segmentation is done. By using this supervised learning methodology, we can again the exhaustive value based on data whiting method which is trained based on the thresholding, which has the ability of self-adjusting to environmental conditions common in outdoor agricultural environments [6]. In Fig. 5, where clustering algorithm will make the center point after initializing the ZCA whitening, and the object is detected, which is good for the feature enhancement and image extraction process. The algorithm steps are as follows

Select K number of points as centroid of cluster;
Set a threshold constant C
While centroid fail to converge {
For(i = 0; i < k; i ++){

For(j = 0; j < number of points in point cloud; j++) {
Calculate the distance between point and the centroid;
If(d < C)
Assign the point to cluster;
Else Discard the point; }
} Recalculate the centroid of cluster by averaging all the points in cluster;
}

4. **Greenness Detection**

Color-based process is used for indices the image from fixed size to resized image (R, G, B components) are these are the early stage in greenness detection where the images are detected in green color component when compared to background pixels. It is difficult to achieve a 100% segmentation rate, especially without manual thresholding, due to uneven data size of image; by this thresholding, we can gain the color variation among different crop and weed species. The main aim of this process is to eliminate 99% of soil and other residuals from the image; from this, we can retain

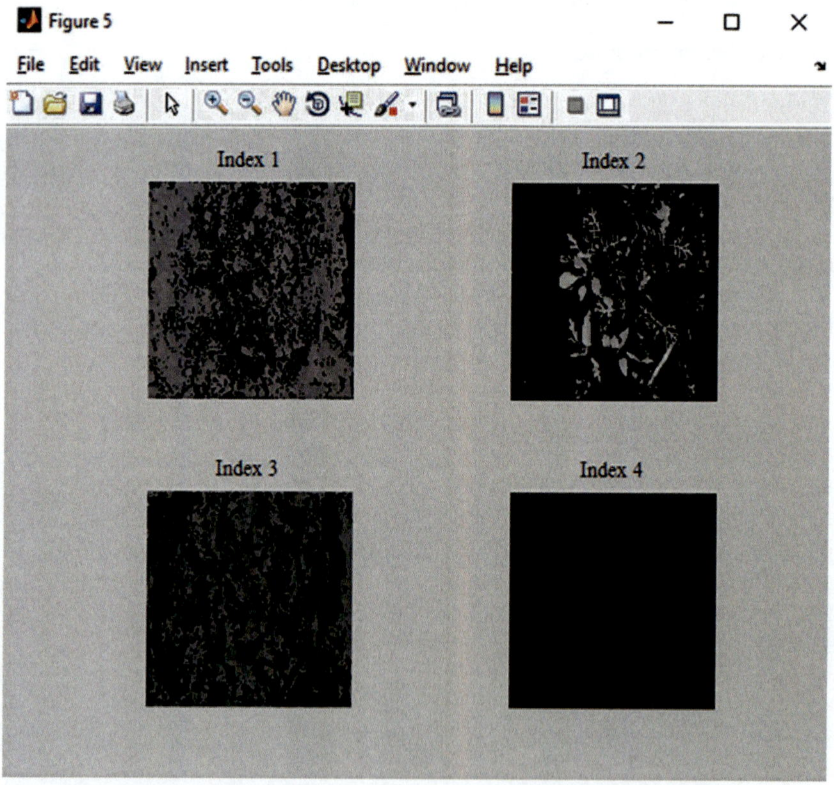

Fig. 6 Object identified image

the pixels of texture properties which will get spoiled. A color index methodology is used to eliminate the blue segmentation which is utilized in the image segmentation approach [7]. This index is mainly concentrated on the pixel value of the each image where the green area is identified in the plant and the rest is background which is shown in black shade color. By using this segmentation algorithm, the pixels are distributed in the three-dimensional that represents the RGB color coding algorithm (where red pixels belong to soil and residual class and blue pixels belong to vegetation class), respectively [8]. From this, we can view the counter-segmentation which is given in Fig. 4; here the greenness area is alone displayed (Fig. 7). The greenness detection algorithms were evaluated by measuring the percentage of the correct segmentation rate (CSR) calculated using the given formula. And also a pseudocode for greenness detection is given below.

$$\mathrm{CSR} = \frac{\text{Number of pixel correctly segmented by algorithm}}{\text{Actual number of pixel of object}}$$

segment image horizontally and partition into four boxes
then for each segmentation (1....N) {
classify different colors
then segmentation of colors is applied to red, green, yellow, and blue.
cluster the object by using K-Means Clustering algorithm
Do {
i. Find the large cluster size
} while cluster is error cluster
return maximum cluster color.

4 Results and Discussions

The performance of CNN algorithm for classification of the weed plant is 94.38% when 30% of images from gained dataset are used for testing the image data set and remaining for greenness detection. When 15% of images are used for testing purpose, and then the remaining set of data are used for efficiency calculation; based on the given dataset, it is reached up to 93.3%. This is clearly explained in Table 1

As a result, by using the CNN classification the number of data set is very high, and these data are trained previously by comparing with other trained data which takes long time. And the use of morphological image segmentation algorithm the detection of crop and weed in agriculture field gave 96% correct predictions. Out of the 108 images present in the data set, weeds in 55 images were identified correctly.

Fig. 7 Greenness detection image

Table 1 Performance of binary classification

S. No	Number of test images used	Number of trial images used	Efficiency (%)
1	12	108	94.38
2	24	96	93.3

5 Conclusion

Shape features of plants are mainly used for the identification of the weed in a field based on the discrimination at early stages of plant growth. Due to this difference in the feature of each plant, there may appear a difference in the main crop; for that purpose, a simple k-means and CNN algorithm are used for classification purposes by using thresholding. In this study, two types of methods are used in linear pattern recognitions, i.e., CNN and *k*-means clustering, and were employed to discriminate the weeds from carrot plants based on shape features. Both CNN and *k*-means were able to recognize the shape and patterns and classify the weed among the plants.

References

1. Weis, M., Gerhards, R.: Detection of weeds using image processing and clustering. Bornimer Agrartechnische Ber. **69**, 138–144 (2009)
2. Khattab, A., Abdelgawad, A., Yelmarthi, K.: Design and implementation of a cloud-based IoT scheme for precision. In: 2016 28th International Conference on Microelectronics (ICM), Giza, 201–204 (2016). https://doi.org/10.1109/ICM.2016.7847850
3. Bossu, J., Gee, C., Truchetet, F.: Development of machine vision system for real time precision sprayer. Electron. Lett. Comput. Vis. Image Anal. **7**(3), 54–66 (2008)
4. Mekala, M.S., Viswanathan, P.: A Survey: Smart agriculture IoT with cloud computing. In: 2017 International conference on Microelectronic Devices, Circuits and Systems (ICMDCS). Vellore (2017), 1–7. https://doi.org/10.1109/ICMDCS.2017.8211551
5. Rani, K.A.A., Supriya, P., Sarath, T.V.: Computer vision based segregation of carrot and curry leaf plants with weed identification in carrot field. In: 2017 International Conference on Computing Methodologies and Communication (ICCMC), Erode, pp. 185–188 (2017). https://doi.org/10.1109/ICCMC.2017.8282671
6. Tang, J.L.: Weed identification based on K-means feature learning combined with convolutional neural network. Comput. Electron. Agric. Elsevier. 0168-1699/2017
7. Bakhshipour, A., Jafari, A.: Evaluation of support vector machine and artificial neural networks in weed detection using shape features. Comput. Electron. Agric. **145**, 153–160 (2018)
8. Sa, I., et al.: weedNet: Dense Semantic Weed Classification Using Multispectral Images and MAV for Smart Farming. IEEE Robotics and Automation Letters **3**(1), 588–595 (Jan. 2018). https://doi.org/10.1109/LRA.2017.2774979
9. Sabzi, S., Abbaspour-Gilandeh, Y.: Using video processing to classify potato plant and three types of weed using hybrid of artificial neural network and partincle swarm algorithm. Measurement **126**, 22–36 (2018)
10. Bakhshipour, A., Jafari, A.: Seyed Mehdi Nassiri, Dariush Zare, Weed segmentation using texture features extracted from wavelet sub-images. Biosyst. Eng. **157**, 1–12 (2017)

Edge Detection Using Modified Third Order Edge Mask for Grayscale Medical Image Segmentation

A. Rijuvana Begum and S. Jana

Abstract Image segmentation, which partitions image into distinct regions, plays a significant role in digital image processing. Edge detection is one of the discontinuity-based image segmentation approaches that segment the images based on abrupt changes in its intensity or gray level. Edges of an object are characterized by pointed change in intensity. Edge detection is widely used in image processing applications, namely object recognition, segmentation, and matching. This paper introduces a new edge detection technique called Modified Third Order Differential Edge Mask (MTODEM) based on third-order differential equation. MTODEM is tested on real-time image samples obtained from patients. Experimental results show that the new edge detector, MTODEM, performs better with higher peak signal-to-noise ratio, lower mean square error, and larger entropy over other edge detection operators like Sobel, Prewitt, Laplacian, and TODEM.

Keywords Image segmentation · Edge detection · Medical images · Third-order differential edge mask · TODEM · MTODEM · PSNR · MSE · Entropy

1 Introduction

Basically in all object recognition tasks, a perceptual unit is extracted from a scene based on edge and region information [1]. Edge of an object constitutes the pixels in the boundary which separate regions of different luminance in an image [2, 3]. The significance of edge is that important physical, geometrical, and visual information of an image is contained in it. Edge, being a vector variable, contains both magnitude, which is a representation of intensity difference of neighborhood pixels, and orientation information which represents the direction of the greatest change in

A. R. Begum (✉)
Prist Deemed to be University, Thanjavur, Tamil Nadu, India
e-mail: arbbegum@gmail.com

S. Jana
Vel Tech Rangarajan Dr. Sagunthala R&D Institute of Science and Technology, Chennai, India
e-mail: drsjana@veltech.edu.in

P. Suresh et al. (eds.), *Advances in Smart System Technologies*,
Advances in Intelligent Systems and Computing 1163,
https://doi.org/10.1007/978-981-15-5029-4_42

intensity [4, 5]. Although numerous researches have been carried out on efficient edge detection techniques in the recent past, an algorithm that is robust to different imaging conditions and that can be used for varied applications has not been developed [6]. The advantage of edge detection is that it makes image analysis simpler with less data to be processed. The limitation of edge detection is that its performance is reduced in noisy images, due to the presence of high-frequency noise and edge components [7]. Edge detection involves 'smoothing, filtering, enhancement, detection, and localization' [4]. This paper presents a novel method to detect edges for grayscale medical image segmentation based on modified third-order differential equation.

2 Review of Related Work

Ilkin et. al [8] proposed a method of enhancing canny edge detection using Prewitt, Robert, and Sobel kernels and showed that with Sobel the detection was more accurate compared to other two kernels. Priyadarshini et.al [9] proposed an edge detection method which involves addition and division operation for values obtained after thresholding. Sample images of size 124 × 124 are considered for experiments. The authors showed that their method is computationally efficient when compared to edge detection using Sobel operator. Acharjya et.al [10] presented a review on edge detection algorithm using various operators such as Prewitt, Sobel, LOG. Real-life images are taken for analysis to measure the parameters such as PSNR, MSE, and Entropy. Statistical analysis shows that edge detection using Prewitt and Sobel operator gives better performance over edge detection using Laplacian and LOG, since these operators produce high MSE, lowest entropy, and low PSNR. Laplacian-based methods offer thick, spotty, and disconnected edges. Shrivakshan et.al [11] presented a comparative analysis of different edge detection filters for the classification of SHARK fish. Filters like Roberts, Prewitt, Sobel, Laplacian, and Canny are applied to extract edges. Experimental results show that the gradient-based methods are highly sensitive to noise. Laplacian filter is better for some features, but suffers from malfunctioning of false edges. They claimed that Canny edge filter performs better in noisy environments but it is limited in terms of processing speed and cost. Vijayarani et.al [6] discussed the performance comparison of facial image detection techniques for image mining applications. Edge detectors such as Canny and Sobel operators were applied over the set of facial image data. The authors concluded that Canny is the best edge detection algorithm for facial edge detection from their experimental results. Jayakumar et.al [4] discussed the different edge detection techniques and methods for satellite images. Roberts, Prewitt, Sobel, and LOG operators are tested over the image for experimental analysis. LOG operator counters the noise sensitivity behavior of the Laplacian operator at preprocessing stage using Gaussian smoothing before applying the second-order differentiation at preprocessing stage. Joyce Mary et.al [12] derived a new edge detection operator using third-order differential equation. Performance of this edge detector is compared with the well-defined gradient and Laplacian-based edge detectors to validate the results. The authors [13] presented an improved method of edge detection for medical images based on 'modified discrete wavelet transform with morphological thinner' (DWT-T). They experimentally proved that their method gave better results in the presence of impulse noise when compared to other methods.

Review on edge detection discussed above elaborates on the different types of operators for different imaging conditions and applications. It is observed that there is no unique edge detection operator suitable to solve all the requirements. Therefore, a new edge detection operator is required which will outperform all the traditional operators.

3 Materials and Methods

Literature provides numerous edge detection methods [9, 14–20]. In the process of edge detection, changes in gray level are identified using differential operator. Basically, edge detection can be categorized into gradient-based and Laplacian-based methods. In gradient-based methods, the largest and the least value of first-order derivative is used in determining the edges of an image. In the Laplacian method, the second-order derivative of an image is computed and its zero crossing is used in the detection of edges. This section provides a description of the various gradient operators, namely 'Robert, Prewitt, and Sobel' and also provides a description of the second-order Laplacian operator.

3.1 Gradient-Based Edge Detection

Gradient-based edge detection is also referred to as first-order edge detection since the first-order derivative of an image is used in the computation of its gradient. For a two-dimensional image $f(x, y)$, the gradient is given by Eq. (1) [2, 11].

$$\nabla f = \begin{pmatrix} \frac{\partial f}{\partial x} \\ \frac{\partial f}{\partial y} \end{pmatrix} \tag{1}$$

Robert, Prewitt, and Sobel are the well-known gradient-based edge detection operators.

3.2 Laplacian-Based Edge Detection

Laplacian is second-order edge detection operator where the second-order derivative enables to detect if the pixel falls on the dark side of the gray scale or whether it falls on the bright side of the gray scale using the zero-crossing property. The Laplacian

[2] of an image, f (x, y), is given by Eq. (2) as follows,

$$\nabla^2 f = \frac{\partial^2 f}{\partial x^2} + \frac{\partial^2 f}{\partial y^2} \quad (2)$$

The Laplacian provides a single, isotropic (rotation invariant) convolution mask of size 3 × 3 [19].

3.3 Third-Order Edge Detection

In order to overcome the limitations of the gradient first-order edge detection masks, namely Robert, Prewitt, and Sobel, and also, the limitation of the Laplacian second-order mask as discussed in Sects. 3.1 and Sects. 3.2, a new edge detection method, Third-Order Differential Edge Mask (TODEM) based on third-order derivative is proposed and is shown in Fig. 1. TODEM [12] is a 4 × 4 mask which is applied directly on the image pixels in order to obtain the edge map. Edge detection using this mask showed improved performance in terms of high PSNR and low MSE values.

Formulation of TODEM is based on third-order differential equation, and for an image $f(x, y)$, it is given by Eq. (3).

$$\nabla^3 f = \frac{\partial^3 f}{\partial x^3} + \frac{\partial^3 f}{\partial y^3} \quad (3)$$

From Eq. (3), the third-order mask of size 4 × 4 is developed. This is non-isotropic and a rotation-independent mask.

Especially for the case of medical images, the operator TODEM produces poor localization and detection. Thus, to improve the detection ability, the TODEM operator is modified and a new mask known as MTODEM has been proposed and described in Sect. 3.4.

Fig. 1 Third-order differential edge mask

0	0	-1	0
0	0	3	0
-1	3	-6	1
0	0	1	0

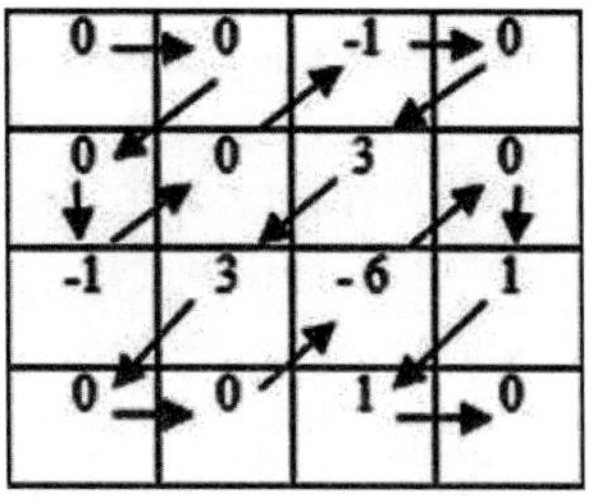

Fig. 2 Zigzag scanning

Fig. 3 Modified TODEM (MTODEM)

0	0	0	-1
0	-1	0	3
3	0	0	-6
0	1	1	0

3.4 Modified Third-Order Differential Edge Mask (MTODEM)

The new mask is formulated from the TODEM mask, such that the mask coefficients are reordered using zigzag scan path as given in Fig. 2.

Therefore, the reordered new mask is obtained and it is shown in Fig. 3.

4 Performance Evaluation

Performance of the MTODEM is implemented using MATLAB® software. Real-time medical images for two different patients were analyzed. The image samples like liver, kidney, and bladder are used for experimental analysis. Pixel-based metrics like SNR, PSNR, ROC, variance, MSE, entropy, and human visual system-based metrics like universal image quality index (UQI) and structural similarity index (SSI) are the two common form of performance evaluation techniques to understand the efficiency of any image processing algorithms. In this paper, the experimental results in terms of PSNR, MSE, and entropy for the image samples are tabulated in Tables 1 and 2 for two different patients. Result of MTODEM is compared with Sobel, Prewitt, Laplacian, and TODEM edge masks. Comparative performance shows that the MTODEM offers better results than the other masks. From Table 1, it is observed that the PSNR of patient 1-Kidney is around 46 dB in case of MTODEM, whereas for the other operators it is around 32 dB only. This situation is reflected for other images also. Similarly from Table 2, the MSE of Patient 2-Liver in MTODEM is 5.177 which

Table 1 Performance comparison of patient 1

Edge masks	Kidney			Liver			Bladder		
	PSNR	MSE	Entropy	PSNR	MSE	Entropy	PSNR	MSE	Entropy
Sobel	32.959	32.898	3.5523	32.409	37.339	4.4233	33.184	31.238	3.3178
Prewitt	32.646	35.353	3.4148	32.114	39.964	4.3123	32.894	32.893	3.2026
Laplacian	31.936	41.635	2.9009	31.291	48.302	3.8315	32.322	38.096	2.7648
TODEM	32.12	39.907	3.088	31.566	45.336	3.9686	32.50	36.566	2.9254
MTODEM	46.084	1.602	5.7461	40.494	5.8023	5.3157	45.885	1.677	4.6162

Table 2 Performance comparison of patient 2

Edge masks	Kidney			Liver			Bladder		
	PSNR	MSE	Entropy	PSNR	MSE	Entropy	PSNR	MSE	Entropy
Sobel	31.602	44.955	3.7311	32.956	32.915	4.062	32.937	33.07	3.0874
Prewitt	31.316	48.019	3.5922	32.651	35.312	3.9683	32.73	34.676	2.9693
Laplacian	30.804	54.031	3.0676	31.778	43.172	3.5315	32.367	37.699	2.7342
TODEM	30.932	52.462	3.2676	32.019	40.846	3.6662	32.574	35.946	2.8994
MTODEM	45.929	1.66	5.1176	40.99	5.177	5.6126	44.753	2.1765	4.5547

are very less comparatively. PSNR and MSE are the pixel-based metrics to determine the efficiency of the image analysis tasks. MSE and PSNR are opposite to each other. Higher PSNR yields lower MSE and vice versa. MSE indicates the pixel difference between the original and edge image. Both MSE and PSNR measures the amount of closeness between the original and processed image, and thereby it ensures the localization property of an edge. The expressions governing MSE and PSNR for an MxN image are given by Eqs. (4) and (5).

$$\text{MSE} = \frac{1}{MN}\sum_{x=0}^{M-1}\sum_{y=0}^{N-1}(f(x,y) - \hat{f}(x,y))^2 \tag{4}$$

$$\text{PSNR} = 10\log_{10}\frac{255^2}{\text{MSE}} \tag{5}$$

where $f(x, y)$ is the original image and $\hat{f}(x, y)$ is the edge image.

In addition to the metrics like PSNR and MSE, the table provides metric called entropy. Entropy is a measure of randomness or uncertainty that can be used to represent the pattern of pixels of an image. Lesser the entropy higher will be the compression, and hence lesser the memory. But for our sample images, the MTODEM offers higher entropy for all the image samples.

Graphical representation of PSNR, MSE, and entropy of all the edge masks and for all the image samples are presented in Fig. 4. It is understood from the plots of PSNR and MSE, and the proposed MTODEM edge detection operator outperforms

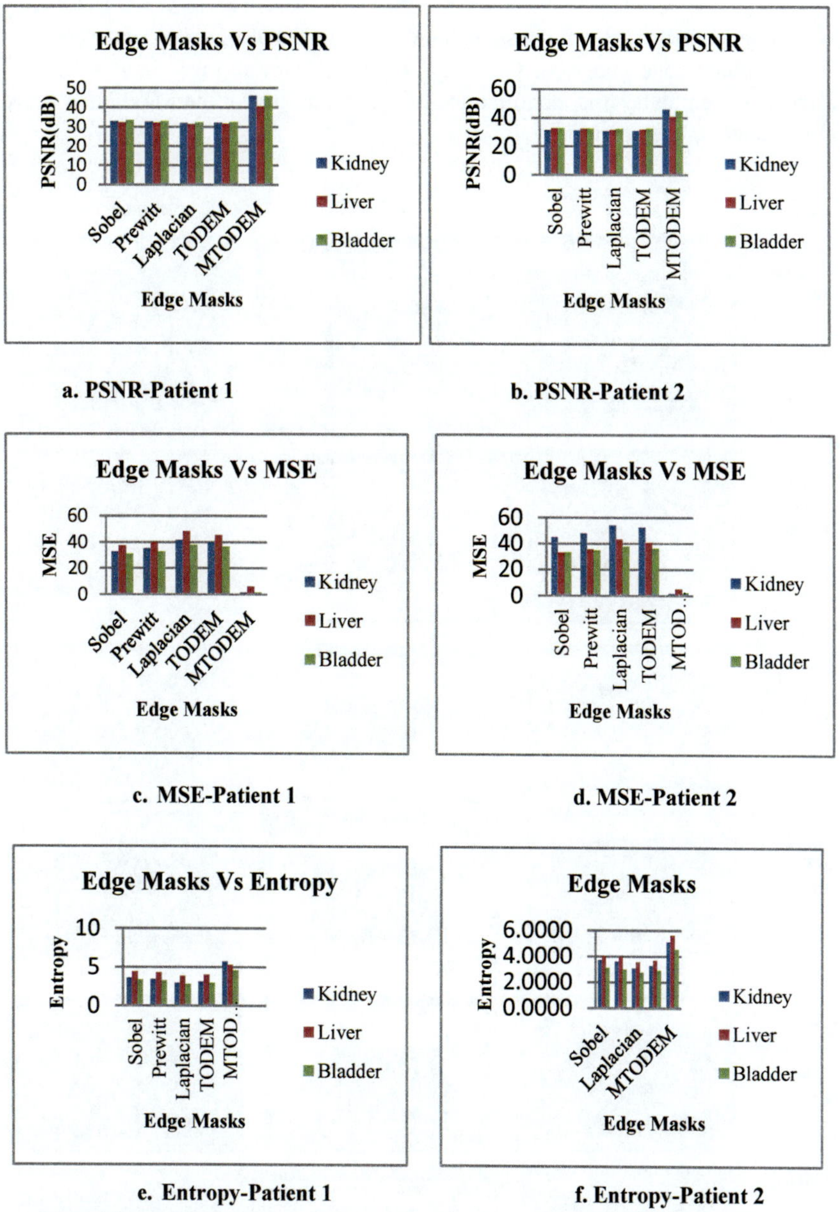

Fig. 4 Graphical representation of the performance of edge masks

the other edge masks. But at the same time modified TODEM offers larger entropy. Also, resultant edge images of the samples are presented in Figs. 5 and 6 for two different patient data. It is identified from Figs. 5 and 6, that the MTODEM mask offers better results comparatively.

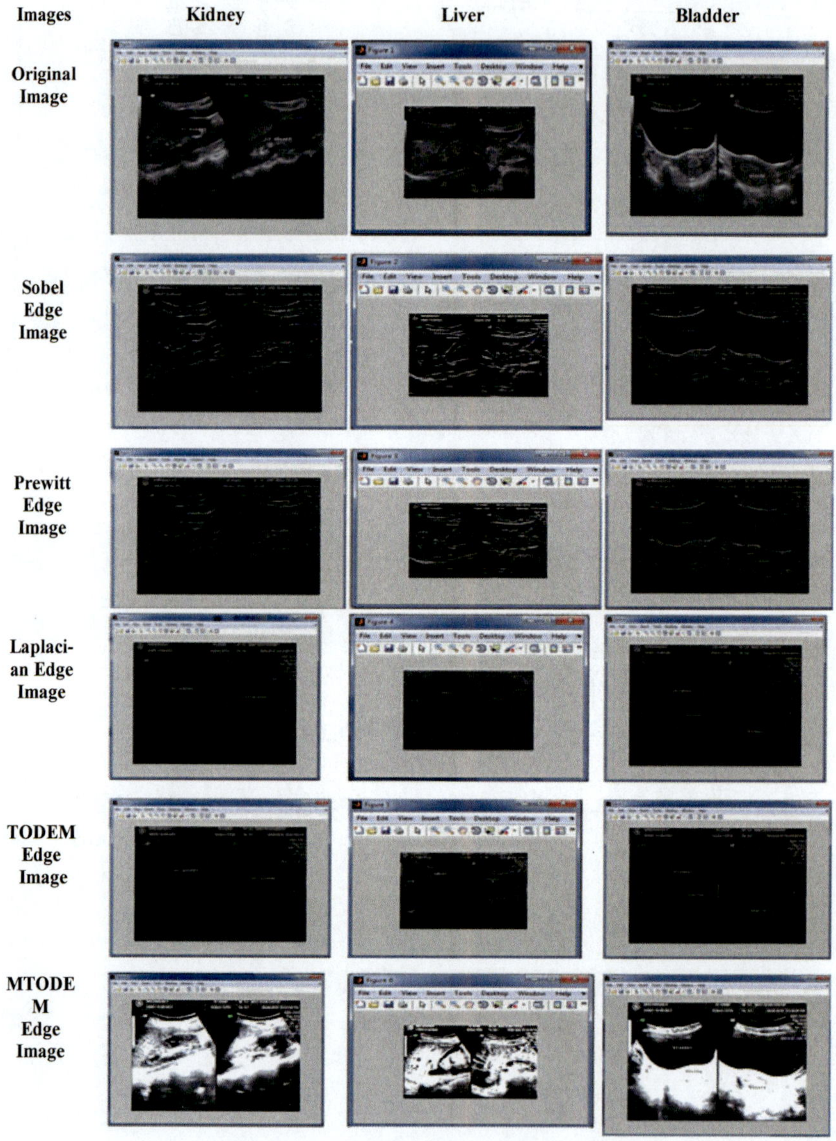

Fig. 5 Edge extracted images -patient 1

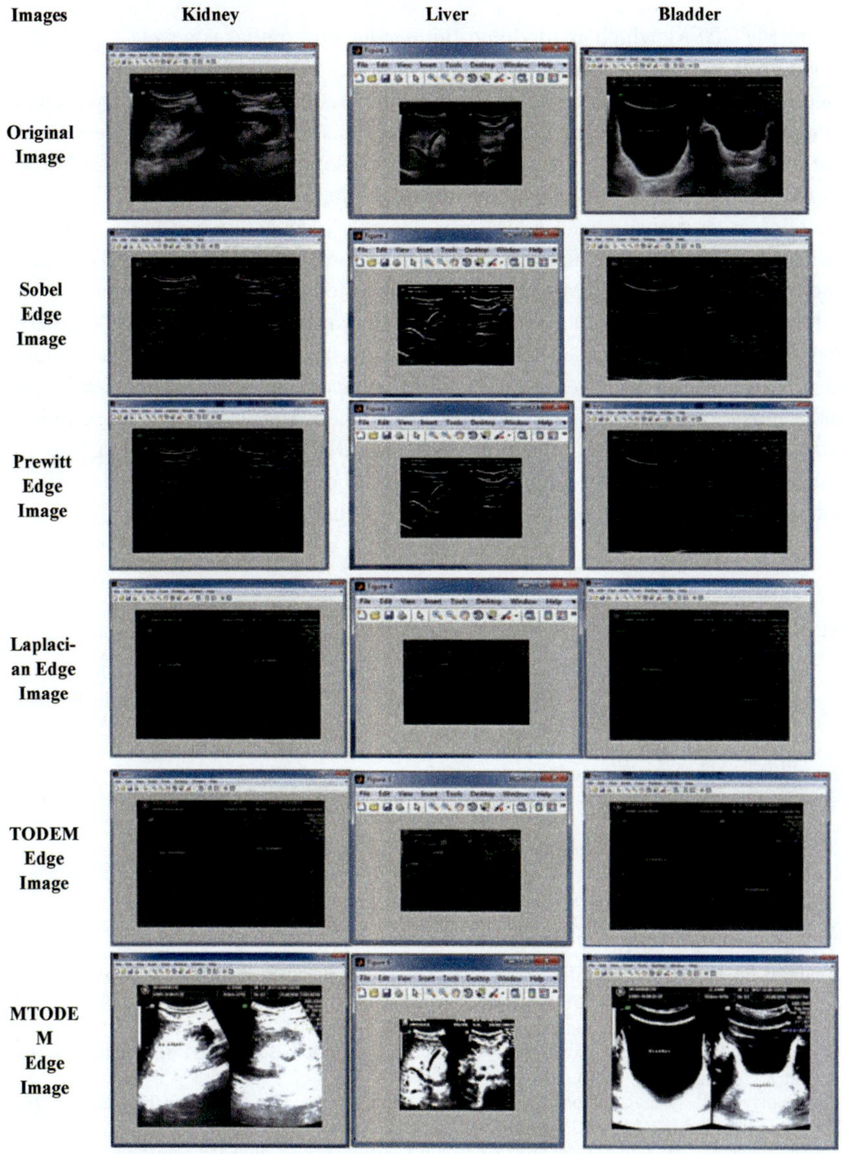

Fig. 6 Edge extracted images -patient 2

5 Conclusion and Future Work

Segmentation in images requires edge detection as the primary step in applications like object detection. Therefore, applications require edge detectors which are more accurate in performance. In this paper, a new modified edge detector called MTODEM which is modified third-order differential edge mask has been proposed. Gradient first-order Robert mask which is highly sensitive to noise is used for applications where simplicity and computation time are significant factors. Whereas, Prewitt and Sobel masks which have the advantage of being less sensitive to noise provide poor edge localization because of thicker edges. Second-order Laplacian mask also results in poor localization in spite of thin edges. TODEM which is a third-order derivative mask also provides poor localization and lower detection ability particularly for the medical images. Therefore, TODEM is modified to obtain improved localization. Proposed MTODEM is tested on six medical images of two different patients. Performance results prove that MTODEM offers good response than TODEM, Laplacian, Sobel, and Prewitt masks in measurement parameters, namely MSE and PSNR values. MTODEM enhances the sharp edges clearly, but under noisy environments its performance is limited. Hence, suitable noise reduction has to be incorporated in the future. Also, MTODEM is useful only for grayscale edge detection. Methods to implement MTODEM directly on color images without grayscale conversion can be taken as the future enhancement.

References

1. Moon, H., Chellappa, R., Rosemfeld, A.: Optimal edge based shape detection. IEEE Trans. Image Proc. **11**(11), 1209–1226 (2002)
2. Gonzalez, R.C., Woods, R.E: Digital Image Processing. Prentice Hall, 3rd edn (2001)
3. William, P.K: Digital Image Processing. Willy-Interscience (1991)
4. Jayakumar, R., Suresh, B.: A review on edge detectionmethods and techniques. IJARCCE **3**(4), 6369–6371 (2014)
5. Shapiro, L., Stockman, G.: Computer Vision (2000)
6. Vijayarani, S., Vinupriya, M.: Performance analysis of canny and sobel edge detection algorithms in image mining. IJIRCCE **1**(8), 1760–1767 (2013)
7. Maini, R., Agarwal, H.: Study and comparison of various image edge detection techniques. IJIP **3**(1), 1–11 (2009)
8. Ilkin, S., Hangisi, F.S., Tafrali, M., Sahin, S.: The enhancement of canny edge detection algorithm using Prewitt, Robert and Sobel Kernel. Proc. Int. Conf. Eng. Technol. 120–124 (2017)
9. Priyadarshini, S., Sahoo, G.: A new edge detection method based on additions and divisions. Int. J. Comput. Appl. **9**(10), 10–13 (2010)
10. Acharjya, P.P., Das, R., Ghoshal, D.: A study on image edge detection using the gradients. Int. J. Sci. Res. Publ. **2**(12), 1–5 (2012)
11. Shrivakshan, G.T., Chandrasekar, C.A.: Comparison of various edge detection techniques used in image processing. Int. J. Comput. Sci. **9**(5, 1), 269–276 (2012)
12. Joyce Mary, G.J., Begum, A.R.: Guided filter smoothing for third order edge mask. Int. J. Comput. Appl. **120**(23), 36–42 (2015)
13. Gupta, S., Sunkaria, R.K.: An improved edge detection algorithm using a modified discrete wavelet transform based on morphological thinner for noisy medical images. IEEE Xplore, (2018)
14. Kalpana, M., Kishorebabu, G., Sujatha, K.: Extraction of edge detection using digital image processing techniques. Int. J. Comput. Eng. Res. **2**(5), 1562–1566 (2012)
15. Masoud, N., Karimi, R., Hariri, M., Malekian, K.: Edge detection techniques in processing digital images: Investigation of canny algorithm and Gabor method. World Appl. Program. J. **3**(3), 116–121 (2013)

16. Siva Vinayaka Pavana Hiranya, P., Venkata Suresh, B., Sabhanayagam, T.: Car license plate detection using structured component analysis. IJEDR **2**(1), 196–200 (2014)
17. Pradhan, R., Pradhan, P., Bhattacharjee, R., Singh, D.: Edge detection using morphological operator: a new approach. Int. J. Adv. Res. Comput. Sci. Softw. Eng. **4**(2), 84–87 (2014)
18. ElifAybar: Sobel edge detection method for MATLAB
19. Mukesh Kumar, R., Saxena, R.: Algorithm and technique on various edge detection: a survey. SIPIJ. 4(3), 65–75 (2013)
20. Kaur, K., Malhothra, S.: A survey on edge detection using different techniques. IJAIEM **2**(4), 496–500 (2013)

Software-Defined Networking and Architecture of IoT with Security, Challenges and Applications: A Survey

Ramesh Pusuluri, K. Kalaiarasan, and A. Yogaraj

Abstract As the IoT incorporates large number of devices with different heterogeneous end systems, the software involved with these systems to run those algorithms and establishing the connections in the network with the help of various protocols is quite complex. With a wide range of diverse devices involved in the system, the design of the architecture should be carefully chosen for providing reliability as well as the scalability. The architecture may be different for various distributed systems located remotely like smart homes, agriculture, enterprise applications, manufacturing, energy management and monitoring of environment. One such architecture is fog. As the information is exchanged among these devices, keeping the information private and providing security by denying the access to un-authorized users is essential. This paper discusses the software-defined networking technologies and basic architecture with security features, challenges involved in it and applications.

Keywords Software-defined networking · Security in IoT · IoT challenges

1 Introduction

IoT is the future communication among diverse devices for control and automation by connecting the various devices in the network. The IoT was a network containing various devices like home appliances, industrial machinery, vehicles and different diverse devices which are equipped with software, electronic circuits,

R. Pusuluri (✉) · K. Kalaiarasan · A. Yogaraj
Department of Electronics and Communication Engineering, Vel Tech Rangarajan Dr. Sagunthala R&D Institute of Science and Technology, Chennai, India
e-mail: rameshpusuluri999@gmail.com

K. Kalaiarasan
e-mail: Kalaiarasan@veltech.edu.in

A. Yogaraj
e-mail: ayogaraj@veltech.edu.in

P. Suresh et al. (eds.), *Advances in Smart System Technologies*,
Advances in Intelligent Systems and Computing 1163,
https://doi.org/10.1007/978-981-15-5029-4_43

network topologies, sensors and actuators which support these devices in order establish the connection and exchanging the data efficiently and effectively. The number of objects connected through Internet of things will be several hundreds or thousands which depends on the infrastructure as well as the computing system embedded with it [1]. IoT technology permits sensing or controlling the objects which are located remotely across the network. IoT integrates the physical objects in the world with computers by reducing the intervention with humans which results in improvement of accuracy as well as economically. The devices in IoT will gather usable data with the help of various sensing and controlling technologies, and this data will be exchanged automatically. Few decades ago, for establishing the network, the infrastructure for networking includes various devices like switches, routers, ASICs and intermediate devices for performing sophisticated operations. These devices are pre-programmed and cannot be modified later to change its functionality. These cannot be pre-programmed with multiple rules for providing optimized networking [2].

2 Software-Defined Networking

The software-defined networking is one of the significant networking architectures, and by using this control over network can be achieved without implementing old techniques. The three layers of software-defined networking are application layer, infrastructure layer and control layer. Multiple application program interfaces are eastbound, westbound, northbound and southbound. One of the architecture for performing the operations of computing, control, storage and networking and distributing the services to users is fog [3]. For improving the experience of the users and tolerant to failures, integration of fog/edge with IoT is necessary. The fog works with both mobile and wire-line devices, and it was compatible with the hardware as well as the software containing data plane and network plane. It supports wide variety of applications in 5G wireless communication devices and artificial intelligence. It was organized by clients or edge devices which provides diverse computational services for the customers [4]. In this type of computing, the data generated by various devices involved in IoT was processed at network edge without sending it to cloud which decreases the consumption of energy as well as the bandwidth.

The systems which effectively combine the cyber technology and the physical components by integrating advanced technologies of communication and a sophisticated computing method are called cyber physical systems. The major purpose of these systems was to change the way of communication methods among physical, human and cyber worlds. These systems monitor and control the devices in more efficient and secured way. These are the systems which are a combination of actuators, sensors, physical components, centres for control and various networks of communication. The function of sensors was to monitor and measure the condition of the various physical components. The actuators operate in a way to achieve the required functionalities on physical components [5]. The communication networks send the data obtained from the sensors and reply to these sensors with some control signals

Table 1 Various types of networking with its features

Type of networking	Features
Edge networking	It provides aggregation of data, flow monitoring, de-aggregation of data, admission control, used to develop SDN-based wireless sensor networks and monitoring of networks
Access networking	Access-core integration, accessing of information, wireless access networking, mobile access network, rule-caching in metropolitan area network, monitors the data flow
Core networking	Providing security to enterprise networks, admission control, classification of packets, information-centric networking, communication among devices in LTE networks, monitoring of data flow, routing in mobile core networks
Data Centre networking	Online resource sharing, traffic engineering, services including virtual network, energy-efficient networking, VM-scheduling and migration

given by control centres. These control centres are utilized to analyse the information and send some feedback for achieving the required mode of operation. There are four types of networking which are edge networking, access networking, core networking and data centre networking. Different types of networking technologies with their features are shown in Table 1.

3 Architecture and Enabling Technologies in Different Layers of IoT

The basic architecture of IoT contains four layers which are perception layer, network layer, service layer and application layer. The functionalities of each layer in IoT were shown in Fig. 1.

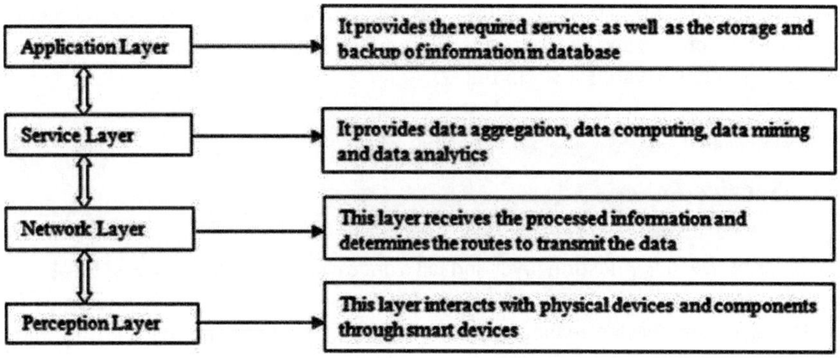

Fig. 1 Functionalities of each layer in IoT

3.1 Perception Layer

This layer was the bottom layer in the architecture of Internet of things which is also called as layer of sensor [6]. The major functionality of this layer was for tracking and identifying the objects. With the help of different smart devices like radio frequency ids, actuators as well as the sensors, this layer communicates with various devices. The RFIDs with unique identification number comprise of antenna, reader, tag which exchanges data with the help of radio signals. The transmission of signals takes place from reader of RFID to tag of RFID with the help of this antenna. The advantages of RFID layer in this technology were less cost, high security, miniature components, rapid scanning, long life, large storage capacity, non-touch operation. Wireless sensor networks are one of the technologies which are having significant role in IoT [1]. These networks perform monitoring and tracking the state of various devices and transmit this data to sink nodes or data control centres. These networks act as interface between cyber world and real world. The advantages of wireless sensor networks are scalable, reconfiguring the network in dynamic way, miniature size, less cost and less power consumption. These networks were mainly used for observing the parameters of real world. This layer transmits the information to upper layers by measuring, collecting and processing the data which is obtained from smart IoT devices.

3.2 Network Layer

Network layer was the prominent layer in IoT as it contains diverse devices like gateways, hubs, switches and diverse technologies of communication like wireless fidelity, LTE and Bluetooth. It is the intermediate layer in the architecture of IoT which is also called as layer of transmission. This layer simultaneously transmits and receives information from various applications via gateways and interfaces [3]. The information sent by the perception layer was received by this layer and sends this data to hubs of IoT by detecting the routes using various routing protocols and achieves reliable and secure forms of communication. Various technologies and protocols in network are shown in Fig. 2.

3.3 Service Layer

The layer between application layer and network layer is service layer, and this layer provides secure and effective to various applications. The various technologies in service layer are service management technology, resource sharing and management, and middleware and interface technology are shown in Fig. 3. The service management technology detects the applications and devices for scheduling reliable services. The interface technology in the service layer is necessary to achieve the

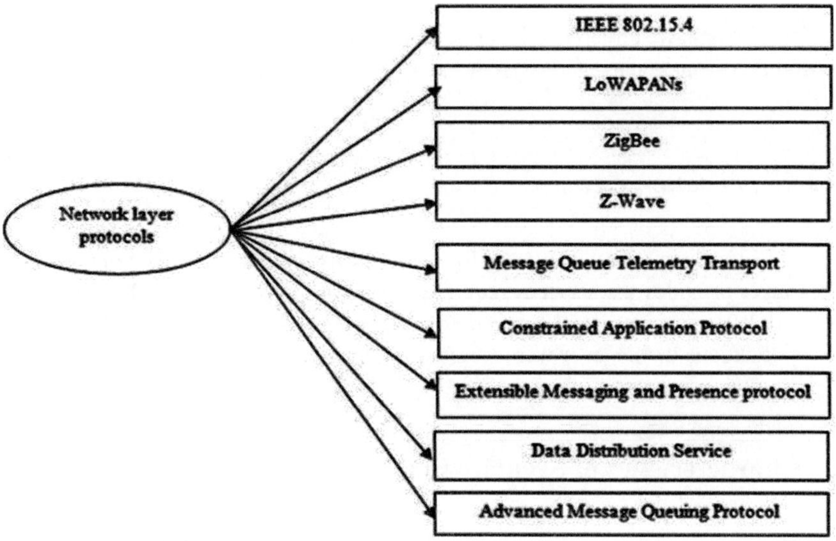

Fig. 2 Various protocols in network layer

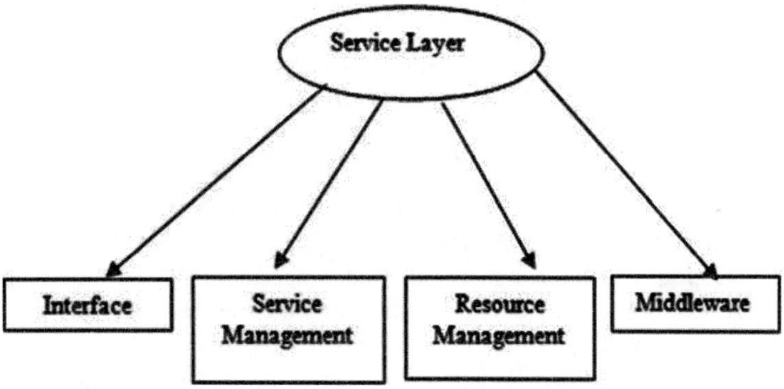

Fig. 3 Various technologies in the service layer

effective exchange of information with high security [2]. The interface maintains the various devices which are interconnected like connection and disconnecting of the devices, operation and communication among the devices. The service management is responsible for discovering the applications and devices and provides reliable and effective services. A simple service contains data collection, analysing and data storage. Middleware is software which gives services between applications of Internet of things. The advantages of middleware are it supports wide variety of applications as well as it can run on different OS platforms. It enhances computing in a distributed way and service interaction in different applications, networks and devices. To achieve effective utilization of resources as well as reducing the cost of resources, resource management plays a very important role.

3.4 *Application Layer*

It was the top layer in architecture of IoT which is also called as the layer of business. This layer provides the services required by the users by receiving the data from network layer, and it stores the data in the database for further analysis [7]. Some of the applications in this layer were smart cities, smart buildings and smart grids.

4 Characteristics of Security in IoT

Security is the main feature for any device for achieving the reliable communication. The information which was sharing among the devices should be confidential and should not be revealed to un-authorized users. For achieving security, various cryptographic algorithms should be incorporated into the devices. The following are some of the security features which should be fulfilled for security.

Privacy: Privacy is one of the important characteristics in IoT, where the authorized users are allowed for accessing the data and un-authorized users are not allowed to access the data. Privacy enables the user to control some part of data, and the user was not allowed to extract additional information [8]. As more number of devices and people are involving in IoT who utilize the common network of communication, privacy is having significance importance.

Authentication: Authentication is achieved, when the devices which send data through network are legitimate and the devices which receive the data are legitimate. As more number of devices operate in IoT, the process of identification and authentication is very difficult. Efficient algorithms and protocols are needed in IoT to achieve authentication.

Integrity: As data has to go through communication networks, there may be a chance of tampering by some intended or un-intended users, and the feature of integrity will avoid this and ensure that accurate data will be provided to authorized users. For normal functioning of the IoT applications, it is a salient feature because if modified data was received, it estimates incorrect operations, and incorrect commands will be sent which could hamper the applications of IoT.

Confidentiality: This is one of the key security features where the information can be given access to the users who are intended to use in the process of communication and this information cannot be eavesdropped by users who are not authorized to use. As more number of devices are connected in IoT, it is required that data gathered by one device should not be known by other devices [9]. To attain complexity, complex and high secure cryptographic techniques are needed. To achieve availability, various efficient and secure routing protocols should be deployed whenever needed.

Availability: This security feature will enable that both information and devices are accessible by users who are authorized to use for their services when they request. As most of the devices in the IoT are real-time systems, the services requested by the users cannot be sent once again if the requested information cannot be reached in time.

5 Challenges Associated with IoT

As most of the devices in IoT uses wireless communication to communication with each other. The data which was shared among these devices are prone to various attacks in different layers. The data may be misused by the attackers or modify the data or send any malware in order to disrupt the functionality of the devices in IoT.

5.1 Perception Layer Attacks

As the perception layer gathers the data in IoT, the major threats to security in this layer were data forging and destruction of IoT devices. The different types of perception layer attacks are shown in the figure.

Malicious code injection attacks: In this type of attack, the attacker will induce malicious code into device's memory. This malicious code will change the functionality of IoT devices; the attacker will be able to access the system, and they can control the entire IoT system.

Node Capturing attacks: In these kinds of attacks, the attackers can take control over the devices or the nodes in IoT by changing the node hardware or replacing the complete node [10]. Once the node was captured by the attackers, the important information in the node will be copied by the attackers like encryption and decryption keys and uses this information to create a fake node; this fake node appears as the actual node and connects to IoT system for stealing the important information. This kind of attack was one of the node replication attacks.

Replay attacks: The attacker or malicious node will send the information to the destination node to appear themselves as the legitimate sender, so that the attacker will gain the trust of IoT devices. These kinds of attacks are induced in the process of authentication to impair certificate authorization.

False data injection attacks: The attacker takes the control of devices in IoT and replaces actual data measured by the sensors or actuators with false data, and the same will be transmitted to the various applications in IoT [11]. These applications will send the wrong commands and services which affect the networks and applications of IoT.

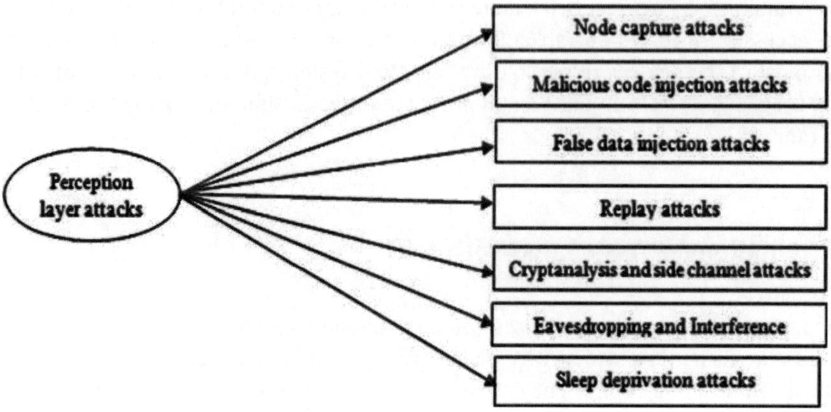

Fig. 4 Various perception layer attacks

Eavesdropping and Interference: The devices in IoT use to send and receive the information via wireless communication networks, and there are chances that the information transmitted through these links will be eavesdropped by the attackers who are not authorized to use. Sometimes, the attackers will send the noisy data to corrupt the actual information. To counteract these issues, complex cryptographic algorithms and key management techniques should be used.

Cryptanalysis and side-channel attacks: The cryptanalysis attack will utilize the generated cipher-text or plaintext to extract the key of encryption utilized in algorithms. The various types of perception layer attacks are as shown in Fig. 4.

5.2 Network Layer Attacks

As the network layer was responsible for transmitting the data which was collected, the resources of the network are more prone to various security issues [8]. All the devices are connected with wireless links, careful design is needed. Various network layer attacks were shown in Fig. 5.

Denial of Service attack: The denial of service attacks will utilize the resources of the network by sending the more traffic and disturbing IoT system services.

Spoofing attacks: The main aim of spoofing attacks was to get the complete access of IoT systems by sending the malicious data in the system. Some of these kinds of attacks in IoT are Internet protocol spoofing and spoofing of radio frequency id [7]. The Internet protocol spoofing attacks is made by obtaining the IP address of the user and utilizes this address for sending the malicious data. In radio frequency id spoofing, the attacker will capture the tag of RF id and utilize this RF Id tag for sending the malicious data.

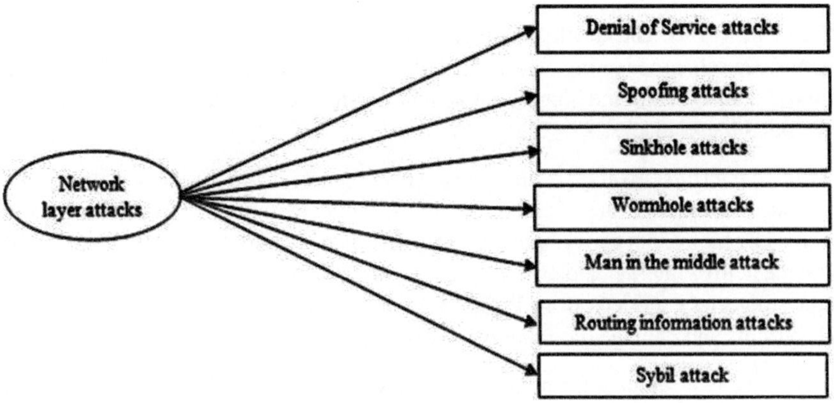

Fig. 5 Various network layer attacks

Wormhole attacks: These attacks are initiated by two nodes in a cooperative manner. These two nodes located in different geographical locations will share the information of routing with the private links for achieving false transmission.

Man in the middle attack: In this attack, malicious nodes are placed between the two nodes which are communication with each other in the IoT. These malicious nodes stole the information regarding the identity of the communicating nodes and forward the data as well as receive the data by appearing as the valid nodes [11]. By deploying highly secured protocols of communication, these attacks can be prevented by not leaking the node's identity to attackers.

Routing Information attacks: These attacks are mainly intended for routing information in systems of IoT. In these attacks, the information regarding routing was modified for creating route loops which are utilized for transmission of data, and this leads to unnecessary source paths and thereby increasing the delay. To counteract these attacks, routing protocols with high degree of security should be implemented in the network and measures should be taken to prevent disclosure of the IP information.

5.3 *Application Layer Attacks*

The major functionality of application layer was to provide the services which are needed by the user. The majority of the attacks in the application layer were software kind of attacks.

Phishing attacks: In these kinds of attacks, the attacker will get the users confidential data like user ids and passwords by sending the users with infecting emails and links for phishing websites.

Malicious worms: The attackers will infect the applications of IoT with Trojan horse, warms in a self-propagating way and acquire the important information [4]. Suitable firewalls, antiviruses and some high standard software need to be implemented to withstand these attacks.

Malicious Scripts: To impair the system functionality of IoT, these scripts will be appended to the software, get changed in the software, and finally deleted thereby decreasing the system performance. As majority of the IoT applications are connected with the help of Internet, attacker will easily make customer to run these scripts while requesting these services from Web. These scripts steal the confidential data from the systems and damage the functionality of the system.

Privacy: In IoT, the data from various devices were gathered through various steps like (1) collection of data; (2) aggregation of data; and (3) mining and analysis of data. The collection of data was done by sensing the condition of the objects which are connected in IoT. The aggregation of data combines the data which was closely related in useful information, and mining and analysis of data will obtain the value of the information for in IoT for the applications specified [8]. Lack of privacy in IoT will lead to loss of property, and it is even unsafe for humans. Various security mechanisms are needed in IoT to ensure privacy.

6 Applications of IoT

6.1 Smart Grid

By combining the IoT and cyber physical systems, smart grid will be developing by replacing the normal power grid for providing consumers with the best energy service. In this system, distributed generators for the energy were introduced in a dispersed manner for improving the usage of resources of energy. To improve the energy storing capability as well as reducing the carbon dioxide emission in the atmosphere, vehicles which run on electricity were introduced. Smart meters and communication networks which communicate in two easy were adapted to obtain for providing interactions among the consumers and the providers of service [7]. By using the IoT technology, more smart meters will be installed in homes and office buildings which are connected with smart grid. These meters will be monitoring the generation of energy, storage of energy and consumption of energy as well as they communicate with providers of service and inform information regarding energy requirement of consumers, and they will charge for the electricity consumption accordingly. By using the infrastructure of fog/edge computing, data which was collected from these smart meters will be stored and processed. For improving the effective energy resource utilization as well as reducing the cost, the providers of the service will send the energy which was dispatched from the grid and consumers will reduce their energy consumption. As these smart meters which were installed in smart grid communicate among each

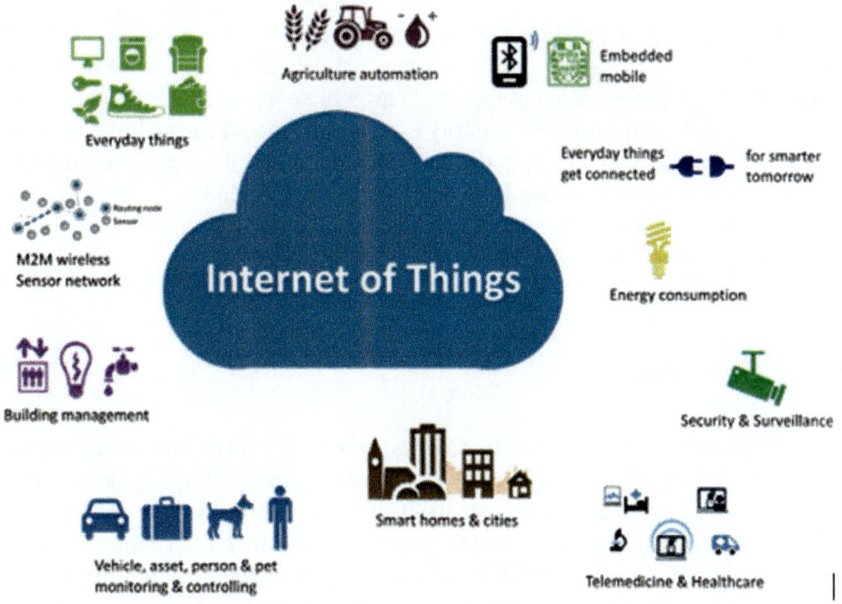

Fig. 6 Some applications of IoT in diverse fields

other with the help of wireless links, there may be possibility that the attackers will extract the information and they will change the content in it. The consumers will get the false information regarding energy consumption, and providers of service will be estimating fault energy supply; it will create disturbance the major functionalities in the grid system. So, highly sophisticated security mechanisms should be employed in the networks so that it will ensure the privacy and confidentiality of the information. Some of the diverse applications of IoT are shown in Fig. 6.

6.2 Smart Transportation

It was also called as intelligent transportation systems which were one of the applications of IoT. For achieving the reliable and effective transportation systems, the control systems, networks used in communication and various techniques for computing were integrated for providing security as well. To achieve this, more number of smart vehicles were included and they are connected with the help of wireless networks. These vehicles will effectively share the traffic data and prepare the timings for the drivers on when to start. These vehicles will identify objects near to them and adjust the velocity of the vehicles in the journey without the intervention of the drivers with high safety [12]. These vehicles are equipped with various controlling units which control and monitor the vehicle's subsystems electronically.

They function as an internal network to share the data which was collected inside the vehicle. They were equipped with communication interfaces for connecting with other vehicles. Using these interfaces, vehicles will communicate with other vehicles as well as the infrastructure. Security mechanisms should be employed in these vehicular networks with complicated data encryption and decryption techniques; otherwise, the attackers will steal the data, modify it and send this information to other vehicles which leads to serious hazards. It will lead to roads with high traffic and increased travel time.

6.3 Smart Cities

Smart cities will utilize the public resources in much efficient manner which results in improving user's QoS and decreases the operational costs for public administrators [9]. The IoT in cities will provide the distributed database for the condition of historical as well as commercial buildings in the cities which can be monitored by deploying deformation and vibration sensors which monitors building stress. The atmospheric conditions in the surroundings can be monitored with sensors which measure the quality of air by computing oxygen levels, air pollution, humidity and temperature. Utilizing this database will reduce the cost for testing the condition of the building manually. This data combined with seismograph readings can be used for studying the effect of small earthquakes on the building in the cities. IoT can be used for managing the garbage waste in large cities which results in saving the manpower as well as maintaining the ecological balance. Used in smart health for monitoring the condition of the patient remotely so that doctor will prescribe the necessary medicines to the patients. IoT can be utilized in smart street lighting which varies the intensity of lights according to different timings, density of people in that location as well as conditions of the weather.

7 Conclusion

In this paper, a brief review of architecture of IoT and technologies was presented with security, challenges and applications associated with it. The basic functionality of the software-defined networking in IoT was discussed. It discusses the computing using fog for supporting wire-line as well wireless applications with the basic relation of IoT with cyber physical system. The architectures of IoT with four layers which are perception layer, network layer, service layer and application layer have been discussed in detail with the technologies used in each layer. For providing security to the devices in the IoT, the issues related to the security and privacy which affect the performance and efficiency of IoT with available solutions were presented. The various security attacks in perception layer, network layer and application layer were presented with the possible solutions. Finally, some of the applications like smart grid, smart transportation, and smart cities were discussed to present the ways for implementing the computations based on fog/edge in real-world applications. In future, billions of devices will be connecting with IoT for providing reliable services to humankind and fulfil the needs by saving human efforts as well as the operational costs by developing the nations economically and ecologically.

References

1. Ahmed, S.H., Kim, G., Kim, D.: Cyber physical system: architecture, applications and research challenges. In: Proceedings of IFIPWireless Days, Nov 2013 (2013)
2. Al-Fuqaha, A., Guizani, M., Mohammadi, M., Aledhari, M., Ayyash, M.: Internet of things: a survey on enabling technologies, protocols, and applications. IEEE Commun. Surv. (2015)
3. Pan, J., Jain, R., Paul, S., Vu, T., Saifullah, A., Sha, M.: An internet of things framework for smart energy in buildings: designs, prototype and experiments. IEEE Int. Things J. (2015)
4. Andrea, I., Chrysostomou, C., Hadjichristofi, G.: Internet of things: security vulnerabilities and challenges. In: Proceedings of IEEE Symposym Computer Communication (ISCC) (2015)
5. Atzori, L., Iera, A., Morabito, G.: The internet of things: a survey. Comput. Networks (2010)
6. Athreya, A.P., Tague, P.: Network self-organization in the internet of things. In: Proceedings of IEEE International Conference on Sensor and Communication Network (SECON) (2013)
7. Atzori, L., Iera, A., Morabito, G., Nitti, M.: The social internet of things (SIoT)—When social networks meet the internet of things: concept, architecture and network characterization. Comput. Networks (2012)
8. Kreutz, D., Ramos, F., Esteves Verissimo, P., Esteve Rothenberg, C., Azodolmolky, S., Uhlig, S.: Software-defined networking: a comprehensive survey. In: Proceeding of the IEEE, Jan 2015 (2015)
9. Cardenas, A.A., Amin, S., Sastry, S.: Secure control: towards survivable cyber-physical systems. In: Proceedings of 28th International Conference on Distributed Computing System, June 2008 (2008)
10. Bressan, N., et al.: The deployment of a smart monitoring system using wireless sensor and actuator networks. In: Proceedings of 1st IEEE International Conference on Smart Grid Communication, Gaithersburg, Oct 2010 (2010)
11. Bonomi, F., Milito, R.A., Zhu, J., Addepalli, S.: Fog computing and its role in the Internet of Things. In: Proceedings of 1st Edition MCC WorkshopMobile Cloud Computing, Aug 2012 (2012)
12. Botta, A., de Donato, W., Persico, V., Pescapé, A.: On the integration of cloud computing and Internet of Things. In: Proceedings of International Conference on Future Internet Things Cloud (FiCloud), Aug 2014 (2014)

Design and Implementation of Secured Cloud-Based Remote Home Access Through an Application for Smart Infrastructure

Annadanam Subbarathinam Shashank Karrthikeyaa, Ramakrishnan Priyadarshini, Gunasekaran Revathi, and Santhanam Sakthivel Murugan

Abstract Internet of things (IoT) is one of the most swiftly growing technologies predominantly and prevalently existing, creating a profound impact on the quality of living with each passing day. It encircles newer and better technologies such as smart cities, smart electric grids, remote smart homes. These technologies are emanating from the integration of the existing systems with the physical world that creates much better ones. Remote home access (RHA) is one of the emerging concepts of IoT enabled smart homes. Remote home access service is a system by means of which a user can remotely monitor the status of his devices and control them as well even from outside his/her home. This paper deals with the design and implementation of a system in which the requests of the users generated within the home or outside can intelligently be managed and controlled by cloud-based services. The functionalities of the system that is proposed here are classified into various service groups such as monitoring, controlling, security, authenticating, authorizing, managing, logging, and user groups. As a part of the RHA architecture, the connectivity between the cloud and the user application is encrypted using Secure Sockets Layer (SSL) certificates to prevent the infringement of user's privacy and the authentication is afforded based on Open Authorization 2.0 (OAuth 2.0) framework for better security.

Keywords Internet of things · Remote home access · Secure sockets layer · Open authorization 2.0

A. S. S. Karrthikeyaa · R. Priyadarshini · G. Revathi
Department of ECE, SSN College of Engineering, Chennai, India
e-mail: shashank16997@ieee.org

R. Priyadarshini
e-mail: rpriyadarshini15091@ece.ssn.edu.in

G. Revathi
e-mail: revathy15089@ece.ssn.edu.in

S. S. Murugan (✉)
Department of ECE, Under Water Acoustic Research Lab, SSN College of Engineering, Chennai, India
e-mail: sakthivels@ssn.edu.in

P. Suresh et al. (eds.), *Advances in Smart System Technologies*,
Advances in Intelligent Systems and Computing 1163,
https://doi.org/10.1007/978-981-15-5029-4_44

1 Introduction

The various applications of the Internet of things (IoT) find use across all kinds of markets and user groups. Smart homes or automated homes is one such part of the Internet of things which constitute a group of networking enabled devices that can regulate themselves according to the priorities of the user which are controlled by a user interface(application), providing comfort and convenience to the people who owns them. Such a system can make all the differences in the quality of our lives not only in terms of sophistication but also in terms of the efficiency including expenditure. This paper presents the smart home automation system, which could be accessed from a remote place, unlike the routine in-house access as highlighted by earlier researchers [1], using a remote production environment as a backend. Protocols like Hypertext Transfer Protocol Secure (HTTPS) and Message Queuing Telemetry Transport (MQTT) are used for communication.

2 Remote Home Access

Remote home access (RHA) is an integral part of IoT which aims to customize our personal abodes making it more optimized, efficient, and easily accessible. Currently, a lot of smart home devices are available in the market, but all of them are derived from and are made to support a specific vendor. The aim of the present work is to create a general interface that could be customized later on, according to the requirements of the user by developing an android application (user interface) using a production environment at the backend with multiple in-built functionalities that are integrated into a group of networking enabled devices—as done earlier in IOT-based home applications [2] , in addition to which user privacy and security are ensured keeping in mind the risk associated with today's IoT-based devices [3].

3 Remote Home Access Features

The following are the features proposed in the architecture.

1. Monitoring: The ON/OFF state of the devices such as lights, fans, or blinds is monitored through the user interface, gateway, and the backend interface.
2. Controlling: The ON/OFF state of the devices such as lights, fans, or blinds is controlled through the user interface, gateway, and the backend interface.
3. Authenticating: User authentication is enabled to afford system access to the users by using the OAuth 2.0 framework [4] that implements a well-constructed authentication system.
4. Authorizing: Users are divided into three groups [5] such as admins, who can add or remove users and devices, primary users who can both view the status

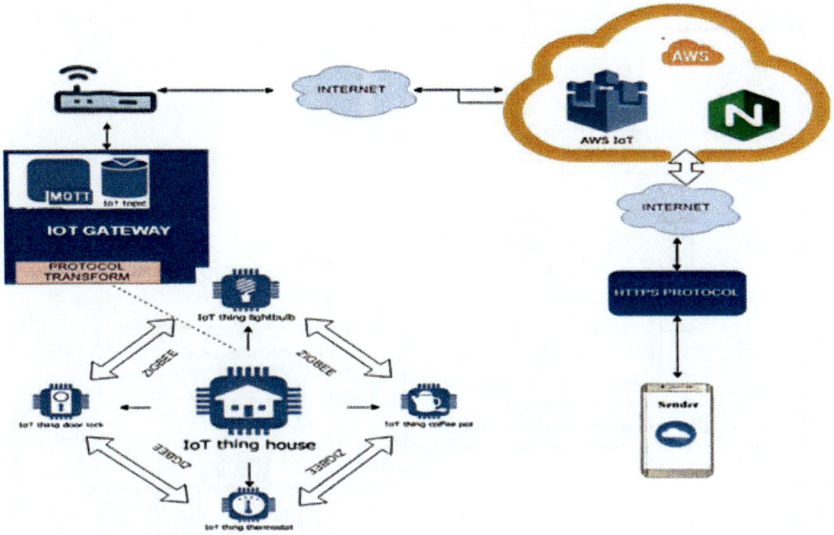

Fig. 1 Proposed system

of and control the devices and secondary users who can view the device status only. The functions and access rights of users are provided according to the user group.
5. Managing: Adding and removing users and reset password.
6. Logging: The various operations performed by the users are stored in a file along with the time stamp for future reference.
7. Security: Secure Sockets Layer (SSL) [6] certificates are used to encrypt the application to server connectivity, serving as a basic layer of security to take care of user privacy and data safety (Fig. 1).

4 System Architecture

4.1 Proposed Network Architecture

The architecture proposed in this paper is novel since it caters the needs of all kinds of devices along with the features described above. Unlike any other system, we have two types of devices.

1. That are enabled for IOT and
2. Another set of devices that are not enabled for the same.

A general interface has been developed for monitoring and controlling these devices. Hence, this architecture provides a better versatility with respect to the devices that can be controlled. A pictorial representation of the architecture is shown in Fig. 2.

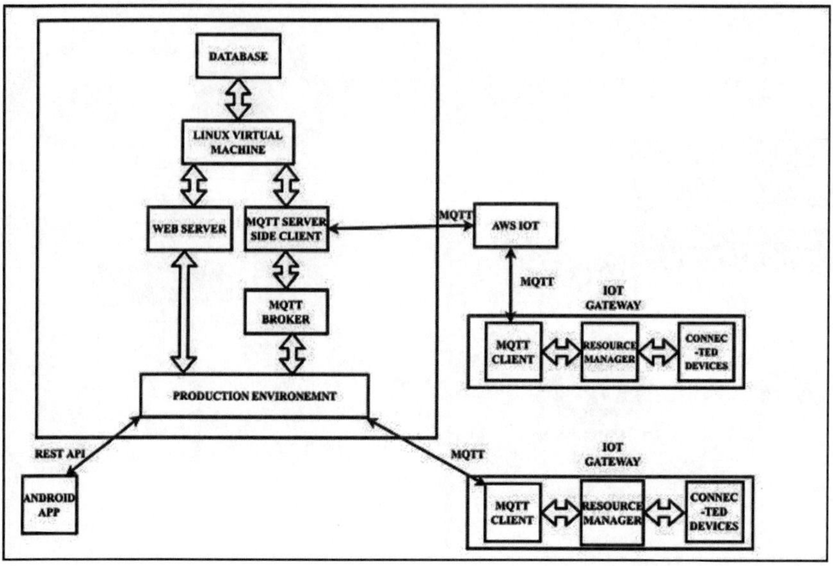

Fig. 2 Flow architecture

MQTT has been used for communication between the gateway and the cloud in view of its lightweight data-centric nature and publishes/subscribes model that is suitable for resource-constrained devices like gateways built using Broadcom chips or a simple Node MCU. It has a very low response time compared to HTTPS when sending multiple messages using the same connection [7]. REST API using HTTPS is used for communication between the android application and the cloud.

It is clear from the communication diagram in Fig. 3 that there is no direct communication between the Web and MQTT part. Both these separate entities are interfaced using databases.

A production environment with enhanced traffic handling capabilities is used for entire networking activities. Extended support has also been provided for devices that have the capability to communicate with AWS IoT.

The response to the request from the application and device is illustrated by various steps in the flowcharts shown in Figs. 4 and 5, respectively.

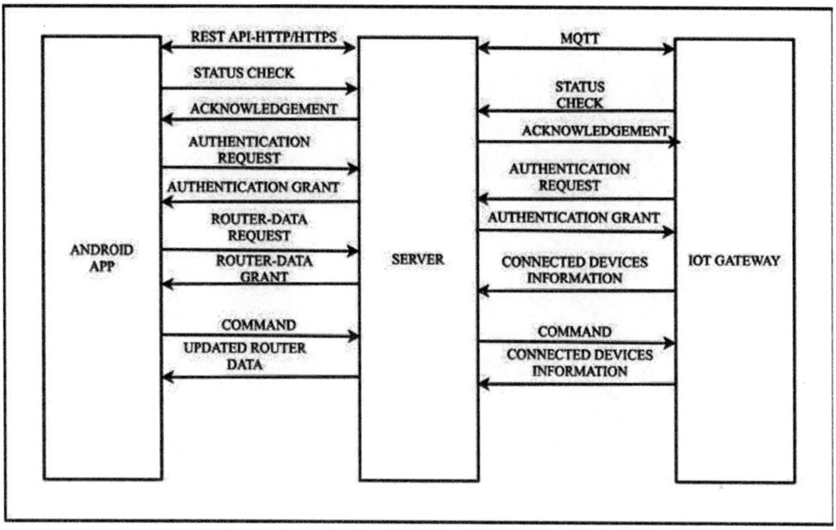

Fig. 3 Communication diagram

4.2 Proposed Security Mechanisms

In today's world, data security is very important [8]. In this proposed architecture, the following types of security measures are used.

1. OAuth 2.0 for REST API's [9].
2. Standard Authentication using credentials for MQTT.
3. SSL-based encryption for REST API's.
4. SSL-based authorization and encryption for MQTT.

OAuth 2.0 for REST API's

This involves randomly generated strings using user's ID and an encryption key to identify the user called as tokens. The generated tokens are used for further communication by the user after successful authentication. The access tokens are used to authenticate the users for further communication, and refresh tokens are used to regenerate access tokens. In the case when both the access and refresh token get expired, the user has to authenticate again. When the user log's out, the tokens are blacklisted to prevent any misuse. A new set of tokens are generated for the next login. The generated tokens are shown in Fig. 6.

Standard Authentication Using Credentials for MQTT

A database is created with randomly generated usernames and passwords. The MQTT broker is configured to only accept requests from the client with proper credentials. The access to publish and subscribe topics is also limited by using the correct configuration to prevent interference of various devices and gateways.

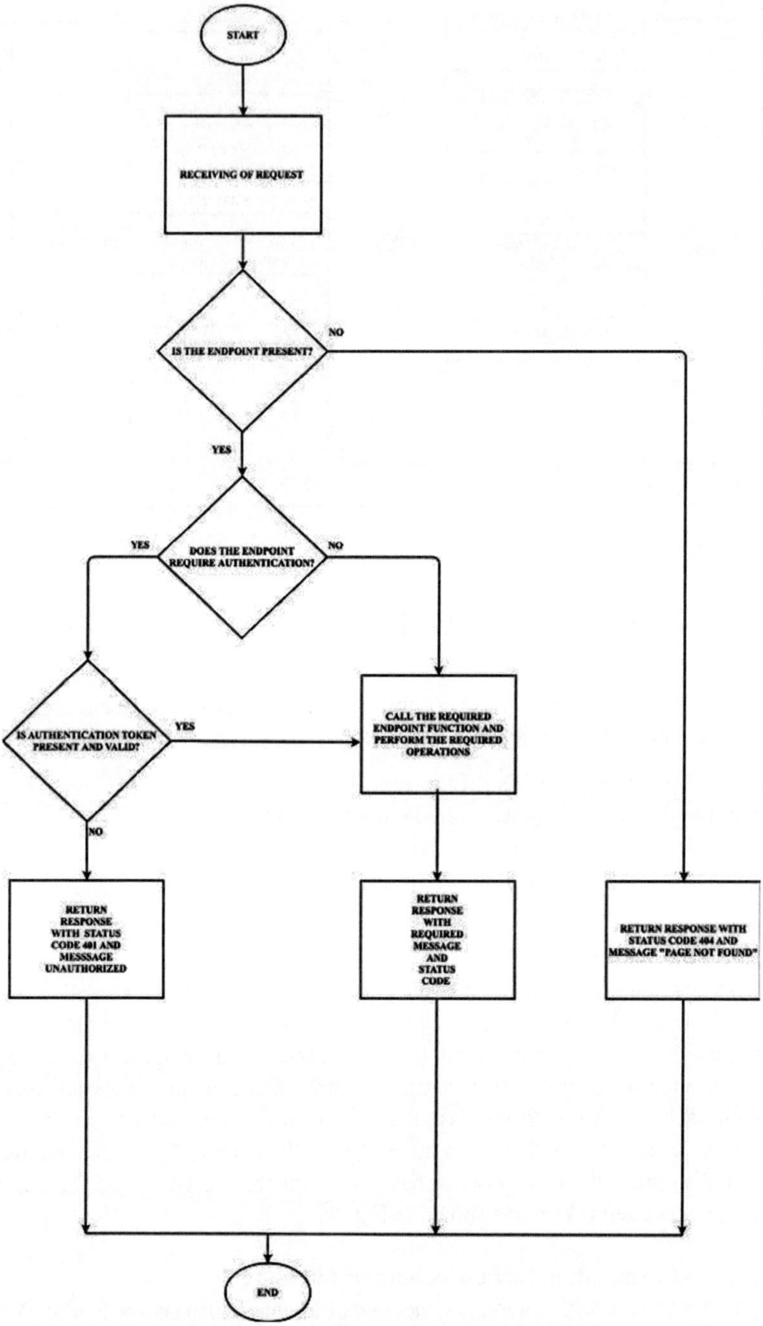

Fig. 4 Flow diagram for REST API communication

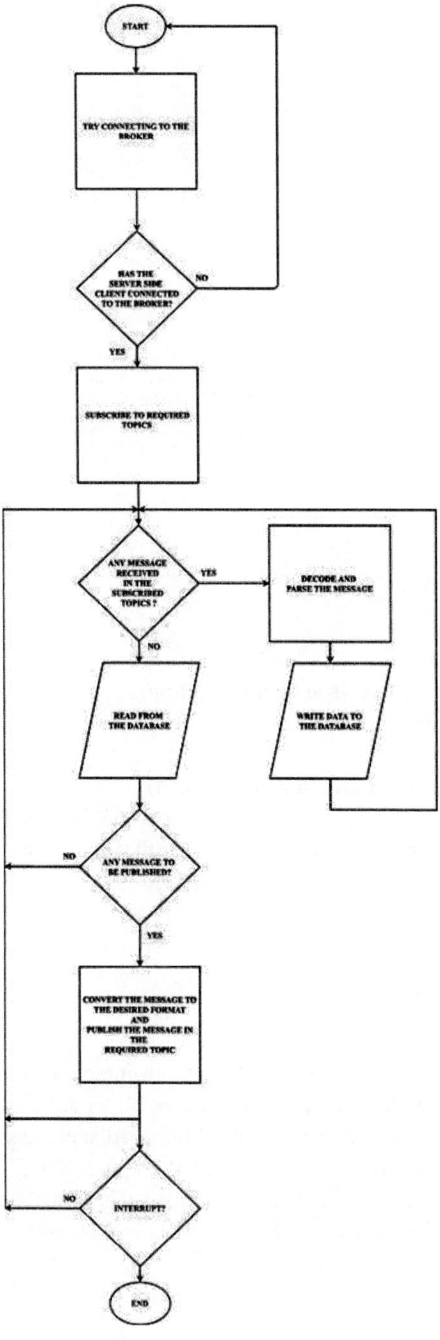

Fig. 5 Flow diagram for MQTT device communication

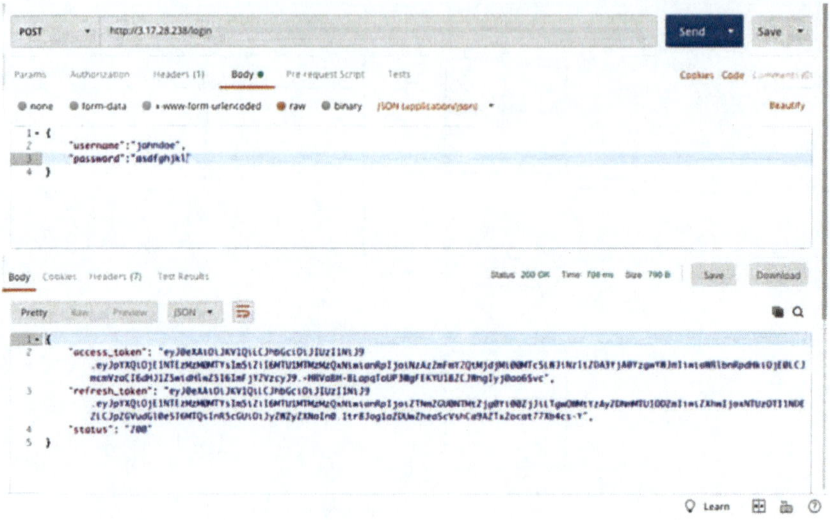

Fig. 6 Generation of access and refresh tokens

SSL-Based Encryption for REST API's

An SSL certificate obtained from a certification authority is used to encrypt the communication between the application and the server.

SSL-based authorization and encryption for MQTT

A pool of SSL certificates is generated for gateways and devices. Each SSL certificate is used to identify the gateway or device and at the same time provides encrypted communication between the broker and the device [9].

4.3 Proposed Hardware Architecture

As described at the beginning of this section, an interface to control devices not enabled for IOT has been developed. A Node MCU is used as communication as well as a control module. A channel relay of required specifications can be used to control n devices of that specification. A 220 V AC to 5 V DC transformer is used to power the Node MCU. The hardware architecture is shown below (Fig. 7).

Existing IOT-enabled devices are made to communicate using a snippet that converts messages to the required format.

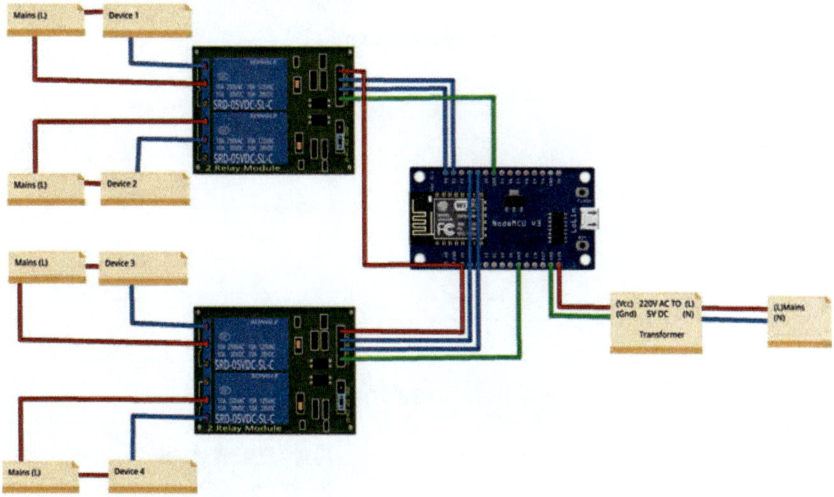

Fig. 7 Hardware architecture

4.4 Proposed Features

The proposed features like managing, controlling, grouping are achieved by performing required operations using databases.

5 Implementation Results

A remote home access system (RHA) with various functionalities such as authentication, authorization, managing, controlling, monitoring, logging has been. An android-based application has been developed for enabling the user to avail the above-mentioned functionalities. The login is aided using the OAuth 2.0 standards, and hence, a good user authentication is ensured. Various features such as adding users, removing users, adding gateways, removing gateways have also been provided to fulfill the stated functionalities (Figs. 8 and 9).

A stable production environment is used in the backend to cater to the needs of many users and devices. With respect to the hardware, the system developed at the frontend and the backend has been tested and verified using Node MCUs and IOT enabled devices. The working prototype is shown in Figs. 10, 11 and 12.

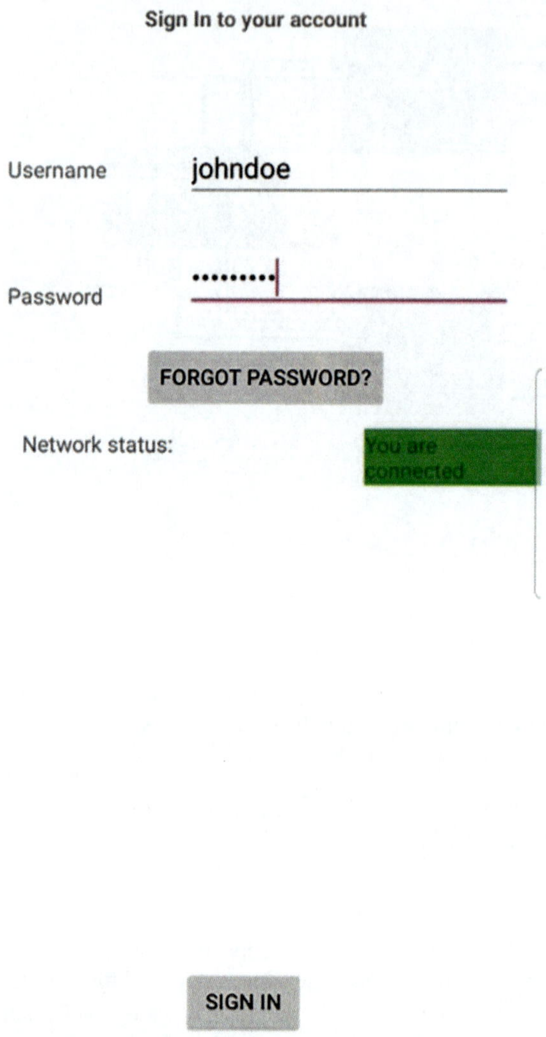

Fig. 8 Sign in page on the application

6 Conclusion

In a nutshell, this paper presents an application-based remote home access system that is integrated with authentication, authorization, monitoring, controlling, managing, logging, and security as its' features. A general interface is developed, and it could further be customized according to the needs of the consumer. These seven functionalities have been implemented, keeping in mind, the need for easy access, privacy, and security of the users and overall good performance and management of the system.

Fig. 9 Various options and devices displayed on the application

A well-furnished system, which can address all these needs, has been designed, developed, and implemented herewith.

7 Future Work

The features can further be extended with some more add-ons, such as reporting using push notifications in the lock screen to inform the user with the current device state without actually entering into the application and informing fault detection to

Fig. 10 A working prototype consisting of 4 bulbs

Fig. 11 Real-time implementation setup

Fig. 12 Real-time Implementation—Devices

the users in case of any appliance mal-functioning, low voltage, short circuit, and other abnormalities. Features such as commands to the device, such as brightness and color control for lights, speed for fans, the temperature for air conditioners can be added. This system can be automated by using sensors and giving them appropriate priorities of control. It can further be integrated with machine learning by monitoring the device data and its flow continuously which makes the user experience a more customized and user-specific system. The data from the device can be used for data analytics and thus can be used for various other purposes like refraining of defects in product and understanding user behavior.

Acknowledgements The authors extend their thanks to SSN College of Engineering, Aricent Technologies, for allowing the use of their hardware, Amazon Web Services whose free tier service is powering the entire backend and Underwater Acoustics Research Laboratory at SSN College of Engineering where work was done.

References

1. Mandula, K., Parupalli, R., Murty, C.H.A.S., Magesh, E., Lunagariya, R.: Mobile-based Home Automation using Internet of Things(IoT). In: 2015 International Conference on Control, Instrumentation, Communication and Computational Technologies (ICCICCT) (2015)
2. Wang, D., Lo, D., Bhimani, J., Sugiura, K.: AnyControl—IoT based Home appliances monitoring and controlling. In: 2015 IEEE 39th Annual International Computers, Software & Applications Conference (2015)
3. Jacobsson, A., Davidsson, P.: Towards a model of privacy and security for smart homes. In: 2015 IEEE 2nd World Forum on Internet of Things (WF-IoT) (2015)
4. https://oauth.net/2/
5. Dey, S., Roy, A., Das, S.: Home automation using internet of thing. In: 2016 IEEE 7th Annual Ubiquitous Computing, Electronics & Mobile Communication Conference (UEMCON) (2016)
6. https://www.verisign.com/en_IN/website-presence/online/ssl-certificates/index.xhtml
7. https://cloud.google.com/blog/products/iot-devices/http-vs-mqtt-a-tale-of-two-iot-protocols
8. https://www.purdue.edu/business/Security/pdf/printoutwebctmaterial.pdf
9. Sarker, S., et al.: An application for end to end secure messaging service on Android supported device. In: 2017 8th IEEE Annual Information Technology, Electronics and Mobile Communication Conference (IEMCON) (2017)

Development of Smart Electricity Energy Management System Using IoT

Senthilkumar Nikhitha, Younus Zohra Noori Mohsina, V. Sneha, and Santhanam Sakthivel Murugan

Abstract This paper aims to develop a system to have smart control of street lights. It is proposed with an aim to reduce the energy utilization of the existing system, to alleviate the maintenance of lights, and monitor the system using an android application. These objectives are accomplished using sensors that allow an automatic mode of operation to switch ON lights when both motion is detected, and light intensity falls below a certain threshold level. The system was implemented in the institution's Underwater Acoustic Research Laboratory on a small-scale basis. This paper also involves the application of an algorithm to find the shortest route from all possible paths in a mesh network. In a mesh network, there is a possibility of repetition of data being resent twice. To prevent this, we opt for Dijkstra algorithm which checks for the optimal path [1, 2] without flooding messages.

Keywords Internet of things · Smart control system · Firebase · Mesh optimization

1 Introduction

IoT is a network of connected objects or things. Here, application from a personal gadget and data from sensors communicates wirelessly over the Internet. The proposed system is developed owing to the extensive use of power in the present

S. Nikhitha · Y. Z. N. Mohsina · V. Sneha
Department of ECE, SSN College of Engineering, Chennai, India
e-mail: snikhitha15109@ece.ssn.edu.in

Y. Z. N. Mohsina
e-mail: zohra15129@ece.ssn.edu.in

V. Sneha
e-mail: sneha15107@ece.ssn.edu.in

S. S. Murugan (✉)
Department of ECE, Under Water Acoustic Research Lab, SSN College of Engineering, Chennai, India
e-mail: sakthivels@ssn.edu.in

P. Suresh et al. (eds.), *Advances in Smart System Technologies*,
Advances in Intelligent Systems and Computing 1163,
https://doi.org/10.1007/978-981-15-5029-4_45

lighting systems that consume most of the world's electricity. The efficient and intelligent use of resources is feasible through IoT. Smart street lighting system controls street lights based on movement detection. Also, mesh network has evolved over the years and is being considered a primary option in the industry for wireless communications [3]. Due to the preference given to wireless networks with the least human disruption, we opt for mesh networks. The above reasons pave the way for the development of this system.

2 Background of Study

The existing system has several practical difficulties, one of them being the uncertainty in turning ON and OFF during changes in the climatic conditions [4, 5]. This is one important difficulty existing in the current system. Traditional systems turn ON for most of the day even in the absence of people. This results in wastage of an immense amount of power unnecessarily.

The system aims at reducing the energy wastage due to the impractical use of lights on the street. It provides automatic ON/OFF with the help of a system embedded along with sensors which monitor the motion and brightness levels.

To eradicate the operator error and thereby minimize the energy utilized by the system, mesh network is established in street lights. Maintenance also becomes easier. In a mesh network, every node is connected to every other node. The algorithm presented in the paper checks for the shortest path from all the possible routes and selects the shortest path between the receiver and the transmitter, considering traffic as the primary deciding parameter. This allows faster data transmission between the nodes [6].

3 Proposed System

IoT is a system of interrelated objects that establish communication with each other to allow exchange of data through Internet. This communication can be through several wireless technologies, some of which are Bluetooth/BLE4.0, **Radio** RF, Wi-**Fi**, GSM/GPS [7]. Here, since we use ESP8266, Wi-Fi is chosen (Fig. 1).

4 ESP8266 Wi-Fi Module

Espressif system developed this Wi-Fi **module** (an SoC—system on a chip module) with a complete TCP/IP stack and capabilities of a microcontroller. Several IoT-based applications make use of this Wi-Fi module. Due to its low cost and high **speed** (computational capabilities), the module is easy to use. It can be either used

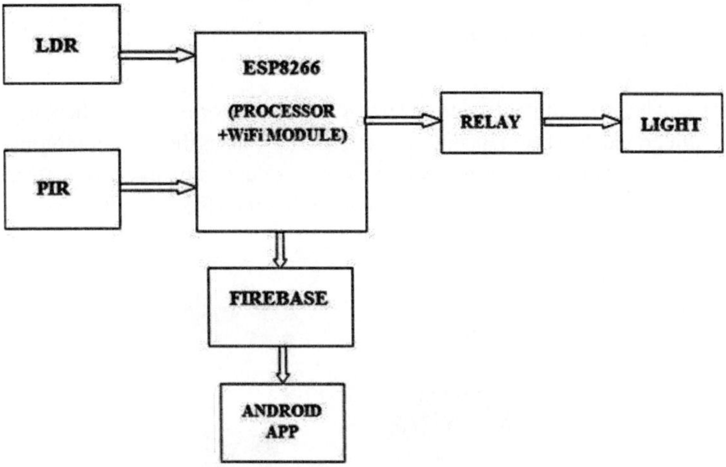

Fig. 1 Blockdiagram

as a microcontroller or a Wi-Fi board or both. Hence, it is easier to control sensors and transfer sensor data without any external influence. Below figure represents the ESP8266 (Fig. 2).

Fig. 2 ESP8266

4.1 *Configuring ESP8266*

ESP8266 can be programmed using several software platforms, one of them being the Arduino integrated development environment (IDE). This is the most commonly used open-source software which programs the Wi-Fi module using its set of instructions. The following steps are followed to configure ESP8266 (Figs. 3 and 4).

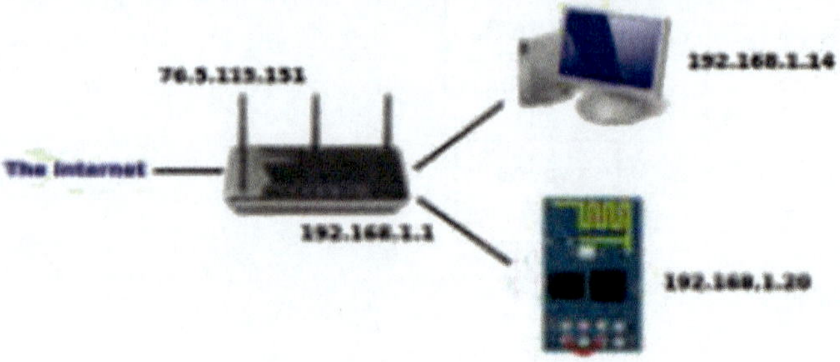

Fig. 3 EPS8266 connection by IP address

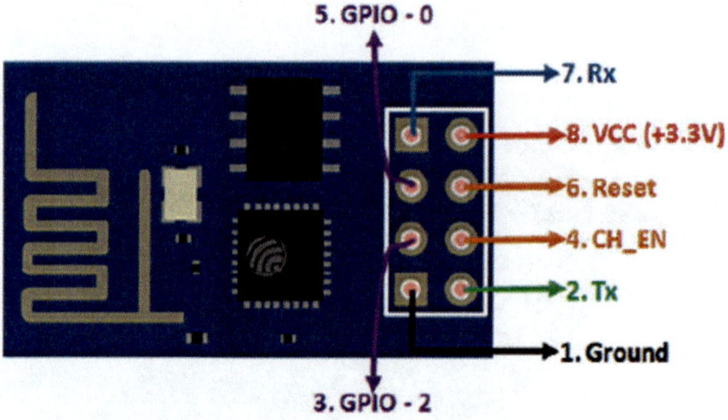

Fig. 4 Pin Diagram

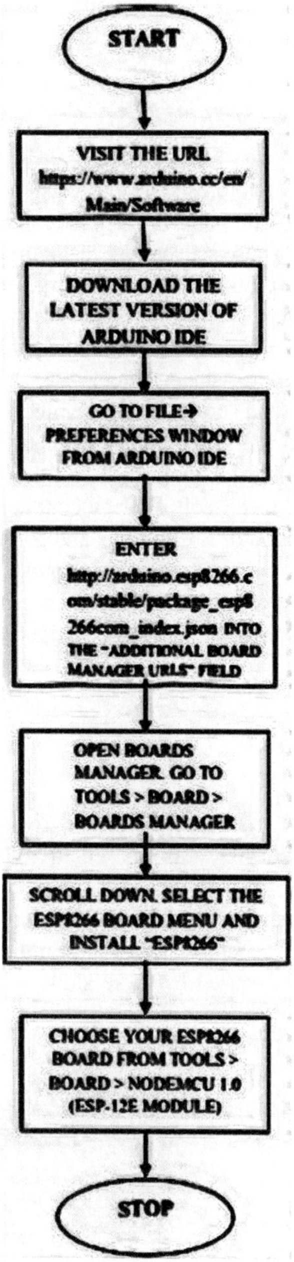
START
VISIT THE URL https://www.arduino.cc/en/Main/Software
DOWNLOAD THE LATEST VERSION OF ARDUINO IDE
GO TO FILE→ PREFERENCES WINDOW FROM ARDUINO IDE
ENTER http://arduino.esp8266.com/stable/package_esp8266com_index.json INTO THE "ADDITIONAL BOARD MANAGER URLS" FIELD
OPEN BOARDS MANAGER. GO TO TOOLS > BOARD > BOARDS MANAGER
SCROLL DOWN. SELECT THE ESP8266 BOARD MENU AND INSTALL "ESP8266"
CHOOSE YOUR ESP8266 BOARD FROM TOOLS > BOARD > NODEMCU 1.0 (ESP-12E MODULE)
STOP

5 Sensors

5.1 Light-Dependent Resistor(LDR)

An LDR is a light-sensitive photoresistor, the resistance of which changes when light is incident on it. As the light intensity increases, the resistance value decreases over a range of magnitudes. These sensors are made up of semiconductors (mostly, cadmium sulphide, CdS). Naturally, LDRs have a large resistance value, considering the fact that a very small number of electrons move freely. A huge number of electrons are confined within the lattice structure and are incapable to allow free movement. Henceforth, LDRs have a high resistance value.

When light is incident on the surface of the semiconductor, a part of the energy of photons is transmitted to the semiconductor electrons. Few of these electrons find this energy to be adequate for them to break the bonds from the crystalline structure, allowing them to move freely thereby conducting electricity. As a result, there is a drop in the semiconductor's resistance, and consequently, this decreases the resistance of the photoresistor on the whole. This course of action continues as light continues to fall and new electrons become free, increasing the overall conductivity. This contributes to further decrease in resistance (Figs. 5 and 6) [8].

5.2 PIR Motion Sensor

Passive infrared sensor (**PIR**) is made up of a pyroelectric sensor. It senses the presence of people near its sensing area. Its sensitivity ranges from 5 m to 12 m, average value being 10 m.

This sensor can distinguish between varying amounts of infrared radiations. It consists of 2 slots which are receptive to infrared radiations. In the normal state, equal amount of radiations is sensed by these slots. If a human being walks past

Fig. 5 A typical LDR

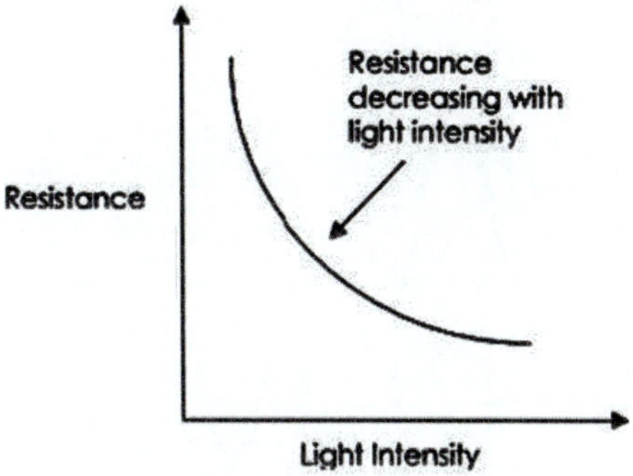

Fig. 6 LDR resistance versus light Intensity graph

the sensor, positive differential pulses are generated and the opposite occurs when the person leaves the area of detection. The detection is based on the pulse changes (Figs. 7, 8, 9 and 10).

Here, the light glows when motion is detected in dark surroundings. Its ON period extends for 10 s when continuous motion is detected. When no motion is detected, it turns OFF after a fixed time.

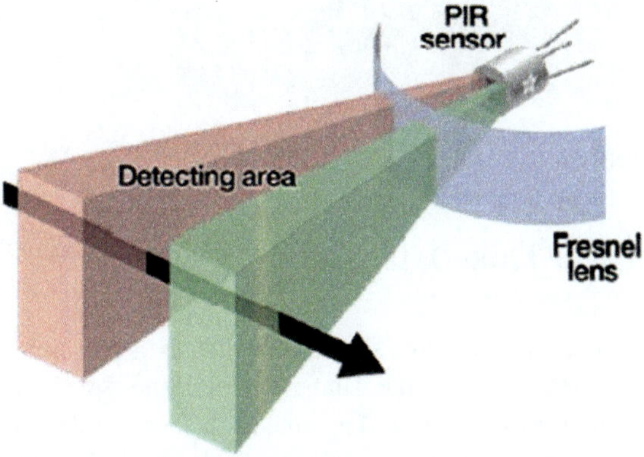

Fig. 7 Detection area of PIR

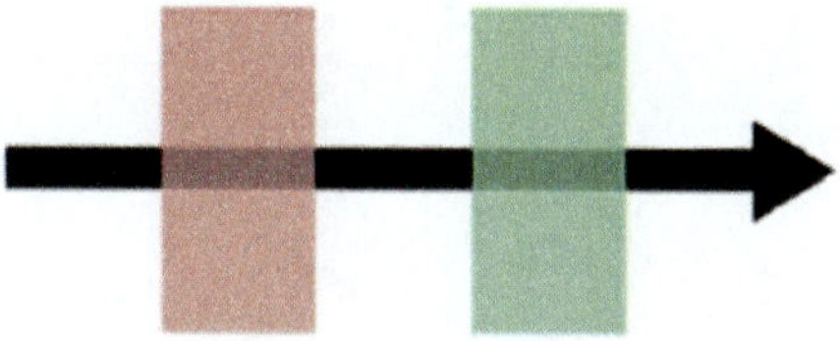

Fig. 8 Heat source movement

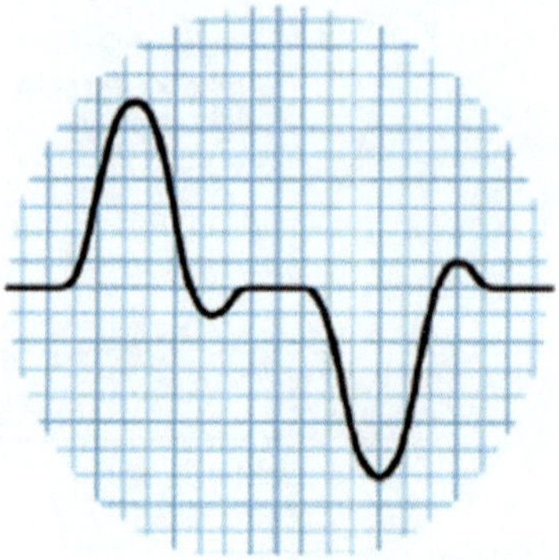

Fig. 9 output signal

Fig. 10 PIR with ESP8266

6 Firebase (Real-Time Database)

The application requests data from Firebase, which is a real-time database. The mobile application is connected to the database with the help of Web sockets, unlike other databases which involve HTTP. This makes the sync of data quite faster. The statuses of the bulbs are communicated to the application via Firebase. Also, the number of units consumed per day is updated to the database.

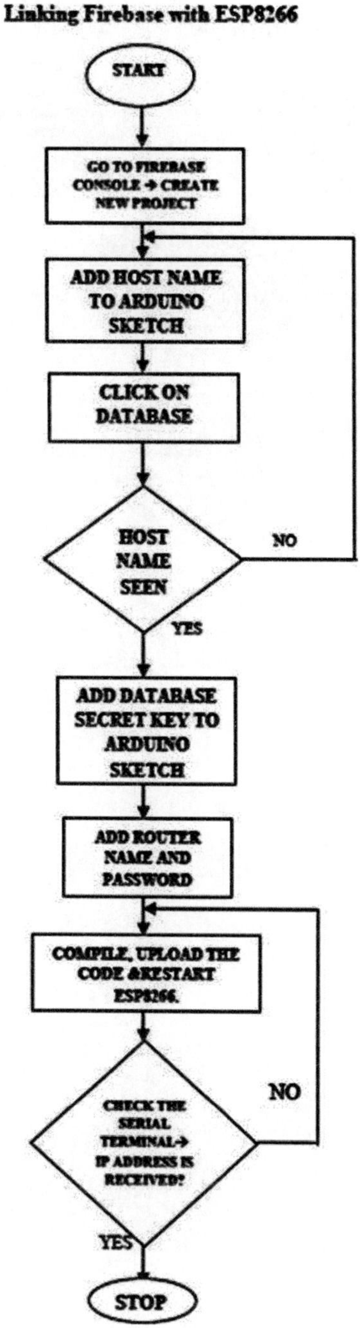

Fig. 11 Firebase console

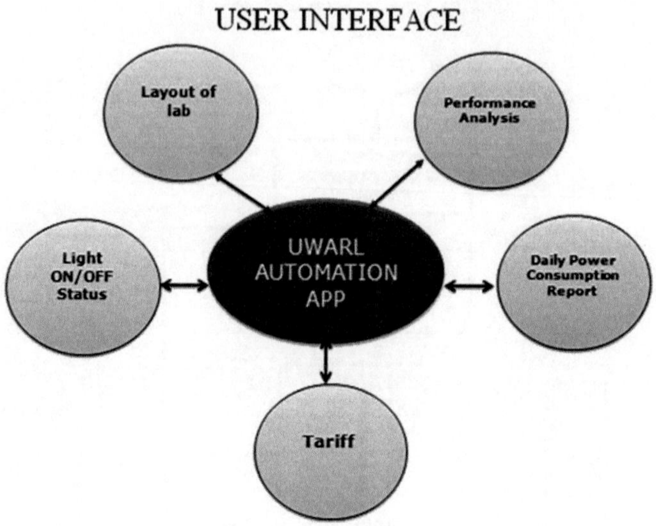

Fig. 12 Features of the application

Linking Firebase with ESP8266

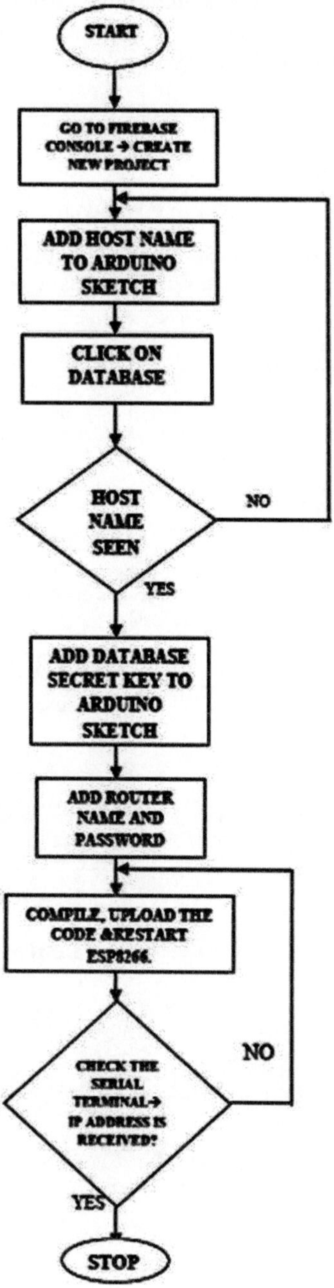

By following the above steps, the application is developed and connection is established between the database and the application. When values are varied, the changes are viewed on the console (Refer Fig. 11).

7 Android Application

We monitor the proposed system using an android application, created using MIT App Inventor 2. The application has several features which is shown in the Fig. 12 [9].

Screenshots of the Application

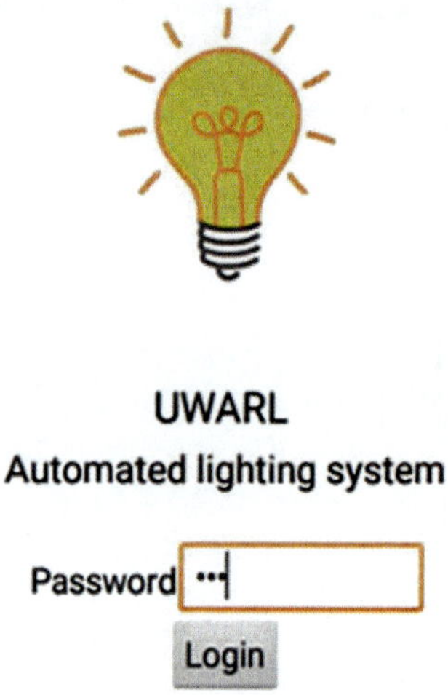

1. Authentication

2. Hardware features

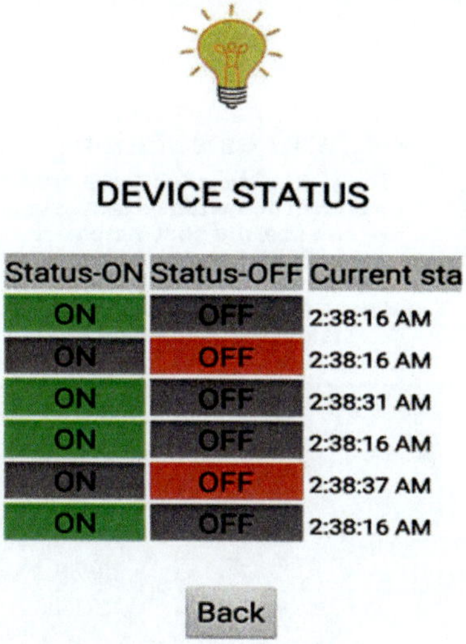

3. Physical layout

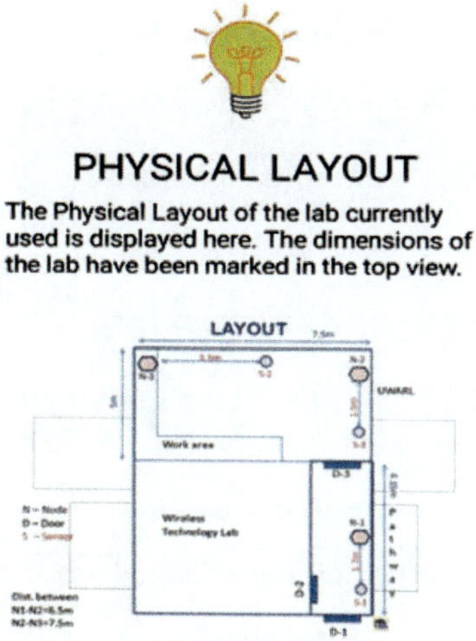

4. Status of the bulb

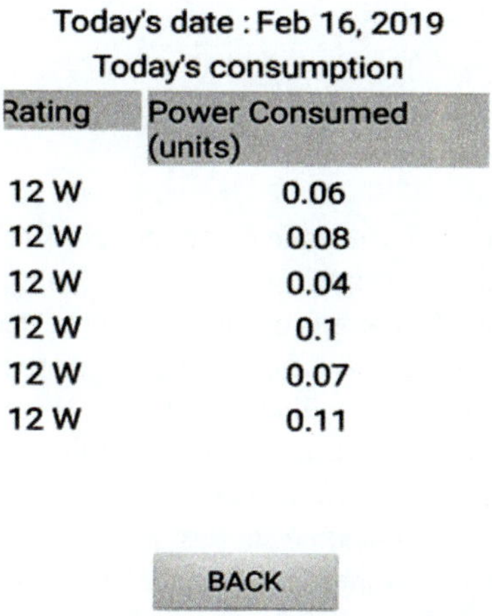

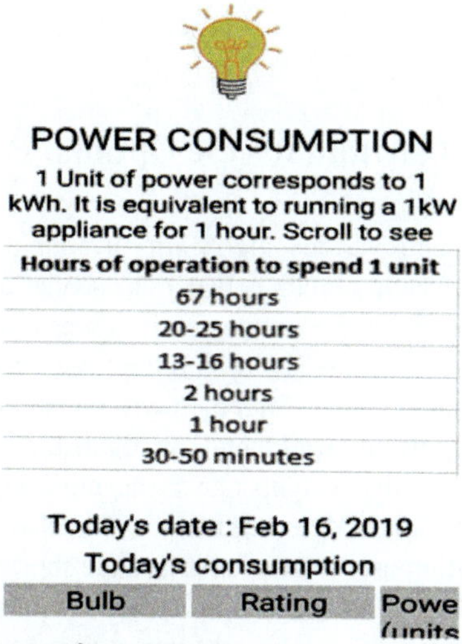

5. Power consumption report

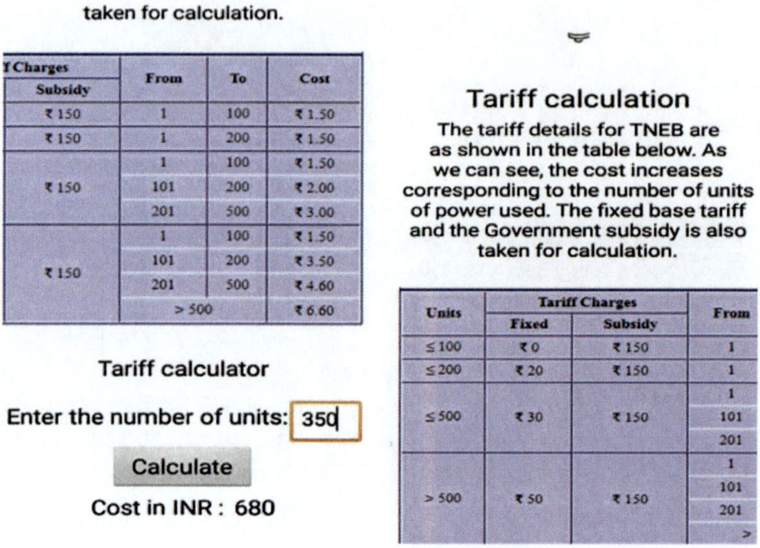

6. Tariff calculation

8 Algorithm for the Optimal Path

There are various algorithms available to find the efficient route or path for transmission of data between the sender and the receiver.

The lights are connected in a mesh network where every node is connected to every other node. So to find the fastest path we go for different algorithms such as Bellman–Ford, A*search, Floyd–Warshall, Dijkstra, flooding algorithm and breadth-first algorithm. Bellman–Ford algorithm is used to solve the problem with the single-source if edge weights may be negative. Dijkstra's algorithm solves the problem with non-negative edge weights which is also for single-source shortest path. A* search algorithm uses heuristics to find the shortest path to speed up the search mechanism. Floyd–Warshall algorithm solves all pairs of shortest paths and finds all possible shortest paths which are unnecessary. So we go for link-state routing algorithm which is a dynamic algorithm that finds the shortest path during the run-time and one such link state algorithm is Open Shortest Path algorithm. We also consider the traffic and assign weights unlike OSPF where the path with the shortest distance is selected. So we go for Dijkstra's shortest path algorithm which even considers the negative weights.

The positions of nodes are static, which is already fixed and not varied. All the other algorithms will work in directed graphs only. A mesh network in which all nodes are connected is assumed, and the number of nodes is taken at input in the form of a value. A traffic matrix is automatically generated randomly that various every time a new message is sent [10]. The matrix is generated randomly here because the algorithm is not embedded in the node, so the traffic cannot be detected by the algorithm.

$W(i, j) = 0$ (Weight), if $i = j$.
$W(i, j) =$ Edge/sweight, if V_i and V_j are connected directly (the outer link).
$W(i, j) =$ Edge/sweight, if V_i and V_j are connected directly (the inner link).

To find the shortest distance from the source vertex to the destination, all possible paths to travel between the two are traversed and the distance is calculated and the data is stored in an array, say len[], that contains the distances between the source and destination. The information about the nodes is stored in another array, node[] which is to be accessed later to find the path through which the message hopped in order to reach the destination. Every time when the loop is repeated, the same process is done and the nodes get updated and stored in the array node[]. There is another two-dimensional array a[][], which is used to form the input graph of vertices and a traffic matrices traffic[][]. Eventually, the shortest distance is calculated from source to destination (Fig. 13).

So the algorithm computes the shortest path between the two nodes, i.e. between the transmitter and the receiver, and checks all the possible ways and also the traffic between the path and chooses the optimal path.

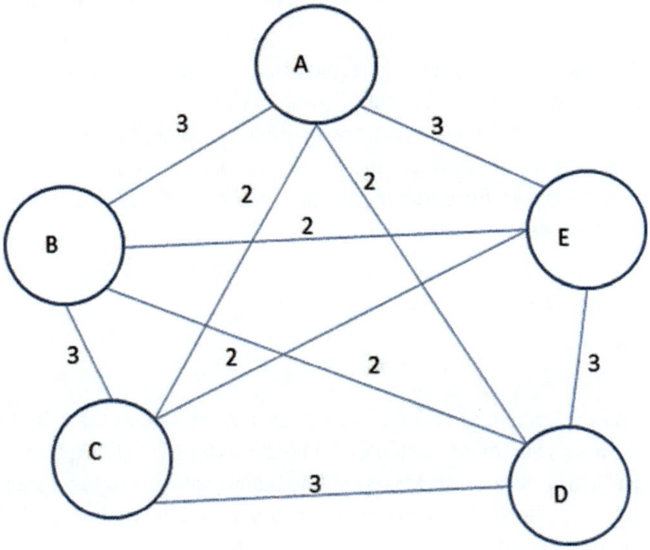

Fig. 13 Mesh network

8.1 Optimal Path Selection with Traffic

The Algorithm is Stated Below

- Enter the value of n-the no. of nodes to be present in the mesh network, the delay of the outer consecutive link -u and the inner direct link-v.
- Generate a random traffic matrix traffic[][] of size $n \times n$.
- Enter the node value of the sender-Vs and the receiver-Vr.
- Repeat step 5 until $i = 1, 2, \ldots. n$ to consider all nodes.
- Repeat loop for $i = 1, 2, \ldots, n$. Start from sender and traverse to every node in matrix and store the distance and the node no. in the corresponding matrix len[][] and node[][].
- Now the path is repeated to find the shortest value. Store the value in a matrix a[][].
- Now add the matrix a[][] with the randomly generated traffic matrix traffic[][] and save as $a[i][j] = .a[i][j] + \text{traffic}[i][j]$.
- [End of loop]
- Set the value of $a[i][j]$to 0 if i ==j and reset the value len[vs] to 0. The value found is the dist for communication between the two.
- Repeat loop while(vd is not included in the set[]). Set j = search_min(len,) [This function to find the shortest dist] and then backtrace the node path in the matrix node[][].
- Print the shortest cost and then the shortest path by backtracking.
- Exit (Figs. 14 and 15).

In the O/P of the algorithm, the loop is run thrice with the source and the destination in all 3 cases as node 4 and node 2 and the number of nodes is 5.

So in the first case, for the random traffic matrix, we get the shortest path to be traversed is via 1 or 4 out of which one is selected. In the second case, the shortest changes to via 2 depending on the traffic, and in the third case, it is similar to the first case where the path is via 1 and 4.

9 Results

The design is developed and is found that it is more proficient than the traditional system. Three nodes are implemented in the Underwater Acoustic Research Laboratory, in which each node controls two bulbs and their respective statuses are displayed on the developed application screen. An algorithm to find the optimal path in a mesh connected lighting system which has traffic as its main constraint is developed.

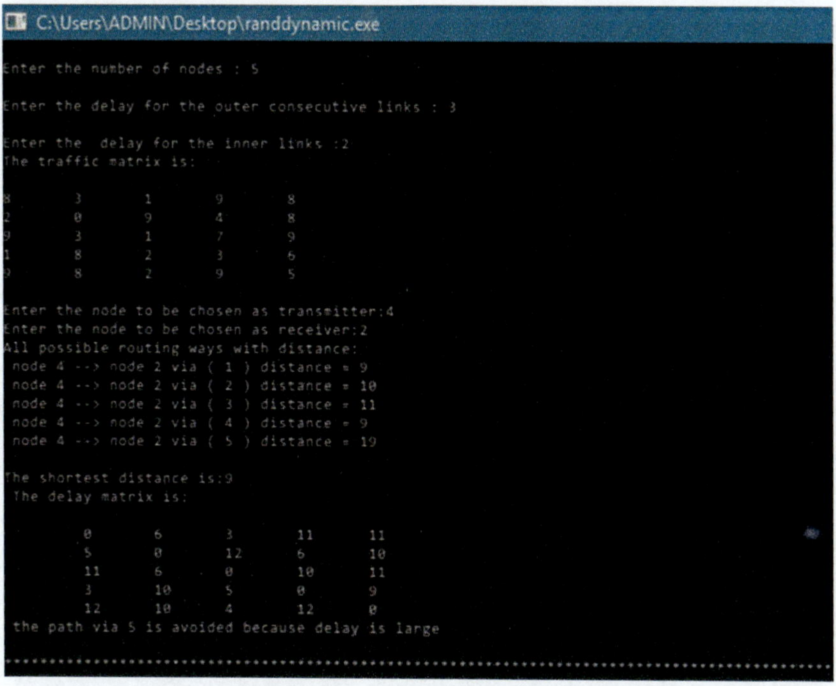

Fig. 14 a Screenshot of output window

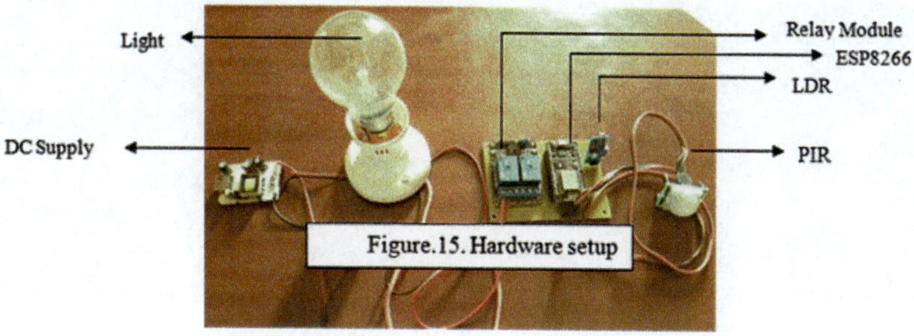

```
The traffic matrix is:

5       4       3       5       9
8       2       9       5       2
6       6       2       6       8
5       6       1       1       5
0       3       6       8       6

Enter the node to be chosen as transmitter:4
Enter the node to be chosen as receiver:2
All possible routing ways with distance:
 node 4 --> node 2 via ( 1 ) distance = 14
 node 4 --> node 2 via ( 2 ) distance = 8
 node 4 --> node 2 via ( 3 ) distance = 13
 node 4 --> node 2 via ( 4 ) distance = 8
 node 4 --> node 2 via ( 5 ) distance = 13

The shortest distance is:8
 The delay matrix is:

        0       7       5       7       12
        11      0       12      7       4
        8       9       0       9       10
        7       8       4       0       8
        3       5       8       11      0
 the path via 1 is avoided because delay is large

*****************************************************************************

The traffic matrix is:

4       1       9       1       4
8       9       7       2       6
2       8       5       9       9
2       4       1       3       3
8       0       5       4       7

Enter the node to be chosen as transmitter:4
Enter the node to be chosen as receiver:2
All possible routing ways with distance:
 node 4 --> node 2 via ( 1 ) distance = 8
 node 4 --> node 2 via ( 2 ) distance = 6
 node 4 --> node 2 via ( 3 ) distance = 15
 node 4 --> node 2 via ( 4 ) distance = 6
 node 4 --> node 2 via ( 5 ) distance = 8

The shortest distance is:6
 The delay matrix is:

        0       4       11      3       7
        11      0       10      4       8
        4       11      0       12      11
        4       6       4       0       6
        11      2       7       7       0
 the path via 3 is avoided because delay is large
--------------------------------
Process exited after 19.2 seconds with return value 0
Press any key to continue . . .
```

Fig. 15 b Screenshot of output window

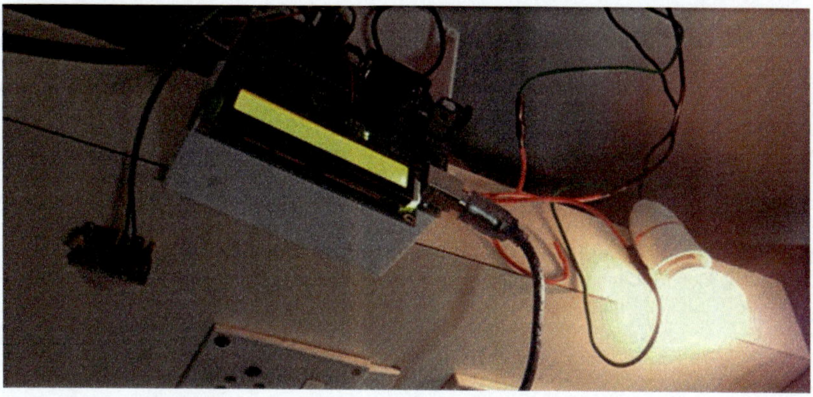

10 Conclusion

The developed system is used for smart control and monitoring of lights. It is evaluated in a real-time scenario within an indoor location for five lights. The system works efficiently, and there is a decrease in the consumed power. At the same time, the required conditions are satisfied. The system is less complex and lowers the overall costs required for maintenance because of its periodic checks on the lights using the application. The future work includes implementation of the algorithm and expanding the number of nodes for an easier execution.

Acknowledgements The authors extend their thanks to SSN College of Engineering, Aricent Technologies for their guidance and Underwater Acoustic Research Laboratory at SSN College of Engineering where the work was done.

References

1. Kumaravel, K., Mincy Sabu, D.A., Marimuthu: Optimal routing algorithm used in wireless mesh networks for energy efficiency. Int. J. Eng. Sci. Res. Technol. **3**(1), (2014)
2. Goldberg, A.V., Kaplan, H., Werneck, R.: Reach for a*: efficient point-to-point shortest path algorithms. In: Proceedings of the Workshop on Algorithm Engineering and Experiments (ALNEX). 129–143 (2006)
3. Akyildiz, I.F., Wang, X.: A survey on wireless mesh networks. IEEE Commun. Mag. **43**(9), S23–S30 (2005). (Pubitemid 41638080)
4. Abinaya, B., Gurupriya, S., Pooja, M.: IoT based smart and adaptive lighting in street light'. In: International Conference on Computing and Communications Technologies (ICCCT) **4**(2), (2017)
5. Archana, G., Aishwarya, N., Anitha, J.: Intelligent street lightweight system. Int. J. Recent Adv. Eng. Technol. **3**(4), (2015)

6. Srivatsa, D.K., Preethi, B., Parinita, R., Sumana, G., Kumar, A.: Smart street lights. In: Texas Instruments India Educators' Conference
7. Deo, S., Prakash, S., Patil, A.: Zigbee-based intelligent street lighting system. In: 2nd International Conference on Devices, Circuits and Systems (ICDCS), Coimbatore (2013)
8. Harikiran, G.C., Menasinkai, K., Shirol, S.: Smart Security Solution for Women based on IoT International Conference Electrical. Electronics, and Optimization Techniques (ICEEOT)
9. Sudhakar, K.S., Anil, A.A., Ashok, K.C., Bhaskar, S.S.: Automatic street lightweight system. Int. J. Rising Technol. Adv. Eng. **3**, (2013)
10. Rudrawar, O., Daga, S., Chadha, J.R., Kulkami, P.S.: Smart Street Lighting System with Light Intensity Control Using Power Electronics. Technologies for Smart-City Energy Security and Power (ICSESP), Bhubaneswar

IoT-Based Automatic Library Management Robot

V. S. Prabhu, R. M. Abinaya, G. Archana, and R. Aishwarya

Abstract Robots are engaged in various activities in our day-to-day life. The aim of this paper is to mitigate the need of a librarian for collecting books from the library counter and arranging them in respective racks one by one. The main principle of the Library automation robot is based on the following: Radio Frequency Identification Technology (RFID), dynamics of the robotic arm and Line Following. The software module allows the user to reserve books that are available in the library. This robot in the library will be helpful in eliminating the problem of misplacement of books. This line follower robot scans the RFID tag in the book and finds the place of the rack where the book needs to be placed. The proposed robot will pick and place the books in its respective racks. This project also provides the database of books available for reservation purposes regardless of the place. At the same time, it sends a reminder message to the user about the due dates. The aim of this robot is to replace the manual work in the library.

Keywords Radio Frequency Identification · Line Follower

1 Introduction

Robots are engaged in various activities in our day-to-day life. Considering this, the paper intends in reducing the efforts of a librarian, who arranges books in a library and this project also facilitates to reserve books over Internet of Things. The data about the books and its respective racks will be previously fed into the robot so that the robot can pick and place the books in the respective racks [1]. The robot will reach the respective rack following the white or black lines made on the floor and places the books into the shelves. The software interfaced with the robot provides details about

V. S. Prabhu (✉)
Department of ECE, Faculty, R.M.D. Engineering College, Chennai, Tamil Nadu, India
e-mail: vsp.ece@rmd.ac.in

R. M. Abinaya · G. Archana · R. Aishwarya
Department of ECE, R.M.D. Engineering College, Chennai, Tamil Nadu, India

P. Suresh et al. (eds.), *Advances in Smart System Technologies*,
Advances in Intelligent Systems and Computing 1163,
https://doi.org/10.1007/978-981-15-5029-4_46

the availability and reservation of books. This robot eradicates the misplacement of books, considered as a major problem in the library. This robot also fetches the book from the shelf and gives it to the user at the front desk [2]. The Line Follower Robot is an automated guided vehicle that follows the black or white lines that are embedded on the floor or ceiling. The robotic arm picks the required book and returns to its initial position to deliver the book. The information regarding all the books in a library is fed in the system database [3].

The name and author information of all the books will be stored in the database. The working principle of the proposed model states the returning of books to the predestinated location in the library. Each book will be attached with a distinctive RFID tag. By scanning these individual distinctive tags on the books, respective destination locations are traced by the sensors [4]. Shelf locations are labelled by a stop pit of the line tracer. The total books that are available in the library will be shown on the webpage that is created and from that the books can be reserved by the user from anywhere. Using sensor-operated motors the robot is designed, to trace the path of bookshelf arrangements in the library. The path to the respective rack is tracked with the help of database information which is already fed into it.

2 Existing System

The usual and current procedure of most of the libraries is, in order to get some books for our exams or for some research purposes. The following issues will be faced. Once we arrived at the library, we would face an unexpected issue: There were at least 50 or 60 other students waiting in a line at the librarian's desk, all of them wanting the same thing as we wanted: quick access to information. After half an hour it was becoming obvious that the entire process of serving a book to a student was too slow. The librarian had to take several time-consuming steps for each student. Formerly, every student had to be identified. Then they have to make their request. Later, the librarian will guide the student where to find the book. Sometimes, the student will not be able to find the book. Then, they will return back to the librarian to seek additional information. Several times, the librarian will go away from his desk to find the book. At times the book will not be available in the library itself. All these will happen while 50 other people were waiting in line. The problem here is obvious: This part of our life has not been developed by modern technology. Probably, the librarians will have a PC in which they search for the student identity. But still, the entire process consumes more time. Thus, to overcome this book sorting issue, an automated pick and place robot model are proposed.

3 Proposed System

The three main units of this system are as follows:

- RFID Reader.
- Line follower arm robot.
- Software Interfacing (Fig. 1).

1. ***Actuation***: Actuators give the power supply of a robot, the parts which change over put away power into development. An actuator responds by receiving a control signal and convert them into mechanical motion. Electric motors control wheels and straight actuators control robots.
2. ***Microcontroller***: The architecture used in the robot is Atmega-328P. Microcontroller controls the entire system which acts as the heart of the project. ATMEGA328P is a 28-pin chip with 8-bit AVR processor. Two 8-bit counters with Separate Prescaler and compare mode, one 16-bit counter with Separate Prescaler, compare mode and capture mode. It contains RAM of 2 KB Internal

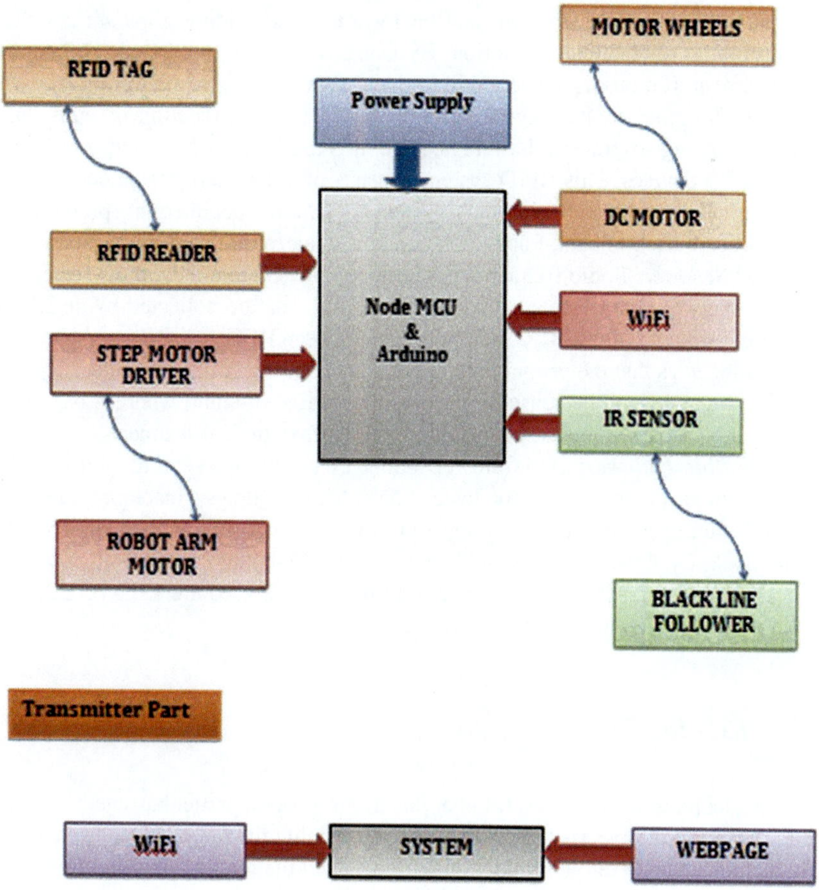

Fig. 1 Block diagram depicting the components involved in designing the automatic library management robot

SRAM, programmable flash memory of 64 KB Flash, 3 Timers (Programmable Watchdog Timer with Separate On-chip Oscillator), 2 external interrupts, 32 GPIO's, ISP programming support, etc. Microcontroller is programmed using KEIL IDE coding done in Embedded C.
3. ***The Base Shaft***: The base shaft is directly connected to the rotary base. It gives height to the device. Also, optional movement can be provided to the base shaft so that it can be tilted forward or backward with the help of motors.
4. ***RFID Module***: RFID (Radio Frequency Identification) module is used to read the identification number in RFID Tags, sends data regarding the ID number serially through UART.
5. **Moving arm**: It is attached to the DC gear motor with fixed body as a platform for support. The arm motion can be controlled in both vertical and horizontal direction to an extent of 90°.
6. ***Gripper***: At the front part of the arm, Gripper is sited. With the help of DC gear motor, they can slide forward and backward. These sliding grips are coupled to arm end for facilitating motion. By using worm gear, the angular (circular) movement of motor is converted in to linear movement of sliding grip.
7. ***RFID Tagging***: RFID tagging system is used for identification of individual things among several similar things. Also, it is used for tracking purposes using small RF devices. This RFID system consists of a tag itself, for device read or write, and also provided with a host system application for collecting, processing, and transmitting of data. Each book contains a tag for unique identification.
8. ***RFID Scanner***: Radio frequency Identification refers to a technology whereby RFID tags or smart labels encoded with digital data are collected by an RFID reader through radio waves. A comparison between RFID scanning and barcode scanning says that the primary has some advantages over the later one. Like the RFID tag, data can be sensed even outside the line-of-sight, whereas barcodes cannot be accessed in case of misalignment with an optical scanner.
9. ***Black Line Follower***: Line follower robots are used in the semi as well as fully autonomous plants. This robot detects the path and follows a black or white line that is made on the floor or ceiling. Here the robot will follow the path towards the book. On a white surface, a black line is made on the floor or ceiling which acts as a path for the robot. The sensors that can be used are LED or LDR or Photodiode or Phototransistor pairs (Fig. 2).

3.1 RFID Reader

RFID is a technology that is useful in applications where a particular object is to be found over several objects. It does the work by sensing the radio frequency waves. This includes a tag fixed with individual objects and the reader reads and finds the particular object we request for [8].

Library Management System (LMS) using RFID technology helps to find the books faster over several books and it is highly secured. This system gives a large

Microcontroller	ATmega328P - 8 bit AVR family microcontroller
Operating Voltage	5V
Recommended Input Voltage	7-12V
Input Voltage Limits	6-20V
Analog Input Pins	6 (A0 - A5)
Digital I/O Pins	14 (Out of which 6 provide PWM output)
DC Current on I/O Pins	40 mA
DC Current on 3.3V Pin	50 mA
Flash Memory	32 KB (0.5 KB is used for Bootloader)
SRAM	2 KB
EEPROM	1 KB
Frequency (Clock Speed)	16 MHz

Fig. 2 Specifications of various components used in designing the automatic library management robot

benefit to the library and to the librarian. Library also consists of information, books, magazines, etc.

In addition to books, many libraries are now also depositories and access points for maps, prints, or other documents. The media storage such as microform like microfilm or microfiche, audio recorded tapes, Compact Disks, LPs, cassettes, videotapes, and Digital Versatile Disks. Libraries have materials arranged in a specified order according to a library classification system, so that items may be located quickly, and collections may be browsed efficiently [9]. There is also Reference stacks available which are different and have only book references. Only selected members can access it.

CDAC is the largest library having 17,000 books. It also has both paraprofessionals and professional librarians in it.

It maintains the user account, book issue and return and arranging of books on right shelf.

This library maintains collection of books, ordering books and maintains library budgets.

Technical Services work behind the scenes cataloguing and processing new material and deaccessioning weeded materials.

Basic processes involved in a library are planning about the materials present in it, proper arrangement of books based on the predefined classification, storing the details of books borrowed and returned. This system will automate the following tasks using RFID technology.

Gives access to many number of books at a time.
To check the presence of a particular book among several books.
Identifying the correct area of location of the book.
Verification of the materials stock and accounting can be easily done.

3.2 Line Follower Robot

Nowadays, Robotics is an emerging technology in the field of research. By developing new robots, it can be used for various practical purposes in various fields. Many robots are designed in such a way that it can perform various works that are hazardous to people such as defusing bombs, mines and exploring shipwrecks.

The library management robot reduces the problem of searching books manually by using Dewey classification and also places the book in its respective rack. It identifies each book using the RFID tag attached to it and the data or information about the books will be already fed into the robot.

The Library Management Robot is developed with a base that has two motors with two degrees of freedom and the column has four motors with four degrees of freedom. These motors are operated with the help of—ATmega328P –8 bit AVR family microcontroller. The robot uses IR sensors for detecting the path which is interfaced with Arduino [9]. This microcontroller is programmed in such a way that the robot performs the function based on our requirements.

The Line Follower Robot is an advanced robot with reliability. This robot follows a particular path based on the information already fed into it. The path made on the floor or ceiling will be a black line on the white surface. Black Line Followers are the basic principle through which the robot tries to take one step ahead to the bookshelf. It is used for both domestic and industrial purposes, etc.

Since the Line follower robot is designed with the pick and place functionality, it could be able to pick the respective book we require and also place the book in the correct shelf. The robot can be cylindrical which can move horizontally and vertically. The spherical robot can provide two rotational and one linear movement. An articulated robot is a fixed robot with three vertical axes rotatory arms. We use this technology to provide the functionality to robot that it picks the book and place it where it is directed (Fig. 3)

Fig. 3 The library management robot picking the requested book from its respective rack

3.3 *Software Interfacing*

Arduino is an open-source electronic prototyping platform allowing the users to develop interactive electronic objects by interfacing both hardware and software. The Arduino reads input in the form of light on a sensor or a push button or a message which is then converted into a hardware operation such as activate a motor or switching on an LED or uploading information in a webpage. Arduino (Uno)consists of 14 digital pins that can be used as an input or output using commands such as pinMode (), digitalWrite () and digitalRead (). They operate on a power supply of 5 volts. Each pin can transmit and receive a maximum of 40 milli Amps and consists of an internal pull-up resistor of 20–50 kΩ.

The board operates between 6 and 20 volts which are given through an external power supply. In ATmega328, 0.5 KB is used for bootloader from 32 KB. It consists of a 2 KB SRAM and 1 KB EEPROM which can read and write with the help of EEPROM library. The Arduino used consists of a serial monitor. This allows a simple data transfer between the microcontroller. When the data is transmitted to the desktop through the USB cable to the serial chip, the USB wire connected to the receiver and transmitter will make the LEDs on the board to glow.

We need to create a system that can handle the related operations for book reservations in the library. The system should integrate with an existing system in the library that manages the visitor information as well as the system should have a web interface that displays all books and their reservation status. The software that has

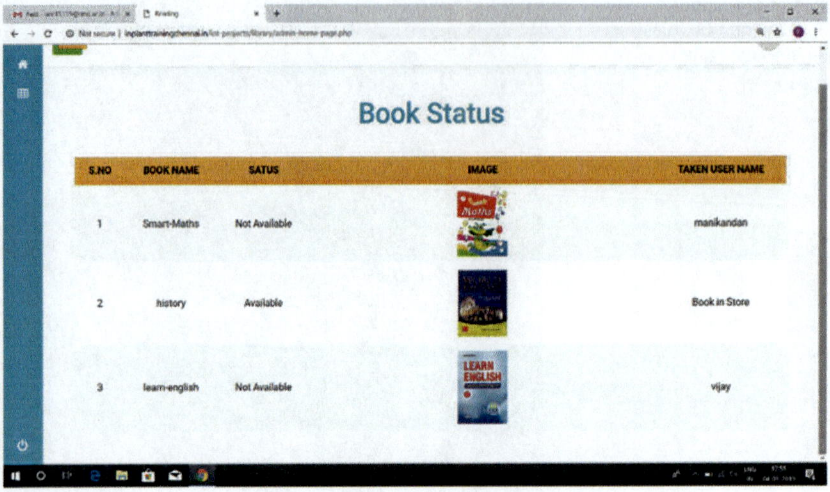

Fig. 4 The webpage displaying the status of availability of books in the library

been interfaced consists of certain features such as various functionalities, security, performance and better efficiency. The software is developed with the help of PHP language (Fig. 4).

A web component will be used to display the web page that prints the books reservation status and also will expose a function to reserve a book providing the visitor name who will reserve the book and due date for the reserved book. This system ensures the person of stating precisely the availability and if available remaining copies left are also provided. Thus, this approach is an advanced and most effective library automation system.

4 Conclusion

The librarian's work can be reduced with the help of this library management robot which arranges the books in its respective racks. This robot in the library will be helpful in eliminating the problem of misplacement of books. This robot provides quality results such as localization, identification and extraction of books. The intended aim of this robot is to facilitate the pick and place operation of the specific book and provide the book to the respective client. Thus, the library can be automated with the help of this Library Management Robot using line follower technique. At the same time, time and manpower can be reduced and the work efficiency can be increased.

5 Future Work

Similar to this project, it can also be implemented in medical shops for picking and placing the medicines. It can also be implemented in supermarkets and book stores.

References

1. Harada, K., Torea, F., Tsuji, T.: Pick and place planning for dual-arm manipulators. IEEE Int. Conf. Robot. Autom (2012)
2. Ramanan, N.V., Manoj Senthil, K.: Library management system to issue and retrieve books from user using autonomous robot. Asian J. Appl. Sci. Technol. (AJAST), **2**(2) 2018
3. Arumugaraja, M., Gugapriya, B., Soundarya, M.: The library management robot. Int. J. Eng. Comput. Sci. (IJECS), **3**(3) (2014)
4. Jose, M., Kayyalakam, C.S., Naveen, J.R., Kuriyakose, T.T.: Library assisting robot. Int. J. Eng. Res. Technol. (IJERT), **4**(4) (2015)
5. Sarvaiya, P., Patel, H., Shah, U., Patel, R., Holia, M.S.: Smart cart for accessing the library catalogue. Int Conf Res Innovations Sci. Eng. Technol. (ICRISET), **1** (2017)
6. Meghadas, K., Jose, S.K., Athira, M.N.: Library assistant service robot using QR code technology. Int. J. Eng. Sci. Comput. (IJESC), **6**(6) (2016)
8. Behan, J.: The development of an autonomous library assistant service robot (2008)
9. Katode, U., Mandavkar, P., Kudu J., Lad, S.: Wireless pick and place library management robot. Int. J. Comput. Appl. (2014)
10. Baby, A., Augustine, C., Thampi, C., George, M., Abhilash, A.P., Jose, P.C.: Pick and place robotic arm implementation using arduino. IOSR J. Electr Electr Eng (IOSR–JEEE), **12**(2) Ver III (2017)
11. Thambi, D.V., Patil, A.A., Shah, P.V., Ahmed, T., Mishra, M.: Raspberry Pi and Arduino based book management system. IOSR J. Eng. (IOSRJEN), **1** (2018)
12. Fujisaki, K.: An RFID-based system for library management and its performance evaluation. IEEE (2015)
13. Feng, C.: Research for application of RFID in library. IEEE (2010)
14. Vandana, C.P., Bhattacharjee, A., Gupta, A.: Library management system based on IOT. J. Comput. Sci. Eng. (IJRDO), **3**(4) (2017)
15. Abdullah, A.T.B., Ismail, I.B., Ibrahim, A.B., Noor, M.Z.H.B.: Library shelf management system using RFID technology. In: 2011 IEEE International Conference System Engineering Technology (ICSET) (2011)

An Efficient Method for Moving Vehicle Detection in Real-Time Video Surveillance

S. Sri Jamiya and P. Esther Rani

Abstract Surveillance systems are important to monitor and control critical situations for the field of traffic flow. Our aim is to enhance the ITS system about highway traffic and retrograde vehicle detection from surveillance video. We review the literature about vehicle detection in various countries and major highways, day to day vehicle transportation face critical situations then instant accident from different locations. This paper proposes an efficient method of detecting traffic surveillance in moving vehicles for various road conditions. Kalman filtering predicts the area where to search for each corner feature. This system is used to detect moving vehicles effectively in complicated situations such as pedestrians crossing, weather conditions, roadside trees, etc. The tracking algorithms are carried out over a various range of vehicles. It is to determine the regions where significant motion has occurred.

Keywords Intelligent transport system · Traffic flow · Video surveillance

1 Introduction

The forthcoming technological advances in various fields are growing rapidly for vehicle transportation, namely Intelligent Transport System (ITS). The method of traffic arrangements to the road-based transport system has been applied to the developed countries and developing countries [1]. This system uses CCTV Cameras for the purpose of surveillance. The vehicle transportation and decision-making are used to calculate the number of vehicles. This task mainly depends on foreground extraction. Normally the natural background includes detecting large objects such as buildings, plants, roads, apartments, etc., it has similar intensity values but intensities differ considerably with each other [2]. The foreground method is used to extract the

S. Sri Jamiya (✉) · P. Esther Rani
Electronics and Communication Engineering, Vel Tech Rangarajan Dr. Sagunthala R&D Institute of Science and Technology, Avadi, Chennai, India
e-mail: vtd655@veltech.edu.in

P. Esther Rani
e-mail: drpestherrani@veltech.edu.in

P. Suresh et al. (eds.), *Advances in Smart System Technologies*,
Advances in Intelligent Systems and Computing 1163,
https://doi.org/10.1007/978-981-15-5029-4_47

background image which captures from optical flow model. It carries out a separate foreground from a background [3] by learning and capture of the vehicle background image. This algorithm is used to detect particular moving objects to make robustness and also various changes have been considered to detect abnormal behavior of a vehicle [4]. Depending upon the frame-level and pixel-level specification, true positive detection rate and false-positive detection rates are calculated by detecting abnormally moving vehicles. It is considered to be positive when detected events are abnormal and are said to be negative when detected events are normal. The incremental framework of background modeling is relatively closer to our proposed work and the number of vehicles is counted [5]. The optical flow and thresholding model is applied to each pixel on the vehicle to determine the abnormal behavior of vehicles. The thing is to achieve the removal of noise. The different types of noises are median noise, impulse noise, Gaussian noise, mean noise, bilateral noise. The drawback of mean filter is, not suitable for more number of connected. The median filter is helpful for lane detection [6], for removing the background subtraction and then taken difference for motion vectors. The edges of the detected object are protected by image processing [4] techniques while removing noise. To overcome all these drawbacks we use morphological operations to reduce noise. The median filter improves robustness in the presence of noise and partially occluded objects. Morphological image processing technique is useful for removing these imperfections by considering the form and structure of video. Finally here comes vehicle identification and counting of vehicles in the video. With variations of vehicle tracking system [7], insurance companies can locate the stolen vehicle and stop. It is mainly used to prevent smuggling, knowing the exact location of the cargo, storing the information and data about routes of trucks, providing information to the customer. The application here we use is mainly of security for theft and to provide emergency services to customers.

2 Methodology

Automatic segmentation of a vehicle from a background and other vehicles so there can be a unique track associated with each vehicle. It deals with a variety of vehicles such as motorcycles, passenger cars, buses, construction equipment, trucks, etc. and deals with a range of traffic conditions like light midday traffic, rush-hour congestion, varying speeds in different lanes [8]. It also deals with a variety of lighting conditions (day, evening, night, sunny, overcast, rainy days). It gives strong robustness for the real-time operation of the system. Vehicle tracking approaches are mainly 3D Model-based, Region-based, Active contour-based, and Feature-based. 3D Model-based region is used for easy vehicle classification like Bus, Truck, and Bike. The drawback of this is the memory and processing consuming approach, unrealistic to expect to be able to have detailed models for all vehicles that are found on the roadway. In region-based tracking, every connected region in the image is "a blob" associated with each vehicle and then tracked over time using cross-correlation [9]. The "blobs" are found by the means of background extraction. It works well in free-flowing traffic conditions. The drawback is partial occlusion under congested traffic conditions leads to a grouping of several vehicles into one "blob". Active contour-based is used for representing the vehicle by bounding the contour of the object and dynamically update it during the tracking. It reduces the computational complexity compared to region-based approach. It is not suitable for partial occlusion. In feature-based tracking, [10] tracking is not an object as a whole, but sub-features such as distinguished points or lines on the object. Partial occlusion is not a problem: some of the sub-features remains visible. In motion-based grouping, congested traffic vehicles are constantly changing their velocities to adjust to nearby traffic, thus giving the grouper the information it needs to perform the segmentation [11]. For free-flowing traffic, vehicles more likely to maintain a constant speed with almost no lane drift, making "common fate" grouping less useful, but there is more space between vehicles.

3 Multiple Moving Object Tracking

The proposed algorithm tracks different types of vehicles such as bus, truck, bike and also detects pedestrians. The figure of proposed module is shown below in Fig. 1.

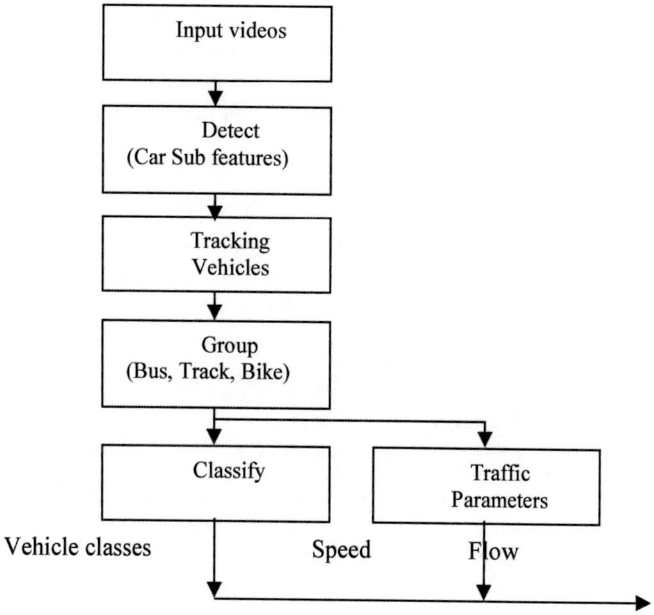

Fig. 1 Flowchart of vehicle tracking

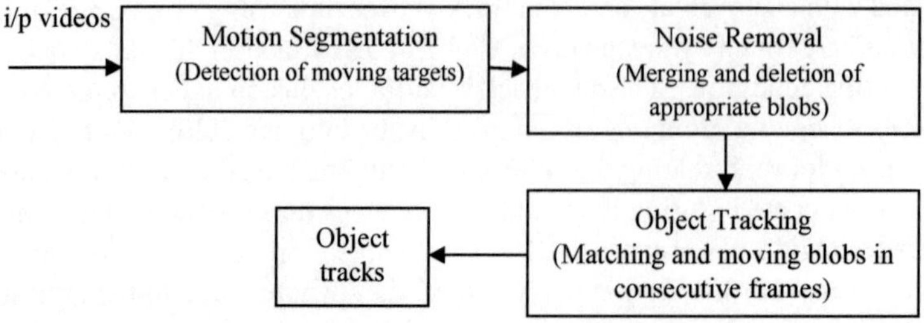

Fig. 2 Architecture of moving object

Estimation of moving vehicles
Instead of computing the entire video, the initial video frames are performed in a series of steps where the features are extracted and segmented. The motions of vehicles are detected in every frame by applying algorithms [7]. The sequences of steps have been introduced to process and analyze the video. In traffic surveillance videos, the motion of the vehicle is detected accurately in a certain number of tested frames. The true detected rate and false-positive detected rate are compared to give efficient moving vehicle detection (Fig. 2).

Identify Movable Objects in the Test Video Frame
The processed test videos are imperfect and it has random noise. The features are extracted and motion of vehicles is estimated by thresholding operations. It is used to divide foreground objects [1] and background objects. Hence, the distinct images are produced by thresholding. Vehicle tracking is used to track the moving vehicles on the road and highways from which information is taken for future use like an accident caused by traffic congestion, security for theft.

Analyze the Remaining Video Frames
Finally, we analyze the remaining video frames in which the moving vehicles are detected. An object is successfully detected by processing the frame by frame in the test videos. Vehicles are tracked by detecting the rectangular box in our videos. For each frame, it shows the number of counted cars.

4 Experimental Results

This section determines the experimental results for real-time videos captured from traffic surveillance cameras. The experimental parameter settings like true positive rate and false-positive rates are evaluated for every lane. The accuracy rate is improved compared with other existing results. Accuracy is calculated by ratio of sum of the true positive and true negative to the sum of the true positive, true negative, false-positive, and false-negative.

True match: A one to one match between ground truth and a group.
False-negative: An unmatched ground truth.
Over segmentation: A ground truth that is matched by more than one group.
False-positive: An unmatched group.

Over grouping: A group that matches more than one ground truth. Output video displays the detected object by detected boxes and produces more reliable segmentation results by the use of these videos.

Experiment 1: For Occlusion

It takes one of the video frames and the estimated moving vehicles are observed by the vehicle detector algorithm. The moving vehicles are recognized and followed even in the presence of occlusion by our proposed tracking algorithms. From test videos, the motion of vehicles is found and estimated (in the presence of obstructed trees, in the dark sky) (Fig. 3).

Experiment 2: Results of multiple vehicle tracking

From the second output video, it not only exhibits the square or rectangular boxes all over the cars but also detected cars in the second test video are viewed in the analyzed frame. The motion vector estimation by the use of median filter is more efficient in detecting vehicles even in the state of being occluded conditions. The following images show the second video for vehicle detection. From this analysis of the second video, the number of vehicles detected and tracked from the processed video frame is 5. This is viewed in top corner of the frame. The counted number of cars will vary for frame to frame (Figs. 4 and 5; Table 1).

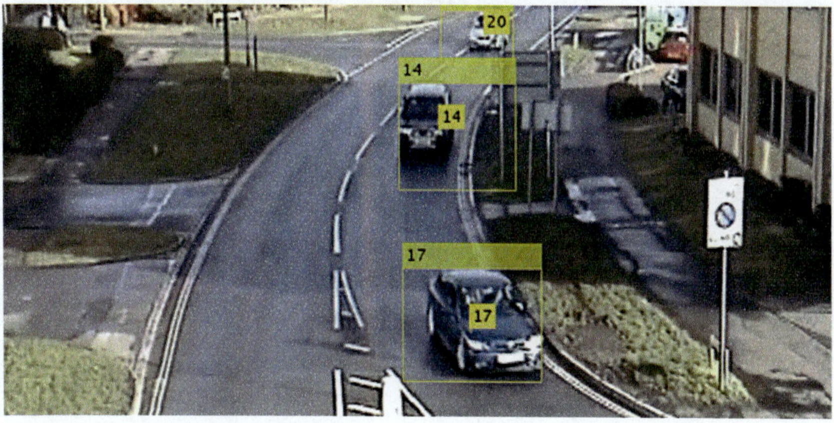

Fig. 3 Predicted results of tracked vehicles

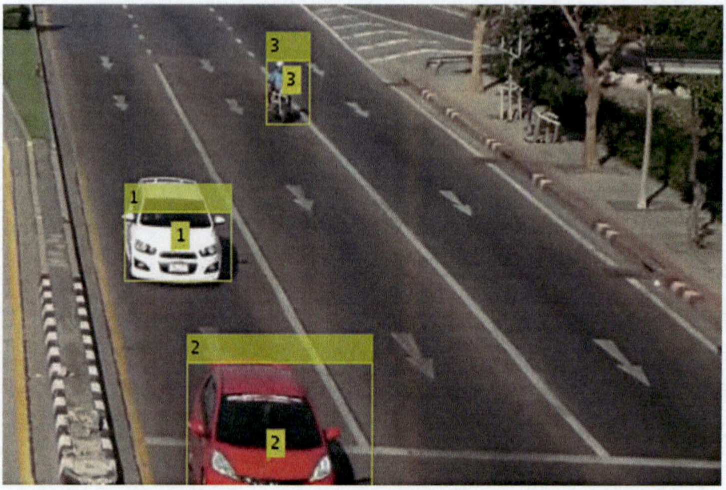

Fig. 4 Predicted results of targeted vehicles (Bike, Car)

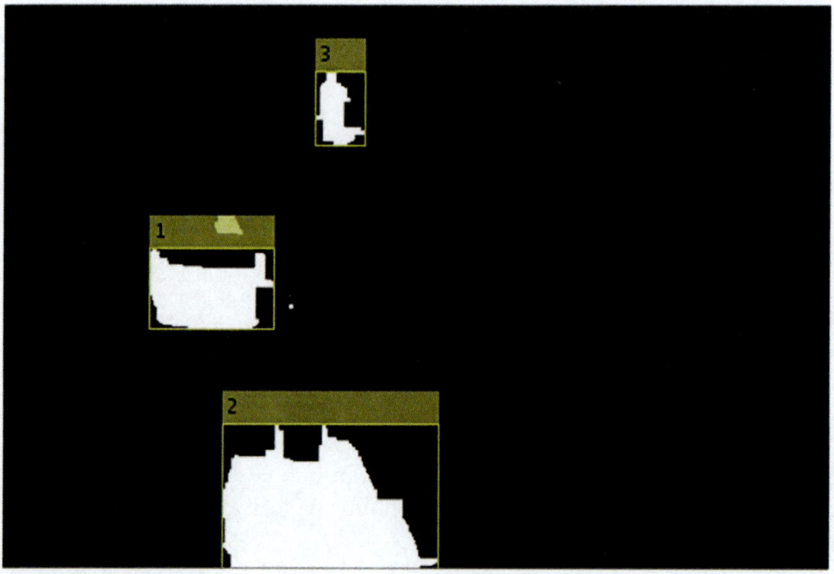

Fig. 5 Thresholding of moving object tracking

Experiment 3: Results for single object tracking

The optical flow tracker algorithm successfully detects and tracks the moving object effectively. The detected moving object is 1 which is shown in the test video frames (Figs 6 and 7; Table 2).

Table 1 Result for multiple object tracking

Video sequences	Miss rate (%)	False positive (%)	Miss matches (%)	MOTA (%)
Lane 1	3.3	2.3	6.8	88
Lane 2	5.2	3.4	7.2	81.1
Lane 3	6.1	2.2	11.8	84.9
Lane 4	3.4	2.1	10.5	89.9

Fig. 6 Predicted results for pedestrian

5 Conclusion

This paper proposes a new vehicle detection algorithm combined with Bayesian and extended Kalman filters, it aims that the background and foreground detection are simple by thresholding and Kalman filters. The moving vehicles can be detected from the original image effectively. Finally, vehicle detection and tracking system are developed by this algorithm by giving strong robustness even in the presence of noise. Our proposed system can detect and track moving vehicles in different climatic conditions. It plays a vital role in traffic congestion relief, road accidents prevention, safer and more comfortable for the public. In the future, the detection system of tracking and classification involves pedestrians behavior that will be considered for further research and practical work.

Fig. 7 Results for single vehicle tracking (Car, Bike)

Table 2 Computation time (s)

Methods	Execution time	
	1	2
Morphological processing	52.34254	18.62341
Particle filter (100 particles)	66.64213	19.46423
Extended Kalman filter	48.86345	17.65473

References

1. Nowosielski, A., Frejlichowski, D., Forczmanski, P., Gosciewska, K., Hofman, R.: Automatic analysis of vehicle trajectory applied to visual surveillance. In: Proceeding Image Processing Communication Challenges, pp. 89–96 (2016)
2. Liu, S., Liu, Y., Ni, L., Li, M., Fan, J.: Detecting crowdedness spot in city transportation. IEEE Trans. Veh. Technol. **62**(4), 1527–1539 (2013)
3. Paragios, P., Deriche, R.: Geodesic Active contours and level sets for the detection and tracking of moving objects. IEEE Trans. Pattern Anal. Mach. Intell. **22**(3), 266–280 (2000)
4. Hai-Feng, S., Hui, W., Dan-Yang, W.: Vehicle abnormal behavior detection system based on video. In: Proceedings IEEE Computational Intelligence Design (ISCID), pp. 132–135 (2012)
5. Le, T., Combs, M., Yang, Q.: Vehicle tracking based on Kalman filter algorithm. Technical Reports Published by the MSU Department of Computer Science, ID: MSU-CS-2013-02 (2013)
6. Siang Teoh, S., Bräunl, T.: A reliability point and Kalman filter based vehicle tracking technique. Int. Conf. Intell. Sys. 134–138 (2012)
7. Iketani, A., Nagai, A., Kuno, Y., et al.: Real time surveillance system detecting persons in complex scenes. Real-Time Imaging **7**(5), 433–446 (2001)
8. Li, Y., Guo, T., Xia, R., Xie, W.: Road traffic anomaly detection based on fuzzy theory. IEEE Access. **6**, 40281–40288 (2018)
9. Indrabayu, B., Achmad, A., Nurtanio, I., Mayasari, F.: Blob modification in counting vehicles using gaussian mixture models under heavy traffic. Asian Res. Publ. Netw. ARPN, **10** (2015)
10. Nagamalai, D., Renault, E., Dhanuskodi, M.: Implementation of LabVIEW based intelligent system for speed violated vehicle detection. In: First International Conference on Digital Image Processing and Pattern Recognition, pp. 23–33 (2011). ISSN: 1865-0929
11. Tan, H., Zhai, Y., Liu, Y., Zhang, M.: Fast anomaly detection in traffic surveillance video based on robust sparse optical flow. In Processing IEEE International Conference on Acoustics Mar. 2016, pp. 1976–1980 (2016)

FPGA-Based Reconfigurable Architectures for DSP Computations

J. L. Mazher Iqbal and T. Manikandan

Abstract With extensive usage of Field Programmable Gate Arrays (FPGAs), a reconfigurable computing platform enhances the wide variety of Digital Signal Processing (DSP) applications using Lookup Table (LUT), Flip-Flop, and multiplexers. Reconfigurable computation using memory becomes a subject of current research. FPGAs can be reprogrammed an unlimited number of times in which various DSP algorithms can be executed on a single hardware device, just as many different software algorithms can run on a conventional processor. The general architectures for DSP consist of a reconfigurable interconnection, Computational Elements (CEs), and memory. This article discusses on the CE architectures based on Memory (CEMs) and Serial arithmetic (CESs). Reconfigurable computing based on memory is much faster, reliable, and, hence the computation using memory requires less delay and power. The CEMs perform multiplication and addition using LUT and the CESs perform computation using serial arithmetic's. The results of the CESs, CEMs, and the computational reuse processing elements are compared. The architectures design based on CEMs significantly improves the performance of the previously reported results in terms of speed, area, and power using Xilinx Virtex-II FPGA, and Xilinx Vivado software.

Keywords FPGA configuration · Computational element architecture · On-Chip memory · DSP specific architecture

J. L. Mazher Iqbal (✉)
Vel Tech Rangarajan Dr. Sagunthala R&D Institute of Science and Technology, Avadi, Chennai, India
e-mail: drmazheriqbaljl@veltech.edu.in

T. Manikandan
Rajalakshmi Engineering College, Chennai, India
e-mail: manikandan.t@rajalakshmi.edu.in

P. Suresh et al. (eds.), *Advances in Smart System Technologies*,
Advances in Intelligent Systems and Computing 1163,
https://doi.org/10.1007/978-981-15-5029-4_48

1 Introduction

Nowadays, consumer appliances are considerably progressive. Consumer appliances are more functional and portable. The important issues in developing Very Large Scale Integrated Circuit (VLSI) architectures for communication systems and Digital Signal Processing (DSP) applications are power, speed, area, and cost [1]. The key devices to resolve these problems are microprocessors, microcontrollers, DSPs, and Field Programmable Gate Array (FPGAs). There are various methods in conventional computing for the execution of the DSP algorithm, Application Specific Integrated Circuit (ASIC), Microprocessors, and FPGAs [1]. ASICs are very efficient and fast because they are designed to perform a specific computation, but it is not possible to alter the circuit after fabrication. As the need for complex compilation for DSP surges, reconfigurable computing remains an active area of research. Many DSP operations for various DSP applications are implemented using enormous logic blocks present in FPGA. Reconfigurable computing using FPGAs outperforms the DSP processor and microprocessor in terms of performance, massive parallelism for low volume applications [2]. Specialized DSP circuitry implementation using reconfigurable computers is more beneficial. The functional resources of reconfigurable computers may be effortlessly altered subsequently configuration are deployment with changing data sets and functioning constraints [3]. The computational unit for reconfigurable computing systems is FPGA. Figure 1 shows a basic island-style architecture, where the functionality of logic elements and switches interconnection are modified to implement various DSP algorithms. FPGA is highly flexible and suitable to implement FSM control circuits such as the Mealy and Moore machine. The current objective is to implement various DSP algorithms with enhanced performance. The implementation is a solution for the problems of power, area, cost, and speed. At the architectural level, the system composed of computational elements (CEs), LUT, interconnections, and control units.

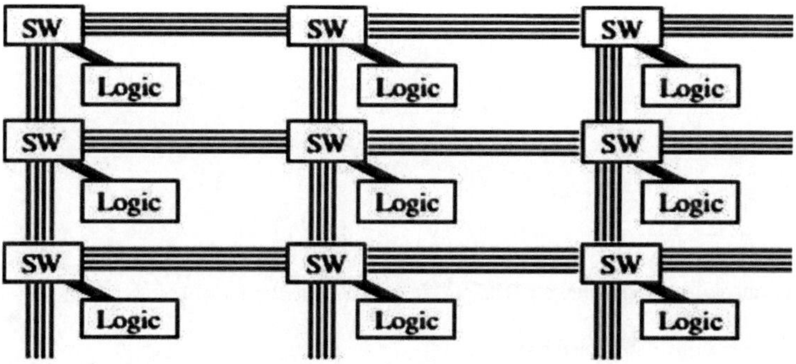

Fig. 1 The structure of FPGA

2 Re-configuration on Programmable Devices

Reconfigurable computing configurations are created either by using hardware description languages (HDL) or schematic entry tools. HDL describes the behavior of electronics circuit or system from which a physical circuit or system can then be attained (implemented) [3]. The transformation of HDL description or RTL to the netlist is called synthesis. The netlist is a machine-readable circuit description. The uncaught errors by behavioral simulation can appear during synthesis. The transformation of the netlist to an FPGA configuration bitstream is called implementation. LUT of the CLBs in FPGA is programmed based on the desired out of the given circuit. DSP circuits are implemented using CLBs. For the combinational circuit, LUT and multiplexers are configured. The output of the LUT is selected directly using multiplexer bypassing the flip-flops. For configuration of the sequential circuit, LUT and multiplexers are configured, such that the output of the LUT is passed through the Flip-Flops and selected through multiplexers. Interconnections can be established using either of the programming techniques such as SRAM, anti-fuse, and flash. The combinational and sequential circuits are mapped unto FPGA [4]. The placement phase allocates physical locations of the circuits on the FPGA while the routing phase interconnects all the sub-circuits by blowing the fuses in the routing channel. Place and route of the design are done using the FPGA vendor proprietary tools, which take into account the devices' architecture and logic block structure. The improvement paths are represented as back arrows which depend on whether the design meets the specified area or timing constraints. The bitstream configuration is the output of the design. This is loaded into the target FPGA and tested. The implemented design is tested in the real-time environment taking into account the switching and propagation delay. Numerous studies show that the implementation of DSP architectures such as Finite Impulse filter (FIR) using reconfigurable hardware establishes that the architectures are appropriate for DSP algorithms [5–8]. Precision word width requirement must be balanced in DSP applications, so as to perform the significant number of operations using a single reconfigurable device. Numerous input streams can be manipulated at a time using external SRAM. FPGA is well suited for DSP applications such as FIR filter because of high parallelism.

3 DSP Processor Architecture on FPGA

The DSP techniques used in the human-computer interface are cellular phone, sound cards, video cards, and speech recognition systems. The DSP performs some common operations on signals such as filtering, transforms, multiplexing and demultiplexing, simple time-domain operations such as addition and multiplication of signals, amplification, attenuation, integration, differentiation, convolution, and signal generation. The combinations of the above operations are performed to eliminate noise from signals. The increased system performance and reduced component count are

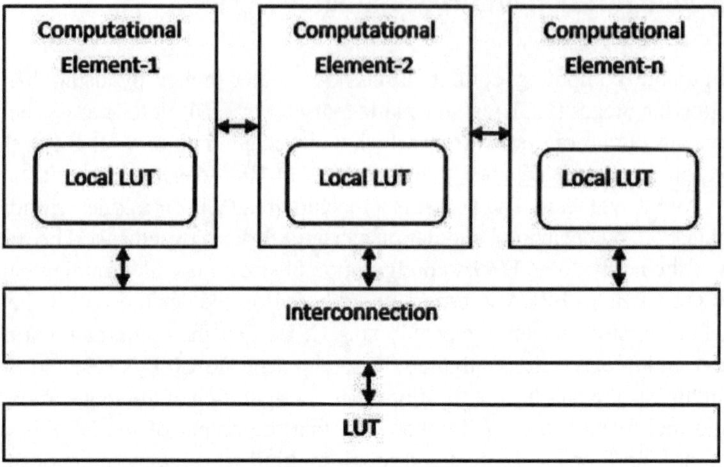

Fig. 2 Memory-based VLSI architecture for DSP

achieved using FPGA based DSP hardware for various communication systems and image processing applications. Computation less architecture using Lookup Table is very effective for signal processing applications. The general diagram of memory-based VLSI architecture for DSP application is shown in Fig. 2. Memory is partitioned into blocks so that all the processing elements get the input simultaneously. Memory-based processing elements perform various DSP operations on the input data and produce the desired output with less complexity. Each computation element acquires access to the memories by a rotational access scheme. The computational element can either writes or reads the data to or from memory. The important blocks to carry out numerous DSP are multiplier, accumulate, and shifter. The proposed method discuss about the serial and LUT based computational element architectures. This computational architecture built using LUT and serial arithmetic procedure with registers, memory, and interconnection on FPGA compromise of worthy balance in terms of chip area, power consumption.

3.1 Computational Element Based on Memory

Computational Element based on Memory (CEMs) is the low complex LUT multiplier. The CEMs gets input data and accomplish computation using LUT. The CEMs do not perform computation as such it just retrieves a value stored in LUT, hence the speed of the CEMs is much higher. The proposed CEMs consumes less area and power compared to existing memoryless based methodology. The computational element is used to develop a number of DSP computations. The CEMs composed of memory to store the product of DSP computations. The low complex memoryless computation elements are shown in Fig. 3. The CEMs composed of encoder, LUT,

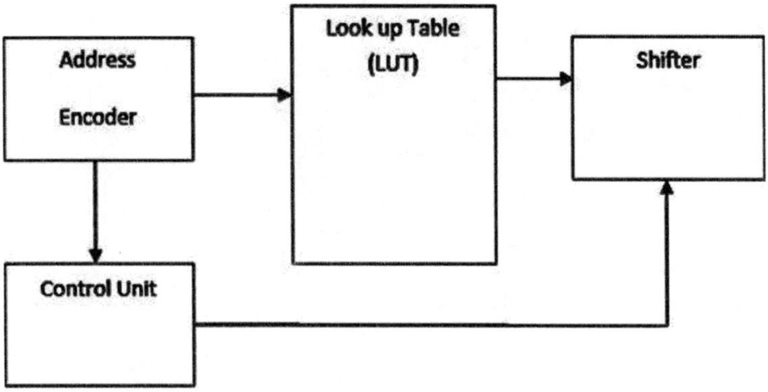

Fig. 3 Low complex memory less computation elements

control unit, and shifter. Memory elements are created using RAM and registers to attain suitable control operation in the CEMs. The registers are used to shift data with high speed. Interconnection provides communication between the CEMs and supplies the computational output into the memory.

3.2 Computational Element Based on Serial Arithmetic's

Computational Element based on Serial Arithmetic's (CESs) based serial arithmetic's shown in Fig. 4. The CEs in bit-serial arithmetic's are processed using serial carry skip adder. The serial carry skip adder with D flip-flop is shown in Fig. 5.

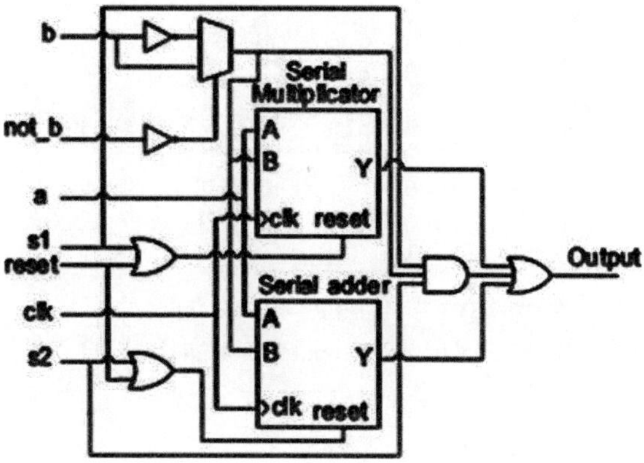

Fig. 4 Serial computation elements

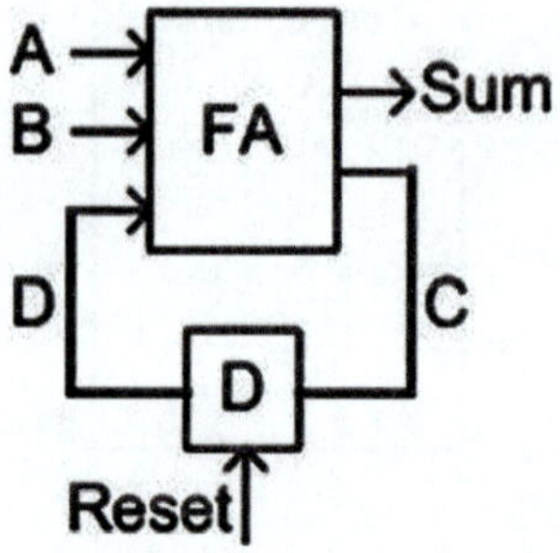

Fig. 5 Serial adder

4 Hardware Implementation

FPGA based computational element for DSP computation is implemented on Xilinx's Virtex-II family 2vp2fg256-6 using Xilinx Vivado software tool. The test bench and suitable stimuli are generated using HDL to test the design. Simulation is critical in verifying developed design behavior. The low complex memory-based LUT processing element and bit-serial arithmetic processes element design has been synthesis and observed. Figure 6 shows the comparison results of various computational elements for with bit-serial architecture and memory-based computational element architectures for DSP algorithm. Figure 7 shows the simulation results of the

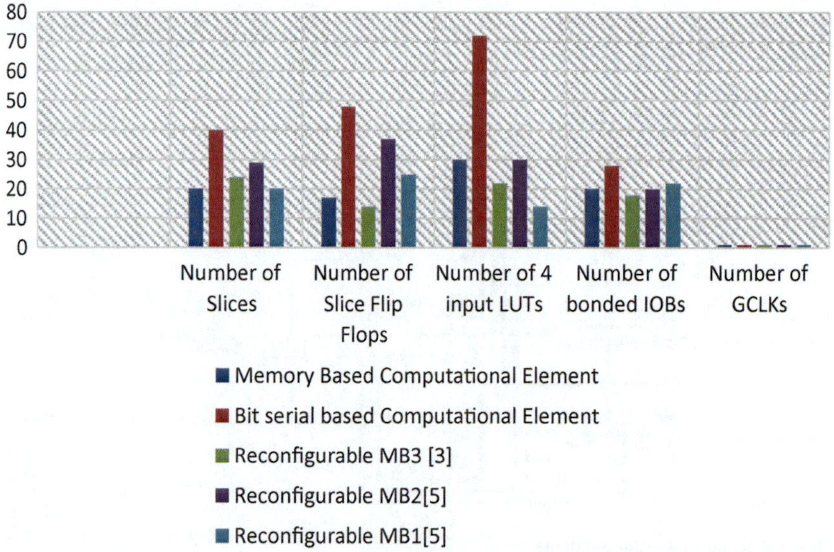

Fig. 6 Comparison results of various computational element for DSP algorithm

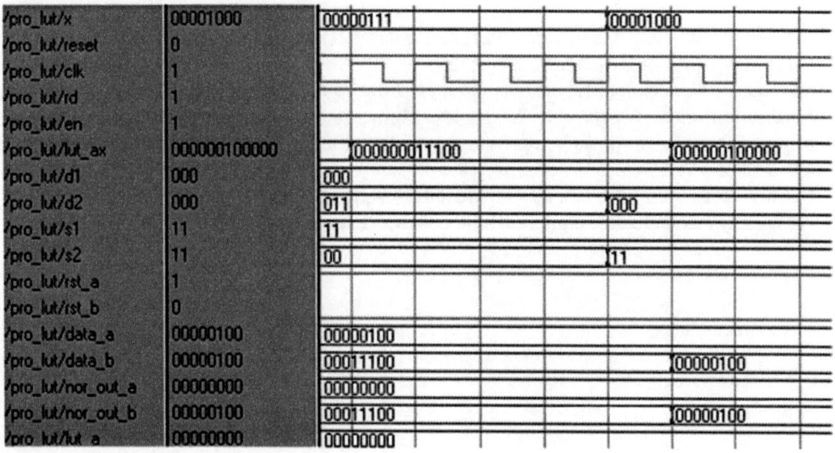

Fig. 7 Simulation result for the CEMs

CEMs. The results show that the CEMs consumes less area and suitable for various DSP implementations.

5 Conclusion

Reconfigurable FPGA attains very complex design with enhanced flexibility for the implementation of various DSP algorithms. Indeed, the latest FPGAs have sophisticated features such as LUT, multiplexers, DSPs, Multipliers, adders, enormous input and output pins, etc. In this paper, we discuss the computational elements based on memory and serial arithmetic using addition, multiplication, and memory. The memory-based architecture implemented on FPGA can be used as a general computational unit for various DSP algorithms implementation.

References

1. Brown, S., Rose, J.: FPGA and CPLD architectures: a tutorial. IEEE Des. Test Comput. **13**(2), 42–57 (1996)
2. Compton, K., Hauck, S.: Reconfigurable computing: a survey of systems and software. ACM Comput. Surv. (CSUR) **34**(2) (2002)
3. Tessier, R., Burleson, W.: Reconfigurable computing for digital signal processing: a survey. J. VLSI Sig. Proc. **28**, 7–27 (2001)
4. Mazher Iqbal, J.L., Varadarajan, S.: High performance reconfigurable balanced shared memory architecture for embedded DSP. Int. J. Comput. Sci. Inf. Secur. **8**(4), 198–206 (2010)
5. Mazher Iqbal, J.L., Varadarajan, S.: A new algorithm for FIR digital filter synthesis for a set of fixed coefficients. Eur. J. Sci. Res. **59**(1), 104–114 (2011)

6. Mazher Iqbal, J.L., Varadarajan, S.: Performance Comparison of Reconfigurable Low Complexity FIR Filter Architectures vol. 250, pp. 637–642. Springer-Verlag Berlin Heidelberg, Springer LNCS-CCIS, (2011). https://doi.org/10.1007/978-3-642-25734-6_151
7. Mazher Iqbal, J.L., Varadarajan, S.: High performance reconfigurable FIR filter architecture using optimized multipliers. Circ. Syst. Signal Proc. **32**, 663–682 (2013). https://doi.org/10.1007/s00034-012-9473-3
8. Mazher Iqbal, J.L., Varadarajan, S.: Memory based and memory less computation for low complexity reconfigurable digital FIR filter. WSEAS Trans. Syst. **12**(3), 142–153 (2014)

Classification by Learning of Wavelet and Texture Features

C. Bagavathi and O. Saraniya

Abstract Image classification is one among the significant tasks that can benefit industry through advancement in science and technology. Texture images possess a specific pattern that can be utilized to uniquely identify a texture class. Classification of Texture images can be done through proper selection of parameter that uniquely identifies an image in a group containing images of different classes. Machine learning approaches are currently the choice of implementation to provide accuracy and robustness in various fields like automatic image recognition, registration, and analysis. This work addresses machine learning methods to classify learnt texture samples through feature fusion of parameters obtained through wavelet and texture analysis.

Keywords Support vector machines · K-Nearest neighbor classifier · Gray level co-occurrence matrix · Haar wavelets · Texture analysis

1 Introduction

Image Classification process is easy to perceive through human perception. Computer vision models, based on classification, need proper parameter group to uniquely identify an image. Human accuracy in image classification approximately crosses 90%. Modern Intelligent networks like convolutional neural networks (CNNs) designed through deep learning networks perform classification through the learning of images and provide accuracy more than that of humans [1]. Image analysis deals with ways to infer image traits through mathematical expressions in equivalence with human observation. Machine learning and artificial intelligence techniques aim to reduce the extent of human intervention in analyzing images for vision-based applications.

C. Bagavathi (✉) · O. Saraniya
Department of Electronics and Communication Engineering, Government College of Technology, Coimbatore 641013, India
e-mail: bagavathi.vlsi@gmail.com

O. Saraniya
e-mail: saranya@gct.ac.in

P. Suresh et al. (eds.), *Advances in Smart System Technologies*,
Advances in Intelligent Systems and Computing 1163,
https://doi.org/10.1007/978-981-15-5029-4_49

Texture is the immediate interpretable quality from an image and it is one of the prominent features noted for image classification. Features such as texture, contrast, edges, contours, shape contexts tag a character to an image. Several feature descriptors have been designed: SIFT [2], SURF [3], HOG [4], binary pattern [5], color-based parameters like correlogram [6], coherence vectors [7], indexing [8].

Textural images contain information regarding a pattern of image and analysis of such images applied in the fields like Quality Assurance of industrial products. The texture feature extraction of end products can detect deformations and abnormalities resulting in an efficient way to assess the reliability and quality of a material. In image classification application with two classes, machines can be trained as a two-state classifier. Training requires a feature to be selected to efficiently represent an image. In this work, Statistical texture methods and wavelets are selected as feature descriptors [9]. This work performs classification of texture images as recognized or unrecognized based on the combined feature descriptors given to a machine classifier. The classifiers selected for learning are Support Vector Machines (SVM) and K-Nearest Neighbor (kNN).

2 Texture Classification

Texture can be regarded as a pattern of picture elements representing a character such as fineness, regularity, brightness, and contrast. Textures are a significant cue in identifying an image. The textural properties derived from an image can be used to identify and segregate a specific pattern of pixels. The complex nature of machine texture perception has motivated researchers to uncover diverse schemes for texture analysis: statistical, signal-based, model-based, geometric, transform-based, and spatial distribution based techniques [9]. Texture primitive sizes are similar to the size of pixels, it is better to employ statistical methods such as filters, domain transforms, matrices, fractals, wavelets to estimate the textural properties. Numerous texture descriptors have been given in literature among which Gray Level Co-occurrence Matrix (GLCM) is the most sort parameter for obtaining allied parameters such as contrast, energy, entropy [10]. Wavelets [11] for image processing pave a path for compaction of the energy of image in fewer wavelet coefficients. Wavelets become more attractive as being a lossless transform. This work utilizes both GLCM and wavelets for texture feature extraction.

2.1 *Wavelet Transform*

Wavelet transform is known as its benefits such as compressing nature, lossless transform, minimum distortion, precise description of signal, time, and frequency information are available and best suited for memory-constrained high-performance applications. Haar is the simplest form of wavelets that preserves energy, requires low computation, and provides approximated and detailed coefficients. Haar wavelets are well suited to represent the image in a compressed and compacted form with reduced dimensionality and enhanced data containment. The ease of computation and suitability make Haar wavelets a wise choice. It also sends integers to irrational numbers and for lossless image compression.

2.2 *Gray Level Co-occurrence Matrices (GLCM)*

Gray Level Co-occurrence Matrices (GLCM) is a histogram-based parameter (a square matrix) of order two that defines the frequency of occurrence of a pair of pixels in the same order in an image [12]. $P(i, j| \Delta x, \Delta y)$ gives the count of occurrence of pixels of intensity '*j*' following a pixel of intensity '*i*' in the direction of Δy with a distance of Δx between the pixels. The matrix has a dimension of $H_g \times H_g$, where H_g is the highest gray level in the image. H_g can be a maximum of 255 in a gray image.

A compact GLCM parameter can be obtained with the pixel distance Δx more than one. Depending on the required level of accuracy in texture calculation, Δx can be fixed where fine texture needs Δx to be small resulting in more samples and larger GLCM. In general, $P(m, n)$ is given as

$$P(m, n) = (i, j)(k, l)\epsilon S | f(i, j) = m,\ f(k, l) = n. \tag{1}$$

GLCM is used to provide 14 parameters relevant to texture description [13]. GLCM has been applied to a variety of applications including medical imaging [14, 15] and seismic studies [16].

3 Learning Methodology

Intelligent machines are developed based on the processes defined to accomplish a task with minimum human intervention. Machine learning methods are best suited when the process expects better accuracy compared to that of humans and when the volume of process exceeds the capability of existing methods. Machine learning methods can be classified based on the type of application for which the algorithm has been designed and the method of building a learning method. Supervised methods are built through a careful set of rules defining different classes and parameters. When machine learning is executed in predicting a class of an object from the guidance available from known similar objects, it is a supervised classification problem. Supervised Classification methods include Support Vector Machines (SVM) [17], Logistic Regression [18], Decision Trees [19], Random Forest [20], Neural Networks [21], Naïve Bayes [22], and Nearest Neighbor [23]. This work analyzes the fittingness of

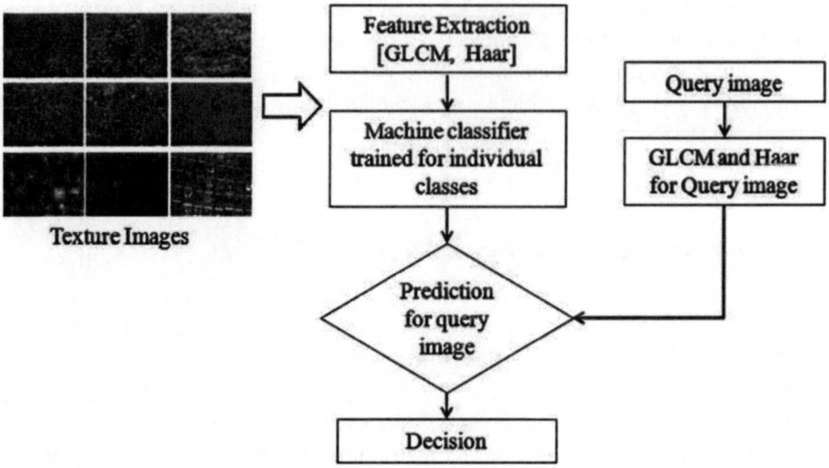

Fig. 1 Methodology of feature fusion for texture classification

support vector machines and nearest neighbor in the classification of textural images as mentioned in Fig. 1.

The texture images are given for analysis for statistical parameter GLCM, detailed and approximate coefficients of Haar wavelets. The computed features are concatenated to form a fused feature set to be trained in the classifier. A query image is selected and its features are calculated. The query parameters are computed and sent for prediction for its class. This method is run for support vector machines and K-Nearest neighbor classifier separately.

A simple support vector machine [17] is a two-class supervised machine learning method most suitable for the classification of extreme cases. Support vectors are the boundary spots at the margins of the group which on a slight variation of the value results in the adjacent class. It is sufficient to train the classifier only with these boundary points as it is the basic condition to properly classify the test case. The main advantage of SVM is that it works efficiently for a higher dimension, provides easy interpretation, and requires few significant training samples. k-Nearest Neighbor (kNN) [23] is a distance-based method that selects k-neighbors of the chosen point and declares the class as that of the class of maximum number of elements in the selected neighbors. It has been used in the recommender system in providing search results, concept search, and computer vision recognition systems. kNN algorithms are best suited for problems that are bounded by simple rules, low prediction speed, and less memory.

4 Results and Discussion

All experiments are evaluated in MATLAB 2017b on an Intel Core i3 processor running at 1.9 GHz. The Δx and Δy for GLCM are fixed as 1 and 1 respectively. This work deals with two datasets: UMD and UIUC datasets. 90 images are trained in SVM with 45 positive images and 45 negative images. The UMD dataset contains high-resolution images of size 1280 × 960 pixels. The entire dataset contains 25 classes with 40 samples/class. This dataset provides samples of the floor, fruits, and mosaic images. The UIUC dataset includes a thousand images, each of dimension 640 × 480 pixels. The images are under varied occlusion, focus, and brightness that are samples of fur, wool, carpet, bricks. Images are under different rotation and texture samples are of varied visible quality. Sample images of the datasets are shown in Fig. 2.

Two different datasets are chosen to prove the efficiency of the proposed method under different image sizes and resolutions. The execution time and classification accuracy are taken as the performance metrics in the proposed work. The results are analyzed and tabulated in Table 1. The support vector classifier is trained with 45 positive-trained images and the efficiency of classification is around 98% and the time required for classification is more than half a minute for two datasets.

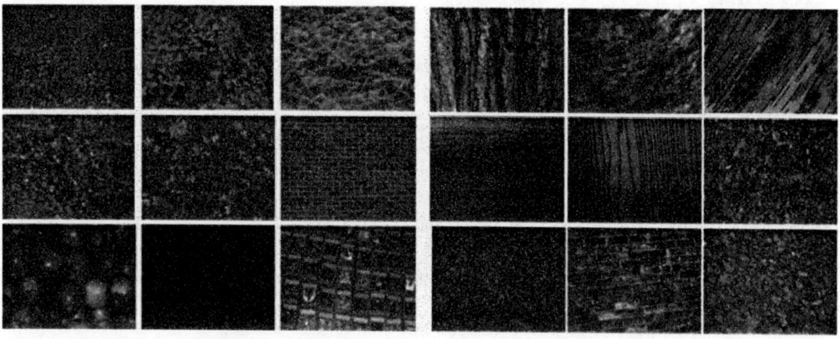

Fig. 2 UMD dataset and UIUC dataset

Table 1 Details of execution run with UMD and UIUC dataset for texture classification and classified through SVM and KNN classifiers

Category	UMD dataset	UIUC dataset		
	SVM	KNN	SVM	KNN
Positive-trained images	45	45	45	45
Correct classification	44	40	43	42
Efficiency (%)	97.78	88.89	95.56	93.33
Average execution time	38.71	10.05	43.25	10.06

The kNN is trained with k value as 5 and the efficiency is 88%. Texture samples of two different datasets are verified for correct classification and SVM performs better for both the sample sets. Number of neighbors is chosen as five following various implementations in existing work [23]. Comparing UMD and UIUC datasets, the accuracy is more for UMD dataset with SVM classification.

5 Conclusion

Texture classification through learning of fused parameters resulted in improved accuracy for SVM with linear kernel. Machine learning forms the basic technique in designing an automated system and the proposed method can be used in industrial quality analysis. Efficiency of classification can be improved by employing other wavelets or texture features. This work can be extended by including non-linear kernels for hyperplane design in SVM and also compare the results for various count of neighbors.

References

1. Dodge, S., Karam, L.: A study and comparison of human and deep learning recognition performance under visual distortions. In: Proceedings of 26th IEEE International Conference on Computer Communication and Networks (ICCCN), pp. 1–7 (2017)
2. Heymann, S., et al.: SIFT implementation and optimization for general-purpose GPU (2007)
3. Bay, H., et al.: Speeded-up robust features (SURF). Comput. Vis Image Underst. **110**(3), 346–359 (2008)
4. Dalal, N., Triggs, B.: Histograms of oriented gradients for human detection. In: Computer vision and pattern recognition, IEEE computer society conference, vol. 1, pp. 886–893 (2005)
5. Ojala, T., Pietikäinen, M.: Unsupervised texture segmentation using feature distributions. Pattern Recogn. **32**(3), 477–486 (1999)
6. Huang, J., Ravi Kumar, S., Mitra, M.: Combining supervised learning with color correlograms for content-based image retrieval. ACM Multimedia **97** (1997)
7. Roy, K., Mukherjee, J.: Image similarity measure using color histogram, color coherence vector, and sobel method. Int. J. Sci. Res. (IJSR) **2**(1), 538–543 (2013)
8. Huang, J., et al.: Spatial color indexing and applications. Int. J. Comput. Vision **35**(3), 245–268 (1999)
9. Materka, A., Strzelecki, M.: Texture analysis methods–a review. Technical university of lodz, Institute of electronics, COST B11 report, Brussels, pp. 9–11 (1998)
10. Antonik, P.: Real-time automated tissue characterisation for intravascular OCT scans. In: Application of FPGA to real time machine learning, pp. 137–160. Springer, Cham (2018)
11. Van de Wouwer, G., Scheunders, P., Van Dyck, D.: Statistical texture characterization from discrete wavelet representations. IEEE Trans. Image Process. **8**(4), 592–598 (1999)
12. Kekre, H.B., et al.: Image retrieval using texture features extracted from GLCM, LBG and KPE. Int. J. Compute. Theory Eng. **2**(5), 695 (2010)
13. Harlick, R.M., Shanmugam, K., Dinstein, I.: Texture feature classification. IEEE Trans. SMC **3**(11) (1973)
14. Zulpe, N., Pawar, V.: GLCM textural features for brain tumor classification. Int. J. Comput. Sci. Issues (IJCSI) **9**(3), 354 (2012)

15. Mohanty, A.K., Beberta, S., Lenka, S.K.: Classifying benign and malignant mass using GLCM and GLRLM based texture features from mammogram. Int. J. Eng. Res. Appl. **1**(3), 687–693 (2011)
16. Yenugu, M., Marfurt, K.J., Matson, S.: Seismic texture analysis for reservoir prediction and characterization. Lead. Edge **29**(9), 1116–1121 (2010)
17. Guler, I., Ubeyli, E.D.: Multiclass support vector machines for EEG-signals classification. IEEE Trans. Inf. Technol. Biomed. **11**(2), 117–126 (2007)
18. Qian, Y., Ye, M., Zhou, J.: Hyperspectral image classification based on structured sparse logistic regression and three-dimensional wavelet texture features. IEEE Trans. Geosci. Remote Sens. **51**(4) 2276–2291 (2013)
19. Otukei, J.R., Blaschke, T.: Land cover change assessment using decision trees, support vector machines and maximum likelihood classification algorithms. Int. J. Appl. Earth Obs. Geoinformation **12**, S27–S31 (2010)
20. Moosmann, F., Nowak, E., Juri, F.: Randomized clustering forests for image classification. IEEE Trans. Pattern Anal. Mach. Intell. **30**(9), 1632–1646 (2008)
21. Park, S.B., Lee, J.W., Kim S.K.: Content-based image classification using a neural network. Pattern Recogn. Lett. **25**(3), 287–300 (2004)
22. Liu, Y.-F., Guo, J.-M., Lee, J.-D.: Halftone image classification using LMS algorithm and naive Bayes. IEEE Trans. Image Proc. **20**(10), 2837–2847 (2011)
23. Hastie, T., Tibshirani, R.: Discriminant adaptive nearest neighbor classification and regression. Adv. Neural Inf. Proc. Syst. 409–415 (1996)

Sensible Autonomous Machine Using Deep Learning and Convolutional Neural Networks

Marialouis Diviya, Sankar Koushik Raghav, Ravichandran Parthiban, and Shanmugam Udhayakumar

Abstract In recent times, automation has achieved improvements by quality, accuracy, and precision. We introduce SAM-Sensible Autonomous Machine, performing two operations. Interfacing SAM to follow back an object by crossing obstacles and to reach the destination by sensing the response signal from Smartphone through Reinforcement Learning. Monitoring entire surrounding using Recurrent Convolutional Neural Networks Algorithm to identify the colour patterns of traffic signal by processing images through TensorFlow and performing turn operations by getting trained in predicting static/dynamic models. The goal is to initiate a self-driving machine, observing the surroundings across a transport region and act accordingly by the providers' instructions. This initiative brings many real-world things to autonomous creature and the main purpose is to save time from user's point of view. We implement a setup to make the machine act to various traffic scenarios and to track the user's location via Latitude and Longitude provided by the user to reach the desired location.

Keywords Deep learning · Recurrent convolutional neural networks · Multi-layer perceptron · Raspberry Pi · Pi camera

M. Diviya · S. Koushik Raghav (✉) · R. Parthiban
Department of Computer Science and Engineering, Rajalakshmi Engineering College, Chennai, India
e-mail: raghavan.sankar@gmail.com

M. Diviya
e-mail: diviyalouis@gmail.com

R. Parthiban
e-mail: rparthiban729@gmail.com

S. Udhayakumar
Saveetha School of Engineering, Saveetha Institute of Medical and Technical Sciences, Chennai, Tamil Nadu, India
e-mail: mailtoudhay@gmail.com

P. Suresh et al. (eds.), *Advances in Smart System Technologies*, Advances in Intelligent Systems and Computing 1163,
https://doi.org/10.1007/978-981-15-5029-4_50

1 Introduction

In recent times, automation has achieved improvements by quality, accuracy, and precision [1]. Recurrent networks are trained with the backpropagation algorithm. At each step, we compare the activations of output units with the desired activations and propagate errors backward through the network. As Artificial Intelligence (AI) [2] progressed and techniques for handling larger amounts of world knowledge were developed, some progress was made on the tasks and new tasks could be attempted. AI with robots involve operations that can think like a human, act like a human, think rationally, and also act rationally.

2 Software Process Steps

2.1 Processing Real-Time Data

Pi camera is integrated with Raspberry Pi. Image processing is done in real-time scenario by recognizing object patterns and detecting traffic light using Recurrent Convolutional Neural Networks implementing with Multi-layer Perceptron (MLP) [3]. Each layer is linked with the neurons from the previous and the next layer which are made to classify different object patterns and undergoing a comparison of real-time data with a loaded dataset.

2.2 Calibration of the Machine

Absorption and Reflection of black and white signals to make the machine sense the road patterns for moving forward and turn operations (left and right). Measuring the distance between the machine and other objects on-road by sensing frequency through Ultrasonic Sensor.

2.3 Manipulation of the Machine

L293D Motor IC is interfaced with Raspberry Pi to perform start, stop, and left-right turning operations at 5 V and 12 V supply from Raspberry Pi [4]. Enable pins are connected to Raspberry Pi to establish communication between IC and Motor, respectively (Fig. 1).

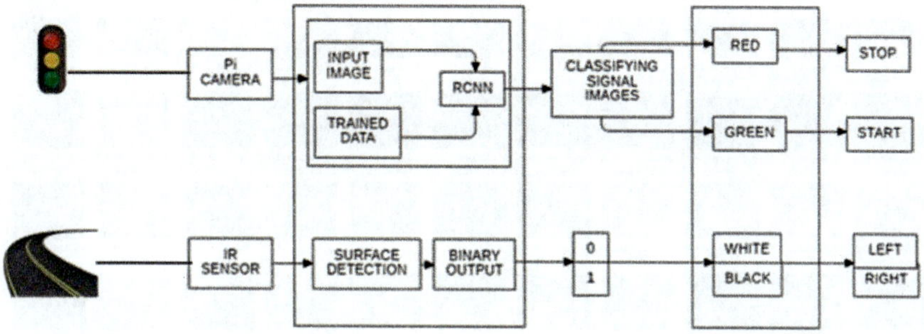

Fig. 1 Architecture diagram

Fig. 2 Day images

3 Training and Testing with Datasets

3.1 Gathering Images

Two different objects are needed to be detected (red, green). Then, it took about another 169 pictures with multiple images in the picture (Figs. 2 and 3).

3.2 Labelling Images

Images are gathered as clusters and labelled to produce as a dataset input for training the machine. The machine classifies image patterns via Pi Camera by comparing real-time data and the dataset that is given in the form of trained data (Fig. 4).

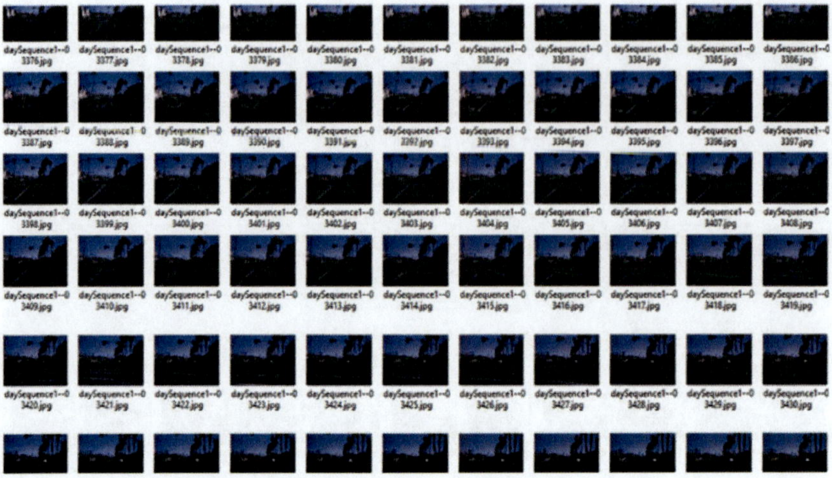

Fig. 3 Night images

Fig. 4 Green signal is labelled by selection

These .xml files will be used to generate TFRecords, which are one of the inputs to the TensorFlow trainer (Fig. 5).

3.3 Generating Training Data

With the images labelled, generate the TFRecords that serve as input data to the TensorFlow training model. First, the image .xml data will be used to create .csv files containing all the data for the train and test images.

Fig. 5 Red signal is labelled by selection

Table 1 Images trained with TensorFlow model

Image type	No. of Images	No. of steps trained	Accuracy (%)
Red signal	82	1089	67
Green signal	87	1125	71

3.4 *Configure Training Data*

The object detection training pipeline is configured and it defines which model and what parameters will be used for training (Table 1).

The progress of the training job by using TensorBoard, which provided information and graphs that show how the training is progressing. The loss graph showed the overall loss of the classifier over time (Fig. 6).

The inference graph is generated and it is made to classify the traffic signal patterns. *X*-axis represents loss and *Y*-axis represents the time taken for data to get trained. The training periodically saves checkpoints. The checkpoint at the highest number of steps used to generate a frozen inference graph [5].

3.5 *Testing Images*

Trained Images are given as input to Pi Camera and it showed outputs as below (Fig. 7).

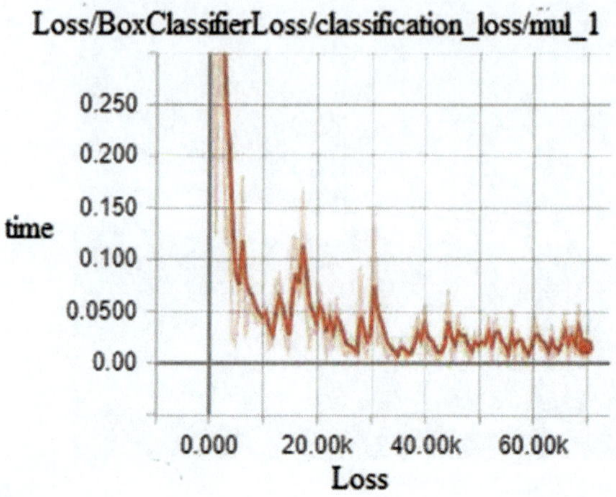

Fig. 6 Tensor board graph of training data

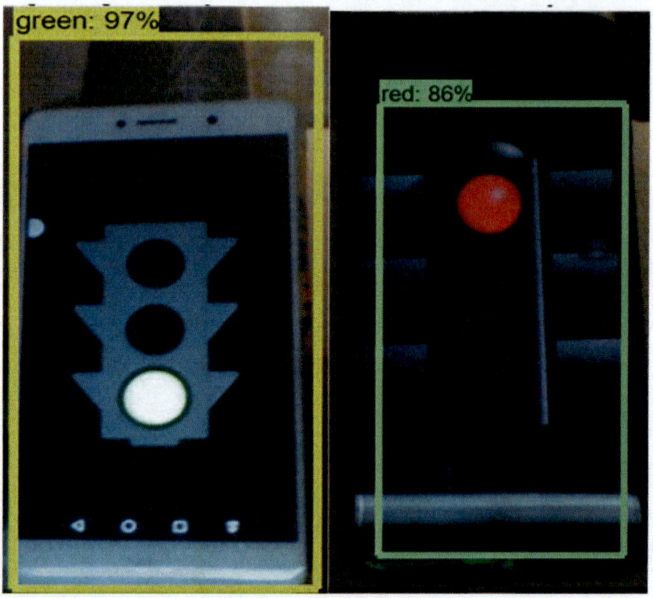

Fig. 7 Pi camera displaying signals

4 Working with Machine

Producing control systems capable of analyzing sensory data in order to provide accurate detection of other vehicles and the road ahead fusing data from multiple sensors. Data Replication between sensory data and real-time camera [6]. Monitoring entire surroundings to identify the colour patterns of traffic signal and performing right/left turn by getting trained in predicting static/dynamic models. Maintain ratio of frames captured and running status of machine.

4.1 Sensor Testing

IR sensor sends binary values by receiving and decoding signals [7]. The led sends and receives live data of road patterns and communicate the output value with the motor IC. Absorption and Reflection of black and white signals to make the machine sense the road patterns for moving forward and turn operations (left and right) (Fig. 8).

4.2 Motor Testing

L293D Motor Driver IC consists of four input pins, four output pins and two enable pins. When enable A is high, left motor (input 1 and input 2) works [8]. When enable B is high, right motor (input 3 and input 4) works (Table 2).

A GPS aerial is mounted at the rear for location information featuring gyroscopes, altimeters, and a tachometer to avoid a minute miscalculation of the machine's positions [9] (Fig. 9).

The machine can safely drive itself under specific conditions such as expressway merging, high-speed cruising, low-speed traffic jam, closed-campus operations by classifying object patterns, and traffic signals. Infrared Sensor performs identification of road patterns during self-parking scenarios by absorbing and reflecting Infrared signals [10]. Pi camera is fixed on top of the machine to detect signal patterns by classifying colours from comparing with the trained data at a frequent measure of time

5 Conclusion and Future Work

The proposed work uses different modules for providing automation to the robotic machine. There are certain systems providing security in existence but with less efficiency but our project deals with live data in the form of continuous images which gets automatically deleted after some point of time. This improves the efficiency of the system. Although the machine senses traffic signals and proceeds moving operations, installation of higher-grade cameras and motor power shall improve efficiency, performance, and speed control of the machine respectively. More complicated images are given as input to train the machine act according to emergency situations by predicting and analyzing.

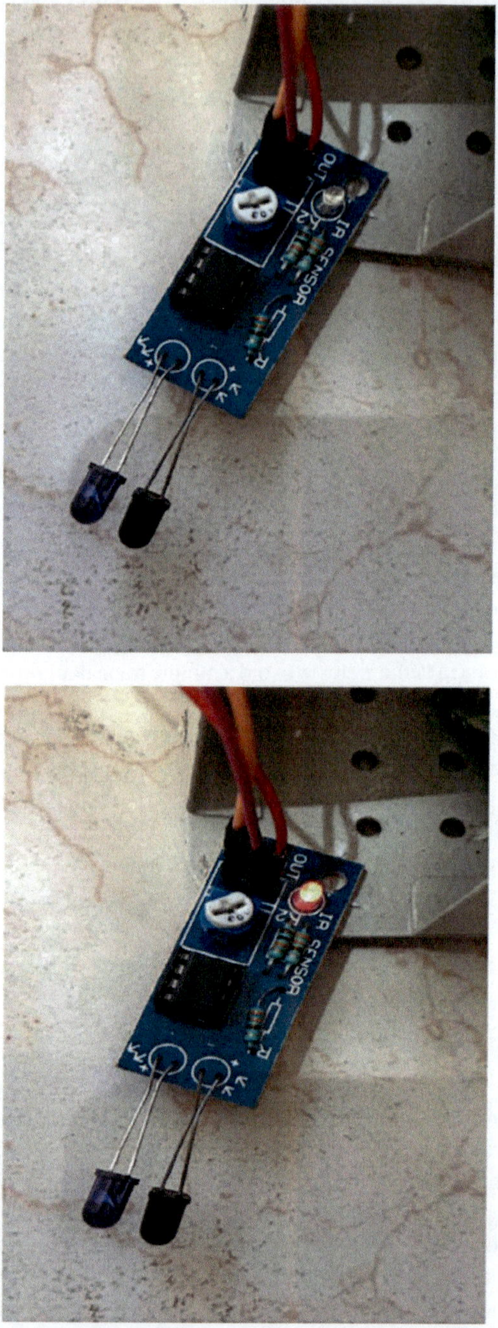

Fig. 8 Sensing the environment for moving/stopping

Table 2 Binary value of input 1 and input 2

Input 1	Input 2	Result
0	0	Stop
0	1	Anti-clockwise
1	0	Clockwise
1	1	Stop

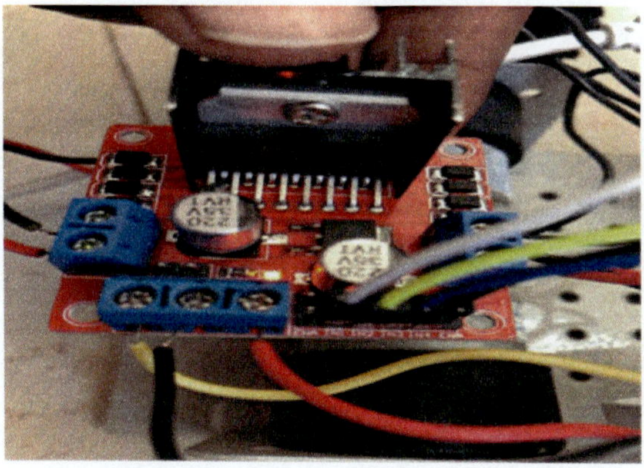

Fig. 9 L293D Motor IC

References

1. Poovam Raj, T.T., Udhayakumar, S., Silviya Nancy, J.: Smart city based mobile application for seamless communication of differently-abled. Int. J. Mult. Ubiquitous Eng. **12**(1), 177–190 (2017)
2. Hemanth, T.A., Hemanth, P.T., Hanumant, G.S.: Prototype of automated cars. Int. J. Eng. Res. Electron. Commun. Eng. **5**(4) (2018)
3. Wang, S., Du, X.: Overview of deep learning. In: 31st Youth Academic Annual Conference of Chinese Association of Automation (2016)
4. Durr, O., Pauchard, Y., Browarnik, D., Axthelm, R., Loeser, M.: Deep Learning on a Raspberry Pi for Real Time Face Recogniton. Zurich University, pp. 11–12 (2015)
5. IsacTornberg: Real time object tracking on Raspberry Pi 2. Thesis (2016)
6. Mustafa, V., Jilani, S.A.K.: Raspberry Pi based real time people detection, tracking and counting system. Int. J. Eng. Technique **3**(6) (20177)
7. Borenstein, J., Koren, Y.: Obstacle avoidance with ultrasonic sensors-robotics and automation. Int. Conf. Adv. Syst. (2015)
8. Li, M., Zhao, C., Hou, Y., Ren, M.: A new lane line segmentation and detection method based on Inverse perspective mapping. Int. J. Digital Content Technol. Appl **5**(4) (2011)
9. Udhayakumar, S, TamilSelvan, L., Umanandhini D., Dhinakaran, K., RajKumar, U.: Power aware zone based routing in a pervasive irrigation management system. In; IEEE International Conference On Computing Communication & Networking Technologies (2012). https://doi.org/10.1109/icccnt.2012.6395953

10. Abed, A.A., Rahmanb, S.A.: Computer vision for object recognition and tracking based on raspberry Pi. In: International Conference on Change, Innovation, Informatics and Disruptive Technology ICCIIDT'16, London (2016)

Speaking Mouth System for Dumb People Using Hand Gestures

Mirna Sarah Daniel, Nithin Prince John, R. Prathibha Devkar, Renu Abraham, and Ritty Elsa George

Abstract One of the beautiful creation of God is human beings who have been blessed with vision, sound, and speech. But there are some human beings who are deprived of this valuable blessing. Since speech is the major mode of communication which provides a way of exchanging information, perspective and utterance among different people, in both verbal and non-verbal manner, dumb people are mostly affected. In this framework, hand gesture recognition system for dumb people is used for conveying their thoughts to others. These people use sign languages to communicate with the world and people finds it difficult to understand what they are really speaking. So this project uses gestures which are an important part of such people and make use of these gestures to convert it into speech. True communication occurs when others understand one's message and they can respond in kind. This system smoothens the communication gap between dumb and normal people and provides an effective conversation. This artificially intelligent system make use of keras as platform over which gestures that are captured in real time via Webcam and trained using convolutional neural network (CNN) are converted into text and further to speech output.

Keywords Gestures · CNN · Artificial intelligent system · Recognition · Keras · Sign languages

M. S. Daniel · N. P. John · R. Prathibha Devkar (✉) · R. Abraham · R. E. George
Department of Computer Science, Saintgits College of Engineering, Kottayam, Kerala, India
e-mail: prathibhadevkar@gmail.com

M. S. Daniel
e-mail: mirnadaniel01@gmail.com

N. P. John
e-mail: nithin.pj@saintgits.org

R. Abraham
e-mail: renu.kk61@gmail.com

R. E. George
e-mail: rittyelsageorge@gmail.com

P. Suresh et al. (eds.), *Advances in Smart System Technologies*,
Advances in Intelligent Systems and Computing 1163,
https://doi.org/10.1007/978-981-15-5029-4_51

1 Introduction

The statistics from the World Health Organization (WHO) shows that about 280 million people are blind, 340 million are deaf, 1 million are dumb, and many others are physically challenged. Science and technology has developed much higher and human life is made more easy and comfortable within a short period of time. During the last few decades, we have come across a wide range of technologies where man power is not required anymore or the human did not have to move his body to do even a small task. But when we are thinking of the advances coming to us; we more often forget the people who are deprived of these advancements and those are the disabled section of the society [1]. We often fail to notice that these new technologies have not been able to provide the comfort and easiness to those who deserve it most. They need to be given the touch of the advancements to make them feel that they are part of the community too. Hand gesture interactions have advanced with gaming as in PlayStation as well as VR systems. Beyond this, it can be used in application aiding the physically challenged community like the dumb people. Hand gesture recognition is the primary requirement for the conversion of sign language to voice. This paper aims at solving the problem of limited communication abilities of the dumb people who know the sign language by transforming it into the form of verbal and vocal communication [2].

There are different methods that are used to interpret gestures. For example, in the glove-based method, one has to wear gloves and some sensors are attached with the gloves for detecting hand gestures [3]. But this requires more maintenance. The proposed system requires very less maintenance as it does not require sensors or accelerometer. In this era, most of the laptops are combined with Webcam. In our work, we are creating a software that is capable of recognizing the hand gestures through the Webcam. The proposed method extracts features from the video frame and converts into text and in turn into speech.

2 Related Works

In the paper by Islam et al. [4], a new hand gesture recognition system is introduced for the low cost color video using Webcam. They have used a deep convolutional neural network (DCNN) for extracting the useful features of hand and to recognize the ASL using those gestures. Finally, a multi-class support vector machine (MCSVM) is used to identify the captured signs. The machine is trained up using the CNN's extracted features and provides an accuracy of 94.57%.

Gurav and Kadbe's [5] model is focused on one of the most important mode of communication, i.e., hand signs to control robots or for household applications. The system proposed in this paper uses AdaBoost-based hand pose detectors. These detectors are then trained with the reduced Haar-like feature set. It also uses the convex hull algorithm to detect the fingertips.

The paper by Padmanabhan and Sornalatha [6] proposes a system that makes use of motion sensor in the form of a glove. Here, all the messages or any templates that have been used are stored into a database which is given to a PIC microcontroller and the motion sensor ed glove is fitted into the hand. When gestures are performed, the sensors get activated and are given to the microcontroller. The microcontroller matches the motion with the already-stated database and produces the speech signal.

The system in the paper presented by Yamunarani and Kanimozhi [7] acts as an interpreter which reduces the communication gap between a normal person, deaf, and a dumb person wherein the English like sentences as text by normal person for the deaf people and the sign language given by the dumb person interprets its meaning as audio to the normal person. The image is captured through Web camera and they are preprocessed to remove the noises using Wiener filter. The obtained filter is given to histogram of oriented gradients (HOG) which extracts out the features. While testing, the image is compared with the trained images done in the classification stage using the minimum Euclidean distance and the PC is interfaced with arduino which gives out a text output in the LCD display and voice through the speaker attached.

Other similar work ideas include: the paper by Tripathy et al. [3] builds a system that captures real-time images of signs and converts it into voice with text as an intermediate output. Signs are detected by using a classifier based on Haar-like features and a speech synthesizer is used for speech modules. This system makes use of Webcam, OpenCV modules, and Microsoft visual studio as IDE.

In the paper [8, 9], a classifier based on CNN for hand shape recognition is trained using a process of transfer learning. This system in-cooperates six static and eight dynamic gestures and for dynamic gestures; segmentation is carried out using HSV skin color algorithm. A CNN architecture named VGG16 is used as the pretrained model.

The paper by Nikam and Ambedkar [10] proposed a system prototype that can recognize sign languages and help the deaf and dumb to communicate easily with others. Recognition of both pattern and gestures are the developing fields of research in this paper. The hand gesture recognition system provides a more interactive way of communicating with computer. Being a significant part in non-verbal communication, hand gestures are playing key role in our daily life.

The software presents a real-time hand gesture recognition based on extracting the shape features of the hand-like orientation, calculating the centroid, status of fingers, and thumb positions.

3 Proposed System

The basic idea of the proposed model is to identify the various gestures (sign languages) and convert it into text to speech. The algorithm been used is implemented using OpenCV and a high-level neural network API, i.e., keras is incorporated with tensor flow.

The system can work by capturing the skin color of the person who's is using it or he can use any color glove to get more accuracy from the system as the skin color segmentation largely depends on lighting condition. Initially, the gestures are captured through the Webcam with the person wearing the glove or without wearing the glove and choosing the appropriate region of interest (ROI). Then, we create a dataset based on the captured image. The image is stored with a dimension of 200 × 200 pixels. Likewise, we create ten gestures and store it into the database. The gestures are trained using CNN and they are recognized in real time with the input from the Webcam. The recognized gesture or the sign is converted to meaningful word and further converting the obtained text to speech. Figure 1 depicts the working of the proposed system.

The system is divided into three modules: Gesture training, Gesture Recognition, and Text to speech.

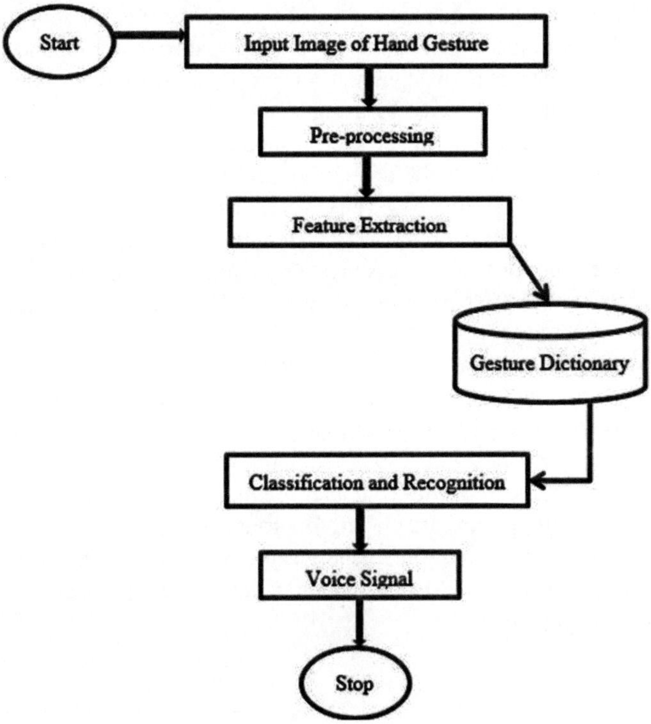

Fig. 1 Flowchart of the overall system working

3.1 Gesture Training

It deals with training the gestures collected from the dumb people. The system trains the captured gestures using convolutional neural network (CNN). The biggest revolution in computer vision area is the development of convolutional neural network (CNN) [9]. It not only have improved the image classification accuracy but also played an important role for generic feature extraction such as scene classification, object detection, semantic segmentation, image retrieval, and image caption [11].

The three layers used in convolutional neural network are the convolutional layers, pooling layers, and full connection layers as shown in Fig. 2 [12].

The system used makes use of 32 convolutional filters of size 3×3 over the image of size 200×200 with an activation function ReLu, followed by MaxPooling. The ReLu or the rectified linear unit is a linear activation function having a threshold at 0 and therefore, the model can take less time to train or run. The purpose of using MaxPooling is to reduce the spatial dimensions of the activation maps and the number of parameters used in the network, thereby reducing the computational complexity. After this, a dropout of 0.5 is used to prevent the model from over fitting and to create some kind of image noise augmentation. The last stage of convolutional neural network is an artificial neural network classifier which is a dense layer. The flattening in CNN converts the output of CNN layer into single long continuous linear vector [13, 6] (Figs. 3 and 4).

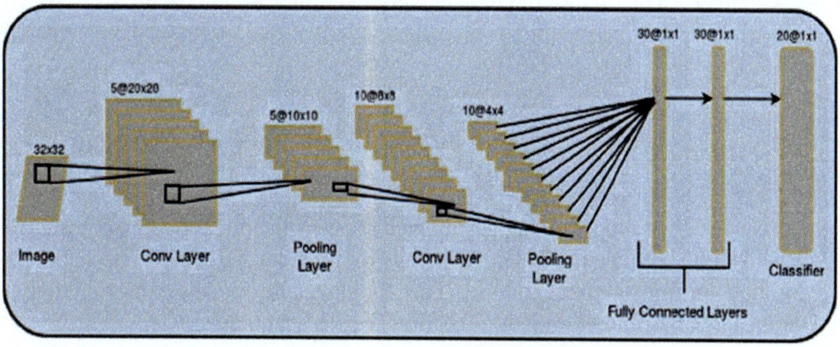

Fig. 2 Sample architecture of CNN

Fig. 3 Flowchart depicting the training phase

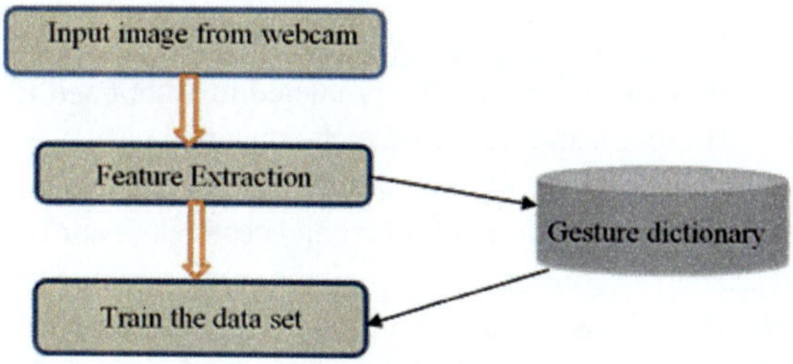

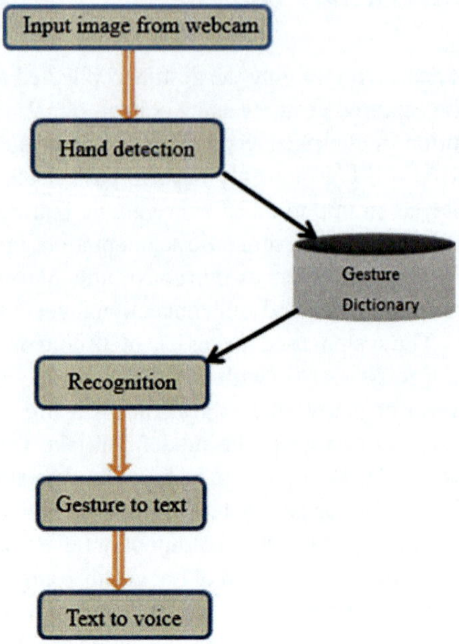

Fig. 4 Flowchart showing the recognition phase

The model uses an optimizer algorithm called Adam which helps us to update the weights in the network, a loss function called categorical_crossentropy, and a list of metrics. Then, it is trained with 15 epochs with a batch size of 128.

3.2 Gesture Recognition

This phase deals with recognizing gesture and converting corresponding gestures to text as shown below. The hand is detected by cropping the ROI and is checked with the available gestures in the dictionary. The recognized gestures is converted to text and then using an offline Python library for text-to-speech conversion.

3.3 Text to Speech

It deals with converting the predicted text obtained from the gesture recognition to voice output using text-to-speech conversion.

In this proposed system, dumb person needs to show the hand gestures in front of the Webcam and this particular hand gesture (gesture for the word 'NICE' is shown in Fig. 5) is recognized and display the corresponding word on the screen itself. (Here, 'NICE' is the word for that particular gesture shown). After the word corresponding to that gesture is identified, then that word is converted into speech.

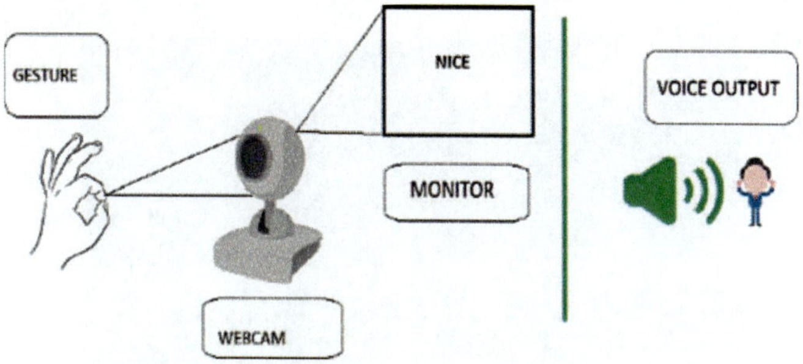

Fig. 5 Text-to-speech conversion of a gesture

4 Result and Discussion

The output of the system is shown below where we capture gestures in real time through Webcam. The 30% dataset is used as training set and the 70% as testing set. The accuracy of the system is nearly 96% when we capture the gestures while wearing the glove; otherwise, it is 90%. The dataset includes the following gestures as Fig. 6.

As mentioned earlier, the region of interest of the hand is selected first and processed. Figure 7 shows the cropping of hand color (particularly the glove color) for recognition and obtained the grayscale HSV image of the same.

The gestures are shown in front of the Webcam shown in Fig. 8 and are correctly recognized and the corresponding words are displayed as text. Finally, the texts are converted to speech.

Fig. 6 Dataset of the project

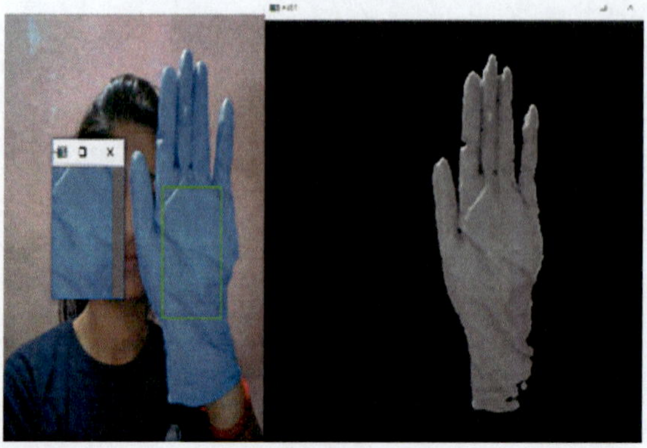

Fig. 7 Cropping the region of interest of the hand and obtaining the grayscale image

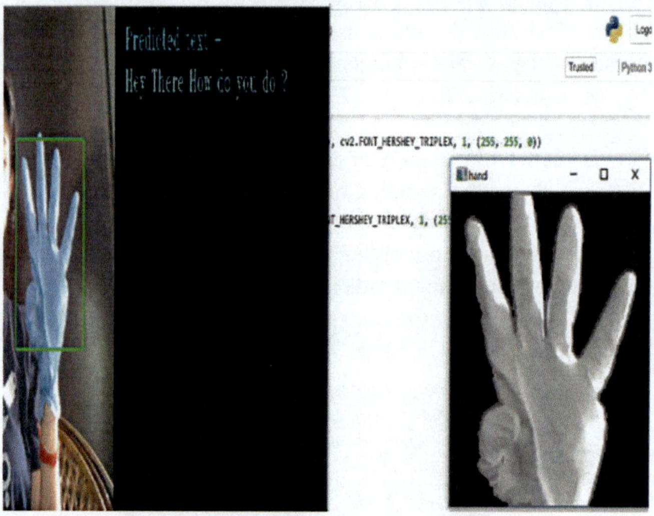

Fig. 8 Recognized gesture are converted into text and then to speech

5 Conclusion and Future Scope

A notable challenge in real-life application is to recognize the hand gestures on the basis of its accuracy. Hand gestures recognition and conversion of these gestures into text and then into speech is presented in the paper. The input hand gestures are captured using the Webcam and processed through CNN. The trained gestures when shown by the dumb person is recognized and produced as speech output. We would like to develop this whole system into a wearable gestural device that has a camera, micro controller, and a speaker.

Acknowledgements This idea was supported by Kerala State Council for Science Technology and Environment (KSCSTE) and provided funding through INNOVATE contest.

References

1. Harish, N., Poonguzhali, S.: Design and development of hand gesture recognition system for speech impaired people. May 28–30 (2015)
2. Dawane, S.P., Ali Sayyed, H.G.: Hand gesture recognition for deaf and dumb people using GSM module. IJSR (2015)
3. Tripathy, A.K., Jadhav, D., Barreto, S.A., Rasquinha, D., Mathew, S.S.: Voice forthe mute. Feb 04–06, 2015 Mumbai India. In: Hussain, S., Saxena, R., Han, X., Ahmed Khan, J., Shin, H. (eds.) Hand Gesture Recognition Using Deep Learning
4. Islam, Md. R., Mitu, U.K., Bhuiyan, R.A., Shin, J.: Hand gesture feature extraction using deep convolutional neural network for recognizing Amercian sign language. 2018. In: 4th International Conference on Frontiers of Signal Processing (ICFSP)
5. Gurav, R.M., Kadbe, P.K.: Real-time finger tracking and contour detection for gesture recognition using opencv. In: 2015 International Conference on Industrial Instrumentation and Control
6. Padmanabhan, V., Sornalatha, M.: Hand gesture recognition and voice conversion system for dumb people. Int. J. Sci. Eng. Res.**5**(5) (2014). ISSN 2229-5518
7. Yamunarani, T., Kanimozhi, G.: Hand gesture recognition system for disabled people using arduino. **4** (2018)
8. Sriram, N., Nithiyanandham, M.: A hand gesture recognition based communication system for silent speakers
9. Wankhade, K.A., Zade, G.N.: Sign language recognition for deaf and dumb people using ANFIS. IJSETR **3**(5) (2014)
10. Nikam, A.S., Ambekar, A.G.: Sign language recognition using image based hand gesture recognition techniques. In: Online International Conference on Green Engineering and Technologies (IC-GET) 2016
11. Tavari, N.V., Deorankar, A.V.: Implementation of neural network based hand gesture recognition system **3**(6), 6395–6401 (2014)
12. Chen, X., Yang, X., Wang, M., Zou, J.: Convolution neural network for automatic facial expression recognition. IEEE-ICAST-2017
13. Albawi, S., Abed Mohammed, T., AL-ZAWI, S.: Understanding of a convolutional neural network. ICET (2017)

Improvement in Performance Attributes of Multivariable System Using RTDA Controller

C. Febina and D. Angeline Vijula

Abstract Recent scientific and technological progress is demanding the development of efficient control systems for industrial processes. Majority of the industrial systems are nonlinear and multivariable in nature and hence, it becomes insistent to develop a superior controller satisfying the performance requirements under various constraints and limitations. Though conventional controllers are widely accepted in many industries, it evokes some limitations when applied to multivariable systems. In this paper, a simplified next-generation unconventional robustness, set-point tracking, disturbance rejection, aggressiveness (RTDA) controller is designed for controlling the quadruple tank system (QTS) which is an eminent benchmark Multi-input multi-output (MIMO) system. RTDA controller with its enhanced features tunes each design parameter separately to obtain the optimum performance attributes, where a conventional controller fails to do so. A comparative study is done between Internal model control (IMC) tuned PI controller and RTDA controller for the decoupled process. From the simulation results and analysis, RTDA controller stands to be preeminent than conventional controller in overcoming the control challenges that are associated with multivariable systems and ensuring set-point tracking, robustness, and disturbance rejection.

Keywords Quadruple tank system · Multivariable control · Decoupler · IMC-PI · RTDA controller

1 Introduction

Process industries have historically acknowledged the importance of process control and its needs. Several developments and process control researches have been made

C. Febina (✉) · D. Angeline Vijula
PSG College of Technology, Coimbatore, India
e-mail: cfebina3@gmail.com

D. Angeline Vijula
e-mail: dav.ice@psgtech.ac.in

P. Suresh et al. (eds.), *Advances in Smart System Technologies*,
Advances in Intelligent Systems and Computing 1163,
https://doi.org/10.1007/978-981-15-5029-4_52

in industry as well as in academia for several decades. A conservative PID controller is a widely accepted control methodology in industries because of its advantages like ease of implementation, simple control structure, and robustness to modelling errors. But it is not suitable for controlling the multi-input multi-output (MIMO) processes and processes having large time constants, substantial time delays, inverse response, nonlinearity, etc. Several researchers have implemented various techniques and used PID algorithm for controlling MIMO systems. Nyuyen Truong Luan vu et al. designed a multiloop PID controller using generalized IMC-PID method for multiloop systems [1]. An optimized PI controller using particle swarm optimization technique and model reference adaptive control for a multivariable system was developed for improving the efficiency of the control system [2, 3]. The tuning parameters of PID controller are not directly related to the performance attributes, so achieving set-point tracking and disturbance rejection simultaneously can be difficult [4] and hence a next-generation controller, namely RTDA controller was developed and it was proposed that RTD Acontroller outperforms classical PID controller [5]. Ogunnaike et al. [6] developed a RTDA controller that combines the simplicity of conventional PID controller and predictive capabilities of model predictive controller (MPC).

The RTDA controller for SISO system was developed as an alternative to PID controller and tuning rules were proposed for a broad class of processes including delay dominant and inverse response processes [7]. Srinivasan et al. characterized the performance of MPC controller with RTDA controller for FOPDT and SOPDT processes [8] and performance of RTDA controller was compared with other controllers for SOPDT processes for minimum zero and non-minimum zero operating conditions [9]. Taking into account all the uncertainities and nonlinearities involved in PEM fuel cell, a disturbance rejection controller was developed with fast set-point tracking [10].

In this paper, a highly interactive multivariable control problem, i.e., Quadruple tank system (QTS) is considered which contains a transmission zero. Reza Katebi et al. proposed multivariable predictive PID controller for quadruple tank system and compared its efficiency with multiloop PI controller [11]. The RTDA controller with its prominent features is implemented in QTS and its tuning parameters, θ_R (robustness), θ_T (set-point tracking), and θ_D (disturbance rejection) can be tuned autonomously. In this approach, efforts taken to improve the set-point tracking ability do not decline the robustness and disturbance rejection capabilities of the control system. The outline of the paper is as follows: Sect. 2 describes briefly the mathematical model with its description and two conditions are developed in order to analyze the responses of the process in both minimum and non-minimum phase operating conditions. The minimum loop interactions are estimated using relative gain array (RGA) method in Sect. 3 and decoupler is designed to correct or compensate the interactions between process variables in Sect. 4 while followed by design of PI controller using IMC based tuning method in Sect. 5. RTDA algorithm is discussed in Sect. 6. Performance analysis and simulation results are in Sect. 7. Lastly, this paper is concluded in Sect. 8.

2 Mathematical Model

A quadruple tank system proposed by Johansson et al. [12] has been used in chemical engineering laboratories to exemplify the performance margins for multivariable systems posed by ill-conditioning, model uncertainties, and right half plane transmission zeros. Quadruple tank process was intended to demonstrate the importance of multivariable zeros and how do they change due to variations in the process.

2.1 *System Description*

QTS consists of two pumps and four interconnected tanks, where V_1 and V_2 are the two input voltages to the pumps, and Y_1 and Y_2 are the two output voltages from the level transmitters. The aim is to control the levels of two lower tanks by regulating the inlet flow rates. The three-way valves are positioned in such a way that the output is split into two, one to the lower tank and other to the upper tank which is diagonally opposite to the lower tank. One of the main features of QTS is that the dynamics of the process can be adjusted to be either in minimum phase or in non-minimum phase by changing the position of three-way valves. For the linearized model, both the location and direction of the zeroes have direct physical interpretations.

The schematic of the quadruple tank equipment is presented in Fig. 1. Parameter γ is determined based on how the three-way valves are set. If γ_1 is the ratio of inlet flow to the tank 1, then $(1-\gamma_1)$ will be the outlet flow to the tank 4. Likewise, if

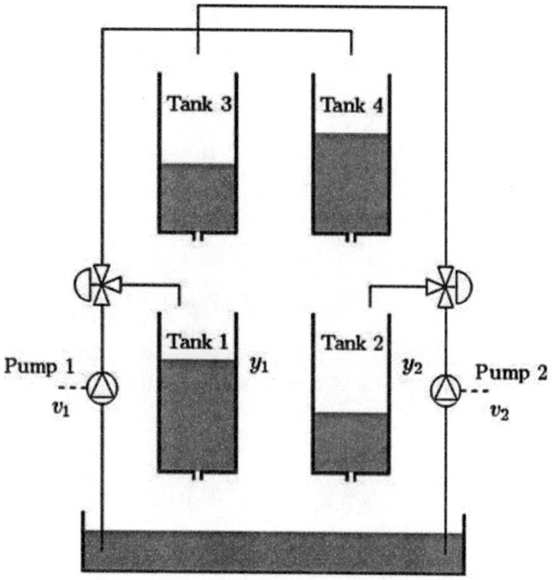

Fig. 1 Schematic diagram of quadruple tank system

γ_2 is the ratio of inlet flow to the tank 2, then $1 - \gamma_2$ will be the outlet flow to the tank 3. The parameters $\gamma_1 \gamma_2 \varepsilon(0,\ 1)$ are determined from the position of valves that are set prior to the experiment. The flow to tank 1 is $\gamma_1 k_c V_1$ and the flow to tank 4 is $(1 - \gamma_1)k_c V_1$ and similarly, for tank 2 is $\gamma_2 k_c V_2$ and tank 3 is $(1 - \gamma_2)\, k_c V_2$, where k_c is the calibration constant.

The measured level signals are $Y_1 = k_m\, h_1$ and $Y_2 = k_m\, h_2$, where k_m is the calibration constant for level measurements. Mathematical model of quadruple tank system is designed using analytical methods with all known process parameters. The state equations of the four tank system based on Bernoulli's law and mass balances are given in Eqs. (1) to (4)

$$\frac{dh_1}{dt} = -\frac{a_1}{A_1}\sqrt{2gh_1} + \frac{a_3}{A_1}\sqrt{2gh_3} + \frac{\gamma_1 k_c}{A_1} v_1 \tag{1}$$

$$\frac{dh_2}{dt} = -\frac{a_2}{A_2}\sqrt{2gh_2} + \frac{a_4}{A_2}\sqrt{2gh_4} + \frac{\gamma_2 k_c}{A_2} v_2 \tag{2}$$

$$\frac{dh_3}{dt} = -\frac{a_3}{A_3}\sqrt{2gh_3} + \frac{(1-\gamma_2)k_c}{A_3} v_2 \tag{3}$$

$$\frac{dh_4}{dt} = -\frac{a_4}{A_4}\sqrt{2gh_4} + \frac{(1-\gamma_1)k_c}{A_4} v_1 \tag{4}$$

where

h_i is water level in Tank 'i'.
a_i is cross-section of outlet hole of Tank 'i'.
A_i is cross-sectional area of Tank 'i'.

The linearized state space model is given by Eqs. (5) and (6).

$$\frac{dX}{dt} = \begin{bmatrix} -\frac{1}{T_1} & 0 & \frac{A_3}{A_1 T_3} & 0 \\ 0 & -\frac{1}{T_2} & 0 & \frac{A_4}{A_2 T_4} \\ 0 & 0 & -\frac{1}{T_3} & 0 \\ 0 & 0 & 0 & -\frac{1}{T_4} \end{bmatrix} X + \begin{bmatrix} \frac{\gamma_1 k_1}{A_1} & 0 \\ 0 & \frac{\gamma_2 k_2}{A_2} \\ 0 & \frac{(1-\gamma_2)k_2}{A_3} \\ \frac{(1-\gamma_1)k_1}{A_4} & 0 \end{bmatrix} U \tag{5}$$

$$Y = \begin{bmatrix} k_c & 0 & 0 & 0 \\ 0 & k_c & 0 & 0 \end{bmatrix} \tag{6}$$

The time constants T_i are calculated using Eq. (7)

$$T_i = \frac{A_i}{a_i}\sqrt{\frac{2h_i^0}{g}}, i = 1, 2, 3, 4. \tag{7}$$

Table 1 Quadruple tank process parameters

Area of tank 1 and 3 (A_1, A_3)	28 cm^2
Area of tank 2 and 4 (A_2, A_4)	32 cm^2
A_1, A_3	0.071 cm^2
A_2, A_4	0.057 cm^2
k_c	0.5 V/cm

Table 2 Steady-state operating points

Steady-state parameters	Minimum phase	Non-minimum phase
h_1^0, h_2^0 (cm)	(12.4, 12.7)	(12.6,13)
h_3^0, h_4^0 (cm)	(1.8, 1.4)	(4.8,4.9)
v_1^0, v_1^0 (V)	(3.00, 3.00)	(3.15, 3.15)
k_1, k_2 (cm^3/Vs)	(3.33, 3.35)	(3.14, 3.29)
γ_1, γ_2	(0.70, 0.60)	(0.43, 0.34)

Equations (8) and (9) are the transfer function matrices for minimum phase and non-minimum phase conditions which are obtained using the process parameters and operating points as given in Tables 1 and 2, respectively.

$$G_{-}(s) = \begin{bmatrix} \frac{2.6}{1+62s} & \frac{1.5}{(1+23s)(1+62s)} \\ \frac{1.4}{(1+30s)(1+90s)} & \frac{2.5}{1+90s} \end{bmatrix} \tag{8}$$

$$G_{+}(s) = \begin{bmatrix} \frac{1.5}{1+63s} & \frac{2.5}{(1+39s)(1+63s)} \\ \frac{2.5}{(1+56s)(1+91s)} & \frac{1.6}{1+91s} \end{bmatrix} \tag{9}$$

The transfer matrix G has two zeroes where one of the zero is always in the left half of s-plane and the other can be located either in right half or left half of s-plane based on the position of three-way valves. If the values of γ_1 and γ_2 satisfy the condition ($0 < \gamma_1 + \gamma_2 < 1$), then the system is in minimum phase, and the system is in non-minimum phase, if the values of γ_1 and γ_2 satisfy the condition ($1 < \gamma_1 + \gamma_2 < 2$).

3 Relative Gain Array Technique

In Multivariable control design, an appropriate operating practice is to determine suitable input and output pairs so that the respective manipulated variable has the maximum effect on the output variable. The relative gain array (RGA) is a widely used method for establishing the best input–output pairings for multivariable process control systems. RGA concept quantifies the change in steady-state gain that occurs when other control loops are closed. According to the proposal given by Bristol [13], RGA ($\wedge$) is given by Eq. (10).

$$\wedge = \begin{bmatrix} \lambda_{11} & \lambda_{12} \\ \lambda_{21} & \lambda_{22} \end{bmatrix} \tag{10}$$

where λ_{ij} is the relative gain between ith input and jth output. For minimum phase condition, λ_{11} is obtained as 0.63 which falls in the range $0.5 < \lambda_{11} < 1$, so the suitable pairing is determined as $y_1 - u_1$ and $y_2 - u_2$. But for the non-minimum phase condition, λ_{11} is obtained as 0.375 which falls in the range $0 < \lambda_{11} < 0.5$, so the pairing is found as $y_2$with $u_1$and $y_1$with u_2.

4 Design of Decoupler

Multivariable control system suffers with undesirable interactions between the control loops and requires a reciprocal system to isolate the loops from each other. Such a design is called as decoupling control. In this study, an ideal decoupler is used to eliminate loop interactions in which diagonal elements are assigned as unity so that the decoupler is insensitive to forward path controller gains. This enables the designer to avoid redesign of decoupler during online controller tuning. The block diagram of the process with decoupler is shown in Fig. 2.

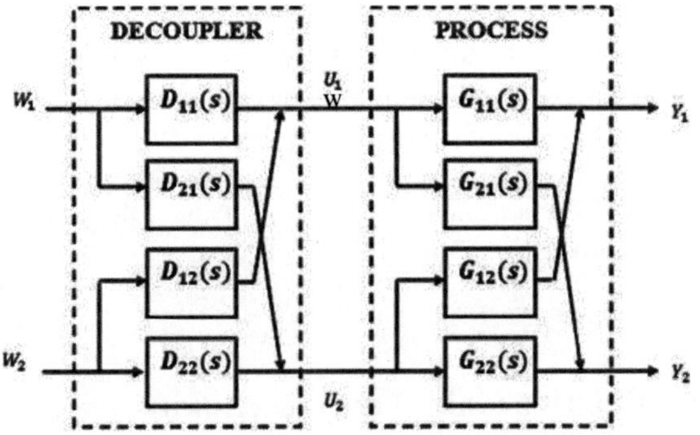

Fig. 2 Architecture of system with decoupler

The ideal decoupler is designed by the method proposed by Zalkind [14] and Luyben [15] and it is as given in Eq. (11).

$$D(s) = \begin{bmatrix} D_{11} & D_{12} \\ D_{21} & D_{22} \end{bmatrix} \tag{11}$$

where the diagonal elements (for ideal decoupler), $D_{11} = D_{22} = 1$ and the off-diagonal elements are given below

$$D_{12} = -\frac{G_{12} D_{22}}{G_{11}} \tag{12}$$

$$D_{21} = -\frac{G_{21} D_{11}}{G_{22}} \tag{13}$$

The designed decoupler matrices for QTS in minimum phase and non-minimum phase operating conditions are given in Eqs. (14) and (15), respectively

$$D_{-}(s) = \begin{bmatrix} 1 & \frac{-0.577}{(1+23s)} \\ \frac{-0.5}{(1+30s)} & 1 \end{bmatrix} \tag{14}$$

$$D_{+}(s) = \begin{bmatrix} 1 & \frac{-1.667}{(1+39s)} \\ \frac{-1.5625}{(1+56s)} & 1 \end{bmatrix} \tag{15}$$

5 IMC Tuned PI Controller

It is well known that despite many sophisticated control techniques and theories that have been devised in last few decades, proportional–integral–derivative (PID) controllers are still the most adopted in practical applications. Due to their simple structure, PID controllers are quite easy to tune and their use is well explicitly understood by majority of automatic control designers and industrial practitioner. In this paper, a decentralized PI controllers is designed which could be tuned simultaneously. In QTS, liquid level is the variable of interest and hence, PI controller is used. The control structure for PI controller with decoupler is designed for both minimum and non-minimum phase configurations. The IMC-based tuning method has many features such as time delay compensation and offset free response at steady state. Shaping the set-point tracking and regulatory responses using filter is also the advantage of IMC tuning method. The generalized PI controller transfer function for IMC-based tuning structure is given in Eq. (16)

Table 3 Controller parameter using model-based tuning

Controller parameters	Minimum phase		Non-minimum phase	
	Loop 1	Loop 2	Loop 1	Loop 2
Proportional gain (K_c)	1.03	0.756	0.292	0.313
Integral gain (K_i)	0.0167	0.018	0.00138	0.001307

$$G_{\mathrm{PI}}(s) = \frac{\tilde{G}_{\mathrm{p}}^{+}(s)^{-1} G_{\mathrm{f}}(s)}{1 - \tilde{G}_{\mathrm{p}}^{-}(s) G_{\mathrm{f}}(s)} \tag{16}$$

$$G_{\mathrm{f}}(s) = \frac{1}{\lambda s + 1}, \tilde{G}_{\mathrm{p}}^{+}(s) = \frac{K_{\mathrm{p}}}{\tau_{\mathrm{p}} s + 1} \text{ and } \tilde{G}_{\mathrm{p}}^{-}(s) = 1 \tag{17}$$

Substituting Eq. (17) in Eq (16), we get,

$$G_{\mathrm{PI}}(s) = \frac{\tau_{\mathrm{p}}}{K_{\mathrm{p}}\lambda} + \frac{1}{K_{\mathrm{p}}\lambda s} \tag{18}$$

The PI controller parameters are inferred from Eq. (18) as follows:

$$K_{\mathrm{c}} = \frac{\tau_{\mathrm{p}}}{K_{\mathrm{p}}\lambda} \tag{19}$$

$$K_{\mathrm{i}} = \frac{1}{K_{\mathrm{p}}\lambda} \tag{20}$$

Using the IMC design formulae, the controller parameters are calculated and are shown in Table 3.

6 RTDA Controller

The key attributes of the overall performance of any control system design are robustness, set-point tracking, disturbance rejection, and aggressiveness. The primary objective of any control system design is to arrive at a practical point of success that would result in good set-point tracking and disturbance rejection without sacrificing robustness. In conventional PID controllers, the tuning parameters are not directly related to the performance attributes, and designing a perfect controller to achieve the desired performance is not easy. Using RTDA control strategy, a simplified model prediction with transparent tuning parameters has been implemented for a class of multivariable system. It is possible to tune each design parameter independently to obtain the optimum performance. Block diagram of RTDA controller is shown in Fig. 3.

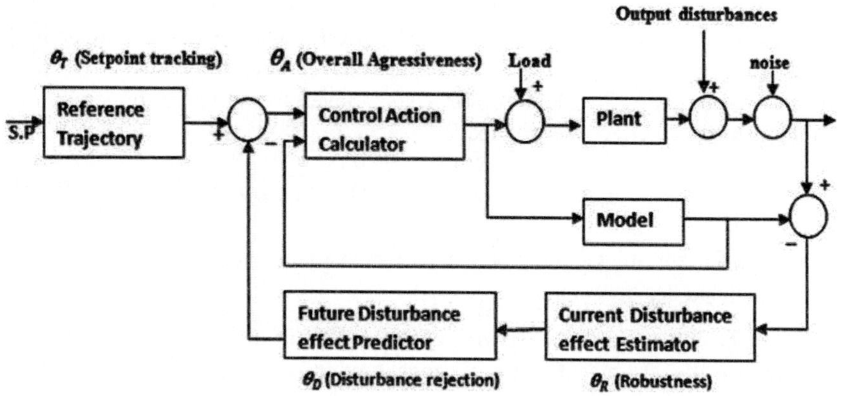

Fig. 3 Block diagram of RTDA controller

The RTDA controller has four tuning parameters (θ_R, θ_T, θ_D, and θ_A) which are directly related to the performance attributes of the closed loop control system. The robustness parameter θ_R is used to improve the controller performance even in the presence of model uncertainties and plant/model mismatch. The parameter θ_D is used to improve the regulatory performance without affecting the servo performance that is rejecting the disturbances without influencing the other parameters. The parameter θ_T is used to improve the servo response, and the aggressive parameter θ_A is used to improve both servo and regulatory responses.

The actual dynamics of the process is approximated to first-order model. The transfer function of the model is given in Eq. (21).

$$y(s) = \frac{K}{\tau s + 1} u(s) \tag{21}$$

The control action $u(k)$ remains the same for the whole prediction horizon, i.e.,

$$u(k + i) = u(k) \tag{22}$$

$$\hat{y}(k + i) = a^i \hat{y}(k) + b_i \eta u(k) \quad \text{for} \quad 1 \le i \le P \tag{23}$$

where, $\eta_i = \frac{1-a^i}{1-a}$.

This prediction must be updated to include the effect of unmeasured disturbances and other sources of modelling errors. The set of P equations are used to compute the control input $u(k)$ which can make the error closer to zero.

Error Decomposition and Model Update: The predicted model output has deviation from actual process output due to plant-model mismatch and prediction needs to be updated. The model mismatch is given by,

$$e(k) = e_m(k) + e_d(k) \tag{24}$$

where $e_m(k)$ and $e_d(k)$ are the estimates which represent the inherent modelling uncertainties and effects of unmeasured disturbances, respectively.
Current Disturbance Effect Estimation: Using Bayesian estimation principle, the current disturbance effect is estimated as

$$\hat{e}_d(k) = \theta_R \hat{e}_d(k-1) + (1-\theta_R)e(k) \tag{25}$$

where θ_R is the robustness tuning parameter which affects the stability of the closed loop system and lies between 0 and 1.

Future Disturbance Effect Prediction: With the current disturbance estimate, the future error is then estimated to update model prediction.

$$\hat{e}_d(k+i) = \hat{e}_d(k) + \frac{1-\theta_D}{\theta_D}[1-(1-\theta_D)^i]\Delta\hat{e}_d(k) \quad \text{for } 1 \le i \le P \tag{26}$$

$\Delta\hat{e}_d(k) = \hat{e}_d(k) - \hat{e}_d(k-1)$, the difference between two consecutive errors.θ_D is the disturbance rejection tuning parameter, when $\theta_D \sim 0$: linearly increasing disturbances (aggressive) and when $\theta_D \sim 1$, constant disturbance.

Updated N-step Model Output Prediction: The future prediction for P-step prediction horizon is given by

$$\tilde{y}(k+i) = \hat{y}(k+i) + \hat{e}_d(k+i) \qquad \text{for } 1 \le i \le P \tag{27}$$

Desired Output Trajectory: The desired trajectory $y_t(k)$ for a given setpoint $S_p(k)$ is represented as

$$y_t(k) = \theta_T\, y_t(k-1) + (1-\theta_T)S_p(k) \tag{28}$$

$$y_t(k+i) = \theta_T^i y_t(k) + (1-\theta_T^i)S_p(k) \quad \text{for } 1 \le i \le P \tag{29}$$

θ_T is the trajectory tracking tuning parameter.
Final Controller Equation: Control action is based on least square minimization technique. The objective function of the controller is given as

$$\min_{u(k)} = \Sigma_{i=1}^{P}(y_t(k+i) - \tilde{y}(k+i))^2 \tag{30}$$

The control action $u(k)$ is the minimization of the difference between model predicted output and reference trajectory for P-step

$$u(k) = \frac{1}{b}\frac{\Sigma_{i=1}^{P}\eta_i\ \psi_i(k)}{\Sigma_{i=1}^{P}\eta_i^2} \tag{31}$$

$\psi_i = y_t(k+i) - a^i\hat{y}(k) - \hat{e}_d(k+i)$ where ψ_i is the stipulated error.

The aggressiveness parameter θ_A depends on prediction horizon P is given by Eq. (32),

$$P = 1 - \frac{\tau}{t_s}\ln(1-\theta_A) \quad (32)$$

where t_s is the sampling time.

7 Simulation Results

SIMULINK models for multiloop control of quadruple tank system in minimum and non-minimum operating conditions for PID and RTDA controller are developed. The responses for all these cases are analyzed and compared using simulation. Through analysis, it is observed that non-minimum phase system is difficult to control when compared to minimum phase system. The analysis also shows that the control loop interactions are completely eliminated with the help of decoupler. The IMC-based PI structure uses the process model in implicit manner that is PI tuning parameters are adjusted based on the transfer function model of the process.

Servo and regulatory responses are obtained for both minimum phase and non-minimum phase for QTS system using IMC-PI controller and RTDA controller. Various performance metrics like rise time, peak overshoot, settling time and Integral of Squared Error (ISE) are considered to compare the performance of the controllers. Servo and regulatory response for minimum phase operating condition using IMC tuned PI controller are shown in Fig. 4 in which interactions between the loops were completed eliminated. Moreover good servo response was achieved and controller was able to reject the disturbances for the applied load change for both the outputs. Similar effects were observed for non-minimum phase condition and it is shown in Fig. 5 but it suffers with inverse response initially.

Servo and regulatory responses for minimum phase operating condition using RTDA controller are shown in Figs. 6, 7, 8 and 9 for different values of tuning

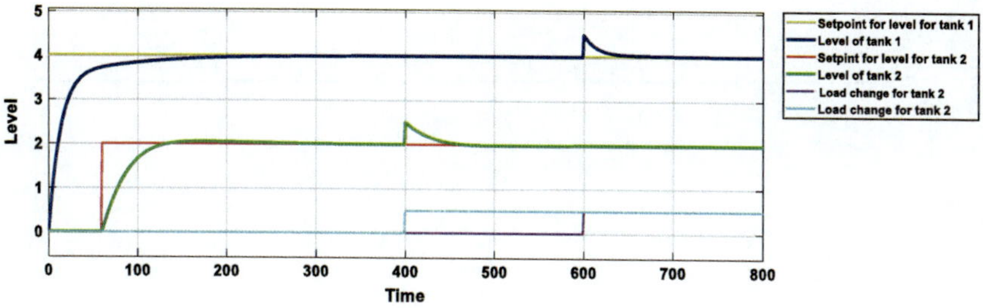

Fig. 4 Servo and regulatory response of IMC-PI controller for minimum phase

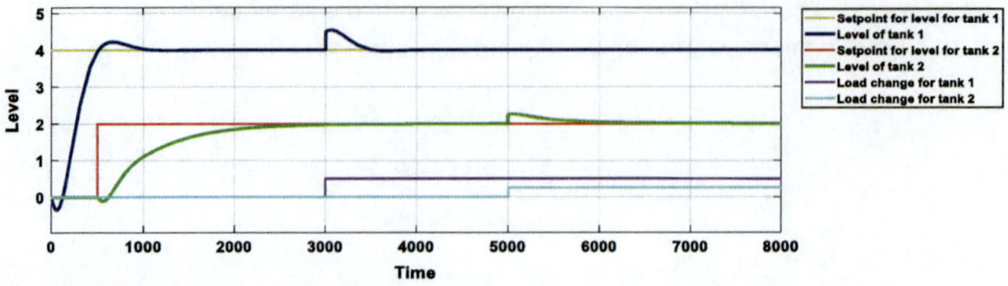

Fig. 5 Servo and regulatory response of IMC-PI controller for non-minimum phase

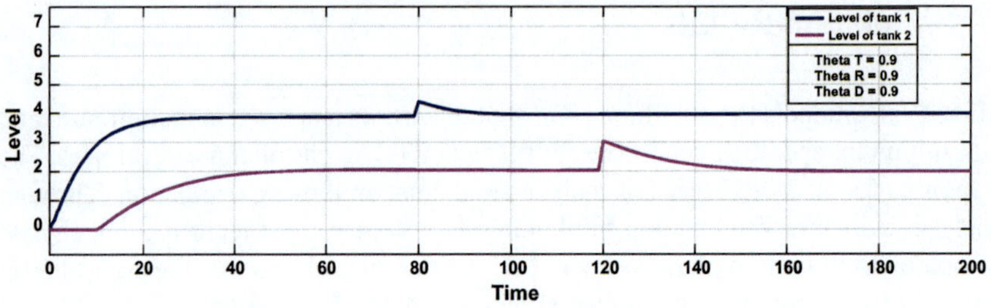

Fig. 6 Response of the system using RTDA controller in minimum phase ($\theta_T = 0.9$, $\theta_R = 0.9$, $\theta_D = 0.9$)

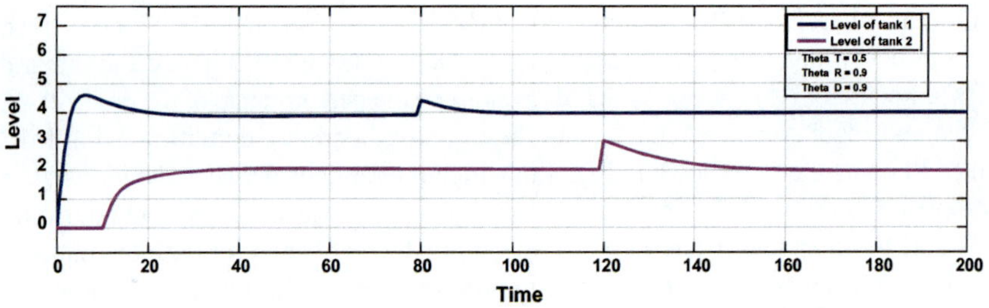

Fig. 7 Response of the system using RTDA controller in minimum phase ($\theta_T = 0.5$, $\theta_R = 0.9$, $\theta_D = 0.9$)

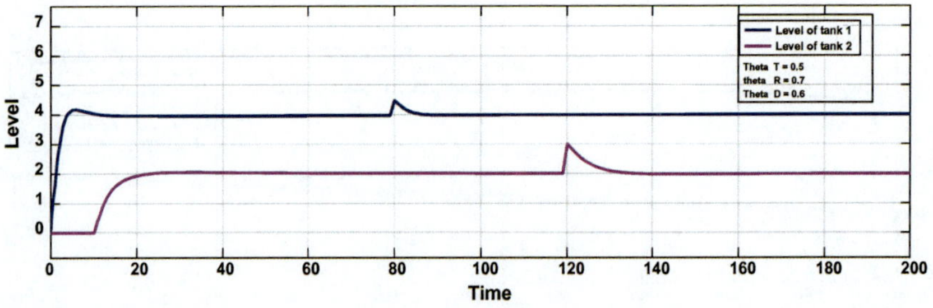

Fig. 8 Response of the system using RTDA controller in minimum phase ($\theta_T = 0.5$, $\theta_R = 0.7$, $\theta_D = 0.6$)

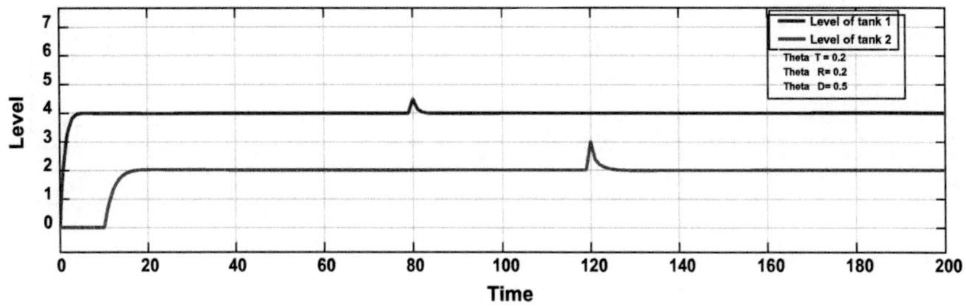

Fig. 9 Response of the system using RTDA controller in minimum phase ($\theta_T = 0.2$, $\theta_R = 0.2$, $\theta_D = 0.5$)

parameters to analyze their effects on closed loop response. Here, the decoupled transfer function is used as model, thereby plant-model mismatch is considered.

The RTDA controller aims to achieve robust performance even in the presence of small modelling errors. In minimum phase condition, the tuning parameters θ_T and θ_D were independent of each other which enabled set-point tracking and disturbance rejection simultaneously. From Figs. 7 and 8, it is observed that system yields better tracking response if θ_T is lowered but overshoot is increased and thereby reducing the robustness of the controller. It is also observed from Fig. 8 that reducing θ_R reduces the overshoot. In Fig. 9, θ_T and θ_R are reduced to get satisfactory responses without producing overshoot. Same effects are obtained for non-minimum phase conditions and are presented in Figs. 10 and 11. The performance of RTDA controller can be improved by reducing the value of θ_T, as the value of time constant is large in both minimum phase and non-minimum phase conditions.

The performance comparison for IMC-PI and RTDA controller is given in Tables 4 and 5 and it is inferred that RTDA controller gives better results in all the performance metrics for both minimum phase and non-minimum phase compared to IMC-PI controller.

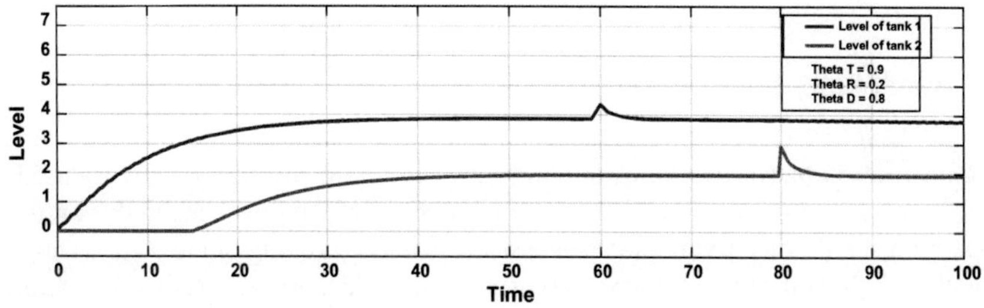

Fig. 10 Response of the system using RTDA controller in non-minimum phase ($\theta_T = 0.9$, $\theta_R = 0.2$, $\theta_D = 0.8$)

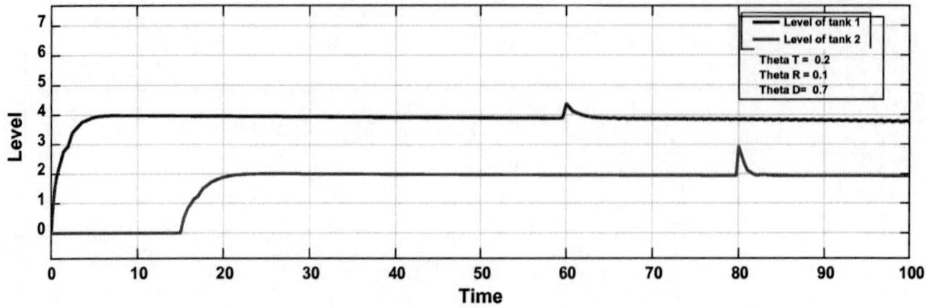

Fig. 11 Response of the system using RTDA controller in non-minimum phase ($\theta_T = 0.2$, $\theta_R = 0.1$, $\theta_D = 0.7$)

Table 4 Performance metrics of PI-IMC controller

Minimum phase			Non-minimum phase	
Parameters	Level 1	Level 2	Level 1	Level 2
Settling time (s)	320	280	1400	3700
Peak overshoot (%)	0	0	25	0
Rise time (s)	288	252	1260	3300
ISE	52.78	133.29	416.52	151.68

Table 5 Performance metrics of RTDA controller

Minimum phase			Non-minimum phase	
Parameters	Level 1	Level 2	Level 1	Level 2
Settling time (s)	8	21	5	22
Peak overshoot (%)	20	0	0	0
Rise time (s)	7.2	18.9	4.5	19.8
ISE	5.9934	2.589	6.2764	2.4121

8 Conclusion

Conventional controllers, which are known for its simplicity, yield better transient and steady-state responses if the system parameters remain unchanged during the operating conditions. But in practical cases, it is not so. Tuning of the parameters is another challenging task for conventional controllers. In this study, a next-generation controller named RTDA controller is designed for a multivariable system and simulated using MATLAB. Simulation results proved that RTDA controller is able to tune all the parameters independently whereas by using PI controller, tuning of the parameters affects the performance attributes. Thus, RTDA controller has better performance in set-point tracking, disturbance rejection, and robustness for multivariable system. Inverse response that is exhibited by IMC-PI controller is completely discarded by RTDA controller in non-minimum phase condition. The main advantage is that RTDA controller has ability to rule out the margins encountered by the conventional controller with transparent tuning parameters. Therefore, RTDA controller will find its place in industrial collaboration to enhance its job to meet current industrial challenges and will be beneficial to the practitioners. The only limitation of RTDA controller is that it cannot be applied for non-self regulatory systems in its present form. In the future, this can be overcome by suitable modification in the algorithm.

References

1. Vu, T.N.L., Lee, J., Lee, M.: Design of multiloop PID controllers based on the generalised IMC-PID method with Mp criterion. Int. J. Control Autom. Syst. **5**(2), 212–217 (2007)
2. Dhanraj, A.V., Nanjundappan, D.: Design of optimised PI controller with ideal decoupler for a non linear multi-variable system using particle swarm optimization technique. Int. J. Innovative Comput. Inf. Control **10**(1), 341–355 (2014)
3. Dhanraj, A.V., Nanjundappan, D.: Design of decentrilized PI controller using model reference adaptive control for quadruple tank process. Int. J. Eng. Technol. **5**(6), 5057–5066 (2014)
4. Mukati, K., Rasch, M., Ogunnaike, B.A.: An alternative structure for next generation regulator controllers, Part I: Basic theory for design, development and implementation. J. Process control, **16**, 499–509, June (2006)
5. Mukati, K., Rasch, M., Ogunnaike, B.A.: An alternative structure for next generation regulator controllers, Part II: II: stability analysis, tuning rules and experimental validation. J. Process Control **19**, 272–287 (2009)
6. Mukati, K., Ogunnaike, B.A.: Stability analysis and tuning strategies for a novel next generation regulatory controllers. In: Proceedings of American Control Conference. Boston, MA (2004)
7. Sendjaja, A.Y., Ng, Z.F., Si, S.: How and vinay kariwala: analysis and tuning of RTD-A controllers. Ind. Eng. Chem. Res. **50**(6), 3415–3425 (2011)
8. Srinivasan, K., Singh, J., Anbarasan, K., Paik, R., Medhi, R., Choudhury, K.D.: MPC and RTDA controller for fopdt & sopdt process. J. Inf. Syst. Commun. **3**(1), 109–113(2012)
9. Anbarasan, K., Srinivasan, K.: Design of RTDA Controller for Industrial Process using SOPDT Model with minimum or non-minimum Zero. ISA Trans. **5**, 231–244 (2015)
10. Mani, G., Pinagapani, A.K.: Design and implementation of a preemptive disturbance rejection controller for PEM Fuel cell air-feed system subject to load changes. J. Electr. Eng. Technol. **11**(4), 1921–718 (2016)
11. Saeed, Q., Uddin, V., Katebi, R.: Multivariable predictive PID controller for quadruple tank. In. J. Mech. Mechatron. Eng. **4**(7), 1178–1183 (2010)
12. Johanson, K.H.: The quadruple tank process: a multivariable laboratory process with an adjustable zero. IEEE Trans. Control Syst. Technol. **8**(3), 456–465 (2000)
13. Bristol, E.: On a new measure of interaction for multivariable process control. IEEE Trans. Autom. Control, **11**(1), 133–134 (1966)
14. Zalkind, C.S.: Practical approach to non interacting control, Part I and II, instrumentation and control systems, **40**(3 and 4), 8–13 (1967)
15. Luyben: Distillation Decoupling, AIChE J. **16**(2), 198–203 (1970)

A Survey on Crack Detection Algorithms for Concrete Structures

M. J. Anitha, R. Hemalatha, and S. Radha

Abstract Crack detection is gaining much attention nowadays due to localization and identification of cracks in the civil structures accurately. The aim of crack detection on images is to automatically detect the cracks or defects considering the properties of civil structures. Among various civil structures, concrete structures are the most important one which is having both domestic and industrial applications. Detection of cracks in concrete structures will be very important in present scenario. Some of the prevailing crack detection methods are manual inspection method, acoustic and vibration-based method, electrical and magnetic methods, visual and optical methods. This survey paper jointly summarizes the various methodologies and issues of automatic crack detection. Performance metrics like accuracy, sensitivity, and specificity are used to analyze the experimental evaluation of different papers effectively. Finally, various challenges and the future research directions of the crack detection techniques are discussed.

Keywords Concrete surface inspection · Defect detection · Crack detection · Artificial neural network · Deep neural network · Defect classification · Image processing · Parameter estimation

M. J. Anitha (✉)
Department of ECE, DOTE, Chennai, India
e-mail: mj_anitha@yahoo.co.in

R. Hemalatha · S. Radha
Department of ECE, SSN College of Engineering, Chennai, India
e-mail: hemalathar@ssn.edu.in

S. Radha
e-mail: radhas@ssn.edu.in

P. Suresh et al. (eds.), *Advances in Smart System Technologies*,
Advances in Intelligent Systems and Computing 1163,
https://doi.org/10.1007/978-981-15-5029-4_53

1 Introduction

Crack is a line on the surface or something along which it has split in the surface. All civil structures degrade over time because of the formation of cracks. The crack is formed mainly due to the usage of different materials, design, construction loads, and exposure conditions. The civil structures are to be checked for cracks periodically, either manually or by utilizing many crack detection methods. Traditionally, the concrete structures are verified by manual inspection at regular intervals resulting in number of accidents due to lack of inspection. For example, the collapse of the four-story residential building in Mumbai on 2017 killed 24 peoples and injured many peoples. Thus, early crack detection using crack detection techniques is gaining more importance nowadays, as there is a need to preserve our ancient buildings and maintain its structural quality. In general, manual inspection is a time-consuming process and it is not possible to detect cracks in the not reachable areas by human beings. The accuracy of manual inspection depends on the knowledge of the specialist. Automating this process will overcome these issues by detecting cracks at the early stage thereby reducing severe damage and failures. Therefore, researchers started working toward automatic image-based crack detection techniques [9]. The major objective of crack detection is to identify the crack in the structure by applying some kind of processing techniques. Generally, crack detection techniques are classified into two types. They are destructive testing and non-destructive testing (NDT). In this, destructive testing methods are used to identify the reason for failure, performance, and material behavior under different loads of structures by using crash test and hardness test. In non destructive testing methods, the cracks are detected without affecting the concrete or steel structures. So, NDT methods are more popular than destructive testing methods. NDT techniques are based on radiography, infrared thermography, ultrasonic testing, and magnetometry. The attention toward image-based crack detection techniques for non-destructive testing methods is increasing drastically due to its simplicity and the ability to detect larger cracks within the minimal duration of time. The main issues that affect the performance of the image-based crack detection techniques are computational complexity, non-uniform illumination, blurred images, low contrast images, and accuracy of the crack detection algorithm. The crack detection algorithm tackling the above-mentioned issues can also be applied to concrete structure. In this paper, the crack detection algorithm applied on concrete structures are reviewed and categorized into following two groups:

- Crack detection techniques based on image processing.
- Crack detection techniques based on deep convolutional neural networks.

This paper provides an extended survey of the crack detection algorithms for concrete structures by analyzing the different works of various researchers. This paper is organized as follows. Section 2 begins with the general architecture of crack detection process. Section 3 depicts the various crack detection techniques using image processing methods used to identify defects on concrete structure. Section 4 describes about different crack detection algorithms using artificial neural network applied on concrete structure. Section 5 discusses about various challenges faced in dealing with crack detection algorithm and Sect. 6 illustrates the experimental evaluation. Section 7 explains the inferences made and highlights the future research directions followed by conclusion in Section 8.

2 Architecture of Crack Detection Process

The general architecture of crack detection process is depicted in Fig. 1. The first stage of crack detection is image acquisition, which acquires the images of defect and non-defect structures. Mostly, this is done by using camera. The factors that affect the quality of image under natural light are climate, seasons, and time. A natural light shield and artificial lighting may be used to reduce the effects of above factors [6]. The resolution of the camera plays an important role in crack detection technique [22]. The different types of cameras used to acquire images are digital camera, DSLR camera, or IR camera depending upon the type of technique used. The next stage is the preprocessing stage which is necessary to remove noises from the cracked image. Some of the popular preprocessing techniques used are Gaussian filters [1, 18], filters based on percolation method [26], wavelet transform [15], Canny edge detector [19], and morphological operations [10, 14]. Preprocessing technique must be selected in such a way that it requires minimal computation time for processing. Some of the crack detection techniques are binarization method based on genetic algorithm with wheel selection strategy [10], morphological feature extraction algorithm [15], and clustering k-means algorithm [14, 12]. Automated crack detection algorithms mostly use convolutional neural network [7], back propagation neural network [18], etc. The final stage is the parameter estimation which is used to measure particular parameters from the crack. Generally, width, location, length, depth, and direction of propagation [9] are measured.

2.1 Image Acquisition

In order to get good quality of crack surface image, illuminating the surface with required light is necessary. The light source is also important to provide fluctuation free-light on the cracked surface. Different types of light sources are wide spectrum tungsten, sunlight, fluorescent tubes, halogen, xeon, LED, and sunlight. They are used to illuminate the crack surfaces of concrete and steel civil structures. Mostly,

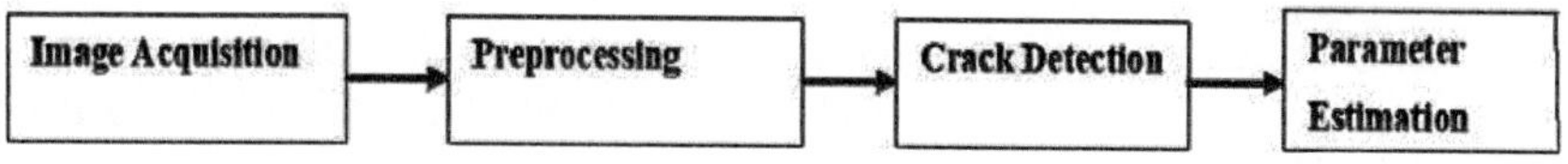

Fig. 1 General architecture of crack detection

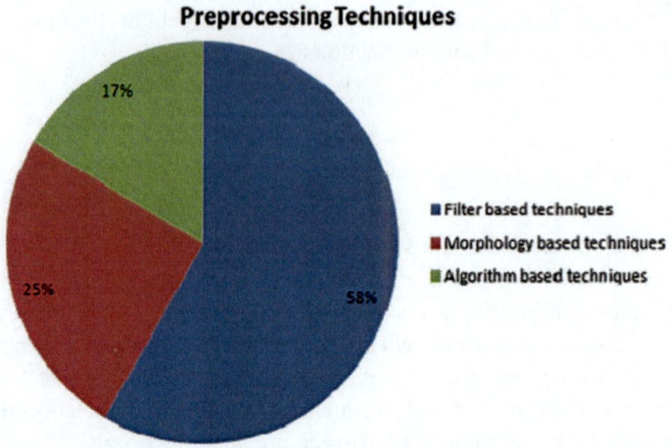

Fig. 2 Types of preprocessing techniques

digital camera is used in concrete structures [12, 19, 22]. The resolution of digital camera and lighting available isused to determine the accuracy of image.

2.2 Preprocessing

Preprocessing techniques are used to separate crack features from the background and they are mostly based on filtering techniques. Muti-sequential image filter is used to eliminate noise and identify indistinct cracks [19]. Smoothing filter removes the irregularly illuminated conditions and shading effects by subtracting the input concrete image with the smoothed image [10]. The input image is obtained from inspection robot and the noise present in the image is removed by using line emphasis filters [7]. Other important preprocessing techniques are morphology-based and algorithm-based techniques. The different types of preprocessing techniques are plotted in pie chart format and are shown in Fig. 2.

3 Crack Detection Techniques Applied on Concrete Structure

This section describes in detail about the crack detection for concrete structure using image processing techniques and neural networks-based algorithms and it is summarized in Table 1.

Table 1 Comparison of crack detection techniques of concrete

References	Dataset size	Camera type	Preprocessing techniques	Crack detection techniques	Parameter estimation
[3]	40 K	DSLR camera	–	Convolutional neural network	Overcome the effects of lightning and shadow changes
[1]	101	Digital camera	Image stitching algorithm	Skeleton transform, fast Fourier transform, neural network	Measurement of crack length, width, and prediction of crack depth
[12]	Clustered images	Digital camera	Otsu's thresholding, morphological operator	Binary distance trans- form, iterative contour evolution	Extraction of crack from shadow images and clustered background images
[19]	2000	Digital camera	Multi-sequential image filter	Re-extraction method, genetic programming	Measure the crack width, length and orientation
[10]	50	–	Smoothing Filter, Hessian matrix filter	Otsu's thresholding	Detection of crack on irregularly illuminated conditions, shading, and blemishes
[9]	60	SONY cyber- shot camera	Hessian matrix with line filter	locally adaptive thresholding	Decrease effects of varying stains, blebs, and shadings
[22]	3500	Digital camera with UAV	–	Convolutional Neural network	Provide a safe and fast crack detection technique
[15]	420	–	Morphological operation	Neural network Using LDA technique	Calculate crack thickness
[18]	2250	–	Gaussian filtering, improved subtraction method	Morphological operation, back propagation neural network	Automatic detection of crack from camera images

(continued)

Table 1 (continued)

References	Dataset size	Camera type	Preprocessing techniques	Crack detection techniques	Parameter estimation
[27]	50	Digital camera	Filter using percolation model	Improved percolation algorithm	To extract crack from concrete blebs, stain, insufficient contrast, and shading
[26]	–	Digital camera	–	the termination procedure	Increase the computation speed for detecting crack in mega pixels images
[21]	10	Samsung Galaxy mobile phone camera	Otsu algorithm	BFS traversal algorithm	Calculate the crack width
[20]	1000	DSLR camera	Gradient method	SVM	To measure crack density
[25]	10	Digital Camera	Sobel filter, Canny Filter	Percolation process	To reduce computational cost
[17]	603	Digital camera	ResNet classifier	Deep residual network	Detect different types of defect in input image
[28]	100	Digital camera	Morphological operations	Percolation model	To measure crack width, length, and orientation
[6]	250	Digital camera	Statistical filter	Morphological operations	Detect minor, major and multiple cracks in buried concrete pipes

3.1 Crack Detection Techniques Based on Image Processing

Cracks appearing on concrete surface are the first-step deterioration of the structure. The regular disclosure of crack will head to the rigorous destruction to the structure. So, it is vital to recognize the cracks at the earliest stage to eradicate these problems. However, crack detection on a concrete image is cumbersome due to many factors such as delaminations and voids. This section deals with various methods that involve crack detection on concrete surface. The cracks on the concrete structures can be detected using various image processing techniques. The crack detection techniques are classified into three categories based on the type of technique used to detect crack on concrete structures.

- Thresholding-based techniques.
- Model-based techniques.
- Pattern-based techniques.

Thresholding-Based Techniques: Generally, the basic idea behind crack detection is that the non-cracked pixels are lighter than pixels in the crack region. So, thresholding methods are used to classify the pixels as crack and noncrack. Fujita et al. [10] fostered preprocessing techniques for accurate crack detection in concrete structures. This method first removes the irregularly illuminated conditions and shading effects by subtracting the input concrete image with the smoothened image. Then, Hessian matrix filters are applied to sharpen the line structures associated with the crack which is followed by thresholding to extract the cracks from background. The main advantage of this system is it handles irregularly illuminated conditions, shading, and blemishes. But, the system's sensitivity and specificity are based on the threshold value. Hence, the system is extended [9] to handle the threshold value using locally adaptive thresholding. This system spots coarse cracks by using probabilistic realization method and fine cracks by locally adaptive thresholding. The major benefit obtained from this system is that it does not require optimization while fixing the threshold value for binarization. Yet, this method is based on optimizing the size parameter.

The difficulties in detection of cracks in concrete buried pipes are randomness and irregularity of cracks. Sinha and Fieguth [23] developed a statistical filter to identify cracks by processing the image along the four directions using windows of increasing size and it is preceded by cleaning and linking operations. Minor cracks can be detected with smaller window size and major cracks can be detected with larger window size. Even though it detects multiple cracks, it fails to detect mushroom cracks. In general, traditional methods provide good crack detection result for a particular filter size with an exact image. Then thresholding step is used to separate cracks from the background which needs manual adjustments resulting in non-automatic solution. Thus, they are not suitable for real-time applications. Hence, Dinh and La [7] developed an automatic computer vision-based method to sense concrete cracks. Here, the input image is obtained from inspection robot, and the noise present in the image is removed by using line emphasis filters. The drawbacks of traditional methods are subjugated using histograms, which provide dynamic parameters for thresholding. Histogram thresholding is used to select cracks from the background. The biggest peak in the histogram of an image is used to boost the threshold value of image binarization. The crack present in low contrast images and uniform images is also detected with high accuracy. Since, it fails to achieve accuracy independent of boundary conditions, Asdrubali et al. [2] developed an algorithm to detect contours of thermal bridges by using infrared thermographic images. This helps to study the energy performance of buildings. The kantorovich operators are used to enhance the thermographic images. Then, by analyzing the histogram, suitable thresholding is used to detect the contours. The accuracy of threshold-based technique is dependent upon the selection of threshold value to detect crack. Generalization of threshold value is difficult. Because of this, the thresholding techniques may be inapplicable to the real-time scenario. Also pixels in shadow part of image will also have same intensity as crack pixels, which will lead to inaccuracy.

Model-Based Techniques: Yamaguchi et al. [27] introduced a new approach, based on the assumption that noises are composed of thin interconnect textures. This method is based on percolation model. The brightness connectivity and shape of the percolated region are used to extract continuous texture. The extraction of continuous texture depends on the length criterion rather than the window criterion. This method works effectively for images with unclear cracks and curtails noise due to insufficient contrast, shading, concrete blebs, and stain. But this system does not work in real-world scenario. There forth, the system is extended to promote an effective and fast-crack detection technique [26] using percolation model. The main parameter that affects the system performance is its computation time. Reducing the computational complexity is important for applying the crack detection in real world due to increase in the size of digital images up to 10 mega pixels. Accordingly, the termination procedure is used to reduce the computation time. Moreover, the percolation technique may skip to the successive pixel based on the circularity of the neighboring pixels. Reduction in computational cost and high accuracy is achieved using this method of crack detection. This is obtained by considering the amount of crack pixels, instead of size of the image. The limitation of this method is that its performance depends on the chosen threshold value at the termination procedure.

Yamaguchi and Hashimoto [25] developed a new algorithm to detect crack using edge information and percolation processing. Horizontal and vertical edges are detected by sobel filter followed by Canny filter for edge detection. It eliminates the repetition of the iterative process starting from every pixel and reduces the computation time while maintaining the precision. But it fails to automate few parameters. The evaluation of safety of post-earthquake structures is done manually and consumes more time and also costly. The traditional techniques only aim at locating crack points not cracks width, length and orientation. Zhu et al. [25] introduced a novel method to gather properties of crack. The crack detection method depends on percolation and is used to detect crack points. The crack properties like width, length, and orientation are measured by using crack skeletons and distance field. This method fails to provide high accuracy and automated parameter selection.

Hiasa et al. [13] introduced a novel method to identify subsurface damage identification in concrete bridges by using infrared thermography. Finite element (FE) model simulations are used to find threshold for infrared data processing. The temperature threshold of non-affected areas is got from the IR image and the affected area is obtained from FE model simulation, respectively. By visual inspection, it is not possible to detect delaminated concrete areas. This difficulty can be overcome by infrared thermography, and delaminated areas of concrete slabs in centimeter depths could be detected. This model-based approaches mainly depend on input from user to start the seed pixels and not scalable due to dependency on user input. Hairline cracks cannot be detected because of the difficulty to identify cracks [16].

Pattern-Based Image Processing Techniques: Nishikawa et al. [19] introduced a digital image processing technique to identify cracks in concrete structures automatically. Genetic programming is used for extracting cracks on concrete surface. In this method, a multi-sequential image filter is developed to detect major cracks by eliminating residual noise. The indistinct cracks are detected by repeatedly applying the image filter to the local regions neighboring the crack. The widths of the crack can be detected using spatial derivatives derived from brightness patterns. Nowadays, the increase in the use of android mobile phones is growing drastically. Henceforth, Shuang-rui et al. [21] introduced an automatic crack detecting system to measure the width of crack using android mobile phones. By using high portability of android mobile phones, images were captured by mobile phone and the image is processed using binarization, filtering, edge recognition, and width calculation techniques. The Otsu thresholding algorithm is used for binarization. Edge recognition is achieved by breadth-first search (BFS) traversal algorithm and the width of crack is measured using Euclidean distance calculations. The main limitation of utilizing this method is automatic optimization of distance between the concrete crack and the camera. Jahanshahi et al. [15] enhanced the system to develop a vision-based crack detection method by extracting the whole crack from its background based on regular edge-based approaches. The purpose is to measure crack thickness more accurately. The drawback is the slow computational speed of detecting cracks. Powerful decision-making tools are required to develop fast analysis which is the major drawback of the system. The system also fails to detect cracks when the length of crack is less than 15 mm. So, they [14] developed a remote-sensing, connection-less crack detection method. In this work, depth perception is availed to identify cracks and ordain thickness of crack images captured at any resolution and from different distance. This damage detection approach allows capturing of the thinnest cracks. The drawback of this method is that it craves independent light source to increase performance and accuracy of the system. Robust classifiers based on pattern recognition methods relay upon training data. Training and validation data require the knowledge of trained experts. It is a labor-demanding and error-prone process.

4 Crack Detection Techniques Based on Neural Networks

The crack detection techniques using image processing techniques are custom made for certain images in the studied datasets. The techniques may not give exact results for new images and datasets collected from various lightning conditions and presence of shadows. These drawbacks are eliminated by using neural network. Artificial neural network is a technique of computing, created by interaction of many connected processing elements. This is a powerful tool used for real-time applications. The network will improve its performance by learning from its experience. The trained network may give complete and correct output from incomplete and partial input information. ANN consists of multiple nodes similar to the biological neurons of human brain. These multiple nodes are organized in layers. The input data is connected to ANN through input layer. Input layer provides input to one or more hidden layers. Hidden layers are responsible for actual processing of neural network through a system of weighted connections. The hidden layer is connected to the output layer. A neural network is initially well trained. Training is a process of providing input and indicating the details of the output to the network.

4.1 Crack Detection Techniques Based on Basic Artificial Neural Networks

Moon et al. [18] tested the concrete surface and picturized the cracks effectively using two steps of processing. In the first step, the filtering is used to extract the crack from the background followed by improved subtraction method and morphological operation. The next step identifies the presence of cracks by back propagation neural network. The system has the limitation of not working properly in existence of fog, hue of concrete surface, and shape of structures. Various optimization techniques are required to define the optimal parameters. As identifying cracks at the pixel level is not possible, an integrated model is developed by an Adhikari et al. [1] to represent the cracks numerically using image processing. This model consists of four stages. In the first stage, crack length is measured by using the perimeter of the skeleton of a crack. The second stage uses Fourier transform of digital images which eliminates the requirement of image registration as required in the traditional methods. In the third stage, neural networks are used to predict crack depth followed by last stage 3D visualization, which is gained by using crack density. The drawback of this model is largely dependent on camera resolution and imaging criteria. The networks here discussed gave good performance to both identification and classification of cracks. Prasanna et al. [20] introduced new crack detection algorithm to maintain structural health and reliability of concrete bridges by detecting cracks on bridge decks. Manual threshold parameter tuning is removed by machine learning classification.

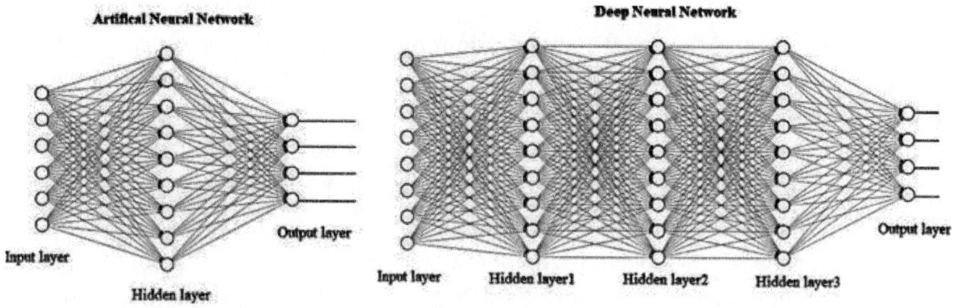

Fig. 3 Artificial neural network and deep neural network

Potential crack regions are spatially identified by robust curve fitting and multiple visual features are calculated. Feature extraction uses scale-space of the local feature and crack density map is also calculated. Even though it is completely automated, the processing time is high. The limitation of ANN is that the performance of ANN is affected by the selection of good input features.

4.2 *Crack Detection Techniques Based on Deep Convolutional Neural Networks*

ANN can have single hidden layer. Hidden layers are responsible for learning features. It does not have multiple hidden layers. So, it cannot learn complex features. So, deep neural network and CNN are used. The deep neural network consists of multiple layers where each layer transforms input data into more abstract representation. These series of layers between input and output, select input features and generate a series of new features depending on the input data similar to human brain. It is an unsupervised learning method and it does not have a desired output and information about output is not required. If more layers are present, then the level of features learnt by deep learning neural network will be higher. It requires a massive dataset to train itself on. In deep learning, it has more hidden layers, generates more features, and requires more training dataset. Some of the important activation functions used in deep learning network are sigmoid or logistic function, hyperbolic tangent function (Tanh), and rectified linear units (ReLu) function. Figure 3 illustrates artificial neural network and deep neural network. This section shows the deep learning-based techniques used to detect and classify crack in concrete structures. The reasons for using deep learning algorithms to identify cracks on concrete structure are as follows. The convolutional neural networks (CNNs) require very less preprocessing compared to other image classification algorithms. Feature extraction is one of the most time consuming and complex process in traditional automatic crack detection techniques. This limitation can be overcome by convolutional neural networks. The CNNs are easily adapted to new problems compared to traditional methods. Thus, the concrete crack detection techniques based on CNN is getting popular. Cha and Choi [3] proposed a novel method adopting a deep CNN for varying lighting and shadow changes. This method does not require image processing techniques for extracting features because convolutional neural networks are able to learn image features automatically. Large size images are scanned by combining both CNN and sliding window method. The limitation of this method is unable to detect internal features like voids and delamination in concrete structures due to photographic image. Silva and Lucena [22] proposed a deep neural network with classification algorithm. It detects the cracks present in concrete surface under different illuminations, surface finish, and humidity conditions. The number of layer in CNN is tuned to produce better performance. The main issue that influences the performance of the system is that it identifies the cracks at pixel level. Most of the traditional techniques identify only specific type of defects in concrete and steel structures. In order to overcome this limitation, Cha et al. [4] broached a semi real-time technique for concurrent detection of multiple defects in various types of images and Faster R-CNN is used to achieve it. This network can reveal crack in concrete, corrosion, and delamination in steel and bolt corrosion. The database used for testing was generated from images which comprehend location of damage and the label corresponding to it. It affords better computation efficiency. The frame work also extended to video-based defect detection also. The limitation of this method is that the accuracy and robustness are reduced when the distance between structural damage and the camera is increased. Supervised learning methods are used to detect and classify the defects in civil structures effectively with enough labeled images. In real-time applications, it is hard to collect labeled images due to the limited availability of experts for identifying and labeling defects in images. To overcome this defect, Feng et al. [17] developed a deep residual neural network, which is used for extraction and categorization of crack in the image. This network is initially trained with a set of images with labels for cracks, then it is applied to choose a subset of new images to retrain the network. This method is used to achieve a detection accuracy of 87.5%. AlexNet, GoogleNet and ResNet are some of the architectures of deep learning neural networks. Among these networks, AlexNet is used to detect cracks in concrete most effectively than any other networks [8]. AlexNet architecture is described in Fig. 4.

5 Experimental Evaluation

Various parameters like accuracy, computation time, error rate, sensitivity, and specificity are adopted to evaluate the performance of various crack detection techniques. Sensitivity and specificity are calculated using the possibilities of binary classifier such as True Positives (TP), True Negatives (TN), False Negatives (FN), and False Positives (FP). The accuracy is measured using sensitivity and specificity metrics given in Eqs. 1 and 2.

$$\text{Sensitivity} = \text{TP}/(\text{TP} + \text{FN}) \tag{1}$$

$$\text{Specificity} = \text{TN}/(\text{TN} + \text{FP}) \tag{2}$$

The sensitivity of crack detection is the ratio of correctly detected cracks to all crack samples where a defect truly occurs. The specificity is the ratio of undetected defect cases to the number of samples where defects do not occur truly. The neural network using LDA technique gives specificity of 91.36 and sensitivity of 97.6 in percentage [15]. The comparison bar chart is plotted with two values such as methodologies adapted for detection of crack and the accuracy in Fig. 5. From the performance comparison chart, AlexNet deep neural network-based crack detection techniques achieved highest accuracy, comparing the accuracy of different crack detection techniques used.

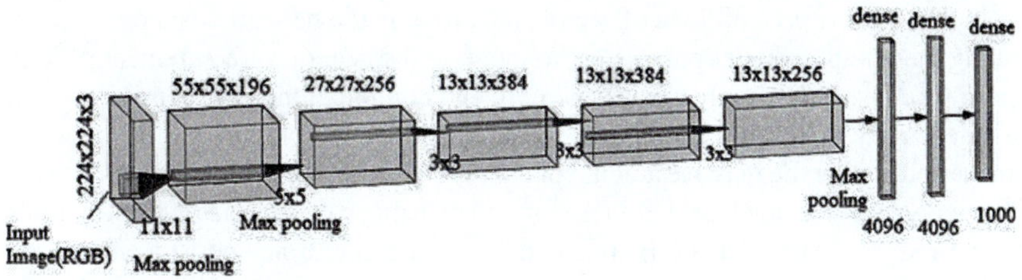

Fig. 4 Architecture of AlexNet

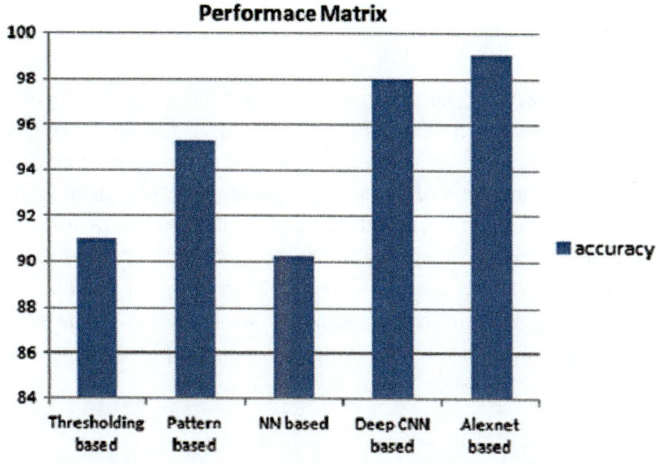

Fig. 5 Performance accuracy of different algorithms

6 Challenges and Future Research Directions in Crack Detection

The main issue that affects the performance of the various crack detection techniques in concrete is the distance between concrete crack and camera [2, 24], camera resolution that may produce blurred images, and low contrast images, and hard to detect thin objects [9]. Most of the existing crack detection techniques adopts photographic images that fails in detecting internal defects like voids, delaminations [13], environmental conditions such as same weather, hue of the concrete surface, existence of fog, the shape of structures [18], the optimization of threshold value for termination procedure [26], blurred and low contrast images which will reduce the accuracy of crack detection algorithm [9]. For neural network-based techniques, the arising problems are it needs millions of sample data to tune the network, data over fitting, tuning the complex hyper parameters to increase the accuracy, requirement of high speed processor like graphic processor unit, and computationally expensive [5, 11]. It is necessary to develop a system to locate, quantify, and classify various types of cracks and handle the massive amount of data acquired by such autonomous systems. Fast analysis of the crack detection performed by using robust decision making tools is required [15]. The effect of lighting condition on accuracy and performance of the crack detection techniques is another important research scenario [14].

7 Conclusion

This survey has analyzed the various crack detection techniques, different methodologies adapted on concrete civil structures. The goal of this survey is to bestow an extensive survey and analyze various techniques and approaches of crack detection methods applied on concrete structures. The challenges while dealing with crack detection on concrete structures and future research directions are also enlightened in this survey. In summary, the survey in crack detection technique illustrates the major breakthrough in various aspects. Though, these works have not met numerous challenges like dealing with different camera resolution, there is a trade off still exists in between the system accuracy and its computational complexity which needs to be improved further.

References

1. Adhikari, R., Moselhi, O., Bagchi, A.: Image-based retrieval of concrete crack properties for Bridge inspection. Autom. Constr. **39**, 180–194 (2014)
2. Asdrubali, F., Baldinelli, G., Bianchi, F., Costarelli, D., Rotili, A., Seracini, M., Vinti, G.: Detection of thermal bridges from thermographic images by means of image processing approximation algorithms. Appl. Math. Comput. **317**, 160–171 (2018)

3. Cha, Y.J., Choi, W.: Deep learning-based crack damage detection using convo- lutional neural networks. Comput.-Aided Civil Infrastruct. Eng. **32**(5), 361–378 (2017)
4. Cha, Y.J., Choi, W., Suh, G., Mahmoudkhani, S.: Autonomous structural visual inspection using region-based deep learning for detecting multiple damage types. Comput.-Aided Civil Infrastruct. Eng. **33**(9), 731–747 (2018)
5. Chen, F.C., Jahanshahi, M.R.: Nb-cnn: deep learning-based crack detection using convolutional neural network and naive bayes data fusion. IEEE Trans. Industr. Electron. **65**(5), 4392–4400 (2018)
6. Chen, J.H., Su, M.C., Cao, R., Hsu, S.C., Lu, J.C.: A self-organizing map optimization based image recognition and processing model for bridge crack inspection. Autom. Constr. **73**, 58–66 (2017)
7. Dinh, T.H., Ha, Q., La, H.M.: Computer vision-based method for concrete crack detection. In: 2016 14th International Conference on Control, Automation, Robotics and Vision (ICARCV). pp. 1–6. IEEE (2016)
8. Dorafshan, S., Thomas, R.J., Maguire, M.: Comparison of deep convolutional neural networks and edge detectors for image-based crack detection in concrete. Constr. Build. Mater. **186**, 1031–1045 (2018)
9. Fujita, Y., Hamamoto, Y.: A robust automatic crack detection method from noisy concrete surfaces. Mach. Vis. Appl. **22**(2), 245–254 (2011)
10. Fujita, Y., Mitani, Y., Hamamoto, Y.: A method for crack detection on a concrete structure. In: Pattern Recognition, 2006. ICPR 2006. 18th International Conference on, vol. 3, pp. 901–904. IEEE (2006)
11. Geethalakshmi, S.: A survey on crack detection using image processing techniques and deep learning algorithms. Int. J. Pure and Appl. Math. **118**(8), 215–220 (2018)
12. Hao, M., Lu, C., Wang, G., Wang, W.: An improved neuron segmentation model for crack detection–image segmentation model. Cybern. Inf. Technol. **17**(2), 119–133 (2017)
13. Hiasa, S., Birgul, R., Catbas, F.N.: A data processing methodology for infrared thermography images of concrete bridges. Comput. Struct. **190**, 205–218 (2017)
14. Jahanshahi, M.R., Masri, S.F.: Adaptive vision-based crack detection using 3d scene reconstruction for condition assessment of structures. Autom. Constr. **22**, 567–576 (2012)
15. Jahanshahi, M.R., Masri, S.F., Padgett, C.W., Sukhatme, G.S.: An innovative methodology for detection and quantification of cracks through incorporation of depth perception. Mach. Vis. Appl. **24**(2), 227–241 (2013)
16. Koch, C., Georgieva, K., Kasireddy, V., Akinci, B., Fieguth, P.: A review on computer vision based defect detection and condition assessment of concrete and asphalt civil infrastructure. Adv. Eng. Inform. **29**(2), 196–210 (2015)
17. Luo, H., Xiong, C., Fang, W., Love, P.E., Zhang, B., Ouyang, X.: Convolutional neural networks: computer vision-based workforce activity assessment in construction. Autom. Constr. **94**, 282–289 (2018)
18. Moon, H., Kim, J.: Intelligent crack detecting algorithm on the concrete crack image using neural network. In: Proceedings of the 28th ISARC, pp. 1461–1467 (2011)
19. Nishikawa, T., Yoshida, J., Sugiyama, T., Fujino, Y.: Concrete crack detection by multiple sequential image filtering. Comput.-Aided Civil and Infrastruct. Eng. **27**(1), 29–47 (2012)
20. Prasanna, P., Dana, K.J., Gucunski, N., Basily, B.B., La, H.M., Lim, R.S., Par-vardeh, H.: Automated crack detection on concrete bridges. IEEE Trans. Autom. Sci. Eng. **13**(2), 591–599 (2016)
21. Shuangrui, C., Zheng, S., Quan-sheng, Y.: Concrete crack width detecting system for android platform. Open Civil Eng. J. **9**, 846–851 (2015)
22. Silva, W.R.L.d., Lucena, D.S.d.: Concrete cracks detection based on deep learning image classification. In: Multidisciplinary Digital Publishing Institute Proceedings, vol. 2, p. 489 (2018)
23. Sinha, S.K., Fieguth, P.W.: Automated detection of cracks in buried concrete pipe images. Autom. Constr. **15**(1), 58–72 (2006)

24. Wang, P., Huang, H.: Comparison analysis on present image-based crack detection methods in concrete structures. In: 2010 3rd International Congress on Image and Signal Processing (CISP), vol. 5, pp. 2530–2533. IEEE (2010)
25. Yamaguchi, T., Hashimoto, S.: Automated crack detection for concrete surface image using percolation model and edge information. In: 32nd Annual Conference on IEEE Industrial Electronics, IECON 2006, pp. 3355–3360. IEEE (2006)
26. Yamaguchi, T., Hashimoto, S.: Fast crack detection method for large-size concrete surface images using percolation-based image processing. Mach. Vis. Appl. **21**(5), 797–809 (2010)
27. Yamaguchi, T., Nakamura, S., Saegusa, R., Hashimoto, S.: Image-based crack detection for real concrete surfaces. IEEJ Trans. Electr. Electron. Eng. **3**(1), 128–135 (2008)
28. Zhu, Z., German, S., Brilakis, I.: Visual retrieval of concrete crack properties for automated post-earthquake structural safety evaluation. Autom. Constr. **20**(7), 874–883 (2011)

A Comparative Analysis of AlexNet and GoogLeNet with a Simple DCNN for Face Recognition

P. Kalaiarasi and P. Esther Rani

Abstract Deep neural networks have achieved great success in many fields like bioinformatics, computer vision, automatic machine translation, etc. DCNN plays vital role in face recognition and outperforms human performance. In recent years, many pre-trained DCNNs were developed by researchers for object detection and recognition. In this work, a simple DCNN with very few layers is compared with two pre-trained networks: AlexNet which has eight layers and GoogLeNet which has 22 layers. Faces of the given image are detected using all three networks, training and testing performances are also studied. At the end of the study, it is observed that GoogLeNet learns the features and get trained very soon with less iteration, AlexNet achieved 100% accuracy after 30th iteration and simple DCNN also achieves 100% accuracy but goes nearly 42 iterations. For analysis, faces94 dataset is used.

Keywords Face recognition · Deep convolutional neural networks · AlexNet · GoogLeNet

1 Introduction

1.1 Deep Learning

Deep Learning is a subset of machine learning, based on learning representations of data. Deep learning architecture is basically similar to artificial neural networks. This artificial neural network was actually designed from the inspiration of human brain. Like biological neuron system, an artificial neuron system was developed in the late 1940s. An artificial neural network is a group of neurons. These neurons are

P. Kalaiarasi (✉) · P. Esther Rani
Department of Electronics and Communication Engineering, Vel Tech Rangarajan Dr. Sagunthala R&D Institute of Science and Technology, Chennai, India
e-mail: vtd603@veltech.edu.in

P. Esther Rani
e-mail: drpestherrani@veltech.edu.in

P. Suresh et al. (eds.), *Advances in Smart System Technologies*, Advances in Intelligent Systems and Computing 1163,
https://doi.org/10.1007/978-981-15-5029-4_54

called nodes and are interconnected to transmit the information from one layer to the other. The connections between the layers are assigned some weights randomly between 0 and 1. Artificial neural network (ANN) structure is constructed with one input layer, hidden layer, and one output layer. When the network is trained with some labeled dataset say images of a cat, it learns from the training and identifies whether image is cat or not. Prior to deep learning, the features are extracted by hand-crafted descriptors and given to the network for training. But now, in case of deep learning, a raw input image is given and the network learns the data by extracting features and classifies them. The deep neural network is same as ANN but the only difference is it has input layer and output layer along with many layers in hidden layer. Since, there are many layers in the DNN, deep manipulations are carried out and the network is trained well to extract more important features and classify. The hidden layer consists of convolution layers which perform convolution operation, pooling layers to reduce the feature maps and fully connected layers.

Convolutional layer The network can have as many as convolutional layers. In this layer, a filter with size n x n is slide over the image with a particular stride value to read each and every pixel. When there are many neurons present, the parameters will also be more. This convolution process reduces the parameters as the network goes deeper.

Pooling layer This layer reduces the computations and parameters. There are two types of pooling: Max pooling and average pooling. Max pooling performs picking the max value and average pooling calculates the average of the particular slide.

Fully connected layer All the neurons in the previous layer are linked to this layer. This layer detects the high level features from the image. The number of nodes in classification layer is equal to the number of classes in the dataset.

When the input is given, weights to the connections between the nodes are also assigned randomly between 0 and 1. The value in the node at the first layer and the weight are multiplied and produced at the node in second layer. Now, the activation function or transfer function is applied to this value to introduce nonlinear properties. The activation function may be any one of the following: Sigmoid, Tan h or ReLu. ReLu is the most commonly used transfer function in today's networks. If the network did not recognize accurately during training, then the error between the expected output and obtained output is calculated and reduced using back propagation method which automatically adjusts the weights until the error becomes zero. Once the error becomes zero in training, testing is carried out and the accuracy or efficiency of the network is obtained.

The learning tasks in deep learning are of three methods:

- *Supervised learning*: The datas given for the training are labeled. When the network is trained, it learns the features for each label and also learns the variations between the features of each label.
- *Unsupervised learning*: The given datas are not labeled and the network is just trained. The network learns and distinguishes the variances between the classes and classifies according to what it has learnt.

- *Semi supervised learning*: Few datas are labeled and the others are unlabeled for training.

The modes of learning generally employed for deep learning are: Stochastic gradient descent, Mini-batch gradient descent, and Batch gradient descent. Of these, the most commonly used is stochastic gradient descent and Mini-batch gradient descent. For training deep neural network, a large set of data are required. Only then, it learns deeply and produces an accurate output which may sometime outperform humans. Deep learning is nowadays applied in many fields like speech recognition, image recognition, language processing, handwriting recognition and generation, automatic machine translation, bioinformatics, military, drug discovery and toxicology, healthcare and more.

In this work, few of the deep convolutional neural networks used for face recognition are analyzed and compared. For face recognition, there are some architecture pretrained and produces good results. The deep learning architectures include deep neural network, deep belief network, and recurrent neural networks. The architectures vary depending on the number of layers and parameters used. The pre-trained networks are AlexNet [1], VGG Face [2], GoogLeNet [3], Squeeze Net [4], and ResNet [5]. In this work, AlexNet and GoogLeNet are compared with a simple deep convolutional neural network.

2 Literature Survey

In last few years, many researches have been done for face recognition. Though the work started around the mid twentieth century, it groomed only after the development of neural network by Fukushima [6] and CNN with back propagation by Lecun [7]. Later, in 2006, deep neural network was introduced by Hinton Lab [8]. Until 2012, it was a quite tough work of detecting and recognizing more number of images. In 2006, Krizhevsky [1] introduced AlexNet which is the first network to classify more than 1.2 million images with an error rate of 16.4%. Later, VGG Face in 2014 [2], GoogleNet in 2014 [3], and ResNet in 2015 [4] were developed. After the evolution of AlexNet, face recognition became much easier that even a face image affected by various factors like noise, illumination, occlusion, contrast, and low resolution can also be detected and recognized. The recently developed ResNet achieves ever less error rate of 3.57% which is lesser than human error rate. In Lu et al. [9] used ResNet model for recognizing a low resolution face image. They used trunk network and branch networks trained with both the high resolution and low resolution images. Using the coupled mapping loss method in branch network, the test images are matched with the HR images and are paired exactly. This work achieved 98.7% accuracy.

In Tikoo et al. [10] used back propagation algorithm to recognize a face which is detected using Viola Jones algorithm. While propagating the network in backward direction, weights are updated to reduce to the error between the expected

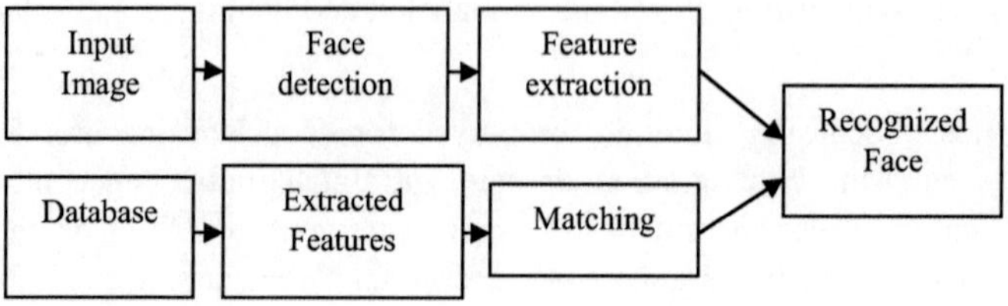

Fig. 1 Face recognition process

output and obtained output. Sigmoid function is applied as an activation function for obtaining the output of the neuron. In Ferrari et al. [11] investigated the pros and cons of various DCNNs. Datasets such as IJB-A, YouTube Faces, UMD Faces and MegaFace was used to study AlexNet, VGG Face, and ResNet networks. For IJB-A dataset, ResNet alone performs little less than the other two; all the three networks perform well on YouTube Faces; VGG Face obtains better result for UMD Faces and ResNet performs well on MegaFace. Grm et al. [12] analyzed the strength and weakness of four deep learning models: AlexNet, SqueezeNet, GoogLeNet, and VGG Face. These networks were experimented on datasets having various noise, blur, brightness, occlusion, and contrast factors for image quality covariates and model-related covariates. At the end of the study, they concluded that SqueezeNet and AlexNet performances seem to be almost similar, and VGG Face was robust to noise but performs less for low brightness. GoogLeNet was robust to all factors except noise and blur and on the whole, all DCNNs perform well in some aspects and less in few factors. Almisreb et al. [13] used AlexNet for recognizing ear in both image and video using transfer learning method. In their work, the last few layers of the AlexNet architecture were modified and obtained 100% accuracy on testing with 50 images of ten subjects and four video clips of two subjects. In Bussey et al. [14] used AlexNet and VGG Face to incorporate humanoid robot for face recognition. Here, NAO is trained with some datasets and tested by making the robot to capture the image by itself. It captures the image of person in front of it, crops it, and feeds to AlexNet and VGG Face for feature extraction and then compress the result. Finally, the matched face is again given back to the NAO. For short distance, VGG Face and AlexNet achieve 100% accuracy and reduce while increasing the distance between the camera and face. In Xie et al. [15] used GoogLeNet to remove artifacts in CT images. Along with ResNet, filtered back projection reconstruction method was used and the time taken for reconstruction is recorded as 4.1 s and PSNR was 49.67% which was better when compared to previous works.

3 CNN in Face Recognition

The general process of face recognition includes face detection, feature extractio, feature matching and recognizing as shown in Fig. 1. A neural network designed

Fig. 2 Neural network

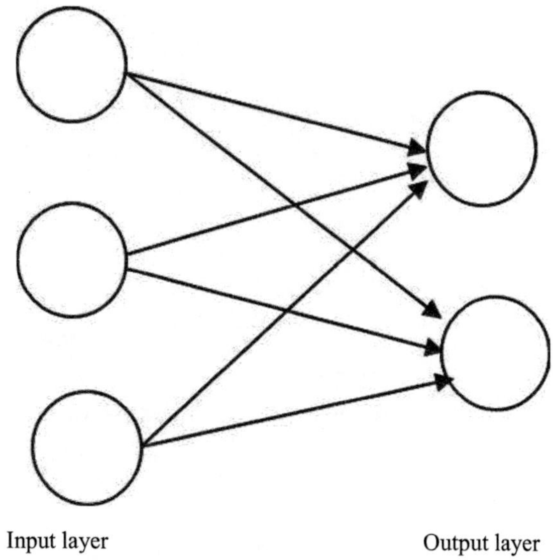

initially had one input layer and one output layer as shown in Fig. 2 [16]. Then hidden layer was introduced between the input layer and output layer as shown in Figs. 3 and 4. In the hidden layer, convolution process was carried out by sliding the filter over the image and convoluting the values obtained for each set of pixels. This network is called convolutional neural network [17]. More number of convolutional layers extracts important features from the image. The number of neurons in the input layer depends on the input; the neurons at the output layer is selected based on the number of classes in the dataset. The neurons in each layer are linked to all the neurons in next layer along with some bias. Each connection has some weights. Both the input value and the weight are multiplied and activation functions such as sigmoid/tanh are applied. This process is carried out till the last layer. If the expected output and obtained output are not same, then the same procedure is carried out from the output toward the input. This process is repeated till the error becomes zero.

Previously, face recognition was done by extracting the features and classifying manually using various methods like SIFT, SURF, GIST, RANSAC, HOG [18], LBP [19] for feature extraction and SVM, RF, PCA [20], Kernel PCA, LDA [21], FDA for classification. Later, deep neural networks were developed and made all the process automatic. The deep convolutional neural network [16] extracts the features with convolutional layers and classifies using softmax classifier. In this work, a comparative analysis is carried out on simple DCNN with pre-trained DCNN models, AlexNet and GoogLeNet using same datasets for all the three networks (Figs. 3 and 4).

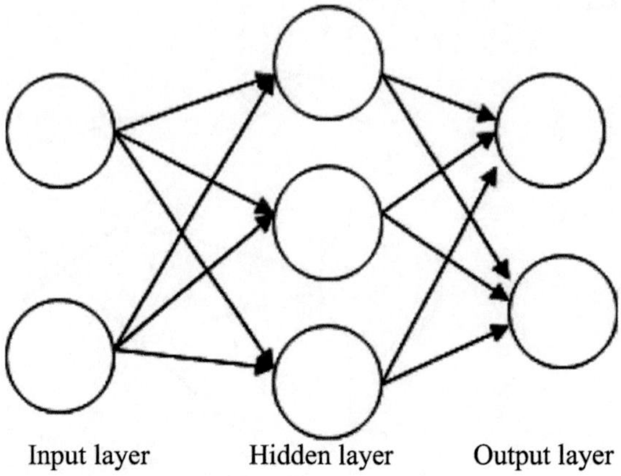

Fig. 3 Neural network with hidden layer

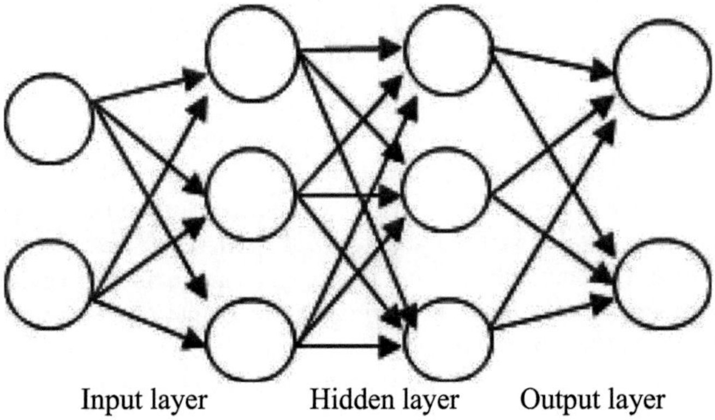

Fig. 4 Convolutional neural network

3.1 *Simple DCNN*

A simple DCNN designed for this work shown in Fig. 5 has three convolutional layers with filter size 5 × 5 and stride 2 followed by ReLu and maxpooling with size 2 × 2 and a stride value of 2. Also one fully connected layer, followed by a softmax layer for classification is used.

3.2 *Alexnet*

It is a pre-trained convolutional neural network which was developed by Krizhevsky et al. [1] The network has eight layers, of which five are convolutional layers and 3

Fig. 5 Architecture of simple DCNN

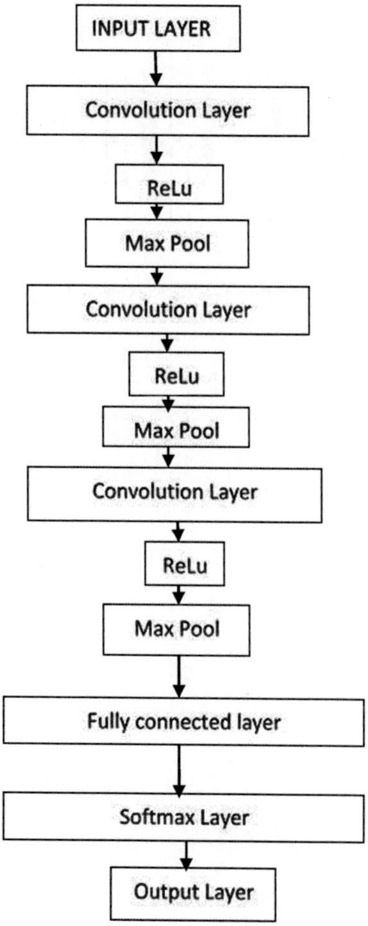

are fully connected layers along with input layer and output layer. The AlexNet takes the input image of size 227 × 227 × 3. The convolutional layer and fully connected layers are followed by ReLu layer as an activation function. After the second and fifth layer, max pooling is used to reduce the dimensions of the feature maps. The fully connected layers contain 4096 neurons and their outputs are passed to the 1000 way softmax classifier for classification. In this architecture, ReLu is used for nonlinear function. A technique called drop out was also used to avoid over fitting. When any neuron is dead and dropped out, they do not involve in further propagation [22]. For learning, this model uses stochastic gradient method and the initial weights are assigned using zero-mean Gaussian distribution having standard deviation 0.01 and learning rate is set to be 0.01. The architecture of AlexNet was shown in Fig. 6.

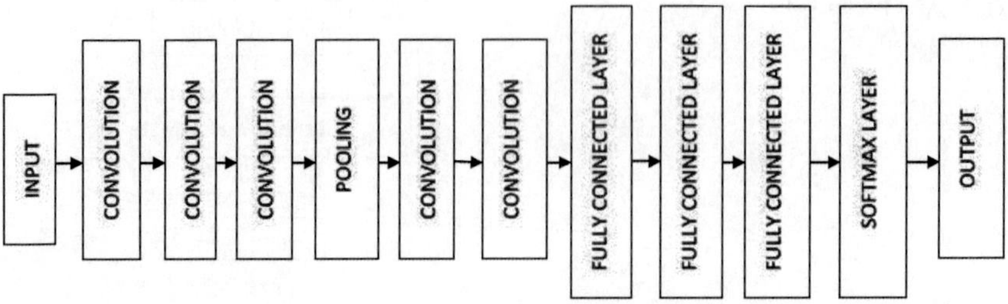

Fig. 6 Architecture of AlexNet

Fig. 7 Inception module of GoogLeNet

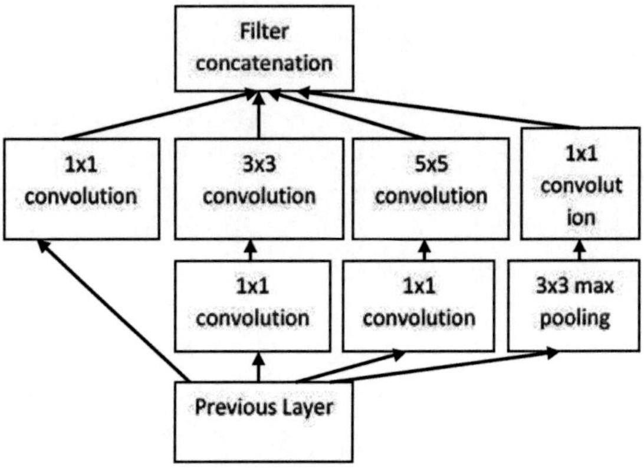

3.3 Googlenet

It is a pre-trained network developed by Google in 2014 for computer vision challenges. This network is nowadays used by many researchers and applications. It has 22 layers with 4 million parameters [3]. Inception module technique was used to cover larger area of the image. It applies three filters on a same patch of image to learn even a very small pixel and inception modules are stacked one on another and then concatenated [23]. Unlike in AlexNet, average pooling was used to avoid over fitting. Compared to other networks, GoogLeNet has very less parameters and computational cost is also 2 times lower than AlexNet. The network achieved an error rate of 6.67% [3] which is very close to human. The architecture of GoogLeNet and inception module is shown in Figs. 7 and 8, respectively.

4 Comparison Between DCNNS

For analyzing the performance of three architectures, Faces94 [24] dataset is used which has 20 images of 150 classes. For all the three networks, value of learning rate, epochs, and batch size is set as same. The ratio of training and testing images

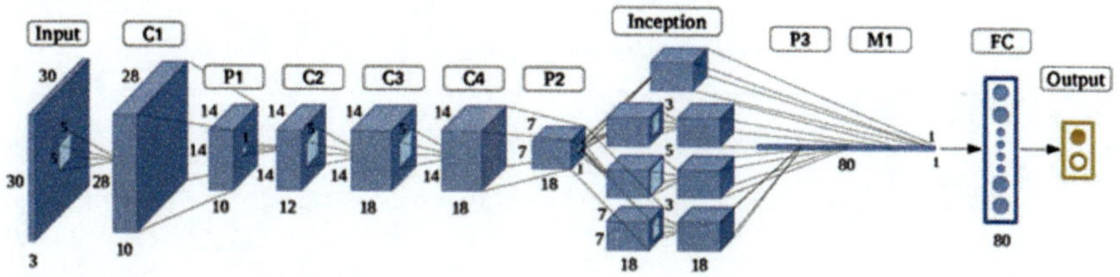

Fig. 8 Architecture of GoogLeNet [12]

on dataset is divided as 70% and 30%, respectively. During training, the networks extract the features and learn to classify the image. The networks are evaluated also for noise and blur images and their impact on the network is also studied. Salt and pepper noise is injected in different values and different levels of blurring are applied. The accuracy and loss for training and validation were plotted as shown in Figs. 9, 10 and 11. When the performances of the networks are compared by providing the datset, all the three networks recognize the image and classifies accurately. But, as the layers and parameters of AlexNet and GoogLeNet are more than the simple DCNN. Pre-trained networks learn the features in lesser epochs. For the given dataset, GoogLeNet achieves 100% accuracy after its 2nd epoch, AlexNet achieves 100% after its 5th epoch but simple DCNN takes nearly 7 epochs for 100% accuracy because of its less parameters. The number of iterations per epoch is 6. All these networks use stochastic gradient descent method or learning. All the three networks detect and recognize the faces exactly as mentioned 100% in Table 1. The detected faces are shown in Figs. 12 and 13. When noise and blurring are injected, accuracy starts decreasing. Figures 14 and 15 show the corresponding performance.

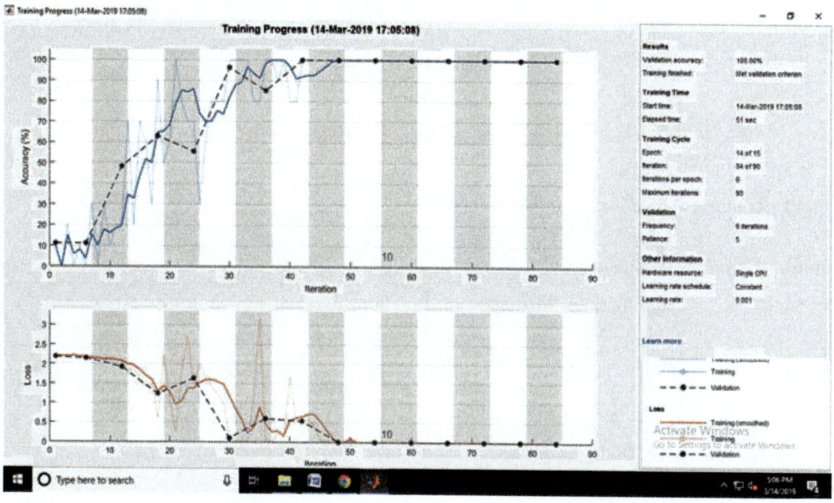

Fig. 9 Training and validation of simple DCNN

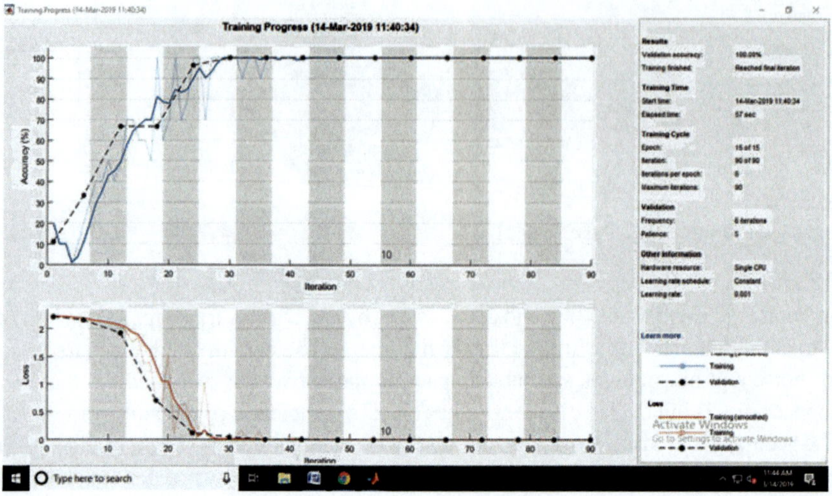

Fig. 10 Training and validation of AlexNet

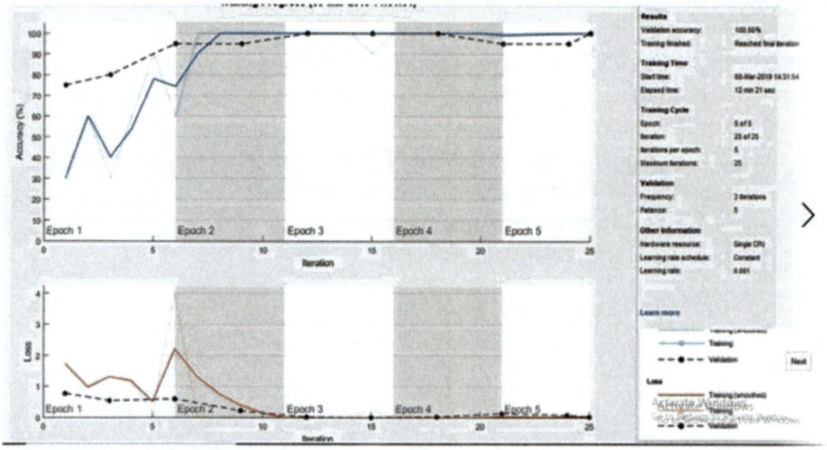

Fig. 11 Training and validation of GoogLeNet

Table 1 Comparison of networks

Network	Learning rate	Max epochs	Iterations to reach 100%	Time in sec	Accuracy (%)
Simple DCNN	0.001	20	42	113	100
AlexNet	0.001	20	30	662	100
GoogLeNet	0.001	20	8	463	100

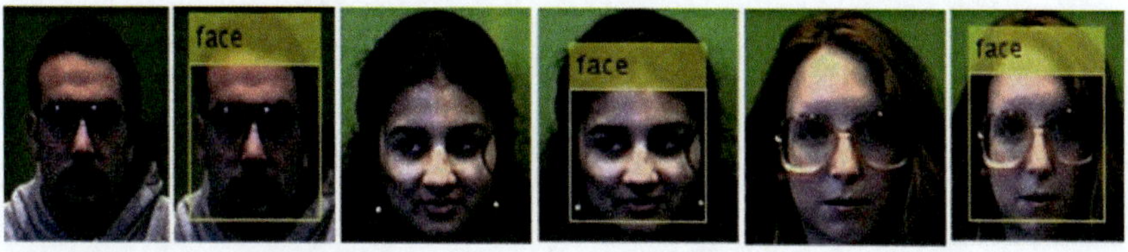

Fig. 12 Faces detected from online database

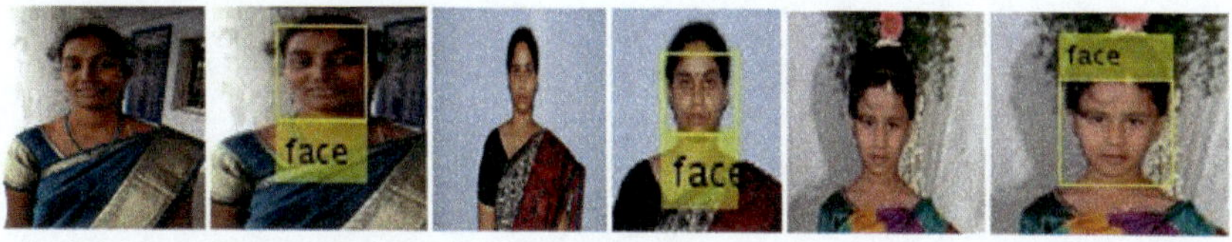

Fig. 13 Faces detected in real-time images

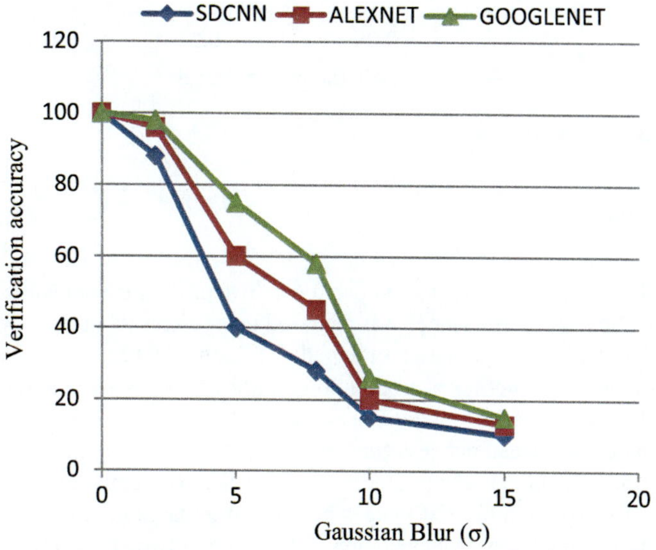

Fig. 14 Effect of blurring on DCNNs

In the simple DCNN, when noise and blurring are injected, the accuracy starts decreasing but the time taken to process the image remains same. After training and validation, testing is done with both database images and real-time images. The network recognizes the correct class of the image and shows the detected face. The faces detected from both the dataset and real-time images by simple DCNN are shown in Figs. 12 and 13.

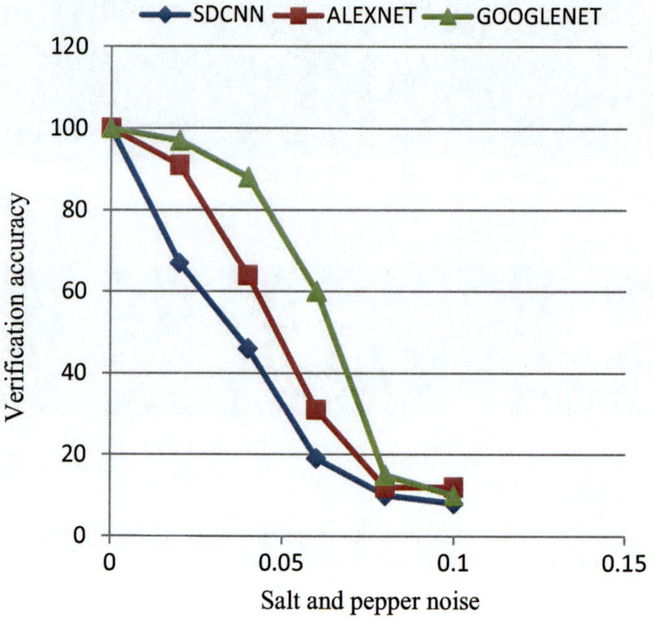

Fig. 15 Effect of noise on DCNNs

5 Conclusion

In this work, a simple DCNN was compared with two pre-trained networks AlexNet and GoogLeNet. The performance was analyzed using Faces 94 dataset. At the end of the study, the following points are observed: (1) GoogLeNet gets trained soon with only 8 iterations, AlexNet takes around 30 iterations to learn the features, and the simple DCNN takes nearly 40 iterations to achieve 100% accuracy. (2) The network with less parameter consumes less time and the network having more parameters need more time. (3) GoogLeNet outperforms AlexNet and simple DCNN for clear image, noisy image, and blurred images because of its large number of layers but the time taken is much more when compared to the other two networks. Finally, for recognizing the image from a dataset having 3000 images, the simple DCNN with minimum layers are enough as they consume less time and give 100% accuracy for clear images. But the only drawback of the simple DCNN is for learning the data, it goes for more iteration and when noise or blurring is added, its performance becomes less. This can be rectified in the future works to make the network robust to noise and blur.

References

1. Krizhevsky, A., Sutskever, I., Hinton, G.E.: Imagenet classification with deep convolutional neural networks. In: Processing of Advances Neural Information Processing Systems, pp. 1097–1105 (2012)
2. Simonyan, K., Zisserman, A.: Very deep convolutional networks for large-scale image recognition. In: Proceedings International Conference Learning Representations (2015)
3. Szegedy, C., Liu, W., Jia, Y., et al.: Going deeper with convolutions. In: Proceedings IEEE Conference Computer Vision and Pattern Recognition, pp. 1–9 (2015)
4. Iandola, F.N., Han, S., Moskewicz, M.W., et al.: SqueezeNet: AlexNet-level accuracy with 50 × fewer parameters and
5. He, K., Zhang, X., Ren, S., Sun, J.: Deep residual learning for image recognition. In: IEEE Conference on Computer Vision and Pattern Recognition (CVPR) (2016)
6. Fukushima, K.: Neocognitron: a hierarchical neural network capable of visual pattern recognition. Neural Networks **1**(2), 119–130 (1988)
7. LeCun, Y., et al.: Gradient-based learning applied to document recognition. Proc. IEEE **86**(11), 2278–2324 (1998)
8. Hinton, G.E., Osindero, S., Teh, Y.-W.: A fast learning algorithm for deep belief nets. Neural Comput. **18**(7), 1527–1554 (2006)
9. Lu, Z., Jiang, X., Kot, A.: Deep coupled ResNet for low-resolution face recognition. IEEE Signal Process. Lett. **25**(4), 526–530 (2018)
10. Tikoo, S., Malik, N.: Detection of face using Viola Jones and recognition using back propagation neural network. Int. J. Comput. Sci. Mob. Comput. **5**(5), 288–295 (2016)
11. Ferrari, C., Lisanti, G., Berretti, S., Del Bimbo, A.: Investigating nuisances in DCNN-based face recognition. IEEE Trans. Image Process. **27**(11), 5638–5651 (2018)
12. Grm, K., Štruc, V., Artiges, A., Caron, M., Ekenel, H.K.: Strengths and weaknesses of deep learning models for face recognition against image degradations. in IET Biometr. **7**(1), 81–89 (2018)
13. Almisreb, A.A., Jamil, N., Din, N.M.: Utilizing AlexNet deep transfer learning for ear recognition. In: 2018 Fourth International Conference on Information Retrieval and Knowledge Management (CAMP), Kota Kinabalu, pp. 1–5 (2018)
14. Bussey, D., Glandon, A., Vidyaratne, L., Alam, M., Iftekharuddin, K.M.: Convolutional neural network transfer learning for robust face recognition in NAO humanoid robot. In: 2017 IEEE Symposium Series on Computational Intelligence (SSCI), Honolulu, HI, pp. 1–7 (2017)
15. Zheng, X., Yang, C., Xie, L., Liu, J., Zhang, Y., Yan, J., Zhu, H., Hu, Y.: Artifact Removal Using Improved GoogLeNet for Sparse-view CT Reconstruction. Published in scientific Reports (2018)
16. Schmidhuber, J.: Deep learning in neural networks: An overview. Neural Networks http://arxiv.org/abs/1404.7828 [cs.NE], **61**, 85–117 January (2015)
17. LeCun, Y., Bengio, Y.: Convolutional networks for images, speech, and time series. Handbook of Brain Theory and Neural Networks, **3361**(10), 1995 (1995)
18. Dalal, N., Triggs, B.: Histograms of oriented gradients for human detection. In: Proceeding IEEE Computer Society Conference CVPR, pp. 886–893 (2005)
19. Ahonen, T., Hadid, A., Pietikäinen, M.: Face description with local binary patterns: application to face recognition. IEEE Trans. Pattern Anal. Mach. Intell. **28**(12), 2037–2041 (2006)
20. Chan, T., Jia, K., Gao, S., Lu, J., Zeng, Z., Ma, Y.: PCANet: a simple deep learning baseline for image classification? IEEE Trans. Image Process. **24**(12), 5017–5032 (2015)
21. Yu, H., Yang, J.: A direct LDA algorithm for high-dimensional data—with application to face recognition. Pattern Recognit. **34**(10), 2067–2069 (2001)
22. Agarwal, P., Prakash, N.: An efficient back propagation neural network based face recognition system using haar wavelet transform and PCA. Int. J. Comput. Sci. Mob. Comput. **2**(5):386–395 (2013)

23. Lin, M., Chen, Q., Yan, S.: Network in network [Online]. Available: http://arxiv.org/abs/1312.4400. (2014)
24. Faces94 dataset from online. Available: https://cswww.essex.ac.uk/mv/allfaces/faces94.html

Time-Varying Analysis of Channel Estimation in OFDM

Rajeshbabu Chitikena and P. Estherrani

Abstract The benefits of MA-signaling schemes need the fading radio channel tracking in the OFDM wireless systems. Here, we discuss about the errors in the channel identified based on time domain statistics. These estimations are based on fading models (slow and fast fading), we give the LS & MMSE estimators. It provides the tradeoff between the complexity and the performances. A sixteen QAM system symbol error rate is given by simulated results. Based upon estimators complexity, MMSE gives 4 dB SNR over the LSE.

Keywords Channel estimation · OFDM · LMS

1 Introduction

Presently, the OFDM [1] system is subjected to the significant investigations. There upon the techniques have adopted to the European DAB systems [2], then the OFDM signaling obtained a broad interest in the channel fading environment. An example of OFDM system is thoroughly investigated by the digital TV broadcastings [3]. A benefit of DPSK system is that it avoids the time-varying channel tracking. Yet, this shall limit the number of bps and result in the loss of 3 dB SN ratio [4]. If the R_x contains, error MA-signaling schemes are used. In [5, 6], 16 QAM modulations in an OFDM systems are verified.

In the OFDM-based MCM system, considered channel is generally impulse response of finite length. A cyclic extension, greater than this impulse response, avoids the interference between consecutive block and maintains the orthogonality based on sub-carrier frequency [7]. The proposal discusses about the channel estimation technique in a wireless OFDM system that uses the properties of the channel

R. Chitikena (✉) · P. Estherrani
ECE Department, Veltech University, Chennai, India
e-mail: rajeshchitikena@gmail.com

P. Estherrani
e-mail: drpestherrani@veltech.edu.in

P. Suresh et al. (eds.), *Advances in Smart System Technologies*,
Advances in Intelligent Systems and Computing 1163,
https://doi.org/10.1007/978-981-15-5029-4_55

impulse response. Hoeher [6] and Cio [8] addresses that properties. In Sect. 2, we describe the system models. Section 3 discusses the minimum mean-square error (MMSE) & least-square (LS) channel estimators. The performance of MMSE estimator is better than the LSE but complexity is more for MMSE rather than LSE. In Sect. 4 shows how the error is estimated using a 16 QAM signalling schemes. The performances are presented in term of MS-error and SE-rate.

2 Basic System Model

Considering the systems from Fig. 1, x_k is the transmitted symbol, the impulse response is denoted circularly symmetric gaussian random variable is denoted with $\tilde{n}(t)$ and the received symbol data is denoted with y_k The Tx'd symbol x_k is taken from a multiamplitude signal constellation. The D2A andA2D converters contain ideal LPF with bandwidth $1 = T_s$ where $1/T_s$ is the sampling duration. A cyclic length T_G for interference reduction and kept the orthogonality between the sub carriers. $g(t)$ can be represented in the following form:

$$g(t) = \sum_m \alpha_m \delta(t - \tau_m T_s) \tag{1}$$

where α_m is amplitude and complex-valued that satisfies the following in equality $0 \le \tau_m T_s \le T_G$ (Fig. 2).

$$g(t) = \delta(t - 0.5T_s) + \delta(t - 3.5T_s)$$

The proposed system can be modeled as the N-point discrete time Fourier transform (DFTN) as

$$y = \left(\mathrm{DFT}_N(\mathrm{IDFT}_N(X)) \otimes \frac{g}{\sqrt{N}} + \tilde{n} \right) \tag{2}$$

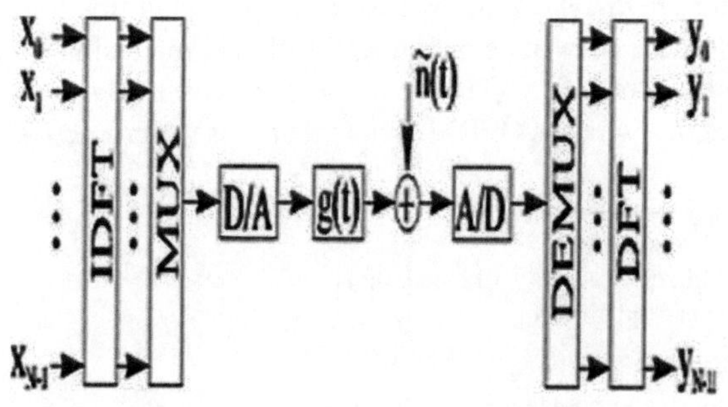

Fig. 1 Base-band OFDM system

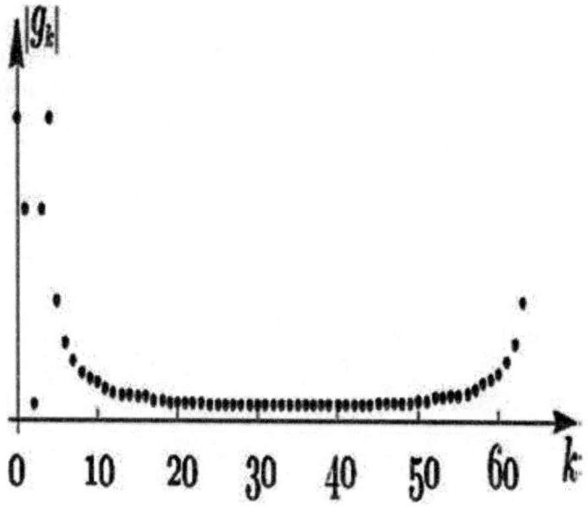

Fig. 2 Leakage between taps for the continuous channel

where ⊗ denotes cyclic convolution,

$$\begin{matrix} X = \left[x_0 x_1 \ldots x_{N-1}\right] \\ y = \left[y_0 y_1 \ldots y_{N-1}\right] \end{matrix}, \tilde{n} = \left[\tilde{n}_0 \tilde{n}_1 \ldots \tilde{n}_{N-1}\right] \quad (3)$$

is a circularly symmetric with iid of Gaussian random variables, and is denoted with $g = [g_0 g_1 \ldots g_{N-1}]$. The $g(t)$ can be expressed as

$$g_k = \frac{1}{\sqrt{N}} \sum_m \alpha_m \, \mathrm{e}^{-j\frac{\pi}{N}(k + (N-1)\tau_m)} \frac{\sin(\pi \tau_m)}{\sin\left(\frac{\pi}{N}(\tau_m - k)\right)} \quad (4)$$

If the delay is τ_m, then a mapping is possible between α_m and $g_{\tau m}$ In case of non-T pulse, energy will leak to all taps g_k which represents τ_m is not an integer indicates the power leakage problems case. From Fig. 3, the N independent Gaussian channels

$$y_k = h_k x_k + n_k \quad (5)$$

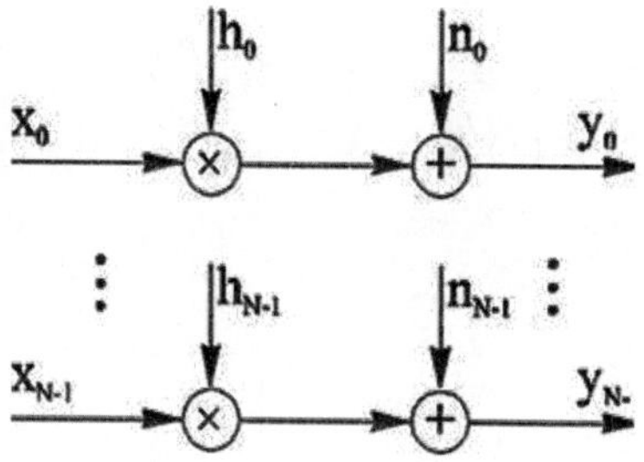

Fig. 3 Parallel Gaussian channels

where h_k is the channel-attn and denoted as $h = [h_0 h_1 \ldots h_{N-1}] = \mathrm{DFT}_N(g)$ and $n = \left[n_0 n_1 \ldots n_{N-1}\right] = \mathrm{DFT}_N(\widetilde{n})$ is an iid. CGM—Gaussian noise.

For our purpose, we write noise vector in the following manner

$$y = X + F_g + n \tag{6}$$

DFT-matrix is represented as $F = \begin{bmatrix} W_N^{\infty} \ldots & W_N^{0(N-1)} \\ W_N^{(N-1)0} \ldots & W_N^{(N-1)(N-1)} \end{bmatrix}$

$$W_N^{nk} = \frac{1}{\sqrt{N}} \mathrm{e}^{-j2\pi \frac{nk}{N}}$$

3 Channel Error Estimation

In previous cases, different error estimation methods are discussed and basic structure is represented in Fig. 4. The Tx'd symbols from the modulation output is X_k, and appeared as an error in the channel and is based on either training sequence or quantized random DM-estimator.

3.1 MMSE and LS Estimators

The noise in the channel is **n** and the symbol information vector **g** is interrelated; the MMSE error eqn is [9]

$$\tilde{g}_{\mathrm{MMSE}} = R_{gy} R_{yy}^{-1} y \tag{7}$$

where

$$R_{gy} = E\{g y^H\} = X F R_{gg} F^H X^H$$
$$R_{yy} = E\{y y^H\} = X F R_{gg} F^H X^H + \sigma_n^2 I_N$$

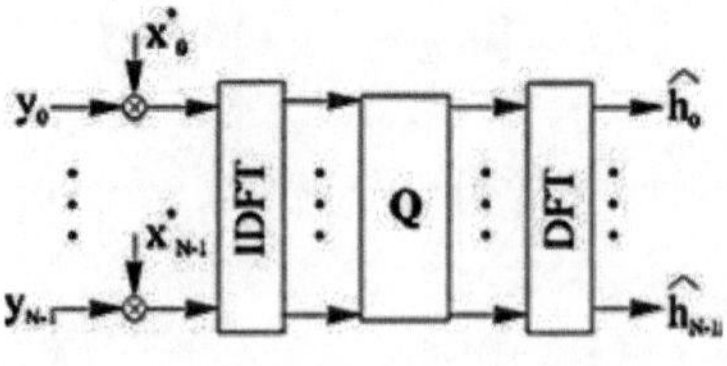

Fig. 4 General estimator structure

R_{gy} is cross covariance matrix and R_{gg} is the auto-covariance matrix of g, and σ_n^2 is the noise variance $E\{|n_k|^2\}$. $\hat{g}_{\text{MMSE}}$ generates the frequency domain of MMSE estimate $\hat{h}_{\text{MMSE}}$ by

$$\tilde{h}_{\text{MMSE}} = F\tilde{g}_{\text{MMSE}} = FQ_{\text{MMSE}}F^HX^Hy$$

where Q_{MMSE} can be shown to be

$$Q_{\text{MMSE}} = R_{\text{gg}}\left[\left(F^HX^HXF\right)^{-1}\sigma_n^2 + R_{\text{gg}}\right]^{-1}\left(F^HX^HXF\right)^{-1} \tag{8}$$

From Fig. 4, the corresponding MMSE is: Based on g is either Gaussian or not, $\hat{h}_{\text{MMSE}}$ is not a MMSE estimator. The better linear estimator as $\hat{h}_{\text{MMSE}}$

The LSE for the CIR g minimizes $\left(y - XF_g\right)^H\left(y - XF_g\right)$ and generates $\hat{h}_{\text{LS}} = FQ_{LS}F^HX^Hy$ where $Q_{LS} = \left(F^HX^HXF\right)^{-1}$

Therefore, higher complexity is a basic problem in MMSE; whereas MSE is more for LSE. So, our main aim is how to rectify the drawbacks in the proposed error estimation methods.

3.2 M-MMSE and MLS Methods

For MM-SE calculations, we consider an $N \times N$ matrix Q_{MMSE}, which leads to more complexive case and reduced by decreasing the size of Q_{MMSE}.and is represented in Fig. 2, g contains the overall energy, $L = \left\lceil \frac{T_G}{T_s} \right\rceil$ taps. The modified MMSE estimator becomes $\tilde{h}_{\text{MMSE}} = TQ'_{\text{MMSE}}T^HX^Hy$

where

$$Q'_{\text{MMSE}} = R'_{gg}\left[\left(T^HX^HXT\right)^{-1}\sigma_n^2 + R'_{gg}\right]^{-1}\left(T^HX^HXT\right)^{-1} \tag{9}$$

Equation (9) is represented in Fig. 5:

L is finite value compared to size of N and because of this, M-MSE complexity reduces drastically Therefore, LSE not depends upon the time-varying channel statistics. g decrease with respect to L size and noise is uniform over entire range [6, 8].

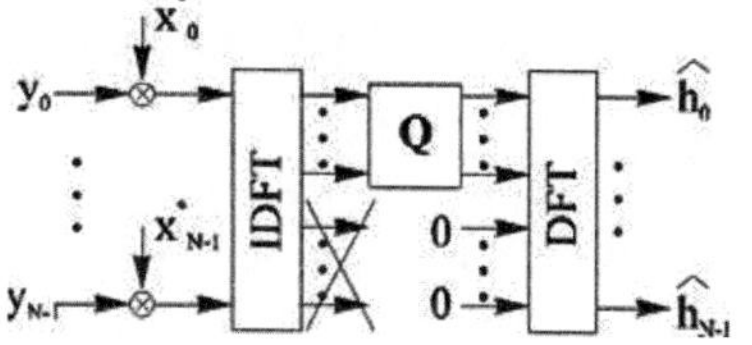

Fig. 5 Model-based M-estimator

Therefore, the modified LS is denoted as

$$\hat{h}_{\text{LS}} = T Q'_{\text{LS}} T^H X^H y$$

where

$$Q'_{\text{LS}} = \left(T^H X^H X T\right)^{-1} \tag{10}$$

Equation (10) is the MLSE and its basic view is shown in Fig. 5.

4 Simulation Result

4.1 Simulation Setup

Parameters:

BW = 500 kHz,
Number of tones = 64
Tot. Symbol Period = 138 μs
Cyclic prefix length = 10 μs
$F_s = 500$ kHz rate
$L = 5$.

The following estimators are used:

Estimator	Notation	Taps used	Size Q_0
MMSE	MMSE	0…63	64 × 64
LS	LS	0…63	NA
Modified MMSE	MMSE_0 MMSE_5 MMSE_10	0…4 0…9, 59…63 0…14, 54…63	5 × 5 15 × 15 25 × 25
Modified LS	LS_0 LS_5 LS_10	0….4 0…9,59…63 0…14, 54…63	5 × 5 15 × 15 25 × 25

4.2 Mean-square Error

Figure 6 indicates the MSE versus SNR for the M-MSE, LSE, MM-SE0, MM-SE5 and MM-SE10. For low SNRs, it effects less compared to the channel noise, while it was more effect on large SNRs.

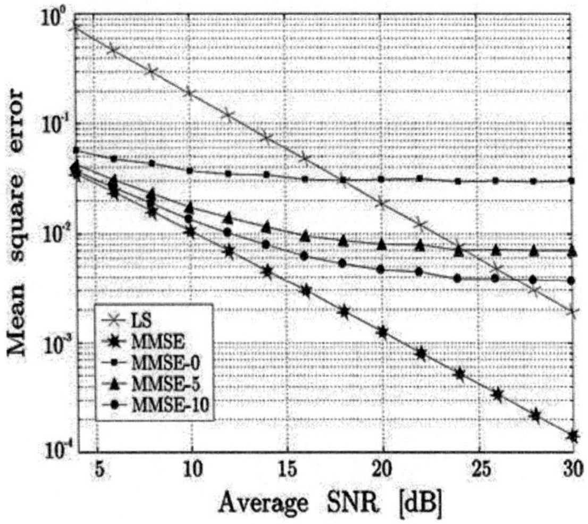

Fig. 6 Mean-square error for three modified MMSE estimators

If the size of $[Q'_M]$ will give lower MSE for every SNRs.

Figure 7 indicates the MMSE versus average SNR for the MM-SE, LSE, LS0, LS5 and LS10. Compared to M-MMSE, M-LSE is better in MSE reduction for a range of SNRs.

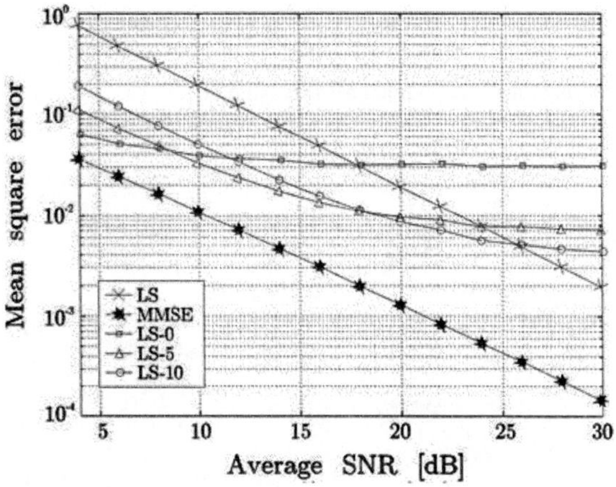

Fig. 7 Mean-square error for three LS estimators

4.3 Symbol Error Rate

The SE-Rate graphs represent the MSE of the channel estimations discussed in the previous section. For SE-Rate, calculations [10].

From Fig. 8, there was a significant improvement in SNR up to 4 dB in M.MMSE over LSE estimator. From Fig. 9, even though the matrix size Q changes, the SNR will be same as M-MMSE compared to conventional MMSE, and graphically represented for MMSE10 and LSE10 in Fig. 10. The mean difference between these two methods is 2 dB in SNR

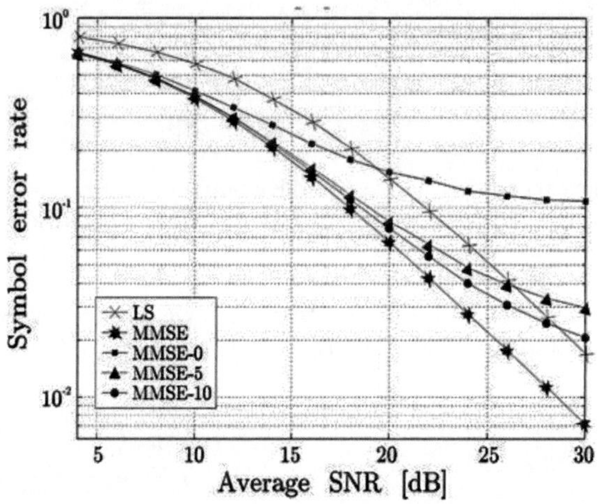

Fig. 8 Symbol error rate for three modified MMSE estimators

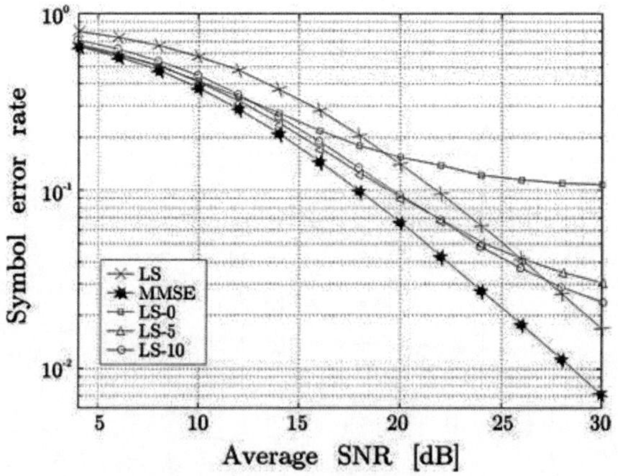

Fig. 9 Symbol error rate for three modified LS estimators

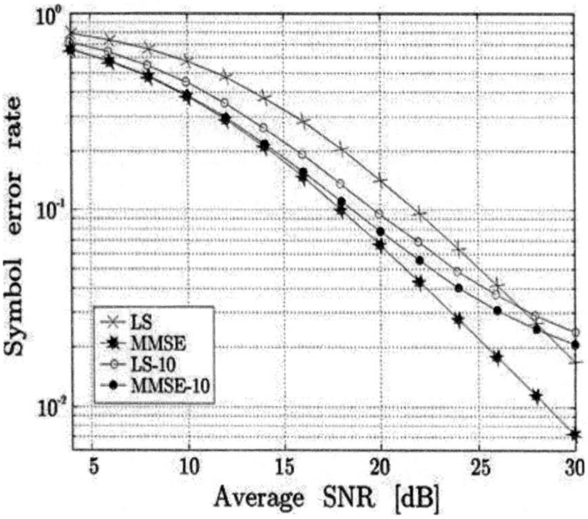

Fig. 10 Comparison of SER between a modified MMSE estimator and the corresponding modified LS estimator

5 Conclusions

These error calculated methods are discussed in this study are greatly estimating the offsets in the OFDM system with base knowledge of channel statistics. By knowing the noise variance and channel covariance, MMSE estimators are working but complexity is more than LSE. LSE is reliably working for high SNRs, due to its simplicity. M-MMSE and M-LSE are reliably working for low SNRs, these modifications provides the trade of between complexity and performance. By using 16-QAM, there was an improvement in the SNR up to 4 dB over the LS estimator.

References

1. Cimini, L.J., Jr.: Analysis and simulation of a digital mobile channel using orthogonal frequency-division multiplexing. IEEE Trans. Commun. **33**(7), 665–675 (1985)
2. Alardand, M., Lassalle, R.: Principles of modulation and channel coding for digital broad casting for mobile receivers. EBU Rev. **224**, 3–25 (1987)
3. Martietal, B.: European activities on digital television broad casting from company to cooperative projects. EBU Tech. Rev. **256**, 20–29, 1993
4. Proakis, J.: Digital Communications. McGraw-Hill (1989)
5. Wilson, S.K., Ellen Khayata, R., Cio, J.M.: QAM modulation with orthogonal frequency division multiplexing in a Rayleigh fading environment. In: Proceedings of VTC 1994, pp. 1660–1664, Stockholm, Sweden, June 1994
6. Hoeher, P.: TCM on frequency-selective land—mobile fading channels. In: Proceedings of the 5th Tirrenia International Workshop on Digital Communications, Tirrenia, Italy, September 1991

7. John, A.C.: Multi carrier modulation for data transmission: an idea whose time has come. IEEE Commun. Mag. **28**(5), 5–14 (1990)
8. Cio, J.M.: Stanford University, Private communication
9. Scharf, L.L.: Statistical Signal Processing. Addison, Wesley (1991)
10. Wilson, S.K.: Digital audio Broad casting in a fading and dispersive channel. Ph.D.-thesis, Stanford University, August 1994

Autonomous Vehicle Sunshade Using Aluminum Panels

Saad Bin Abul Kashem, Fawad Shahid, Uvais Qidwai, P. Suresh, Jubaer Ahmed, and U. Saravanakumar

Abstract The heat build-up inside of a vehicle cabin due to the direct composure of thermal energy when parked outside during the day can be a huge dilemma, the intensity of the problem can be decreased by covering the windows of the automobile, but drivers are reluctant to do it manually. There are several semi-automatic car sunshades available in the market but none of them provide a hard shield to protect the car from smashing the window. The aim of this project is to design, develop, and construct a smart autonomous automobile sunshade that not only prevents the thermal heat caused by the sun to penetrate the automobiles but also provides the vehicle with mild security. Adding the fact that this project contributes to the automobile industry applications and takes a step forward to autotomize yet another gadget used in our everyday life. This was done using an aluminum frame consisting of wheel sliders and aluminum composite material panels to cover the windows, and sensors to avoid a collision, this device was able to cover three sides of the automobile as 4 × 4 vehicles have accessories installed on the rear window to provide the boot with some extra strength. A test prototype has been constructed and the legitimacy of the mechanical design was tested. The design can be improved by using pneumatic pistons as a substitute for the linear actuators.

S. Bin Abul Kashem (✉)
Qatar Armed Forces—Academic Bridge Program, Qatar Foundation, Doha, Qatar
e-mail: skashem@qf.org.qa

F. Shahid · J. Ahmed
Swinburne University of Technology, Sarawak, Malaysia
e-mail: fshahid@gmail.com

J. Ahmed
e-mail: jahmed@swinburne.edu.my

U. Qidwai
Qatar University, Doha, Qatar
e-mail: uqidwai@qu.edu.qa

P. Suresh · U. Saravanakumar
Vel Tech University, Chennai, India
e-mail: sureshp@ieee.org

P. Suresh et al. (eds.), *Advances in Smart System Technologies*,
Advances in Intelligent Systems and Computing 1163,
https://doi.org/10.1007/978-981-15-5029-4_56

Keywords Sunshade · Aluminum composite material · Wheel sliders · Heat reflection efficiency · Linear actuator · Arduino

1 Introduction

In countries like Malaysia which possess hot weather, when a car is parked outside on a hot sunny day, the temperature levels inside the cabin of the car can reach up to 20 °C above the ambient temperature. When the visible light from the sun penetrates the car, the plastic area of the car dashboard starts to cark due to the UV damage. On top of that, the electrical components also get damaged and the fabric on the car seats could decolorize on top of that it represents an uncomfortable atmosphere for the driver and the passengers. For the past decade, many advancements have been done to protect the car from this thermal radiation build-up inside a car cabin but all of them have some flaws. This build-up is caused by the phenomenon called the greenhouse effect and due to this, countries like Malaysia and the countries in the middle east where the population of automobiles is high, people's daily routines are affected by this problem.

The greenhouse effect is when the thermal radiation passes through one medium to another it changes its waveform. In this case, when the radiation passes through the car window it decreases its wavelength. Whereby, the radiation cannot pass back through the car window to go out, hence, causes the radiation to be trapped inside the car cabin. Moreover, an increase is seen in the car theft of 4 × 4 vehicles, and in most cases, the side windows are smashed to break into the vehicle. There are many devices in the market to block the thermal radiation sourcing from the sun but they all require manual installation whenever the car is parked in the sun.

1. The device should be constructed of a material having a reflective efficiency high enough to prevent thermal radiation to enter the car
2. The device should mount on the roof of a 4 × 4 truck and use an infrared remote and receiver to open the sunshade wirelessly for the convenience of the operator
3. The material chosen should cover the car from the sides and the front to prevent window smashing.

2 Literature Review

When a vehicle is parked outside in the sun the temperature of the car cabin can rise to 65 °C. Various research and experimental analysis have been conducted by Al-Kayiem et al. [1] on the thermal accumulation and distribution inside a parked car cabin. In his study, both experimental and numerical analyses were conducted. The experimental results were obtained from measurements on a parked car in an unshaded area. The temperature was measured at 12 different locations inside the car for six different cases. Two cases worth researching were, when all of the car windows were closed, and when a sunshade was kept beneath the front windshield.

In the first case it was found that the surface temperature of the dashboard can reach up to 70 °C, in the second case after placing the sunshade the temperature of the surface of the dashboard was decreased by 27%, Furthermore, the temperature of the air inside the car cabin had a 26% decrease to it.

To reduce the heat build-up in automobiles, several devices have been developed over the past years. An awning assembly for parked cars was documented in the United States Patent No. 5,230,545 issued on July 27th, 2008 [2]. It consists of a pair of frames and a driving mechanism fixed to the roof of the vehicle, A pair of powerful actuators which retracts two bars which are connected to the driving mechanism. The awning roller assemblies are fixed to the frame. The problem which this design is that it does not extend and cover the whole side windows, but it does provide passengers with useful protection when they are getting on or off the car in rainy days. Another invention was documented by Rhea et al. on the 7th of March 2006, US patent No. 7,008,002 B2 [3]. It consists of one viper motor that is attached to a piece of cloth with the help of sting when the motor rotates it extends or folds the cloth depending on the direction of rotation. The problem with this invention was that the cloth flies away and changes its position when the weather is windy.

TefsaGuma documented his invention in the United States, patent [4]. It is an exterior cover that automatically extends and covers the car. Furthermore, the cover also includes flaps that automatically extend outwardly to the sides of the vehicle to provide protection for the sides. The assembly of this invention took a lot of valuable space and when in function exhibits masses of noise. A window shade device was documented in the United States Patent No. 1,289,281, issued on December 31, 1918, to Shaft [5]. This device is a pleated shade for the window which is changeable by cords at the bottom and top of the window. The interior sunshade solution is still the most commonly used today since it is the cheapest solution, but these sunshades usually cover only one of the four sides of the car, it was documented in the United States Patent No. 4,775,180 on 4th of October 2005 [6] Another similar patent documented in the United States Patent No. 4560245 on 24th of December 1985 [7] Another invention documented is US patent No. 7,140,662 B1 documented in 2006 [8] The unit uses a pleated shade section with a solar reflective coating on at least the outside facing portion of the shade. It is motorized and operated by pulleys. But unlike the first one which can be removed when required and used on other sides this one must be installed on the front windshield only, and it extends vertically rather than horizontally. US patent No. 5,468,040 [9] uses one motor to lift a mechanical transmission with is attached to the sunshade cloth. The interesting thing about this sunshade is that it is controlled by a remote controller to extend and retract the sunshade. Inventor Bing JyeCherng documented his design US patent No. 5,752,56 [10–17] this electric sun shield for automobiles consists of a tube. A motor mounted at the intermediate portion of the tube. A pair of screw rods connected with two ends of the motor. A pair of threaded sleeves. Connected with the two screw rods a pair of supporting rods connected with the two threaded sleeves. Us patent No. 5,291,934 invented by Gaston Ouvrard. The inventor took account of the fact that the rear screen is curved. It has a winder roller at the end of the cloth and the other end is connected to the pull bar. The pull bar is hinged at the other end to the sliders which is parallel to the axis of the winder roller. A drive device synchronously moves the carriages, and hence the ends of the arm in opposite directions on the slider to thereby extend the cloth. The pull bar is elastically deformable in a direction perpendicular to the cloth to support the curved screen. Another interesting invention is the US patent No. 7,475,935 B2. This design was the most convenient solution to this problem. The comfort factor provided by the inventor to the occupants of the motor vehicle was the arrangement of a sunshade on the side of Windows. US patent 5,344,206 documented by Roy J. This invention relates to also a retractable sunshade designed to be secured in the interior of an automotive windshield. US patent No. 2,723,714 was documented by Murray R. Moore. This is another type of interior car sunshade which only covers the front screen of the car. United States Patent No. 5,291,934, issued in 1994 to Ouvrard et al. [13], United States Patent No. 5,468,040, issued in 1995 to Peng Hsieh et al. [14] and United States Patent No. 5,752,560, issued in 1998 to Cherng [15] revealed sunshades devices controlled by power electronics for vehicles. These devices use extension rods which are placed within the vehicle. The rods can be extended automatically to cover the vehicle when it is parked in the sun but the user needs to give command using a switch or remote. All of these devices have the downside that they don't have a pleasant look or they have a lack of slick design and cannot cover the side windows. As these devices place inside the vehicle, it consumes valuable space within the vehicle. The devices also failed to reduce the temperature inside the car adequately as there is no protection or cover for the vehicle roof.

3 Design and Methodology

This section will describe the material selection and mechanical design respectively. A prototype has been built. The video of the test run can be seen at https://youtu.be/sl1qryZorkE.

3.1 Material Selection

An imperative step conducted in this research was the material selection. Selecting the right material was the absolute key for the sunshade to obtain the best results possible as material selection can directly affect many of the projects attributes, the selection was done based on several characteristics which are

1. Weight
2. Heat reflective efficiency
3. Cost-effectiveness
4. Strength.

Table 1 Aluminum alloy 6061 properties

Properties	Aluminum alloy 6061
Elastic modulus (N/m^2)	6.9×10^{10}
Mass density (kg/m^3)	2700
Tensile strength (N/m^2)	124.1×10^6
Yield strength (N/m^2)	55.1×10^6
Specific heat capacity (J/k K)	1300
Thermal conductivity (cal cm/s-cm^2-°C)	1160

Aluminum composite material was used in our application to cover the windows of the vehicle. It is a sandwich formed by two thin skins of metal bonded to a plastic core. The plastic core is thermoplastic material with an outer skin of aluminum sheet. The 3 mm thick panel weights 3.8 kg per sqm and has a modulus of elasticity of 2400 Mpa. Another prominent feature is that it has an impact strength of 50 kg/cm^2. This material also has a high reflectivity.

The major part of the design is the frame, which decides the stability, strength, and durability of the invention. After reviewing several types of metal, it was short-listed to aluminum alloy 6061, It has a visible light reflectivity of around 80% making it ideal for our application. Its properties are shown in Table 1.

3.2 *Mechanical Design*

The mechanical design of the project was composed using SOLIDWORKS. Solid works motion analysis was used to get the speed, acceleration, and position at any given time.

The base of this design is a custom-built roof frame. The frame has all the components mounted on it. And this aluminum frame has the capability to fit most of the 4×4 trucks. 3 linear actuators are used to extend the aluminum composite covers resting on the frame from its initial position to an extended position, the design also consists of rolling sliders and a C section channel. The aluminum sheets cover three sides of the car when in an extended position, front screen, and the two sides. The sunshade does not cover the back window of the vehicle since all the trucks have an assembly installed.

The dimensions of the frame shown in Fig. 1a were inspired from the dimensions of a jeep XJ Roof rack. This roof rack is claimed to be the most aerodynamically stable racks in the current market. The black colored region in Fig. 1b is the slope region designed to make the frame more aerodynamically stable, this region is facing the front of the car when installed on the roof.

As seen in Fig. 2b, the frame is fitted with two C channels with a rolling wheel slider inside which are connected to the panel, the wheel will roll along the path of the channel, moving the aluminum composite panels. The sliders are used to have a

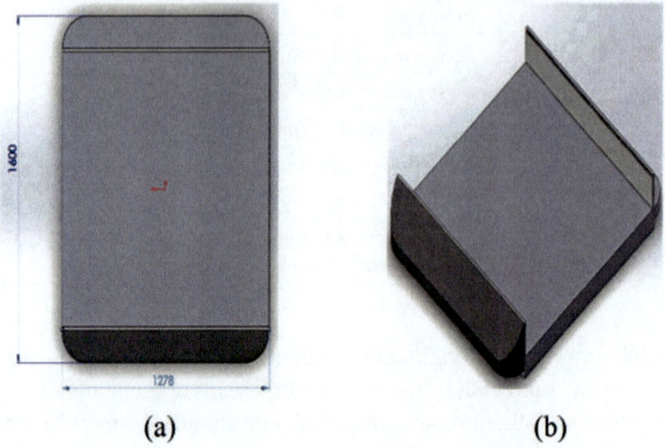

Fig. 1 Sunshade frame. Top view (**a**). Isometric view (**b**)

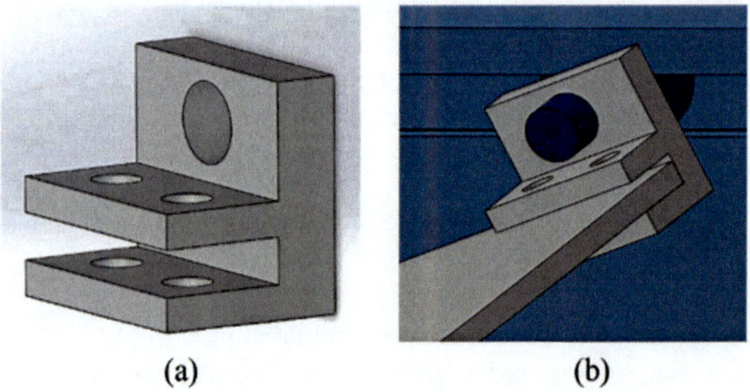

Fig. 2 **a** Shows the plate holder, **b** shoes the panel connected to panel holder

minimum friction movement of the panels within the channel, the rolling sliders are connected to panel holders which are screwed to the panels. There are many types of sliders that could be used, these sliders vary depending on the number of wheels it contains. The chosen slider has four wheels, this is sufficient to reduce the friction of the motion to a point that it no longer stresses the actuator.

The panel holder seen in Fig. 2a is connected to the rolling slider and can rotate freely. This will enable the panel to have the second axis of motion and allow it to tilt downwards when the slider moves along channel.

The panel is a cut out from a 3 mm thick sheet of aluminum composite material with a height of 635 mm and a width of 1260 mm for the side cover and height of 750 mm and a width of 1158 mm. The aluminum composite sheet will be riveted to an aluminum frame. The frame for the sheet (cover) is needed to add strength to the design. If the sheet was directly connected to the slider the panel would've deformed, therefore, to provide extra support and rigidity to the design an aluminum frame was riveted to the aluminum composite sheet. The panel is connected to the sliders on both ends. It can be seen in Fig. 3a, when the linear actuator extends, the slider will be forced to slide, and the plate will slowly tilt downwards due to gravity. As seen in Fig. 4. When retracted, the side panels stack on top of the frame uncovering the windows of the car the front panel rests beneath the frame at the retracted position. In addition, when extending the panel covering the front screen opens first followed by the left cover and then the right one. The panels to the side have a displacement of 450 mm to reach its extended position and are the same for the panel covering the front screen of the car. As seen in Fig. 3b the side panels tilt at 68° from the horizontal axis. The front panel has a lesser tilt since the front screen of the car has a lesser slope than the side windows.

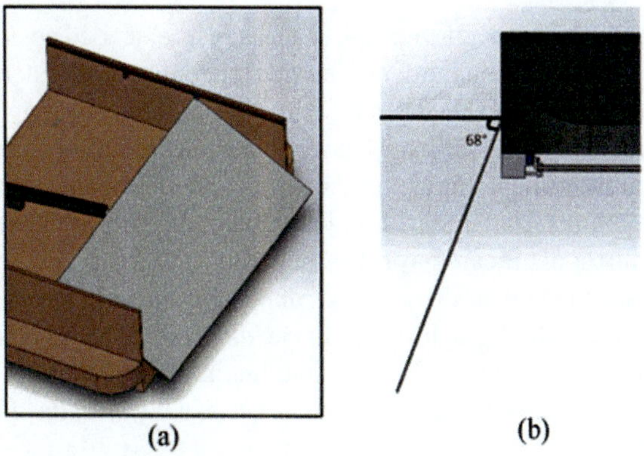

Fig. 3 **a** Panel assembled with the C channel of the frame and connected to the linear actuator; **b** The angle of the panel after the extension

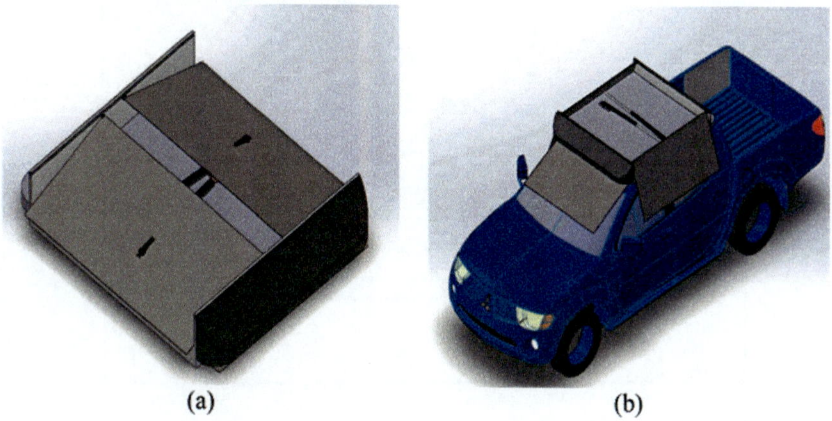

Fig. 4 Folded Sunshade (**a**), extended sunshade (**b**)

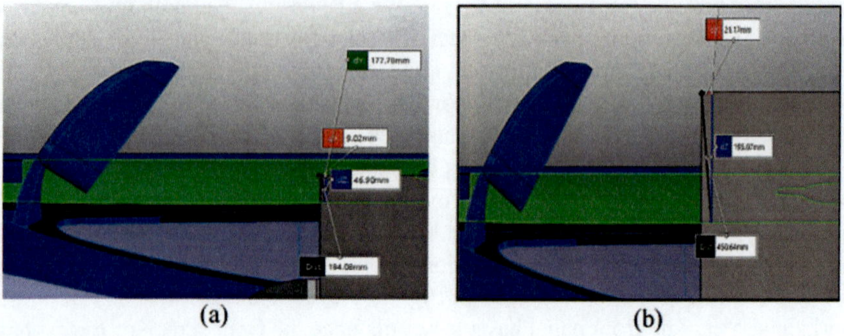

Fig. 5 Top view in extended position (a), top view at maximum displacement (b)

Figure 5a illustrates that even when the shade is in its extended position it is 4.6 cm away from the furthest point on the door. The side mirror has more extension outwards than the sunshade covers. Hence, when the sunshade is deployed after parked outside neighboring vehicles could still park beside this car without the sunshade being a hindrance to their movement. Figure 5b shows that when the panel is at its maximum displacement in the horizontal axis, it is just 19 cm apart from the vehicle but still behind the side mirror of the car this proves that even when a car is parked close beside the vehicle the sunshade would be able to deploy itself without any obstructions. Figure 6 shows the plotted graph of the linear displacement in X-axis of the side panel from the edge of the frame, it has a maximum displacement of just 326 mm, it should be noted that the edge of the frame aligns exactly with the edge of the roof of this car (Fig. 7).

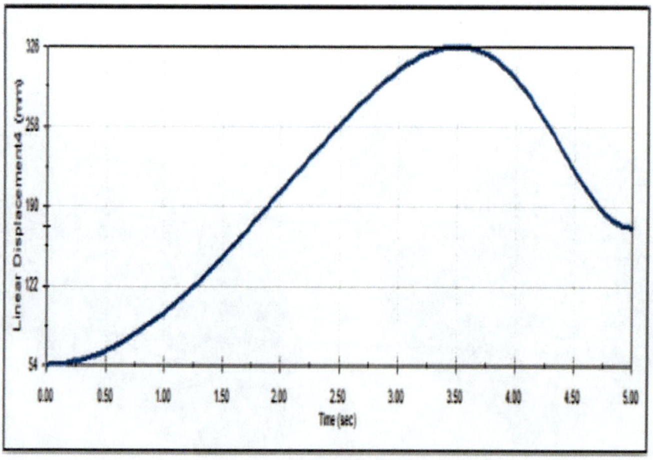

Fig. 6 Displacement graph

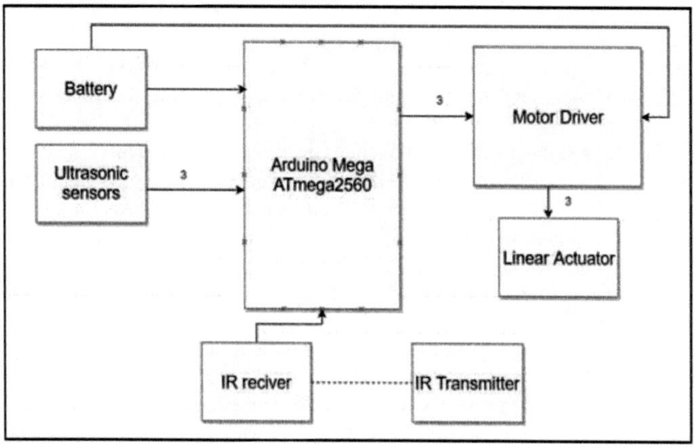

Fig. 7 Electrical design block diagram

The sunshade works using a combination of linear actuators, ultrasonic sensors, and infrared remote control. The scroll down window on the left of the MS Word Formatting toolbar. These components are controlled via a microcontroller.

4 Conclusion

This research is a new opportunity for the automobile industry. There are several semi-automatic car sunshades available in the market but none of them provide a hard shield to protect the car from smashing the window. If this idea adopted by a real automobile manufacturing corporation, a practical system can be produced. This paper has shown the capability of making such a thing. Due to the limitations of finance and equipment, better results could not be obtained. However, the current results are promising for future work. In the end, this research and development project offers a real solution to a very common problem that people face on a daily basis regardless of the obstacles that appear. The design can be improved by using pneumatic pistons as a substitute for the linear actuators.

References

1. Al-Kayiem, H.H., Sidik, M.F.B.M., Munusammy, Y.R.A.L.: Study on the thermal accumulation and distribution inside a parked car cabin. Am. J. Appl. Sci. **7**(6), 784–789 (2010)
2. Huang, C.Y., Wang, Y.F.: Convenient awning assembly for parked cars (Google Patents, ed.) (1993)
3. Rhea, T.C, Johnson, J.R., Troublefield, M.S.: Automated covering for an automobile (Google Patents, ed.) (2006)
4. Guma, T.: Portable automatic carport (Google Patents, ed.) (1984)

5. Shaft, W.S.: Window-shade (Google Patents, ed.) (1918)
6. Phillips, A.J.: Automatic retractable shade (Google Patents, ed.) (1988)
7. Sarver, P.I.: Vehicular windshield curtain for inhibiting heat transfer (Google Patents, ed.) (1985)
8. Wilkinson, K.L., Wilkinson, T.L.: Retractable sunshade (Google Patents, ed.) (2006)
9. Hsieh, S.F.P., Lee, C.L.: Power-operated automobile sunshade (Google Patents, ed.) (1995)
10. Kashem, S.B.A., Ektesabi, M., Nagarajah, R.: Comparison between different sets of suspension parameters and introduction of new modified skyhook control strategy incorporating varying road condition. Veh. Syst. Dyn. **50**(7), 1173–1190 (2012)
11. Kashem, S., Nagarajah, R., Ektesabi, M.: Vehicle Suspension Systems and Electromagnetic Dampers. Springer, Singapore (2018)
12. Kashem, S.B.A., Roy, S., Mukharjee, R.: A modified skyhook control system (SKDT) to improve suspension control strategy of vehicles. In: 2014 International Conference on Informatics, Electronics & Vision (ICIEV), pp. 1–8. IEEE (2014)
13. Kashem, S., Sufyan, H.: A novel design of an aquatic walking robot having webbed feet. Int. J. Autom. Comput. **14**(5), 576–588 (2017)
14. Kashem, S.B.A., Sheikh, M.I.B., Ahmed, J., Tabassum, M.: Gravity and buoyancy powered clean water pipe generator. In: 2018 IEEE 12th International Conference on Compatibility, Power Electronics and Power Engineering (CPE-POWERENG), pp. 1–5. IEEE (2018)
15. Kashem, S.B.A., Tabassum, M., Sufiyan, H.: A novel design of an amphibious robot having webbed feet as duck. In: Proceedings of IEEE Xplore 2017, International Conference on Computer and Drone Applications, 9–11 Nov 2017
16. Kashem, S.B.A., Jawed, S., Ahmed, J., Iqbal, A.: An experimental study of the amphibious robot inspired by biological duck foot. In: 2018 IEEE 12th International Conference on Compatibility, Power Electronics and Power Engineering (CPE-POWERENG), pp. 1–6. IEEE (2018)
17. Cherng, B.J.: Electric sunshield for automobiles (Google Patents, ed.) (1998)

Design and Evaluate Low-Cost Wireless Sensor Network Infrastructure to Monitor the Jetty Docking Area in Rural Areas

Jiang Hong Chang, Mujahid Tabassum, Uvais Qidwai, Saad Bin Abul Kashem, P. Suresh, and U. Saravanakumar

Abstract Applications of wireless communication technologies are still on the rise since its introduction, known as the Internet of things. These applications often focus on monitoring applications and real-time services. Jetty monitoring becomes one of the suitable situations to implement the Internet of things. For rural areas, implementing a jetty monitoring system can be difficult due to limitations of power supply, wireless access and external environmental conditions. This project proposes a low-cost smart jetty docking monitoring system that is able to conduct smart monitoring for the jetties in rural areas. The system helps to monitor sudden changes in the water stream level. If the water level is too high, the system will alert the administrators, so they can take immediate action. These results are stored in a database for other future environmental research purposes and a web server is used to display the results graphically. Before the system is actually implemented, a model is created to replicate an actual jetty docking area to test the system's performance. Data correlation algorithms are introduced to improve the system's sustainability by reducing its power consumption and byte transmission volume. Sensor fault detection algorithm is added to ensure if any damaged sensors in the system is reported immediately for replacement.

J. H. Chang
Swinburne University of Technology, Sarawak, Malaysia
e-mail: bernardcjh95@gmail.com

M. Tabassum
Higher College of Technology, Muscat, Oman
e-mail: mujahid.tabassum@hct.edu.om

U. Qidwai
Qatar University, Doha, Qatar
e-mail: uqidwai@qu.edu.qa

S. Bin Abul Kashem (✉)
Qatar Armed Forces—Academic Bridge Program, Qatar Foundation, Doha, Qatar
e-mail: skashem@qf.org.qa

P. Suresh · U. Saravanakumar
Vel Tech University, Chennai, India
e-mail: sureshp@ieee.org

P. Suresh et al. (eds.), *Advances in Smart System Technologies*,
Advances in Intelligent Systems and Computing 1163,
https://doi.org/10.1007/978-981-15-5029-4_57

Keywords Wireless sensor networks · Jetty docking monitoring · Wireless applications · Zigbee

1 Introduction

Wireless sensor networks (WSNs) have been applied to many applications especially monitoring applications. Jetty monitoring is known as one of the suitable situations to apply wireless sensor network technologies. However, not all jetties have proper electrical supply and wireless access. Implementation of a jetty monitoring system will be challenging due to those factors and external environmental conditions that may damage the system. A typical WSN should consist of sensor nodes, gateways or sinks and a base station to monitor the entire system [1]. Each sensor node in the system consists of sensors, a microcontroller and a radio transmitter. The gateways should redirect the data received from sensors and transmit to the base station, where data is processed and uploaded to the server. In this research project, a low-cost smart jetty docking monitoring system is proposed. The system is expected to measure the water stream level to warn the administrators immediately if the water level is too high. Also, it measures temperature, humidity and rainfall quantity that can be used to discover how these parameters affect the water stream level and the collected data can be used for other future environmental researches.

During the research project, a few project objectives are created such that the proposed system must be low-cost and able to perform basic monitoring such as observing the water level. The system must be sustainable to reduce maintenance cost and prolong its operational time.

2 Literature Review

2.1 Related Studies for Wireless Sensor Network Applications

2.1.1 Forest Environmental Monitoring

In this case study, a smart WSN system is created to detect any possible flames in the forest [2]. The sensors used in the system are mainly flame detection sensors and other applicable environmental sensors. Filter algorithms are used in the nodes to improve data accuracy whilst saving power. Gateways are established to collect data sent by the nodes and allow remote monitoring from the base station. Figure 1 shows the structure of the entire proposed system.

In the figure, routers are added to further expand the network inside the system. The proposed system is expected to filter unnecessary data to reduce the packet amount to be sent from the nodes. The system uses Zigbee for data transmission as

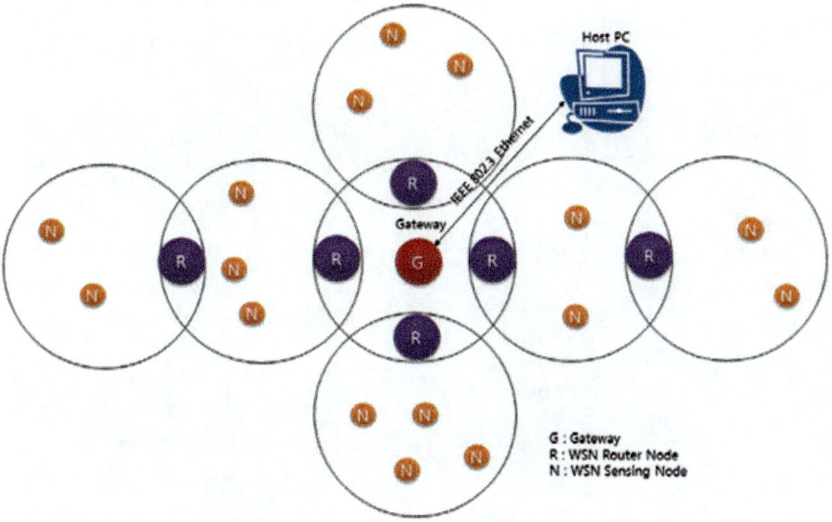

Fig. 1 Real-time forest environment monitoring system structure

Zigbee allows long transmission range, up to 100 m depending on power output and environmental characteristics.

2.1.2 Green Borderline Surveillance

Wireless multimedia sensor networks (WMSNs) are used to monitor forest trails and other green areas that are not in the line of sight [3]. Every sensor node has a sensing module, communication module, power module, processing module and a SIM card as shown in Fig. 2.

The sensing unit involves a thermal camera, scalar sensor and an analogue-to-digital converter that collects data, whereas the processing unit has a coordination subsystem, location management, motion controller and network synchronization to process the collected data. SIM card allows direct communication from the host to the sensor module. Data is collected and processed in the sensor node before transmission through the communication module. Transmitted images are collected and shared in the nearest police stations and central monitoring room to detect any trespassing in borderline areas and deploy patrol units accordingly.

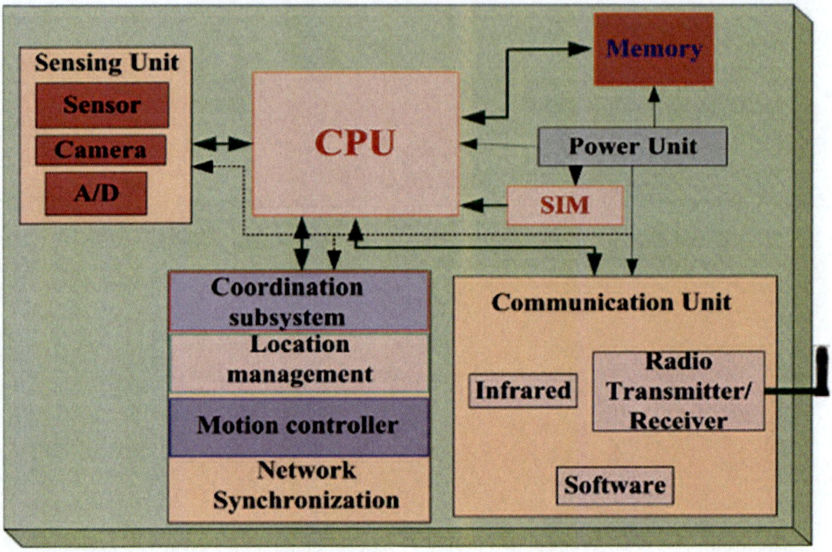

Fig. 2 System architecture of sensor node

2.2 Additional Features of Wireless Sensor Networks (WSNs)

2.2.1 Security Enhancement

A telecare system is integrated with bilinear pairing and advanced user authentication methods [4]. User public keys are created using the user's personal information using bilinear mapping. This is done by pairing characters into function that creates a unique bilinear map pattern. For user authentication methods, conventional login identifier (ID) and passwords are used, which the patient's login attempt is further verified by assigned trust authorities.

2.2.2 Power Efficiency and Data Congestion Reduction

To improve power efficiency and reduce data congestion, the byte transmission rate in the system must be reduced [5]. Spatial and temporal correlation can be used for data that has common relationships. Collected data will be identified whether it has any relationship compared to its previous samples. The sampling rate will be halved until its minimum threshold if the current data is the same as its previous data.

3 System Components

The proposed low-cost smart jetty monitoring system consists of a sensor node and a gateway that can be remotely controlled by a host computer. This will be explained in the diagram in Fig. 3.

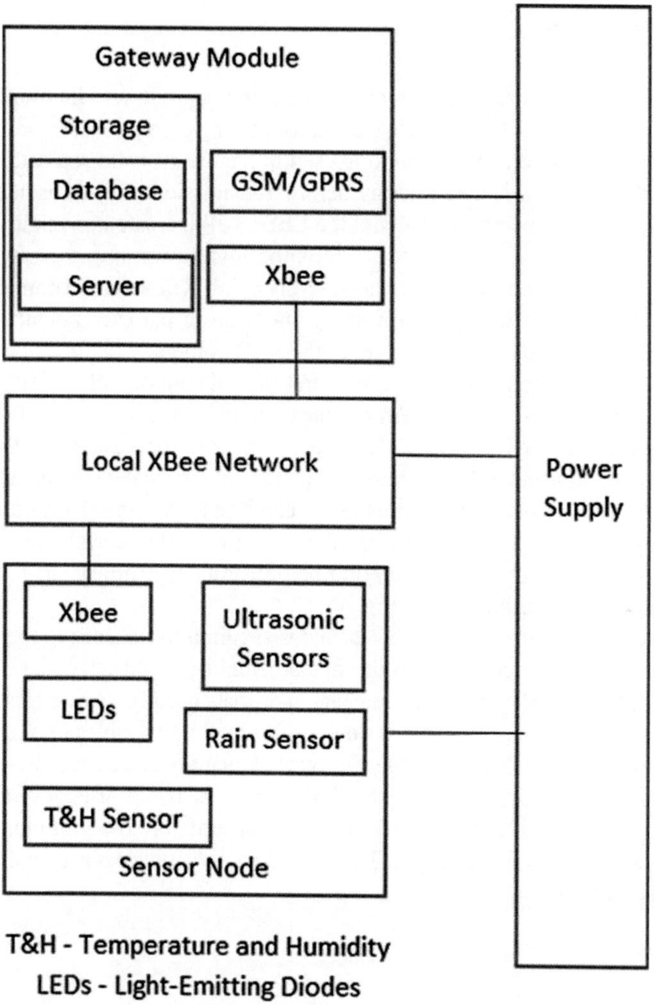

Fig. 3 Block diagram of the proposed system

Table 1 LED colour table

LED colour	Water level status
Green	Safe
Yellow	Caution, please monitor the jetty docking area
Red	Unsafe, please act immediately

3.1 Sensor Node

The sensor node consists of ultrasonic sensors, temperature and humidity (T&H) sensor, rain sensor, an XBee module and light-emitting diodes (LEDs) with colours of red, yellow and green. An Arduino is used to handle all collected data from the sensors and conduct basic data filtering before transmission. The ultrasonic sensors will monitor the water stream level and the LEDs will indicate how high the water level is to nearby users. If the water level is too high, a warning message will be transmitted from the sensor node to the gateway using XBee as the communication protocol. The gateway will send a warning message to the user's device through email. Table 1 will explain the meaning of the LED colours.

For the T&H sensor, it collects the temperature and humidity of the jetty docking area. The rainfall sensor will determine the rainfall quantity at the jetty docking area. The data collected by these two sensors can be used for weather prediction and discover the relationship between these parameters and the water stream level. Also, they can be used for other environmental research purposes in the future. All data collected by the sensor node will be filtered using a simple data correlation algorithm to reduce the number of bytes required to be transmitted. A flow chart is drawn to describe how the algorithm works (Fig. 4).

With these algorithms, the number of bytes required to transmit to the gateway can be significantly reduced. Using the fundamental type table, 1 byte of memory is usually allocated for 1 character, integers and floats usually require four bytes to store its value. In normal cases, the system sends two float variables for temperature and humidity and two integer variables for rainfall quantity and current water stream level that is around 16 bytes of data excluding characters that are used for formatting. With the proposed algorithm, the system only transmits up to a minimum of four bytes, which are only four characters. This can effectively reduce power consumption and data congestion [6].

3.2 Gateway Module

The gateway module, which is the Raspberry Pi, will receive data transmitted by the sensor node through XBee. Further data processing is conducted to identify the received data. The data is then sent to the MySQL database and an Apache server web page is used to access the database to display the information using line graphs. A flow chart for the gateway module will be shown (Fig. 5).

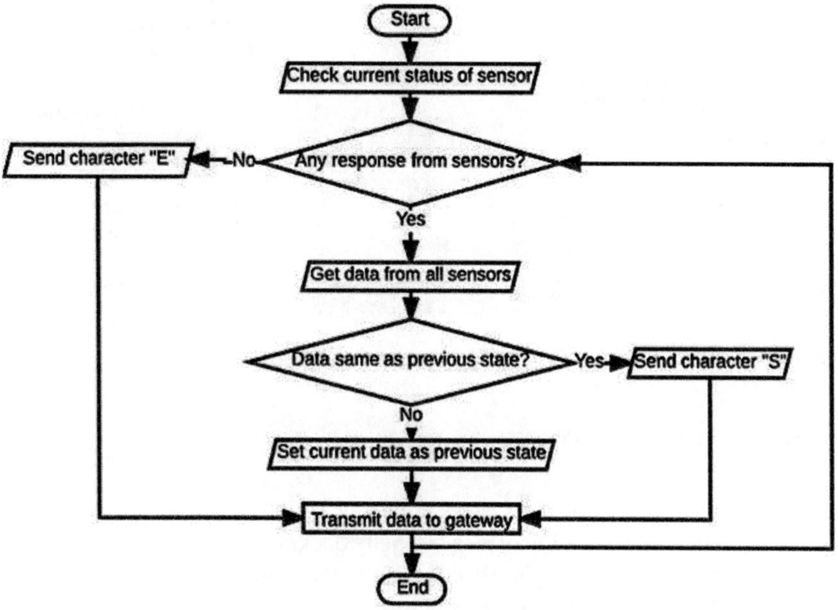

Fig. 4 Flow chart of overall sensor module with data correlation algorithm

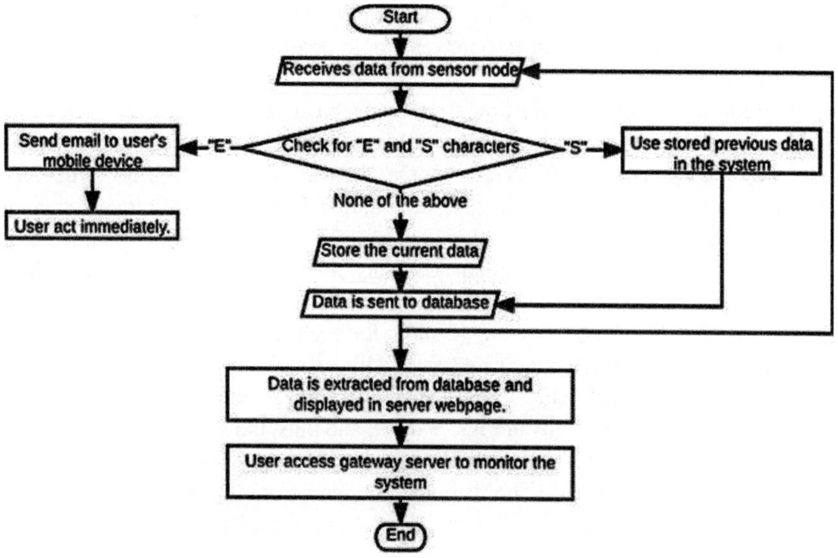

Fig. 5 Flow chart of the gateway module

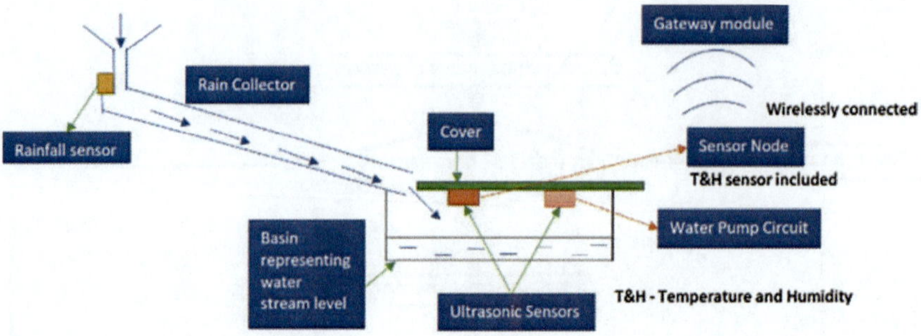

Fig. 6 Flow chart of water stream level algorithm used by ultrasonic sensors

4 Results and Discussion

To test the effectiveness of the system, a temporary jetty docking area model is created to replicate the situation of an actual jetty docking area. The system will run for a few days to demonstrate the effectiveness of the system. A simple diagram of the jetty docking area model is shown in Fig. 6.

Based on Fig. 6, the blue arrows represent the rainwater flow collected by the rain collector. The rain sensor is attached near the rain collector to obtain rainfall quantity readings. The basin will collect the water that represents the water stream level. A cover is attached to ultrasonic sensors to determine the water level in the basin. A water pump circuit is connected to one of the ultrasonic sensors to ensure the collected rainwater does not overflow from the basin. The other ultrasonic sensor is used by the sensor node to gather results. All sensors will be connected to the sensor node, which transmits data wirelessly to the gateway module at 30-min interval. The user then remotely accesses the gateway module through Wi-Fi connection, and power adapters are used in the model to supply sufficient power to the system.

The algorithm for the water stream level has been adjusted to fit in the model, which is shown in Fig. 7.

Three batches of results showing 12 plots for each measured parameter in 2-D line graphs are obtained from the web server will be shown in Figs. 8, 9 and 10.

Based on the results above, it can be proved that the system is working fine. No rain was detected for the results in Figs. 8 and 9, and thus, the water level in the jetty docking area model remains the same. For Fig. 10, light rain was detected at the end of the monitored time period. The readings obtained from the system may not be accurate due to the constraints of the jetty docking area model.

5 Conclusion

Based on the results obtained from the model, it can be concluded that the system is feasible and able to perform for long hours. The system will require sufficient

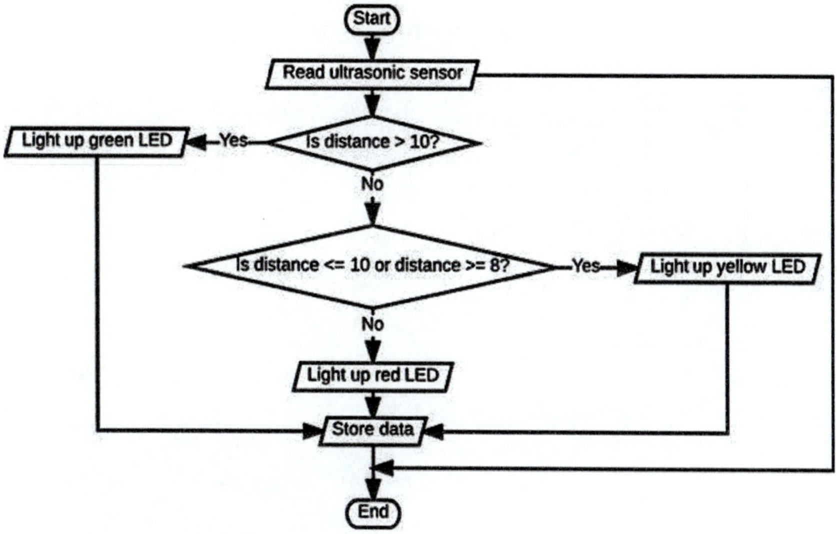

Fig. 7 Flow chart of water stream level algorithm used by ultrasonic sensors

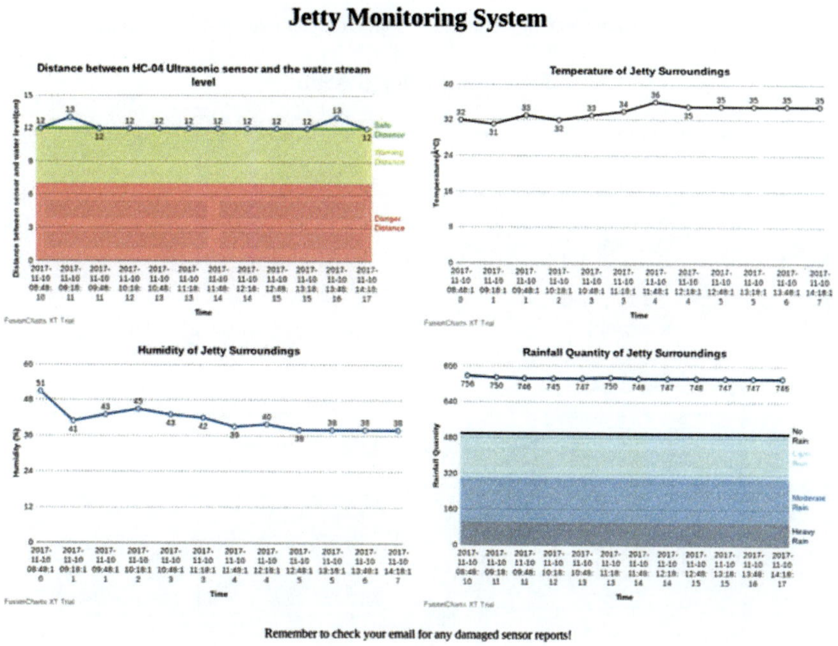

Fig. 8 Status of jetty docking area model from 8:48 a.m. to 2:18 p.m. on 10 November 2017

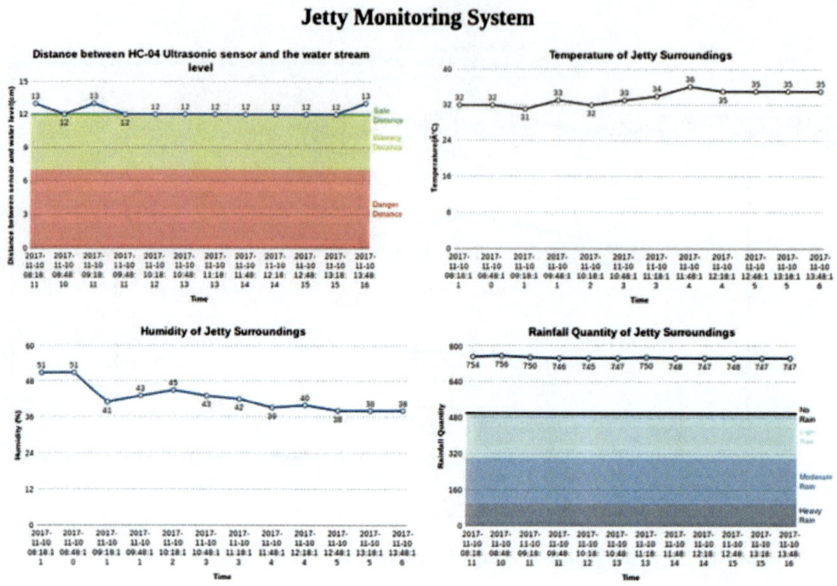

Fig. 9 Status of jetty docking area model from 8:25 a.m. to 1:55 p.m. on 11 November 2017

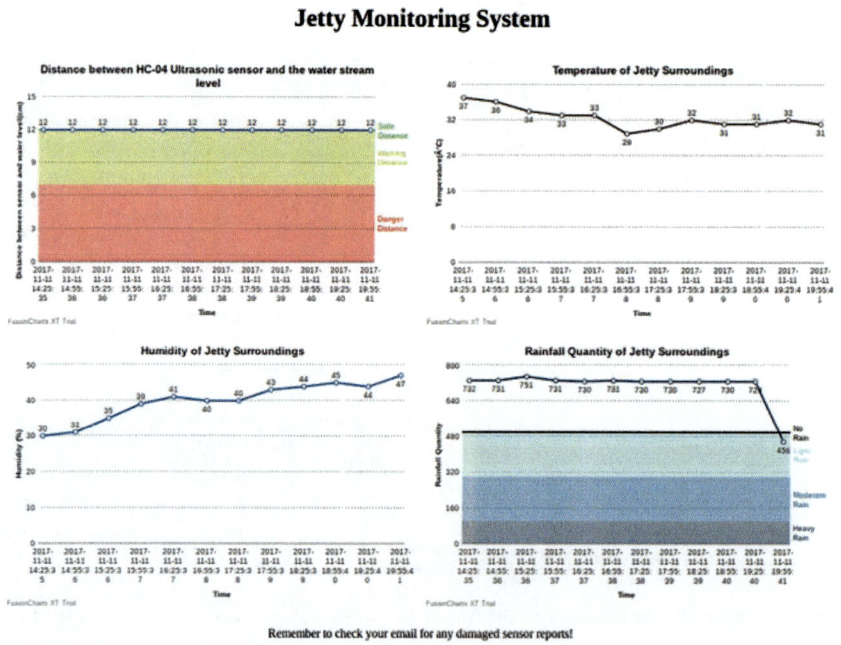

Fig. 10 Status of jetty docking area model from 2:25 p.m. to 7:55 p.m. on 11 November 2017

protection such as proper plastic casings for each component to withstand extreme environmental conditions in the actual jetty docking area. The power supply for the system will be replaced with durable batteries. Wi-Fi connection will be replaced with service-oriented resource architecture (SORA). Apache web server is relatively slow when multiple users are accessing the server as it needs to handle all pending requests at the same time. Deployment of the system at an actual jetty is required to further test its reliability and sustainability in actual environmental conditions.

Appendix

SORA is a network architecture that encapsulates all web service providers and customers in a network that allows them to interact and share common information with each other in a pre-defined network interface whilst maintaining their own way of data processing methods [7–14]. It represents a platform where all end-users, application providers, service portals, Internet service provider (ISP) and cloud-computing provider can collaborate with each other using common system communication methods. All information of applicants such as functionality, home address or even recommended system requirements is preserved in the system and organized by categories [15–18]. This reduces the redundancy for all involved users in the architecture to recreate their integration methods each time a change is proposed in one of the roles inside SORA.

References

1. Akyildiz, I., Vuran, M.: Wireless Sensor Networks. Wiley, The Atrium, Southern Gate, Chichester, West Sussex United Kingdom (2010)
2. Seo, J.H., Park, H.B.: Forest environment monitoring application of intelligence embedded based on wireless sensor networks. KSII Trans. Internet Inf. Syst. **10**(4), 1555–1570 (2016)
3. Hulaj, A., Shehu, A., Bajrami, X.: Application of wireless multimedia sensor networks for green borderline surveillance. Ann. DAAAM Proc. **27**, 845–853 (2016)
4. Liu, C.-H., Chung, Y.-F.: Secure user authentication scheme for wireless healthcare sensor networks. Comput. Electr. Eng. **59**(no. Supplement C, 2017/04/01/), 250–261 (2017)
5. Shabna, V.C., Jamshid, K., Kumar, S.M.: Energy minimization by removing data redundancy in wireless sensor networks. Int. Conf. Commun. Signal Process. **2014**, 1658–1663 (2014)
6. Fundamental Types (C++), Msdn.microsoft.com, 2017. [Online]. Available: https://msdn.microsoft.com/en-us/library/cc953fe1.aspx. Accessed: 11 Nov 2017
7. Huan, X., Yongqiang, X.: SORA: a network based service oriented resource architecture. in: 2013 15th IEEE International Conference on Communication Technology, pp. 407–412 (2013)
8. Kashem, S.B.A., Ektesabi, M., Nagarajah, R.: Comparison between different sets of suspension parameters and introduction of new modified skyhook control strategy incorporating varying road condition. Veh. Syst. Dyn. **50**(7), 1173–1190 (2012)
9. Kashem, S., Nagarajah, R., Ektesabi, M.: Vehicle Suspension Systems and Electromagnetic Dampers. Springer, Singapore (2018)

10. Kashem, S.B.A., Roy, S., Mukharjee, R.: A modified skyhook control system (SKDT) to improve suspension control strategy of vehicles. In: 2014 International Conference on Informatics, Electronics & Vision (ICIEV), pp. 1–8. IEEE (2014)
11. Kashem, S., Sufyan, H.: A novel design of an aquatic walking robot having webbed feet. Int. J. Autom. Comput. **14**(5), 576–588 (2017)
12. Kashem, S.B.A., Sheikh, M.I.B., Ahmed, J., Tabassum, M.: Gravity and buoyancy powered clean water pipe generator. In 2018 IEEE 12th International Conference on Compatibility, Power Electronics and Power Engineering (CPE-POWERENG), pp. 1–5. IEEE April, 2018
13. Kashem, S.B.A., Tabassum, M., Sufiyan, H.: A novel design of an amphibious robot having webbed feet as duck. in: Proceedings of IEEE Xplore 2017, International Conference on Computer and Drone Applications, 9–11 Nov 2017
14. Kashem, S.B.A., Jawed, S., Ahmed, J., Iqbal, A.: An experimental study of the amphibious robot inspired by biological duck foot. In: 2018 IEEE 12th International Conference on Compatibility, Power Electronics and Power Engineering (CPE-POWERENG), pp. 1–6. IEEE (2018)
15. Tabassum, M., Gabr, M., Mohanan, S., Mathew, K.: Development of smart vehicle security and entertainment system (SSES) using raspberry pi. Int. J. Eng Adv. Technol. **9**(3), 4077–4083 (2020)
16. Tabassum, M., Perumal, S., Ab Halim, A.H.: Review and evaluation of data availability and network consistency in wireless sensor networks. Malays. J. Sci. Health Technol. **4**(Special) (2019)
17. Tabassum, M., Zen, K.: Evaluation and improvement of data availability in WSNs cluster base routing protocol. J. Telecommun. Electron. Comput. Eng. (JTEC) **9**(2–9), 111–116 (2017)
18. Tabassum, M., Zen, K.: Signal interference evaluation of Eko wireless sensor network. In: 19th International Conference on Transformative Research in Science and Engineering, Business and Social Innovation, SDPS 2014 (2014)

IoT-Based Energy Analytic Platform for Foundry Units

M. Thillai Rani and R. Sivakumar

Abstract Energy efficiency practices also help in reducing the energy cost of a manufacturing plant. Energy is one of the top three cost components of any manufacturing facility. In the last decade, the general awareness level in energy efficiency has improved very rapidly and as a result, metering system for tracking energy consumption is improved as compared to the last decade. Energy Data Analytics means the processing of energy, production, and operational data to report energy performance of the industry. It saves the time and resources required to analyze data and prepare reports and help to discover the operational inefficiencies and their impact, hidden energy-saving potential.

Keywords Specific energy consumption · Project management unit (PMU) · Internet of things (IoT) · Machine to machine (M2M)

1 Introduction

Energy is one of the top three cost components of any manufacturing facility. In the last decade, the general awareness level in energy efficiency has improved very rapidly and as a result, metering system for tracking energy consumption is improved as compared to the last decade. Energy efficiency practices also help in reducing the energy cost of a manufacturing plant. Not only energy but also fuel consumption and other process parameters are also being logged. Data analytics is the science of analyzing raw data to make an inference from that information. Data can be uploaded to the server or it can be manually entered in the web portal. Processing and analysis of data are done by the help of intelligent algorithms and returns intelligent reports and analytics in the web portal. The Internet has boomed out to make the possibility

M. Thillai Rani (✉)
Department of ECE, Dr. N.G.P. Institute of Technology, Coimbatore, India
e-mail: mthillaisiva@gmail.com

R. Sivakumar
GEF-UNIDO-BEE Project, CO INDIA, Coimbatore, India
e-mail: siva.raja076@gmail.com

P. Suresh et al. (eds.), *Advances in Smart System Technologies*,
Advances in Intelligent Systems and Computing 1163,
https://doi.org/10.1007/978-981-15-5029-4_58

for infinite connectivity between services and more. In the age of IoT, everything that generates data has the potential to be linked to the vase web of information. In some industries, like critical facility management, IoT connectivity offers enhanced system monitoring that can save energy. The application of such IoT computing makes energy management easier and the industry can allocate their resources to manage energy use and implementation of energy conservation projects.

1.1 Energy Data Analytics

Energy Data Analytics means the processing of energy, production, and operational data to report energy performance of the industry [1]. It saves the time and resources required to analyze data and prepare reports and help to discover the operational inefficiencies and their impact, hidden energy-saving potential. For taking advantage of Energy Data Analytics, a proper metering and data acquisition system is required to collect accurate errorless energy and production information [2]. Figure 1 provides the pictorial representation of the need for energy management.

Energy consumption reduction has been topping the priority list for organizations all around the world. Central administrations all around the world have started to take note and this has led to several policy decisions as well; perform, achieve, and trade scheme being a prominent one in the Indian context. Conventional methods like energy audits do not cut it for organizations anymore. Data is the only way to uncover hidden energy potential savings, opportunities for optimization, and monitoring processes and equipment at a cellular level. The inability to use primary data obtained from metering systems have restricted organizations for long. It is time to use the data, analyze it, and draw meaningful inferences so as to reinforce your business with information and data-driven decisions.

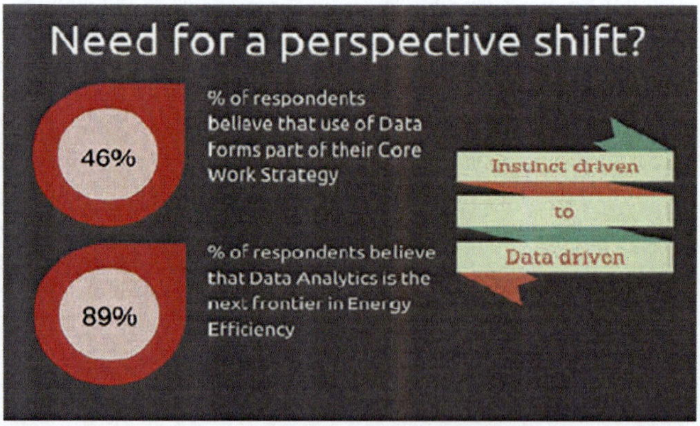

Fig. 1 Need for energy management

2 IoT-Based Energy Management

IoT computing is the activity of store and access data and programs over the Internet instead of local computers. Access to data or the programs is limited to that one computer and other computers connected to the local network. IoT-based Energy Information Management and Analytics System is to help the industry in assessing, reporting, and managing energy performance in a faster, better, and cost-effective way [2]. It is developed to capture energy performance data and provide periodic analysis of the same for the foundry units. Figure 2 depicts the need for energy management.

2.1 *Benefits on IoT-Based Energy Management*

Easy data access: Energy consumption can be measured online easily from the data collected and stored in IoT.

Identify energy-saving opportunities in real-time: Off schedule or excess energy consumption can be identified using circuit-level real-time energy monitoring.

Energy initiative prioritization: The specific inefficiency sources across all systems and all sites using benchmarks is identified thus making informed energy efficiency retrofit decisions and increasing ROI.

Drive green branding and behavioral change: Energy accountability and green awareness can be increased by making energy insights visible to all stakeholders within the organization.

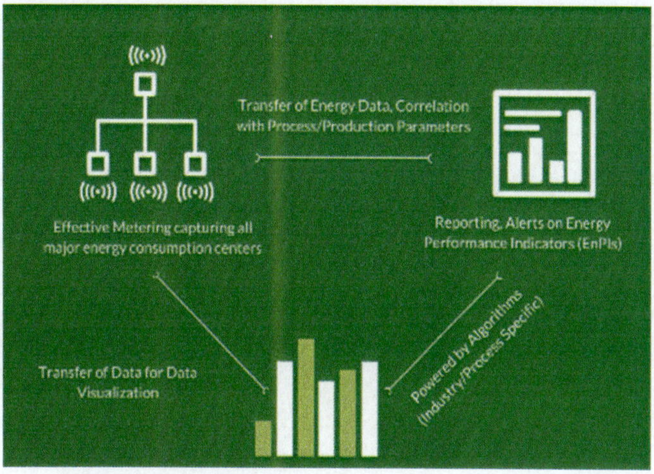

Fig. 2 IoT-based energy management system

Reduce energy costs Real-time alerts and customized reports result in cost reduction.

Benchmark and report on level basis.

3 IoT-Based Energy Management in Foundry Units

Each industry has certain drivers/parameters that define the production quality and in turn the production cycle and energy consumption pattern. So for the same amount of production, the parameters may be different and hence the Specific Energy Consumption (SEC). For sustained energy efficiency improvements, it is imperative that the unit has an established and working towards energy data strategy. Energy data can be combined with process/production data it provides actionable information based on which energy consumption behavior could be analyzed and reduction assessments can be done [3]. The typical energy consumption pattern in a foundry unit is shown in Fig. 3.

Average Specific Energy Consumption of Melting Cast Iron
Induction Furnace: 620 kWh/Tone or 1 Gcal/Tone
Cupola Furnace (Coke based): 13 kgs of Coke/Tone or 0.9 Gcal/Tone).

Depending on the installed power density and the melting practice, the thermal efficiency of the induction furnace can exceed 80%, but usually, it is in the range of 60–78%. The theoretical requirement of energy for melting iron is only 340 kWh per ton whereas the actual power required is around 700 kWh. This difference is due to two factors namely.

Inherent in the principle of melting in an induction furnace which includes the inefficiency in electrical bus bar losses, eddy current losses, refractory losses, and cooling water losses, etc.

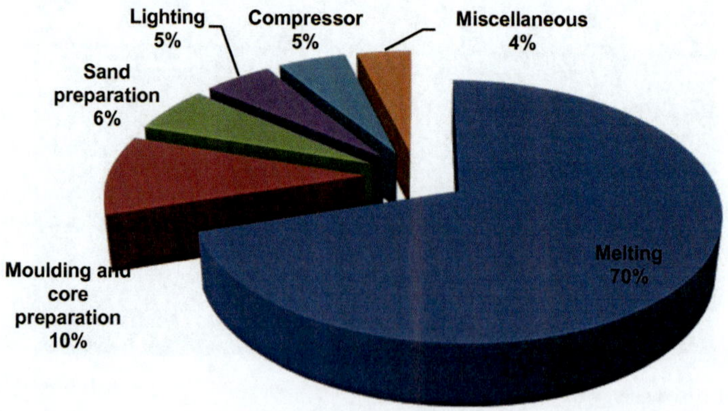

Fig. 3 Energy consumption trend in foundry

Foundry Name	Product Type	Production in MT	Energy consumption in kWh	SEC in kWh/MT
Foundry A	FG 200	10 MT	8500	850
	SG 400/7	10 MT	9300	930
Foundry B	FG 200	10 MT	7800	780
	SG 400/7	10 MT	8950	895

Fig. 4 Production versus energy consumption comparison

Operational losses are largely due to unnecessary and excessive holding of liquid metals in the induction furnace.

Consider the example in the table given below. In foundry A and B product types FG 200 and SG 400/7 have different energy consumption for the same production quantity. In many of the cases we have different SEC, so how does a verifier account for the impact of energy-saving measures on the SEC?

A comparison chart on Production vs energy consumption is shown in Fig. 4. In this case, study takes the case of casting process (foundry sector) and look at identify a process in which variation in SEC could be accounted to some of the parameters, which in turn will help identify the effect of energy-saving measures on the Gate to Gate (GTG) specific energy consumption.

One of the widely adopted and acknowledged methods is to establish a correlation between the type of castings (FG grade/SG grade) and corresponding electrical energy consumption, which can be done by developing a statistical model. The statistical model takes the production quantity as input and in return provides the expected electrical energy consumption.

The government is taking many initiatives towards promoting energy efficiency and renewable energy in India. As a part of the initiatives, the PMU had implemented the IoT-based energy analytics software in 16 foundry units. In this regard, M/s E Cube Energy Trading Private Limited tie-up with the project and implemented the software as per the individual foundry requirements [2]. Figure 5 shows the structure of IoT-based management system.

Table 1 refers to the energy consumption versus production taken from the two

Fig. 5 Structure of IoT-based energy management system

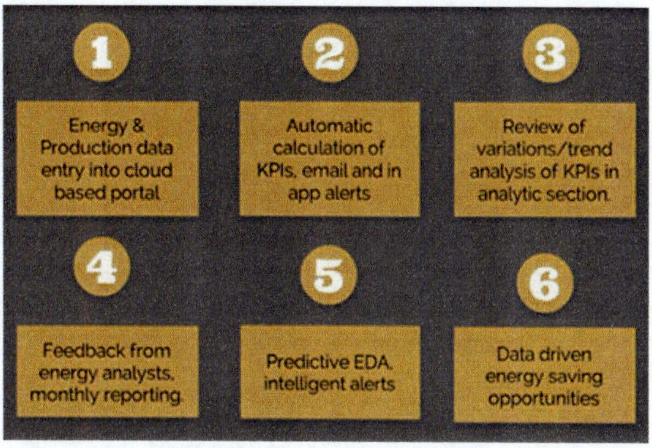

Table 1 Energy consumption versus production

Months	Total raw material for FG260 Grade to induction furnace 1 (Tons)	Total raw material for SG400/15 Grade to induction furnace 1 (Tons)	Molten metal produced for FG260 Grade from induction furnace 1 (Tons)	Molten metal produced for SG400/15 Grade from induction furnace 1 (Tons)	Electricity cons by induction furnace 1 (kWh)	Total raw material for FG260 Grade to induction furnace 2 (Tons)	Total raw material for SG400/15 Grade to induction furnace 2 (Tons)	Molten metal produced for FG260 Grade from induction furnace 2 (Tons)	Molten metal produced for SG400/15 Grade from induction furnace 2 (Tons)
Sep'17	98.51	4.81	99.64	4.66	78,900	159.12	8.63	147.85	18.50
Oct'17	119.43	1.53	81.67	1.48	61,740	147.70	10.82	145.16	10.19
Nov'17	150.77	0.97	146.75	0.91	101,710	127.23	2.91	122.21	2.81
Dec'17	208.05	0.88	204.87	0.93	141,130	114.45	2.49	114.72	2.40
Jan'18	134.36	0.00	134.82	0.00	94,080	116.56	0.00	114.78	0.00
Feb'18	190.12	0.00	187.72	0.00	129,880	116.63	0.00	118.20	0.00
Mar'18	162.37	0.00	164.32	0.00	110,870	125.27	0.00	123.07	0.00

Months	Electricity consumption by induction furnace 2 (kWh)	Electricity consumption from Grid (kWh)	Good Casting produced (Tons)
Sep'17	123,820	271,057	161.86
Oct'17	115,680	235,456	142.91
Nov'17	85,470	250,079	163.60
Dec'17	78,380	293,332	193.74
Jan'18	75,480	222,476	149.75
Feb'18	82,300	277,465	183.54
Mar'18	83,460	258,477	172.42

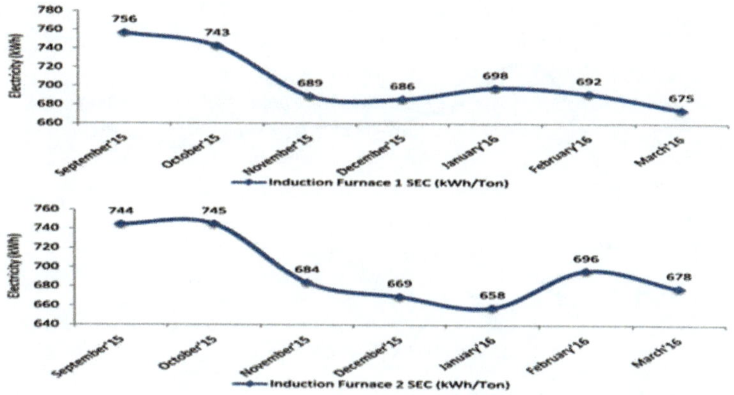

Fig. 6 SEC trend for furnace 1 & 2

induction furnaces in a particular foundry unit. Based on the production plan and casting grades the two furnaces have been used.

3.1 Specific Energy Consumption by Furnace (kWh/Ton)

The graph shown in Fig. 6 refers to the specific energy consumption for the induction furnace 1 and 2. Depending on the furnace utilization and casting produced from the individual furnaces the SEC is varied. This graph will give a clear indication of the individual furnace performance irrespective of production plan, type of casting produced, and raw material mixture [4]. The SEC value is fluctuated from month to month and based on this value the foundry unit will analyze the raw material mixture and plan.

3.2 Furnace Efficiency Index

Figure 7 shows the furnace efficiency index. The ideal furnace efficiency index is around 32% and the foundry unit can compare this value. Corrective actions can be taken to improve the furnace utilization and it leads to an increase the efficiency index.

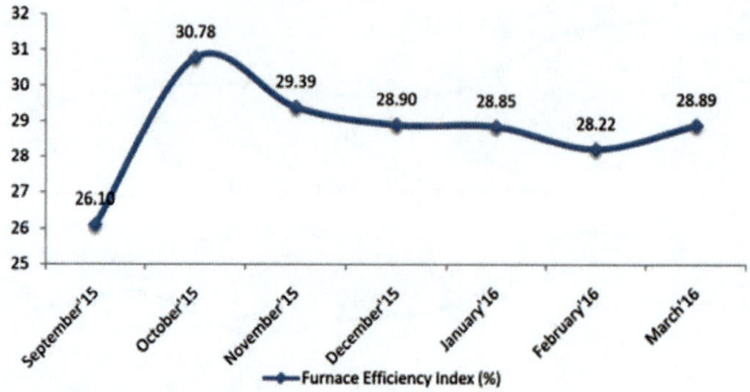

Fig. 7 Furnace efficiency index trend

3.3 Production Profile

Inner circle shows FG-260 and outermost circle show SG-400/15 for all the graphs. Furnace wise production details can be taken for analyzing the performance and efficiency index. From the chart, the foundry unit is able to visualize the product matrix taken by month wise can be analyzed. Month-wise production profile is shown in Fig. 8.

4 Analytical Observations

4.1 Specific Energy Consumption (SEC)

The expected SEC of Induction Furnace 1 and 2 over the period is shown in Figs. 9 and 10. The benchmarking values were arrived based on the past data entry and history of records entered in the IoT-based energy management analytic software. Now the foundry units are able to identify the gap in the SEC and the same can be analyzed.

4.2 Potential Energy Efficiency Gap

There has been a potential energy efficiency gap, which is a big reason for concern. The below table refers to the various reasons to improve the energy efficiency in the particular unit. In this case, most of the time the unit losses the energy in the heat time. If proper corrective action has been taken and tries to reduce the heat time than there is high-energy savings that can be achieved [3, 4]. Similarly, large amount of energy was lost in the furnace heat loss and the unit identified the corrective measures

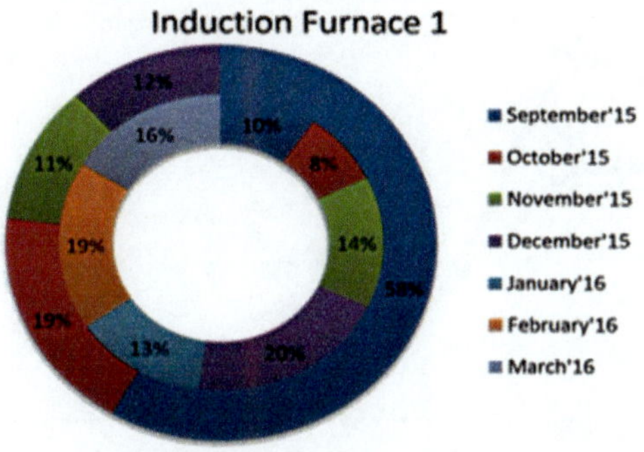

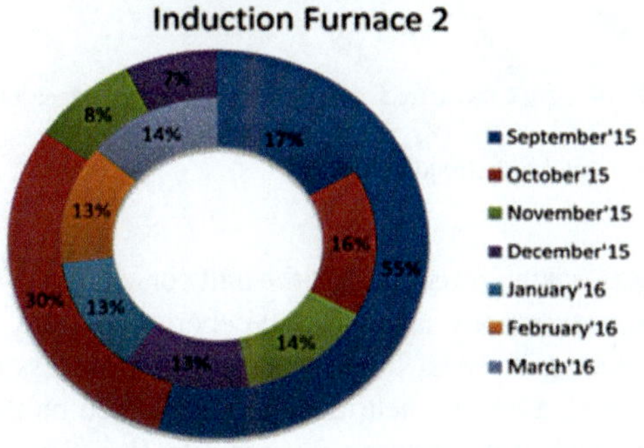

Fig. 8 Month wise production trend

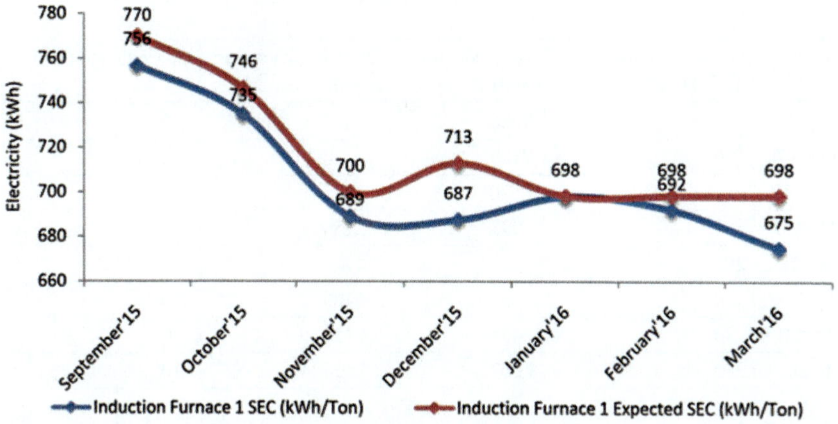

Fig. 9 Month wise furnace 1 SEC trend in kWh/MT

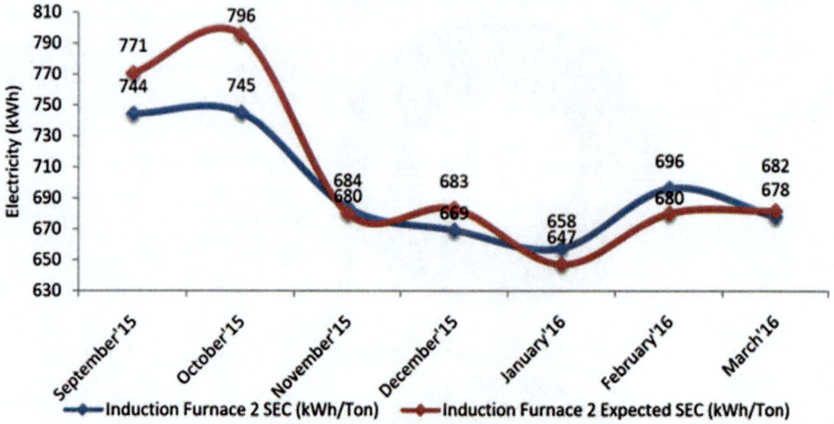

Fig. 10 Month wise furnace 2 SEC trend in kWh/MT

as there is huge energy-saving possibility. If the unit concentrates on the high-energy loss area then there is enormous savings in the energy possible. Corrective action can be taken for reducing the heat time by installing a process control timer and the timing can be adjusted by the melting supervisor based on the casting process and grade of iron melted in the furnace. Proper refractory lining and crucible lid implementation will help to reduce the heat losses in the furnace.

	Reasons			
Months	Due to molten metal YIELD (kWh)	Due to heat time (kWh)	Due to furnace loss (kWh)	Unaccounted (kWh)
Sep'17	0	9686	0	0
Oct'17	0	7155	0	0
Nov'17	4	0	3295	110
Dec'17	0	938	2754	0
Jan'18	202	39	4946	0
Feb'18	0	35	6406	0
Mar'18	30	454	433	0
Total	236	18,307	17,834	110

5 Conclusion

Increasing cost and demand for energy has led many organizations to find smart ways to monitor and save energy. Emerging technologies like IoT and M2M can be utilized to manage energy consumption. These technologies have gained widespread popularity across multiple filed that they are deployed to address challenges in energy. Hence, the customers now demand to know the energy consumption and the usage pattern in real-time so that they can have control over it. This shift in customer attitude has forced the end-users and enterprises to feel the need to use predictive analysis to effectively analyze the huge volume of data and predict future usage and potential problem.

References

1. www.eetpl.in
2. www.en-view.com
3. http://foundry.en-view.com/
4. http://impt.en-view.com/

Energy Efficient Multifocus Image Fusion in DCT Domain

G. Sreeja and O. Saraniya

Abstract Fusion of multifocus images in discrete cosine transform (DCT) domain has been extensively exploited in recent years due to low complexity. In particular, the fusion by DCT is useful in visual sensor networks (VSN), where the images have to be transferred in coded format. Many research works have been done by combining spatial domain methods in DCT domain. In the proposed work, the energy of correlation coefficient in DCT domain is chosen as fusion criteria. The method works by evaluating the focus measurement between the input images and Laplacian-based sharpened images in DCT domain. The results obtained by the proposed method are compared in terms of objective metrics and the results show that the proposed work avoids inappropriate block selection which exists in available methods.

Keywords Multifocus image fusion · DCT · Image enhancement · Energy of correlation coefficient · Consistency verification

1 Introduction

Visual sensor networks (VSN) capture the same scene with different focal depths. Objects within the defined focal point appear to have high resolution but images go blurry beyond the focal lengths. In order to combine different images obtained with these cameras, image fusion is employed in VSN. Application of VSN includes military surveillance, remote sensing and object tracking [1]. Image fusion takes multiple images as inputs and then combines them into single full focused image by considering its significant information. The need for less computation time and memory had forced to employ DCT instead of multiscale transforms for image fusion in VSN. In recent years, many DCT-based image fusion algorithms have been evolved

G. Sreeja (✉) · O. Saraniya
Department of ECE, Government College of Technology, Coimbatore 641013, India
e-mail: sreeja2592@gmail.com

O. Saraniya
e-mail: sarandas2000@yahoo.com

P. Suresh et al. (eds.), *Advances in Smart System Technologies*,
Advances in Intelligent Systems and Computing 1163,
https://doi.org/10.1007/978-981-15-5029-4_59

for VSN since the images have to be processed in compressed format [2]. Existing DCT-based fusion algorithms are based on variance, contrast, by selecting maximum AC coefficients [3, 4], by combining energy of Laplacian (EOL), variance of Laplacian (VOL) [5], using singular value decomposition (SVD) [6], and sum-modified Laplacian (SML) [7]. The methods with variance and contrast are highly suffered from blocking artifacts. The rest of the methods have the impact of inappropriate block selection.

In this manuscript, a novel energy efficient multifocus image fusion is proposed in DCT domain. DCT is performed by dividing images into 8 × 8 blocks. For each block, energy of correlation coefficient is calculated as fusion measure and decision map is formed by selecting image blocks with higher energy value. Through consistency verification (CV), the final fused image is obtained with better quality and it is measured using performance metrics such as $Q^{AB/F}$ [8], feature mutual information (FMI) [9] and mutual information (MI). The efficiency of proposed work is verified by comparing the work with existing DCT-based fusion algorithms [6, 7, 10].

2 Proposed Energy-Based Fusion Work

Multifocus image fusion method combines relevant information from differently focused images to form single full focused image. Energy reflects the contrast of an image and it is chosen as focus measure for performing fusion. The proposed energy-based multifocus image fusion in DCT domain is depicted in Fig. 1. Edge enhancement by applying Laplacian kernel to the image and adding the resultant image obtained back to the original image is the preprocess for proposed work. Laplacian filtering is more sensitive to noise, and to reduce the impact of such artifacts, Gaussian smoothing is applied to the images before applying Laplacian kernel. After the process of enhancement, for implementing DCT, all these four images are level shifted with the range [−127 to 128] and divided into blocks with size of 8 ×

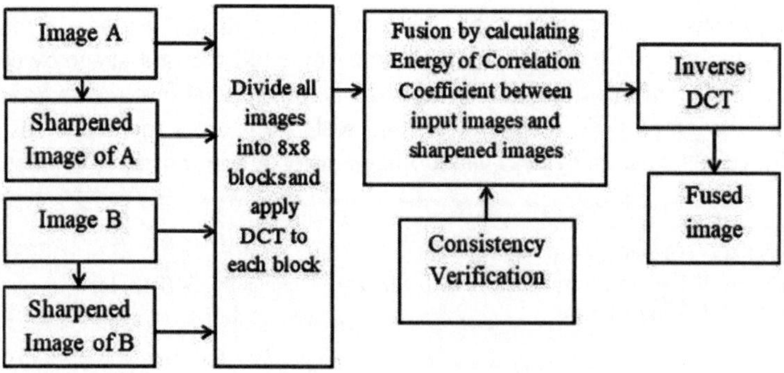

Fig. 1 DCT-based image fusion using energy as focus measure

8. The correlation energy is calculated for each DCT block and the one with highest energy value is selected. With the obtained decision map, the focused regions are observed, and finally, inverse DCT is implemented to get the fused image.

DCT separates the given image into spectral subbands according to its visual quality. The DCT has been exploited in image processing due to its energy compaction property [11]. Before applying DCT transformation, the image is level shifted and the divided into 8×8 blocks. For each block, the two-dimensional DCT is applied and the coefficient is obtained.

Energy is one of the focus measures that reflect the contrast of an image. The energy is calculated by the formula

$$\text{Energy}_{\text{DCT}} = \sum_{i} \sum_{i} \text{image}A(i, j)^2. \tag{1}$$

The energy is calculated for input images and sharpened images. The correlation energy is a useful criterion to measure the focused region. The product of [5] energies of original image and sharpened image scaled by (1 − Correlation) is the deciding parameter for image fusion. The energy of correlation coefficient is calculated using the equation

$$\text{En}_A \times (1 - \text{Correlation}(\text{image } A, \text{sharp } A)) \times \text{En}_{\text{sharp } A}. \tag{2}$$

The fusion is performed by selecting block with high energy value. The initial decision map based on correlation energy is given by

$$\text{Decision map} = \begin{cases} \text{Dmap} = +1 \text{ for image } A \text{ if Energy } A > \text{Energy } B \\ \text{Dmap} = -1 \text{ for image } B \text{ if Energy } B > \text{Energy } A \end{cases} \tag{3}$$

2.1 Consistency Verification

After fusion, the quality of fusion performance is improved by applying consistency verification (CV) method [6]. It reduces the errors caused by wrong selection of image blocks. Using majority filter, the fusion process is improved. The final decision map by CV method is represented by

$$\text{Fusion} = \begin{cases} \text{image } A \text{ if Dmap} \geq 0 \\ \text{image } B \text{ if Dmap} < 0 \end{cases} \tag{4}$$

3 Performance Metrics

The quantitative analysis of image fusion algorithms is performed by exhibiting fusion objective metrics. In the proposed work, three non-reference fusion performance metrics namely Petrovic's method ($Q^{AB/F}$), FMI and MI are employed for validation.

Mutual information defines the information quantity of input images that is retained in the fused image after fusion. Feature mutual information (FMI) uses gradient operator and calculates the fused image quality by averaging the mutual information.

The level of edge information transferred from the original source to the resultant image is measured using Petrovic's method specified in [8].

4 Analysis of Simulation Results

The proposed work is simulated using MATLAB 2017b software using Intel core i5 processor with 4 GB RAM. LytroDatasets [12] are employed for evaluating the proposed method. The dataset consists of 20 pairs of differently focused image of size 520 × 520. The efficiency of proposed work is validated through three non-reference fusion metrics, namely Petrovic's metric ($Q^{AB/F}$), feature mutual information (FMI) and amount of mutual information (MI). These objective metrics obtained with proposed method are compared with the relevant DCT-based techniques like DCT + SML [7], DCT + SVD [6] DCT + Sharp and DCT + Corr [10]. The original codes of all these methods are simulated on same software and the values obtained are tabulated in Table 1. In Fig. 2, for visual comparison, the resultant image obtained by proposed work is presented along with all other methods and a graph is depicted in Fig. 3.

The results of LytroDataset reported in Table 1 shows that the proposed method produces best results when compared to competing DCT-based fusion methods in three performance metrics $Q^{AB/F}$, FMI and MI by scoring 0.7521, 0.8747 and 8.2547, respectively. In terms of FMI, DCT + SVD method achieves better score. The overall performance of the proposed work seemed superior to other existing DCT-based methods.

Table 1 Comparison of fusion metrics for non-referenced datasets

Methods	$Q^{AB/F}$	FMI	MI
DCT + Sharp	0.7444	0.8736	8.0165
DCT + Corr	0.7516	0.8745	8.2508
DCT + SML	0.7449	0.8737	8.1139
DCT + SVD	0.7337	0.8748	7.3916
Proposed	0.7521	0.8747	8.2547

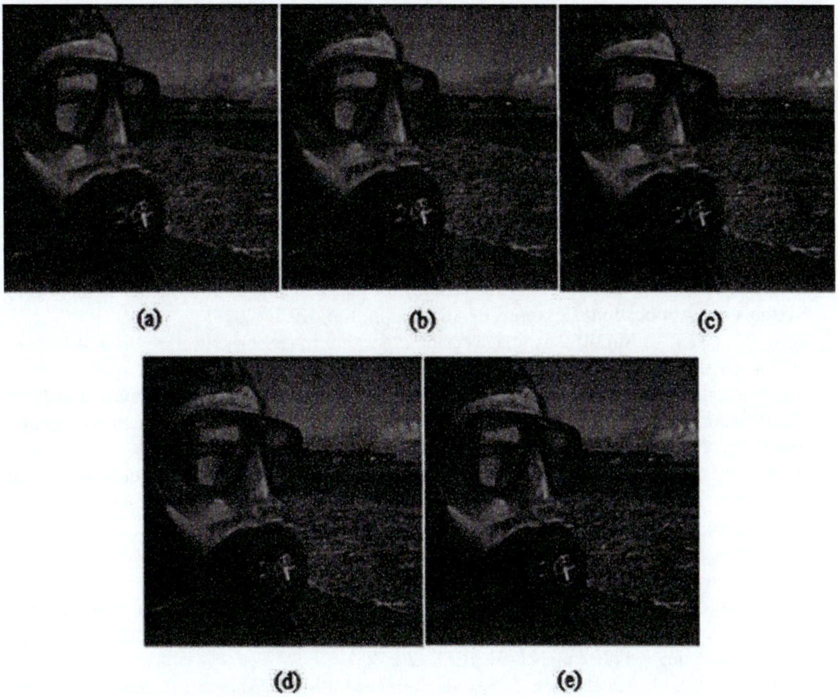

Fig. 2 Visual comparison of fused image obtained by **a** DCT + Sharp, **b** DCT + Corr, **c** DCT + SML, **d** DCT + SVD and **e** proposed work

Fig. 3 A graphical representation of result comparison

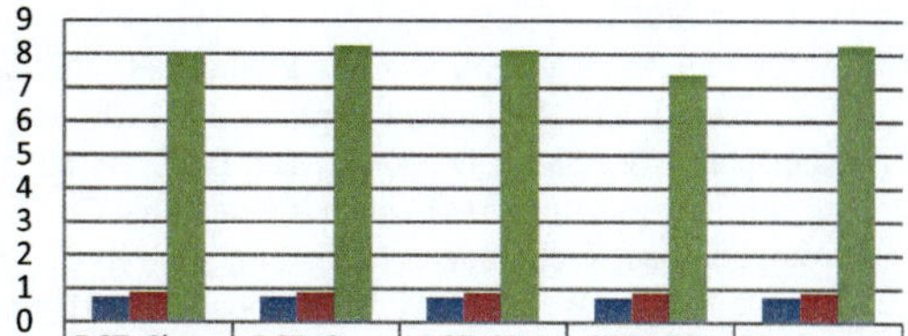

	DCT+Sharp	DCT+Corr	DCT+SML	DCT+SVD	Proposed
QAB/F	0.7444	0.7516	0.7449	0.7337	0.7521
FMI	0.8736	0.8745	0.8737	0.8748	0.8747
MI	8.0165	8.2508	8.1139	7.3916	8.2547

5 Conclusion

A new image fusion technique based on energy of correlation coefficient for different focus images is proposed in this paper. For both the source image and its corresponding sharpened image, the energy is calculated. By choosing the block with highest energy value, fusion is performed. The method is compared with four DCT-based fusion methods, and results show that the problem of blocking artifacts is removed by the proposed method. It is evident from the resultant graph that the proposed method outperforms the competing fusion algorithms in terms of edge information and visual quality.

References

1. Oral, M., Sevgi Turgut, S.: A comparative study for image fusion. In: Innovations in Intelligent Systems and Applications Conference (ASYU), pp. 1–6. IEEE (2018)
2. Tong, Y., Chen, J.: Multi-focus image fusion algorithm in sensor networks. IEEE Access **6**, 46794–46800 (2018)
3. Paramanandham, N., Rajendiran, K., Narayanan, D., Anand, M.: An efficient multi transform based fusion for multi focus images. In: International Conference on Communications and Signal Processing (ICCSP), pp. 0984–0988. IEEE (2015)
4. Albuquerque, H.R., Ren, T.I., Cavalcanti, G.D.: Image fusion combining frequency domain techniques based on focus. In: 24th International Conference on Tools with Artificial Intelligence, vol. 1, pp. 757–762. IEEE (2012)
5. Amin-Naji, M., Aghagolzadeh, A., Ezoji, M.: Ensemble of CNN for multi-focus image fusion. Information Fusion (2019)
6. Amin-Naji, M., Ranjbar-Noiey, P., Aghagolzadeh, A.: Multi-focus image fusion using singular Value decomposition in DCT domain. In: 10th Iranian Conference on Machine Vision and Image Processing (MVIP), pp. 45–51. IEEE (2017)
7. Abdollahzadeh, M., Malekzadeh, T., Seyedarabi, H.: Multi-focus image fusion for visual sensor networks. In: 24th Iranian Conference on Electrical Engineering (ICEE), pp. 1673–1677. IEEE (2016)
8. Kalaivani, K., Phamila, Y.A.V.: Analysis of image fusion techniques based on quality assessment metrics. Indian J. Sci. Technol. **9**(31), 1–8 (2016)
9. Jagalingam, P., Hegde, A.V.: A review of quality metrics for fused image. Aquatic Procedia **4**, 133–142 (2015)
10. Amin-Naji, M., Aghagolzadeh, A., Ezoji, M.: CNNs hard voting for multi-focus image fusion. J. Ambient Intell. Humanized Comput., 1–21 (2019)
11. Mahapatra, S., Sa, K.D., Dash, D.: DCT Based Multifocus image fusion for wireless sensor networks. In: 2018 Second International Conference on Inventive Communication and Computational Technologies (ICICCT), pp. 871–875. IEEE (2018)
12. Zhang, Y., Wei, W., Yuan, Y.: Multi-focus image fusion with alternating guided filtering. Sig. Image Video Process., 1–9 (2018)

Automatic Speed Control System in Vehicles Using VANET

A. Sathyapriya, K. Sathiya, T. M. Sneha, D. Rohit Raja, and T. Manikandan

Abstract Intelligent Transport System is the need of the hour technology which paves the easy way for smart city implementation in the future. As much as the road rules and safety regulations are formulated, there is still no assurance that it will all be followed by everyone. Road accidents are increasing every day and mortality is reaching hundreds of thousands around a year. So, it is necessary to implement technologies to prevent them from happening. Vehicular ad hoc network commonly known as the VANET technology can be used at several advantages. Controlling over speed vehicles is the main objective of this paper. It can be done by sending the speed limit data over the network to the engine control unit (ECU) in the vehicle. Taking this data, ECU controls the speed of the vehicle according to the speed limit in that region. Single broadcasting of the data from a central control location is enough to trigger the vehicle to vehicle communication. A vehicle first receiving the signal passes it to the nearby vehicles and the chain goes on. VANET restricts the disadvantage of signals being lost due to the speeding of vehicles. This approach can easily control the vehicle speed automatically. V2V transmission is efficient than prior methods like GPS transceiver systems, which needs transceivers along the road at regular intervals. This signal might get lost if the vehicle is moving at high speed. With VANET, these disadvantages are overruled.

Keywords VANET · Speed control system · Engine control unit · Intelligent transport system · Location tracking · Networking · V2V

1 Introduction

Safety is the highest priority when it comes to human life. According to Ministry of Road Transport and Highways data, by 2016, 66.5% (about 268,341) of the total road accident deaths are the consequence of over-speeding. And 61.0% (about 73,896) of them has lost their lives [1].

A. Sathyapriya (✉) · K. Sathiya · T. M. Sneha · D. Rohit Raja · T. Manikandan
Rajalakshmi Engineering College, Chennai, Tamil Nadu, India
e-mail: sangfroidshakthi@gmail.com

P. Suresh et al. (eds.), *Advances in Smart System Technologies*,
Advances in Intelligent Systems and Computing 1163,
https://doi.org/10.1007/978-981-15-5029-4_60

Intelligent Transportation System (ITS) is the emerging application of implementing intelligence systems in vehicles to achieve safety and automation [2–6], [7], and [8]. Road transportation is an important means for people to reach places. But the increased population and vehicles have reduced and safety of lives on road. Both pedestrians and drivers get affected by road accidents. The directive pronounced by the European Union 2010/40/EU defines Intelligent Transportation System as the information and communication technologies which are used in road transport (mainly) for traffic management and other interfaces with other modes of transport [9].

GPS enabled speed control systems [2] also established. In this paper, speed limit data is fed to the vehicle through GPS, and according to this data, ECU compares and alters the speed of the vehicle. In addition to ECU, cruise control system is also used [3] for the limiting of vehicle's speed. But this method faces the limitation in no network coverage regions. Smart zones are established in paper [4], and in these smart zones alone, obstacle detection is established using RF technology. But using RF technology has the disadvantage in high-speed vehicles. Another method is the cooperative adaptive cruise control (CACC) [8] which is the improvement of adaptive cruise control system.

These two limitations can be overcome by using VANET systems. Different routing protocols are available for VANET [10] applications. Accident prevention system is established in this system. Nazmus S Nafi discusses in his model [5] about using VANET for intelligent road traffic signaling system. In this system, live information about traffic flow and trip time is fed to sensors on road unit and is received by the receivers in vehicles. This system establishes a wide are traffic control system connecting several nodes. Similarly, traffic flow data is also sent to roadside units (RSUs) [7]. The combination of RF, GPS, and GSM [6] has also been realized for emergency and accident detection application. The security issue in VANET is not left unattended. A motion model has been created [11] to reduce the security issues in VANET using advanced on-demand routing protocol (AODV).

Yet there is an improvement needed in adopting VANET for automatic speed control systems in vehicles. The proposed method here overcomes the limitations in using GPS and RF technology for speed control and combines VANET technology for the same as explained further.

2 VANET Technology

ITS exclusively uses wireless communication methods [6, 7, 10, 11], as it is the most efficient and highly exploited technology. There are several ITS techniques already applied in various aspects of transportation. There are several technologies to achieve these applications. But VANET is efficient of them all.

Wireless ad hoc network (WANET) is a decentralized mesh-type network. In this networking, there is no centralized unit to monitor the nodes and the data that is being transmitted among them. The nodes can enter and leave a WANET according to its requirement, and as long as the node stays in that network, it can transmit and receive data among the other nodes in the network. This wireless network is the pioneer for VANET technology.

VANET is based on the WANET technology and is applied for vehicle to vehicle (V2V) communication. In this network, any number of nodes (vehicles) can be connected. And, the nodes can join the network at any time. This system is used in many live smart vehicle system applications. In this paper, we are using it for automatic speed control of the vehicle.

3 Proposed Work

The proposed work in this paper is based on VANET. I2V, V2V communication is done through DSRC. DSRC is suitable for transmitting data to high mobility vehicles. The data transmitted from the central system is broadcasted to the vehicles using suitable technology. The data received by a node (vehicle) is transmitted among other nodes through DSRC. On receiving the data, it is sent to the vehicle's engine control unit which in turn compares the current speed with the set speed and alters it to match the speed limit.

3.1 Dedicated Short-Range Communication (DSRC)

The basis for dedicated short-range Communication (DSRC) is IEEE802.11p. IEEE802.11 is the standardization used for vehicle-based communication. The ultimate aim is to provide communication between vehicle to vehicle or vehicle to other roadside infrastructure. It provides V2V as well as V2I communication which is known as hybrid V2X communication. This is an enhanced standard of IEEE802.11 (WI-FI). This technology is useful in high mobility vehicles with low latency. It is not sensitive to environmental changes as it can withstand even harsh climatic conditions and it is not affected by changes in the topology of the vehicle (Fig. 1).

Possible applications include:

- Obstacle detection and alert
- Accident warning system
- Blindspot vehicle awareness.

3.2 GSM/GPRS

Global system for mobile communication (GSM) is a standard that is being followed worldwide for various applications. GSM module establishes a communication

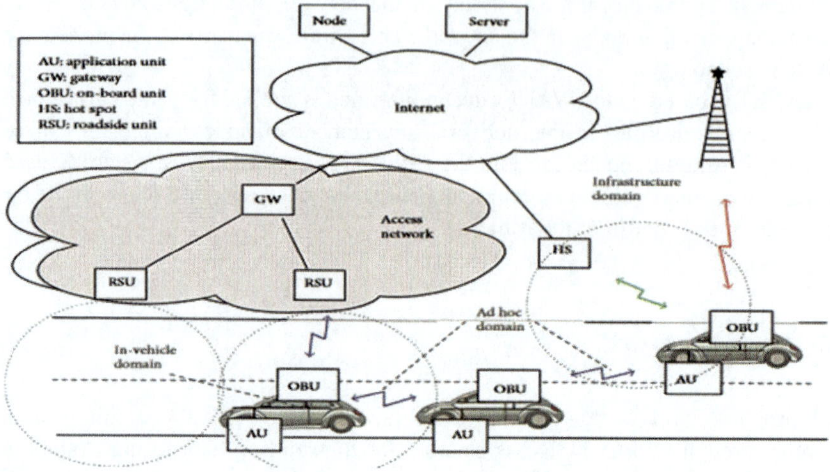

Fig. 1 V2V architecture

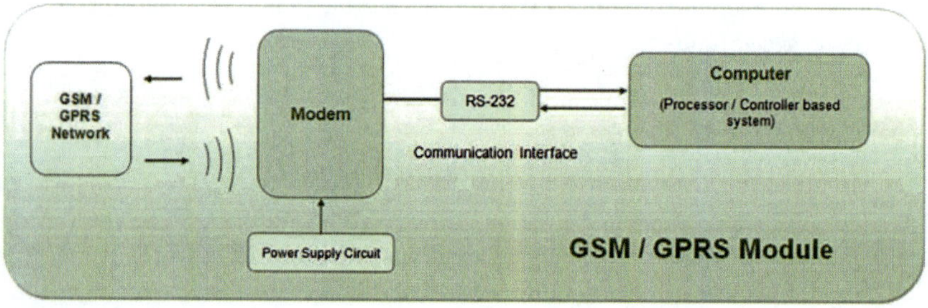

Fig. 2 GSM/GPRS module

between devices. It is basically a chip that is powered by a supply and transmits data among nodes (Fig. 2).

3.3 Engine Control Unit (ECU)

Engine control unit is the electronic brain of a vehicle which controls all the process of an engine. It performs various operations such as controlling of air–fuel mixture, controlling of ignition timing, and controlling an idle speed of the vehicle. ECU monitors the speed of the vehicles by reregulating air–fuel mixture reaching the engine. Depending on the signals that ECU receives, it controls the amount of the fuel mixture reaching engine. Thereby, the speed of the vehicle is controlled (Fig. 3).

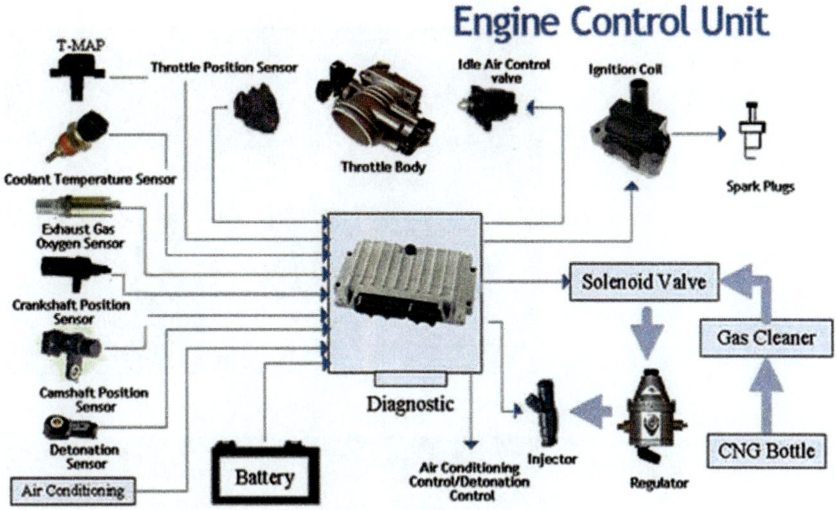

Fig. 3 Engine control unit

3.4 ESP8266 12E NodeMCU Wi-Fi Module

ESP8266 ESP8266 is one of the Wi-Fi chips built upon TCP/IP layers. It is first developed by Espressif Systems in China and is the cheapest available microchip. NodeMCU is the open-source firmware used for ESP8266 and it can be programmed with Arduino IDE itself. It is the cheap and efficient Wi-Fi modules available in the market (Fig. 4).

Fig. 4 NodeMCU

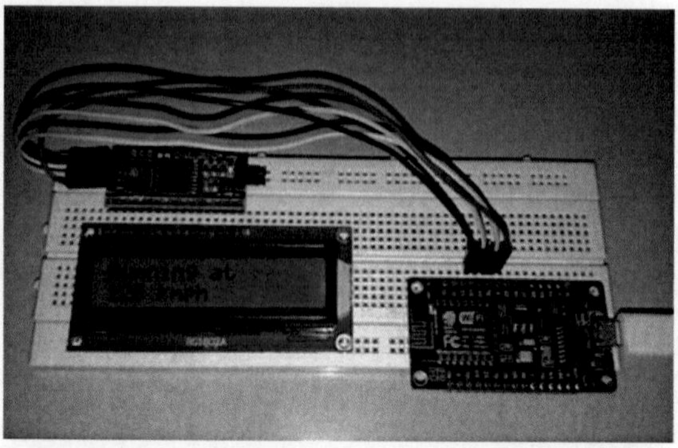

Fig. 5 Hardware output before receiving Wi-Fi signal

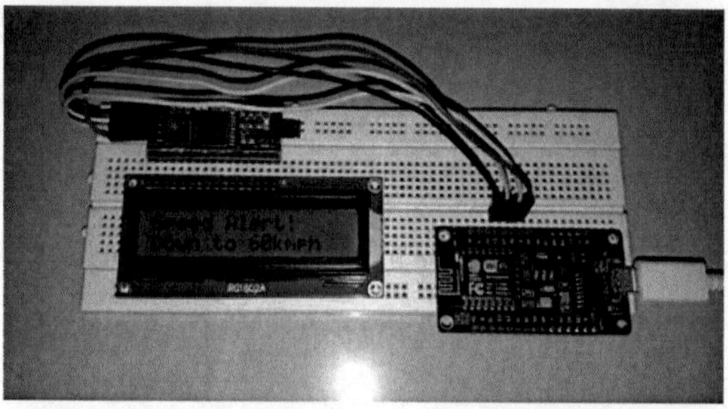

Fig. 6 Hardware output after receiving Wi-Fi signal

3.5 Prototype Model

To realize how the proposed system works, a prototype has been developed using NodeMCU ESP8266 12E module. The real-time system can be realized by similar methods with suitable technologies as defined above.

The motor runs in maximum speed (Fig. 5). A signal is transmitted from a mobile hotspot to NodeMCU and on receiving this signal; NodeMCU controls the speed of the motor that is connected to it (Fig. 6). It can be understood that the NodeMCU unit defines the signal transmission unit in VANET and the speed control in motor is synonymous to vehicle speed control in real-time system.

4 Results and Discussions

The proposed method can achieve the concerned application well enough. The usage of GPS requires network coverage throughout the journey. But the system fails in the no network regions. Also, RF technology has a disadvantage when it comes of high mobility vehicles. These disadvantages can easily be overcome by implementing the system in VANET.

The proposed is explained in the prototype using NodeMCU. Hotspot signal from a mobile is being sent and received by NodeMCU. Upon receiving the signal, the microcontroller reduces the motor speed (Fig. 6). This prototype depicts the work of VANET and ECU combined system.

VANET is used to transmit the valid speed limit data across a wide area and several nodes in a short time and without loss. The distance covered by VANET is expandable through the roadside units, but it's ideally 50–100 km. The signal and data transmitted are received by the on-board unit (OBU) in the vehicle. The received signal by the vehicle sends a command to engine control unit. This command triggers ECU to reduce fuel injection to the vehicle's engine gradually. The speed limitation is thus achieved.

The shortcoming of using VANET is that there is a possibility of anonymous node to transmit corrupt data. There is no strong security since the messages does not include the identity of the person/vehicle sending the information. To solve this issue, the message has to be sent from an authorized source. And, the identity of that authorized source has to be included as a token while transmitting the data. Research is going on in making the messages encrypted. Future works are to be done in this area before widening the technology to several more applications.

5 Conclusion

Thus, from this proposed system, it is possible to achieve automatic speed control in vehicles. This system can be successfully and effectively utilized in speed-critical zones and as well as other regions. Using VANET has paved way for the upcoming 5G system adaptation for this model. This model is adaptable for future implementation of 5G systems, thereby.

6 Future Work

Smart cities are the next goal everyone is working toward. Putting forward, realization of that goal, our proposed model can be extended in that direction. A whole set of applications include; vehicle guidance, traffic management, pathfinding, obstacle detection, live monitoring of the path, updating the routes, and etc.

There is an immense possibility of extending this idea toward smart cities. There is no restriction on what vehicles could be connected as long as a portable connected mobile device is available. On further extending this idea, even pedestrians can be alerted of their path. This can help visually challenged people more.

References

1. Directive 2010/40/EU of 7 July 2010 on framework for ITS deployment in road transport
2. Jahan, A.S., Hoq, I., Westerfield, D.: GPS enabled speed control embedded system speed limiting device with display and engine control interface. In: IEEE Long Island Systems, Applications and Technology Conference (LISAT) (2013). https://doi.org/10.1109/lisat.2013.6578244
3. Afify, M., Abuabed, A.S.A, Alsbou, N.: Smart engine speed control system with ECU system interface. In: IEEE International Instrumentation and Measurement Technology Conference (I2MTC) (2018). https://doi.org/10.1109/i2mtc.2018.8409871
4. Vinod Rao, S., Saketh Kumar, P., Anil Kumar, N., Saleem Yusuf, M.D., Krishna, B.: Smart zone based Vehicle Speed control using RF and obstacle detection and accident Prevention. Int. J. Emerg. Technol. Adv. Eng. (IJETAE) (2014). 10.1.1.637.7540
5. Nafi, N.S., Khan, J.Y.: A VANET based intelligent road Traffic Signaling system. In: Australasian Telecommunication Networks and Applications Conference (ATNAC) (2012). https://doi.org/10.1109/atnac.2012.6398066
6. Dahat, D., Thakur, M., Sahare, R., Ramtekkar, R., Barve, V., Khanwe, V.: VANET for emergency services and accident detection. Int. J. Eng. Sci. Comput. (2017)
7. Kumar, R.: Methods to resolve traffic jams using VANET. Int. J. New Innov. Technol. (2015)
8. Milanés, V., Shladover, S.E., et al.: Cooperative adaptive cruise control in Real traffic situations. IEEE Trans. Intell. Transp. Syst. (2014). https://doi.org/10.1109/tits.2013.2278494
9. Report on road accidents in India—2016 by Ministry of road transport and highways, Transport research wing, Government of India
10. Dutta, R., Thalore, R.: A review of various Routing protocols in VANET. Int. J. Adv. Eng. Res. Sci. (2017)
11. Patil, K., Mahajan, G., Patil, D., Maharajan, C.: Implementation of motion model using VANET. Int. J. Comput. Eng. Res. Trends (2016)

Design and Development of a New Microstrip Patch Antenna for ISM Band Applications

V. Venkataramanan, T. Manikandan, M. Sathish, R. Srilakshmi, and K. Kalaiarasan

Abstract The design and simulation of the hexagonal-shaped microstrip patch antenna and its array was proposed in this paper. The structure is capable of operating in the 2.4 GHz ISM band frequency. The 2×2 designated hexagonal patch antenna array acts as a radiating element. The designated structure was analysed for the series-fed and corporate-fed manner, and various antenna performance parameters like gain, return loss and bandwidth were measured. The simulated results show that the proposed work significantly enhances the bandwidth and gain than the existing microstrip patch arrays for the ISM band applications.

Keywords Microstrip patch antenna · Hexagonal shape · ISM band · Series-fed · Corporate-fed

1 Introduction

To meet the demand of distant communication, the wireless systems operating in various frequency bands like L, S, C and X needs appropriate antenna array with good radiation characteristics. The microstrip patch antenna array is mostly preferred for different applications because of its characteristics like less profile, inexpensive, low weight and easy fabrication methods. Due to the necessity of higher gain, the antenna elements increases which results in the complication of the feeding network.

Antennas are found to be the significant components in the wireless communication system. The antenna in the wireless system is used for both transmission and

V. Venkataramanan (✉)
Communication Systems, Rajalakshmi Engineering College, Chennai, India
e-mail: venkataramanan32@gmail.com

T. Manikandan · M. Sathish · R. Srilakshmi
Department of ECE, Rajalakshmi Engineering College, Chennai, India
e-mail: manikandan.t@rajalakshmi.edu.in

K. Kalaiarasan
Vel Tech Rangarajan Dr. Sagunthala R&D Institute of Science and Technology, Chennai, India

P. Suresh et al. (eds.), *Advances in Smart System Technologies*,
Advances in Intelligent Systems and Computing 1163,
https://doi.org/10.1007/978-981-15-5029-4_61

reception of the signals. The emerging development towards ease and effectiveness led to the thought of combining microstrip patch antenna into an array of solitary element which could attain radiating tasks and higher gain concurrently.

2 Literature Survey

In [1], microstrip patch antenna with hexagonal shape for Wi-Fi application was outlined and they have implemented a suitable technique in designing the filter. In [2], the approach of integrating microstrip patch antenna array with the Chebyshev filter of order three has been discussed [2] and they have showed the rejection of unwanted frequency signals. Three-element Yagi antenna [3] was designed and resulted in the improved performance characteristics by introducing frequency-domain-based filtering approach. In [4], the miniaturized slot dipole antenna with second-order BPF has been designed and experimented. The printed and upturned L-shaped antenna and the parallel coupled microstrip line sections were introduced [5] which resulted in high bandwidth and flat gain response. The Γ-shaped antenna and the miniaturized coupled line resonator were designed [6] which showed an optimum bandwidth and gain response.

In [7], the metal posts installed into the horn antenna generate the filtering function, and in [8], the coupled cavities were created into a leaky wave antenna which generates filtering performance on the antenna. However in both [7] and [8], the designs have not been implemented using a systematic approach and also not considered much on the specifications of filter and antenna. The additional resistance transformation structure has been configured [9] in between the filter and antenna, and hence, this arrangement correctly helps to integrate both band-pass filter and the antenna. In recent times, many designs of filtering antennas were implemented using the filter synthesis approach [10] and [11]. In [12], the third resonator and output feed line were substituted by the patch antenna and the characteristics of the filter were preserved by considering the external quality factor. In [13], the microstrip patch antenna replaces the final resonator and the load impedance of the BPF.

3 Proposed Work

3.1 Specifications of Single Microstrip Patch Antenna

The schematic layout of single hexagonal-shaped microstrip patch antenna is shown in Fig. 1.

The hexagonal patch dimensions were designed for frequency (f) of 2.4 GHz. Let S be the sides of the hexagonal patch, $\varepsilon_{r(\text{effective})}$ as the effective relative permittivity and b added with c are the assumed value of length of the feed as shown in Fig. 1.

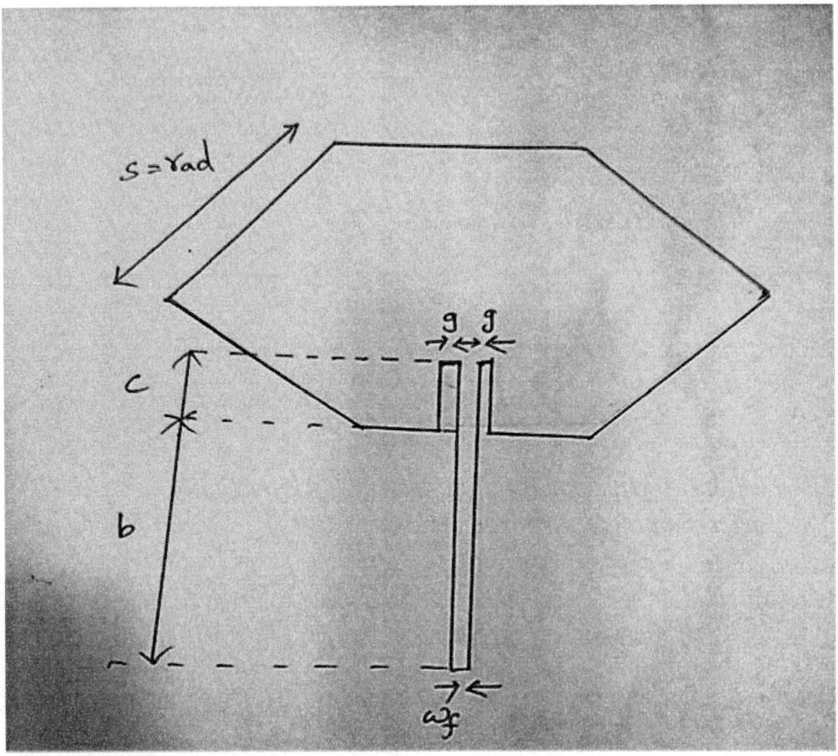

Fig. 1 Hexagonal patch antenna

$$f_r = \frac{1.8412C}{2\pi S\sqrt{\frac{2.598}{\pi}}\sqrt{\varepsilon_r}} \tag{1}$$

$$f_r = \frac{c}{3.1033S\sqrt{\varepsilon_r}} \tag{2}$$

$$S = \frac{c}{3.1033f_r\sqrt{\varepsilon_r}} \tag{3}$$

Based on the above formulas, different parameters of an antenna are given in Table 1.

Table 1 Specifications of microstrip patch antenna

Symbol	Quantity	Values (mm)
Material: FR4_Epoxy		
ε_r	Dielectric constant	4.4
W_f	Width of the feed	2.528
S	Sides of the patch antenna	19.2026
H	Substrate thickness	1.6
G	Gap between feed line and patch	0.3
B	Length of the feed	31.25
C	Inset depth_50 Ω	13.93
a, e	Effective length and width	4.8,4.8

3.2 Design of Microstrip Patch Antenna Array Using Series-Fed and Corporate-Fed

Microstrip patch antennas are popular in arrays as well as the single element. Antenna arrays are highly useful in the enhancement of gain and efficiency. The feeding arrangement of the microstrip patch antenna array can either be done by single elements or multiple lines. Depending on the type of feeding, the array can be classified as series-fed and corporate-fed [8].

The input signal which is fed from one end of the feed network is coupled serially to all the antenna element which was done in series-type feeding network. This method is compact, and therefore, the line losses related with this type of array are less compared to the corporate-fed array. The corporate-fed type antenna array consists of multiple feed lines, and it controls the feed of every element. The layout of series-fed and corporate-fed antenna array is shown in Figs. 2 and 3. An optimum value of distance ($\lambda/2$) is maintained between the two patches.

The length of the feed should be maintained properly because a minute variation may also cause issues in radiation. In Fig. 2, width of the feed is considered as

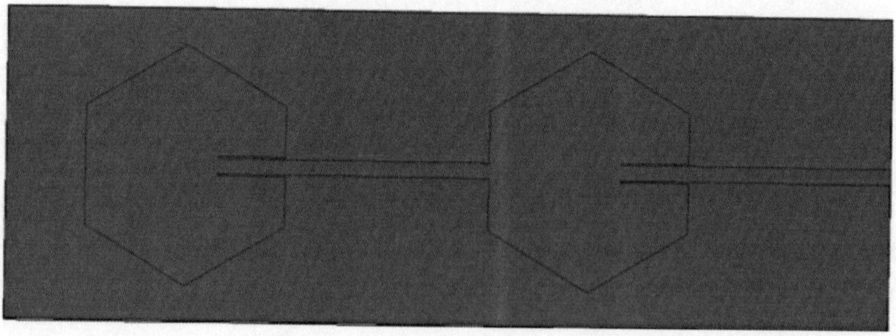

Fig. 2 Microstrip patch antenna array using series-fed

Fig. 3 Microstrip patch antenna array using corporate-fed

2.528 mm in order to maintain the input impedance as 50 Ω. To achieve enhanced results, the gap g between the inset feed line and the patch antenna is considered to be 0.3 mm. In parallel-fed type, to match radiating elements' input impedance, logic of network theory is used. Now, we are going to design the parallel-fed array of hexagonal microstrip patch antenna as shown in Fig. 3. It can be designed either by using the same formula for single hexagon or may be duplicated using the software option, and the input is fed parallel as shown in Fig. 3.

In HFSS, to design the hexagon, the centre position should be (0, 0, 0) and the start position should be (rad, 0, 0) where rad is the radius of the equivalent circle of desired hexagon. In our design, radius that is side of the hexagon is chosen as 19.2026 mm. Number of segments is chosen as 6.

The position of feed should be calculated by,

$$p = \left(-\frac{wf}{2}\right), (\text{rad} - c), 0$$

and total length of the feed is $(c + b)$.

The length of the feed to unite is calculated by the below equation,

$$\begin{aligned} b &= \frac{\lambda}{4} \\ &= \frac{c}{4f} = \frac{3 \times 10^8}{4 * 2.4\,\text{GHz}} \\ &= 31.25\,\text{mm} \end{aligned}$$

The above calculated value 31.25 mm is the optimum value of distance ($\lambda/2$) between the two patches.

4 Results and Discussions

The high-frequency structure simulator (HFSS) was used for the analysis. The various antenna parameters like return loss, gain and radiation pattern were analysed. It is observed that from Figs. 4 and 5 the single microstrip patch antenna is resonating at a frequency of 2.4 GHz with a return loss of −14 dB. The gain and bandwidth are 2.59 dB and 56.5 MHz. As shown in Figs. 6 and 7, the antenna array connected in series-fed is resonating at a frequency of 2.52 GHz and a return loss of −20 dB, which is marginally lesser than the single patch antenna. The gain and bandwidth are 3.34 dB and 58.8 MHz, which are marginally higher than single patch antenna.

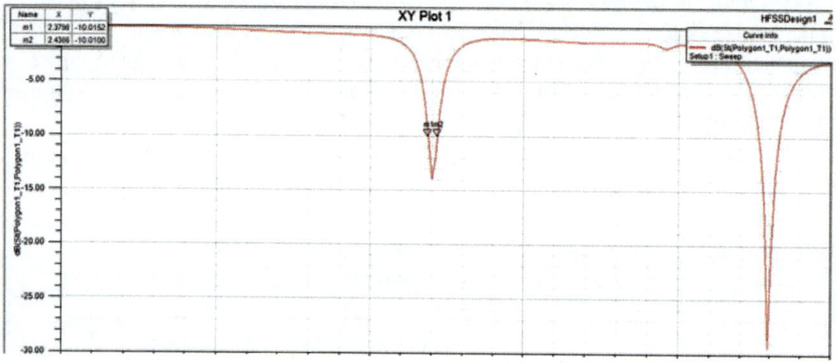

Fig. 4 Return loss of single patch antenna

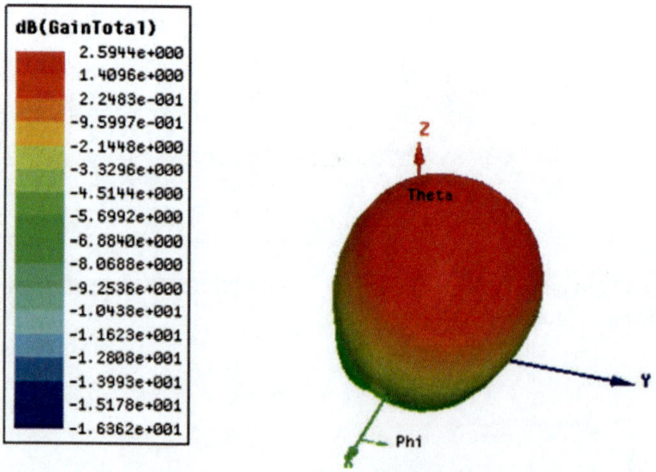

Fig. 5 Gain of single patch antenna

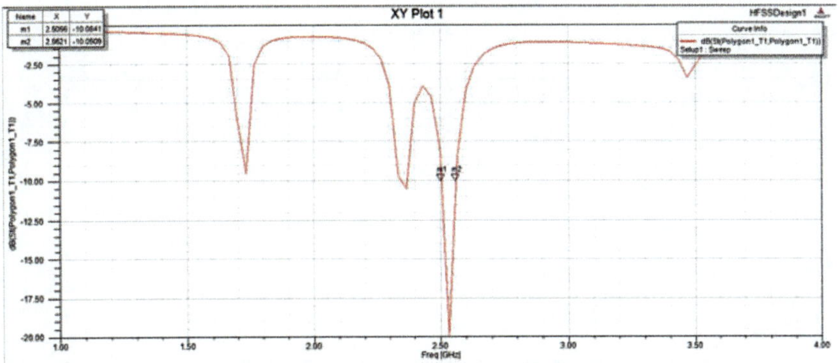

Fig. 6 Return loss of series-fed patch antenna array

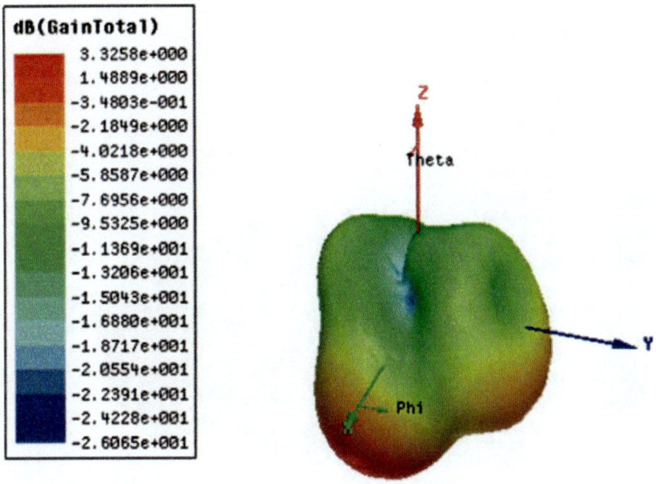

Fig. 7 Gain of series-fed patch antenna array

The antenna connected in parallel-fed manner is resonating at a frequency of 2.37 GHz which is shown in Fig. 8 with a return loss of −34 dB, which is significantly lesser than both single patch antenna and series-fed patch antenna array.

The gain of the corporate antenna array is 3.54 which is given in Fig. 9. The gain is 1.4 times more when compared to the single patch antenna. Further, an optimum bandwidth value of 139.2 MHz is achieved when the antenna array is connected using corporate-fed type.

From the above results, it is clear that corporate-fed array yields reduced return loss, improved gain and bandwidth than the single patch antenna and patch antenna connected in series-fed manner.

The far-field radiation patterns of all three antenna types such as single patch, series-fed antenna array and corporate-fed antenna are given in Fig. 10a, b, c,

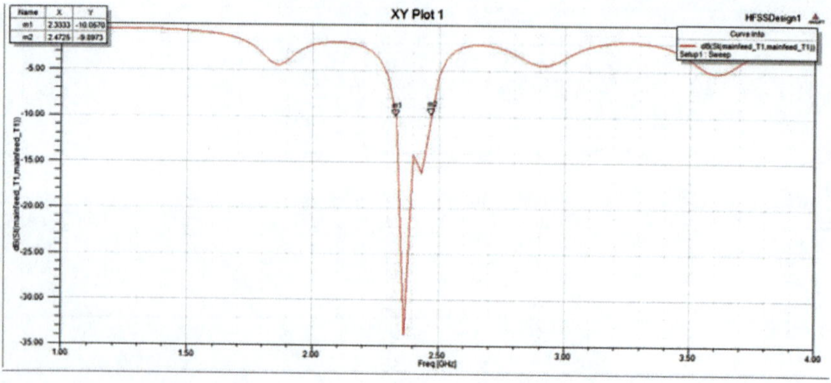

Fig. 8 Return loss of corporate-fed patch antenna array

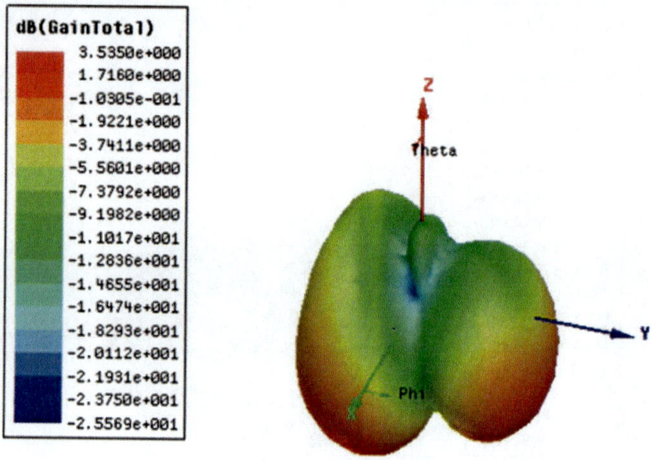

Fig. 9 Gain of corporate-fed patch antenna array

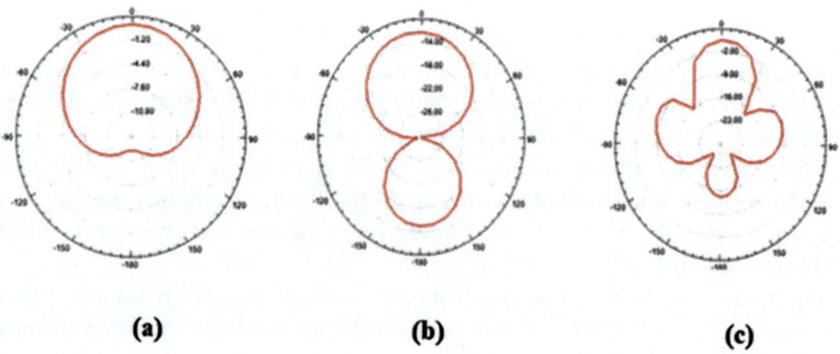

Fig. 10 Radiation pattern

Table 2 Summary of results

Antenna type	Gain (dB)	Return loss (dB)	Bandwidth (MHz)
Single patch	2.59	−14	56.5
Series-fed array	3.33	−20	58.8
Corporate-fed array	3.54	−34	139.2

respectively.

Single patch (b) Series-fed antenna type (c) Corporate-fed antenna array The results obtained from the above analysis summarised below for single patch element, series-fed and corporate-fed patch antenna array in Table 2.

5 Conclusion

This paper clearly explains the design procedure and various performance parameter analysis of single microstrip hexagonal patch antenna and array of hexagonal microstrip patch antennas on FR4 substrate. Based on the simulation results, the antenna parameters such as gain, return loss and bandwidth helps us to identify the effective antenna array which is mostly suitable for ISM band and S-band frequency applications. Thus, the optimum return loss, gain and bandwidth values of −34, 3.54 dB and 139.2 MHz are achieved, respectively, when the hexagonal patch element is connected using corporate-fed array. In future, the band-pass filter may be incorporated with corporate-fed patch antenna array which may yield higher bandwidth and radiation efficiency.

References

1. Anusha, S., Devi, P.: Hexagonal shaped micro-strip patch antennaf or Wi-Fi application. Int. J. Innov. Res. Comput. Commun. Eng. **5**(3), 5655–5658 (2017)
2. Chuang, C., Chung, S.: Synthesis and design of a new printed filtering antenna. IEEE Trans. Antennas Propag. **59**(3), 1036–1042 (2011)
3. Sharipov, Z., Guine, F.: Microstrip frequency Selective surface for use in horn filtenna. In: IEEE Radar Methods and Systems Workshop, pp. 107–109 (2016)
4. Chuang, C.-t., Chung, S.-j.: New printed filtering Antenna with selectivity enhancement. In: Proceedings of the 39th European Microwave Conference New, Issue October, pp. 747–750 (2009)
5. Filtenna, R., Farzami, F.: Reconfigurable linear/circular polarization rectangular waveguide filtenna. IEEE Trans. Antennas Propag. **66**(1), 9–15 (2018)
6. Chuang, C.-T., Chung, S.-J.: A new compact filtering antenna using defected ground resonator. In: Proceedings of Asia-Pacific Microwave Conference 2010, pp. 1003–1006 (2010)

7. Kayser Azam, S.M., Ibrahimy, M.I.: A compact bandpass filter using microstrip hairpin resonator for WLAN applications. In: 7th International Conference on Computer and Communication Engineering (ICCCE), pp. 313–316 (2018)
8. Zhang, S., Zhu, L.: General Synthesis method for Symmetrical Even-order Chebyshev bandpass filter. In: Proceedings of APMC, pp. 667–669 (2012)
9. Yousefi, M., Aliakbarian, H.: Design and integration of a high-order hairpin Bandpass Filter with a Spurious suppression circuit. In: Loughborough Antennas & Propagation Conference (LAPC), pp. 1–4 (2015)
10. Velidi, V., Kumar, M., Subramanyam, A.V.G.: Uniplanar microstrip patch antenna with reduced harmonic radiation using multi-stub filter unit. In: IEEE International Conference on Antenna Innovations & Modern Technologies for Ground, Aircraft and Satellite Applications (IAIM), pp. 1–5 (2017)
11. Srilakshmi, R., Manikandan, T., Banu, S.T.: Performance comparision of S-band antenna array with series fed and corporate fed microstrip array. Int. J. Eng. Technol. **7**(2.33), 1036–1039 (2018)
12. Tawk, Y., Costantine, J., Christodoulou, C.: A varactor-based reconfigurable filtenna. IEEE Antennas Wirel. Propag. Lett. **11**, 716–719 (2012)
13. Chuang, Chao-Tang, Chung, Shyh-Jong: A compact printed Filtering antenna using a ground-intruded coupled line resonator. IEEE Trans. Antennas Propag. **59**(10), 3630–3637 (2011)

Smart Connected Street Lighting System

K. Srinidhi and N. Krishna Prakash

Abstract Streetlights demand for most of the electricity that is generated. An efficient street lighting provides wider opportunities for energy savings. Smart street lighting forms the important and a valuable asset for the city's infrastructure that improves the efficiency of the lighting systems and reduces the overall energy consumption in the city. This paper discusses the design and development of traffic-based street lighting system test bed for smart cities. In addition to vehicle detection-based streetlight control, the setup includes a cloud-based remote monitoring system for fault detection. The system developed is tested assuming a highway environment, and the results show 30% energy saving compared to the conventional system.

Keywords Street lighting system · LED-based streetlights · Traffic adaptive control · Smart cities · Internet of things (IoT) · Cloud

1 Introduction

Street lighting systems are very much essential for every aspect of day-to-day life. Street lighting plays a critical role in the contribution of traffic as well as pedestrian safety. Lighting systems in city environment consume about 50% of the total electricity [1]. Statistical data shows around 27 million streetlights are being lit up the whole night in India and most of them are CFL or sodium vapor lamps. Nowadays, energy demand being comparatively higher than the energy production, the overall energy consumption has to be reduced. An efficient street lighting provides wider opportunities for energy savings in a smart city [2]. There is a greater interest in reducing lighting energy usage through more efficient lighting systems and by introducing smartness to it. Therefore, the government is looking forward to adopt LED technology in order to reduce the energy demand.

K. Srinidhi (✉) · N. Krishna Prakash
Department of Electrical and Electronics Engineering, Amrita School of Engineering, Amrita Vishwa Vidyapeetham, Coimbatore, India
e-mail: ksrinidhi23@gmail.com

P. Suresh et al. (eds.), *Advances in Smart System Technologies*,
Advances in Intelligent Systems and Computing 1163,
https://doi.org/10.1007/978-981-15-5029-4_62

Streetlights that are connected in a network significantly improve the city's lighting system by controlling electricity consumption and reduce the overall operational costs and maintenance costs. Optimization in street lighting is done by several ways like minimizing the overall cost of the system, luminance level, and operational constraints. In addition to camera, sensors such as passive infrared sensors(PIR) and Light-dependent resistors (LDR) are also used for vehicle detection and sensing the surrounding environmental conditions [3–5]. Embedded systems and wireless sensor networks (WSN)-based remote monitoring of street lighting lamps provide savings in terms of the electric power and maintenance costs [6, 7]. Several current balancing techniques and driver circuits are designed for high-power LED lamps to achieve better energy efficiency [8–10].

In a smart city, along with the introduction of IoT (Internet of things), information and communication technologies have also emerged to develop new services. A completely automated system along with remote monitoring of the street poles without any manual intervention based on IoT is designed and found to be more easy and accurate [11, 12]. A smart solution that will collect accurate data of the streetlights, and a detailed analysis of the faults will go a deeper way in making streetlights efficient. Remote monitoring of each and every street pole and identification of faults in the system has to be properly addressed in an Internet of things platform [13, 14]. This paper discusses the development of an efficient LED lighting system which consumes lesser power in comparison with other lighting systems and use of emerging communication technologies in developing a network of streetlights.

2 System Overview and Methodology

The proposed idea of smart connected systems includes vehicle detection, communication between the poles, and identification of faults. Street poles consist of PIR sensors which can detect obstacles up to a maximum distance of about 12 m. It is placed at a certain distance from the ground facing of the road such that it can detect both the sides. Pedestrians are not considered in this design as the system is proposed for highway. Being a Wi-fi-enabled smart city, each pole communicates with the other through the available Wi-fi. The role of embedded system is to measure the overall current that the LEDs consume and calculate the power consumption by the streetlights and prevent its malfunctioning.

In Fig. 1 P1, P2, P3, and P4 mentioned are the poles. The design overview gives a clear description about the street poles placed at the center of the road with the pole controllers and a data concentrator for a set of poles to collect the faulty information from the poles and upload it in the streetlight management system. Light-dependent resistor (LDR) and real-time clock (RTC) together are used to detect the surrounding environmental conditions. During daytime, all the poles are in the sleep mode and once when the surroundings become dark, the poles start glowing at a minimum intensity level. If suppose the vehicle reaches the first pole, the PIR sensor placed at pole 1 senses the motion of the vehicle and glows at the maximum intensity. It then

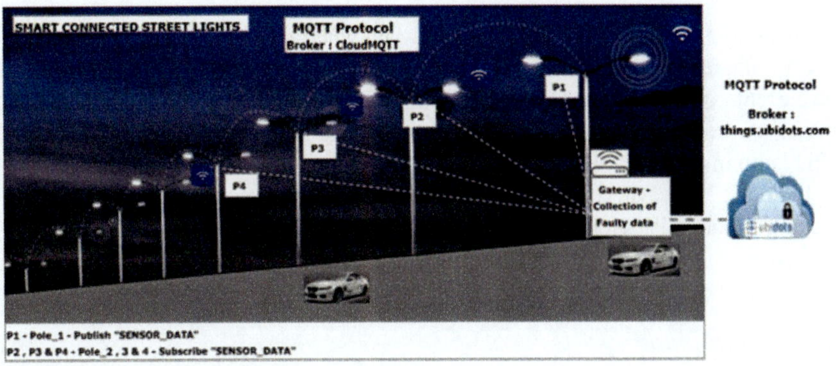

Fig. 1 Smart connected street lighting system

communicates to its set of poles to glow at the same intensity level. Once the vehicle crosses the particular pole, the poles go back to their original intensity level. The sensor value remains high for any number of vehicles that pass by, and so this logic can be applicable even when the vehicle density is high. There can be any number of faults that can occur in a pole. So, the faults that occur in the poles are detected and communicated to the management system through gateway. This prevents the manual intervention.

2.1 *Pole Design*

Individual pole consists of a light-dependent resistor (LDR), real-time clock (RTC), PIR sensor, microcontroller, and a lamp with the driver circuit as shown in Fig. 2. LDR and RTC are used in sensing the surrounding environmental conditions, and PIR sensor is used for motion detection. Once when the vehicle is detected, the lamps

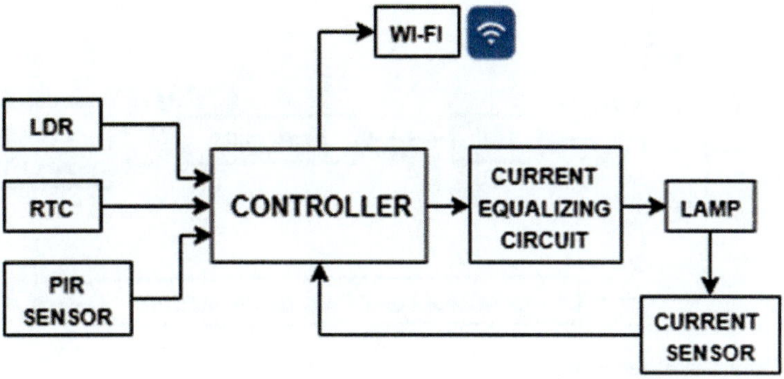

Fig. 2 Pole design

are made to glow at a maximum intensity. And when the vehicle leaves, it is made to glow at a reduced intensity level. The lamp is connected to the controller through the current-equalizing circuit which equalizes the current flow through all the LEDs in a string. Current sensor is used to measure the overall current consumption by the LED lamp in comparison with the other conventional lamps. The controller used in each pole is Node MCU and Raspberry pi is used as a gateway through which the fault information is uploaded into the Ubidots cloud platform.

2.2 Lamp Node

The LED driver shown in Fig. 3 balances the current that flows to a string of LEDs. Without proper driver, there is a chance of LEDs getting overdriven and so lifetimes of the LEDs may be deteriorated causing poor performance or failure. It can also cause uneven luminous outputs from the strings. Different circuits consisting of passive elements like resistors, inductors, and capacitors are available for current balancing in LED strings [8]. The efficient and cost-effective circuit is designed using active devices [10].

The current through MOSFET develops a voltage drop V_a across the source terminal resistance R_S. The voltage V_a is converted into a smooth dc voltage V_b using a low-pass filter. A non-inverting amplifier is used to amplify the voltage V_b

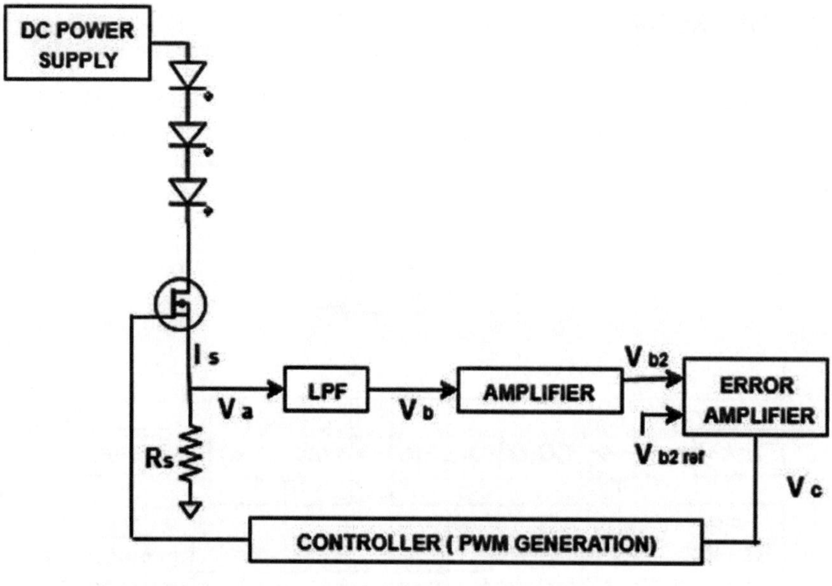

Fig. 3 Current equalizer circuit

and is given to the error amplifier for system stability. The control signal V_c generates PWM signal which is given to MOSFET forming a closed loop.

2.3 Communication Node

In a smart city, Wi-fi facilities are readily available and so it is preferred to use Wi-fi that is already available in the city. The network architecture is shown in Fig. 4, and it follows a star topology. Researches prove that MQTT protocol (also known as publish/subscribe protocol) is one of the most reliable among other protocols and best suitable for IoT-related applications. Hence, it is decided to use MQTT protocol in our proposed design. MQTT Broker is the heart of the protocol which is responsible for exchanging the messages between the client and the server nodes. The broker used here is Cloud MQTT. The presence or absence of vehicle is sensed by the first pole and is communicated to all the other poles in its set.

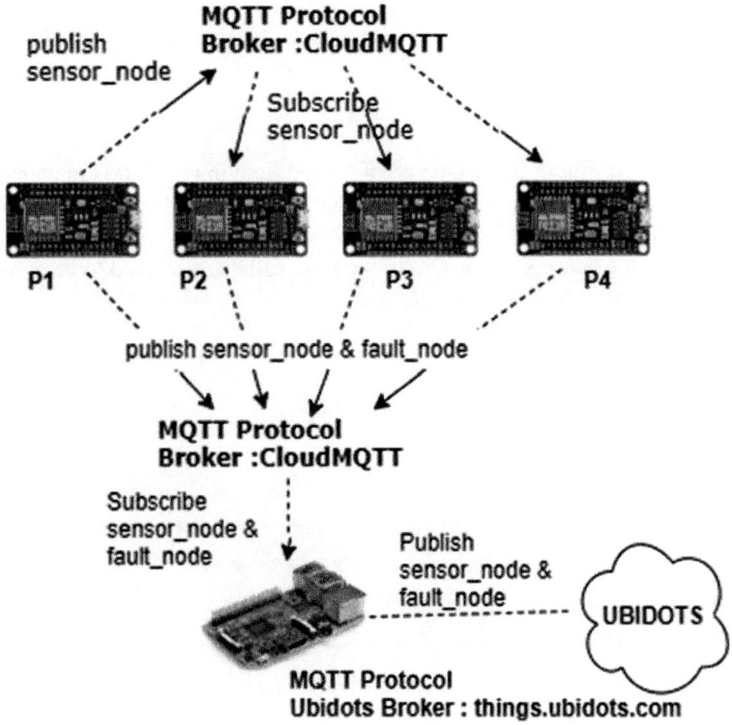

Fig. 4 Network architecture

2.4 *Fault Detection*

For any real-time system, it is essential to ensure the proper functioning of each module. There are many possibilities for occurrence of faults in a system like sensor fault, failure of the hardware components, aging of lamps, communication loss, and improper functioning of the controller. Hence, any type of fault that can occur in the system has to be identified and informed to the management system for the necessary actions to be taken which will become a tedious process and time-consuming when human intervention is involved.

The pole performance is continuously monitored in an IoT platform. Ubidots platform is used to update the fault information. Ubidots IoT platform merges the physical world and the digital world in order to convey the working condition of any number of sensor nodes. It also provides visualization tools for easy understanding. Here, the controls of unlimited number of luminaries are done in a single platform.

3 Implementation Results

3.1 *Driver Circuit*

A scaled-down version of the prototype of a street lighting system is implemented and evaluated for its performance. A 3-W LED lamp is designed for use in light poles. The current equalizer circuit is designed for the lamp which maintains a constant current flow in the lamp. The current through the LED is measured to be constant, and Fig. 5 shows the simulation result.

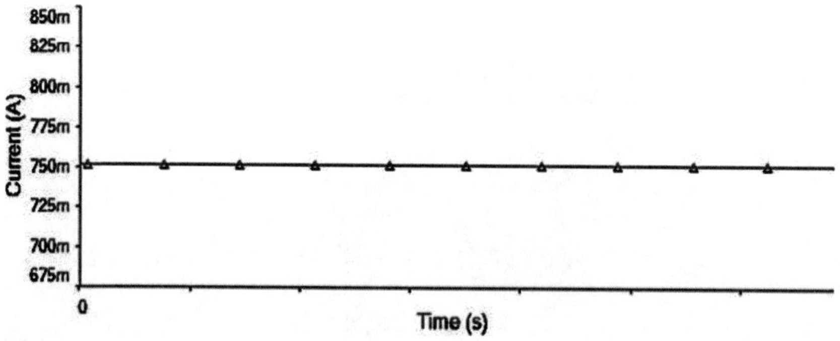

Fig. 5 Simulation of current equalizer circuit

3.2 Development of Test Bed

A complete scaled-down pole implementation is shown in Fig. 6. The test bed of street lighting system consists of four individual poles as shown in Fig. 7. The third pole is the faulty node that shows the fault information being logged into the management system.

When LDR senses the environment to be bright and RTC shows current time as daytime, then the nodes are in sleep mode. When LDR senses the environment to be dark and RTC shows current time nighttime, then the nodes are in active mode. Pole-to-pole communication has two different cases:

Case 1: When no vehicle is present, all the lights remain in dimming condition.

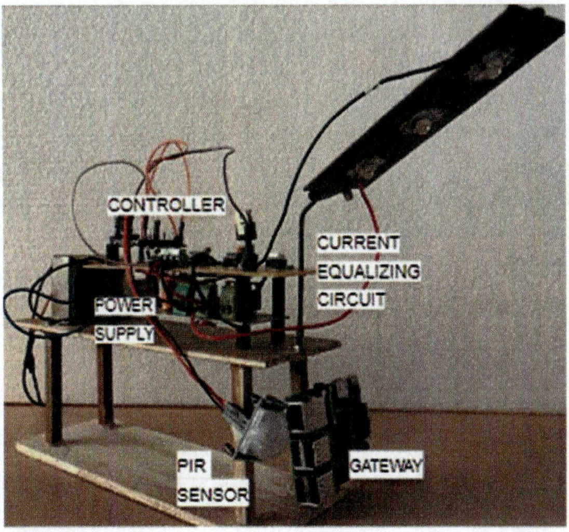

Fig. 6 Pole implementation

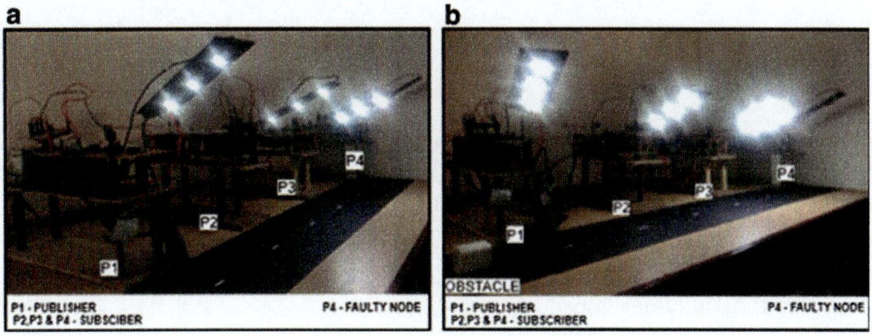

Fig. 7 Streetlight test bed with **a** no vehicle detected, **b** vehicle detected

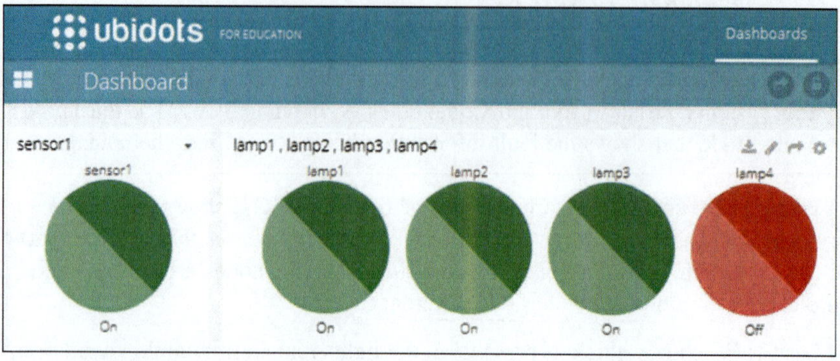

Fig. 8 Ubidots platform

Case 2: The lights glow with a maximum intensity when a vehicle is detected, and it comes back to the dimming state after a particular time interval. This saves the overall power consumption by the streetlights. Figure 7 shows two conditions when vehicle detected and when no vehicle is detected. A highway environment is assumed, and the system is evaluated for different test cases. The results show that around 30% of energy can be saved with the proposed methodology.

3.3 Cloud Platform

Gateway node known as data concentrator collects the fault data from all the poles. Here, Raspberry Pi subscribes for the lamp node fault from the poles and publishes it in the Ubidots platform. The sensor node status and the lamp node status have been collected by the gateway node and are being displayed in Ubidots platform. In the test bed, the fourth pole is faulty and is indicated as red in Fig. 8. This makes the maintenance easier.

4 Conclusion

A prototype that mimics a smart street lighting system in a city is implemented and evaluated for different test cases. This system is extremely efficient under the condition of cost-effective installation and also serves the city infrastructure to reduce overutilization of conventional power. The proposed system works efficiently and economically in all the aspects and can be deployed in any parts of the country which paves way for a fully developed smart city.

References

1. Novak, T., Pollhammer, K.: Traffic-adaptive control of LED-based streetlights [industry Forum] public lighting in urban environments accounts for about 50% of the electricity con. IEEE Ind. Electron. Mag. **9**(2), 48–50 (2015)
2. Yogavani, D., Natarajan, K.P.: Implementation of wireless sensor network based multi-core embedded system for smart city. IJCTA **10**(02), 119–123 (2017)
3. Shahzad, G., Yang, H., Ahmad, A.W., Lee, C.: Energy-efficient intelligent street lighting system using traffic-adaptive control. IEEE Sens. J. **16**(13), 5397–5405 (2016)
4. Mahoor, M., Salmasi, F.R., Najafabadi, T.A.: A hierarchical smart street lighting system with brute-force energy optimization. IEEE Sens. J. **17**(9), 2871–2879 (2017)
5. Gagliardi, G., Casavola, A., Lupia, M., Cario, G., Tedesco, F., Lo Scudo, F., Gaccio,F.C., Augimeri, A.: A smart city adaptive lighting system. In: Third International Conference on Fog and Mobile Edge Computing (FMEC), pp. 258–263. IEEE, Barcelona (2018)
6. du Toit, P., Kruger, C., Hancke, G.P., Ramotsoela, T.D.: Smart street lights using power line communication. In: IEEE AFRICON, pp. 1581–1586. Cape Town (2017)
7. Bhosale, S.A., Ankalkote, T.: Modern street lighting system with intensity control based on vehicle movements and atmospheric conditions using Zigbee. In: International Conference on Information, Communication, Engineering and Technology (ICICET), pp. 1–4. IEEE, Pune (2018)
8. Zhang, R.,Chung, H.S.: Paralleled LED strings: an Overview of current-balancing techniques. IEEE Ind. Electron. Mag. **9**(2), 17–23 (2015)
9. Nair, D.R., Prakash, N.K.: Design and implementation of a wireless sensor network based efficient LED lighting system. In: 6th International Conference on Computing, Communication and Networking Technologies (ICCCNT), pp. 1–6. IEEE, Denton, TX (2015)
10. Lin, C., Hung, T., Wang, C., Pai, K.: A balancing strategy and implementation of current Equalizer for High Power LED backlighting. In: 7th International Conference on Power Electronics and Drive Systems, pp. 1613–1617. IEEE, Bangkok (2007)
11. Sikder,A.K., Acar, A., Aksu, H., Uluagac, A.S., Akkaya, K., Conti, M.: IoT-enabled smart lighting systems for smart cities. In: IEEE 8th Annual Computing and Communication Workshop and Conference (CCWC), pp. 639–645. Las Vegas, NV (2018)
12. Amrutur, B., Rajaraman, V., Acharya, S., Ramesh, R., Joglekar, A., Sharma, A., Lele, A., Mahesh, A., Sankaran, S.: Poster abstract: an open smart city IoT test bed. In: IEEE/ACM Second International Conference on Internet-of-Things Design and Implementation (IoTDI), pp. 323–324. Pittsburgh, PA(2017)
13. Elejoste, P., Angulo, I., Perallos, A., Chertudi, A., Garcia Zuazola, I., Moreno, A., Azpilicueta, L., Astrain, J., Falcone, F., Villadangos, J.: An easy to deploy street light control system based on wireless communication and LED technology. Sensors **3**(5), 6492–6523 (2013)
14. Daely, P.T., Reda, H.T., Satrya, G.B., Kim, J.W., Shin, S.Y.: Design of smart LED streetlight system for smart city with web-based management system. IEEE Sens. J. **17**(18), 6100–6110 (2017)

Identity Management Using Blockchain Network for Fail-Safe e-Governance

J. N. Benedict, S. Udhayakumar, B. R. Vikram, and C. Vignesh

Abstract One of the major problems in today's society is the tampering and altering of government records such as police case information, Aadhar card information, driving license and using such false information for obtaining various benefits such as government jobs, passports, visas, and other organizations requiring trust. This project proposes a solution to this problem using blockchain implementation. The idea is to create a block for each individual and branch their various information in these blocks and chain them together. These blocks use hashing and digital signatures as a primary security requirement. Hash values of these blocks are created using the entire contents of the block, and each block contains the hash of previous block, thereby creating a chain. The various hashing algorithms used are SHA-256, and the public-key cryptography is using RSA or ECC algorithm. If someone tries to tamper with the contents of the block, the hash values change and get disconnected from the chain. Every operation on the each of the blocks will be recorded with a time stamp, and it cannot be altered. Hence, this permissioned blockchain network of all the individuals can be used to prevent the tampering and altering of various government records such as police case records, passport records, and PAN and help various organizations such as visa verification and income tax department. This blockchain

J. N. Benedict · B. R. Vikram · C. Vignesh
Department of Computer Science and Engineering, Rajalakshmi Engineering College, Chennai, India
e-mail: mailtoudhay@gmail.com

J. N. Benedict
e-mail: benedict.jn@rajalakshmi.edu.in

B. R. Vikram
e-mail: vkrmbr13@gmail.com

C. Vignesh
e-mail: vigneshchand19@gmail.com

S. Udhayakumar
Saveetha School of Engineering, Saveetha Institute of Medical and Technical Sciences, Chennai, India

P. Suresh et al. (eds.), *Advances in Smart System Technologies*,
Advances in Intelligent Systems and Computing 1163,
https://doi.org/10.1007/978-981-15-5029-4_63

network provides a solution for all the existing problems in tampering and altering of government records.

Keywords Blockchain · Smart contracts · Identity management · Secure e-governance · Ethereum

1 Introduction

All the traditional information systems use a database in the back-end of its application. Apparently, it is quite easy for people to change or alter that information without leaving traces of those changes made. Most of these traditional information systems are centralized. They have a client–server relationship model. Therefore, it is easy for offenders to attack these centralized systems. Offenders just have to attack this centralized server to change or alter their records and hide their tracks. This happens to be an easy task with all the latest attacking tools and technologies. Even with a good non-breakable security firewall for the information system, there is always the problem that occurs because of corrupted employees who are working inside the organization. Offenders can just bribe the employee to change their records.

Most of the governmental organizations are still using paper-pen-based records. Some have evolved to the digital age, but still they have poor security mechanisms. Adopting to good security mechanisms is costly and time-effective. Most of the systems are centralized, and therefore, there is always the problem due to 'Single point of failure.' And there is always the problem of a corrupted employee inside the organization. Therefore, what is required is a system that will not allow offenders to change or alter the records and even if they manage to alter those records, the change of records should be transparent. They must not be able to hide their tracks. And also the system must be cost-effective and feasible. This project proposes a solution to this problem by using a distributed application with blockchain as the back-end. The proposed system is distributed, and therefore, there will be no single point of failure and offenders cannot break the system by attacking at a single server.

The proposed system uses the new evolving technology—'Blockchains' to replace the traditional databases as the back-end of applications. Traditional databases are easily breakable and tamperable whereas, blockchain uses a wide variety of security mechanisms to provide effective security as well as proving to be very cost-effective. Blockchain is a distributed ledger consisting of an increasing list of records and blocks which are associated to its adjacent blocks using cryptographic mechanisms. Each block in the blockchain consists of a cryptographic hash of the preceding block, a time stamp, and the transaction information. The cryptographic hash is computed by using various efficient hashing algorithms such as SHA-256 and Keccak-256. The hash is computed for the entire contents of the block, and the next block in the chain contains this hash value. Therefore, even the slightest change in the block results in abundant change in the hash value which will not be the same as stored in the next block. Hence, the attacker should change the hash values in all the blocks in order to tamper and get away without any traces. In a blockchain with thousands of blocks or records, it is not feasible to change all the block hashes in order to tamper one record. Therefore, blockchain technology is very much resilient and strong to alteration of data of any sorts. It is like a distributed ledger that is open to all and that allows to record transactions between two entities in an capably and in a provable and a permanent way.

Every block or record in the blockchain is associated with a time stamp. Hence, the time of creation of every block will be known to the system. The time stamp is also included when computing the hash of a block. Therefore, changing the time stamp will also result in removing the block out of the block chain. Blockchain consists of transaction data. We will store the required data in the form of these transaction data. In order to change or alter the transaction data, we need to create a new block to store those new transaction data. The old data will never be removed from the block chain. Only then there will be a link of some sorts between the old and new transaction data.

Blockchain is a peer-to-peer network that allows for inter-node communications, and it is decentralized. Every node in the block chain network has the copy of all the blocks in the blockchain, and the blocks are distributed throughout the network. Therefore, the offenders cannot attack a single node and alter the data. Whenever a new data or alteration of some sorts is being done, all the nodes involved in the blockchain network must provide consensus to add the block to the blockchain. Here, no centralized 'official' copy of the blockchain exists and no node in the blockchain is trusted more than any other nodes. Blocks in the blockchain have valid transactions that are held in batches which are encoded into a Merkle tree. A Merkle tree is nothing but a hash tree which is a tree structure with root node, intermediate nodes, and leaf nodes. The root node consists of a hash value which is obtained by hashing all the nodes in its branches. Whenever a change in hash is made in a particular leaf node, only the nodes in the path from that leaf node to root node need to update their hash values.

The average time that takes for a network to generate or add a new block to the blockchain is termed as the block time. If block times are shorter, then faster the transactions in the network. Block time can be specified in most of the existing block chain technologies. Most of the popular applications of blockchain use a greater block time so that attackers would not be able to attack and recreate the blocks instantly. Larger block time is the main reason that makes attackers infeasible to alter all the blocks in the blockchain.

There are three types of blockchains, viz, (1) blockchains with only cryptocurrency, (2) blockchains with cryptocurrency and business logic, and (3) blockchains with only business logic. There are three categories of blockchain, viz (1) public, (2) private, and (3) permissioned. This project uses blockchain only with business logic with no cryptocurrency, and it is a private blockchain. It uses Ethereum blockchain network to implement the distributed application. Ethereum blockchains use smart contracts to execute the business logic.

2 Literature Review

Blockchain is an evolving technology. It was first used by Sakoshi Nakamoto in the Bitcoin technology which is an electronic currency [1]. Bitcoin is an e-cash or a pure peer–to–peer version of cash in electronic form that allows transactions to make online without the need for an intermediate organization such as the bank. Bitcoin or the cryptocurrencies will disrupt the use of banks in the future. Bitcoin is the first to use the blockchain technology. This paved the way for many other innovations and ideas [2]. Bitcoin is purely for cryptocurrency where many other technologies were developed for implementing business logic along with the cryptocurrency and also for implementing only business logic without cryptocurrency.

The leading blockchain technologies are Ethereum and the IBM Hyperledger [3]. Ethereum uses smart contracts to implement the business logic which are deployed on the blockchain using the popular Web3 standard. Hyperledger provides a variety of frameworks and tools for various businesses to implement their own private blockchain. Since blockchain is an evolving technology, most of these frameworks and tools are in incubation stage with some out for production and beta testing [4]. Most of these frameworks and tools get regular updates and patches. Hyperledger allows only business logic without any cryptocurrencies. The smart contract in Hyperledger is called chain code [5].

3 Proposed Fail-Safe e-Governance Model

This project uses the Ethereum blockchain platform to store the data of various individuals and to run the smart contracts, which are applications that run as exactly programmed by the programmer. These are the logics that will be used to store, retrieve, and perform operations on the data stored in the blocks. The model consists of a Web user interface that allows the organization to enter the details of the individuals. The various details that are used in this project are Aadhar card number, name of the person, and phone number of the person. These details are entered by the organization in order to give each individual a digital identity by creating a block for each individual (Fig. 1).

The logic for storing the details of the individuals is written as smart contracts [6]. Smart contracts are a piece of code that is deployed on the blockchain nodes. These smart contracts are written using the programming language Solidity, and these smart contracts are deployed in the block chain using Web3 JavaScript Ethereum API's [7].

The model consists of another Web user interface that allows the police organization to search for an individual by entering the Aadhar number and then adding the criminal cases that are filed on that particular individual. It also has a feature that allows the organization to enter the Aadhar number and retrieve all the transactions that were performed on the blocks that correspond to that particular Aadhar number.

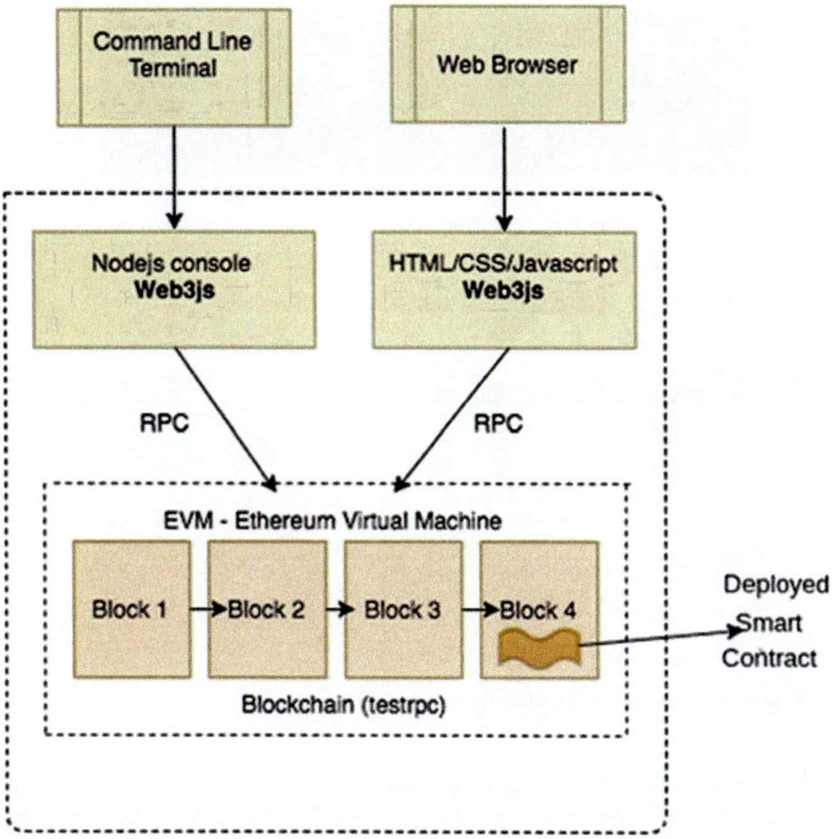

Fig. 1 Architecture of distributed application using Ethereum blockchain network

In laymen terms, it retrieves all the cases that are filed on the individuals with that particular Aadhar number.

3.1 *Ganache-Based Ethereum*

This project uses Ganache tool to run a blockchain in the local system. Ganache allows us to instantly create a personnel Ethereum blockchain network in our local system which can be used for testing our smart contracts, execute the commands, and inspect the state of the blocks in the blockchain. Ganache shows the number of blocks in the network, the list of all the transaction, and all the logs in the network. It provides ten accounts with 100 ethers which can be used in our distributed application. Ganache provides a port number which can be used to connect the Ethereum blockchain that in running using Ganache to connect to the Web user interface (Fig. 2).

Fig. 2 Ganache tool running Ethereum blockchain

The transactions section shown in the Ganache tool shows all the transactions in the blockchain along with the hash value. It shows the time stamp in the particular transaction along with the block number. This helps us when we are testing out code and logic. The logs section shows all the logs from starting of the Ganache tool. It is very helpful in tracking the events and debugging the logic in our smart contracts. The blocks section shows all the individual blocks with block number and the associated hash values.

This project uses Truffle IDE; truffle manages the smart contract life cycle management by taking care of managing the smart contract artifacts, custom deployments, and linking of libraries. The truffle framework allows to initialize a project, which creates all the folders and files required for a distributed application, compile the smart contracts, test the smart contracts, and migrate the smart contracts which deploys the smart contracts on the blockchain. It also provides many other commands and features that can be used for developing a distributed application.

The various truffle commands are:

1. truffle init
2. truffle develop
3. truffle compile
4. truffle test
5. truffle compile.

Truffle develop runs a local blockchain network in the system. It provides various keys and neumonics that can be used when writing smart contracts. It also provides a port number that can be used to integrate with a Web user interface.

Truffle compile is used for compiling the smart contracts that are inside the contracts folder which are written using solidity. After compiling, the artifacts are written to the build folder in JSON format which will be used for executing the code.

Truffle test allows to write programs using solidity or JavaScript to test the smart contracts. Truffle migrate is used to deploy the smart contracts in the blockchain using Web3 JavaScript.

```
1_initial_migrations.js :
    const Migrations = artifacts.require("Migrations");
    module.exports = function(deployer) {
      deployer.deploy(Migrations);    };
2_deploy_contracts.js :
    var Identity = artifacts.require("./Identity.sol");
    module.exports = function(deployer) {
    deployer.deploy(Identity);
    };
```

These two JavaScript programs are used to deploy the smart contract in the blockchain. Web3 JavaScript Ethereum API's are used for building the distributed application using Ethereum blockchain. Web3.js is a collection of libraries that allow the application to interact with the Ethereum blockchain network. Web3 js is used to deploy the smart contract in the blockchain. The smart contracts use solidity code to store the data on the blockchain. Then, we use Web3 js to interact with the blockchain from the Web user interface to retrieve the blocks and display them in the Web page (Figs. 3 and 4).

The following code is used to connect to the Web provider:

```
if (typeof web3 !== 'undefined') {
        web3 = new Web3(web3.currentProvider);
    } else {
        // set the provider you want from Web3.providers
  web3=newWeb3(new Web3.providers.HttpProvider("http://localhost:7545")); }
```

The various Web3 js APIs that we will use are web3, eth, net, utils., etc. These APIs provide a rich set of features for admin, debugging, mining, etc.

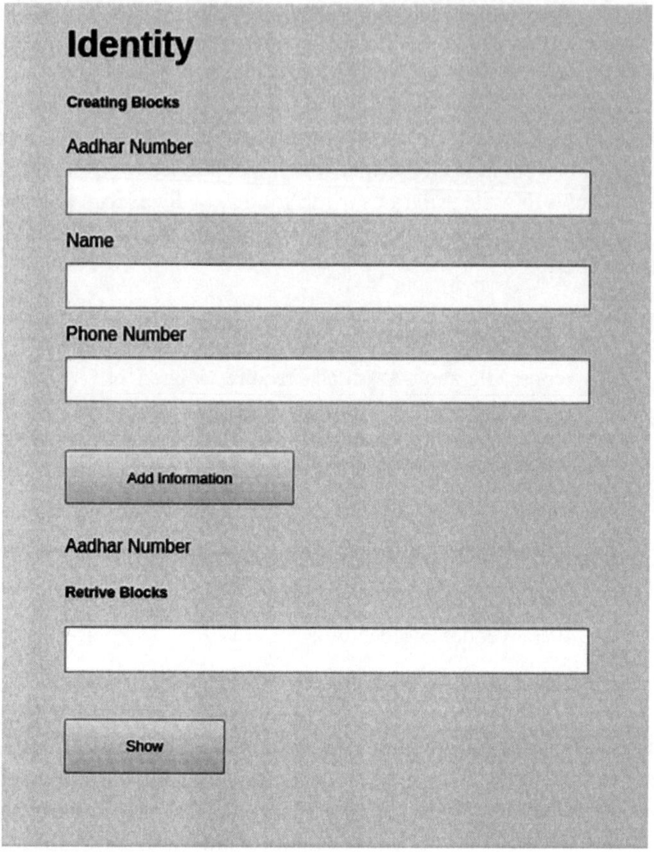

Fig. 3 Web user interface for collecting Aadhar details

3.2 Algorithm

The following algorithm is used by the smart contracts to get the details from the organization and store them in the blockchain:

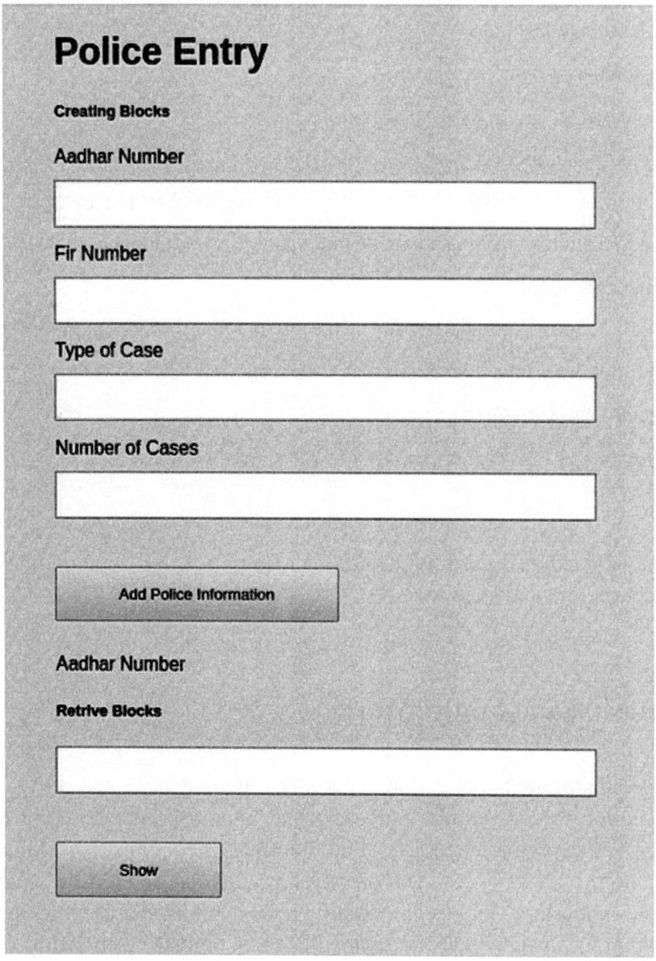

Fig. 4 Web user interface for entering police case details

```
struct Identity {
  string aadhar;
  string name;
  string phonenumber;  }
maapping(uint => Identity) identityInfo; // Mapping integers to structure
function getDetails( string _aadhar, string _name, string _phonenumber){
if(_aadhar is unique) {
  if(length(_aadhar)=12 and length(_name)>0 and length(phonenumber)=10 ){
hash_the_details_using_keccuk256;
store_details_in_structure; }
else revert(); }
else revert(); }
```

The same algorithm is used by the contract for police department logic with the appropriate structure and state variables.

4 Conclusion and Future Work

The proposed distributed application using blockchain as back-end proves to be an effective solution for preventing tampering and alteration of records or case file. Blockchain is an evolving technology, and therefore, the application of this revolutionary technology is very vast. The proposed model can be used by other similar organizations that want to replace the existing traditional databases with more secure and transparent mechanism that would not allow any tampering or frauds. Examples of where the proposed model can be used are income tax department, visa department, driving license department, and many more. This project only tries to solve the problems related to police cases and individual identities. But the proposed model can be applied to solve a wide range of existing problems ranging from markets to banks. Block chain can do for business what the Internet did for communication. It is called the next generation of the Internet. In future, every application will communicate and coordinate using blockchains.

References

1. Nakatomo, S.: Bitcoin: a peer-to-peer electronic cash system. https://bitcoin.org/bitcoin.pdf (2009)
2. Christidis, K., Devetsikiotis, M.: Blockchains and smart contracts for the Internet of things (2016)

3. Wang, S., Yuan, Y., Wang, X., Qin, R.: An overview of smart contract: architecture, applications and future trends. In: IEEE Intelligent Vehicles Symposium (2019)
4. Cai, W., et al.: Distributed applications: the blockchain empowered software system. IEEE Access (2018)
5. Wohrer, M., Zdun, U.: Smart contracts: security patterns in the Ethereum ecosystem and solidity. In: International Workshop on Blockchain Oriented Software Engineering (IWBOSE) (2018)
6. Alharby, M., van Moorsel, A.: Blockchain-based smart contracts: a systematic mapping study. arXiv preprint arXiv:1710.06372 (2017)
7. Zhang, P., White, J., Schmidt, D.C., Lenz, G.: Applying software patterns to address interoperability in block chain based healthcare apps. arXiv preprint arXiv: 1706.03700 (2017)

Predictive Education—from Idea to Implementation

V. Sathya Durga and J. Thangakumar

Abstract Predictive education is a new term coined, which means using machine learning methods and building prediction models that can be used in the educational sectors. Using datasets collected from schools and colleges, prediction model is built and implemented. These structures can be used in the day-to-day activities of an educational institute. By using these prediction structures in routine, quality of the education process will improve definitely. This paper discusses the predictive education, and few experiments on educational datasets are performed.

Keywords Prediction · Performance prediction · Predictive education

1 Introduction

Prediction is an advanced form of data analysis activity, which involves forecasting future events with data available in hand. Almost all industries use prediction for their effective functioning. Health industry, telecommunication, retail, and finance sectors are some industries to name few. Predictive education is a new concept which is about using machine learning and prediction models to forecast the performance of the students [1].

Machine learning is an important category of algorithms which allows computers to learn by themselves, take intelligent decisions, act in accordance to the protocol, correct their actions, and learn from their mistakes. By using machine learning in educational sector, we can revolutionize and pave way for smarter and intelligent educational system.

V. Sathya Durga (✉) · J. Thangakumar
Hindustan Institute of Technology and Science, Padur, India
e-mail: sathyadurga.v@gmail.com

P. Suresh et al. (eds.), *Advances in Smart System Technologies*,
Advances in Intelligent Systems and Computing 1163,
https://doi.org/10.1007/978-981-15-5029-4_64

2 Related Works

Srinivas and Ramaraju [2] implement an educational model. They use naïve bayes algorithm in their work. In their work, they use student's dataset with 257 students' record. They use two feature selection techniques in their work. The accuracy of their work is 84% [2]. Okubo et al. work and develop an academic prediction system. They use neural network-based prediction in their work. With data from online learning system, they predict the grade of students. The accuracy of their work is 90% [3]. Karishma and Swati [4] uses decision tree algorithm in their work. They use the decision tree algorithm ID3 and C4.5 algorithm. They work on 200 academic datasets. The accuracy of their research work is 98.5% [4].

3 Materials and Methods

In this paper, we use academic datasets of students. Using the dataset, we plot the confusion matrix, ROC curve, and scatter plot. Weka and MATLAB2018a were used in this research work.

4 Results and Discussions

Using Weka 3.8 and linear regression classifier, four types of error were calculated. The results show that for the selected academic dataset mean absolute error is 4.84 and root-mean-square error is 6.90 (Fig. 1).

Correlation coefficient	0.9287
Mean absolute error	4.8499
Root mean squared error	6.9057
Relative absolute error	34.1793 %
Root relative squared error	36.8191 %
Total Number of Instances	200

Fig. 1 Results of linear regression

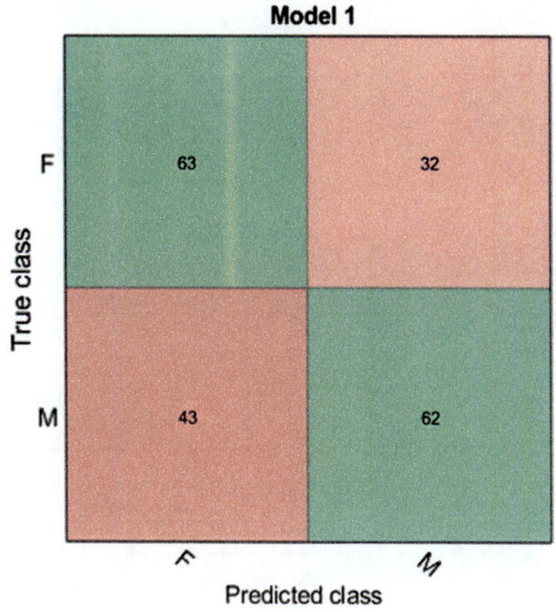

Fig. 2 Confusion matrix

4.1 Confusion Matrix

The confusion matrix is one of the important metrics. Confusion matrix of the dataset is plotted. In this dataset, we obtain higher results for true positive values. If true positive is higher, it is meant that prediction values are correct (Fig. 2).

4.2 ROC Curve

The ROC curve is plotted. The AUC value of the ROC Curve is 0.68. From the ROC value, it can be inferred that the prediction model is good in predicting performance (Fig. 3).

4.3 Architecture of the Prediction Model

The detailed architecture of the prediction model is shown in Fig. 4.

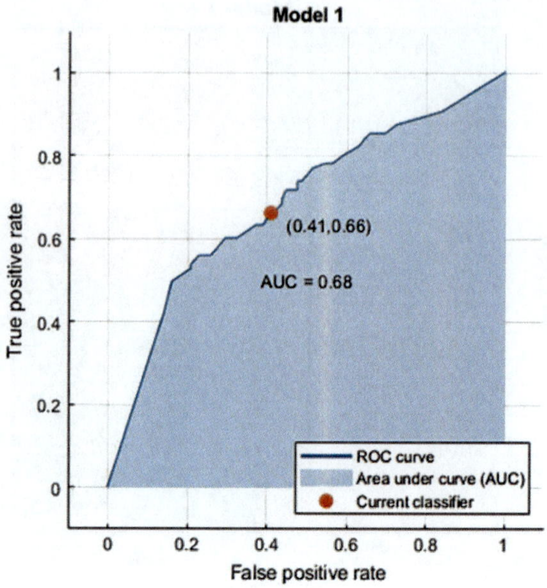

Fig. 3 ROC curve

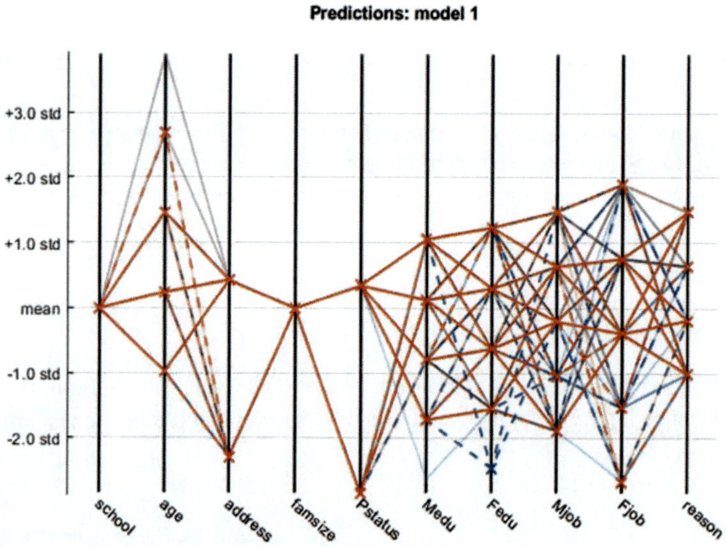

Fig. 4 Architecture of the model

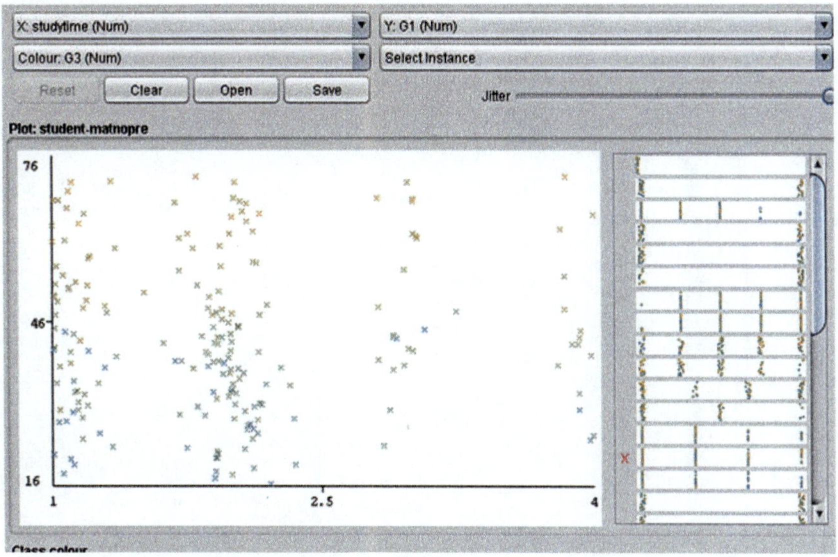

Fig. 5 Scatter plot of the model

4.4 *Scatter Plot of the Dataset*

The scatter plot of the dataset can be seen in this diagram.*X*-axis is the study time, and *y* axis is the grade obtained by the students. It can be inferred that students who studied more than 2.5 hours achieved higher grades (Fig. 5).

4.5 *Heat Map of the Dataset*

Figure shows the heat map of the dataset (Fig. 6).

5 Conclusion

In this paper, we have taken educational dataset and performed experiments on those datasets. We have used linear regression classifier, and the root mean square of the dataset is 6.90. We have plotted the confusion matrix and ROC curve of the dataset. Scatter plot and heat map of the dataset are plotted. Finally, it can be concluded that if sincere effort is taken to implement prediction models in educational domain, without any doubt we can redefine the education domain as a whole. Predictive education will pave way for smarter systems and will benefit the students, teachers all those integrated with the educational sector.

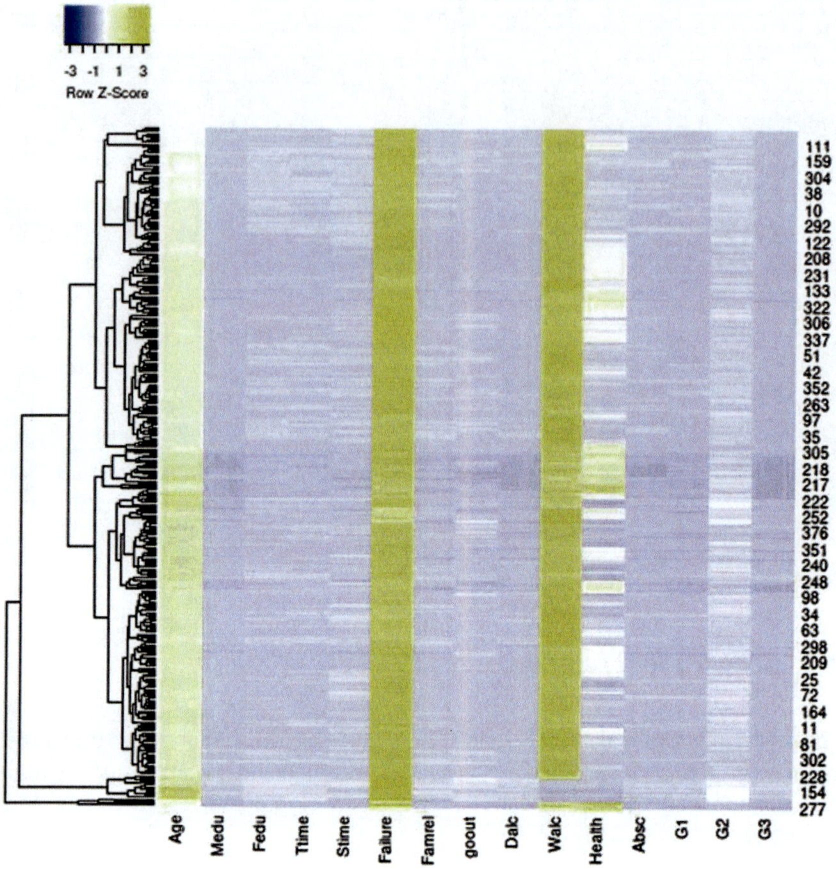

Fig. 6 Heat map of the model

References

1. Why education is important in life? https://www.reference.com/education/education-important life2c3b80038e953b9f. Last accessed 20 Mar 2019
2. Srinivas, A., Ramaraju, V.: Detection of failures by analysis of student academic performance using Naïve Bayes classifier. Int. J. Comput. Math. Sci. **7**(3), 277–283 (2018)
3. Okubo, F., Yamashita, T., Shimada, Ogata, H.: A neural network approach for students' performance prediction. In: Proceedings of the Seventh International Learning Analytics & Knowledge Conference, pp. 598–599, Canada (2017)
4. Karishma Bhegade, B., Swati Shinde, V.: Student performance prediction system with educational data mining. Int. J. Comput. Appl. **146**(5), 32–35(2016)

IoT-Based Intelligent Healthcare Module

Meenakshi Saji, Madhusudhan Sridhar, Anitha Rajasekaran, Rohan Akshai Kumar, A. Suyampulingam, and N. Krishna Prakash

Abstract Technology advancements in healthcare give us a platform to address ever-changing patient requirements and healthcare reforms. Remote Health Monitoring Systems, which work in real time, have been identified as a solution to address the issue that current healthcare facilities may not be able to handle huge crowds of people in an efficient manner. This paper discusses the design and development of a system for real-time monitoring of health and important body parameters like blood pressure, body temperature and pulse rate, using sensors connected to a system on chip. The data is stored in cloud using Internet of things, and the collected along with preliminary analysis is communicated to the user, as well as healthcare providers, and next of kin, alerting them to any possible emergencies.

Keywords Real-time health monitoring system · Internet of things · Intelligent healthcare unit · Remote patient monitoring · Cloud

1 Introduction

The rapid rise in population calls for better healthcare facilities, and the current systems might not be feasible to treat the horde of people who will need immediate healthcare. Technology advancements in healthcare give us a platform to address ever-changing patient requirements and healthcare reforms. A real-time health system (RTHS) represents a care delivery system, which will help providers implement collection of relevant information from different sources (e-records, devices, applications), analysis on the data and low-level diagnosis. Remote Health Monitoring Systems, which work in real time, have been identified as a solution to address the issue that current healthcare facilities may not be able to handle huge crowds of people in an efficient manner. The monitoring of different health

M. Saji · M. Sridhar (✉) · A. Rajasekaran · R. A. Kumar · A. Suyampulingam · N. Krishna Prakash
Department of Electrical and Electronics Engineering, Amrita School of Engineering, Amrita Vishwa Vidyapeetham, Coimbatore, India
e-mail: madhus.sri@gmail.com

P. Suresh et al. (eds.), *Advances in Smart System Technologies*, Advances in Intelligent Systems and Computing 1163,
https://doi.org/10.1007/978-981-15-5029-4_66

parameters doesn't necessarily need to be a process which needs a hospital or the presence of a medical professional. It would ease the workload on hospitals and medical professionals to automate, or make remote, the process of measuring health parameters.

The monitoring of general health parameters such as Blood Pressure, ECG, Body Temperature, Pulse Rate, Respiratory Rate, EMG, Weight, Height, Oxygen Saturation and Galvanic Skin Response do not need supervision as they do not involve invasive methods of measurement. Measurement of parameters like sugar level in blood (glucose) and haemoglobin content involve invasive methods which need the presence of authorized personnel. Parameters which involve non-invasive methods, when measured locally in the comforts of one's home, can help eradicate some of the issues the common people face today when it comes to healthcare, like availability to healthcare, affordability and accessibility. The technology used for that is commonly called remote health monitoring system (RHMS) or remote patient monitoring (RPM). RPM will also reduce healthcare delivery costs, which would be a boon to the people who aren't able to afford healthcare now. This is due to the use of sensors, control devices and cloud technology. The patients monitored will also have the additional comfort that they will get the necessary support immediately if any problem arises. RPM will also reduce the number of emergency hospital visits, and the duration of hospital stay, as some problems can be diagnosed beforehand. IoT provides low-cost solutions to remote health monitoring, and it can help detect signs of deteriorating health earlier, enabling earlier responses and treatment [1].

Various patient monitoring systems were implemented for measuring parameters like temperature, pressure, heart rate, ECG etc. [2–4]. Remote monitoring systems were also implemented using wireless sensor networks and IoT [5–7].

This paper discusses the development of a system designed for real-time monitoring of health and important body parameters (blood pressure, body temperature, and pulse rate), storage of the data in cloud using Internet of things, and analysis of the stored data for low-level diagnosis and plotting.

2 Background Study

2.1 Temperature Measurement

The temperature sensor used by most RHMS is dependent on the measurement of skin temperature. This temperature has significant variation from core body temperature, as skin temperature has direct correlation to surroundings and environmental factors. The core temperature measurement requires invasive methods. The sensor requires to be ingested for accurate measurement, which is not a safe option for many. Human body temperatures span a wide range. Even though skin temperature under extreme environmental conditions may drop to 0° Celsius, healthy core temperature is around 37 °C [8]. While many attempts have been made to derive the core body temperature from the skin temperature, it is not practical for continuous monitoring. Sensors like DHT11 and LM35 have been used extensively in projects for measurements for skin temperature [2, 5]. But measurement of core temperature under the tongue requires a probe. This can be achieved using a DS18B20 sensor.

2.2 *Blood Pressure Measurement*

The most common blood pressure measurement device is a sphygmomanometer, consisting mainly of a cuff, which is inflated via a manual bulb, and deflated by way of a valve. In the system implemented by Zhang Jin-ling et al. [9], the cuff control is done by using an MCU. A filter circuit was designed to capture the cuff pressure signal and the pulse wave signal from the noise signals, to adjust them to the appropriate signal's levels and input to the A/D conversion circuits [10, 11]. The systolic and diastolic pressures were computed from the peak amplitude values [12, 13].

2.3 *ECG Measurement*

ECG is commonly measured by use of electrodes which record the electrical activity generated by heart muscle depolarizations, which propagate in pulsating electrical waves towards the skin [14, 15]. The most common implementation of ECG consists of 3 electrodes (neutral, positive and negative) which are called Right Arm, Left Arm, and Right Leg, which are placed appropriately on the body. AD8232 is a heart rate monitoring board which can be used for integrating our ECG module with the control module.

3 Proposed Methodology

The proposed system aims to create a remote health monitoring system that consists of data acquisition using sensors, graphical user interface, Internet of things (IoT) integration, and analysis of data for low-level diagnosis and plotting. Sensors are interfaced to a microcontroller/microprocessor and required parameters are obtained from the user as shown in Fig. 1. The health parameters acquired through the module is sent to cloud storage, which is followed by analysis of the data, and generation of relevant statistics and reports. With this analysis, possible critical conditions are identified, and concerned contacts are notified, so that the correct course of action can be taken.

The required sensors are connected to a Raspberry Pi through an external ADC. User interface can be facilitated either via a touchscreen and display, or a module and display, which provides easy interaction for the user with the sensors and enables him/her to view the generated results. The report may also be available to the user on their iPhone/Android devices via an application or mail. Sensor data is acquired using sensor modules designed for the purpose. The data acquired is send to cloud, as historic data is required for a thorough analysis. Certain files are also stored locally, to prevent loss of data.

4 Hardware Implementation

Figure 2 shows the blood pressure measurement circuit. The cuff is inflated and deflated with the help of motor and valve controls. The appearance and disappearance of oscillations signify systolic and diastolic pressures respectively. MPX2040 is chosen as the pressure sensor for this module, with the output voltage linearly related to the pressure measured. A filter is imposed, to allow only signals within frequency range of 0.338 Hz to 6.63 Hz. The DC component indicates the pressure inside the handcuff and the AC component indicates the pumping of heart.

A waterproof probe-like sensor, DS18B20 is used to measure temperature under the tongue. Its one-wire bus enables ease of use, and 12-bit Celsius temperature is obtained as default output from the sensor.

The electrocardiogram assesses the electrical and muscular functions of the heart. The electrocardiogram can measure: Rate (refers to how fast the heart beats) and

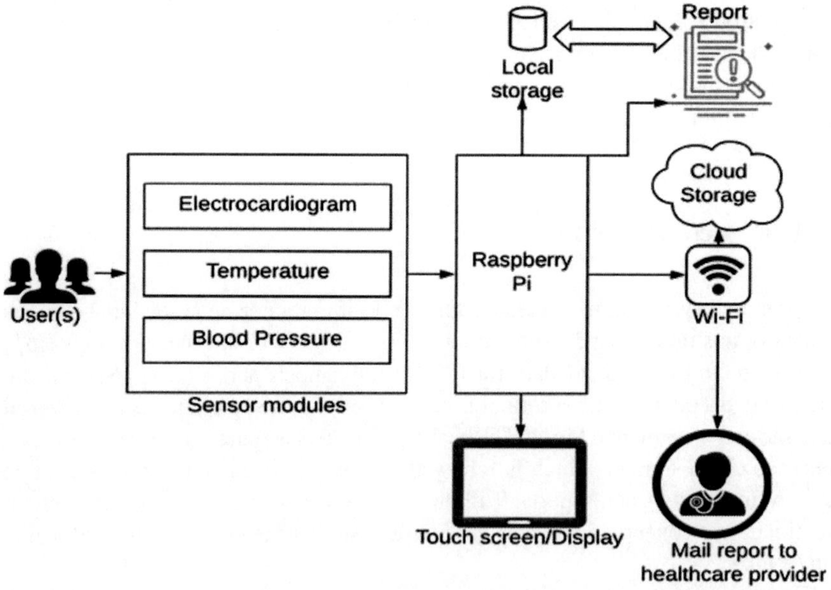

Fig. 1 Block diagram of proposed methodology

Rhythm (refers to the type of heartbeat). The AD8232 is a 9-pin chip used for this application, with a three-lead electrode system as shown in Fig. 3. MCP3008 is used as an external ADC since the Raspberry Pi does not have one inbuilt. The sensor pad placement is in accordance with the Einthoven's Triangle, for best possible results. The ECG signal is obtained from 3 lead ECG sensor. ECG signal is plotted and QRS Complex of signal is detected. The time between the peaks are used to identify the heart rate in bpm. This can be done by simply measuring distances, or by using 6 s method, for irregular heart rates.

4.1 User Interface and Cloud Storage

Raspberry Pi with a touch screen is used as a means of communication with the user. Hence, the GUI has been designed to work with the Raspberry Pi and the touch screen (Figs. 4 and 5). This is done by using the Kivy library for python, which is open source and contains a lot of features to develop various kinds of GUI. Google offers a suite of public cloud services, known as Google Cloud Platform, or GCP. Google Cloud Storage, which supports IoT applications through IoT Core, Pub/Sub for real-time messaging and so on is used as main service. After creating a new project, it is assigned a project ID. API's should be enabled and the credentials are generated. These credentials are locally stored as a json object that can be called at the time of authorization. Google libraries are used to transfer the files from Raspberry Pi to cloud service. The layout of visual application components and user interface generally starts with a Mockup. The Pencil application being open source, was used to create mock-ups for tentative screens required for the GUI of a Health Monitoring System.

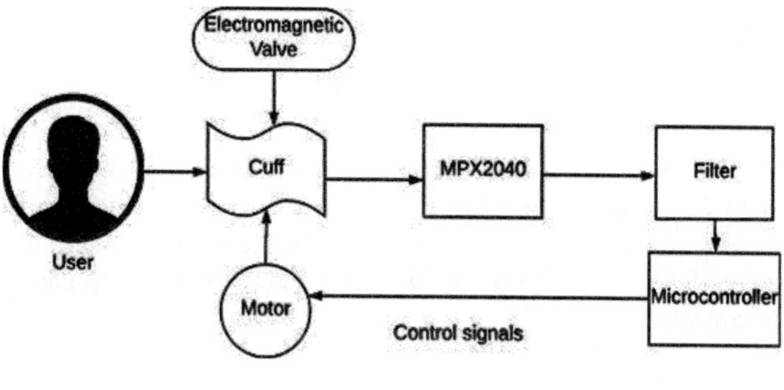

Fig. 2 Digital sphygmomanometer design

Fig. 3 ECG module block diagram

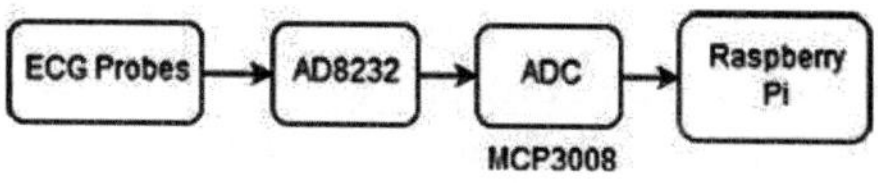

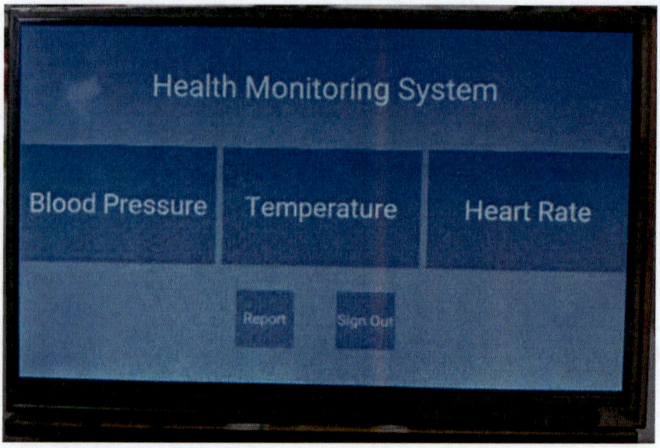

Fig. 4 GUI implemented on the touch screen

The GUI prototype created by the designer aids the developer in creating a functional GUI. This makes the development easy, as the design is already available. A library that supports multi-touch is essential for GUI development for a Raspberry Pi touchscreen, which serves as the display unit. Libraries like Kivy and Tkinter are available for GUI development. Kivy is a modern graphical user interface toolkit, which is also open source. It allows designer to easily develop natural interfaces for a wide selection of devices, while also supporting multi-touch. The designs made using Pencil were recreated using Python and the Kivy library and functionality was added to the GUI like switching between screens and other back-end logic.

4.2 Results Obtained

For a single user, the parameters were measured from the prototype and from existing modules for temperature, blood pressure and pulse rate measurement. Blood pressure and pulse rate were measured using their respective modules for the e-Health sensor shield and temperature was measured using a standard digital thermometer. The values measured from the 2 setups are compared and the accuracy is calculated and tabulated in Table 1.

5 User Health Report

The GUI of the HMS provides an option to generate a report with the help of python-docx library. The Report contains the date and time of generation, as well as key

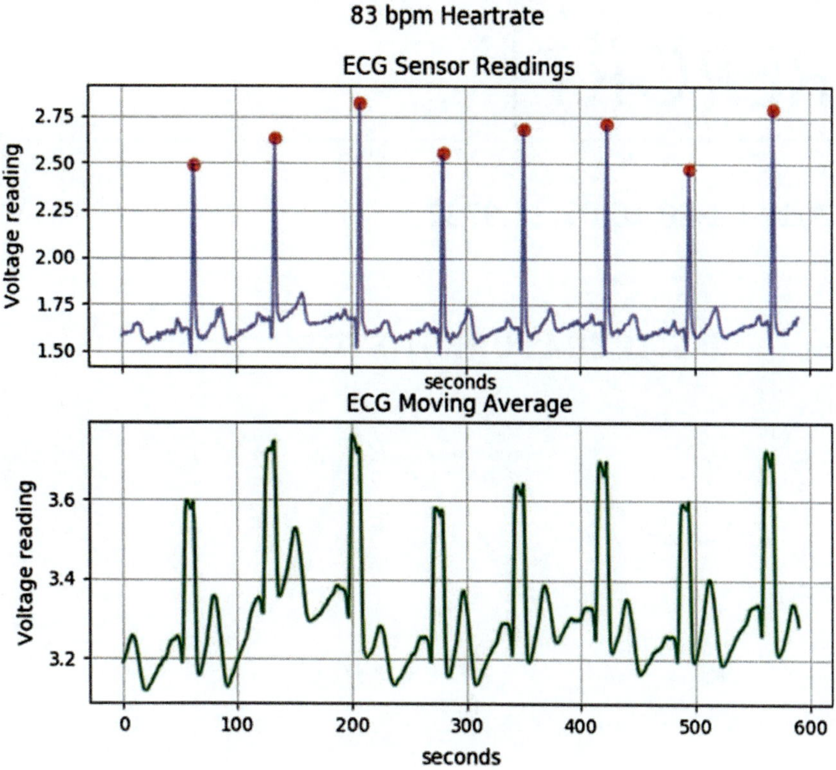

Fig. 5 ECG analysis displayed on the GUI

Table 1 Comparison of measurements with standard measuring kit

S. No.	Parameters	Measurement from standard module	Measurement from prototype	Accuracy (%)
1.	Body temperature (F)	98.70	98.10	99.4
2.	Blood pressure (mmHg)	117/81	127/79	99.15
3.	Pulse rate (bpm)	80	83	96.25

graphs and values from the three measured parameters, namely: Temperature, ECG, and Blood Pressure (Fig. 6). Temperature consists of the last recorded temperature, as well as a graph representing historical data over the past 10 days. ECG contains the values, as well as its analysis graphs, which extract the heart rate from the measured ECG voltages, as well as the RR-peak graphs of the last measured instance. BP contains the systolic and diastolic pressures as measured by the module (Fig. 7).

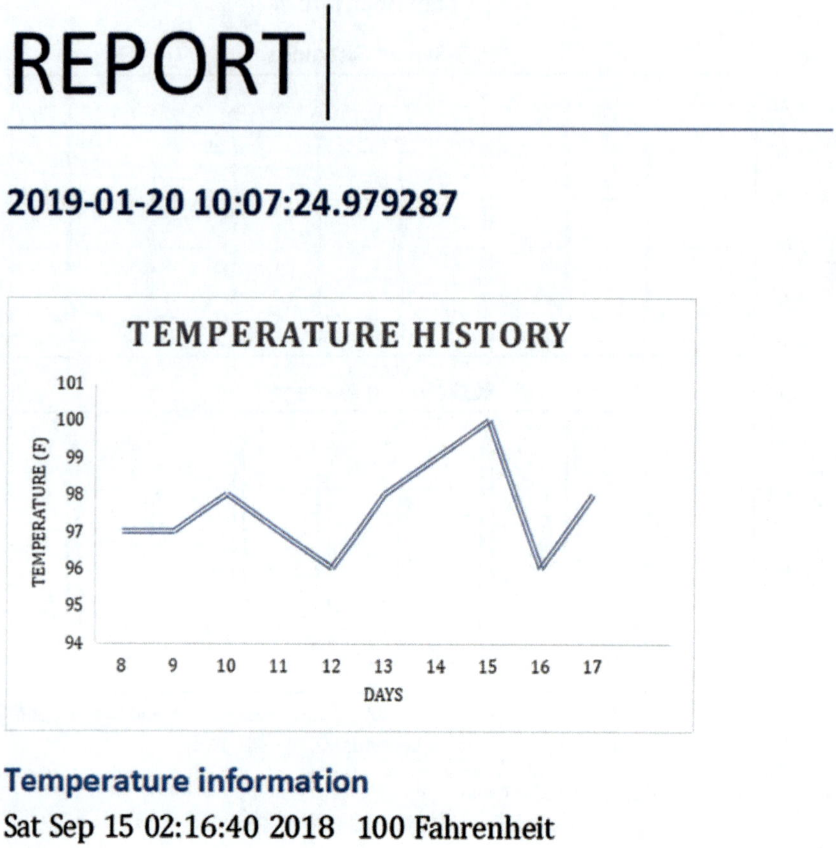

Fig. 6 Page 1 of user health report (word document)

The generated report can be conveyed to the healthcare provider, if the user wishes so, or if there is a critical condition. The user is required to sign-in into their mailing accounts via the GUI and enter the ID of the healthcare provider. The latest report and measured parameters are then sent to the mail ID provided for a basic overview, and any more processing that is required. The mail is sent using python's smtplib module, which uses simple mail transfer protocol, through RFC 821 protocol. A secure connection is made with the help of smtplib's TLS encrypted connection, which validates the host name and its certificates and optimizes the security of the connection. Since it is not secure to store passwords, the user inputs them at the time of sending, and the password is lost as soon as the code block ends.

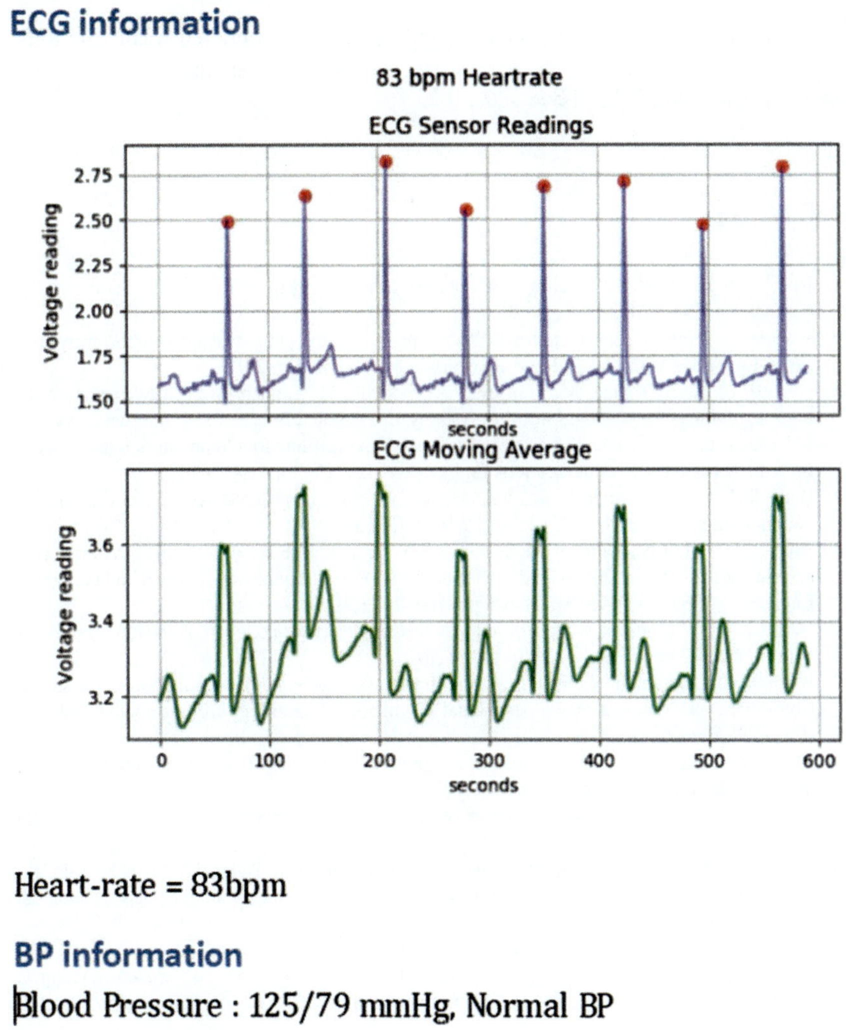

Fig. 7 Page 2 of user health report (word document)

6 Conclusion

The aim of the IoT-based Intelligent Health is to create a system that enables easy healthcare amenities to those with constraints when it comes to accessing healthcare services. All sensors and sensor systems are calibrated in comparison with existing kits to reduce any chance of errors to the best of our ability. The ability to communicate health reports to providers and making historic data available would enable easier processing of data, and faster diagnostics. The developed prototype been tested on 5 test subjects, and health data monitored was found to be at par with data measured using standard measurement techniques and average accuracy was found to be above 98%.

References

1. Runyon, B, Lefebure, S.: The Stages of Maturity in the Journey to the Real-Time Health System. Gartner, ID: G00263098, August 2016
2. Abdulkadir Ibrahim, A., Muhammad, Y., Zhuopeng, W.: IOT patient health monitoring system. Int. J. Eng. Res. Appl. vol. 7, Issue 10, (Part-7), pp. 01–03, Oct 2017
3. Ahmed, M.U., Björkman, M., Čaušević, A., Fotouhi, H., Lindén, M.: An overview on the internet of things for health monitoring systems. In: Mandler, B. et al. (eds), Internet of Things. IoT Infrastructures. IoT360 2015. Lecture Notes of the Institute for Computer Sciences, Social Informatics and Telecommunications Engineering, vol. 169. Springer, Cham
4. Baig, M.M., Gholamhosseini, H.: Smart health monitoring systems: An overview of design and modeling. J. Med. Syst. **37**(2), article 9898 (2013)
5. Ahmed, S., Millat, S., Rahman, M.A., Alam, S.N., Zishan, M.S.R.: Wireless health monitoring system for patients. In: 2015 IEEE International WIE Conference on Electrical and Computer Engineering (WIECON-ECE), Dhaka, pp. 164–167 (2015)
6. Garbhapu, V.V., Gopalan, S.: IoT based low cost single sensor node remote health monitoring system. Proc. Comput. Sci. **113**, 408–415 (2017)
7. Krishna, C.S., Sampath, N.: Healthcare monitoring system based on IoT. In: 2nd International Conference on Computational Systems and Information Technology for Sustainable Solution (CSITSS), Bangalore, pp. 1–5 (2017)
8. Uth, M.F., Koch, J., Sattler, F.: Body core temperature sensing: Challenges and new sensor technologies. Proc. Eng. **168**, 89–92 (2016)
9. Ogoina, D.: Fever, fever patterns and diseases called 'fever'–a review. J. Infection Public Health **4**(3), 108–124 (2011)
10. Chandrasekhar, A., Kim, C.S., Naji, M., Keerthana, Natarajan, Hahn, J.O., Mukkamala, R.: Smartphone-based blood pressure monitoring via the oscillometric finger-pressing method. Sci. Trans. Med. **10**(431), 07 (2018)
11. Fajar, M.R.Z., Hadiyoso, S Rizal, A.: An interfacing digital blood pressure meter with arduino-GSM module for real-time monitoring. In: 2017 International Conference on Control, Electronics, Renewable Energy and Communications (ICCREC), Yogyakarta, pp. 98–102 (2017)
12. Alisinanoglu, F.: Wireless Blood Pressure Monitoring System. Unpublished. https://doi.org/10.13140/RG.2.1.2104.5609
13. Jin-Ling, Z., Yue, L., Jia-Bao, W.: Design of electronic blood pressure monitoring system based on mobile telemedicine system. In: 2013 ICME International Conference on Complex Medical Engineering, Beijing, pp. 145–149 (2013)
14. Biel, L., Pettersson, O., Philipson, L., Wide, P.: ECG analysis: a new approach in human identification. IEEE Trans. Instrum. Meas. **50**(3), 808–812 (2001)
15. Krishna Prakash, N., Reseeda Banu, S.M., Haseena Banu, S.M.: Denoising of ECG by statistical adaptative thresholding and detection of T-wave alternans using principal component analysis. In: International Conference on Advances in Computing, Control, and Telecommunication Technologies, pp. 778–780 (2009)

A Survey on Methodologies for Handling Imbalance Problem in Multiclass Classification

S. Sridhar and A. Kalaivani

Abstract An imbalanced class distribution introduces the difficulty in classification tasks and exists in various real-world applications. Imbalanced class distribution happens when minority class has fewer instances than the majority class which has a large number of instances. The misclassification problem classifies the minority class into wrong class as compared to the majority class. Much work and solution are proposed for binary or two-class imbalance problem whereas the class imbalanced datasets that contain more than two-class labels with varying degree of imbalance have been received limited attention. Other than smaller representation of instances, factors like noise, overlapping classes, multidimensional classes, and data instances with multiminority and multimajority classes may affect the predicting rate of traditional classifiers on minority instances. Several solutions have been proposed for the multiclass imbalance problem that includes data sampling techniques, class decomposition techniques, cost-sensitive ensemble methods, etc. This paper presents a detailed survey of the approaches to handle multiclass imbalance problem.

Keywords Multiclass data imbalance · Multimajority · Multiminority · Classification · Boosting · Sampling

1 Introduction

Many real-world problems have more than two classes. Protein family classification, biomedical diagnosis, protein fold prediction, weld flaw classification, detection of fraudulent bank account transactions, etc., are some examples of multiclass problems. Multiclass classification is used in predicting the labels of instances of a dataset with more than two classes. The problem in multiclass classification arises when some

S. Sridhar (✉)
SRMIST, Vadapalani Campus, Chennai, TamilNadu, India
e-mail: sridhars2@srmist.edu.in

A. Kalaivani
SIMATS, Chennai, TamilNadu, India
e-mail: kalaivani.sse@saveetha.com

P. Suresh et al. (eds.), *Advances in Smart System Technologies*,
Advances in Intelligent Systems and Computing 1163,
https://doi.org/10.1007/978-981-15-5029-4_67

classes have smaller instances compared to other classes. Many conventional machine learning algorithms suffer from these uneven distribution of data instances among classes and make less effective in predicting instances of minority class.

Most of the existing imbalanced data handling techniques are designed only for two-class scenarios. They have been identified as less or not effective in dealing with multiclass scenarios. Also, in multiclass scenario, any classifier aims to achieve high classification accuracy by favoring the majority class which leads to misclassifying the minor samples into majority class. In general, multiclass classification problem can be approached either by directly applying the dataset involving multiple categories of data on a multiclass classifier or combining the results of several two-class classifier results. Due to imbalance nature among the samples in different classes, we need to focus on several methods to mitigate the imbalance problem. Some of the widely used methods include data leveling methods by using sampling techniques to make the training data balanced one before the learning of a classifier takes place [1], constructing new learning models that can understand the data distribution among classes [2], assigning or adjusting weights based on the misclassification rate thereby improving the classification performance [3], and finally combing the data leveling techniques and ensemble methods [4, 5]. In this paper, the imbalance problem in multiclass classification is discussed and presents the several effective methods taken by various researchers to tackle the data imbalance scenarios.

2 Multiclass Imbalance Problem

Traditional learning algorithms focus only on underlying class distribution and attempts to increase the accuracy rate and do not consider the underlying distribution of data among classes. Consequently, instances in majority class are well classified whereas instances in minority class have a tendency to be misclassified and this imbalance scenario is almost a part of all classification problems. For example, in the case of medical data analysis, false positive analysis can cause more expensive tests rates and anxiety of patients, whereas the false negative analysis can lead to loss of life due to less number of minor disease samples. This imbalance nature must be mitigated by classifying the minority samples correctly. Most of the existing methods [6] associated with solving class imbalance problem in multiclass scenario are discussed in the following Sect. 3. The general methods to handle multiclass imbalance data classification are shown in Fig. 1.

Fig. 1 Multiclass imbalance data classification handling methods

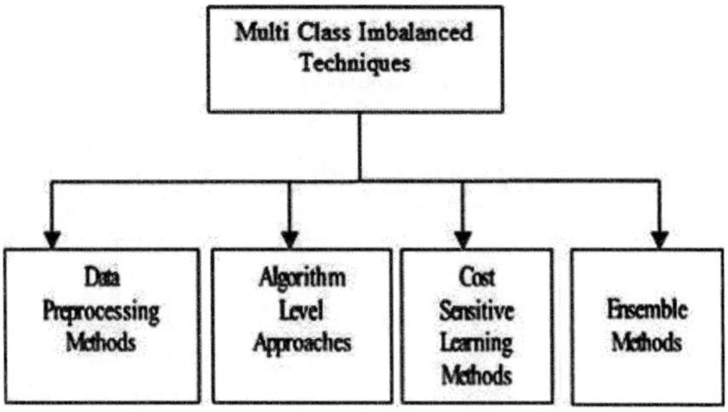

3 Methodology

3.1 Data Preprocessing Techniques

The methods discussed here directly deal with the data itself rather than the model, which include the data preprocessing techniques that try to balance the given samples before we train the classifier. Several sampling methods which are a part of data preprocessing technique help us to either add data samples in the minority class or to remove some data samples from the majority class.

Oversampling Techniques. Synthetic minority oversampling technique (SMOTE), the broadly used oversampling method aims to create new minority class examples (synthetic examples) by comprising a number of minority class instances that lie together. The weakness is it results in great chance of class mixture and too much sampling percentage leads to overfitting. SMOTE and its variants are listed in Table 1 which effectively handle the multiclass imbalanced datasets.

Chawla et al. [1] in his paper suggested that SMOTE shows a significant improvement in generating samples than random oversampling method. The main advantage of oversampling technique is generating a balanced data distribution without compromising the information of majority class.

Undersampling Techniques. Since the likelihood of over fitting happens in oversampling methods, many researchers suggests to perform under sampling techniques which tries to balance the data by randomly eliminating the samples from majority class. Again the downside is it can remove the essential data samples, needed for the training process [5]. Some of the widely used undersampling techniques are listed in Table 2.

Nguyen et al. [18] studies showed that undersampling outperforms oversampling technique when the training dataset is small. Chawla et al. [1] suggested that the bias of the learner toward class with major samples can be mitigated by combining the random undersampling and SMOTE techniques. The strength of undersampling techniques lies in its reduced training time on larger imbalanced datasets [19]. Sudarsun

Table 1 SMOTE and its variants

SMOTE boost (SMOTE + Boosting)	Concentrates more on difficult samples of minority class by giving weightage to synthetic samples (Chawla lal 2003) [7]
MSMOTE (Modified SMOTE)	Focuses on the minority class data distribution and thereby eliminates the noise (Shengguo Hu 2009) [8]
LNSMOTE (Local neighborhood-SMOTE)	Avoids the overgeneralization issue by carefully choosing the nearest neighbor to create the synthetic samples (Maciejewski 2011) [9]
BDLSMOTE (Borderline-SMOTE)	Tackles misclassification by considering only the borderline and the nearby samples to create synthetic samples as they are more error-prone (Han et al. 2005) [10]
MWMOTE (Majority weighted minority oversampling technique)	Assigns weights to the minority class samples with respect to its distance to the nearest samples of majority class and forms the cluster of minority class samples. New concrete synthetic samples can be created from this cluster (Barua et al. 2014) [11]
G-SMOTE (Geometric SMOTE)	Extends SMOTE by safely selects a radius around each class with minor samples instance and generates sample data within a safe boundary and his experiments revealed that this method performs well than the other oversampling methods (Georgios Douzas 2017) [12]

Santhiappan [20] proposed a new method namely TODUS, which aims at minimizing the information loss that usually happens while performing during random undersampling.

3.2 *Algorithm-Level Approaches*

The goal is to make the learning classifier to adjust the learning toward the minority class by creating or modifying the typical machine learning algorithms [21] instead of favoring the majority class which requires intensive knowledge on classifier's behavior and the underlying data distribution.

Support Vector Machines. Support vector machines (SVMs), basically a two-class classifier, use a hyperplane to separate the data into binary groups that works fine when there are only two-class labels, but to handle more than two categories, several approaches have been suggested, but widely known two most popular divide and conquer approaches are described below.

One-versus-one (*OVO*): A pairwise classification method where a classifier is trained to discriminate samples of one class from the samples of another class one by

Table 2 Types of undersampling techniques

Condensed nearest neighbor decision rule (CNN rule)	Creates a subset of examples using a one nearest neighbor rule, which can correctly classify the original dataset (Hart 1968) [13, 14]
Edited nearest neighbor rule (ENN)	Removes examples for those having class label different from the class of minimum half of its k-nearest neighbors (Tomek 1976) [15]
Neighborhood cleaning rule (NCL)	Modifies the edited nearest neighbor method and increases the role of data cleaning by finding examples for those having class label different from the class of minimum half of its k-nearest neighbors. If this example belongs to majority class remove it. Otherwise, remove its nearest neighbors which belong to the majority class (Laurikkala 2001) [16]
Cluster-based undersampling	Focuses on dividing the training examples into clusters and representative data for majority examples are selected from each cluster based on the number of minority examples and thereby increasing the prediction accuracy (Yen and Lee 2009) [17]
NearMiss undersampling	Performs undersampling of majority class examples with respect to its distance to other points in the minority class (Zhang 2003) [2]

one. For a new observation, classification is done based on maximum voting, where each classifier votes for one class. Space consumption and time consumption are two drawbacks of this method.

One-versus-all (*OVA*): This approach constructs one classifier per class which is trained to discriminate samples of one class from the samples of remaining classes. For a new observation, classification is done based on maximum output among all classifiers. The drawback of OVA is when the training data of each subproblem is unbalanced, which will affect the performance of each binary classifier.

The multicategory problem can be transformed to several binary category problems using either OVO or OVA and then we can apply a traditional algorithm, to produce a classifier for each of the simplified binary classification problems. Finally, the results are aggregated to form a complete solution. Thammasiri et al. [22] experiment results reveal that the support vector machine along with SMOTE sampling technique achieves a good performance on classification. Hanaa [23] proposed a new hierarchical model as given in Fig. 2, to rebalance the class samples based on grouping algorithm and uses SVM and multiclass SVM for effective classification.

Logistic Regression. It is a predictive analysis method used when we have a categorical outcome (2 or more categories). This technique can be used to analyze and predict variables that are "discrete," "nominal," and "ordered." The different types of logistic regression are binary, multinomial, and ordinal. The different forms of logistic regression can be used to model many real-world scenarios with a relatively

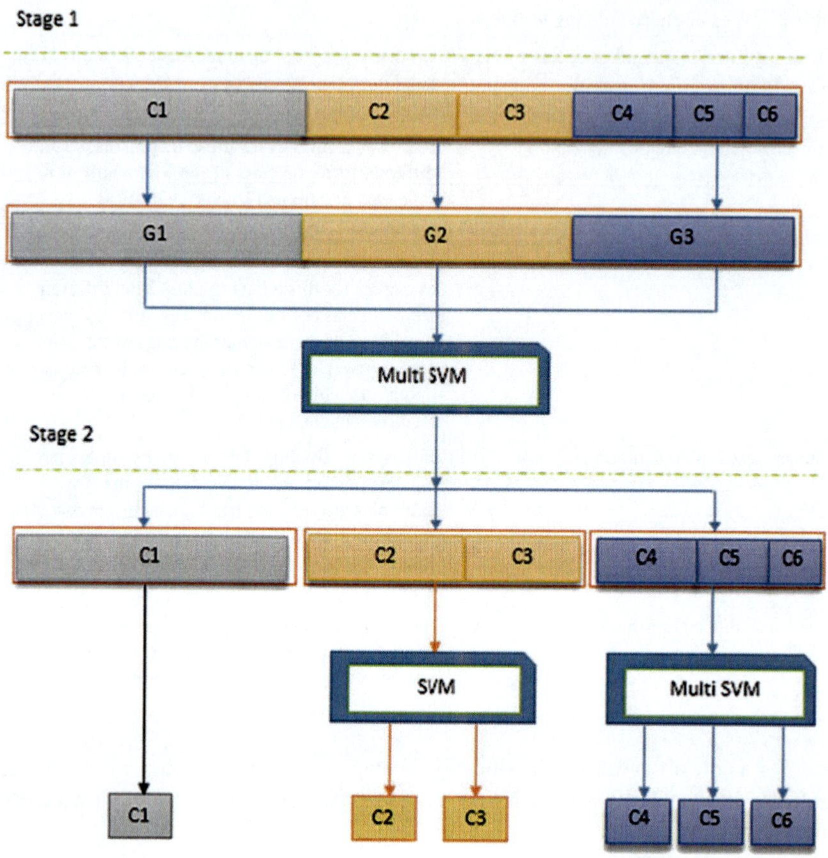

Fig. 2 Hierarchical model SVM

easily interpretable outcome as shown in Fig. 3. The drawback is performance is based on the number of outliers.

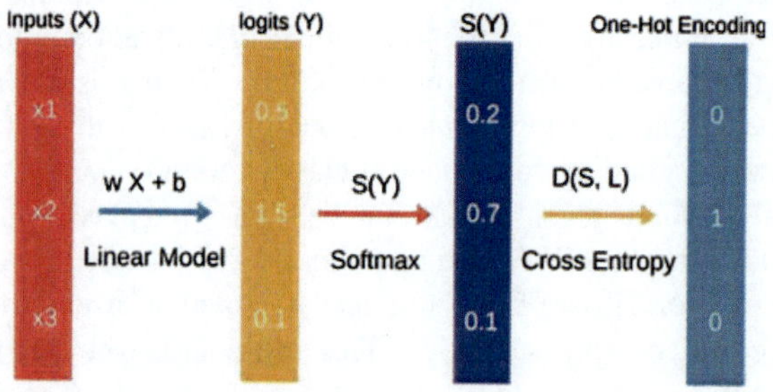

Fig. 3 Multinomial logistic regression model

Multinomial logistic regression methods classify the given data sample x based on their probability $p_k\ (x)$ of being in class C_k and it is expressed as:

$$p_k(x) = \Pr\{y = C_k|x\}. \tag{1}$$

These traditional methods use linear scoring functions to find $p_k\ (x)$. Miho Ohsaki [24] suggested in his work that KLOGR meant for multiclass classification performs well than other regression techniques especially on imbalanced biomedical datasets as it creates nonlinear boundaries based on softmax function as defined in Eq. 2

$$p_k(x) = \exp(y_k(x)) \tag{2}$$

$$k$$

$$\sum \exp(y_l(x))$$

$$l = 1$$

where
$y_k\ (x) = w_k{\cdot}\Phi\ (x) + \theta_k$, with $\Phi\ (x)$ as the kernel function that determines the closeness of a sample to a given class and θ_k as the bias term.

Radial Basis Function Networks. Radial basis function network (RBF), a type of feed-forward artificial neural networks, has three layers in its structure, an input layer, only one hidden layer, and an output layer as given in Fig. 4. It uses radial basis function which is a nonlinear function as an activation function. Generally, it uses Gaussian function as radial basis function and experiments revealed that RBF performance degrades as the imbalance nature in the dataset increases [25]. Some suitable data preprocessing techniques may be adopted to overcome this problem.

The radial basis function that activates the neurons in the hidden layer is mostly expressed as Gaussian function as defined in Eq. 3.

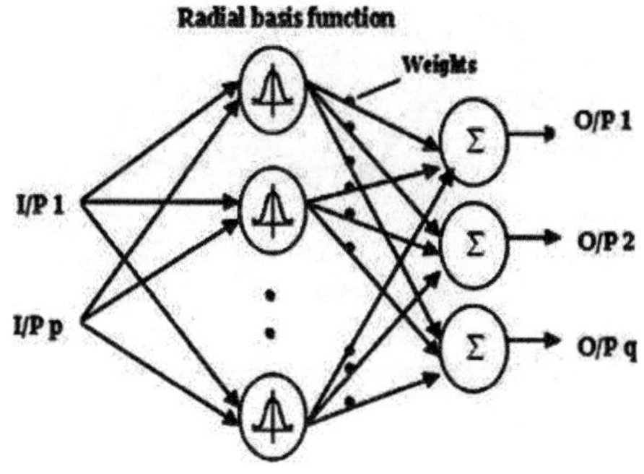

Fig. 4 RBF network architecture

$$\phi_i(\vec{x}) = \phi_i\left(e^{-(\|\vec{x}-\vec{c}\|/d_i)^2}\right) \tag{3}$$

where c, d represent center and width of basis function and the nodes in the output layer use the function as defined in Eq. 4.

$$f_j(\vec{x}) = \sum_{i=1}^{m} w_{ij}\phi_i(\vec{x}) \tag{4}$$

To train the weights, methods like least mean squares (LMS), Cholesky decomposition, orthogonal least squares (OLS), and singular value decomposition (SVD) were widely used [26].

Pérez-Godoy (2014), in his work, suggested that when imbalance ratio is high in the dataset, the LMS methods perform well than the other methods. The adaptability of RBF to different training sets gets affected due to the prior assignment of hidden neurons. To overcome this issue, HuiWen (2016) proposed an incremental hybrid method namely RBF-BP (ILRBF-BP) to incrementally create the number of hidden neurons based on the training sample size [27]. His experiments revealed that the hybrid method shows a significant improvement in classification performance.

Multilayer Perceptron. Multilayer perceptron (MLP) is a feed-forward artificial neural work which uses backpropagation to train the networks. The model is composed of one input layer, arbitrary number of hidden layers, and an output layer as shown in Fig. 5.

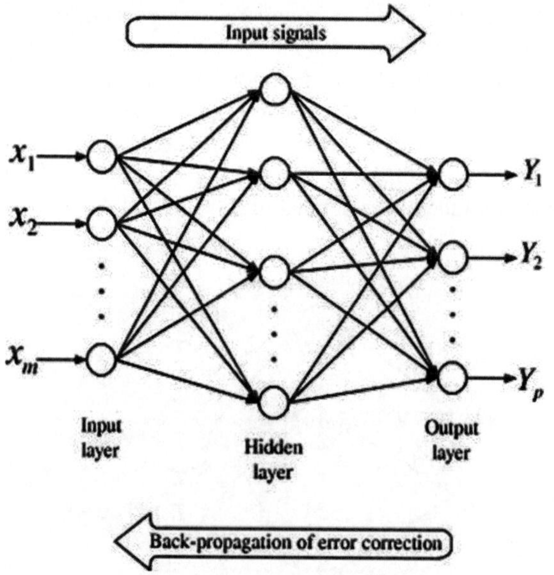

Fig. 5 Multilayer perceptron model

The learning process occurs in two stages namely forward and backward pass. In the forward pass, the model tries to learn the correlation between the input and output pairs of data, and in the backward pass, it uses backpropagation method to adjust weights and bias relative to the error. The above two passes of the model are repeated until the network converges or its predictions become accurate. The downside of this approach is it may converge at local minimum and its structural complexity and recent studies revealed that evolutionary artificial neural networks (EANNs) overcome these problems [6].

Shenfield (2017) proposed multiobjective ANN optimization approach for fetal cardiotocograms analysis and he showed that his approach significantly recognizes the minority class. Yildirim (2017) in his research on predicting kidney disease with the imbalanced datasets concluded that multilayer perceptron model which achieves 99.8% precision is the best suited model for disease prediction [28].

Decision Trees. The decision tree algorithms construct the tree in a top-down manner, which starts with a root node containing all the training data. Nodes are recursively split until some termination condition is met as shown in Fig. 6. Decision tree [21], a simple and an efficient supervised learning technique, works based on the splitting criteria function which decides how the given data should be split in order to increase its used classification accuracy. Studies revealed that to handle multiclass imbalanced datasets the most common and successfully used splitting criteria functions are gain ratio (C4.5) and Hellinger distance (HDDT) [29]. Marcellin et al. [30] used an asymmetric entropy measure for classifying imbalanced data and proposed that decision rules resulting from a decision tree constructed with asymmetric entropy as split criteria are more precise for classifying the minority class.

Cost-Sensitive Learning Methods. Here, the methods integrate both data-level alterations by adding costs to data samples and algorithm-level changes by modifying the learning process to handle the costs [31]. The challenge of this approach is how to devise a cost matrix which is not available with the dataset and how to associate

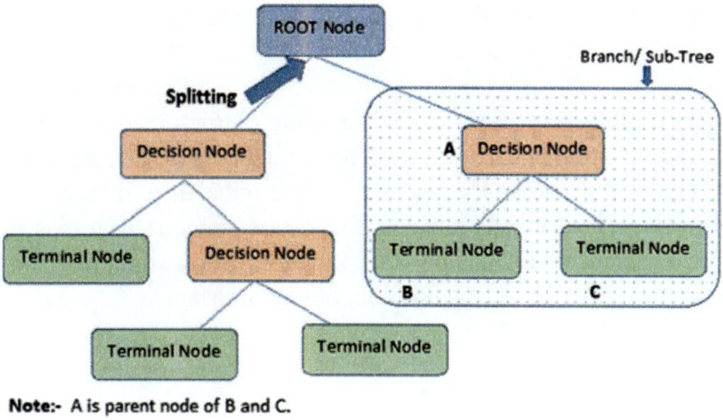

Fig. 6 Decision trees

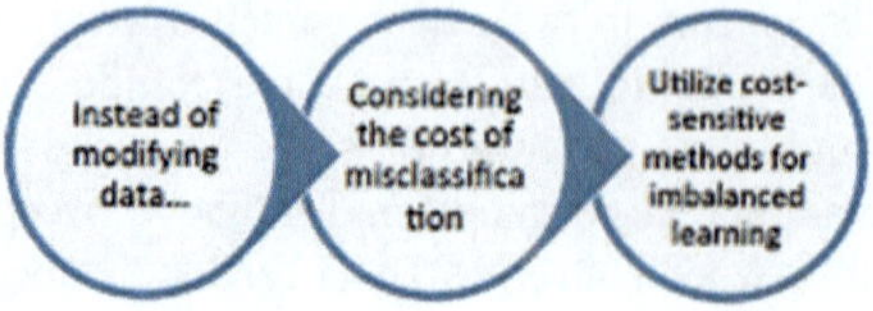

Fig. 7 Cost-sensitive methods approach

it with the algorithm. Cost-sensitive learning methods consider the variable rate of a misclassification of the diverse classes as shown in Fig. 7. These methods attempt to reduce the amount of high cost errors and the total error involved in misclassifying the instances [1]. The cost associated with misclassification can be of two types: sample-dependent cost and class-dependent cost. Most of the works concentrate on class-dependent cost. The most widely used cost-sensitive learning approach is rescaling approach which exists in different forms like assigning weights to training samples in ratio to misclassification costs, resampling the training samples before the classifier gets trained and threshold moving after the classifier gets trained [32], and these rescaling approaches attempt to convert the cost blind classifier to a cost-sensitive classifier.

Ensemble Methods. Ensemble-based methods have been recognized as more effective than sampling approaches [33]. These methods try to improve the single classifiers' performance by constructing several classifiers and combining them to obtain a new classifier and can be applied to imbalanced datasets [34, 35] to increase the classification accuracy. Random forest (a series of decision trees) is an ensemble model that can be used for classification as well as regression problems as shown in Fig. 8. It works well with both categorical and continuous features [36]. Studies revealed that it is more robust to outliers [5] and handles multiclass imbalance datasets effectively.

Several proposals in the literature reveal that hybrid techniques like combining ensemble learning with imbalanced data learning techniques such as sampling methods help us to overcome the class imbalance problem in a better way. Similarly, classic undersampling when combined with ensemble approaches [4, 37], like UnderBagging, attain good results. However, in learning imbalanced datasets, choosing more useful instances should help ensemble structure to perform multiclass imbalanced classification in a better way. Yoav et al. [7] presented adaptive boosting (AdaBoost) algorithm to solve the problem of data imbalance. It focuses mainly on data points that have been mostly misclassified by the previous weak classifier. Chris et al. [10] presented a combined approach called RUSBoost (sampling and boosting algorithms) algorithm to lessen the problem of data imbalance.

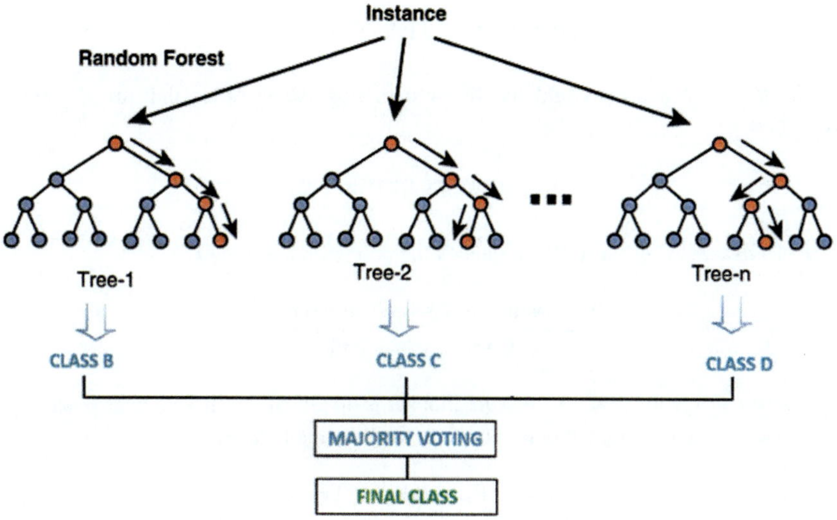

Fig. 8 Ensemble methods

4 Performance Metrics

The widely used measures for multiclass classification are discussed in this section. The first step before evaluating the classifier performance is to generate the multiclass confusion matrix against the actual target class and classifiers predicted class for the given samples as shown in Fig. 9.

In the above matrix, TP in the diagonal position refers to ***true positives*** means the classifier correctly classifies the given samples. The non-diagonal elements tell us the count of misclassified samples which can be either false positive (FP) or false negative (FN).

False positive. means the classifier wrongly classifies the given sample as positive samples which actually belong to negative class.

False negative. means the classifier misclassifies the given sample as negative samples which actually belong to positive class.

Precision: A measure to find the positive samples that are correctly predicted from the total predicted samples in a positive class.

Predicted / Actual	Class A	Class B	Class C	Class D
Class A	TP			
Class B		TP		
Class C			TP	
Class D				TP

Fig. 9 Confusion matrix

$$\text{Precision} = \text{TP}/(\text{TP} + \text{FP})$$

Recall: A measure to find the fraction of positive samples that are correctly classified

$$\text{Recall} = \text{TP}/(\text{TP} + \text{FN})$$

F-Measure: A measure that balances the precision and recall

$$\frac{\text{F-Measure} = 2 * \text{precision} * \text{recall}}{\text{Precision} + \text{recall}}$$

Microaverage method: A measure that sums up all true positives, false positives, and false negatives for each class and then the average is taken.

$$\frac{\text{Precision}_{\mu} = \text{TP}_1 + \text{TP}_2 + \text{TP}_3 + \cdots \text{TP}_k}{(\text{TP}_1 + \text{TP}_2 + \text{TP}_3 + \cdots \text{TP}_k) + (\text{FP}_1 + \text{FP}_2 + \text{FP}_3 + \cdots \text{FP}_k)}$$

$$\frac{\text{Recall}_{\mu} = \text{TP}_1 + \text{TP}_2 + \text{TP}_3 + \cdots \text{TP}_k}{(\text{TP}_1 + \text{TP}_2 + \text{TP}_3 + \cdots \text{TP}_k) + (\text{FN}_1 + \text{FN}_2 + \text{FN}_3 + \cdots \text{FN}_k)}$$

Macroaverage method: The overall performance of a model across different samples can be measured by averaging the precision of classifier on different classes

$$\text{Precision}_{\text{M}} = \text{PR}_1 + \text{PR}_2 + \text{PR}_3 + \ldots.\text{PR}_k/\text{K}$$

Where PR_1 refers to precision measure with respect to class 1. Similarly, the performance of a model across different samples can be measured by averaging the recall of classifier on different classes [38].

Both the microaverage and macroaverage techniques can be used when there is an imbalance in multiclass data. Studies revealed that when the microaverage value of a model is greater than macroaverage value, then minority classes must be examined for poor metric performance and for model having the microaverage value lesser than macroaverage value, the majority classes must be examined for poor metric performance.

5 Applications

Various real-world applications require accurate class prediction especially when it involves more than two classes. Again the challenge arises when the application faces imbalanced data within the multiple classes. The following table summarizes how the multiclass classification is carried out on imbalanced datasets in medical and biological domain (Table 3).

Table 3 Medical and biological applications

Domain	Area of research	Researchers	Method used	Dataset
Medical domain	Automated analysis of cardiotocograms	Shenfield [6]	Multiobjective ANN optimization approach	Fetal CTG dataset
	Predicting albendazole adverse event outcomes	Yildirim [39]	ID3 classifier with resampling technique	Albendazole dataset
	Chronic kidney diseases prediction	Drall [40]	Multilayer perceptron algorithm	Chronic kidney disease dataset
Biological domain	Yeast data classification	Hanaa [23]	Grouping algorithm and MSVM	Yeast dataset
	Brugada syndrome of 148 Exome dataset	Farid [41]	Clustering-based data rebalancing method	Exome dataset
	Multiclass classification of biological data	Begum [42]	greedy hierarchical binary classifier (MLP)	Biological dataset (Begum and Aygun 2012)

6 Challenges

The key challenge in tackling multiclass imbalance dataset classification lies in identifying the portion to start with. Be it in selecting the best sampling techniques, grouping the classes to rebalance the classes before the classifier comes to play with the data, selecting the appropriate model with enough consideration on minority samples, and selecting appropriate misclassification cost or using hybrid techniques varies from application to application. Apart from this, the following are the challenges identified in the survey.

- Depending on sampling strategies may either result in overfitting and noise in the case of oversampling techniques or loss of precise information in the case of undersampling techniques [29, 35, 43, 11].
- When using divide and conquer approach like converting multiclass problems into several two-class problems, the challenge lies in selecting the best classifier, the number of classifiers, and training cost [22].
- Cost-sensitive learning approaches are not steady in yielding classification results as it is hard to get the precise classification cost and diverse misclassification cost [41, 44].
- Selecting features for ordinal classification especially in biological classification problems [45].
- Selecting the suitable evaluation metrics to analyze the performance of a classifier in precisely predicting the minor class samples [23].

- Making the model to focus more on the underlying data distribution rather than the class distribution [7].

7 Inference from the Survey

Much of the survey works done in this paper concentrated on multiclassification of medical and biological datasets under imbalance scenario. The inference from this survey is summarized as follows:

- Identifying new data preprocessing techniques, defining new cost-sensitive and hybrid methods can be proposed to handle the multiclass imbalance nature of data especially for medical datasets
- Understanding the order of relationship among various classes [10]
- Handling outliers and noise in multiclass environment effectively [35, 11]
- Developing cost learning models without binarizing the problem for inconsistent costs [43]
- Group-based rebalancing the datasets can be tested with different classification models [23].

8 Conclusion

In this paper, we conducted a detailed study on various ways to tackle data imbalance problem in multiclass classification. It covers how the imbalance data can be mitigated using data preprocessing methods before the classifier gets trained, algorithm level modifications to handle the multiclass imbalance data before and after the classifier gets trained and also discussed the ensemble methods to tackle the multiclass imbalance problem in an effective way. Hybrid approach may seem to work well with multiclass imbalanced data. Also, we discussed the various performance measures for multiclass imbalanced data classification. Future research can be concentrated on developing new data preprocessing techniques, defining new cost-sensitive and ensemble methods that can handle imbalance nature distributed among multiclasses effectively.

References

1. Chawla, N.V., Bowyer, K.W., Hall, L.O., Kegelmeyer, W.P.: SMOTE: Synthetic minority oversampling technique. J. Artif. Int. Res. **16**, 321–357 (2002)
2. Zhang, J., Mani, I.: KNN approach to unbalanced data distributions: A case study involving information extraction. In: Proceedings of the ICML'2003 Workshop on Learning from Imbalanced Datasets, Washington, DC, USA, 21 Aug 2003

3. He, H.B., Garcia, E.A.: Learning from Imbalanced Data. IEEE Trans. Knowl. Data Eng. **21**, 1263–1284 (2009)
4. Blaszczyński, J., Stefanowski, J.: Neighbourhood sampling in bagging for imbalanced data. Neurocomputing **150**, 529–542 (2015)
5. Galar, M., Fernandez, A., Barrenechea, E., Bustince, H., Herrera, F.: A review on ensembles for the class imbalance problem: bagging-, boosting-, and hybrid-based approaches. IEEE Trans. Syst. Man Cybern. Part C Appl. Rev. **42**, 463–484 (2012)
6. Shenfield, A., Rostami, S.: Multi-objective evolution of artificial neural networks in multi-class medical diagnosis problems with class imbalance. In: 2017 IEEE Conference on Computational Intelligence in Bioinformatics and Computational Biology (CIBCB)
7. Chawlalal, N.V.,Lazarevic, A.,Hall, O.: SMOTEBoost:Improving prediction of the minority class in boosting. In: The 7th European Conf on Principles and Practice of Knowledge Discovery in Databases.Berlin, Springer, pp.107–119 (2003)
8. Hu, S., Liang, Y.: MSMOTE: Improving classification performance when training data is imbalanced. In: 2009 Second International Workshop on Computer Science and Engineering
9. Maciejewski, T., Stefanowski, J.: Local neighbourhood extension of SMOTE for mining imbalanced data. In: IEEE Symposium on Computational Intelligence and Data Mining (2011)
10. Han, H., Wang, W.Y., Mao, B.H., Smote, B.: A New Over-Sampling Method in Imbalanced Data Sets Learning. In: International Conference on Intelligent Computing, ICIC 2005: Advances in Intelligent Computing, pp. 878–887
11. Barua, S., Islam, M.M., Yao, X., Murase, K.: MWMOTE–Majority Weighted Minority Oversampling Technique for Imbalanced Data Set Learning. IEEE Trans. Know. Data Eng. (2014)
12. Georgios, D., Fernando B.: Geometric SMOTE: Effective oversampling for imbalanced learning through a geometric extension of SMOTE (2017)
13. Wilson, D.L.: Asymptotic properties of nearest neighbor rules using edited data. IEEE Trans. Syst. Man Cybern. **3**, 408–421
14. Hart, P.E.: The condensed nearest neighbor rule (corresp.). IEEE Trans. Inf. Theory **14**(3), 515–516 (1968)
15. Tomek, I.: An experiment with the edited nearest-neighbor rule. IEEE Trans. Syst. Man Cybern. **6**, 448–452
16. Laurikkala, J.: Improving identification of difficult small classes by balancing class distribution. In: AIME'01 Proceedings of the 8th Conference on AI in Medicine in Europe: Artificial Intelligence Medicine, pp. 6–66 (2001)
17. Yen, S.J., Lee, Y.S.: Cluster-based under-sampling approaches for imbalanced data distributions. Expert Syst. Appl. **36**(3), 5718–5727 (2009)
18. Nguyen, H.M., Cooper, E.W., Kamei, K.: A comparative study on sampling techniques for handling class imbalance in streaming data. SCISISIS 2012, Kobe, Japan, 20–24 Nov 2012
19. Fernńdez, A., López, V., Galar, M., del Jesus, M.J., Herrera, F.: Analysing the classification of imbalanced data-sets with multiple classes: Binarization techniques and ad-hoc approaches. Knowl. Based Syst. **42**, 97–110 (2013)
20. Santhiappan, S., Chelladurai, J., Ravindran, B.: A novel topic modeling based weighting framework for class imbalance learning. In: CoDS-COMAD' 18: The ACM India Joint International Conference on Data Science &Management of Data, 11–13 Jan 2018
21. Lin, Y., Lee, Y., Wahba, G.: Support vector machines for classification in nonstandard situations machine learning. **46**(1–3), 191–202
22. Thammasiri, D., Delen, D., Meesad, P., Kasap, N.: A critical assessment of imbalanced class distribution problem: The case of predicting freshmen student attrition. Expert Syst. Appl. **41**, 321–330 (2014)
23. Hanaa. S.A., H.S., Saeed, F.A.: New hierarchical model for multiclass imbalanced classification. J. Theoret. Appl. Inf. Technol. **95**(16) 31 Aug 2017
24. Ohsaki, M., Wang, P., Matsuda, K., Katagiri, S., Watanabe H., Ralescu A.: Confusion-Matrix-Based Kernel Logistic Regression for Imbalanced Data Classification, pp 1806–1819 (2017)

25. Haddad, L., Morris, C.W., Boddy, L.: Training radial basis function neural networks: Effects of training set size and imbalanced training sets. J. Microbiol. Methods **43**(1), 33–44 (2000)
26. Pérez-Godoy, M.D., Rivera, A.J., Carmona, C.J., del Jesus, M.J.: Training algorithms for Radial Basis Function Networks to tackle learning processes with imbalanced data-sets. Appl Soft Comput **25**, 26–39 (2014)
27. Wen, H., Xie, W., Pei, J., Guan, L.: An incremental learning algorithm for the hybrid RBF-BP network classifier. EURASIP J. Adv. Signal Process (2016)
28. Yildirim, P.: Chronic kidney disease prediction on imbalanced data by multilayer perceptron: Chronic kidney disease prediction. In: 2017 IEEE 41st Annual Computer Software and Applications Conference (COMPSAC), Turin, Italy, pp. 193–198 (2017)
29. Fan, X.N., Tang, K., Weise, T.: Margin-based over-sampling method for learning from imbalanced datasets. In: Advances in Knowledge Discovery and Data Mining; Springer, Berlin, Heidelberg, Germany, vol. 6635, pp. 309–320 (2011)
30. Marcellin, S., Zighed, D.A., Ritschard, G.: Evaluating decision trees grown with asymmetric entropies. ISMIS008, pp. 58–67
31. Zhou, Z., Liu, X.Y.: Training cost-sensitive neural networks with methods addressing the class imbalance problem. IEEE Trans. Knowl. Data Eng. **18**, 63–77 (2006)
32. Zhou, Z.H., Liu, X.Y.: On multi-class cost-sensitive learning. In: AAI'06 Proceedings of the 21st national conference on Artificial intelligence, vol. 1, pp. 567–572
33. Mellor, A., Boukir, S., Haywood, A., Jones, S.: Exploring issues of training data imbalance and mislabeling on random forest performance for large area land cover classification using the ensemble margin. J. Photogramm. Remote Sens. **105**, 155–168 (2015)
34. Liu, T.Y.: Easy ensemble and feature selection for imbalance data sets. In: Proceedings of the 2009 International Joint Conference on Bioinformatics, Systems Biology and Intelligent Computing, IJCBS'09, Washington, DC, USA, pp. 517–520, 3–5 Aug 2009
35. Qian, Y., Liang, Y., Li, M., Feng, G., Shi, X.: A resampling ensemble algorithm for classification of imbalance problems. Neurocomputing **143**, 57–67 (2014)
36. Breiman, L.: Random forests. Mach. Learn. **45**, 5–32 (2001)
37. Tahir, M.A., Kittler, J., Yan, F.: Inverse random under sampling for class imbalance problem and its application to multi-label classification. Pattern Recogn. **45**, 3738–3750 (2012)
38. Sokolova, M., Lapalme, G.A systematic analysis of performance measures for classification tasks. Information Processing and Management 45 (2009) 427–437
39. Yildirim, P.: Pattern classification with imbalanced and multiclass data for the prediction of albendazole adverse event outcomes. In: The International Workshop on Data Mining for Decision Support (DMDMS) (2016)
40. Drall, S., Drall, G.S., Singh, S., Naib, B.B.: Chronic kidney disease prediction: A review. Int. J. Manage. Technol. Eng. **8**(5), 2249–7455, p. 288 (2018)
41. Farid, D.M., Nowe, A., Manderick, B.: A new data balancing method for classifying multi-class imbalanced genomic data. In: Proceedings of Benelearn (2016)
42. Begum, S., Aygun, R.S.: Greedy hierarchical binary classifiers for multi-class classification of biological data. Network Modeling Anal Health Inf Bioinf **3**, 53 (2014)
43. Agrawal, A., Viktor, H.L., Paquet, E.: SCUT: Multi-Class imbalanced data classification using SMOTE and cluster-based undersampling, In: Proceedings of the 7th International Joint Conference on Knowledge Discovery, Knowledge Engineering and Knowledge Management (IC3K 2015), KDIR, vol. 1, pp. 226–234
44. Sun, Y., Kamel, M.S., Wong, A.K., Wang, Y.: Cost-sensitive boosting for classification of imbalanced data. Pattern Recogn. **40**, 3358–3378 (2007)
45. Misganaw, B., Vidyasagar, M.: Exploiting Ordinal Class Structure in Multiclass Classification: Application to Ovarian Cancer. IEEE Life Sciences Letters (2015)

Prioritized Congestion for Detection and Avoidance by Tadr Algorithm Toalemorite Throughput in Wireless Sensor Networks

B. Vijayalakshmi, C. Ram Kumar, and S. Anusha

Abstract Wireless sensor network is a rising technology that provides the various applications in defense and industries. Congestion and security are the main parameters in sensor networks. Congestion takes place when several nodes can transmit packets to a single channel. When nodes get closer to the sink, the traffic arises, which leads to congestion. It leads to loss of information, heavy packet loss, packet delay and less energy efficiency. A congestion control scheme is necessary to improve the quality of service (QoS) and system lifetime. The modified LACAS is to avoid traffic from many to one node. This is mainly designed for health care units. The nodes interact with the previous nodes at continuous intervals of time to reduce the congestion. Here, TADR algorithm with buffer management is implemented with MAC layer and transport layer. A TADR algorithm is designed to route packets around the blocking regions and spread out the unnecessary packets in different paths of idle nodes and low weighted nodes. The buffer automatically adapts the sensor's forwarding rates to be nearly optimal without causing congestion. The simulation result shows that the throughput gets increased, and packet loss is reduced drastically.

Keywords TADR · MAC layer · LACAS

1 Introduction

Wireless sensor networks consist of a huge amount of sensor nodes, users and sinks that are squashed, underloaded and battery-powered plans be able to use in almost every location. In practical applications, there are two types of users, namely traditional users and mobile users. In traditional, sensor networks produce traffic by sensing more nodes and burst the messages, and it is shown in Fig. 1. Buffer drops

B. Vijayalakshmi (✉) · S. Anusha
Sri Ramakrishna Engineering College, Coimbatore, India
e-mail: vgaug25@gmail.com

C. Ram Kumar
Dr. N. G. P. Institute of Technology, Coimbatore 641048, India

P. Suresh et al. (eds.), *Advances in Smart System Technologies*,
Advances in Intelligent Systems and Computing 1163,
https://doi.org/10.1007/978-981-15-5029-4_68

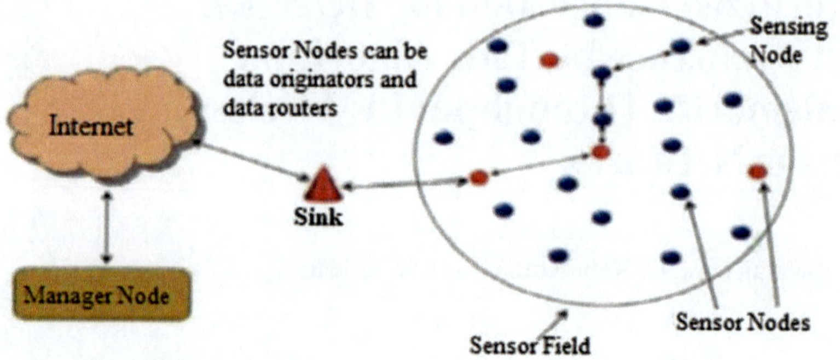

Fig. 1 Intercommunication of sensor nodes

and increase in delay are the main problems for congestion in sensor networks. Many applications require the value collected by the sensors and are stored in the station. Congestion happens mainly in source to sink direction when packets are transmitted. Congestion control technique's functions in different layers in which hop by hop layer, source rate-limiting scheme, priority queuing and scheduling the packets [1]. The congestion arises when the source node transmits the data to another node through multi-hop technique. At the receiving end, the data is received in a single path in which traffic arises. The congestion control will improve the network performance, lifetime and throughput and also reduce the packet loss and delay.

There are two types of congestion: Node-level congestion and link-level congestion. Figure 2 shows the issues that take place in the network. The node-level congestion is depended on the buffer overflow in which the sequence number with network ID takes place. Due to this cause, delay gets increased, thereby packet loss occurs. The link-level congestion is mainly designed for multipath, in which multiple sensors start transmitting the node at the same time [2].

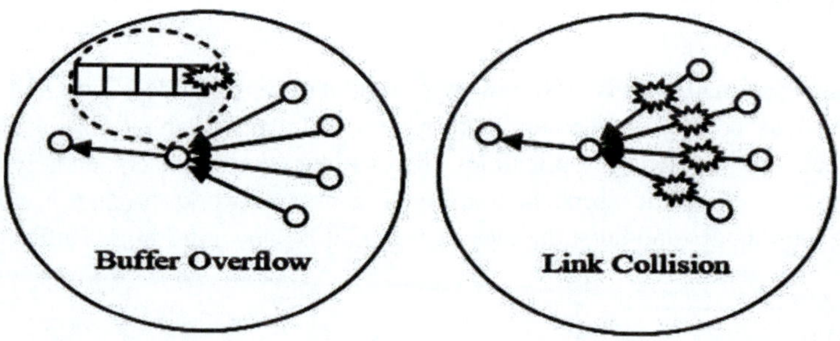

Fig. 2 Types of congestion

2 Related Work

2.1 Congestion Avoidance and Detection (CODA)

CODA has accurate and efficient congestion over the wireless sensor networks. The congestion should prefer the previous channel and the past channel for transmission. The sensor network must share the channel, leads to the traffic on the medium. The channel monitors the entire network to reduce the traffic. This leads to the high throughput and packet loss.

In an open loop, the node broadcasts the backpressure message to detect congestion. In a closed loop, the packet arrival time is measured from source to sink. The feedback is passed through every neighbor node to avoid congestion. Each sensor monitors the fairness and buffer level to improve the network throughput [3].

2.2 TADR

Traffic-aware dynamic routing is the technique in which congestion controls will lead to increase the throughput of the wireless sensor networks. It will be increased by multipath techniques, in which route request (RREQ) and route reply (RREP) are established. When congestion is detected, it will find the alternate routes with the shortest path. To design this network, the idle nodes or sleeping node is used to reduce the network overload [4]. TADR has a large end to end delay, which produce more congestion at the sink. To avoid this drawback, we are fixing the hotspot node to avoid congestion, thereby reducing traffic in the entire network.

2.3 RC-MACand MAODV Routing Protocol

The RC-MAC protocol allows the node to act in the different channel by reducing unnecessary channel switching. Receiver-centric MAC avoids the interference due to improve communication throughput of wireless sensor network [5]. It consists of two main part of RC-MAC: 1. Channel access scheduling. 2. Lost packet retransmission.

MAODV is used for analyzing the multiple routes. Primary routes and secondary routes are the main technique for route failure. In case failure occurs, it finds the alternate path to reach the destination node. The broken link leads to the packet loss and use of more bandwidth. Shortest path algorithm, selection of QoS, RREQ, and RREP for broadcasting are used in modified AODV.

3 Congestion Control Techniques

3.1 Congestion Detection

Congestion plays a vital role in wireless sensor networks. Some of the methods for congestion detection are packet loss, queue length, channel load and transmission delay. The packet gets lost at sender or receiver. ACK is used at sender side and sequence number is used at the receiver side. Every node has a unique buffer in which its size can be used as a threshold. If buffer length exceeds the threshold, the congestion is indicated to the neighbor nodes [6]. The channel load is the part of the interval when the path is busy to the total time. In MAC layer, packets are removed leads to reduce the packet drop. Delay in MAC layer is depended on the packet arrival time. Scheduler defines as how many packets per arrival time schedule the transmission range from nodes to nodes from the queues, forwards the packets to the MAC layer.

3.2 Congestion Notification

Explicit congestion notification (ECN) allows end to end notification from sender to receiver via ECN enabled switch without loss of packets, and it is shown in Fig. 3. It acts in internet layer and transport layer [7, 8]. Instead of dropping packets, IP header is marked in the TCP/IP network. At the destination side, the packet forwards the congestion suggestion to the transmitter, reduces the transmission rate. The main goal of ECN is to reduce packet loss and delay by decreasing the transmission rate without dropping packets [9].

The occurrence of frame relay in the wireless network is realized by frame relay access device (FRDA). The transmission delay is occurred in the frame or packets,

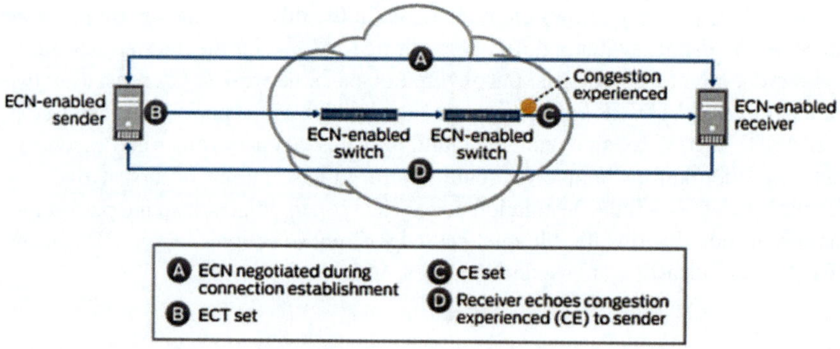

Fig. 3 Explicit congestion notification

and it gets lost. To recover the packets or frames, the devices requesting the retransmission of packets. Transmission delay may increase the congestion. It is used in frame relay LAPF in which link access service takes place [10].

3.3 Congestion Avoidance

When the queue occupancy is less than the congestion threshold, congestion is avoided in the buffer state. The source should send the information to the neighbor nodes in which the alternate path is chosen, leads to congestion avoidance on the links and paths. The congestion occurs due to packet collision and multipath traffic, which reduces the energy efficiency and system performance of the network [11]. To avoid congestion, packets are forwarded to every nodes and packet fairness in the sensor networks. Upstream and downstream nodes are used for queue monitoring to check the congestion level inside all the nodes.

The overall architectural model of congestion control protocols is shown in Fig. 4. The sensors are transmitting the packets to the network. If the node receives the packet, then the packet is checked for congestion. The checked congestion may be in the form of data rate, upstream/downstream, queue length, node delay, etc. If congestion is detected, congestion notification may send either implicit or explicit. Now, congestion is avoided by using AMID, rate control and traffic-aware mechanism. The congestion avoided packet is transmitted to the next node [12, 13].

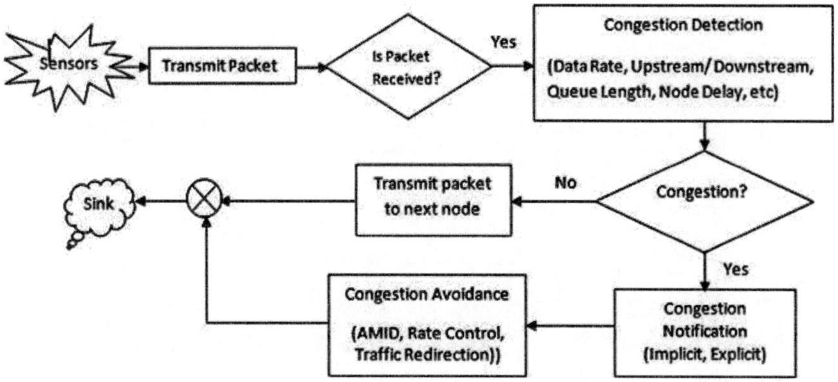

Fig. 4 Overview of congestion control protocols

4 Simulation Results

The simulation is carried out in network simulator tool (Ns-2).The experimental setup contains 40 nodes, and it is in random position. In the implementation, the packet delay and the throughput are measured.

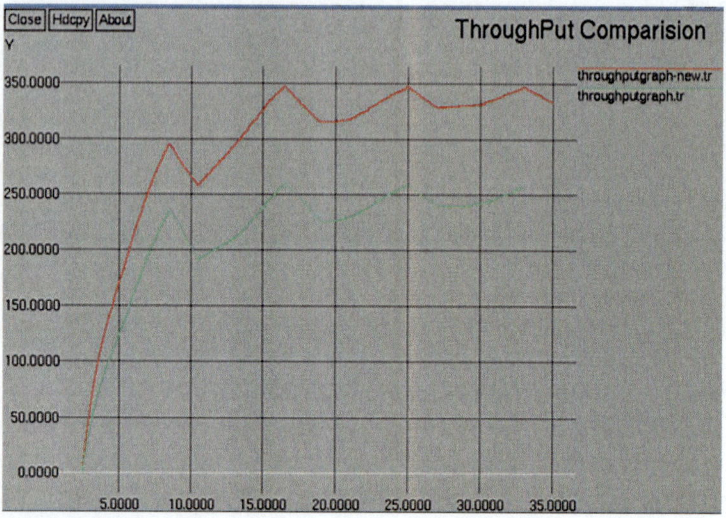

Fig. 5 Throughput comparison

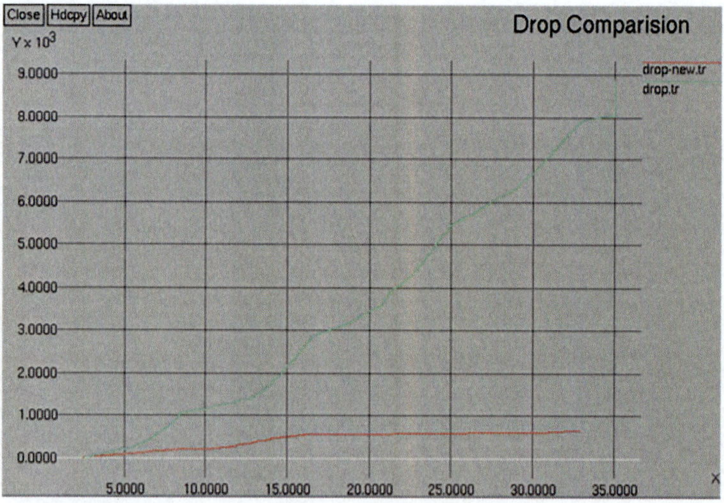

Fig. 6 Packetdrop comparison

In this technique, TCP layer is used to avoid the packet drop even though the transmission speed of the packets is very high [14]. This algorithm appears to be strong to packet errors by keeping a maximum throughput. Throughput is measured in simulation. Throughput is shown in the following graph. Figure 5 shows throughput comparison and it is maximized in the proposed system and thus has better performance than the existing system (Fig. 6).

The parameters of congestion control techniques if described in Table 1. The number of packets that are transmitted in the network is shown in the following Graph 1. Throughput is measured after the link failure occurrence has been rectified. It is shown that the throughput after the link failure has been overcome is high. There are 39%. Packet delivery ratio is defined as the ratio of the number of packets transmitted to the number of packets received. The packet loss reduced by 5.4%. The following Graph 2 depicts the comparison of the two cases.

Table 1 Parameters of congestion control techniques

Protocol	Network topology	Congestion detection	Congestion Avoidance	Priority	Throughput	Packet Loss	Energy
ARC	Bus	Hub placed between nodes	******	Token Message is passed	X	X	✓
SCTP	Linear	Fragment of messages into a number of data chunks	Ethernet Jumbo Frames	Improved Error Detection	X	✓	✓
DCCP	Tree	*******	Tradeoff between timeliness and reliability	Standardized Framework	✓	✓	X
CODA	Tree	Transmission medium is shared	Reduction of energy tax	Responsible for congestion scenarios	X	X	✓
DSDV	Hybrid	Reducing the unnecessary flooding of packets	sequence number is used to distinguish stale routes from new ones	Routing table Management	✓	✓	X
AODV	Hybrid	Reduce the Transmission Rate	******	It will not overload the network	X	✓	✓
PCCP	Ad hoc	Measure Queue Length	Exact Rate Control	Provides better QoS and Network Traffic	✓	X	X
M-AODV	Tree	Broadcast route	*******	Multicasting	✓	✓	X
RC–MAC	Mesh	********	Channel Contention and Radio Collision	Increase Lifetime	✓	✓	X
ECODA	Tree	Dual Buffer and Weighted Buffer	Rate Control	Use flexible Queue Scheduler	✓	✓	✓
HCCP	Ad hoc	Retains the buffer state	allocates or reduces the data rate	Data rate adjustment phase	X	✓	✓

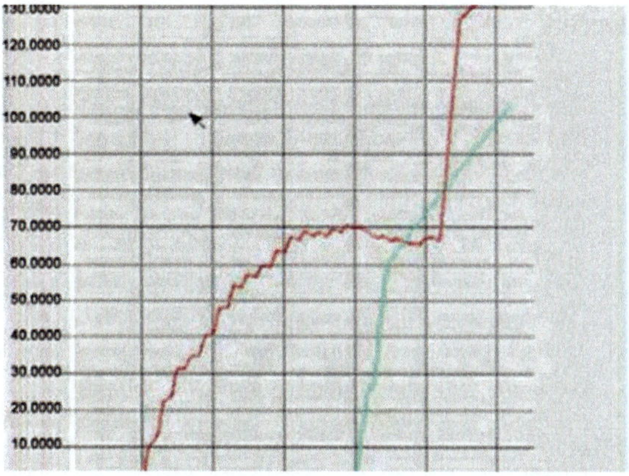

Graph 1 Comparison of throughput after link failure

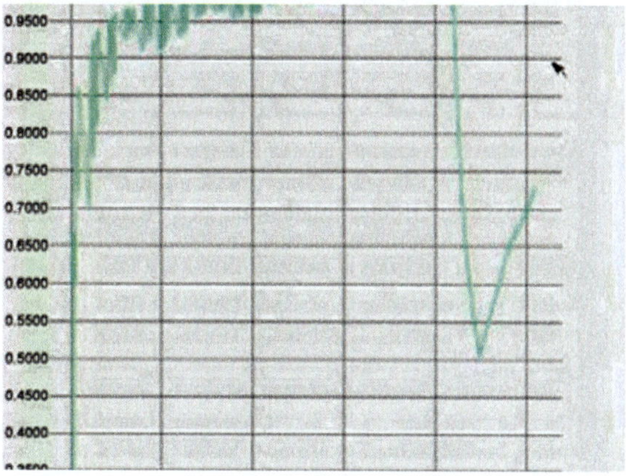

Graph 2 Comparison of packet drop

5 Conclusion

Congestion in wireless sensor networks is the new era to improve the system performance of the network. Congestion and traffic lead to high packet loss and also reduce the energy. This new technique TADR with geographical features will improve the QoS and system performance. TADR algorithm with buffer management is implemented with MAC layer and transport layer. A TADR algorithm in MAC layer is implemented to segment the packets that are being routed around the blocking areas and spread out the unnecessary packets in different paths of idle nodes and underloaded nodes. Various protocols were demonstrated in the network simulator. With the new technique, we have minimized the packet loss and increase in throughput.

References

1. Stann, F., Herdemann, J.: RMST: Reliable data transport in sensor networks. In Proceedings on 1st IEEE Workshop SNPA, Anchorage, AK, pp. 102–112 (2003)
2. Oh, H., Bahn, H., Chae, K.: An energy-efficient sensor routing scheme for home automation networks. IEEE Trans. Consumer Electr. **51**(3), 836–839 (2005)
3. Ram Kumar. C., Jennie Bharathi, M.: Enhancing coding aware routing and handling link failure in WSN. J. Comput. Appl. (JCA) **4**(4) (2011)
4. Wan, C.-Y., Eisenman, S.B., Campbell, A.T.: CODA: Congestion detection and avoidance in sensor networks. In: Proceedings ACMSenSys, Nov 2003
5. Velu, G., Suryatejaa, M.B.: Congestion avoidance based on RC-MAC protocol in wireless sensor networks. Int. J. Comput. Intel. Inf. **4**(4) (2015)
6. Antoniou, P., Pitsillides, A., Blackwell, T., Engelbrecht, A., Michael, L.: Congestion control in wireless sensor networks based on bird flocking behavior. Comput. Netw. **57**(5), 1167–1191 (2013)
7. Denesh Babu, M., Karpagam, M.: Improved method for detection and avoidance of congestion in sensor networks. Int. J. Eng. Res. Technol. (IJERT) **2**(10) (2013)
8. Romer, K., Mattern, F.: The design space of wireless sensor networks. IEEE Wirel. Commun. **11**(6), 54–61 (2004)
9. Chiu. D.M., Jain, R.: Analysis of the increase and decrease algorithms for congestion avoidance in computer networks. Comput. Netw. ISDN Syst. **17**(1), 1–14 (1989)
10. Ram Kumar, C., Raghunath, R.: Enhancing code aware routing by idling methods to improve energy efficiency in wireless networks. Int. J. Appl. Eng. Res. **10**(4), 10731–10741 (2015)
11. Pradeep Kumar, H.R., Krishnaprasanth, S.: EAACM: Enhanced ACK aware clustering mechanism for energy efficient and secure routing in wireless sensor networks. IMPACT: Int. J. Res. Eng. Technol. (IMPACT: IJRET), **2**(4), 53–62 (2014)
12. Kim, H., Song, J., Lee, S.: Energy-efficient traffic scheduling in IEEE 802.15.4 for home automation networks. IEEE Trans. Consumer Electr. **53**(2), 369–374 (2007)
13. Li, Q.T., Feng, Q.Y.: ECODA: Enhanced congestion detection and avoidance for multiple class of traffic in sensor networks. IEEE Trans. Consum. Electron. **56**(3), 1387–1394 (2010)
14. Iyer, Y.G., Gandham, S., Venkatesan, S.: STCP: A generic transport layer protocol for wireless sensor networks. In: Proceedings of IEEE International Conference on Computer Communications and Networks (ICCCN 2005), pp. 449–454 (2005)

Video Captioning for Proactive Video Management Using Deep Machine Learning

S. Om Prakash, S. Udhayakumar, R. Anjum Khan, and R. Priyadarshan

Abstract Inspired by the new advancements in machine learning and search engine technology, it is possible to caption videos and index them for efficient managing in applications like YouTube, Dailymotion, Netflix, etc. This system can generate a set of tags which when put together will give a sensible translation of the video in text format. The key challenge is the semantic gap between identification of objects by the human brain and the computer AI. In this proposed model, we propose to transcribe the video clips based on the contents and construct a pipeline through the convolutional neural network (CNN) which will efficiently and accurately process the video and search for the contents. This would require processing of the video clip and indexing it in a database. The same indexing will enable the use of this technology in content-based image retrieval (CBIR) tasks. The user can search for videos based on the context, and the time spent is seeking through the video to search for the scene is solved. We use deep learning model to search through video content and show that it performs efficiently.

Keywords Convolutional neural networks · Video processing · Content-Based image retrieval · Deep learning

S. Om Prakash · R. Anjum Khan · R. Priyadarshan
Department of Computer Science and Engineering, Rajalakshmi Engineering College, Chennai, India
e-mail: mailtoudhay@gmail.com

S. Om Prakash
e-mail: prakash1996@gmail.com

R. Anjum Khan
e-mail: anjumkhanr@gmail.com

R. Priyadarshan
e-mail: mailmefriends95@gmail.com

S. Udhayakumar
Saveetha School of Engineering, Saveetha Institute of Medical and Technical Sciences, Chennai, India

P. Suresh et al. (eds.), *Advances in Smart System Technologies*,
Advances in Intelligent Systems and Computing 1163,
https://doi.org/10.1007/978-981-15-5029-4_69

1 Introduction

There has been a boom in the rise of social content on the Internet; this social content includes images, videos, and micro-videos called Gifs. Scores of similar kinds of data have encouraged the development of various algorithms to relate the semantics and to analyze the content for better management. The latest breakthrough in establishing this semantic relationship is convolutional neural networks (CNNs). It provides us with ultramodern results in image recognition, image segmentation, and detection, and further with respect to image retrieval also. This is widely used in content-based image retrieval (CBIR) systems. CNNs have proved to accurately detect the features of an image. New techniques in the field of machine learning have emerged to understand the learning process in depth, generally known as "deep learning." This learning environment is nothing but a family of machine learning (ML) algorithms that explore the idea of modeling high-level abstractions in any data, weather it may be text, image, or video, by suitably employing architectures that can get deep into the process of learning, which is composed of multiple nonlinear transformations [1]. The use of CNNs along with deep learning can bridge the semantic gap between detection of objects by the human brain and an artificial intelligence. It is a challenge to train the intelligent machines to identify all the objects in an image or a video like the humans.

The existing systems use the traditional techniques where each frame of the video must be analyzed. This could take a very long time, and it would also need compatible hardware to complement this large amount of data. Inspired by the achievement of deep learning and CNNs, we attempt to implement these techniques with application to large-scale video classification. Since there are no video grouping benchmarks that match with the variety of existing image datasets, videos are significantly more difficult to collect, annotate and store. We propose to summarize a video by skipping the frames and taking periodic breaks. If we make it searchable, the user can simply search for the part of the video using the captions and arrive at it without a delay.

2 Related Works

Bidirectional image sentence mapping. Deep learning using visual and NLP are being introduced into this model to retrieve images and phrases. While the earlier models have mapped numerous images and text sentenced into a shared space, this proposed work by Andrej [2] uses a common area to embed image fragments and sentences as objects and typed dependency tree relationships, respectively. It is used as structured max-margin approach which allows the bidirectional model to associate the different pieces across various modalities. Let us consider a simple query, such as "*a fisherman catching a fish.*" To retrieve a corresponding image, the sentence mapping model identifies the complex visual scene based on the relationship that is present in the sentence and the processes it to retrieve the scene. The literary work gives an approach on the classification of images that is done using the probabilistic model.

Large-scale video classification. Convolutional neural networks (CNNs) is one of the best models for image recognition, there have been numerous research going in this domain, especially an extensive evaluation of large-scale videos are being classified based on the empirical dataset of around 1 million YouTube videos. The approach focuses on developing a time space neighborhood spatial data for foveae design. The work proposes for a multi-resolution environment [3].

Grounded compositional semantics. This method introduces the new dependency tree recursive neural networks (DT-RNN) model [4]. The model proposed to find the dependency trees for sentence embedding into a space vector to retrieve the relevant images which are described by the given sentences.

The DT-RNN model overcame the issues in the previous RN based models by using actions and agents in the sentences and further can abstract the details of syntactic expressions.

Video summary evaluation through text. This method evaluates a video and then reviews and compares for retaining the semantic information present in the actual video [5]. This video summary technique develops a text-based approach for semantic analysis, and it is done by first examining the semantic distance between the text reviews which is measured using NLP metrics generated by humans. A simple case may be for example, in a video, a man named Bob is driving a car, and then, the generated word would be Bob is driving a car. Thus, this work of video set precisely transfers the evaluation of the videos to text domain.

3 Proposed Work for Video Captioning

The traditional search engine works based on parameters like the meta-tags of the content, description of the title given during uploading, and the name of the application. However, it is impossible to manage the content with the above-given information accurately. Let us assume a user wants to search a video on YouTube based on the color of objects and the actual objects in the video. Since YouTube is a generic video-sharing platform, it indexes every video based on its meta-information. Usually when a query is submitted by the user, the search engine will only search the meta-information of that video, and not the content of the video. To overcome this, we introduce video captioning. Here, we first breakdown the video into chunks and send it to the video processor for analysis.

Our video processor will find a stable frame in our video, so that image processing will be effectively done. This means that a permanent structure from a set of frames is taken and the image processing algorithm is run to identify the objects in the frame. We submit these image chunks to a hypothesis that we have modeled. The theory will predict the objects in the images with which we will be able to identify the contents of each frame in the video. Stitching it all together, we get the caption for each frame in the video. These captions will be indexed by the search engine. Whenever a user submits a query, our search engine will find these captions and direct to that particular video. Although the core part of this paper is about video machine learning, we also insist on improving the search functionality of the search engine. This paper is dedicated to provide a sense of ease for users who are searching for a video based on public content. The core layer of this functionality depends on the task of uploading the video by the content owner, and this video must then be processed, assessed, and later stored. After the image is stored, the captions are developed and indexed in a database. Once the captions are indexed, this data helps when a user searches for a video based on captions. This retrieval process is done with the support of a cloud server. We have proposed a model which transcribes video clips into text, based on its contents and then construct a pipeline that can and accurately search for this content by an extension of the ultramodern skip through vectors. This task is a challenge because it requires a detailed understanding of the content of videos. This challenging task is divided into two modules: (1) Captioning and Indexing. (2) Managing and Searching.

3.1 Captioning and Indexing

In this first module, the content owner uploads the video onto the server. Since there would be n number of uploads at any point in time, we have used a distributed server to keep the efficiency of the application (Fig. 1).

The video from the content owner is first sent to the video processing broker which queues up the videos for processing and eventual indexing. Once the video reaches the processor, the input video signals are processed to output video signals where each frame is sent to a convolutional network. Here, the CNNs select the appropriate

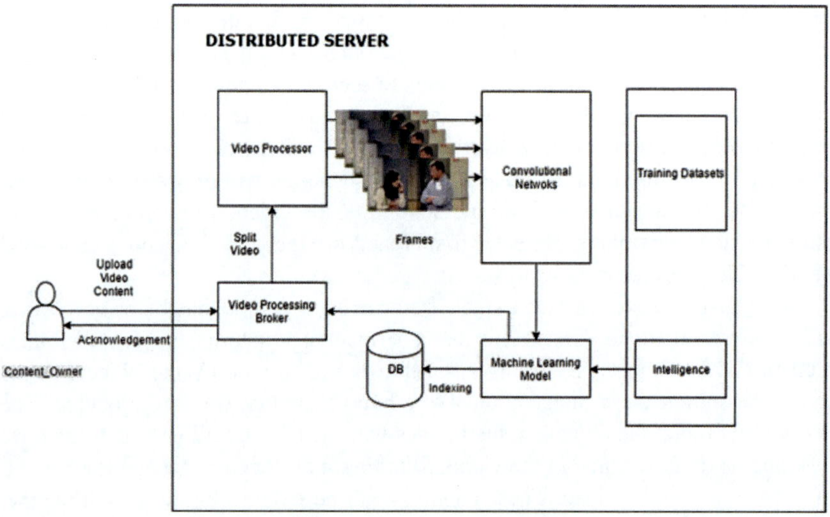

Fig. 1 Architecture of video captioning using deep machine learning

frame from a set of frames, i.e., videos, with frame rates of 24 fps (normal default for high definition video) which are run through the network, and the CNNs select a frame from every second which would contain all the objects from the intended scene. These frames are now raw images. These pictures are sent to an intelligent engine which is trained to detect and identify each object along with all the parameters such as relevance, color, and depth using datasets. Once the objects are identified by the intelligent system, it is sent to a machine learning model where captions are developed and stored in a database for further use.

3.2 *Managing and Searching*

Once the videos are captioned and indexed, they are stored in a database. A user's query is taken in and broken down and queued up in a cloud server. This query is then sent to a text processor where every part of the question is identified, and the annotations are labeled and sent to the database which contains the indexed keywords from the captions. These captions are then mapped to each frame, which is outputted to the user (Fig. 2).

4 Video Processing in a Multi-resolution CNN

The convolutional NN generally takes long time to train a model, even it takes few weeks to prepare and adapt substantial scale datasets even on the quickest accessible GPUs [6]. Since computing equipment with various parameters and hyper parameter settings are accessible, execution time assumes a noteworthy job in the proposed work. This motivates to build a suitable model to increase the efficiency. Thus, it is a challenge to balance between either to improvise the existing hardware's or manage weights, and it can even be achieved using better optimization and initialization techniques. Therefore, one of the approaches is to speed up the network by reducing the no. of layers of neurons; however, this may prove to be a drawback because it

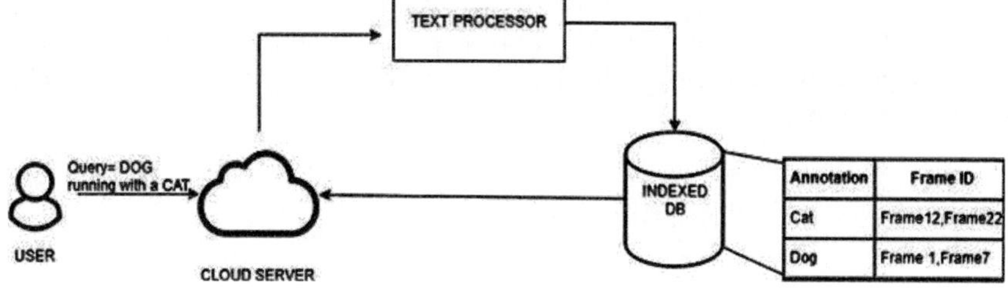

Fig. 2 Architecture of video retrieval process from an indexed DB

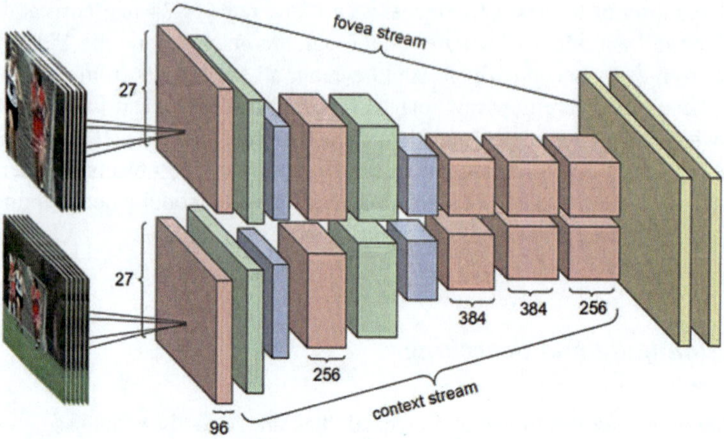

Fig. 3 Multi-resolution CNN architecture

may lower the performance [7]. Hence, to counter it, the resolution of the image can be reduced instead. Thus, the improvement in time is not compromised for reducing the network layer. The proposed architecture uses two separate streams of processing over two spatial resolutions as depicted in Fig. 3.

A 178 × 178 frame is down-sampled to 89 × 89 pixels which creates the context stream. The fovea stream receives the center (89 × 89 pixels) region at the original resolution. By this, the dimension is divided to half. There are two separate streams of processing for the input frames, the first is a context stream which used the low-resolution images and next is the fovea stream that uses the high-resolution center images. Both the streams have convolution layer, normalization layer, and pooling layer. Then, it is converged into two fully connected layers.

5 Learning Process

5.1 Video Optimization

To optimize our models over a computing cluster, downpour stochastic gradient descent algorithm is used. Here, the replicas vary from 10 to 50, and further split into 4–32 partitions.

5.2 *Video Augmentation and Preprocessing*

Further processing is required to reduce the effect of over-fitting; hence, it is necessary to process all the images by cropping to center region and then resizing to 200 × 200 pixels. This is done by sampling in a random process by taking a 170 × 170 area, and randomly flipping horizontally the images with 50% probability. The same steps are applied to all the frames in the video clip. Finally in the last stage of preprocessing, a value of 117 is subtracted as constant from the raw pixel value. This number is nothing but the mean approximate value of all the pixels in the image.

5.3 *Video Summarization Dataset*

This dataset is a key to our training model since it contains video summarizations which were acquired through crowd sourcing. Imagine wedding videos are searched in the YouTube. This is totally time consuming and inefficient, because we need to go through thousands and thousands of results returned. Instead, a summary of essential videos would make the task rather simple. The video summarization dataset contains summaries written by ordinary humans after analyzing the content for more than an hour. Each video includes multiple summaries written by humans. We can pick the best ones through ground truth summary. On the other hand, the summaries are generated through a concatenation of annotations from multiple reviews written by humans.

6 Technical Approaches and Models

As discussed in the previous sections, the task is divided into two different sets:—(1) The task of captioning the videos (2) The task of content-based video retrieval. Each of these tasks requires a convolutional neural network which processes and sends the video frames to a series of interconnected long short-term memory networks of RNN. The LSTM will hold all annotations that were generated by the convolutional network during image recognition process. The annotations in the LSTM are supplied to an affine layer to make a caption. Affine layer here makes the LSTMs in the previous layer connected to each other in the current one. With this, a SoftMax is produced. The LSTMs rearrange the annotations and pass it to the affine layer based on the SoftMax output. The process continues until a lowest SoftMax threshold is hit. The lesser the SoftMax, better is the caption formation.

The video captioning model consists of the following components:

1. We use the convolutional neural network; a state-of-the-art extracts semantic information from the video frames to construct a temporal sequence of features that can be collected and fed into a language model. We supply a video to

this network, frame by frame and thus get ten features (maximum layer) of a trained ConvNet model corresponding to the ten frames per video in our current dataset. For these experiments, we used a pretrained VGG Net and ResNet. These temporal sequences of features are then fed into a bidirectional LSTM.

2. The output of the convolutional network is fed into the bidirectional LSTM, which takes the temporal sequence of the frame features extracted and tries to model the time domain dependency in the video frame features.
3. The output of the bidirectional LSTM is converted into a single feature vector using an affine layer that serves as an entry point to the language model.
4. The language model is an LSTM, which takes as input the concatenation of all the single feature vectors for all frame of a video segment. It then generates a SoftMax output at each time step predicting a word, until an end of sequence symbol is not predicted.
5. The generated caption is fed to a pretrained skip-thoughts model. Given a tuple $(\text{si} - 1, \text{si}, \text{si} + 1)$ of sentences in contiguous order, with *s-i* and the *i*-th sentence of the given book, the phrase or sentence "si" is encoded, then it tries to reconstruct the previous sentence si-1 and next sentence si + 1. We convert our generated captions into vectors by running the encoder part of our model in a semantic space. These vectors are then used for retrieval.
6. The extraction pipeline is very simple. The vectors are returned by increasing the order of magnitude of the Euclidean distance. In $k = \text{argmin}\ (\text{vq}-\text{vi})$, vq is the vector corresponding to the query, vi is the vector corresponding to the *i*-th clip in the dataset, and *k* is the top retrieved clip from the dataset. We can retrieve top-*n* videos, based on the increasing order of the Euclidean distance calculated from the argmin equation.

7 Experiments and Results

The project has two stages—generating captions for video clips and making the video searchable using these subtitles. For the first stage of the pipeline, we tested different combinations of the convolutional networks for extracting features with various temporal layers used to merge the features of consecutive frames.

(1) VGG16—Affine—LM

ConvNet used for feature extraction was a pretrained VGG-16 model which was slightly fine-tuned while training. The features extracted for the batch were combined using an affine (ReLU) layer and fed into the language model.

(2) VGG16—BiLSTM—LM

In this model, the bidirectional LSTM replaces forward LSTM to capture symmetric dependency on previous as well as following frames.

(3) ResNet200—Affine—LM

To improve the feature extraction from the frames, the VGG16 model was replaced with a pretrained ResNet network with 200 layers. The rest of the model was the same as described in 1.

(4) ResNet200—BiLSTM—LM

Same as described in 3, except that VGG16 is replaced by the ResNet200 architecture.

The second stage of the pipeline was initialized using pretrained skip-thought model. This was used to convert the generated captions into vectors. The motivation behind using skip thoughts was that the skip-thought vector in their original paper had been shown to be capable enough to capture the "meaning" of the sentences they encode. When used, without any fine tuning, on tasks like semantic analysis, paraphrase detection, etc., they work very well. So, we believe that the skip-thought vectors would be able to cluster captions that are similar to each other in their meaning. The user query, too, was converted into a vector using the same model. The similarity measurement used to rank the results for the query was Euclidean distance.

7.1 *Evaluation*

We evaluated the quality of the generated captions by using two separate metrics, the cross-entropy loss from the SoftMax layer of the language model and the bilingual evaluation understudy score (BLEU) [8] for the generated captions. The BLEU is a standardized method for evaluating the quality of text that has been translated by machine from one language to another. The entire model is trained using back-propagation on the cross-entropy on the SoftMax output of the language model. We measure performance regarding this using cross-entropy (Table 1).

Table 1 Evaluation of entire pipeline regarding precision, recall, and mean average precision

Model	mAP	P@K = 1	P@K = 10	R@K = 1	R@k = 10
VGG16 Affine-LM	0.492	0.473	0.410	0.0301	0.0289
VGG16 BiLSTM LM	0.691	0.674	0.583	0.0524	0.0507
ResNet200Affine-LM	0.506	0.489	0.433	0.0329	0.0296
ResNet200 BiLSTM LM	0.725	0.699	0.614	0.0561	0.0552

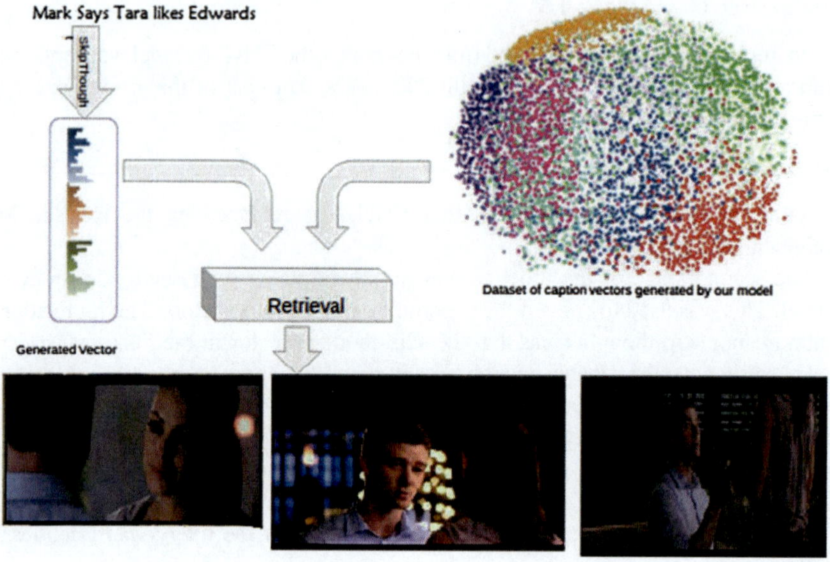

Fig. 4 User query to video retrieval process

7.2 Results

We ran a few queries to evaluate the qualitative performance. For example, "*Mark says Tara likes Edward.*" As described earlier, this query was converted into a vector representation using the skip-thought model, and the set of vectors for the captions were ranked per their Euclidean distance from the query vector. The top five clips corresponding to the closest vectors had four clips from the Castle episode. The second clip among this had the original caption "*Mark asks Tara if she likes Edward.*" The rest had one or more of the three characters in the frame talking to each other. Correspondingly, good results were observed when queries like "*My friends and I were walking in the park,*" "*I cooked in the kitchen.*" were tried. We found in the qualitative results that even without audio input our model could associate the entities with their respective videos. For a user-submitted query, the machine retrieves the desired scene from the set of caption vectors which are mapped against an individual view of a video is presented in Fig. 4.

8 Conclusion

With the improvements in machine learning and neural networks, in this paper, we aim to improve the method of managing the video and introduce new techniques to generate captions for the video using intelligent systems and indexing them in a database. These captions can be utilized by an average user to search for any desired video. Based on these search queries submitted by the user, our system can accurately identify and provide the results. To support the claims, we conduct a series of experiments by combining some convolutional networks and evaluating individual results. The outcome of these combinations is pipelined to convert the captions to vectors. They are then tested, the correctness of the captions is generated, and the meaning is evaluated. The results of all these were compared, and it is found that our model can classify the captions to the video without even using the audio from the video clips.

References

1. Singh, A.V.: Content-based image retrieval using deep learning. Master's thesis, Rochester Institute of Technology (2015)
2. Karpathy, A., Joulin, A., Fei-Fei, L.: Deep Fragment Embeddings for Bidirectional Image Sentence Mapping. Stanford University (2014)
3. Karpathy, A., Toderici, G., Shetty, S., Fei-Fei, L.: Large-scale video classification with convolutional neural networks. In: IEEE Conference on Computer Vision and Pattern Recognition (2014)
4. Socher, R., Karpathy, A., Le, Q.V., Manning, C.D., Ng, A.Y.: Grounded compositional semantics for finding and describing images with sentences. Trans. Assoc. Comput. Ling. **2**, 207–218 (2014)
5. Yeung, S., Fathi, A., Fei-Fei, L.: VideoSET: Video Summary Evaluation through Text, ArXiv (2014)
6. Goodfellow, I., Bengio, Y., Courville, A.: Deep Learning. MIT Press. http://www.deeplearningbook.org (2016)
7. Bergado, J.R., Persello, C., Stein, A.: Recurrent multiresolution convolutional networks for VHR image classification. IEEE Trans. Geosci. Remote Sens. (2018)
8. Papineni, K., Roukos, S., Ward, T., Zhu, W.J.: BLEU: A method for automatic evaluation of machine translation. In: 40th Association of Computational Linguistics, pp. 311–318 (2002)

Experimental Analysis of Gravity and Buoyancy Powered Energy Generation Storage Systems

Saad Bin Abul Kashem, David Tionge Ngambi, Jubaer Ahmed, Uvais Qidwai, and P. Suresh

Abstract The concept of harnessing energy from buoyancy as well as the ability to have underwater energy storage is an area of research that, compared to other renewable energy generation techniques, is relatively unexplored. This study presents an experimental analysis of a buoyancy generation and storage system. Tests were performed under standard laboratory conditions with the primary fluid being chlorinated municipal tap water. The buoyancy systems were designed, fabricated, and tested, with a focus on scalability and the need for results extrapolation. The system charge cycle was analyzed based on the system discharge voltage. Energy losses in the system due to drag or mechanical loss were estimated. The power generation and storage system utilized air transfer between containers to account for buoyancy increase and decrease in the main air vessel. Both systems were designed to function as standalone systems. Key variables of the experiment were the system float and the charge depth. Conclusions are based on experimental observation and compared with

S. B. A. Kashem
Computer Skills Faculty, Qatar Armed Forces—Academic Bridge ProgramQatar Foundation, Doha, Qatar
e-mail: skashem@qf.org.qa

D. T. Ngambi
Department of Mechanical Engineering, Swinburne University of Technology, Sarawak, Malaysia
e-mail: 4309448@students.swinburne.edu.my

J. Ahmed
School of Engineering, Computing, and Sciences, Swinburne University of Technology Sarawak, Sarawak, Malaysia
e-mail: jahmed@swinburne.edu.my

U. Qidwai
Department of Computer Science and Engineering, Qatar University, Doha, Qatar
e-mail: uqidwai@qu.edu.qa

P. Suresh (✉)
Department of ECE, Vel Tech Rangarajan Dr. Sagunthala R&D Institute of Science and Technology, Chennai, India
e-mail: sureshp@ieee.org

P. Suresh et al. (eds.), *Advances in Smart System Technologies*,
Advances in Intelligent Systems and Computing 1163,
https://doi.org/10.1007/978-981-15-5029-4_70

expected results from theoretical analysis, with a focus on performance improvement and real word practicality.

Keywords Wireless sensor networks · Monitoring · Wireless applications · Sensors

1 Introduction

In order to find a solution for the growing need for small consumer energy demands, microgeneration technology has become increasingly popular by providing a way to generate sufficient energy for small appliances such as lighting and small- to medium-sized electronic devices. The most common sustainable microgeneration technologies include, but are not restricted to, wind-generated energy, solar photovoltaic energy, and pico-hydro energy generation. While these forms of energy generation continue to be the norm, the study on alternate forms of microgeneration energy continues, with a focus on finding microgeneration technology that does not have the drawbacks of the conventional forms of energy generation. The importance of the study into microgeneration technologies is reflected by its large potential in urban as well as rural areas due to the technological flexibility it affords [1].

With an abundance of resources and energy generation techniques to harvest these resources, the need to integrate this energy into an electric grid to both transfer and store this energy is becoming increasingly important. In order to facilitate the advance of energy generation techniques and electrical grid adaptation, key aspects such as grid demand intermittency would have to be addressed. One approach to addressing both the need for microgeneration techniques and intermittency of the electrical grid is through buoyancy-powered generation and energy storage. While energy generation and storage techniques are varied in their design and functionality, the overall requirements of the system remain the same. Energy generated and stored in the system during periods of low demand and the energy is released during periods of peak demand. There are a variety of energy storage techniques that are currently being developed which include flywheels [2], super capacitors [3], and pumped hydro systems [4], among others.

The aim of developing and testing buoyancy-based system is to gain understanding of the systems and how they operate under standard laboratory conditions and thus assess their capacity for real world application. By testing the system power output against the amount of air put into and taken out of the system, we can get a benchmark on whether or not buoyant force can generate and store sufficient energy. Varying the charge distance and float volume during the testing will be done in order to determine if, and how much, energy can be generated or stored under these parameters. Results of varying the aforementioned parameters in the test will act as a platform on which to base system scalability options for future development. By developing and testing a fully functional buoyancy prototype, their impact on future energy generation and storage techniques can be further realized.

2 Literature Review

2.1 Buoyancy-Based Energy Generation (BBEG)

Buoyancy-based energy generation system is a field of energy generation that is yet to receive thorough research due to the complexity of the system and its apparent unfeasibility. The system involves the use of an object submerged in water with a varying buoyant force depending on the amount of air in the object. The changes in the amount of air in the object should change the buoyancy force and thus generate mechanical motion. By harnessing this mechanical motion via a circular wheel or a gear and rack system, energy can be generated. The main premise of a buoyancy-based energy generation system is to have a higher energy output from the system than the energy input required to cycle the air in the system. While the system would appear plausible, the main drawback of the system is the inability to cycle the air with less energy than the energy that would be produced by the system. Machines that generate more energy than they receive are commonly known as perpetual machines which are not possible because they violate the first and second laws of thermodynamics.

Brad and forest [4] in 2008 developed a motor that would run would use the coupled force of buoyancy and gravity to create an efficient and perpetual source of energy. The prototype suggests the need for a continuous chain to revolve through a tube system with one part containing liquid and the other containing air. The chain would include multiple floats that would be placed along its length. Ideally, these floats would rise through the liquid section of the pipes due to the buoyant force acting on them, and they would fall due to the force of gravity while they are in the section of the pipe containing air. A similar prototype was proposed by Grondahl [5]; however, the prototype was intended to be an improvement in power generators and not a standalone system.

Prototypes similar to the aforementioned have been theorized and designed to harness the coupled effects of the force of buoyancy [6–9]; however, none have proved successful due to the nature of these designs making them perpetual machines. The majority of these concepts have been derived from patents and while they appear to be probable energy generation methods, none of them account for the fundamental laws of thermodynamics and fluid mechanics that they would need to overcome in order for them to be successful.

While the devices mentioned above are not achievable because they are designed to be standalone devices, they would be more viable if they were to utilize excess or wasted energy from different systems. By utilizing waste energy as input energy to the BBEG system, it may be possible to create a feasible energy generation system. In addition to this, if the system is designed for microgeneration, the system may be adapted to have manual energy input from the user that would be converted into electrical energy in an efficient and desirable way.

2.2 Buoyancy-Based Energy Storage (BBES)

The buoyancy-based energy storage system utilizes principles similar to the BBEG system; however, its primary function is the storage of energy rather than generation. By utilizing the buoyant force of an object submerged in water, energy can be stored as potential energy until required for release. The energy from this system can be recovered using the same mechanical energy recovery methods as the BBEG system. The amount of energy that can be stored and discharged in this system is dependent on the cable tension (assuming a pulley system will be used to connect the buoyant object to the generator). The amount of tension in the cable is dictated by the buoyant force of the object. The tension acting on the cable, derived from Archimedes principle, can be in [10, Eq. (1)] as follows;

$$C = \rho g V - mg \tag{1}$$

where C is the cable tension, ρ is the fluid density, g is the gravitational acceleration, V is the float volume, and m is the mass of the float. The energy that the system would generate, expressed in terms of work done, is expressed in [10, Eq. (2)] as follows;

$$E_{\text{discharge}} = C Z_{\text{charge}} \tag{2}$$

where Z_{charge} is the charge depth. Assuming the fluid is ideal, incompressible, and the float volume is constant, the ideal ES capacity can be expressed in [10, Eq. (3)] as follows

$$E_{\text{ideal}} = \rho g V Z_{\text{charge}} \tag{3}$$

The power from the system can be expressed in [10, Eq. (4)] as follows;

$$\text{Power} = E_{\text{ideal}}/3600 \tag{4}$$

By using the aforementioned equations, the idealized energy storage values can be obtained for varying depths and float volumes as shown in Figs. 1 and 2.

Scalability of the systems is an integral part of analyzing whether or not a system will be feasible. BBES and BBEG systems have very good scalability characteristics as shown in Figs. 1 and 2. The main parameters that would need to be scaled are the charge depth and the volume of the float. The implications of the scalability criteria are that the system has lower storage densities than other energy storage systems such as chemical batteries. However, the advantage of using buoyancy-based system is that the maritime real estate required is less demanding than the land-based real estate.

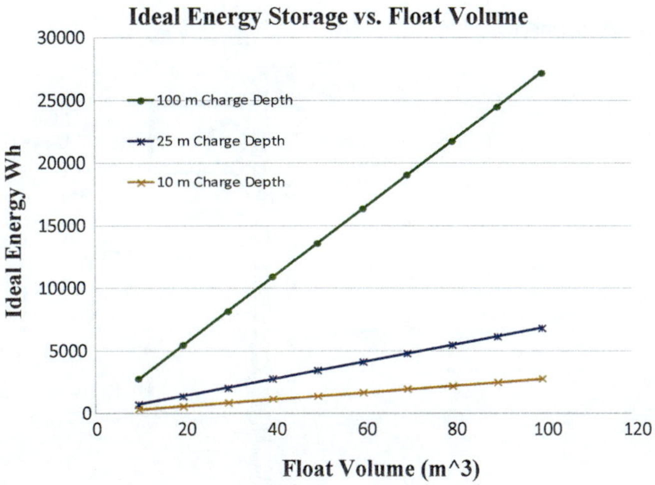

Fig. 1 Ideal energy storage values with varied float volume at different depths

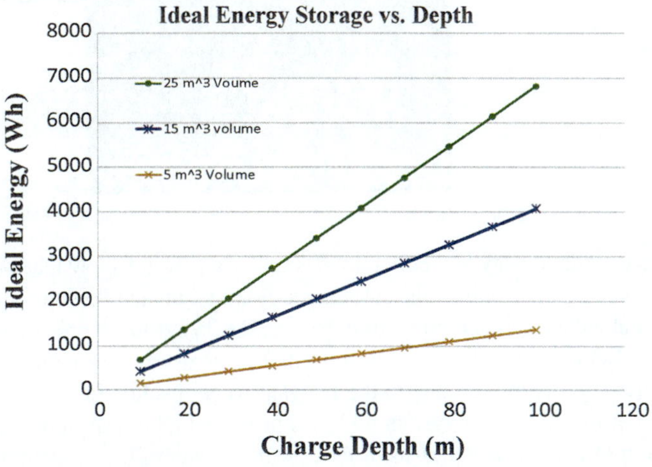

Fig. 2 Ideal energy storage values at varying depths with different float volumes

3 Methodology

3.1 Proof of Concept Testing

In order to test the potential of a BBEG system, a small test rig was made based on a prototype as shown in Fig. 3. However, instead of using an actuator, the air was manually moved between the syringes by a manual operator. While the system

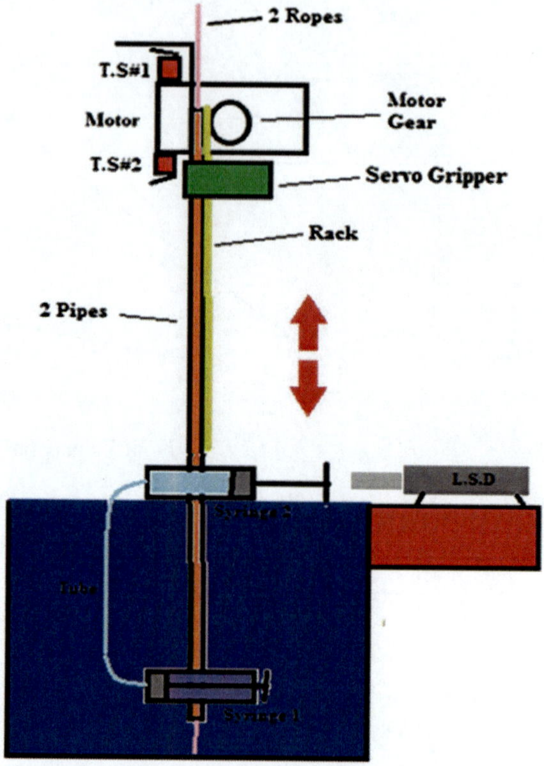

Fig. 3 Buoyancy generator prototype design

was unable to generate any power due to its small scale, it did show that with small changes in the amount of air in the submerged float, an up and down motion could be achieved albeit a small one. In order to test the BBES system, a prototype derived from the BBEG prototype in Fig. 3 was used. The float was submerged in a large pool of water and the motion due to the buoyant force resulted in the rotation of a motor fixed at the end of the system and thus produced a buoyant force. However, the main aim of the experiment was to feed air in and out of the system in order to analyze whether small changes in the amount of air in the system would be sufficient to charge and discharge the system.

For both systems, Eqs. 3 and 4 were used to calculate the ideal amount of energy the system that would be able to produce as per predefined float volume and charge depth specification.

The purpose of the proof of concept testing was to deduce whether the basic overall operation principles of the system would be functional prior to larger scale testing. The system was tested with varying quantities of air being fed into and out of the system to determine whether it would produce sufficient motion for energy generation. While the primary objective of the proof of testing was not to get results, in order consider the test as a success, it would have to create a range of motion that would produce an output current.

3.2 Proof of Concept Results

The proof of concept experiments had results that confirmed the theory but were also, in some cases, inconsistent. For both the BBEG and the BBES systems, there was inconsistency with the results regarding the amount of air that would require an upward or downward motion on the float. This result made it so that it became difficult to find the threshold where the increase in the amount of air would make the object float, and a decrease in the amount of air would make it sink. On the other hand, the proof of concept testing did show that the system would indeed generate and store energy for a specified amount of time and release it as expected.

3.3 Air Exchanger

A key component of the BBEG and BBES system is the air exchange system. This system is required to add or remove air from the float in order for it to rise or fall in the water. For the experiment, the float used was a syringe attached to the side of a container for stability. The system is required to measure and analyze how much energy would be required for the system to rise or fall in the water. The addition of air in the BBES system would act as the charging mechanism and therefore by analyzing how much air would be required to make the float rise, the amount of energy required to charge the system can be determined. In the BBEG system, the constant addition and subtraction of air via the air exchanger would result in an upward and downward motion which would generate electricity via mechanical connection to a generator.

For both the BBEG and the BBES systems, a 100 ml syringe was used to add or remove air from the float. The amount of air was recorded using the pre-manufactured markings on the syringe to ensure accuracy to a reasonable degree. The syringes were also placed under water with the float to ensure that an observable leak would be rectified during the testing.

3.4 Float Design and Anchorage

The dimensions and weight of the float will have significant impact on the performance of a BBEG and BBES system when they are scaled up for offshore usage. Parameters of affecting performance such as drag reduction and system anchorage would have a large impact on the performance of the system.

Testing using a symmetric marine buoy was attempted, and results suggested that during the motion of the buoy in ambient fluid, there would be yawing and pitching of the buoy [10]. Such movements of the float in the water would change the angle of attack with regard to the direction that the float is moving. The same observations were found using a cylindrical buoy in the test below regardless of whether it was a

BBES or a BBEG system being tested. Attempts at stabilizing the floats in the system by using a rod to hold it in place resulted in the float being unable to move freely due to the additional friction of the rod. Due to the severity of the float movement being low, the same float was used further for quantitative analysis.

The anchorage system used in the experiment was not subject to large loading and therefore was held in place with large weights. A pulley was attached to large weights which were placed in water, and during testing, this method of anchorage appeared sufficient. However, it should be noted that for larger scale installations, this method of anchorage would likely be insufficient. The need for adequate float and anchorage systems is one of the major drawbacks of a BBEG and BBES system. The cost of installation and anchorage is likely to be high.

3.5 Large Tank Testing

After reviewing the positive results of the proof of concept testing, the system was up scaled by increasing the charge depth that the system could rise through. The charge depth being one of the key parameters of the experiment was expected to give more concise results. The system was tested in a large body of water with a maximum depth of 2 meters. In order to fully test the system, changes were made to the prototype, and a string was used to attach the system to the motor rather than the road specified in Fig. 3 prototype.

Modifications to the initial prototype were influenced by work done on a prior BBES system [10–20]. By placing a system of pulleys and a cable between the float and the charging system, it meant that static friction would be reduced which was a drawback in the previous design. In addition to this, the newer design allowed for flexibility in the way we would charge the system. Rather than constantly having to charge and discharge the system with air, the string could be connected to a crank and reeled in. A similar design was theorized by in a patent [11, 21–30], and it uses a symmetrical system of floats to generate large amounts of energy.

In order to test the amount of energy that the large tank system would generate, the same procedure as the proof of concept tests were used. As a result of the only variables that changed being the float volume and the reel distance, both of these parameters were accounted for in Eq. 3 and therefore could easily be plotted once the tests were done.

4 Results and Discussion

Both the BBES and the BBEG were tested in a large tank while the energy output was recorded. The values of voltage and current were recorded. The main objective of the testing was to see if and how much power would be generated in the BBEG or stored in the BBES if the main source of air came from the exchange of air via the syringe system.

Results from the experiment demonstrated that, while an increase in the amount of air via the syringes would cause an upward motion due to buoyancy, the force would in some cases be insufficient to generate movement in the system. The amounts of air that would be required to create an up and down motion in the BBEG varied as shown in Table 1. The implications of these results are that the threshold at which a change in the amount of air would cause movement, could not be found. Additionally, repeating of the same tests produces a large amount of disparity and variance, and therefore, conclusions could not be made about how much air exactly would cause motion in the BBEG system.

In the case of the BBES, similar problems as the BBEG system were faced in the test results; however, in order to test the BBES system, it required only upward motion as the charging of the system could be done while it was submerged. However, similar to the BBEG system, the BBES system did not consistently produce results; however, in the times that it did work, the output values were taken, and an average of two successful series of tests are tabulated in Table 2. Analysis of the theorized ideal amount of energy was calculated as per Eq. 3, and the output values are tabulated along with the experimental values in Table 2 and the percentage disparity between the theorized and the final calculated.

The purpose of the test results was to see whether the system could be charged with a small syringe, which in this case provided between 90 and 100 ml of air to the system and thus deduce whether the system would return the same amount of energy during upward motion. Results from Table 2 show that the tested buoyancy systems returned 78% to 67% of the energy they were expected to produce after being charged using the air exchange system. The power results from the tests were tabulated as shown in the graph in Fig. 4. A trend in the results does indicate an increase in power from the system with time. The power plateaus after the initial acceleration has been overcome, and the maximum upward velocity of the float is reached as it rises in the water. This upward acceleration may have an impact on the total amount of energy that the system would generate. The amount of energy generated as a result of the float having to accelerate through the water may be reduced due to the hydrodynamic

Table 1 Results of required amount of air to generate motion in BBEG and BBES system

Test number	Air output required for upward motion (mL)	Time taken for full downward motion (s)	Air input required for upward motion (mL)	Time taken for full upward motion (s)
1	15	25	25	22
2	0	0	25	22
3	0	0	35	20
4	40	27	40	15
5	0	0	40	23

Table 2 Ideal energy and experimental power output values

Test 1					Test 2			
	Volume (m^3)	Charge depth (m)	Power (Wh)	Percentage difference (%)	Volume (m^3)	Charge depth (m)	Power (Wh)	Percentage difference (%)
Experimental	9×10^{-5}	0.6	12.6×10^{-5}	22.3	1×10^{-4}	0.6	9.7×10^{-5}	33.9
Ideal	9×10^{-5}	0.6	16.3×10^{-5}		1×10^{-4}	0.6	14.7×10^{-5}	

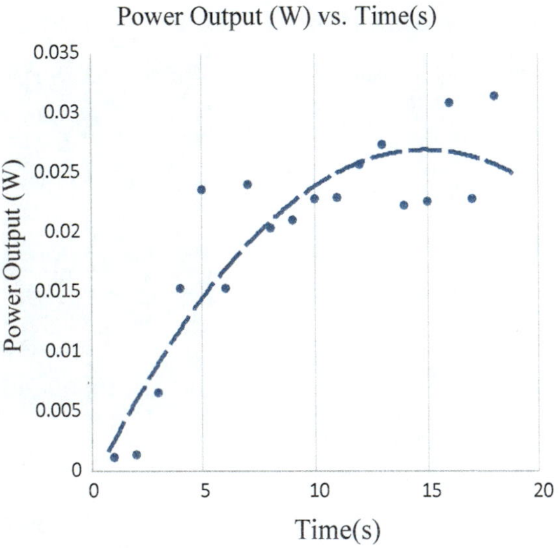

Fig. 4 Power output (W) versus time (s)

force acting on the float; however, there is no evidence to suggest this is the case in this experiment (Fig 5).

5 Conclusions and Recommendation

The experimental testing shown above produces results that show that it is indeed possible to generate power from BBEG and BBES systems under standard laboratory testing conditions. By taking system scalability into account, the experiment shows that the systems have capacity for real world application. The results shown above demonstrate the ability of buoyant force to store and generate energy by altering the key parameters which in this case are float volume and charge depth. Variation of charge volume produced results that supported the theory that scalability would have an impact on the buoyancy system in line with the theory put forward in the literature review. Additionally, the results indicate the ability to use small air transfer containers as charging and discharging methods. The tests above, as shown in Table 1, do also highlight the inconsistency in testing and result gathering. BBE systems are inconsistent in their operation and thus make it difficult to predict the levels and thresholds that would be required for them to function as expected.

Tests done on the BBEG system did not show any conclusive results that would suggest that this method of power generation would feasible within the parameters and conditions set. This is due to the systems inability to produce sufficient upward and downward movement consistently, using the air exchange system. While the BBEG does occasionally generate electrical energy, the inconsistency of its ability to generate power stands as its major flaw. However, it can also be concluded that

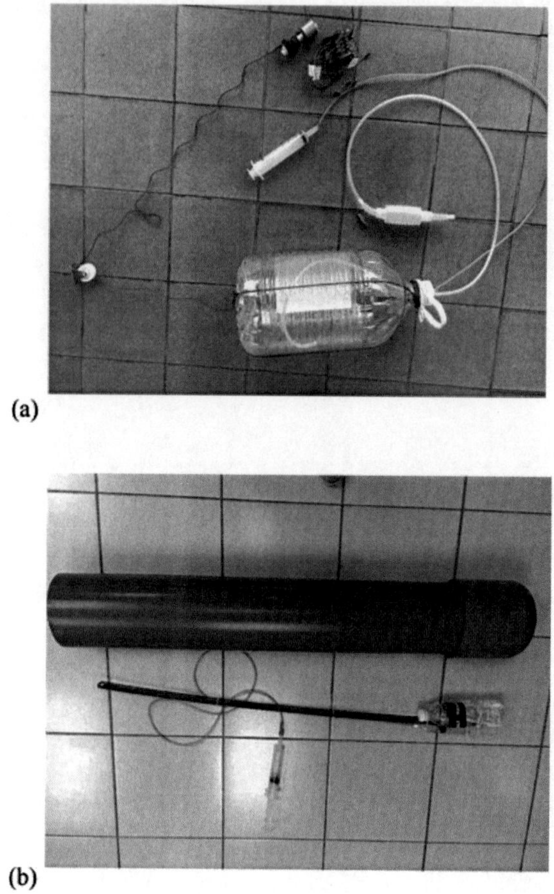

Fig. 5 **a** BBEG/BBES test system; **b** Proof of concept BBEG testing apparatus

should another system of air exchange be used, one that would have a higher air capacity, and the system could generate power more consistently. The BBES system tested did show more positive results, and the experimental results were comparative to the idealized results within a reasonable margin of error. The use of a syringe as the primary charging system was successful and therefore shows that a similar charging system that can be used on BBES systems.

While the tests above were done within reasonable limits of accuracy, there is ample reason to believe that some energy was lost during the experiment. Friction in the pulley system and losses in the generator should be taken into account. Additionally, because the exact amount of air in the float was not measured between tests of the BBEG system, there may be a disparity between the amount of air at the same section between tests. This would account for why in some cases, inserting the same amount of air, in two different tests, would give different results. In addition to this, due to the size of the apparatus for the BBEG and BBES systems, tests were conducted in a body of water that was neither fully at rest, which would mean that the ambient fluid was not at rest during the experiment.

For future study, more accurate air exchange systems with significantly larger capacity would be better suited to produce results that are more easily recorded. In addition, it may likely be easier to find the threshold of air required to create an upward or downward motion if the air chambers were larger. The use of larger charge ratios would also be essential for future study in order to further isolate the system steady-state performance. Recommendations of a charging system alternate to the aforementioned syringe system include the use of a crank to lower the float into the water. Analysis of the system should also be carried out in a smaller laboratory setting in order to understand the float behavior and hydrodynamic response of the system. The following ideal conditions [10] should be taken into account when testing the BBEG and BBES systems

- Ambient fluid is ideal
- Ambient fluid is at rest
- Cable does not stretch
- Change in mass of cable due to length remains constant.

Acknowledgements The composition of this research report has given me insight into what is required to do productive and progressive research. I would like to thank Dr. Lai for providing me with the framework and foundation for this research topic. I would also like to thank Dr. Chong Kok Hing for his invaluable input during the experimentation and composition of this report

References

1. Coles, A.-M., Piterou, A., Genus, A.: Sustainable energy projects and the community: Mapping single-building use of microgeneration technologies in London. Urban Studies **53**(9), 1869–1884 (2016)
2. Cimuca, G.O., Saudemont, C., Robyns, B., Radulescu, M.M.: Control and performance evaluation of a flywheel energy-storage system associated to a variable-speed wind generator. IEEE Trans. Industr. Electron. **53**(4), 1074–1085 (2006)
3. Abbey, C., Joos, G.: Supercapacitor energy storage for wind energy applications. IEEE Trans. Indus. Appl. **43**(3), 769–776, May–June 2007
4. Forrest, B.A.: Buoyancy and gravitation motor. U.S. Patent 5944480 A, Aug 31 1999
5. Grondahl, A.G.: Power Generator. U.S. Patent 2037973 A, 21 April 1936
6. Hastings, D.R.: Vertical gravity/buoyancy power generator. U.S. Patent 8011182 B2, 6 Sept 2011
7. Manoj, V. Gravity-buoyancy object turbine. Patent WO2013128446 A2, 6 Sept 2013
8. Gilmore, R.: Gravity-Buoyancy motor. U.S. Patent 3857242 A, 31 Dec 1972
9. Woolfolk, M.Y.: Motor utilizing buoyancy forces. U.S Patent 4726188 A, 23 Feb 1988
10. Bassett, K., Carriveau, R., Ting, D.S.-K.: Experimental analysis of buoyancy battery energy storage system. IET Renewable Power Generation [Online], 10. Available: http://digital-library.theiet.org/content/journals/10.1049/iet-rpg.2016.0033 (2016)
11. Morgan, J.P.: Buoyancy Energy Storage and Energy Generation System. U.S. Patent 20100107627 A1, 6 May 2010

12. Kashem, S.B.A., Ektesabi, M., Nagarajah, R.: Comparison between different sets of suspension parameters and introduction of new modified skyhook control strategy incorporating varying road condition. Vehicle Syst. Dynamics **50**(7), 1173–1190 (2012)
13. Kashem, S., Nagarajah, R., Ektesabi, M.: Vehicle Suspension Systems and Electromagnetic Dampers. Springer Singapore (2018)
14. Kashem, S.B.A., Roy, S., Mukharjee, R.: A modified skyhook control system (SKDT) to improve suspension control strategy of vehicles. In: 2014 International Conference on Informatics, Electronics & Vision (ICIEV) (pp. 1–8). IEEE (2014)
15. Kashem, S., Sufyan, H.: A novel design of an aquatic walking robot having webbed feet. Int. J. Autom. Comput. **14**(5), 576–588 (2017)
16. Kashem, S.B.A., Sheikh, M.I.B., Ahmed, J., Tabassum, M.: Gravity and buoyancy powered clean water pipe generator. In: 2018 IEEE 12th International Conference on Compatibility, Power Electronics and Power Engineering (CPE-POWERENG) (pp. 1–5). IEEE April 2018
17. Kashem, S.B.A., Tabassum, M., Sufiyan, H.: A novel design of an amphibious robot having webbed feet as duck. In: International Conference on Computer and Drone Applications Proceedings of IEEE Xplore 2017, 9–11 Nov 2017
18. Kashem, S.B.A., Jawed, S., Ahmed, J., Iqbal, A.: An experimental study of the amphibious robot inspired by biological duck foot. In: 2018 IEEE 12th International Conference on Compatibility, Power Electronics and Power Engineering (CPE-POWERENG), (pp. 1–6). IEEE April 2018
19. Chowdhury, M.A., Kashem, S.B.A.: H∞ loop-shaping controller design for a grid-connected single-phase photovoltaic system. Int. J. Sustain. Eng. 1–9 (2018)
20. Mubarak, H., Kashem, S.B.A.: Comparison of different energy saving lights using solar panel. Frontiers Energy **10**(4), 466–472 (2016)
21. Siddique, M.B.M., Kashem, S.B.A., Mathew, K.: Home and water heating using biofuels. In: Proceedings of International Conference on Recent Innovations in Engineering and Technology (2017)
22. Tay, F., Kashem, S.B.M.: Automated miniature greenhouse. Adv. Sci. Lett. **23**(6), 5309–5313. (Indexed: ISI Thomson Reuters, Scopus) (2017)
23. Safe, A.A., Kashem, S., Moniruzzaman, M., Islam, M.T.: Design, fabrication & analysis of twisted blade vertical axis wind turbine (VAWT) and a simple alternator for VAWT. In: 2014 9th International Forum on Strategic Technology (IFOST), (pp. 304–308). IEEE (2014)
24. Tabassum, M., Kashem, S.B.A., Mathew, K.: Distributed energy generation—is it the way of the future? In: Proceedings of the 1st Springer International Conference on Emerging Trends and Advances in Electrical Engineering and Renewable Energy (2016)
25. Sheikh, M. Bilal, I., Kashem, S.B.A., Choudhury, T.: Enhancing solar power generation using gravity and fresh water pipe. In: IEEE International Conference on Mechatronics Proceedings of IEEE Xplore 2017, pp. 266–271 (2017)
26. Tabassum, M., Kashem, S.B.A., Siddique, M.B.M.: Feasibility of using Photovoltaic (PV) technology to generate solar energy in Sarawak. In: 2017 International Conference on Computer and Drone Applications (IConDA) (pp. 11–16). IEEE November 2017
27. Siddique, M.B.M., Kashem, S.B.A., Iqbal, A.: Biofuels in Malaysian perspective: Debates and benefits. In: 2018 IEEE 12th International Conference on Compatibility, Power Electronics and Power Engineering (CPE-POWERENG) (pp. 1–6). IEEE. April 2018
28. Shabrin, N., Kashem, S.B.A., et al.: Investment and construction cost analysis on net-zero energy building technology. Int. J. Mech. Prod. Eng. **5**(4) (2017). ISSN: 2320-2092
29. Shabrin, N., Kashem, S.B.A.: A comprehensive cost benefit analysis of green building. Int. J. Adv. Mech. Civil Eng. (IJAMCE) **4**(2) (2017). ISSN: 2394-2827
30. Kashem, S.B.A., De Souza, S., Iqbal, A., Ahmed, J.: Microgrid in military applications. In: IEEE 12th International Conference on Compatibility, Power Electronics and Power Engineering (CPE-POWERENG) (pp. 1–5). IEEE April 2018

Smart Home Security System Based on Zigbee

Chua Boon Liang, Mujahid Tabassum, Saad Bin Abul Kashem, Zulfiqar Zama, P. Suresh, and U. Saravanakumar

Abstract Sarawak has a high burglary rate. This project aims to create a low-cost smart home security system based on Zigbee that protects Sarawakian homeowners against flood, smoke, and intrusion. The system includes a hub, a battery-powered sensor node, and an Android application. The sensor node sends sensor output to the hub via Zigbee. If intrusion, flood, or smoke is detected, the hub starts an alarm and triggers the Android application to display an alarm in real time through the Internet via Firebase. If an intrusion is detected, a photo of the intruder is taken and processed via Haar Cascades and Histogram of Oriented Gradient (HOG) to emphasize the face and body of the intruder. The system also provides live mobile monitoring. The system is secure, provides a real-time response, Internet-controlled, and low cost but has low battery life.

Keywords Home security · Zigbee · Human detection · Firebase

C. B. Liang
Swinburne University of Technology, Sarawak, Malaysia
e-mail: 4315839@students.swinburne.edu.my

M. Tabassum
Higher College of Technology in Muscat, Muscat, Oman
e-mail: mujahid.tabassum@hct.edu.om

S. B. A. Kashem
Qatar Armed Forces—Academic Bridge Program, Qatar Foundation, Doha, Qatar
e-mail: skashem@qf.org.qa

Z. Zama
Faculty of Information Technology, Higher college of Technology, Muscat, Oman
e-mail: Zulfiqarzaman@gmail.com

P. Suresh (✉) · U. Saravanakumar
Department of ECE, Vel Tech Rangarajan Dr. Sagunthala R&D Institute of Science and Technology, Chennai, India
e-mail: drpsuresh@ieee.org

U. Saravanakumar
e-mail: skashem@qf.org.qa

P. Suresh et al. (eds.), *Advances in Smart System Technologies*,
Advances in Intelligent Systems and Computing 1163,
https://doi.org/10.1007/978-981-15-5029-4_71

1 Introduction

Burglary is the entry of a building to commit theft. It is a big issue in Sarawak, where crime prevention is still inadequate. Kuching has a crime index of 70.23 in the year 2013, which is the 16th highest crime index globally. Hence, security systems are necessary as extra protection against burglars.

A smart home security system is an alarm system that can detect intrusion into a building or area. Its main purpose is to prevent burglary, property damage, and harm to homeowners. Some smart home security systems also include protection against fire, flooding, and smoke. If an intruder is detected, it will trigger alerting devices such as sirens and lights to attempt to scare off burglars and alert the homeowner.

This paper proposes a smart home security system to deter burglars. The objectives are to design and develop a low-cost smart home security system, to develop a Zigbee-based infrastructure to offer smart home surveillance and to enhance security and surveillance with live mobile monitoring.

2 Literature Review

This paper incorporates ideas from other smart home security system projects such as:

- In [1], Hou et al. proposed a Zigbee-based security system that uses infrared, gas, and smoke sensors to detect intrusion and fire. If the sensors are tripped, it will take photos and process them to produce clear photos to be sent to authorities via MMS.
- In [2], Lartigue et al. proposed a Zigbee-based security system that uses motion and door sensors to detect intrusion. The system is controlled using an Android application that shows in real time the status of the sensors. The application also controls an automatic door lock.
- In [3], Noor proposed a security system that takes a photo of an intruder if a passive infrared (PIR) sensor is tripped. The image is then uploaded to a server, and an interface shows the image along with an alarm. It also sends an alert to the owner's Twitter. The advantage of this system is that it can eliminate blind spots and can be hidden, so it can produce better detection and human face recognition.

3 Methodology

The system has one hub and a battery-powered sensor node. The sensor node has a PIR sensor to detect motion, a reed switch to detect the opening of doors, a water level sensor to detect floods, and an MQ-2 smoke sensor to detect smoke. The sensor node sends sensor output wirelessly to the hub via Zigbee. The XBee module is set to sleep to save power when it is not transmitting. To ensure that, the sensor node is functional, and it is set to transmit every minute (Figs. 1, 2 and 3).

The Raspberry Pi at the hub analyzes sensor output sent by the Arduino. If smoke or flood is detected, it triggers the smoke or flood alarm, respectively. Triggering of the intruder alarm depends on the mode. The system has three modes, which are 'Stay', 'Away', and 'Night'. 'Stay' is for when the homeowners are at home. 'Away' is for when the homeowners are not home. 'Night' is for when the homeowners are at home yet sleeping. If the hub does not receive the regular transmission from the sensor node for two minutes, it will trigger a fault alarm. This will prevent faulty sensor nodes from affecting security and protect against intruder jamming. If an alarm is activated, the Raspberry Pi sounds the alarm and commands the Android application to display an alarm.

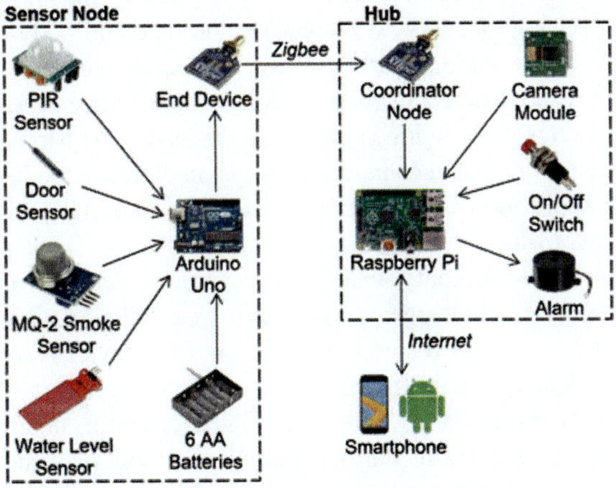

Fig. 1 System block diagram

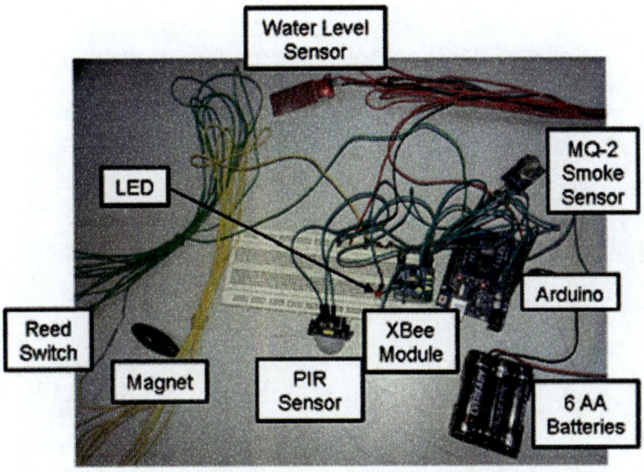

Fig. 2 Sensor node

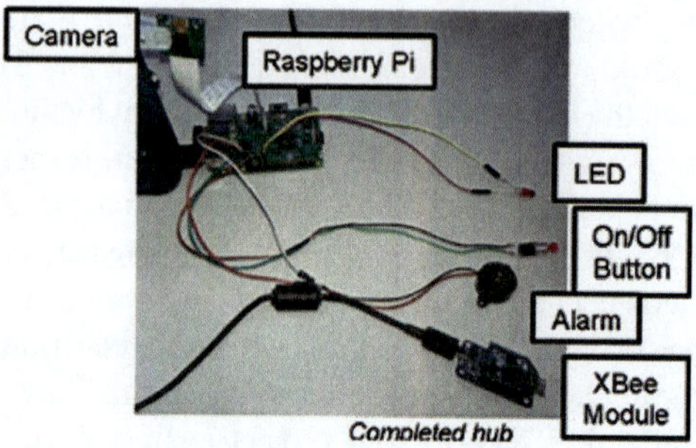

Fig. 3 Hub

If the intruder alarm is triggered, the hub takes a photo of the intruder. The photo will be processed using OpenCV to emphasize the face and body of the intruder as well as attach a timestamp. Haar Cascades is used to detect the face, whereas HOGDescriptor is used to detect the person. A light blue square is drawn around the face, whereas a green square is drawn around the person. This photo of the intruder is saved in Firebase storage and can aid authorities in catching the intruder (Fig. 4).

The Android application is the graphical user interface (GUI) for the security system. It is password-locked (Fig. 5).

Logging in starts a service that listens for an alarm command from the hub. If there is an alarm command, it will trigger a notification that will play notification sounds and vibrate the device. This will alert the user even if the device is screen-locked or the application is closed [4–10] (Fig. 6).

There are four types of alarms, which are: intruder alarm, flood alarm, smoke alarm, and fault alarm. During an alarm, the application will play alarm sounds and show the sensors triggered as well as their location in real time to allow the user to know the location of the disaster. Moreover, if the user wishes to view CCTV feed, the Raspberry Pi will show CCTV feed via MJPG-streamer and Dataplicity.

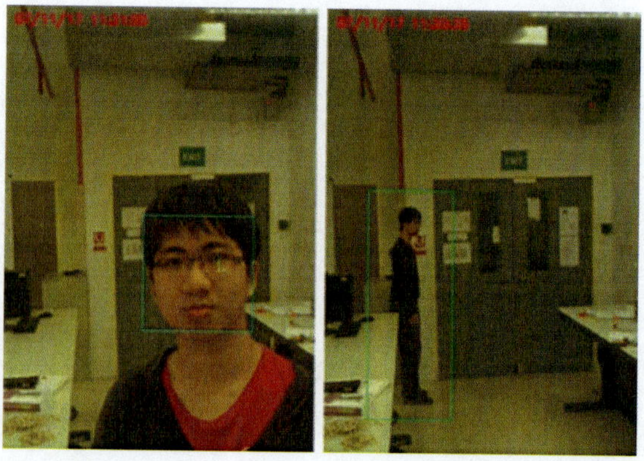

Fig. 4 Face detection and people detection

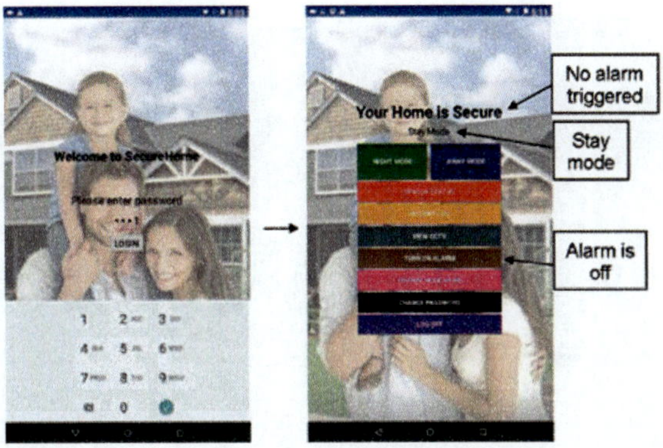

Fig. 5 Password and home screen

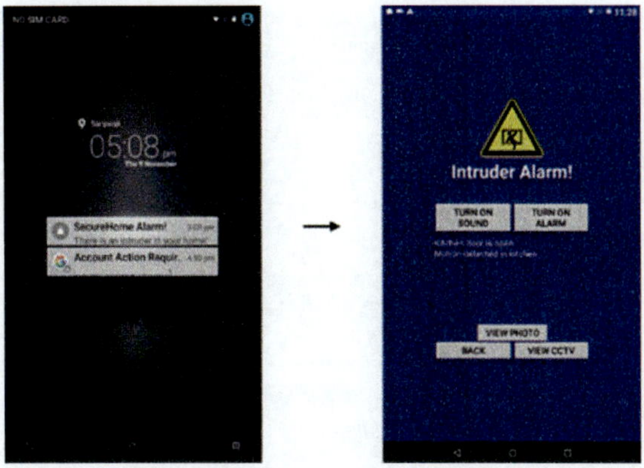

Fig. 6 Alarm will trigger notification

This provides live mobile monitoring. In addition, the user can view the photo of the intruder stored in Firebase storage from the application. Furthermore, the alarm at the hub and the alarm sounds at the application can be turned off (Fig. 7).

The application includes features to facilitate control over the system including changing the mode, checking sensor status, viewing history log, viewing CCTV feed, turning on or off the alarm, changing sensor node name, changing the password, and logging off (Fig. 8).

4 Results

The system was tested for one full day on November 7, 2017, 11.28 am to November 8, 2017, 11.28 am at Swinburne laboratory E504 (Fig. 9, 10 and 11).

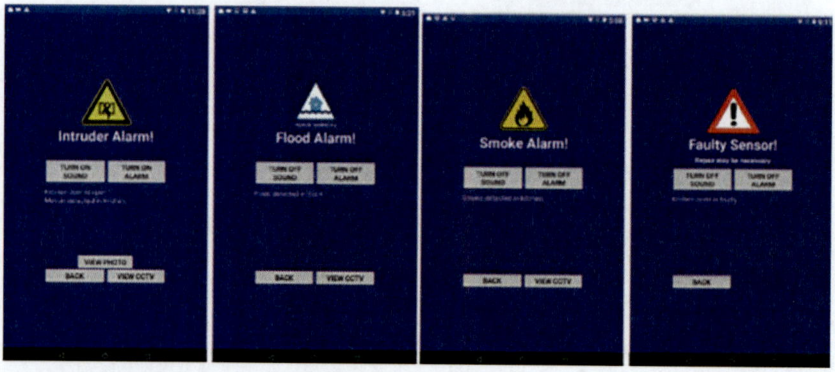

Fig. 7 Intruder, flood, smoke, and fault alarm

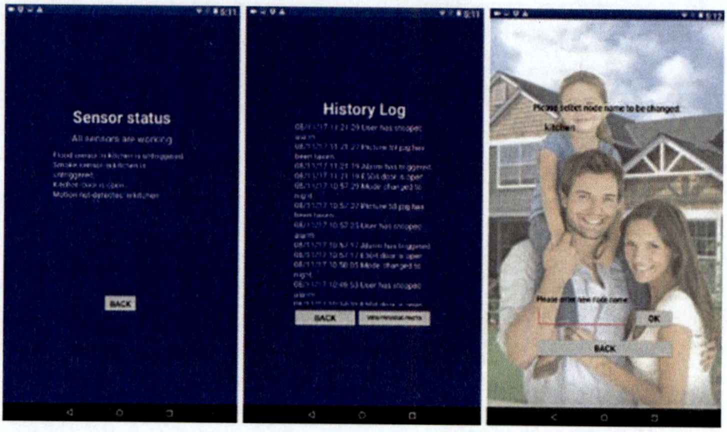

Fig. 8 Sensor status, history log, change node name

The system works properly as opening the door or triggering the motion sensor will trigger the intruder alarm (Fig. 12).

The results show that there are 59 intruder alarms, zero flood alarms, and zero smoke alarms in the testing period. The results are as predicted as it is unlikely to have flood or smoke in the location. The system records the occurrences in the history log. The photos taken are categorized into three types: photos with intruder successfully detected, photos with intruder but not detected, and photos without intruders.

The flood and smoke alarm was tested in simulated testing as it cannot be tested in a real scenario. The water level sensor was put into a container, and water was slowly poured in. The flood alarm triggers when the water level is 1 cm as intended. The smoke alarm was tested by putting a burning paper near the MQ-2 smoke sensor. The alarm triggers immediately. Hence, both flood and smoke alarms are working as intended (Fig. 13).

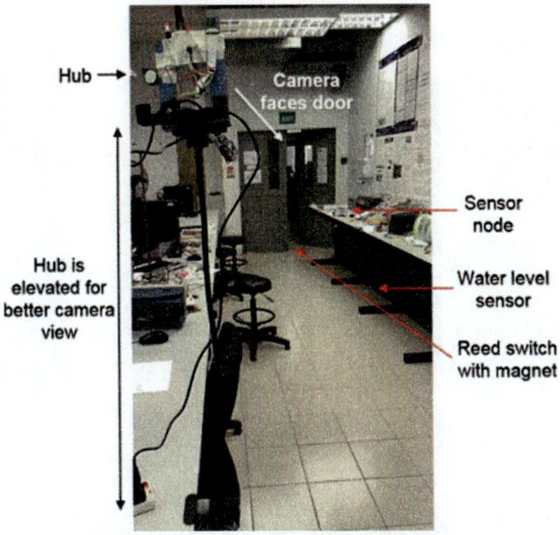

Fig. 9 Photo of testing location

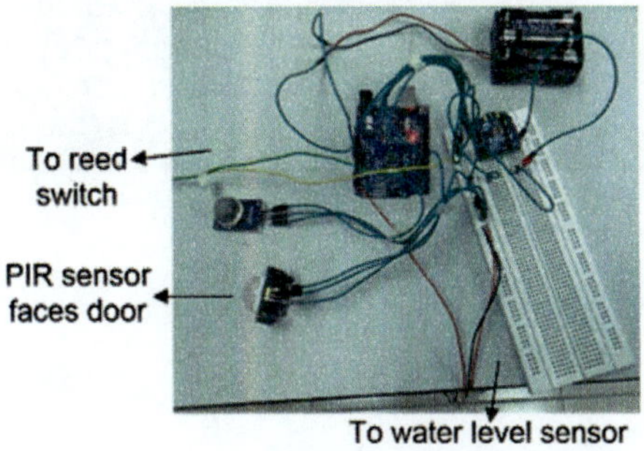

Fig. 10 Sensor node during testing

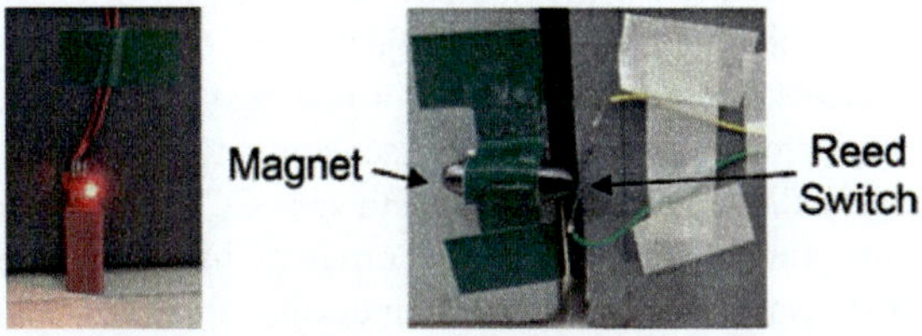

Fig. 11 Water level sensor, reed switch with magnet

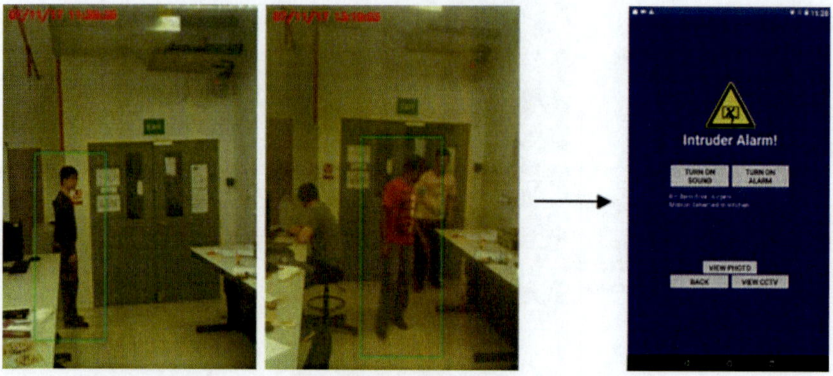

Fig. 12 Motion or opening of the door triggers the alarm

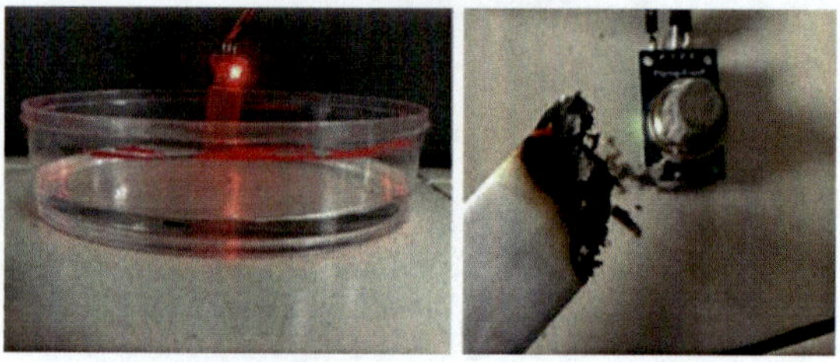

Fig. 13 Flood alarm test and smoke alarm test

4.1 Discussion

The system is fully functional as a smart home security system as it triggers the alarm when it detects intruders, flood, or smoke almost instantaneously. The application features are functional and allow the homeowner to control or check the status of the security system. The total cost of this project is RM612.53. As the cost is less than RM1000, it meets the requirements for a low cost. The system uses Zigbee as a communication protocol, so it has high scalability. The system provides live mobile monitoring by allowing the homeowner to view CCTV footage in the Android application via the Internet to know the occurrences in their home. The system provides a real-time response as it alerts the homeowner almost instantaneously after an intruder, flood, or smoke is detected. The system can be accessed from anywhere if there is an Internet connection, making it part of the Internet of Things (IoT). This allows the homeowner to easily control the system away from home and receive alarm notifications. The system has multiple security features. The Android application is password-locked, so that it is only accessed by the owner. Communication between the hub and the sensor node is encrypted to prevent hacking. Communication between the hub and the application uses Firebase that requires authentication and is hosted on secure sockets layer (SSL), which is a standard security technology. Moreover, the system will trigger the fault alarm if a sensor node does not send transmission within two minutes, preventing jamming.

The system is unable to detect all the intruders from the photos. This is because pedestrian identification requires high-quality photos, but the sample photos are blur due to motion and low resolution. The camera used is of low quality and cannot capture moving objects clearly at high resolution. Another reason is that a portion of the body of the intruder is blocked, preventing HOGDescriptor from identifying them as people. Moreover, the face detection feature did not detect any faces. This is because the intruders are too far from the camera and the photo is of low resolution. Hence, the face is too blur or small to be detected. However, face detection is functional if the target face is near the camera. To solve these problems, a higher resolution camera with fast shutter speed is required. However, the cost is high and goes over the budget constraint. Moreover, a stable location to place the camera is needed to prevent shaking. From the real scenario testing, the battery life for the sensor node is five hours. One method to increase battery life is using low current smoke sensors such as photoelectric smoke sensors. However, this will increase cost. Another method is to replace the linear regulator (70% efficient) on the Arduino UNO with a switching regulator (90% efficient). The Arduino UNO cannot be set to sleep because it would require interrupts to wake up the Arduino but interrupts would affect Zigbee transmission as an interrupt may stop a Zigbee transmission halfway. Finally, higher capacity batteries can be used. In this project, Eveready super heavy duty batteries (carbon–zinc batteries) were chosen because they are easily available and replaceable. This would allow the sensor node to be on at all times without being recharged. AA battery alternatives include alkaline batteries and nickel–metal hydride batteries.

5 Conclusion

In conclusion, the system is fully functional as a smart home security system as it triggers the alarm when it detects intruders, flood, or smoke almost instantaneously. This project meets all of the objectives. The system is functional and provides a low-cost smart home security system alternative to middle-class households. Moreover, the project is very attractive as it suits the contemporary lifestyle. This will attract them to purchase smart home security systems. If a large portion of Sarawakians installs smart home security systems in their homes, the burglary rate of Sarawak will decrease.

The system has a fault alarm to alert the user that the sensor node is not transmitting, which may imply that the battery is low. However, the PIR sensor on the sensor node would start giving wrong output repeatedly before the sensor node stops transmitting. This is because the sensors require 5 V, whereas the Zigbee module requires 3.3 V. Hence, the sensors will fail before the Arduino or the Zigbee module. To remedy this, a separate voltage monitor circuit would be required to monitor battery voltage and alert the user alert if the voltage drops below 6 V. However, this will not be implemented in this project due to lack of time, so it will be done in future works.

Future works include the addition of voltage monitoring circuit to alert when low battery, the addition of high-resolution camera to enable better intruder detection, improvement of battery life by using low-power smoke sensors and the addition of a backup battery and global system for mobile communication (GSM) to enable the system to function during blackouts.

References

1. Hou, J., Wu, C., Yuan, Z., Tan, J., Wang, Q., Zhou, Y.: Research of intelligent home security surveillance system based on ZigBee. Int. Symp. Intell. Inf. Technol. Appl. Workshops **2008**, 554–557 (2008)
2. Lartigue, J.W., McKinney, C., Phelps, R., Rhodes, R., Rice, A.D., Ryder, A.: A tablet-controlled, mesh-network security system: An architecture for a secure, mesh network of security and automation systems using Arduino and Zigbee controllers and an Android tablet application. In: Proceedings of the 2014 ACM Southeast Regional Conference, ACM, 2638500, pp. 1–4 (2014)
3. Noor, M.N.M.: Community based home security system using wireless mesh network. Int. J. Academic Res. Part A **5**(5), 73–79 (2013)
4. Kashem, S.B.A., Ektesabi, M., Nagarajah, R.: Comparison between different sets of suspension parameters and introduction of new modified skyhook control strategy incorporating varying road condition. Vehicle Syst. Dyn. **50**(7), 1173–1190 (2012)
5. Kashem, S., Nagarajah, R., Ektesabi, M.: Vehicle Suspension Systems and Electromagnetic Dampers. Springer Singapore (2018)
6. Kashem, S.B.A., Roy, S., Mukharjee, R.: A modified skyhook control system (SKDT) to improve suspension control strategy of vehicles. In: 2014 International Conference on Informatics, Electronics & Vision (ICIEV), (pp. 1–8). IEEE (2014)
7. Kashem, S., Sufyan, H.: A novel design of an aquatic walking robot having webbed feet. Int. J. Autom. Comput. **14**(5), 576–588 (2017)
8. Kashem, S.B.A., Sheikh, M.I.B., Ahmed, J., Tabassum, M.: Gravity and buoyancy powered clean water pipe generator. In: 2018 IEEE 12th International Conference on Compatibility, Power Electronics and Power Engineering (CPE-POWERENG), (pp. 1–5). IEEE April 2018
9. Kashem, S.B.A., Tabassum, M., Sufiyan, H.: A novel design of an amphibious robot having webbed feet as duck. In: Proceedings of IEEE Xplore 2017 International Conference on Computer and Drone Applications, 9–11 Nov 2017
10. Kashem, S.B.A., Jawed, S., Ahmed, J., Iqbal, A.: An experimental study of the amphibious robot inspired by biological duck foot. In: 2018 IEEE 12th International Conference on Compatibility, Power Electronics and Power Engineering (CPE-POWERENG), (pp. 1–6). IEEE April 2018